Biotechnology in Agriculture and Forestry 17

High-Tech and Micropropagation I

Edited by Y.P.S. Bajaj

With 222 Figures

Springer-Verlag
Berlin Heidelberg New York
London Paris Tokyo
Hong Kong Barcelona Budapest

Professor Dr. Y.P.S. Bajaj
A-137
New Friends Colony
New Delhi 110065, India

ISBN-13:978-3-642-76417-2 e-ISBN-13:978-3-642-76415-8
DOI: 10.1007/978-3-642-76415-8

Library of Congress Cataloging-in-Publication Data. High-tech and micropropagation / edited by Y. P. S. Bajaj. p. cm. -- (Biotechnology in agriculture and forestry ; 17) Includes bibliographical references and index.ISBN-13:978-3-642-76417-2(U.S.: alk.paper)1.Plant propagation--In vitro. I. Bajaj, Y. P. S., 1936- . II. Series SB123.6.H54 1991 631.5'3--dc20 91-22304

Softcover reprint of the hardcover 1st edition 1991

Production Editor: Ingrid Samide, Heidelberg
Typesetting: International Typesetters Inc., Makati, Philippines
31/3145-5 4 3 2 1 0 – Printed on acid-free paper

ALLE ZEIT WACH 1842

Volumes already published

Volume 1: Trees I (1986)
Volume 2: Crops I (1986)
Volume 3: Potato (1987)
Volume 4: Medicinal and Aromatic Plants I (1988)
Volume 5: Trees II (1989)
Volume 6: Crops II (1988)
Volume 7: Medicinal and Aromatic Plants II (1989)
Volume 8: Plant Protoplasts and Genetic Engineering I (1989)
Volume 9: Plant Protoplasts and Genetic Engineering II (1989)
Volume 10: Legumes and Oilseed Crops I (1990)
Volume 11: Somaclonal Variation in Crop Improvement I (1990)
Volume 12: Haploids in Crop Improvement I (1990)
Volume 13: Wheat (1990)
Volume 14: Rice (1991)
Volume 15: Medicinal and Aromatic Plants III (1991)
Volume 16: Trees III (1991)
Volume 17: High-Tech and Micropropagation I (1991)

Volumes in preparation

Volume 18: High-Tech and Micropropagation II (1992)
Volume 19: High-Tech and Micropropagation III (1992)
Volume 20: High-Tech and Micropropagation IV
Volume 21: Medicinal and Aromatic Plants IV

Dedicated to
Gitanjali Bajaj

Preface

Micropropagation of plants is a multibillion dollar industry being practiced in hundreds of small and large nurseries and commercial laboratories throughout the world. At the present, it is the only component of plant biotechnology which has been commercially exploited on such a large scale, especially for the production of ornamentals. Now micropropagation of trees and medicinal plants is assuming great importance. With recent progress made in the propagation of fruit and forest trees, and the immediate need for afforestation and planting of orchards, propagules and plantlets are required quickly and in large numbers. Although the basic methodology of in vitro culture of plant tissues has been fairly mechanized over the last 25 years, conventional micropropagation methods are relatively slow, time-consuming, expensive, and labor-intensive; up to 50–80% of the cost of plant production is attributed to labor. In order to reduce the cost and to increase the efficiency and scale up of plant production, automation is being incorporated. Robots, microcomputers, and other automated systems of artificial intelligence have been employed to surface-sterilize tissues, dispense media, and transfer plants to pots and field. The automatic control of environments in greenhouses to protect cultivation is another area of tremendous commercial importance.

The use of artificial intelligence, image analysis, video cameras, bioreactors, mistifiers, automatic production lines, synthetic seed, storage and cryopreservation of elite genetic stocks, and the computerization of information management will facilitate micropropagation and protected cultivation. Judging from the enthusiasm and interest of biotechnology enterprises, it is envisaged that it will not be long before we see the routine jobs being done by robots for industrial micropropagation.

Taking the above-mentioned points into consideration, it appeared necessary to review the literature and to present the state of the art on micropropagation of plants. Thus, 119 chapters contributed by experts from 30 countries are compiled in a series composed of the following four parts:

High-Tech and Micropropagation I comprises 29 chapters on various aspects of micropropagation and automation, i.e., setting up a commercial laboratory, meristem culture, somatic embryogenesis, graft incompatibility, vitrification, acclimatization, nature and origin of explants, problems with explant exudates, disposable culture vessels, induction of rooting, artificial substrates, rockwool, autotrophic micropropagation, storage and cryopreservation of cultures, quiscent somatic embryos, synthetic seed, automated systems, artificial intelligence, use of robots in transplantation, information management, bioreactors, and computerized greenhouses.

High-Tech and Micropropagation II comprises 27 chapters on American sweetgum, black locust, birches, cedar, cocoa, larches, Persian walnut, poplars, *Quercus*, sandalwood, *Sorbus*, apple, pear, peach, plum, cherry, *Citrus*, raspberry, strawberry, grapevine, avocado, kiwi, Japanese persimmon, papaya, pineapple, date palm, and olive tree.

High-Tech and Micropropagation III comprises 34 chapters on vegetable and tuber crops, medicinal and aromatic plants, fiber and oilseed crops, and grasses. The specific plants dealt with include: asparagus, lettuce, horseradish, cucumber, potato, sweet potato, artichoke, white yam, cardamom, fennel, celery, thyme, leek, mentha, turmeric, lavender, agave, yucca, cotton, jute, sunflower, ryegrass, zoysiagrass, and various species of *Aconitum, Artemisia, Camelia, Centaurium, Digitalis, Dioscorea, Glehnia, Parthenium,* and *Pinella.*

High-Tech and Micropropagation IV comprises 29 chapters on ornamental species of *Alstromeria, Antirrhinum, Begonia, Chrysanthmum, Cornus, Euphorbia, Gardenia, Gladiolus, Hyacinthus, Impatiens, Iris, Lycoris, Nemetanthus, Phalaenopsis, Paeonia, Pelargonium, Rhododendron, Ruscus, Saintpaulia, Senecio, Syringa,* orchids, cacti, roses, and Boston ferns.

These books will be of use to advanced students, research workers, and teachers in horticulture, forestry, tissue culture, and plant biotechnology in general, and also to individuals interested in commercial micropropagation.

New Delhi, September 1991

Professor Y. P. S. Bajaj
Series Editor

Contents

Section I General Aspects, Factors Affecting Micropropagation, Meristem Culture, Somatic Embryogenesis, Grafting, Vitrification, Acclimatization, Disposable Culture Vessels

I.8 Acclimatization of Micropropagated Plants
T. Kozai (With 7 Figures)

I.9 The Effect of the Nature and Origin of Explants on Micropropagation
D. Cachiţă-Cosma (With 9 Figures)

I.10 Problems with Explant Exudation in Micropropagation
J. E. Preece and M. E. Compton (With 1 Figure)

I.11 Woody Plant Micropropagation with Cytokinins
J. W. Einset (With 4 Figures)

I.12 The Effect of the Culture Vessel on Micropropagation
K. K. Kavanagh, A. P. Drew, and C. Maynard
(With 4 Figures)

I.13 Disposable Film Culture Vessels
M. TANAKA (With 17 Figures)

Section II Rooting, Artificial Substrates, Autotrophic Micropropagation, Nursery Planting

II.1 Rooting of in Vitro Cuttings
CH. MONCOUSIN (With 4 Figures)

II.2 Rooting of in Vitro-Produced Shoots in Nonsterile – an Inexpensive and Efficient Technique for en masse Micropropagation
S. B. RAJBHANDARY and Y. P. S. BAJAJ (With 2 Figures)

II.3 Use of Artificial Substrates for in Vitro Propagation
P. BROCHARD (With 6 Figures)

II.6 In Vitro Nursery System for Vegetable Crops – Tomato and Melon
S. Ohki, K. Nasuda, Y. Mori, and H. Katsuta
(With 9 Figures)

Section III Storage/Cryopreservation of Cultures, Quiescent Somatic Embryos, Artificial Seed

III.1 Storage and Cryopreservation of in Vitro Cultures
Y. P. S. Bajaj (With 7 Figures)

III.2 Quiescence and Dormancy in Somatic Embryos
D. J. Gray and A. Purohit (With 3 Figures)

III.3 Artificial Seeds – Encapsulated Somatic Embryos
K. Redenbaugh, J. Fujii, D. Slade, P. Viss, and M. Kossler
(With 9 Figures)

Section IV Automation, Artificial Intelligence, Robots in Transplanting, Information Management, Bioreactors, Computerized Greenhouses

IV.1 Automated Systems
B. Tisserat (With 6 Figures)

IV.2 Artificial Intelligence Technology to Control Plant Environments
Y. Harazono (With 15 Figures)

IV.3 Robots in Transplanting
G. E. Miles and L. J. Kutz (With 15 Figures)

IV.4 Computing and Information Management in Micropropagation
M. Hempel (With 5 Figures)

List of Contributors

ATTREE, S. M., Department of Biology, University of Saskatchewan, Saskatoon, Saskatchewan, Canada, S7H 0WO

BAJAJ, Y. P. S., A-137 New Friends Colony, New Delhi 110065, India

BROCHARD, P., Laboratoire de Génétique Appliquée, Challet-Herault Créations, BP Route de Cholet, 49340 Nuaillé, France

BOT, G. P. A., Institute of Agricultural Engineering (IMAG), Department of Horticulture, Wageningen Agricultural University, Postbus 43, 6700 AA Wageningen, The Netherlands

CACHIŢĂ-COSMA, D., Biological Research Centre, I.C.B. Str. Republicii no. 48, 3400 Cluj-Napoca, Romania

CASSELLS, A. C., Plant Biotechnology (UCC) Ltd., University College, Cork, Ireland

CHALLA, H., Institute of Agricultural Engineering (IMAG), Department of Horticulture, Wageningen Agricultural University, Postbus 43, 6700 AA Wageningen, The Netherlands

CHARLES, A., Ball Floraplant, P.O. Box 335, West Chicago, IL 60185, USA

COMPTON, M. E., Department of Plant and Soil Science, Southern Illinois University, Carbondale, IL 62901-4415, USA

DATE, K., Yamamoto Agricultural Corporation, 532 Yatoutajima Toyoda-Chou, Iwata-Gun Shizuoka-Prefecture, Japan

DICKENS, C. W. S., University of Natal/FRD, Research Unit for Plant Growth and Development, Department of Botany, University of Natal, P.O. Box 375, Pietermaritzburg 3200, Republic of South Africa

DREW, A. P., SUNY College of Environmental Science and Forestry, Syracuse, NY 13210, USA

EINSET, J. W., EniChem Americas, 2000 Princeton Park Corporate Center, Monmouth Junction, NJ 08852, USA (Present address: 151 North Union St., Lambertville, NJ 08530, USA)

Fowke, L. C., Department of Biology, University of Saskatchewan, Saskatoon, Saskatchewan, Canada, S7H 0WO

Fujii, J., University of Chicago, Chicago, IL, USA

Gaspar, T., Hormonologie fondamentale et appliquée, Institut de Botanique (B 22), Université de Liège, Sart Tilman, 4000 Liège, Belgium

Gray, D. J., Central Florida Research and Education Center, Institute of Food and Agricultural Sciences, University of Florida, 5336 University Avenue, Leesburg, FL 34748, USA

Harazono, Y., National Institute of Agro-Environmental Sciences, Division of Agrometeorology, Kannondai 3-1-1, Yatabe-machi, Tsukuba-gun, Ibaraki-ken 305, Japan

Hempel, M., Plant World Explorations, P.O. Box 1210, Bowral, NSW 2576, Australia

Katsuta, H., Laboratory of Horticulture and College Farm, Fukui Prefectural College, 97-21-3, Obatake-cho, Fukui 910, Japan

Kavanagh, K., Department of Forest Science, Oregon State University, Corvallis, OR 97330, USA

Kossler, M., Calgene Inc., 1920 Fifth Street, Davis, CA 95616, USA

Kozai, T., Department of Horticultural Engineering, Faculty of Horticulture, Chiba University, Matsudo, Chiba 271, Japan

Kutz, L. J., Agricultural Engineering Department, Auburn University, Auburn, AL 36849, USA

Maynard, C., SUNY College of Environmental Science and Forestry, Syracuse, NY 13210, USA

Miles, G. E., Department of Agricultural Engineering, Purdue University, West Lafayette, IN 47907-1146, USA

Moncousin, Ch., Centre Horticole de Lullier, Ecole d'ingénieurs (ETS), 1254 Lullier, Switzerland

Moore, R., Department of Biological Sciences, Wright State University, Dayton, OH 45435, USA

Mori, Y., Laboratory of Horticulture and College Farm, Fukui Prefectural College, 97-21-3, Obatake-cho, Fukui 910, Japan

Nasuda, K., Laboratory of Horticulture and College Farm, Fukui Prefectural College, 97-21-3, Obatake-cho, Fukui 910, Japan

Ohki, S., Laboratory of Horticulture and College Farm, Fukui Prefectural College, 97-21-3, Obatake-cho, Fukui 910, Japan

Preece, J. E., Department of Plant and Soil Science, Southern Illinois University, Carbondale, IL 62901-4415, USA

Purohit, A., Central Florida Research and Education Center, Institute of Food and Agricultural Sciences, University of Florida, 5336 University Avenue, Leesburg, FL 34748, USA

Rajbhandary, S. B., National Herbarium and Plant Laboratories, Godawari, Kathmandu, Nepal

Redenbaugh, K., Calgene Inc., 1920 Fifth Street, Davis, CA 95616, USA

Slade, D., University of Seattle, Seattle, WA, USA

Sonneveld, C., Glasshouse Crops Research Station, POB 8, 2670 AA Naaldwijk, The Netherlands

Takayama, S., Department of Biological Science and Technology, Tokai University, 317 Nishino, Numazu, Shizuoka 410-03, Japan

Tanaka, M., Department of Horticulture, Faculty of Agriculture, Kagawa University, Miki-cho, Kagawa 761-07, Japan

Tisserat, B., Fruit and Vegetable Chemistry Laboratory, USDA/ARS, 263 South Chester Ave., Pasadena, CA 91106, USA

Van Staden, J., University of Natal/FRD Research Unit for Plant Growth and Development, Department of Botany, University of Natal, P.O. Box 375, Pietermaritzburg 3200, Republic of South Africa

Viss, P., Plant Research Laboratory, Modesto, CA, USA

Wang, P.-J., Ball Floraplant, P.O. Box 335, West Chicago, IL 60185, USA

Section I
General Aspects, Factors Affecting Micropropagation, Meristem Culture, Somatic Embryogenesis, Grafting, Vitrification, Acclimatization, Disposable Culture Vessels

I.1 Automated Micropropagation for en masse Production of Plants

Y.P.S. Bajaj[1]

1 General Account

Micropropagation of plants is a multibillion dollar industry being practiced in hundreds of small and large nurseries and biotech labs throughout the world. Presently, it is the only component of plant biotechnology which has been commercially exploited on such a large scale. There are about 1000 plant species that can now be micropropagated (Murashige 1989). Micropropagation has been employed for about half a century, especially for ornamental plants (see Ammirato et al. 1989). Millions of plants are produced, sold locally, or exported to different countries (Chu and Kurtz 1989). Presently, interest has been extended to agricultural/plantation crops (e.g., potato, strawberry, oil palm, banana, etc.), medicinal and aromatic plants (Bajaj et al. 1988), and trees (see Bajaj 1986a, 1989a, 1991), (especially those which are difficult to propagate. Micropropagation of fruit and forest trees offers not only means for the mass multiplication of existing stocks of germplasm for biomass energy production, but also for the conservation of important, elite, and rare plants that are threatened with extinction.

For the clonal propagation of such plants, afforestation, and orchard plantations, a large number of plants are needed. Conventional micropropagation is time-consuming, labor-intensive, and sometimes may be unsafe for the workers, e.g., due to use of poisonous mutagens, radioactive chemicals, etc. Moreover, the possibility of contamination, resulting in the high cost of plantlets, is increased.

During the last decade, there has been an increased interest in problems affecting scale-up operations and cost reduction of plant production for commercial micropropagation (Boxus 1985; Sluis and Walker 1985; De Bry 1986; Debergh 1987; Giles and Morgan 1987; Levin et al. 1988; Pierik 1988; Cassells 1989; Chu and Kurtz 1989). To overcome some of the above-mentioned problems and to increase the efficiency and scale up, the production of plants by means of automation has been employed. Robots, microcomputers, and various other mechanized systems are being developed and refined for surface sterilization of tissues/explants, disposable culture vessels, media distribution, inoculation, and transference of plants to pots or to fields. The automatic control of environment and growth conditions in the greenhouse is another area of tremendous commercial importance.

[1]Former Professor of Tissue Culture, Punjab Agricultural University, Ludhiana, India. (Present address: A-137 New Friends Colony, New Delhi 110065, India)

Biotechnology in Agriculture and Forestry, Vol. 17
High-Tech and Micropropagation I (ed. by Y.P.S. Bajaj)

This introductory chapter deals with some of the problems and factors affecting micropropagation, and various aspects of automated micropropagation. The information has been derived mainly from various chapters in this book.

2 Problems and Factors Affecting Micropropagation

The success of a micropropagation program depends on a number of factors, such as the type and origin of the explant, sterilization method, media, and culture conditions. The problems often encountered are browning of the tissue, vitrification, noninduction of rooting, difficulty in acclimatization, and high mortality upon transfer to soil. Some of these aspects were discussed earlier (Withers and Bajaj 1982), and are summarized here.

1. Origin and Nature of Explant. The age, nature, origin, and the physiological state of the inoculum/explant and seasonal variation play a crucial role in the establishment of cultures and subsequent plant regeneration; some of these aspects have been discussed in Chap. I.9, this Vol.). In general, explants from young and juvenile plants are easier to grow than from mature and old trees.

Recently, Pierik (1990) has discussed the problems related to rejuvenation and maturation in micropropagation. Virus-free meristems and explants with shoot buds are used preferably to establish cultures. Lately, young leaves and immature embryos have been the material of choice. Cell suspensions of these tissues enable the production of somatic embryos in large quantities.

2. Browning of the Explants. Browning of the explant immediately after inoculation has been a serious problem, especially with the woody plant species. This has been attributed to phenolics and tannins in tissues (Compton and Preece 1988). This problem can be overcome by using, e.g., antioxidants, ascorbic acid, activated charcoal, PVP, cysteine, incubation in the dark, or frequent transfers to liquid media (see Chap. I.10, this Vol.).

3. Media and Growth Regulators. Various basal medial have been employed, but MS medium (Murashige and Skoog 1962) with minor modifications, e.g., supplemented with vitamins and sucrose, is most commonly used; however, its strength is often reduced to half. Basically, there are two modes of differentiation (1) *direct* formation of buds, plants/somatic embryos from the excised explants, and (2) *indirect* regeneration of plants via callusing. Direct regeneration ensures clonal fidelity, the second mode often shows genetic erosion. The mode of differentiation, i.e., organogenesis, embryogenesis, and rhizogenesis, is controlled by the growth regulators (Reinert et al. 1977). Higher concentrations of auxins generally induce undesirable callusing, whereas cytokinins regulate shoot bud formation. When cytokinins are incorporated into media for micropropagation by shoot multiplication, the artificially high cytokinin concentration produced within shoots stimulates continued growth and also inhibits apical dominance resulting in a branched axis (Einset 1985).

4. Vitrification. Vitrification is a serious problem in micropropagation. This term is generally used to characterize hyperhydric malformations. The vitrified plants appear turgid, watery at the surface and hypolignified, translucent, less green, and are easy to break (Gasper et al. 1987). Vitrification mostly occurs during intensive multiplication. The leaves are thick with large intracellular spaces. Vitrification is reportedly caused by the high level of salts, especially ammonium nitrate as in MS medium (Daguin and Letouzé 1986), liquid medium (Druart et al. 1984), and cytokinins have also been implicated (Lesham et al. 1988). Thus, to avoid vitrification the use of agar and low levels of cytokinins, ammonium nitrate, and dilute MS minerals have been recommended (for details, see Chap. I.7, this Vol.).

5. Induction of Rooting and Acclimatization. Sometimes there is a high mortality rate due to failure of root induction on the in vitro shoots or due to subsequent problems of acclimatization. Various factors affecting adventitious rhizogenesis have been discussed (Németh 1986; Moncousin 1987). To induce rooting the cuttings are often grown aseptically in a medium containing auxins such as IAA, IBA, or NAA. However, this practice can be avoided by adopting a simple and inexpensive method, i.e., by inserting in vitro shoots directly into a nonsterile sand. This method has been successful in a number of plant species, e.g., *Ficus, Tactona,* potato, etc. (see Rajbhandary and Bajaj 1991).

The heavy loss of in vitro plants during transfer from culture vessels to soil has stimulated considerable research on acclimatization. To increase the survival rate of plants, humidity tents, antitranspirants, supplementary light, and CO_2 enrichment have been employed (see Chap. I.8, this Vol.). A microcomputerized acclimatization unit has therefore been developed to provide better environmental control, high survival, and rapid growth of plants (Kozai et al. 1987).

3 Automation – Robots, Microcomputers, and Artificial Intelligence

A robot is a machine which mimics the motions and functions of a living being (De Bry 1986). It is the intelligent control of a system (i.e., a group of procedures, often using machines as tools to increase productivity and reduce costs), whereas *mechanization* is the use of nonintelligent machinery to replace human activities (Tisserat, see Chap. IV.1, this Vol.). There are a number of opportunities to use automation in plant tissue culture operations (Harrell and Simonton 1986; Rowe 1986; Fujita 1989).

Robots have always been an object of interest combined with fantasy. Industrial robots doing routine, and sometimes complicated jobs have been in use for quite some time. The increased development of artificial intelligence and microcomputers (Hempel 1989) now provides robots with "thinking, reasoning, learning and decision making abilities" (De Bry 1986). These abilities have been considerably extended to manipulate biological systems (ASAE 1983), including plant cell and tissue culture.

The history, development, classification, and other general aspects of robots have been discussed by De Bry (1986). According to him, robots can be grouped into three categories: (1) *first generation robots* are immobile and have only an arm attached to a stationary base plus some form of intelligence. They are intended to do a specific job. (2) *Second generation robots* have multitask capabilities. They are more intelligent and have sensory perception, most notably vision; they can detect environmental variation and respond according to their own decision-making faculty. (3) *Third generation robots* are much more complex; they can perceive in three dimensions. Some of these are mobile and can, for instance, pick fruit in an orchard (ASAE 1985).

3.1 Automated Surface-Sterilization System

Efforts are being made to surface sterilize tissues/explant by using automated, computer-controlled systems. Maurice et al. (1985) constructed such a surface-sterilization system (SSS) in which agitation, addition and removal of disinfectant, sterile water rinses, and time of surface sterilization were automatically controlled by programmed instructions in the microcomputer. Comparative tests between manual and automated surface-sterilization procedures with seeds of *Cucurbita pepo* and *Phoenix dactylifera*, and petiole explants of *Anethum graveolens* were conducted and usually no significant difference in effectiveness was found. Depending on the disinfectant employed, the automated method was found to be more effective in the surface sterilization of date seeds in comparison to the manual method. The SSS thus constructed was inexpensive and provided a labor-saving and reproducible method to sterilize large numbers of plant tissues.

3.2 Disposable Culture Vessels

Culture vessels are known to affect the growth and development of in vitro cultures (Kavanagh et al., Chap. I.12, this Vol.). It is required that culture vessels have high light transmittance, low vapour transmittance, and chemical inertness for growth and development of tissues. Glass vessels satisfy these properties; however, they are relatively expensive, less durable, must be thoroughly cleaned/sterilized, and are liable to break during transport. To overcome these discrepancies and problems, Tanaka et al. (1988a,b) developed polymer film (Neoflon film) bags. These films have superior properties, such as thermal stability, chemical inertness, and electrical reliability. They have a melting point of 265–310 °C, and hence can be autoclaved safely. These films also have a low water vapor transmittance and high light transmittance similar to glass. Two types of film culture vessels for micropropagation have been developed (for details, see Chap. I.13, this Vol.):

1. Box-shaped vessel (culture pack);
2. Envelope-shaped vessel (culture bag).

These vessels are successfully used for micropropagation of *Spathiphyllum* and *Cymbidium*. The plants grown in the culture pack can be acclimatized by gradually making slits at the top of the film. Liquid medium is drained off by making several holes at the bottom. The cost of plantlets with this system is also expected to be reduced.

3.3 Automated Plant Culture System and Bioreactors

In vitro cultures require periodic transfer to fresh media. To avoid or delay this and to extend the culture period, an automated plant culture system (APCS) was developed by Tisserat and Vandercook (1985). This system accommodated independent or multiple concurrent growth of cultures. It consisted of silicon tubing, two impeller pumps, two glass medium reservoir bottles, a three-way stainless steel valve, a plant culture chamber, and an interface module containing relay boards. The control of this system was obtained through a microcomputer, which controlled medium introduction, evacuation, and replenishment in a sterile environment. Such an automated system was inexpensively constructed and provided a labor-saving, long-term method to culture plants in vitro. The excised shoot tips of *Aster*, cultured in automated and manual systems, elongated and flowered within 30–45 days after planting. However, both shoots and flowers were larger when cultured in the automated system compared to those growth in culture tubes. Subsequently, a microcomputer-controlled apparatus was constructed (Tisserat and Vandercook 1986) which was used to promote proliferation and growth of orchid shoot tips into whole plants under sterile conditions. Growth of such shoot tips exceeded that occurring using agar medium and manual transference. The culture chamber of the automated system is 10–100 times larger than most culture vessels. The early APCS required gas sterilization (Tisserat and Vandercook 1986), although later models are completely autoclavable (see Chap. IV.1, this Vol.).

In September 1987, at a conference on electronics management of living plants, held in Monte Carlo, Monaco, a number of papers were presented on automation and mechanization in micropropagation. Some of these works are cited here.

A prototype of an intelligent robotic system which can see, and pick up and transfer seedlings has been developed (Deleplanque et al. 1985). Image processing permits this system to discriminate between acceptable and imperfect plantlets. It can transfer plantlets with an efficiency of 1000 plantlets per hour. Rowe et al. (1987) designed a robot work station based on industrial manipulators and a vision system interfaced by specialized software. Such a manipulator could cut and plant a piece of potato tuber in less than 3 s.

Efforts are now being made to substitute human sight by computerized image analysis and video camera. Chee and Cantliffe (1988) and Grand d'Esnon et al. (1987) reported the application of image analysis in the sorting of propagules during artificial seed production. Zryd (1987) used two-dimensional analysis for selection of plant cells and protoplasts. Meyer et al. (1988) compared the effectiveness of image analysis and visual screening methods for quality evaluation of micropropagated plants. Robots have also been developed for transplanting seedlings and in vitro plants (Kutz et al. 1987; Ting et al. 1988; also see Chap. IV.3, this Vol.).

Levin et al. (1988) developed a vitromatic system which is capable of sorting, sizing, and dispensing somatic embryos into compartmentalized culture vessels. The embryos are allowed to grow into plantlets in the culture vessels before transplanting into greenhouse trays by an automated transplanting machine. In this system 12 000 plants can be transferred to greenhouse trays in 1 h.

The automated system developed for *Radiata* pine supports multiplication of meristems, which are then grown into shoot hedges under conditions which allow the automatic control of the liquid medium and the gaseous environments (Aitken-Christie and Jones 1987).

Recently, Miwa (1989) developed a "Mericlone Robot" which transferred the tissues aseptically. It can transplant an in vitro grown plantlet in 1 min. For more information on acclimatization and robotic transplantation, see Section 3.4 (this Chap.).

Bioreactors are being developed primarily with two purposes:

1. Industrial production of secondary metabolites through cell suspension/root culture (Wagner and Vogelmann 1977; Tanaka et al. 1983; Smart and Fowler 1984; Ushiyama et al. 1986; Fowler 1987; Fujita 1988; Hashimoto and Azechi 1988).
2. Large-scale production of somatic embryos/plants (Takayama 1984; Styler 1985; Stuart et al. 1987; Preil et al. 1988; Akita and Takayama 1988).

Different types of bioreactors have been used by various workers, e.g., rotating drum bioreactor (Tanaka et al. 1983), gaseous phase bioreactor (Ushiyama et al. 1984), airlift column bioreactor (Smart and Fowler 1984), spin filter bioreactor (Styler 1985), light-introducing bioreactor, and other models (Margaritis and Wallace 1984). Some examples of the large-scale production of plant cells, include *Coleus blumei* 450 l (Ulrich et al. 1985), *Lithospermum erythrorhizon* 750 l (Fujita 1988), *Panax ginseng* 2000 l (Ushiyama et al. 1986), *Catharanthus roseus* 5000 l (Schiel and Berlin 1987), *Nicotiana tabacum* 20 000 l (Azechi et al. 1985), etc. Recently, Rittershaus et al. (1990) produced cell cultures of *Echinacea purpurea* and *Rauwolfia serpentina* in 75 000-l fermenters. This was the first time that production of plant cells on such a large scale was achieved.

The large-scale production of shikonin, an antibacterial compound, through cell cultures of *Lithospermum* is a well-known example of the application of plant biotechnology in pharmacy. Two sets of culture apparatuses are used consisting of a 200-l tank for cell growth (first stage) and a 750-l tank for shikonin derivative production (second phase; see Fujita 1988). *Panax ginseng*, a source of saponins used as a tonic and aphrodisiac, is another example. Ushiyama et al. (1986) studied the suspension cultures from root tissues in 2000-l bioreactors, and subsequently produced 100 kg callus per month. Takayama (1984), using a small bioreactor (4–20 l), produced 4000 to 20 000 plantlets within 1–2 months. For large-scale bioreactor production of gladioli, lilies, strawberry and begonia plants, see Chap. IV.5 (this Vol.). Akita and Takayama (1988) grew potato microtubers in a bioreactor. Stuart et al. (1987) grew alfalfa somatic embryos in bioreactors. Preil et al. (1988) studied the development of somatic embryos of *Euphorbia pulcherrima* in Biostat M; this bioreactor had silicone tubes for bubble-free aeration. The vitromatic system (Levin et al. 1988) enables the rapid multiplication of meristematic clusters in liquid medium in a bioreactor.

3.4 Artificial Substrates, Acclimation, and Robotic Transplantation

Though agar is used routinely for in vitro cultures, other substitutes and synthetic materials have been applied by various workers, i.e., plastic foam (Hussey 1980), peat plugs (Zimmerman et al. 1987), cotton plugs, glasswool, etc. In *Mimosa dealbata* rooting could not be induced on agar medium, but on the aerated substrate STRATAIR (polyurethane foam with active charcoal) wetted with a nutritive medium, this substrate gave a rooting rate of 40% (see Chap. II.3, this Vol.). Aldrufeu (1987) compared the use of substrates like cellulose, sand, perlite, etc. with agar for rooting of *Pelargo-*

nium, and observed that root hair developed in all the substrates except agar. Moreover, agar creates problems at the time of removal of in vitro plantlets, i.e., washing is difficult, the roots may break, and the contamination rate is also high.

To transfer in vitro plants, vermiculite, sand, perlite, and peat are commonly used in nurseries. In addition, different products such as plastic foam (McCown 1985), textile fibers (Brochard 1987), rockwool (Debergh 1987), and clods of pressed peat (Zimmerman et al. 1987) have been employed. Brochard (see Chap. II.3, this Vol.) used two types of artificial clods:

1. A MILAP plug made of polypropylene fibers.
2. A STRATAIR plug is composed of polyurethane foam, to which a fungicide is added during the preparation. This has been suitable for a number of plant species.

Rockwool is now being widely used as an artificial substrate in many countries, especially in the Netherlands where about 2500 ha of vegetables and flowers are grown on rockwool. Dutch growers harvest about 60 kg of cucumbers, 50 kg of tomato, and 20 kg of sweet peppers per m^2 annually (Sonneveld and Welles 1984, 1988). Rockwool is composed mostly of diabase and limestone. The cubes/slabs are very porous, have a high water retention, and different qualities and sizes are used for horticultural purposes. The Dutch standard system is based on rock wool slabs in bags of plastic film. At the top, the bags are provided with planting holes and at the bottom with slits to drain the surplus water. The young plants are propagated in rockwool cubes or plugs, and placed on top of the slabs, or in planting holes in the slabs.

Acclimatization is the climatic adaptation of organisms, especially the plant, which has been moved to a new environment (Conover and Poole 1987). A high percentage of plants is sometimes lost during transfer from test tube to soil. Thus, acclimatization of in vitro plantlets acquires great importance. Various aspects of acclimatization have been studied, e.g., Brainerd and Fuchigami (1981) and Wardel et al. (1983) studied the effect of humidity. Sutter and Hutzell (1984) used humidity tents and antitranspirants to increase the survival percentage of micropropagated plants in the acclimatization stage. Read and Fellman (1985) developed a controlled environment rooting facility for accelerating the acclimatization of woody ornamentals. Desjardins et al. (1988) showed that supplementary lighting and CO_2 enrichment in the acclimatization stage led to faster autotrophy and growth, and ultimately improved the productivity of asparagus, raspberry, and strawberry plants. Kozai (see Chap. I.8, this Vol.) adopted two approaches to solve the problem of acclimatization: (1) the control of the environments by using a microcomputer-control system, and (2) autotrophic (sugar-free) micropropagation. Micropropagated strawberry plantlets, cultured autotrophically for 28 days at the rooting and acclimatization stages, were about 1.7 times as heavy and more than 4 times as tall as plantlets cultured using conventional tissue cultured conditions.

Kozai and coworkers (Kozai et al. 1987; Hayashi et al. 1988) developed a microcomputer acclimatization unit which provided better environmental control, a higher percentage of survival, and rapid growth of plants. With such a unit, light intensity, relative humidity, air temperature, carbon dioxide concentration, air flow rate, and temperature of the nutrient solution could be controlled. There are several advantages of the acclimatization unit method over the conventional method, i.e., (1) a shorter culture period, (2) no need for transplanting during rooting and acclimatization, (3) more vigorous plants, and (4) no contamination loss during rooting.

Various types of robotic transplantation systems have been developed (Kutz et al. 1987; Ting et al. 1988; Levin et al. 1988). For details, see Chap. IV.3 (this Vol.) by Miles and Kutz.

3.5 Artificial Intelligence and the Climatic Control of Greenhouses

The climatic control of greenhouses for protected cultivation is a complex subject, since many factors such as temperature, light, air flow, humidity, CO_2 concentration, water/nutrient solution, etc. are involved. The regulation of these factors is rather difficult, since different crops with different requirements, involving different factors, have to be manipulated. Thus, automation has been introduced, though expensive it enables off-season production of pest- and pathogen-free vegetables, fruits and flowers, which bring higher prices, thus compensating for the expenditure incurred. Microcomputerized greenhouses are gaining popularity in many countries. Various factors, the control of greenhouse climate (Challa et al. 1988), and optimal greenhouse production under economical constraints have been discussed (Seginer 1989).

Microcomputers were employed to control the greenhouse environments in Japan as early as 1979 (Japan Greenhouse Horticulture Association 1988), and they are now being used routinely (Kozai et al. 1982; Harazono et al. 1984, 1988; Hashimoto and Fukuyama 1985; Kurata 1988) for a variety of plant species, i.e., *Cattleya, Oncidium, Phalaenopsis*, muskmelon, cucumber, and many other vegetables and fruits.

When the air temperature in the greenhouse exceeds the set point of room temperature for ventilation, the roof ventilators are opened first. When the air temperature drops below the set point, the ventilators are closed in reverse order. Furthermore, light, humidity, CO_2 concentration, water sprays, etc. are also automatically controlled (for details, see Chap. IV.2, 6, 7, this Vol.).

3.6 Storage/Cryopreservation of Cultures

In vitro cultures require periodic transfer to fresh media, which not only involves higher cost for manpower, chemicals, and culture vessels, but also the increased possibility of contamination, sometimes resulting in the complete loss of material. Moreover, such cultures are prone to genetic erosion resulting in a change in ploidy, mutations, gene amplification, and translocations. Thus, to avoid periodic transfers various methods have been employed to reduce the growth rate of culture, thus reducing the frequency of the subculture period. Of all these methods low-temperature storage (nonfrozen) has been applied to a large number of species (see Bajaj, Chap. III.1, this Vol.); moreover, such methods enable short-term storage.

For long-term conservation, cryopreservation has been recommended. The principle underlying cryopreservation is to bring the cultures to a zero level of metabolism, which is attained by freezing the cultures in liquid nitrogen (–196 °C). Complete plants have been regenerated from protoplasts, cells, tissues, and organs cryopreserved for various lengths of time. This subject has been reviewed by the author on plant protoplasts (Bajaj 1989b), medicinal and aromatic plants (Bajaj 1988), vegetatively propagated crops (Bajaj 1990a), cereals (Bajaj 1990b; Bajaj and Sala 1991), legumes and oilseed crops (Bajaj 1990c).

Cryopreservation can be of immediate use for (1) storage of germplasm of vegetatively propagated crops, (2) plants with recalcitrant seeds, and (3) storage of selected somaclonal cell lines.

3.7 Cost Reduction in Plant Production and Storage

The most important aspect of commercial micropropagation is the economics involved and the cost of production of a plantlet, which in some cases is exorbitant. The cost of labor involved, especially in the developed countries, varies from 40% (Sluis and Walker 1985), 60% (Donnan 1986), to more than 70% (Anderson and Meagher 1977; Chu and Kurtz 1989). Although exact information is not revealed by the commercial growers, various workers have tried to give a fairly accurate account of the situation (Sluis and Walker 1985; De Bry 1986; Donnan 1986; Kozai et al. 1987; Long 1987; Chu and Kurtz 1989; Harper 1989; Jones 1989; Levin and Vasil 1989). The following presents information extracted from these publications.

A number of measures have been taken and attempts made to reduce the cost of plant production by manipulating methods, culture conditions, media, and adopting mechanical gadgets and resorting to automation. Some of these measures have already been discussed in this chapter. For instance, the use of cheap and disposable culture vessels is an aspect to be considered further. The disposable polymer film culture bags designed by Tanaka et al. (1988b) are durable, thermally stable, and chemically inert. The plants grown in the culture packs can be gradually acclimatized, and the cost of plantlets is expected to be reduced.

Autotrophic micropropagation is an area which needs to be explored to reduce the cost of production, as no sugar is required in the medium. Kozai et al. (1987) have demonstrated that small plantlets are able to change from heterotrophic to mixotrophic, and then to autotrophic provided the CO_2 concentration is increased. In another procedure described by Vanderschaeghe and Debergh (1987) medium can be controlled by adding a layer of liquid medium over the exhausted agar medium (Maene and Debergh 1985), and cooling the bottom of the culture vessel to reduce evaporation of the medium. To save on the cost of media, Weathers and Giles (1988) used "Mistifiers" which regulate the exposure to the medium as well as the gaseous environment by intermittently bathing the culture in a fine mist of liquid medium. The use of liquid media in bioreactors for the large-scale production of somatic embryos/plants has received encouraging response (Akita and Takahashi 1988). Cost reduction at various stages of growth, development, multiplication, and transfer of plantlets to the greenhouse has been dealt with in detail by Levin and Vasil (1989). According to them, two systems, i.e., the vitromatic system (Levin et al. 1988) and the shoot hedge system developed for *Radiata* pine (Aitken-Christie and Jones 1987), provide optimum opportunities for significant cost reductions. According to De Bry (1986), the estimates suggest that a reasonably priced automated production could become almost three times cheaper than conventional techniques.

The direct transfer of in vitro produced shoots to nonsterile sand (Rajbhandary and Bajaj 1991) has been successful in a number of plant species. According to this method, the intermediate step of first transferring the shoots to an auxin-containing medium is skipped, resulting in considerably less cost. Storage of germplasm is an important aspect of micropropagation. Cryopreservation of particular cultures and somaclones (Bajaj 1986b) would avoid unnecessary, periodic transfers, and thus would help to save on labor, cost, etc. Synthetic seed could thus become an accepted method for storage (Redenbaugh et al. 1987).

4 Summary and Conclusions

Conventional micropropagation is labor-intensive and time-consuming, resulting in higher cost of plant production. To overcome this and to produce plants in large quantities automation/mechanization is being introduced in commercial micropropagation laboratories. Various types of robots and microcomputers have been developed for surface sterilization of tissues, media distribution, inoculation, and transfer of plants to soil. Moreover, computerized greenhouses have also helped to control growth environments.

The use of artificial intelligence, image analysis, and video cameras, bioreactors, mistifiers, automatic production lines, synthetic seed, storage, and cryopreservation of genetic stocks as well as computerization of information management will facilitate micropropagation and protected cultivation. However, the cost factor is still the major constraint and considerable efforts are being made to reduce cost. Judging from the enthusiasm and interest in biotechnology enterprises, it is envisaged that routine jobs will be done soon by robots for industrial micropropagation.

References

Aitken-Christie J, Jones C (1987) Towards automation: *Radiata* pine shoot hedges in vitro. Plant Cell Tissue Organ Cult 8:185–196

Akita M, Takayama S (1988) Mass propagation of potato tubers using jar fermentor techniques. Acta Hortic 230:55–61

Aldrufeu A (1987) Rooting and acclimatization of *Pelargonium zonale* plantlets. Acta Hortic 212:361–366

Ammirato PV, Evans DA, Sharp WR, Bajaj YPS (eds) (1989) Handbook of plant cell culture, vol 5. Ornamental species. McGraw-Hill, New York

Anderson WC, Meagher GW (1977) Tissue culture propagation of broccoli, *Brassica oleracea*, for use in F_1 hybrid seed production. Am Soc Hort Sci 102:69–73

ASAE (ed) (1983) Robotics and intelligent machines in agriculture. Am Soc Agric Eng (ASAE) Publ 4–84: 155 pp

ASAE (ed) (1985) Agri-Mation I. ASAE Publ 01–85, 375 pp

Azechi S, Hashimoto T, Nakashizuka M, Murata M, Ishii H (1985) Batch culture of tobacco plant cells in high concentration medium. Jpn Tob Salt Publ Corp 20:61–71

Bajaj YPS (1986a) Biotechnology of tree improvement for rapid propagation and biomass energy production. In: Bajaj YPS (ed) Biotechnology in agriculture and forestry, vol 1. Trees I. Springer, Berlin Heidelberg New York, pp 1–23

Bajaj YPS (1986b) Cryopreservation of potato somaclones. In: Semal J (ed) Somaclonal variations and crop improvement. Martinus Nijhoff, Dordrecht, pp 244–250

Bajaj YPS (1988) Cryopreservation and the retention of biosynthetic potential in cell cultures of medicinal and alkaloid-producing plants. In: Bajaj YPS (ed) Biotechnology in agriculture and forestry, vol 4. Medicinal and aromatic plants I. Springer, Berlin Heidelberg New York, pp 169–187

Bajaj YPS (ed) (1989a) Biotechnology in agriculture and forestry, vol 5. Trees II. Springer, Berlin Heidelberg New York

Bajaj YPS (1989b) Cryopreservation of plant protoplasts. In: Bajaj YPS (ed) Biotechnology in agriculture and forestry, vol 8. Plant protoplasts and genetic engineering I. Springer, Berlin Heidelberg New York, pp 97–106

Bajaj YPS (1990a) Cryopreservation of germplasm of vegetatively propagated crops. Bull Soc bot Fr 137:99–114

Bajaj YPS (1990b) Cryopreservation of germplasm of wheat. In: Bajaj YPS (ed) Biotechnology in agriculture and forestry, vol 13. Wheat. Springer, Berlin Heidelberg New York, pp 669–681

Bajaj YPS (1990c) Cryopreservation of germplasm of legumes and oilseed crops. In: Bajaj YPS (ed) Biotechnology in agriculture and forestry vol 10. Legumes and oilseed crops I. Springer, Berlin Heidelberg New York, pp 49–62

Bajaj YPS (ed) (1991) Biotechnology in agriculture and forestry, vol 16. Trees III. Springer, Berlin Heidelberg New York

Bajaj YPS, Sala F (1991) Cryopreservation of germplasm of rice (*Oryza sativa* L.). In: Bajaj YPS (ed) Biotechnology in agriculture and forestry, vol 14. Rice, Springer, Berlin Heidelberg New York, pp 553–571

Bajaj YPS, Furmanowa M, Olszowska O (1988) Biotechnology of the micropropagation of medicinal and aromatic plants. In: Bajaj YPS (ed) Biotechnology in agriculture and forestry, vol 4. Medicinal and aromatic plants I. Springer, Berlin Heidelberg New York, pp 60–103

Boxus Ph (1985) Big losses due to bacterial contamination can be avoided in mass propagation scheme. Symp In vitro problems related to mass propagation of horticultural crops, Gembloux

Brainerd KE, Fuchigami LH (1981) Acclimatization of aseptically cultured apple plants to low relative humidity. J Am Soc Hort Sci 106:515–518

Brochard P (1987) Use of MILAP substrates for in vitro rooting and coming out of plants obtained by micropropagation. Acta Hortic 212:322–323

Cassells AC (1989) Pitfalls for nurserymen in micropropagation and how to avoid them. Proc Int Plant Prop Soc 38

Challa H, Bot GPA, Nederhoff EM, Van de Braak NJ (1988) Greenhouse climate control in the nineties. Acta Hortic 230:459–470

Chee RP, Cantliffe DJ (1988) Propagule sorting by computerized vision. HortSci 23:173

Chu IYE, Kurtz SL (1989) Commercialization of plant micropropagation. In: Ammirato PV, Evans DA, Sharp WR, Bajaj YPS (eds) Handbook of plant cell culture, vol 5. Ornamental species. McGraw-Hill, New York, pp 126–164

Compton ME, Preece JE (1988) Effects of phenolic compounds on tobacco callus and blackberry shoot cultures. J Am Soc Hort Sci 113:160–163

Conover CA, Poole RT (1987) Acclimatization of indoor foliage plants. Hort Rev 6:120–154

Daguin F, Letouzé R (1986) Ammonium-induced vitrification in cultured tissues. Physiol Plant 66:94–98

Debergh PC (1987) Improving micropropagation. IAPTC Newslett 51:1–10

Debergh PC, Maene LJ (1981) A scheme for commercial propagation of ornamental plants by tissue culture. Sci Hortic 14:335–345

De Bry L (1986) Robots in plant tissue culture: an insight. IAPTC Newslett 49:2–22

Deleplanque H, Bonnet P, Postaire JPG (1985) An intelligent robotic system for in vitro plantlet production. Revisec (IFS Publ) 5:305–314

Desjardins Y, Laforge F, Lussier C, Gosselin A (1988) Effect of CO_2 enrichment and high photosynthetic photon flux on the development of autotrophy and growth of tissue-cultured strawberry, raspberry and asparagus plants. Acta Hortic 230:45–53

Donnan A (1986) Determining and minimizing production costs. In: Zimmerman RJ, Griesbach FA, Hammerschlag F, Lawson RH (eds) Tissue culture as a plant production system for horticultural crops. Martinus Nijhoff, Dordrecht, pp 167–173

Druart Ph, Paques M, Boxus Ph (1984) Morphologie anormale semblable aux symptomes de vitrification reccontres au cours de la micropropagation de *Prunus glaudulosa* Thurn. var. sinensis. Influence de l'AVG, de l'AIBA et d'$AgNO_3$. Arch Int Physiol Biochem 92:V22–V23

Einset JW (1985) Role of cytokinins in woody plant micropropagation. Proc Int Plant Prop Soc 35:608–615

Fowler MW (1987) Process systems and approaches for large scale plant cell culture. In: Green CE et al. (eds) Plant tissue and cell culture. Alan R Liss, New York, pp 459–471

Fujita N (1989) Application of robotics to micropropagation system. In Vitro Cell Dev Biol 25(3):22A

Fujita Y (1988) Shikonin: production by plant (*Lithospermum erythrorhizon*) cell cultures. In: Bajaj YPS (ed) Biotechnology in agriculture and forestry, vol 4. Medicinal and aromatic plants I. Springer, Berlin Heidelberg New York, pp 225–236
Gaspar Th, Kevers C, Debergh P, Maene L, Paques M, Boxus Ph (1987) Vitrification: morphological, physiological, and ecological aspects. In: Bonga JM, Durzan DJ (eds) Cell and tissue culture in forestry, vol 1. Martinus Nijhoff, Dordrecht, pp 152–166
Giles KL, Morgan WM (1987) Industrial scale plant micropropagation. TIBTECH 5:35–39
Grand d'Esnon, Harrell R, Chee R (1987) Propagule sorting by computerised vision during artificial seed production. Proc Moet-Hennessy Conf Electronics management of living plants, Monte Carlo, p 22
Harazono Y, Taenaka T, Yabuki K (1984) Optimizing environmental control of greenhouse climate for cucumber growth by means of "learning control method". Acta Hortic 148:259–265
Harazono Y, Kamiya H, Yabuki K (1988) A control method based on an artificial intelligence technique and its application for controlling plant environment. Acta Hortic 230:209–214
Harper PC (1989) Myth and money in micropropagation. Proc Int Plant Prop Soc 38:232–246
Harrell R, Simonton W (1986) Automation opportunities in plant tissue culture operations. Am Soc Agric Eng Pap 86: p 1596
Hashimoto T, Azechi S (1988) Bioreactors for the large-scale culture of plant cells. In: Bajaj YPS (ed) Biotechnology in agriculture and forestry, vol 4. Medicinal and aromatic plant I. Springer, Berlin Heidelberg New York, pp 104–122
Hashimoto Y, Fukuyama T (1985) Some speaking plant approaches to the synthesis of control system in the greenhouse. Acta Hortic 174:219–226
Hayashi T, Nakayama M, Kozai T (1988) An application of the acclimatization unit for growth of carnation explants, and for rooting and acclimatization of plantlets. Acta Hortic 230:189–194
Hempel M (1989) Microcomputers in micropropagation. Acta Hortic 226:699–702
Hussey G (1980) In vitro propagation. In: Ingram DS, Helgeson JP (eds) Tissue culture methods for plant pathologists. Blackwell, Oxford, pp 51–61
Japan Greenhouse Horticulture Association (1988) Areas of glasshouses, plastic houses etc. for horticulture
Jones JB (1989) Economic considerations in tissue culture propagation of ornamentals. In: Ammirato PV, Evans PV, Sharp WR, Bajaj YPS (eds) Handbook of plant cell culture, vol 5. Ornamental species. McGraw-Hill, New York, pp 165–178
Kevers C, Coumans M, Coumans-Gilles MF, Gaspar Th (1984) Physiological and biochemical events leading to vitrification in plants cultured in vitro. Physiol Plant 61:69–74
Kozai T, Itou T, Inayama M, Kobayashi H (1982) Development of microcomputer-based environment control and data-logging system for greenhouses and its field test. J Agric Meteorol 38:45–50
Kozai T, Iwanami Y, Fujiwara K (1987) Effects of CO_2 enrichment on the plantlet growth during the multiplication stage. Plant Tissue Cult Lett 4(1):22–26
Kurata K (1988) Greenhouse control by machine learning. Acta Hortic 230:195–200
Kutz LJ, Miles GE, Hammer PA, Krutz GW (1987) Robotic transplanting of bedding plants. Trans ASAE 30(3):586–590
Lesham B, Shaley DP, Izhar S (1988) Cytokinin as an inducer of vitrification in melon. Ann Bot 61:255–260
Levin R, Vasil IK (1989) Progress in reducing the cost of micropropagation. IAPTC Newslett 59:2–12
Levin R, Gaba V, Tal B, Hirsch S, DeNola S, Vasil IK (1988) Automated plant tissue culture for mass propagation. Bio/Tech 6:1035–1040
Long RD (1987) Economic and strategic aspects of micropropagation. Proc Ireland Bot Meet 1987, pp 95–102
Maene L, Debergh P (1985) Liquid medium additions to established cultures to improve elongation and rooting in vitro. Plant Cell Tissue Organ Cult 5:127–133
Margaritis A, Wallace JB (1984) Novel bioreactor systems and their applications. Bio/Tech 2:447–453
Maurice V, Vandercook CE, Tisserat B (1985) Automated plant surface sterilization system: Physiol Vég 23:127–133
McCown DD (1985) Plug systems for micropropagules. In: Zimmerman RH et al. (ed) Tissue culture as a plant production system for horticultural crops. Martinus Nijhoff, Dordrecht, pp 53–60
Meyer MJ, McClelland MT, Smith MAL, Spomer LA (1988) Image analysis of growth and development parameters in microcultured whole plants. HortSci 23(3):173

Miwa Y (1989) Robot in tissue culture nursery. In: Measurement, control and information in production of plants. Keisoku Jidou Seigyo Gakkai, Tokyo, pp 202–210 (in Japanese)
Moncousin C (1987) Recent developments in the field of adventitious rhizogenesis of in vitro cuttings. In: Plant micropropagation in horticultural industries. Florizel 87:79–91
Murashige T (1989) Plant propagation by tissue culture: practice with unrealized potential. In: Ammirato PV, Evans DA, Sharp WR, Bajaj YPS (eds) Handbook of plant cell culture, vol 5. Ornamental species. McGraw-Hill, New York, pp 3–9
Németh G (1986) Induction of rooting. In: Bajaj YPS (ed) Biotechnology in agriculture and forestry, vol 1. Trees I. Springer, Berlin Heidelberg New York, pp 49–64
Pierik RLM (1988) Handicaps for the large scale commercial application of micropropagation. Acta Hortic 230:63–71
Pierik RLM (1990) Rejuvenation and micropropagation. IAPTC Newslett 62:11–21
Preil W, Florek P, Wix U, Beck A (1988) Towards mass propagation by use of bioreactors. Acta Hortic
Rajbhandary SB, Bajaj YPS (1991) Rooting of in vitro produced shoots in non-sterile sand – an inexpensive and efficient technique for en masse micropropagation. In: Bajaj YPS (ed) Biotechnology in agriculture and forestry, vol 17. High-tech and micropropagation I. Springer, Berlin Heidelberg New York pp 262–269
Read PE, Fellman CD (1985) Accelerating acclimation of in vitro propagated woody ornamentals. Acta Hortic 166:15–20
Redenbaugh K, Slade D, Viss P, Fujii J (1987) Encapsulation of somatic embryos in synthetic seed coats. HortSci 22:803
Reinert J, Bajaj YPS, Zbell B (1977) Aspects of organization – organogenesis, embryogenesis, cytodifferentiation. In: Street HE (ed) Plant tissue and cell culture. Blackwell, Oxford, pp 389–427
Rittershaus E, Ulrich J, Westphal K (1990) Large-scale production of plant-cell cultures. IAPTC Newslett 61:2–10
Rowe JW (1986) New technologies in plant tissue culture. In: Zimmerman RH et al. (ed) Tissue culture as a plant production system for horticultural crops. Martinus Nijhoff, Dordrecht, pp 35–51
Rowe WJ, Rowe NM, Roepe HP (1987) The application of robotics for an automated tissue transfer system. Symp Innovations and plant production. Sept 1987, Monaco, p 41
Schiel O, Berlin J (1987) Large-scale fermentation and alkaloid production of cell suspension cultures of *Cathranthus roseus*. Plant Cell Tissue Organ Cult 8:153–161
Seginer I (1989) Optimal greenhouse production under economic constraints. Agric Systems 29:67–80
Sluis CJ, Walker KA (1985) Commercialization of plant tissue culture. IAPTC Newslett 47:2–12
Smart NJ, Fowler MW (1984) An airlift column bioreactor suitable for large-scale cultivation of plant cell suspensions. J Exp Bot 35:531–535
Sonneveld C, Welles GWH (1984) Growing vegetables in substrates in the Netherlands. Proc VIth Int Congr Soilless Culture, pp 613–632
Sonneveld C, Welles GWH (1988) Yield and quality of rock wool-grown tomatoes as affected by variations in EC-value and climatic conditions. Plant Soil 111:37–42
Stafford JEH (1985) Experience with a robotic analytical system. Lab Pract 34:19–22
Stuart DA, Strickland SG, Walker KA (1987) Bioreactor production of alfalfa somatic embryos. HortSci 22:800–803
Styler DJ (1985) Bioreactor technology for plant propagation. Basic Life Sci 32:117–130
Sutter E, Hutzell M (1984) Use of humidity tents and antitranspirants in the acclimatization of tissue-cultured plants to the greenhouse. Sci Hortic 23:303–312
Takayama S (1984) Mass propagation of plants through shake culture techniques. Plant Tissue Cult Lett 1:8–13
Tanaka H, Nishijima F, Suwa M, Iwamoto T (1983) Rotating drum fermenter for plant cell suspension cultures. Biotech Bioeng 25:2359–2370
Tanaka M, Goi M, Higashiura T (1988a) A novel disposable culture vessel made of fluorocarbon polymer films for micropropagation. Acta Hortic 226:663–670
Tanaka M, Jinno K, Goi M, Higashiura T (1988b) The use of disposable fluorocarbon polymer film culture vessel in micropropagation. Acta Hortic 230:73–80
Ting KC, Giacomelli CE, Roberts WJ (1988) SCARA robots work cell for seedling transplantation. Acta Hortic 230:221–228
Tisserat B, Vandercook CE (1985) Development of an automated plant culture system. Plant Cell Tissue Organ Cult 5:107–117

Tisserat B, Vandercook CE (1986) Computerized long-term tissue culture for orchids. Am Orchid Soc Bull 55:35–42

Ulrich B, Wiesner W, Arens H (1985) Large-scale production of rosmarinic acid from plant cell cultures of Coleus blumei Benth. In: Neumann KH, Barz W, Reinhard E (eds) Primary and secondary metabolism of plant cell cultures. Springer, Berlin Heidelberg New York, pp 293–303

Ushiyama K, Miyamoto Y, Oda H, Ishida Y (1984) Gaseous phase culture equipment for continuous culture of plant tissues. Jpn Patent Kokai Tokyo Koho 59–45873 (in Japanese)

Ushiyama K, Oda H, Miyamoto Y (1986) Large scale tissue culture of *Panax ginseng* roots. Abstr VI Int Congr Plant tissue and cell culture. Minnesota Univ, 252 pp

Vanderschaege A, Debergh P (1987) Automation of tissue culture manipulations in the final stages. IAPTC Newslett 51:10

Wagner F, Vogelmann H (1977) Cultivation of plant tissue culture in bioreactors and formation of secondary metabolites. In: Barz W, Reinhard E, Zenk MH (eds) Plant tissue culture and its biotechnological application. Springer, Berlin Heidelberg New York, pp 245–252

Wardel K, Dobbs EB, Short KC (1983) In vitro acclimation of aseptically cultured plantlets to humidity. J Am Soc Hort Sci 108:386–389

Weathers PJ, Giles KL (1988) Regeneration of plants using nutrient mist culture. In Vitro 24(7):727–732

Withers LA, Bajaj YPS (1982) The establishment of plant tissue cultures and the regeneration of plants from cultures. In: Withers LA, Williams JT (ed) Crop genetic resources – the conservation of difficult material. IUBS, Ser B 42, FAO Rome, pp 93–96

Zimmerman RH, Yae BW, Fordham I (1987) Comparison of rooting methods for apple cultivars in vitro. Acta Hortic 212:303–310

Zyrd J-P (1987) Image analysis: its uses in the analysis of plant cells and protoplasts. Symp Innovations and plant production. Monte Carlo, Monaco, Sept 1987, p 21

1.2 Setting Up a Commercial Micropropagation Laboratory

A. C. CASSELLS[1]

1 Introduction

A structural analysis of the plant production industry (Jones 1986; Table 1) shows that micropropagation has been integrated into the industry at several levels. Some laboratories (Type 1) sell proliferating cultures or rooted in vitro plantlets to conventional growing-on nurseries while others (Type 2) sell acclimatised young plants. A further strategy (Type 3) is to increase the value-added aspect of production by growing on the plants in-house (Harper 1988).

In carrying out a feasibility study prior to establishing a micropropagation company, it becomes apparent that to satisfy market needs and to overcome adverse product reaction (Cassells 1988a) the micropropagator must be able to produce, directly or via growing-on nurseries, plants of good health status, of good habit and with high genetic stability. To achieve these objectives, it is argued, even a Type 1 laboratory must be equipped to carry out routine plant pathology investigations and progeny testing. Thus, in considering the setting up of a commercial laboratory here, provision for ancillary services to the main micropropagation activity will be included. These latter activities can be regarded as the basis of a quality assurance system.

A prerequisite for the establishment of a viable business is a sound business plan (West 1988), economic aspects of which for micropropagators have been discussed previously (Long 1987). This business plan will determine the capital availability and the relative weighting to be put on each element of the business including the provision for laboratory accommodation and technology acquisition. Here, a state-of-the art approach is taken to micropropagation laboratory technology, i.e. that manual transfer methods are used. It should be appreciated that automation or partial automation of the handling of proliferating cultures (Levin et al. 1988) or a move to production based on bioreactors and somatic embryogenesis (Giles and Morgan 1987) will, through cost reductions, broaden the market for micropropagation (Sluis and Walker 1985). Such advances in the core production technology will not, however, lessen the need for quality assurance.

In discussing the setting up of a micropropagation laboratory, the need for a quality assurance system is considered integral. The key functions to be described are production of axenic cultures, cloning and monitoring for contamination and progeny testing. The latter, for which some glasshouse accommodation is required, includes confirmatory studies on ease of establishment, normality of habit and trueness-to-type (see Cassells 1985, 1988a).

---

[1]Plant Biotechnology (UCC) Ltd, University College, Cork, Ireland

Biotechnology in Agriculture and Forestry, Vol. 17
High-Tech and Micropropagation I (ed. by Y.P.S. Bajaj)

Table 1. Structural analysis of the horticultural plant growing and distribution industry. (After Jones 1986)

	Type	Activity
1.	Micropropagator/shipper with no greenhouse	Sells Stage[a] II and/or Stage III in vitro cultures in vessels
2.	Micropropagator/grower/shipper with greenhouse	Mainly sells Stage IV acclimatised plants, but could sell the same Stages as Type 1
3.	Micropropagator/grower with greenhouse	Sells mature plants, liners, plug trays etc. Uses all production internally
4.	Propagator/grower/shipper with greenhouse, no laboratory	Buys from Types 1 and 2. Sells Stage IV acclimatised plants, liners, plug trays, etc. and mature plants
5.	Propagator/grower with greenhouse, no laboratory	Buys from Types 1 and 2. Sells mature plants
6.	General grower consumer	Buys Stage IV acclimatised plants from Types 1, 2 and 4. Resells products wholesale

[a] Stages as defined by Murashige (1974).

2 The Planning Phase

Given that a capital figure has been agreed, the planning phase involves site acquisition, building and equipment purchase.

In approaching site acquisition there should be recognition in the business plan that following start up, market and other forces (Fig. 1) may cause a change in the business, e.g. from a Type 1 to a Type 2 company (see Harper 1988). The latter would require expansion of the glasshouse facilities and hence more space. If expansion is not possible on the original site there may be a significant cost penalty in relocating to a fresh site (West 1988).

Following site acquisition, or preferably in parallel, on-site discussions should take place with the building contractor. In these planning meetings, inputs from an experienced micropropagator are essential for, while relevant expertise exists in companies familiar with hospital laboratory design and construction, glasshouse erection etc., the precise mix of expertise is not commonly assembled in construction companies. Areas where planning errors can be costly are in the design and provision of services (see Sect. 4.3).

3 Decision on Laboratory Specification

As mentioned above, micropropagation has not developed to the stage where there are established specialist companies offering complete standard packages for building and technology on a turnkey basis. Laboratory specification is usually decided on personal experience, exchange of information through scientific publications (e.g. Broome 1986; Pierik 1987) and visits to micropropagation laboratories. Within this framework the options are considerable and will be discussed further below. However, it should be

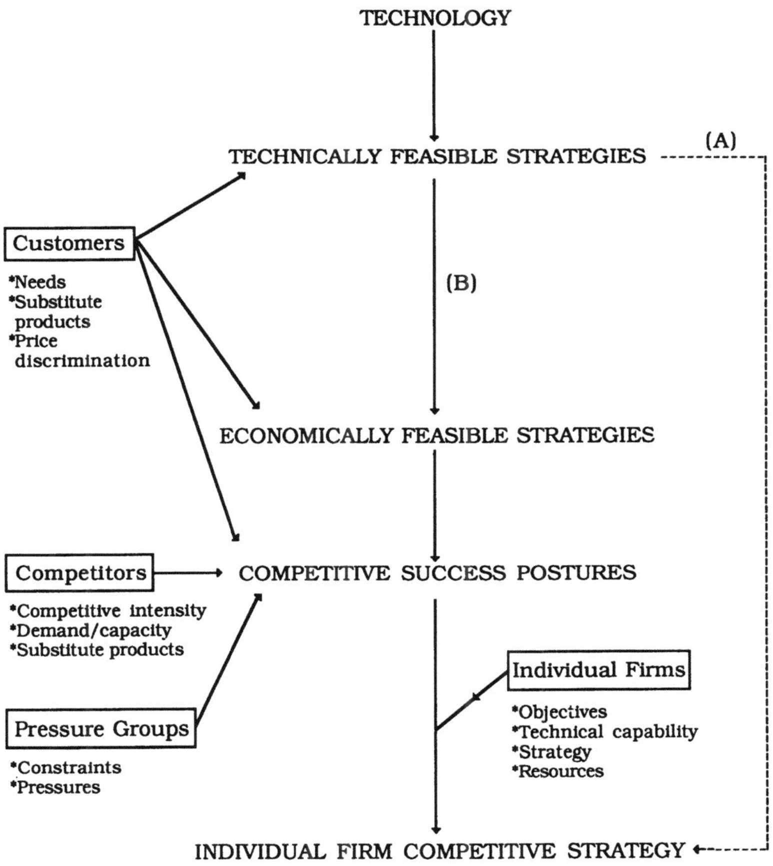

Fig. 1. Factors determining competitive strategy. (Long 1987)

stressed that the capital available and the type and range of product to be sold determine the specific requirements.

Given the above consideration (see also Sect. 1), some basic decisions can be made about size and relative location of functions. Quality assurance requires the provision of plant pathology services and glasshouses. These should be located separate from the "clean air" micropropagation area. Media preparation should also be in the "clean area". Figure 2 shows an example of a micropropagation laboratory designed for Type 1 production according to the criteria discussed in the Introduction.

4 The Micropropagation and Ancillary Facilities

4.1 Media Preparation Area

The media requirements will vary from small quantities for microbiology and tissue culture development studies, to tens of litres for production. Consequently, the equipment required will reflect this demand. Media preparation may start with brought-in

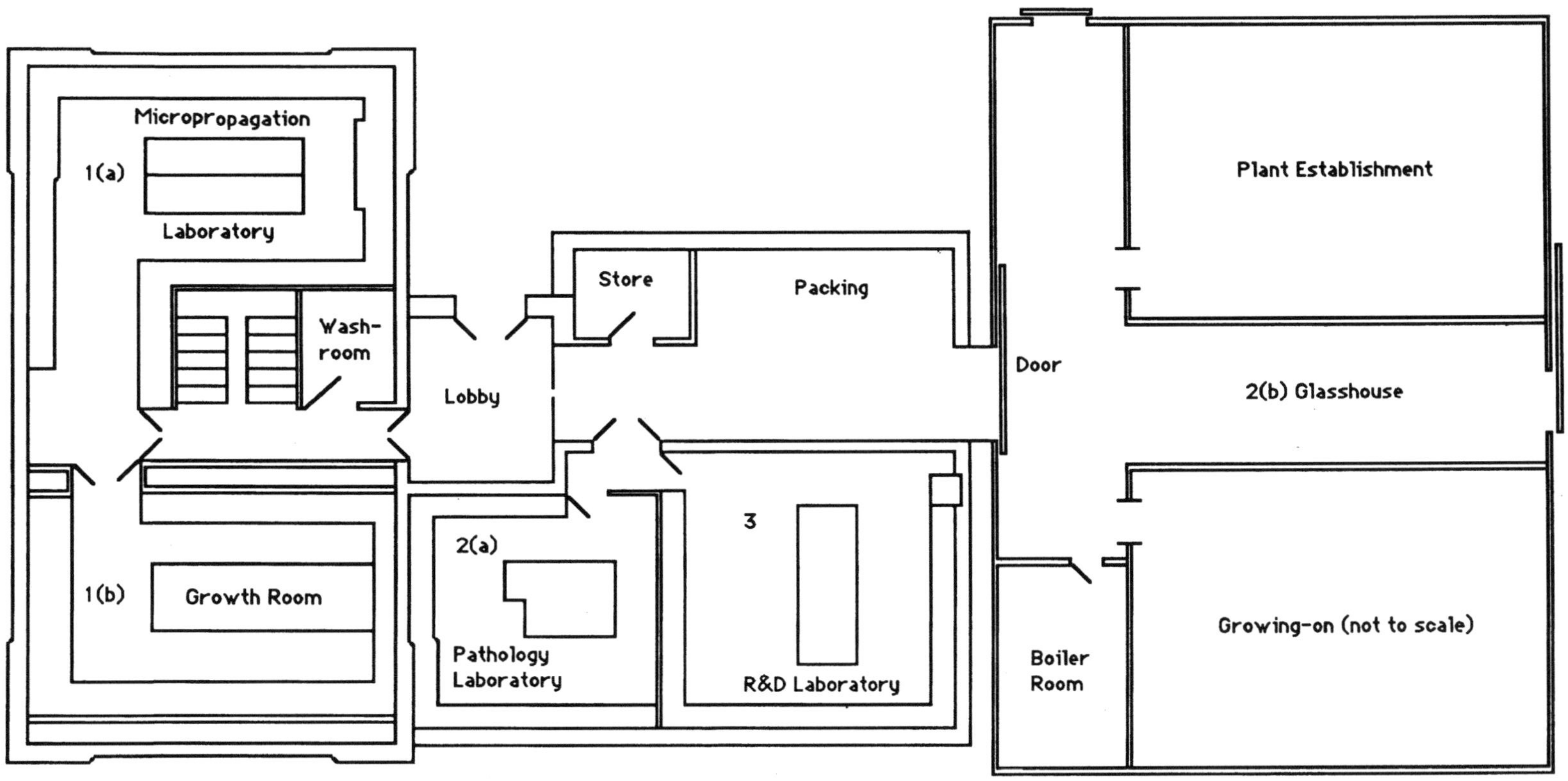

Fig. 2. Laboratory floor plan for a Type 1 micropropagation company with a commitment to a roll-over business plan and a capability to respond to clients offering new genotypes for cloning. *1* Micropropagation area ('clean' area); (a) micropropagation and media preparation; (b) growthroom, *2* Quality assurance; (a) pathology laboratory; (b) glasshouse. *3* Research and Development laboratory

dedicated formulations in preweighed powder form for 1 litre to 50 litre volumes (e.g. from Flow Laboratories, Irvine, UK). Alternatively, the laboratory may choose to make up its own formulations from the basic ingredients (for recipes see, e.g. George et al. 1988). The author in general prefers bought-in basal or complete media, where quality control may be higher than for in-house formulations.

A source of high quality distilled water will be required and here glass or PTFE-coated elements are superior to metal elements which may contaminate the water. In some geographical areas, it may be necessary to double distill the water. It is advisable to send water samples for analysis to a specialist laboratory.

Sterilisation of media on a large scale is achieved by autoclaving and, as mentioned above, to cover the range of activities it may be desirable to have a pressure cooker for small volumes and an autoclave for larger volumes. In purchasing an autoclave, attention should be paid to the cycle time – it may be necessary to compromise, for a short cycle may mean low capacity (Fig. 3a). Uniformity of heating is a problem when using larger volume containers (George et al. 1988). Consequently, it may be worth considering an automatic agar steriliser. This can be linked to an automatic dispenser for Petri dishes or, via a pump, deliver metered volumes to the container of choice (Fig. 3b).

For labile material or where containers are filled with liquid media (e.g. see Debergh and Maene 1981), the liquid may be decontaminated by an on-line 0.2 micron bacterial filter. Microwave ovens have application in melting agar in small volume areas (Cowan and Allen 1985). Depending on the scale of the operation, media may be prepared for immediate use or subject to storage. It is important if the latter strategy is employed to avoid contamination and drying out (see below).

Finally, there will be a need for a washing up area. Again, if volume is high, a domestic or laboratory dishwasher may be required. In all washing up operations the ware must be free of residual detergent.

4.2 Micropropagation Area

The main feature of the micropropagation area is the laminar-flow cabinet (Fig. 3c). These may be individual compartments or linked to form a laminar-flow wall. It is advantageous to pressurise both the micropropagation laboratory and growthrooms with filtered air. It is advisable, for workers entering this "clean area", to wear laboratory coats, overshoes and to coat their hands with antiseptic isopropanol gel when culturing. Before operation, the cabinet walls may be sterilised with UV light (caution, it will react with Perspex and is harmful to eyes) or swabbed with ethanol or preferably isopropanol. Laminar-flow cabinets should be routinely serviced. The instruments may be autoclaved before production starts. Between transfers instruments may be sterilised in an alcohol flame, in a heat gun or in a bead steriliser (Fig. 3d).

The container used for micropropagation is of critical importance in a number of respects. Firstly, it influences the microclimate for culture growth. Secondly, its design influences the ease with which the cultures can become contaminated, assuming that the process starts from axenic stock cultures. The simplest culture container is the Petri dish (Fig. 4a). It has the advantage that it can be bought presterilised and, using a media steriliser/plate pourer, filling can be automated (see Fig. 3b). It is in our experience useful for value-added nodal culture, e.g. of potato virus-free stock (Roca et

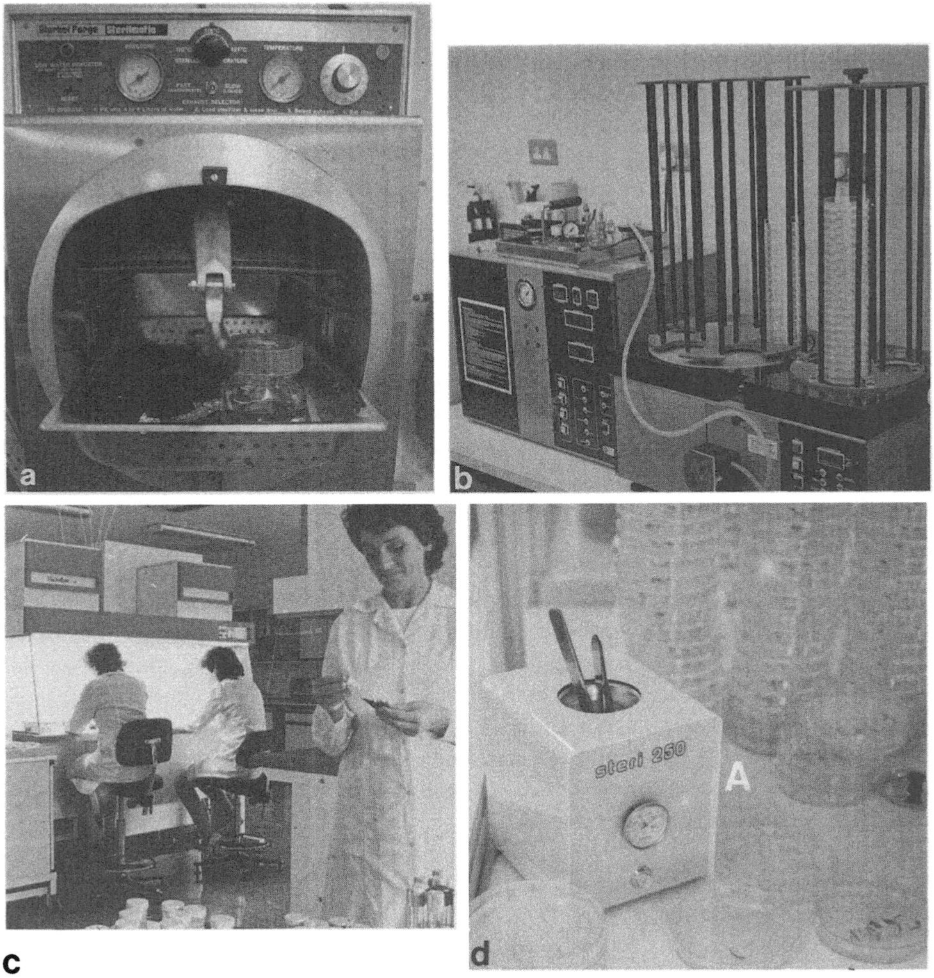

Fig. 3. **a** Short cycle autoclave, note limited capacity. **b** Bulk media steriliser (*left*), capacity 8 l; automatic sterile agar dispenser (*right*). **c** Laminar flow cabinets (by kind permission of Twyford Plant Laboratories Ltd). **d** Glass bead instrument steriliser

al. 1978). A disadvantage is that the plates, if unsealed, may dry out, depending on the ambient growthroom conditions (see below), and are prone to contamination. If sealed, e.g. with parafilm or other high barrier material, vitrification may occur (Keevers et al. 1984).

The commonest containers in use appears to be ca. 500-ml volume glass or plastic jars (Fig. 4b). Various lids are employed from the metal screw cap to glass (as in a preserving or Kilner jar) (Fig. 4b). To control gaseous exchange, lids are left loose or a washer of gas-permeable material is used. A disadvantage of these containers is that they require sterilisation and they are not readily amenable to automated filling. An advance over the latter is the use of sterile transparent stackable food containers (Fig. 4c), or the Plastem system of stackable autoclavable pots (Plastem AG Schwarzenberg; Fig. 4b).

Fig. 4. **a** Petri dishes, note each plate is coded numerically. **b** *Left* Plastem container, note ridge which allows stacking. *Right* Kilner jar. **c** Clear food container in transparent stackable tray. **d** Heat-sealed double-wrapped container (see Fig. 3)

In this laboratory, we have exploited the concept of "biopackaging", in which plastic food quality containers are heat-sealed with membranes which have differential permeability to oxygen, carbon dioxide, ethylene and water vapour (Cassells 1987) (Fig. 4d, also See Sect. 6). Heat sealing has the added advantage that contamination in the growthroom is eliminated.

In summary, the following aspects of the culture container are of economic importance and influence quality and losses due to contamination. Firstly, can the containers be bulk-sterilised, or are they available sterile? Secondly, can filling be automated or semi-automated? Thirdly, can the gaseous atmosphere in the container be controlled? – such, that, on the one hand, the containers do not dry out, or on the other hand, the humidity is high so that vitrification may occur. Fourthly, can the containers be stacked in trays to reduce handling costs? While container design can be customised, the design and mould cost can be significant and, at least at start-up, economic alternatives should be sought. Disposable sterile ware for the food industry should be examined for suitability. A visit to a packing trade exhibition is advised.

Finally, a caution: it is advised that only food-grade plastic materials, or Petri dishes approved for plant tissue culture is used.

4.3 The Growthroom(s) Area

Growthroom design poses several problems and flexibility in design is advisable (see Sect. 2). The basic elements are an air-conditioning system, shelving and lighting (Fig. 5a).

To achieve uniform conditions, i.e. avoiding hot and cold spots, especially when the shelving is installed and possibly moveable, an efficient air circulation system is required. It is advisable, to avoid contamination when using "open" culture containers, that the circulating air is filtered. Further, it is essential that safety features prevent dramatic temperature fluctuations should the air-conditioning unit fail. In addition to locating, where feasible, heat-generating components of the system outside the room, it is essential that the lighting switch off immediately the air conditioning fails. It is recommended that a backup air-conditioning unit be installed with oscillation between the units to avoid excessive wear on the primary unit.

Fig. 5. **a** Simple, readily dismantled shelving, note wall vents for air circulation. **b** Computer-controlled plant growth rooms (Twyford Plant Laboratories Ltd.). **c** Paternoster type shelving (Zippel, Altdorf, West Germany). **d** PAR light measurement

A dilemma facing the laboratory designer is whether, in the interests of capital and operational costs, a single growthroom is built or a number of smaller units. Here again, the business plan indicates the likely species to be propagated and, if their temperature requirements are similar, one temperature regime may suffice. An alternative to constructing rooms to run at different temperatures is the use of temperature-controlled shelving, but this can involve high capital costs.

Shelving ranges from open knockdown shelving which can be easily dismantled (Fig. 5a), e.g. to vary distances between shelves, to horizontally mobile shelving which gives efficient use of floor space (Fig. 5b), or the "paternoster" or vertically mobile shelving which avoids the need to bend or stretch in placing and removing cultures (Fig. 5c).

Lighting, in both quality and intensity, is, like temperature control, problematic. Light quality and intensity are known to influence the growth of cultures (Ammirato 1986; Seabrook 1987) but capital costs (or inherent design faults or lack of foresight in the business plan) introduce operational rigidity. Micropropagators often fail to replace lights as intensity declines with age. A regular monitoring and replacement programme should be followed (Fig. 5d).

5 Research and Development and Quality Assurance

Research and development (R&D) and quality assurance (QA) are conveniently located side by side. The R&D requirement recognises that differences may exist in even closely related genotypes requiring modification of the tissue culture process or, indeed in some cases development of a new process (e.g. Cassells and Morrish 1987). While it is not appropriate to treat this aspect of micropropagation in detail here, the equipment requirements are discussed. For R&D work, the main equipment items are as listed in Table 2. In addition, one or a series of experimental temperature-controlled cabinets,

Table 2. Basic equipment requirements of a micropropagation industry

Balances (0.01 mg to 1 kg)
Water still
pH Meter
Fridge/freezer (–18° to +4°C)
Autoclave
Dishwasher
Drying oven
Laminar flow cabinet(s)
Binocular dissecting microscope
PAR light meter
Optional
Agar head instrument steriliser
Agar steriliser/dispenser
Bar-code printer
Microcomputer with bar-code reader

e.g. modified commercial "cool drinks" cabinets (Fig. 6a) will serve for experimental scale production where the temperature range is extended upwards and, if appropriate, lighting is increased. There will also suffice to provide a slow growth environment for stock storage (e.g. Staritsky et al. 1986).

Quality control has several aspects. Firstly, it is a facet of research and development, i.e. the production of axenic cultures (see e.g. Cassells 1986). Secondly, it is the routine monitoring of cultures for contamination, and here, the possibility of latent contamination should be emphasised (Cassells 1988b). To produce axenic cultures, the usual strategy is to carry out indexing for known diseases of the crop, seeking escapes as tissue culture stock plants (Cassells et al. 1988), or using one of the established tech-

Fig. 6. a Modified 'cool drinks' cabinet for experimental studies on temperature requirements of cultures or stock storage. **b** Use of plug trays (courtesy of Hortus OY). **c** Bar-coding (*left*) printer (*right*) decoder/personal computer

niques, e.g. meristem culture to produce putative axenic cultures (Cassells 1986, 1991). The health status of the latter must be confirmed by appropriate indexing for known pathogens and for cultivable bacteria which may affect production. Further, production should also be monitored for contaminants introduced during transfer or in the growthroom. Especially common contaminants of plant tissue cultures are various species of *Corynebacterium*, *Erwinia*, *Xanthomonas* and *Pseudomonas*, although exotic bacterial species and yeast should not be overlooked (Cassells 1988).

The question arises as to how much of the above work should, or indeed can, be done in-house. Production monitoring can be carried out using commercial bacteriological media formulations (see Cassells 1986, 1991). Specific disease indexing may require the use of more complex bacteriological tests. Alternatively, organisms isolated in pure culture can be sent to university or government laboratories for diagnosis. Increasingly, immunological methods are being exploited for disease indexing, and kits are becoming available commercially (Cassells 1986, 1991), making these techniques available to micropropagators. University and other laboratories will usually provide staff training if required.

The major aspect of quality assurance is to confirm the quality of the product, particularly important for Type 1 companies selling cultures, where failure to establish may cause a loss of customer confidence (see Sect. 6). This operation requires plant establishment facilities and a growing-on area for progeny testing. Acknowledging that plants produced in vitro may be "soft", establishment requirements vary from propagators on heated benches, to misting or fogging systems (Scott 1986; see Sect. 6 and Fig. 7 also). Habit assessment and trueness-to-type evaluation require standard glasshouse or plastic tunnel facilities which may require supplementary heating and lighting.

6 Trends in the Micropropagation Industry

Many factors influence an individual micropropagation company's competitive strategy (Fig. 1). Important comparative advantage for the micropropagator lies in the ability to rapidly clone novel genotypes of high disease status because of the low stock and seasonal independence of the process. Critically important, however, is production cost (Donnan 1986). A discernible change in micropropagators' competitive strategy, in general, has been a move away from the Type 1 activity to Types 2 or 3. Arguably, these reflect consumer resistance to accepting risks associated with the establishment of in vitro material. This factor transfers the significant cost burden of establishing plants to the micropropagator, i.e. Type 2 activity, and again, arguably, has resulted in a shift to the value-added Type 3 activity (see Harper 1989a,b). However, the move away from trade in in vitro material also has implications, for transnational movement of in vivo material poses difficulties not least in relation to quarantine regulations (Kahn 1986). Use of inventions like that described in Fig. 7, which facilitate weaning of microplants, may contribute to the profitability of Type 1 companies and widen international trade.

A major factor limiting market expansion for the micropropagator is production cost, mainly labour cost. The horticulture industry has introduced a high degree of mechanisation/automation in downstream aspects of composting, tray-filling, repotting/transplanting (McCown 1986). The challenge for micropropagation is to auto-

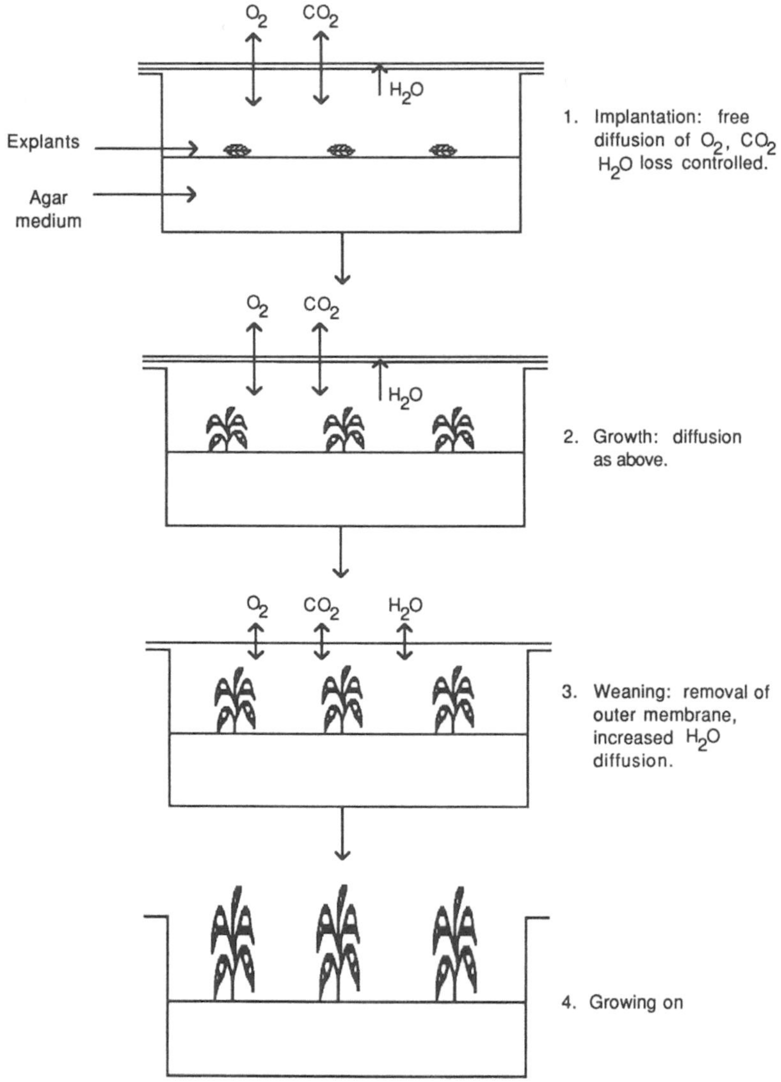

Fig. 7. A double membrane patented system which allows control of the gaseous culture environment during microplant development and weaning. (Cassells 1987)

mate the multiplication and establishment stages to interface with automated horticultural systems.

Approaches to mechanisation of micropropagation must address both of the stages identified above. In achieving this objective, different procedures are likely to be developed for handling meristem cultures, nodal cultures and adventitious cultures. Sluis and Walker (1985) have discussed the problems in applying artifical intelligence to the design of precision cutting equipment for meristem cultures with variable morphology. The problem is less in the case of nodal cultures where increased regularity can be accomplished, while adventitious cultures may be amenable to random cutting. Levin et al. (1988) have described machinery for the mechanisation of meristem cultures which is applicable also to adventitious cultures. Further, Plant Biotechnology Industries Ltd (Israel) have patented machines for both mechanical cutting and modular planting of propagules which suggest how these problems may be approached. Since this aspect of micropropagation is in the commercial area, the patent literature may be more informative than scientific journals.

In crops which reproduce naturally by vegetative propagules, e.g. potato, in vitro production of micropropagules may facilitate automation of planting where the latter can be handled by automatic seeders.

In parallel with automation of production is the need to computerise records. Here, bar-coding/reading of cultures can enable the micropropagator to monitor facets of production, relative growth rates, operative efficiency and stock levels (Fig. 6c). Equally important, such methods can reduce labelling errors, which in an exponential process can result in major problems, including loss of customer confidence.

7 Conclusions

As the individual business plan determines the specification of the product, so the balance between in-house and sub-contracted activity will vary. However, it is argued that in-house, in addition to an efficient production system, ancillary functions are required; these are for the production of axenic cultures, process development and quality assurance. The latter involves monitoring progeny establishment, habit quality, trueness-to-type and confirming the health status. The increasing cost of consequential loss insurance may oblige all Types 1 and 2 companies to raise their quality assurance standards.

In summary, Fig. 8 represents the stages and strategies presented to the micropropagator. Included are the quality assurance steps. An important step is the decision whether to initiate entry into culture of a given variety (Cassells 1988a).

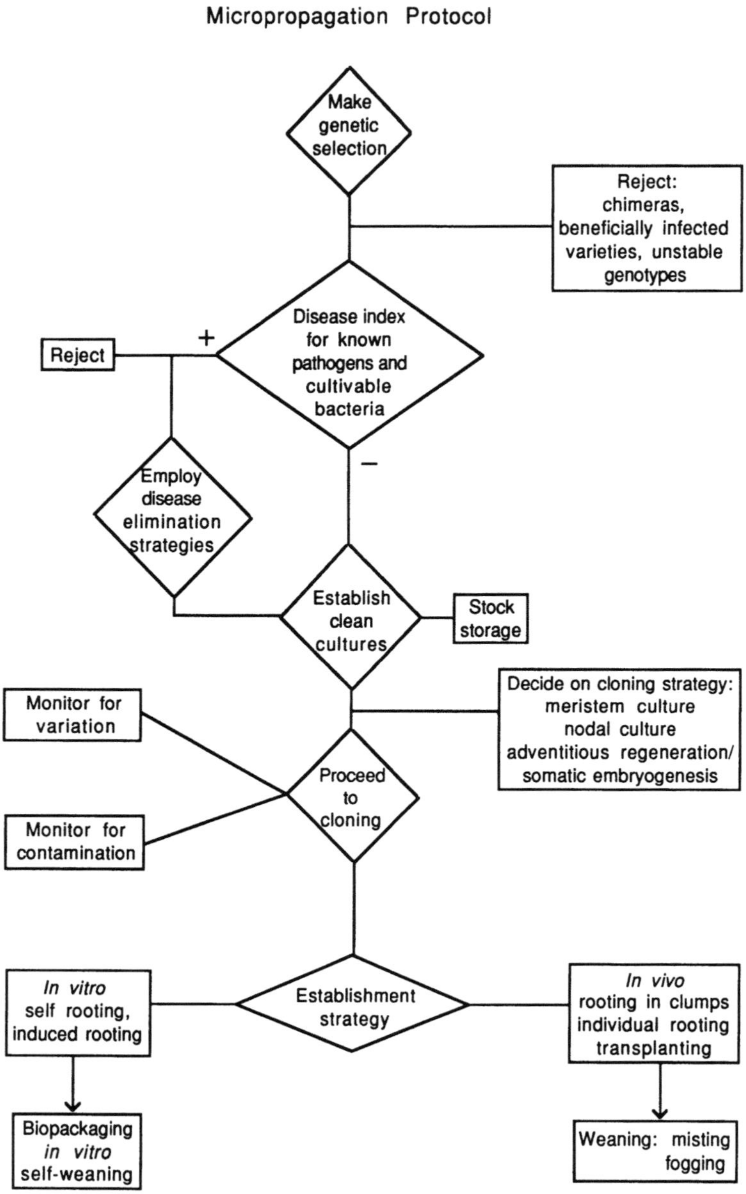

Fig. 8. Factors influencing the production of good quality microplants

References

Ammirato PV (1986) Control and expression of morphogenesis in culture. In: Withers LA, Alderson PG (eds) Plant tissue culture and its agricultural applications. Butterworth, London, pp 23–46

Broome OC (1986) Laboratory design. In: Zimmerman RH, Grisebach RJ, Hammerschlag FA, Lawson RH (eds) Tissue culture as a plant production system for horticultural crops. Martinus Nijhoff, Dordrecht, pp 351–364

Cassells AC (1985) Genetic, epigenetic and non-genetic variation in tissue culture derived plants. In: Schafer-Menuhr A (ed) In vitro techniques – propagation and longterm storage. Martinus Nijhoff, Dordrecht, pp 111–120
Cassells AC (1986) Production of healthy plants. In: Alderson PG, Dullforce WM (eds) Micropropagation in horticulture, practice and commercial problems. Institute of Horticulture, London, pp 53–70
Cassells AC (1987) Patent PCT/GB87/00218. Patent Office, London
Cassells AC (1988a) Pitfalls for nurserymen in micropropagation and how to avoid them. Proc Intl Plant Prop Soc 38:240–246
Cassells AC (ed) (1988b) Symposium on bacterial and bacteria-like contaminants of plant tissue cultures. Acta Hortic 225, 225 pp
Cassells AC (1991) Problems in tissue culture: culture contamination. In: Debergh P, Zimmerman RH (eds) Micropropagation. Kluwer, Dordrecht, pp 31–44
Cassells AC, Morrish FM (1987) Genetic, ontogenetic and physiological factors and the development of a broad-spectrum medium for *Begonia rex* Putz cultivars. Sci Hortic 31:295–302
Cassells AC, Carney BF, Harmey MA, McCarthy E, McHugh A (1988) Problems posed by cultivable bacterial endophytes in the establishment of axenic cultures of *Pelargonium* × *domesticum*: the use of *Xanthomonas pelargonii*-specific ELISA, DNA probes and culture indexing in the screening of antibiotic treated and untreated donor plants. Acta Hortic 225:153–161
Cowan ME, Allen J (1985) Microwave processing of dehydrated culture media. Med Lab Sci 42:156–160
Debergh PC, Maene LJ (1981) A scheme for commercial propagation of ornamental plants by tissue culture. Sci Hortic 14:335–345
Donnan A (1986) Determining and minimising production costs. In: Zimmerman RH, Grisebach RJ, Hammerschlag FA, Lawson RH (eds) Tissue as a plant production system for horticultural crops. Martinus Nijhoff, Dordrecht, pp 167–173
George EF, Puttock DJM, George HJ (1988) Plant culture media, Vols I and II. Exegetics, Edington, UK
Giles KL, Morgan WM (1987) Industrial-scale plant micropropagation. Tibtech 5:35–39
Harper PC (1988) Myths and money in micropropagation. Proc Int Plant Prop Soc 38:232–236
Jones JB (1986) Determining markets and market potential of horticultural crops. In: Zimmerman RH, Grisebach RJ, Hammerschlag FA, Lawson RH (eds) Tissue culture as a plant production system for horticultural crops. Martinus Nijhoff, Dordrecht, pp 175–182
Kahn RP (1986) Plant quarantine and international shipment of tissue culture plants. In: Zimmerman RH, Grisebach RJ, Hammerschlag FA, Lawson RH (eds) Tissue culture as a plant production system for horticultural crops. Martinus Nijhoff, Dordrecht, pp 147–164
Keevers C, Coumans M, Coumans-Giltes MF, Gaspar Th (1984) Physiological and biochemical events leading to the vitrification of plants cultured in vitro. Physiol Plant 61:69–74
Levin R, Gaba V, Tal B, Hirsch S, De-Nola D, Vasil IK (1988) Automated plant tissue culture for mass propagation. Biotechnology 6:1035–1040
Long RD (1987) Economic and strategic aspects of micropropagation. Proc Ir Bot Meet 1987:95–102
McCown DD (1986) Plug systems for micropropagules. In: Zimmerman RH, Grisebach RJ, Hammerschlag FA, Lawson RH (eds) Tissue culture as a plant production system for horticultural crops. Martinus Nijhoff, Dordrecht, pp 53–60
Murashige T (1974) Plant propagation through tissue culture. Annu Rev Plant Physiol 25:135–166
Pierik RLM (1987) In vitro culture of higher plants. Martinus Nijhoff, Dordrecht
Roca WM, Espinoza WO, Roca MP, Bryan JE (1978) A tissue culture method for the rapid propagation of potatoes. Am Pot J 55:691–701
Scott MA (1986) Weaning of cultured plants. In: Alderson PG, Dullforce WM (eds) Micropropagation in horticulture, practice and commercial problems. Institute of Horticulture, London, pp 173–182
Seabrook JEA (1987) Changing the growth and morphology of potato plantlets in vitro by varying the illumination source. Acta Hortic 212:401–405
Sluis CJ, Walker KA (1985) Commercialisation of plant tissue culture propagation. Newsletter Int Assoc Plant Tissue Cult 47:2–12
Staritsky G, Dekkers AJ, Louwaars NP, Zandvoort EA (1986) In vitro conservation of aroid germplasm at reduced temperatures and under osmotic stress. In: Withers LA, Alderson PG (eds) Plant tissue culture and its agricultural applications. Butterworth, London, pp 277–283
West A (1988) A business plan. Pitman, London

I.3 Micropropagation Through Meristem Culture

P.-J. WANG and A. CHARLES[1]

1 Introduction

Isolation and growth of the shoot meristem are of value in recovering pathogen-free plants. When vegetatively propagated plants are systemically infected with disease, the pathogen passes from one vegetative generation to the next. The entire population of a given clonal variety may be infected with the same pathogen. With latent viruses, the symptoms may be as difficult to detect as reduced yield or lower plant quality. Such infections can be difficult to eliminate.

In early work on virus multiplication in vitro, it was discovered that virus infection was not always maintained when the root tip of an infected tomato root was subcultured (White 1934). This was the first work to establish the theory of using meristem culture to eliminate virus pathogens. It was later demonstrated that virus titer decreased not only in root tips but also in the tip of the vegetative buds (Limasset and Cornuet 1949). This work was supported by Morel and Martin (1952, 1955), who recovered virus-free shoots from infected *Dahlia* and potato. However, it was necessary to graft the shoots onto virus-free seedlings to establish plants.

Morel (1960) applied the technique of shoot meristem culture to produce virus-free cymbidiums. He later established the process of clonal propagation for virus-free orchid production (Morel 1963). Since this work, shoot meristem culture has been used extensively for a variety of plants, and forms the basis of the horticulture industry for micropropagation.

2 Shoot Meristem Culture

2.1 Apical Meristem Structure

Shoot meristem culture is a technique in which a dome-shaped portion of the meristematic region of the stem tip is dissected and inoculated on nutrient medium which supports plant growth. In classical plant anatomy, the apical meristem is divided into two regions: the "promeristem", composed of the apical initials and adjacent cells, and below it the "peripheral meristem" in which the protoderm, procambium, and ground meristem can be distinguished (Esau 1960).

[1]Ball Floraplant, P.O. Box 335, West Chicago, IL 60185, USA

Biotechnology in Agriculture and Forestry, Vol. 17
High-Tech and Micropropagation I (ed. by Y.P.S. Bajaj)

Using a common definition, the explant is restricted to the apical dome, the region of the shoot apex distal to the youngest leaf primordium (Cutter 1971). (Fig. 1) The dome region is approximately 0.075 to 0.12 mm in diameter and 0.2 to 0.25 mm in length. The apical dome is not connected through a vascular system to the immature procabium, young leaf primordium, and young axillary buds. This lack of vascular connection provides a basis for using the meristem culture technique for pathogen elimination.

Typically, in meristem culture the apical dome and a few leaf primordia located in the subapical region are dissected. If a portion of the stem beyond the meristematic subapical region is included, the procedure should be referred to as shoot tip culture, not meristem culture (Cutter 1971; Wang and Hu 1980, 1985).

The cells which compose the apical meristem tend to be genetically stable. In a micropropagation system, these cells have the highest potential of generating plants having the same genotypic and phenotypic composition as the explant source.

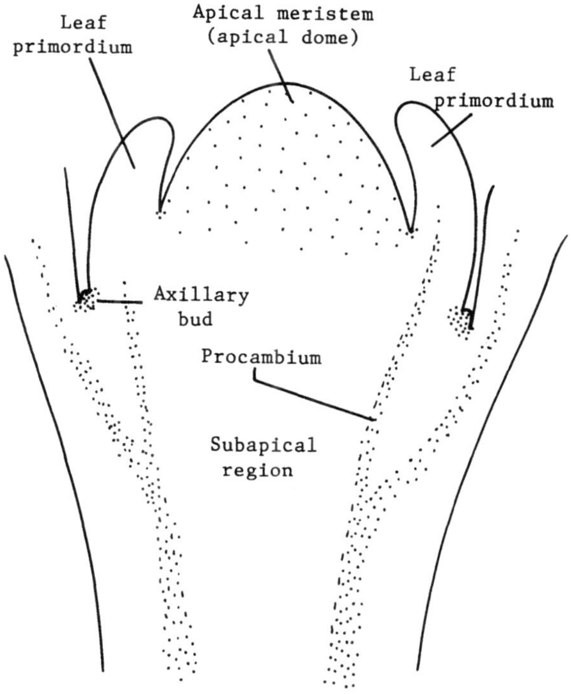

Fig. 1. Diagram of a shoot meristem containing the apical meristem (*apical dome*) and the meristematic subapical region with attached leaf primordia

2.2 Meristem Culture Procedure

2.2.1 Equipment

In our laboratory at Geo. J. Ball Inc., the following supplies have been used:

1. Laminar flow unit (sterile hood)
2. Ultrasonic cleaner
3. Stereomicroscope (8–40x)
4. Glass fiber optic illuminator
5. Long forceps
6. Hypodermic needles with attached handles
7. Bunsen burner or alcohol lamp
8. 75% ethanol
9. Petri dishes

2.2.2 Plant Material Preparation

Actively growing shoot tips are recommended for meristem culture. When shoots are in an active growth phase, the pathogen concentration tends to be reduced and the growth potential after dissection is greater. Actively growing shoots can be obtained by propagating sterilized stem sections, bulbs, corms, or tubers in sterilized artificial soil mix.

The ideal method for the disinfestation of plant material before planting varies with species. For material with tightly covered buds, such as *Narcissus* and *Gladiolus*, surface sterilization by immersing in 10% Clorox for 20 min is usually sufficient. Several rinses in sterile, distilled water should follow. Plants with loosely covered buds, such as potato, carnation, and impatients, should be sprayed with a systemic disinfectant. In our laboratory, a spray mixture composed of 0.1% Benlate (DuPont Co.) and 0.1% Streptomycin sulfate has been used successfully. In addition, it is beneficial to spray stem sections from trees and shrubs having winter buds. Microorganisms are abundant between the bud scales because of the stem's long period of exposure to open air and rainfall. Some segments should be retained on the mother plant and sprayed with systemic disinfectant at the 10th and 5th day before excision.

Although shoot meristems covered by young leaves and leaf primordia are aseptic, surface disinfection before meristem dissection is a typical procedure. Since the tissues of a shoot apex are very tender, mild chemicals and procedures must be used for surface disinfestation. In our laboratory, the following procedures have been adopted. Buds with tightly covered leaves, such as pineapple and chrysanthemum, are immersed quickly in 75% ethanol. Buds with loosely covered leaves, such as carnation and potato, are immersed in 2% Clorox in an ultrasonic cleaner for 10 min. A few drops of Tween 20 should be added to each liter of solution to decrease surface tension, thereby reducing trapped air bubbles in the bud. The solution is removed with two rinses in sterile, distilled water.

2.2.3 Meristem Dissection

The dissecting of meristems is done under aseptic conditions using standard aseptic technique. A stereomicroscope, illuminator, and microscalpels are needed to excise meristems. Most spotlight illuminators on dissecting microscopes release intensive heat energy. In order to prevent heat damage, the meristems should be exposed to light as briefly as possible during dissection. A cool ray illuminator such as glass fiber optic is desirable to reduce heat damage. Hypodermic needles (21 gauge) are used as microscalpels. They can be cut into different lengths and fixed into handles.

To dissect meristems, the shoot tip is held under the microscope with forceps. Short hypodermic needles (2 cm) are used to remove leaves and leaf primordia. It is easier to excise the apical meristems of plants when the primordia are not arranged closely to the apical dome. Special care must be taken when this is not the case. As the dissecting procedure approaches the apex, a different sterile needle should be used to cut each primordium. This technique is a precaution to avoid transferring microorganisms to the sterile apex. It is especially important to follow this procedure when dissecting species having abundant hairs on the primordia, such as strawberry. Longer needles (3 cm) are used for meristem excision to provide a smooth, V-shaped cut below the apical dome. The explant is then transferred to the surface of the culture medium.

Explant orientation does not appear to be critical as long as it is not placed below the medium surface. A thin layer of sterile, distilled water helps to create contact between the explant and the medium surface. In addition, the water may provide a means to disperse toxic metabolic wastes released by the expland.

2.2.4 Culture Media

There is no general-purpose medium available for meristem culture. Most of the commonly used media were originally developed for root culture or other purposes. The optimum concentrations of the ingredients, especially of the micronutrients, have not been critically evaluated for meristem culture. Among the media used, Murashige and Skoog medium with some modifications is the one used most frequently with the greatest success (Murashige and Skoog 1962). The growth additives and salt concentration requirements of the medium vary from species to species and, especially, from one stage of culture development to another. The most frequent additives are growth regulators. However, some species are known to be cultured in media without growth regulators. It is likely that a sufficient quantity of endogenous hormone is already present in the explants.

Since roots are the principal site of cytokinin biosynthesis, the majority of culture media published are supplemented with a cytokinin to support axillary shoot proliferation. Some species are known to regenerate adventitious roots readily, the regenerated roots may act as a new source of cytokinin. Three cytokinins are frequently used to supplement culture media: Kinetin (Kin), N6-benzyladenine (BA) and N6-(2-isopentenyl)-adenine (2-iP). BA is most effective for shoot meristem culture, followed by Kin. 2-iP has been used less frequently. Hu and Wang (1983) have reported that in shoot meristem culture articles, 68% used BA, 23% used Kin and 9% used 2-iP, zeatin (Zea), and SD8339. The low percent usage of the last three cytokinins is partially attributed to their high cost and limited availability. One should be aware of the fact that, although a given cytokinin may not work well in certain species, it may be quite effective in others.

An auxin is required for shoot growth. Since the young shoot apex is an active site for auxin biosynthesis, exogenous auxin is not always needed in Stage I medium. This is especially observed when relatively large shoot tips are the explant sources. Thus, many of the Stage I media are auxin-free. Resting buds and meristems 0.4 mm or less may not produce enough endogenous auxins for shoot growth. In these cases, exogenous auxin needs to be added. Indole-3-acetic acid (IAA), indole-3-butyric acid (IBA), naphthyl-1 (or 2)-acetic acid (NAA), and 2,4-dichlorophenoxyacetic acid (2,4-D) are the most frequently used auxins in plant tissue cultures. IAA is the weakest of the four and is inactivated by light and by IAA-oxidase activity. NAA is the auxin routinely used by most laboratories for meristem shoot tip culture. In Hu and Wang's review chapter (1983) NAA was used in 51%, IBA in 27%, IAA in 22%, and 2,4-D in 6% of the cases. The usual concentration range used in Stage I is between 0.045 to 10 μM.

Gibberellin (GA) has been included in only 17% of the Stage I media listed in Hu and Wang's review chapter (1983). Evidently sufficient quantities of this hormone are synthesized by most explants. The addition of GA to the medium was necessary for meristems of cherry rootstocks that formed rosette plants when cultured in vitro. In GA- and BA-containing medium, axillary buds developed. These buds could be removed and cultured on BA medium, which resulted in internode elongation (Druart 1985).

Liquid medium may yield better results than agar solidified medium. Meristems of *Canna indica* developed better in half strength liquid MS than on agar solidified medium (Kromer and Kukulczanka 1985).

2.2.5 Culture Conditions

Few studies have been directed to revealing the optimum physical incubation conditions, including such factors as light intensity, photoperiod, light quality, temperature, and humidity. Typically, culture conditions are selected arbitrarily.

In general, shoot meristem cultures are incubated under lighted conditions. In potato meristem cultures, it was found that the optimum light intensity for initiating culture is 100 lx. The intensity can be increased to 2000 lx after 4 weeks of incubation. As the shoot reaches a length of 1 cm, the light intensity should be increased to 4000 lx. The high light intensity is advantageous for the survival rate of the cultures when shoots begin growth and after transfer to rooting medium (Wang and Hu 1980). Experiments with *Asparagus, Gerbera, Saxifraga*, and bromeliads identified an optimum light intensity of 100 lx for culture establishment and multiplication, while a higher optimum, 3000 to 10 000 lx is needed to condition the plantlets prior to transplantation into soil (Hasegawa et al. 1973).

Since light stimulates tissue browning for explants with high polyphenol content, it is advisable to reduce light intensity below 1 klx or even incubate in darkness. In *Pelargonium* shoot tip culture, a definite dark period is necessary, possibly to minimize the inhibitory effect of polyphenol oxidase activity in light (Pillai and Hildebrandt 1968). In contrast, dark incubation was found inferior to light incubation for potato. This may be due to an increased polyphenol oxidase activity in continuous darkness (Huth and Bode 1970).

In order to maximize the in vitro growth and prevent induction of dormancy, long photoperiods are used with few exceptions. The most commonly used photoperiod regime is 16/8 h day/night period. In most cases, light is provided by fluorescent tubes of the cool white or Gro-lux type. Small wattage incandescent bulbs are sometimes added to supplement the red and far-red region of the light spectrum.

The optimum incubation temperature can vary with species. The climatic zone where the plant originated offers some clue as to what temperature should be selected. Most workers select a constant incubation temperature ranging between 20 and 28 °C with the majority in the middle of this range. When the incubation temperature is raised above 28 °C, water tends to condense on the plants and vessel walls, and may restrict growth (Lane 1979). A day/night fluctuating temperature may be desirable for some plants, especially for those adapted to temperate and desert climates. Most workers who used a fluctuating day/night incubation temperature did so through arbitrary choice, without experimental support. A 25/20 °C day/night temperature was used for *Betula papyrifera* nodal cultures (Minocha 1980) and *Rheum rhaponticum* meristem cultures (Roggemans and Claes 1979). For *Gladiolus* bud culture, a 24/20 °C day/night temperature was used (Ziv 1979). No advantage of lower night temperature was established for shoot tip cultures of *Asparagus* (Hasagawa et al. 1973).

Relative humidity is not frequently controlled during incubation. Those who do control the humidity usually set it between 60 to 80%, with 70% being the most frequent setting.

3 Factors Influencing Disease Elimination

3.1 General

The value of meristem culture becomes apparent in the production of disease-free plants, in particular virus-free plants. Researchers have reported that the shoot and root apices of virus-infected plants are frequently devoid of pathogens or contain very low viral concentrations (White 1934; Kassanis 1957). It has also been demonstrated that virus-free shoots could be recovered from infected plants using meristem culture technique (Morel and Martin 1952, 1955). In addition, meristem culture can also be used to produce pathogen-free plants from stocks systemically infected with pathogens such as mycoplasma (Fedotina and Kryolva 1976; Jacoli 1978), fungi (Csinos and Hendrix 1977), bacteria (Knauss 1976; Theiler 1977), and seedborne viruses (Kartha et al. 1979). Such a method is of immeasurable value to the horticulture industry.

3.2 Pathogen Distribution

It has been demonstrated that the distribution of viruses in the plant is not uniform. Not all tissues or cells of a diseased plant are necessarily infected by the virus, and the shoot and root apices of virus-infected plants are frequently devoid of viral particles or contain very low viral concentrations (White 1934; Kassanis 1957). Limasset and Cornuet (1949) showed the increase of the TMV concentration distally from the apex of the tobacco plant. Using immunofluorescent and bioassay methods, Mori and Hosokawa (1977) proved that there is no virus in the apical dome and the two youngest leaf primordia of TMV infected tomato, petunia, and tobacco. A high virus concentration was detected in the vascular and pith regions below the third primordium. Using the same method, it was found that no virus was distributed within 0.2 mm of the shoot tips of tobacco and petunia infected with cucumber mosaic virus.

Apical domes are not always free from viral infection. Using the indicator plant, *Chenopodium amaranticolor*, Hollings and Stone (1964) proved the existence of carnation mottle virus in the apical dome of the infected carnation plant. Walkey and Webb (1968) used electron microscopy to observe viral particles in the apical dome of *Nicotiana rustica*. Using the same method, Appiano and Pennazio (1972) detected particles of potato virus X in the shoot meristem of infected potato.

To date, the fact that the apical dome and its adjacent region have a lower viral concentration or are even free from viral infection is not clearly understood. The following possibilities are assumed: (1) lack of vascular system, (2) high metabolic activity, (3) inactivating system in this region, and (4) high auxin concentration (Hollings and Stone 1964, 1965).

3.3 Plant Material and Condition

The following factors may affect the success of meristem culture: the season in which the bud is obtained, the position on the stem of the excised bud, the size of the explant, and treatments applied before and/or after culture.

3.3.1 Growth Season and Condition of Stock Plant

Like the conventional cutting method, the success of shoot meristem cultures is affected by the growth condition of the stock plant and the season during which the explant is obtained. Stone (1963) found a higher survival rate of carnation meristems excised and cultured in early spring and early autumn than those in summer and winter. Van Os (1964) stated that carnation meristems cultured in winter root more easily, whereas those cultured in summer yield a higher percentage of virus-free plants. Mellor and Stace-Smith (1969) and Stace-Smith (1985) found that for most potato varieties, meristems excised after dormancy, i.e., in spring or early summer, root more readily than those taken later in the year. With storage organs in general, the best results may be expected when meristems are dissected at the end of their dormancy period when sprouting has begun.

3.3.2 Bud Location

In a comparison of in vitro survival rates of carnation meristems, Stone (1963) found that 20.1% of the explants from terminal buds developed into plantlets, whereas only 13% of the explants from lateral buds developed. In chrysanthemum meristem culture, Hollings and Stone (1968) noted that the success rate of explants obtained from terminal buds is 32%, whereas from lateral buds it is 18%. The terminal buds presumably have a stronger growth potential than laterals, providing a greater success rate in meristem culture. For some plants the flowering habit is such that the terminal bud differentiates into a flower bud while the adjacent axillary bud develops into the terminal vegetative bud. For these plants, such as tomato, the vegetative axillary bud should be utilized for meristem culture.

3.3.3 Explant Size

In general, the larger the explant the better the chance for its survival. Stace-Smith and Mellor (1968) reported that potato buds 1 mm or more in length root more readily than smaller buds. In *Cassava*, only explants exceeding 0.2 mm in length form complete plants. Those less than 0.2 mm produce either callus or roots (Kartha and Gamborg 1975). However, in order to take advantage of the nonuniform and imperfect distribution of viruses in the host plant body, the size of the excised meristem should be as small as possible. As a compromise, explants of intermediate size should be used. Stone (1963) found that when excising carnation shoot meristems, explants smaller than 0.2 mm are unlikely to root, while those larger than 0.75 mm produce plants that still contained mottle viruses. Tips between 0.2 and 0.5 mm most frequently produce virus-free plants.

Although rare, there are examples using large shoot tips for successful viral elimination. Vine and Jones (1969) obtained virus-free plantlets from 5 mm shoot tips. Leu (1972) eliminated a sugarcane mosaic virus by cultivation of 4-mm shoot tips of sugarcane.

The presence of leaf primordium can determine the capability of a meristem explant to develop. Kassanis and Varma (1967) used 0.1-mm-long potato meristems and found that the explants with one leaf primordium had a more advanced development than those lacking leaf primordium. However, there are exceptions. Some researchers report successful culture of apical dome explants (Smith and Murashige 1970; Riviers 1973; Shabde and Murashige 1977). In addition, Kartha and Gamborg (1975) detected that the presence of leaf primordia is not essential for plant regeneration for *Cassava* meristem culture. For general clonal propagation purposes, meristems with three to four leaf primordia are dissected. If the objective is to eliminate systemic pathogen infections, meristems with only one pair of leaf primordia are cultured.

3.4 Addendum in Culture Media

Data on the distribution of virus particles in the shoot tip frequently disagree with the actual virus elimination data obtained through meristem cultures. For example, Hollings and Stone (1964) excised carnation meristems which could induce lesions of mottle virus types on the indicator plant, *Chenopodium amarauticolor*. However, after a 30-h incubation on nutrient medium, the infection could no longer be detected. This work suggests that the contact of the virally infected tissue with the culture medium may eliminate viruses present in the tissue.

Antiserum or nucleoside analog in culture media can be used to inactivate viruses in shoot tips or cultured tissues. Jordan et al. (1983) eliminated potato virus X through meristem culture in nucleoside analog medium. Inouye (1983) found that the pretreatment of isolated *Cymbidium* meristems with the antiserum of *Odontoglossum* ringspot virus for 2 h to 1 day markedly increased the frequency of virus-free plantlets from shoot meristem culture.

Virus suppression has been reported with the use of chemicals such as Virazole (ribavirine) and Vidarabine (antimetabolites) in the culture medium (Walkey 1980; Kartha 1986; Long and Cassells 1986; Bockholt et al. 1989). Virus-free tobacco and potato plants were produced with the use of Virazole (Cassells and Long 1980, 1982). Five antiphytoviral compounds were tested to eliminate potato viruses X, S, and M from potato cultivars. Virus-free plants were recovered from all treatments (Borissenko et al. 1985). Paludan (1985) found only a limited inactivating effect of ribovirin or amantadine against TMV in *Aeschyranthus* meristems. The use of Virazone (ribavirin) and dyestuffs gave a higher percentage of virus-free plants than petiole explants produced without these chemicals or through meristem culture (Aminuddin and Singh 1985). Hollings and Stone (1965) has reported that the elimination of virus was enhanced by the incorporation of malachite green, 2-4-D, and thiouracil, into the culture medium. The inhibitory effect of these compounds on virus multiplication may be through interference with nucleic acid metabolism.

In addition to chemical controls, heat treatment has been used to inactivate some viruses (Ten Houten et al. 1968; Quak 1977). The ability of the plant to survive the high temperature required is often a limiting factor.

4 In Vitro Clonal Propagation of Disease-Eliminated Plants

4.1 Incubation Conditions and Polyphenol Oxidation Problem

General incubation conditions are similar to those for shoot meristem culture as described in Section 2.2.5.

Many plants are rich in phenolic compounds, especially those plants which are resistant to diseases or insects. In general, the plants requiring shoot meristem culture are mainly the group having disease problems, therefore they tend to have less polyphenol oxidase activity. After tissue injury from dissection, polyphenol oxidase activity will be activated and the tissue can turn brown or black. The oxidation products are known to inhibit enzyme activity in plant body, darken the tissues and culture media, and even kill the explants. Such phenomena impose a serious block on the establishment of primary cultures. Some of the procedures used by various workers to combat this problem are (1) adding antioxidants to culture medium, (2) presoaking explants in antioxidant before inoculating, (3) incubating in reduced light or darkness for the initial period of primary culture, (4) transferring explants to fresh medium whenever browning of the medium is observed, and (5) using liquid media or adding a thin layer of liquid medium on the surface of solid media to dilute toxic metabolites.

The inclusion of antioxidants, such as ascorbic acid, polyvinylpyrrolidone (PVP), diothiothreitol, or bovine serum albumen into Linsmaier and Skoog (LS) agar medium has been reported in *Anigozanthos* bud culture (McComb and Newton 1981). According to Ichihashi and Kako (1977), the browning of *Cattleya* shoot tips was most effectively inhibited by adding an antioxidant into a liquid medium in stationary condition. The inhibitor for polyphenoloxidase activity has little or no effect on the browning process of a solid medium. Stevenson and Harris (1980), however, reported reduction of agar medium discoloration with PVP-10 (0.01%) in *Fuchsia* shoot tip cultures. In *Tactona grandis* bud cultures (Gupta et al. 1981), the medium turned black and all the explants died. In order to reduce blackening, the explants were suspended in a solution of different antibrowning agents in 0.058 M sucrose and agitated on a rotary shaker for 45 min before inoculation onto MS solid medium. The agents tested were H_2O_2 (5%), ascorbic acid (0.28 mM), soluble PVP (0.7%), and polyclar AT (0.7%). Blackening was reduced by all the treatments. Multiple shoots, nevertheless, were formed only from explants treated with polyclar AT, an insoluble PVP, which is known to combine with phenolics by hydrogen bonding. More recently, polyclar AT has been replaced by insoluble polyvinylpyrrolidone (Sigma Chem. Co. USA), which gives similar results.

Since the harmful phenolic oxidation products are formed under illumination, reduction of light intensity at the initial period of Stage I culture incubation should be beneficial. Less than 4% browning of the garlic meristem explants was obtained by Wang and Huang (1974) when the incubation light intensity during the first month was 150 lx, whereas 18% browning resulted with 500 lx. In order to reduce the accumulation of phenolic oxidates, several workers carried out the initial incubation period from 1 to 6 weeks in darkness (Monaco et al. 1977; Adams et al. 1979; McComb and Newton 1981).

The tissue and medium discoloration in thornless blackberry culture were effectively controlled by Broome and Zimmerman (1978) when the shoot tip explants were transferred to fresh medium 1 to 2 days after initial culturing. In shoot tip culture of mountain laurel, Lloyd and McCown (1980) transferred the explants into fresh liquid media 12 and 24 h after the initial culturing, then continued the transfer process on a daily basis for 1 week before medium discoloration was completely under control. Axillary shoots were successfully produced from these explants 1 to 2 months later.

4.2 Multiple Shoot Formation

The main objective of the multiple shoot formation stage is to produce the maximum number of useful propagule units. In culture, a meristem may develop either into single shoots, multiple shoot masses (Bajaj and Dhanju 1979; Bajaj and Gill 1986), or even into rooted plantlets. All of these cultures may be used for micropropagation.

4.2.1 Axillary Shoot Proliferation

Among the methods for achieving propagule multiplication axillary shoot proliferation provides genetic stability and is easily achievable for most plant species. The method has rapidly gained popularity in recent years. For instance, in 1974 only four species that were being multiplied utilized this method. By the end of 1981, there were at least 90 more species that could be added to the list. With many genera, a millionfold enhancement of clonal plant increase in a year's time has been estimated using this method.

In axillary shoot proliferation, cytokinin is utilized to overcome the apical dominance of shoots and to enhance the branching of lateral buds from leaf axils. Shoots obtained from meristem cultures of five grape cultivars were proliferated using half strength MS medium supplemented with BA (Zatyko and Molnar 1985). In general, BA is the most effective cytokinin for stimulating axillary shoot proliferation, followed by, in decreasing order, Kin and 2-iP (Kitto and Young 1981; Papachatzi et al. 1981; Yang et al. 1981).

Exogenous auxins do not promote axillary shoot proliferation; however, their presence may improve culture growth. It was found that the multiplication potential of potato shoots during in vitro layering declined after seven to ten subcultures. This degeneration phenomenon could be completely reversed and eliminated by enriching the medium with 0.54 µM NAA (Wang and Hu 1982). One of the possible roles of auxin in Stage II medium is to nullify the suppressive effect of high cytokinin concentrations on axillary shoot elongation and restore normal shoot growth (Lundergan and Janick 1979). Too high a concentration of auxin may not only inhibit axillary bud branching (Hasegawa 1980), but also induce callus formation, especially when 2,4-D is used.

Gibberellic acid (GA) used in medium is exceedingly low compared with other in vitro culture systems. Its role is essentially for axillary bud elongation (Schnabdrauch and Sink 1979). Wochok and Sluis (1980) observed that a topical treatment of *Atriplex* shoot explants with GA was effective not only in stimulating shoot elongation, but also in enhancing shoot multiplication beyond that of the most effective auxin-cytokinin combination. Lundergan and Janick (1979), however, detected no influence on either proliferation or growth of apple axillary buds when 0.29 µM GA was added. Ancora et al. (1981), working with globe artichoke, suggested that GA was not necessary to maintain shoot proliferation in vitro.

4.2.2 In Vitro Layering

The in vitro layering method was developed by Wang (1977) to mass-clone PVX-free potato plants. The plantlets obtained through meristem cultures are first placed on the medium in a horizontal position. The terminal bud as well as the axillary buds develop into two to three shoots within 3 weeks. Repeated layering of these new shoots results in a multiple-shoot plant mass. The shoots, with roots at their lower nodes, are then harvested 1 cm above the medium surface and laid on the surface of the fresh medium. Two to three new shoots will again develop from each cutting. The shoots can thus be repeatedly subcultured at 3-week intervals and the number doubled or tripled with each transfer.

4.2.3 Adventitious Organogenesis and Embryogenesis

Callus production can be induced in explants derived from meristem culture. From this callus, adventitious shoots, protocorms, or embryoids can be induced to differentiate in large quantities in numerous species. Adventitious organogenesis may enable a substantially faster cloning than axillary shoot multiplication. Nevertheless, the investigators should be warned that chromosome aberrations and ploidy changes frequently occur when an intermediary callus phase has been included in a mass cloning process (Boonnour et al. 1988; Glendening and Sjolund 1988; Kane et al. 1988).

4.2.4 Storage Organ Formation

Storage organs may be induced to form in culture. The production of storage organs is favored over plantlet formation because there is no difficulty in transferring the dormant storage organs out of the aseptic culture and storing them nonaseptically. Wang and Hu (1982) induced in vitro potato tuberization from PVX-freed plantlets.

In addition, in vitro storage organ production has also been induced in *Gladiolus* (Ziv et al. 1970), lily (Robb 1957; Hackett 1969), and onion (Fedotina and Krylova 1976).

4.3 Root Formation

In the root formation stage, adventitious roots regenerate from vascular bundles in the stems of Stage II plantlets. Adventitious root formation can be induced quite readily in many herbaceous species, but it can be recalcitrant in most woody species.

Reducing the cytokinin levels before rooting can be beneficial. Wang and Hu (1984) reduced the BA concentration from 264 to 22 μM to stimulate the elongation of *Sassafras* shoots from Stage II cultures before transferring them to rooting medium. Meristem cultures of some cherry rootstock species yielded shoots that were difficult to root. Preconditioning the shoots in the dark for 9 days and the addition of proline and riboflavin to the medium improved rooting (Druart 1985).

Auxin is essential for root intiation; however, high auxin concentration has been reported to inhibit the root elongation phase (Thimann 1977). James and Thurbon (1979) reported that in order to provide the hard-to-root apple rootstock with a strong root induction stimulus, as well as to avoid callusing, the shoots were first cultured in an auxin-containing "root initiation medium" for 4 to 8 days, then transferred to an auxin-free "root developing medium". This procedure effectively prevented callus formation and resulted in 95% rooting, compared to those in continuous contact with auxin. Similar results have been obtained by Kusey et al. (1980) who predipped the base of in vitro-produced *Gypsophila* shoots in 0.13 mM NAA or 0.12 mM IBA solution for 5 s. Shoots were then transferred into hormone-free White's medium for root initiation and development.

Regardless of the hormones present, the N, P, and K salt concentrations found in MS and B5 (Gamborg et al. 1968) media are often reported to be too high for rooting. When the salt concentration in the medium is lowered to one-half, one-third, or one-fourth of the standard strength, the survival rate of meristem culture and rooting becomes abundant (Kartha et al. 1981; Monteuuis 1987). Although lower salt concentration in a medium may be beneficial to root induction, it sometimes results in poor top growth. Wang (1978) observed that a one-ninth strength of MS medium stimulated 100% rooting in *Cryptomeria japonica*, but resulted in poor shoot growth. A one-third strength of MS medium, on the other hand, resulted in 87% rooting with good top growth. Gupta et al. (1981) successfully rooted *Eucalyptus citriodora* after transferring shoots from MS to White's medium, but the rooted shoots failed to survive on this low salt medium, the leaves gradually turned yellow and dropped off. If the plants were transferred within 2 weeks, just after the emergence of the first roots, back into MS liquid medium, the yellowish leaves turned green and a well-developed root system was formed.

Agar is used to solidify most rooting media. Physiologically, agar is not a completely inert material, but it is a source of various types of substances which may affect growth. Agar, thus, may result in poor root growth in certain sensitive species. In addition to the presence of possible growth inhibitors, the reason for poor rooting in agar medium may also be due to poor aeration and a slow rate of diffusion of the toxic metabolites released from growing tissues. Two methods are commonly used to circumvent the inhibitory effect of agars: (1) supplementing fine powder of activated charcoal (AC) to the agar medium; and (2) using a liquid and filter-paperbridge system in place of agar. AC may absorb toxic substances in the medium, thereby improving root regeneration and development (Wang and Huang 1976). Improved root development was noted for Malling Merton apple rootstocks when shoots were removed from IBA-containing medium and transferred to AC-containing medium (Snir and Erez 1980). AC may also absorb residual cytokinin from the medium. Takayama and Misawa (1980) reported a BA inhibition of *Lilium* root formation. Adding AC to the medium completely reversed the inhibition. Root formation and growth were even better in BA-free medium containing AC than in medium without AC. AC is also capable of shading in vitro roots from light, which may inhibit root growth.

For the forest trees, the age of the plant plays a significant role in the root-regenerating capacity. The root development of *Eucalyptus* nodal cultures was most frequent when explants were obtained from the basal end of the seedlings (de Fossard et al. 1974). Wang and Hu (1984) experienced a sporadic rooting in shoots resulting from bud cultures of a 5-year-old *Sassafras randaiense* tree, whereas the rooting of shoots resulting from embryo cultures of the same species was essentially 100%. It is known that the subculturing process may change the physiological state and gradually rejuvenate the explants. Juvenile characteristics were induced from the bud explants of an adult clone of *Vitis vinifera* after two to three subcultures (Mullins et al. 1979). In bud culture of 20-year-old *Eucalyptus citriodora* trees, Gupta et al. (1981) reported that none of the Stage III treatments resulted in root formation in either the initial explants or the first three subcultures. However, 35–40% of the shoots rooted at the fourth subculture. In the fifth and subsequent passages the percentage of rooting was about 45–50%. Subculturing thus may result in a return of some characteristics associated with juvenility.

4.4 Planting Out

Excessively high water loss from the leaves of plants immediately after transplanting has been reported. Such a high rate of water loss is related to: (1) the reduced quantities of epicuticular wax (Sutter and Langhans 1979), (2) the high volume of mesophyll intercellular spaces and (3) the slowness of stomatal response to water stress (Brainerd et al. 1981). A period of humidity acclimatization is required for the newly transferred plantlets to adapt to the outside environment. During acclimatization, humidity is gradually reduced over a period of two or three physiological adaptations, enabling the development of typical terrestrial plant water control.

5 Long-Term Storage

Considerable interest has been generated in exploring the possibilities of storing plant meristems or vegetative buds in vitro. Meristem and shoot-tip culture are preferred since the probability of maintaining the genetic stability of clonal selections is greater than with other tissue culture systems. An in vitro germplasm preservation method could reduce the labor and space requirements of traditional storage methods.

One approach to long-term storage is to maintain cultures at lower growth temperatures and low light intensity. At the CATIE laboratory, meristem cultures of selected root and tuber crops and shoot-tip cultures of vanilla and *Musa* were stored for up to 18 months at 18 °C. Cultures were maintained on filter paper bridges in liquid media under low light intensity (Jarret et al. 1986). Alleweldt and Harst-Langenbucher (1987) used growth inhibitors for the long-term storage of grapevine.

Another germplasm preservation method utilizes cryogenic technology (Bajaj 1986). Liquid nitrogen is used to provide ultra-low temperatures. At these temperatures, almost all of the cells' metabolic functions are arrested. This state of inactivity enables long-term storage. Shoot development after thawing is essential. Cryoprotectant solutions are used to improve survival of frozen plant material. Vegetative buds, as well as meristems, have been subjected to cryogenic temperatures.

Various reports of the use of cryogenic temperatures for germplasm storage include carnation (Seibert 1976; Uemura and Sakai 1980; Dereuddre et al. 1987), strawberry (Sakai et al. 1978; Kartha et al. 1980), potato (Bajaj 1977a, 1981; Grout and Henshaw 1978; Towill 1981; Manzhulin et al. 1983), cassava (Bajaj 1977b, 1983, 1985; Kartha et al. 1982), pea (Kartha et al. 1979; Haskins and Kartha 1980), tomato (Grout et al. 1978), date palm (Mater 1987), *Arachis* and *Cicer* (Bajaj 1979), *Rubus* (Reed and Lagerstedt 1987), mulberry (Yakuwa and Oka 1988), mint (Towill 1988), and *Pyrus* (Reed 1990).

6 Implications in Commercial Propagation

It is highly probable that in vitro propagation techniques will become the preferred commercial method for cloning desirable genotypes of forest and fruit trees, ornamentals, and select F_1 hybrids. If production systems are cost-effective, in vitro micropropagation can offer space- and time-saving advantages (Wang and Hu 1982).

The main advantage of using meristem culture in a commercial micropropagation system is to attempt to recover pathogen-free plants. It is well known that disease can become a major problem when plant material contaminated with pathogenic bacteria, fungi, viruses, viroids, mycoplasmas, or rickettsia is propagated under in vitro or nursery conditions. With such stock, plant quality can be reduced, or devastating crop losses can occur as noted when Rieger Begonia becomes infected with *Xanthomonas*.

6.1 Production of Disease-Indexed Plants

Before pathogen-free plants became available, there was no practical method of determining the losses in quality and yield of the clonal crops due to pathogen infection.

However, in vitro regeneration of virus-free plants has made such estimations possible. Wang and Hu (1982) eliminated potato virus X and Y through meristem culture. The virus-free plants grew rapidly and displayed greater vigor than those infected. Culture virus-indexed geraniums were reported to exhibit larger blooms in greater numbers, to be more vigorous yet self-branching, to have reduced finishing time, and to be capable of blooming in low light conditions (Oglevee-O'Donovan 1986).

It is important to be aware that meristem cultures do not always produce pathogen-free plants. Large meristem tips are frequently used by commercial propagators. In such cases, if the source plants happened to be infected, and are not pathogen-indexed, the disease would spread to clonal progenies. Tissue culture is considered the primary cause of a devastating virus disease problem in the orchid industry (Lawson and Hearon 1973). Before orchid tissue culture became standard practice, viruses were considered manageable. With the widespread use of micropropagated plants, viruses were transferred throughout the orchid industry.

Since in no case is meristem culture 100% effective in producing pathogen-free plants from infected plants, disease-indexing programs must be used to test the in vitro-regenerated plants. Indexing programs should be incorporated into various stages of in vitro propagation as well as into the large-scale field production. Some indexing systems are rigid, such as for the production of cultured virus-indexed geraniums. Since some pelargonium vascular bacteria and fungal pathogens can severely damage a crop, cuttings are culture indexed to detect the pathogens. Infected cuttings are removed from the system. Culture-indexed plants are then subjected to heat therapy and meristem culture. Plants recovered from this system are grown in isolation and virus-indexed several times before being horticulturally rated and numbers increased for commercial growers (Oglevee-O'Donovan 1986).

Depending on the species and the production requirements it may be possible to maintain the indexed stock in vitro. To maintain controlled disease and insect conditions in the greenhouse is space- and time-consuming. The in vitro system offers a possibility of isolating indexed plants from infection and it can often be cheaper and easier (Pierik 1987). When needed, some of the indexed stock could be propagated in vitro or in the greenhouse.

6.2 Status of Commercial Production

It is difficult to estimate the number of companies currently producing plants using micropropagation techniques. An estimate from 1984 noted 90 laboratories in the US with an additional 115 outside the US involved with the propagation of herbaceous ornamental crops. Approximately 100 laboratories were engaged in the mericloning of orchids (Chu 1986). Holland has shown an increase in the number of micropropagation laboratories since 1984, when 33 produced an estimated 29 million plants to 1988 with 65 to 70 producing over 53 million (Pierik 1988; Misawa pers. commun. 1989). An estimate in 1988 for Western Europe noted 288 laboratories, of which 154 were public and 134 privately owned, while the number in North America was 60 (Misawa pers. commun. 1989).

Private companies throughout the world are propagating plants by tissue culture for internal needs. For example, orchid growers often have a tissue culture facility for sowing seeds or cloning expensive orchids. Ornamental plant growers propagate select lines for their commercial nursery. Larger companies such as seed businesses use micropropagation to produce select breeding lines (Misawa, pers. commun. 1989).

Most establishments undertaking aspects of plant genetic engineering are usually distinct from the plant micropropagation business. However, in many countries government-subsidized laboratories supply research as well as micropropagated plants for commercial use. Laboratories of some botanical gardens have a similar function (George and Sherrington 1984).

Although the majority of commercial laboratories are still concerned with the propagation of herbaceous ornamentals, there is an increasing trend toward the propagation of woody perennials, vegetables, and other crops.

6.3 Limiting Factors for Commercial Production

Despite the growth of micropropagation over the last decade, there are still limiting factors for this technology. Pierik (1988) clearly outlines several reasons for the lack of widespread use of large-scale commercial application of micropropagation. Some of these factors will be highlighted.

A major factor is that the process of transferring culture vessels and harvesting plantlets is highly labor-intensive. Labor can account for up to 70% of the final cost. Producing in countries having low labor cost is only a temporary solution; cost reduction methods are required. It can be difficult to use labor efficiently because production requirements are often seasonal.

Knowledge concerning the in vitro system is often lacking. The control of the physical environment and its influence on plant biochemistry and physiology needs further study. This knowledge will help solve the problems of vitrification, acclimation, and woody plant micropropagation.

Contamination can be a limiting production factor especially when meristem culture followed by an indexing program is not used.

Genetic mutations can often limit the use of a production system. Mutations have been noted to occur when regeneration from callus is a part of the system. For example, an oil palm regeneration system produced a 30% genetical change (Chu 1986). Using meristem culture is no assurance that mutations will not occur in the micropropagation system. Despite this disadvantage, this phenomena can enable the recovery of new plant types. A plantlet originating from an adventitious shoot of a banana meristem culture was found to be tolerant to Panama disease race 4. The tolerant line could be micropropagated (Hwang et al. 1984). For sugarcane, researchers have increased tiller number, stalk length, leaf altitude, and stalk diameter (Lourens and Martin 1987). In ornamental Eustoma, more branched plants were selected through a meristem culture system (Griesbach and Semenluk 1987).

There continues to be a need for micropropagation system development. Propagation techniques developed at universities or institutes are not always economically viable, and may be rejected. Industry and university efforts should be combined to refine production systems and meet economical standards.

References

Adams RM II, Koenigsberg SS, Langhans RW (1979) In vitro propagation of *Cephalotus follicularis* (Australian pitche plant). HortScience 14:512–513

Alleweldt G, Harst-Langenbucher M (1987) The effect of growth inhibitors on long-term storage of in vitro cultures of grapevine. Vitis 26(2):57–64

Aminuddin Singh BP (1985) *Petunia hybrida* Hort. explant culture: effect of virazole and certain dyestuffs on *Petunia* mosaic virus elimination. Indian Phytopathol 38:692–694

Ancora G, Belli-Donini ML, Cuozzo L (1981) Globe artichoke plants obtained from shoot apices through rapid in vitro micropropagation. Sci Hortic 14:207–213

Appiano A, Pennazio S (1972) Electron microscopy of potato meristem tips infected with potato virus X. J Gen Virol 14:273–276

Bajaj YPS (1977a) Clonal multiplication and cryopreservation of cassava through tissue culture. Crop Improv 4:198–204

Bajaj YPS (1977b) Initiation of shoots and callus from potato tuber sprouts and axillary buds frozen at –196 °C. Crop Improv 4:48–53

Bajaj YPS (1979) Freeze preservation of *Arachis hypogaea* and *Cicer arietinum*. Indian J Exp Biol 17:1405

Bajaj YPS (1981) Regeneration of plants from potato meristems freeze-preserved for 24 months. Euphytica 30:141–145

Bajaj YPS (1983) Cassava plants from meristem cultures freeze-preserved for 3 years. Field Crops Res 7(3):161–167

Bajaj YPS (1985) Cryopreservation of germplasm of potato (*Solanum tuberosum* L.) and cassava (*Manihot esculenta* Crantz): viability of excised meristems cryopreserved up to 4 years. Indian J Exp Biol 23:285–287

Bajaj YPS (1986) In vitro preservation of genetic resources – techniques and problems. Int Symp Nuclear techniques and in vitro culture for plant improvement. IAEA, Vienna, pp 43–57

Bajaj YPS, Dhanju MS (1979) Regeneration of plants from apical meristem tips of some legumes. Curr Sci 48:906–907

Bajaj YPS, Gill MS (1986) Micropropagation and germplasm preservation of cotton (*Gossypium* spp.) through shoot tip and meristem culture. Ind J Exp Biol 24:581–583

Bockholt S, Hellwald KH, Buchenauer H (1989) Untersuchungen zum Wirkingsspektrum und zur Wirkungsweise zweier quarternarer Ammoniumverbindungen gegenüber phytopathogenen Viren. J Phytopathol 127:331–340

Boonnour K, Wainwright H, Hicks TGT (1988) The micropropagation of *Lonicera periclymenum* L. (honeysuckle). Acta Hortic 226:183–189

Borissenko S, Schuster G, Schmygla W (1985) Obtaining a high percentage of explants with negative seriological reactions against viruses by combining potato meristem culture with antiphytoviral chemotherapy. Phytopathol Z 114:185–188

Brainerd KE, Fuchigami LH, Kwiatkowshi S, Clark CS (1981) Leaf anatomy and water stress of aseptically cultured Pixy plum grown under different environments. HortScience 16:173–175

Broome OC, Zimmerman RH (1978) In vitro propagation of blackberry. HortScience 13:151–153

Cassells AC, Long RD (1980) The regeneration of virus-free plants from cucumber mosaic virus and potato virus Y-infected tobacco plants cultured in the presence of virazole. Z Naturforsch 35C:350–351

Cassells AC, Long RD (1982) The elimination of potato viruses X, Y, S and M in meristem and explant cultures of potato in the presence of virazole. Potato Res 25:165–173

Chu IYE (1986) The application of tissue culture to plant improvement and propagation in the ornamental horticulture industry. In: Zimmerman RH, Griesbach RJ, Hammerschlag FA, Lawson RH (eds) Tissue culture as a plant production system for horticultural crops. Martinus Nijhoff, Boston, pp 15–33

Csinos A, Hendrix W (1977) Toxin produced by *Phytophythora cryptogea* active on excised tobacco leaves. Can J Bot 55:1156–1162

Cutter EG (1971) Plant anatomy: experiment and interpretation. Part 2. Addison-Wesley London, p 343

deFossard RA, Nitsch C, Cresswell RJ, Lee ECM (1974) Tissue and organ culture of *Eucalyptus*. NZ J For Sci 4:267–278

Dereuddre J, Galerne M, Gazeau C (1987) The effects of sucrose on survival of freezing in liquid nitrogen (–196 °C) of carnation (*Dianthus caryophyllus* L.) apical meristems cultivated in vitro. CR Acad Sci Paris 304(19):485–487
Druart P (1985) True-to-type propagation of cherry rootstocks and varieties by in vitro culture. Acta Hortic 169:319–328
Esau K (1960) Anatomy of seed plants, 1st edn. Wiley Interscience, New York, p 376
Fedotina VL, Krylova NV (1976) Ridding tobaccos of the mycoplasmic infection big bud by the method of tissue culturing. Dokl Bot Sci 228:49–51
Gamborg OL, Miller LA, Ojima K (1968) Nutrient requirements of suspension cultures of soybean root cells. Exp Cell Res 50:151–158
George EF, Sherrington PD (1984) Commercial tissue culture laboratories. In: George EF, Sherrington PD (eds) Plant propagation by tissue culture – handbook and directory of commercial laboratories. Eastern Press, Great Britain, pp 549–551
Glendening TM, Sjolund T (1988) In vitro propagation of kohlrabi from leaf explants. HortScience 23(4):772
Griesbach RJ, Semenluk P (1987) Use of somaclonal variation in the improvement of *Eustoma grandiflorum*. J Hered 78:114–116
Grout BW, Henshaw G (1978) Freeze-preservation of potato shoot tip culture. Ann Bot 42:1227–1229
Grout BW, Westcott R, Henshaw G (1978) Survival of shoot meristems of tomato seedlings frozen in liquid nitrogen. Cryobiology 15:478–483
Gupta PK, Mascarenhas AF, Jagannathan V (1981) Tissue culture of forest trees – clonal propagation of mature trees of *Eucalyptus citriodora* Hook, by tissue culture. Plant Sci Lett 20:195–201
Hackett WP (1969) Aseptic multiplication of lily bulblets from bulb scales. Proc Int Plant Prop Soc 19:105–108
Hasegawa PM (1980) Factors affecting shoot and root initiation from cultured rose shoot tips. J Am Soc Hortic Sci 105:216–220
Hasegawa PM, Murashige T, Takatori FH (1973) Propagation of *Asparagus* through shoot apex culture. II. Light and temperature requirements, transplantability of plants, and cytohistological characteristics. J Am Soc Hortic Sci 98:143–148
Haskins RH, Kartha KK (1980) Freeze-preservation of pea meristems: cell survival. Can J Bot 58:833–840
Hollings M, Stone OM (1964) Investigation of carnation viruses. I. Carnation mottle. Ann Appl Biol 53:103–118
Hollings M, Stone OM (1965) Investigation of carnation viruses. II. Carnation ringspot. Ann Appl Biol 56:73–86
Hollings M, Stone OM (1968) Techniques and problems in the production of virus-tested planting material. Sci Hortic 20:57–72
Hu C, Wang PJ (1983) Meristem, shoot tip and bud culture. In: Evans DA, Sharp WR, Ammirato PV, Yamada Y (eds) Handbook of plant cell culture, vol I. Techniques for propagation and breeding. MacMillan, New York, pp 177–227
Huth W, Bode O (1970) Befreiung Völlig infizierter Kartoffelsorten von Infektionen des Kartoffel-S-Virus durch Meristemkultur. Nachrichtenbl Dtsch Pflanzenschutdienst (Braunschw) 22:37–39
Hwang SC, Chen CL, Lin JC, Lin HL (1984) Cultivation of banana using plantlets from meristem culture. HortScience 19(2):231–233
Ichihashi S, Kako S (1977) Studies on clonal propagation of *Cattleya* through tissue culture method. II. Browning of *Cattleya*. J Jpn Soc Hortic Sci 46:325–330
Inouye N (1983) Effect of antiserum treatment on the production of virus-free *Cymbidium* by means of meristem culture. Nogaku Kenkyu 60:123–133
Jacoli GG (1978) Sequential degeneration of mycoplasma-like bodies in plant tissue cultures infected with aster yellows. Can J Bot 56:133–140
James DJ, Thurbon IJ (1979) Rapid in vitro rooting of the apple rootstock M9J. HortScience 54:309–311
Jarret RL, Fernandez ZR, Salazar S (1986) In vitro conservation at CATIE. Plant Genet Resour Newsl 68:6–10
Jordan M, Corters I, Montenegro G (1983) Regeneration of *Lapageria rosea* plantlets by tissue culture. Gartenbauwissenschaft 48:97–100
Kane ME, Sheehan TJ, Philman NL (1988) A micropropagation protocol using *Fraser photinia* for mutation induction and new cultivar selection. Proc Fla State Hortic Soc 100:334–337

Kartha KK (1986) Elimination of viruses from cultures in presence of antiviral chemicals. In: Withers LA, Alderson PG (eds) Plant tissue culture and its agricultural applications. Butterworth, London, pp 219–238
Kartha KK, Gamborg OL (1975) Elimination of cassava mosaic disease by meristem culture. Phytopathology 65:826–828
Kartha KK, Leung N, Gamborg O (1979) Freeze-preservation of pea meristems in liquid nitrogen and subsequent plant regeneration. Plant Sci Lett 15:7–15
Kartha KK, Leung N, Pahl K (1980) Cryopreservation of strawberry meristems and mass propagation of plantlets. J Am Soc Hortic Sci 105:481–484
Kartha KK, Pahl K, Leung NL, Mroginski LA (1981) Plant regeneration from meristems of grain legumes: soybean, cowpea, peanut, chickpea, and bean. Can J Bot 59:1671–1679
Kartha KK, Leung NL, Mroginski LA (1982) In vitro growth responses and plant regeneration from cryopreserved meristems of cassava (*Manihot esculenta* Crantz). Z Pflanzenphysiol 107:133–140
Kassanis B (1957) The use of tissue cultures to produce virus-free clones from infected potato varieties. Ann Appl Biol 45:422–427
Kassanis B, Varma A (1967) The production of virus-free clones of some British potato varieties. Ann Appl Biol 59:447–450
Kitto SL, Young MJ (1981) In vitro propagation of *Carrizo citrange*. HortScience 16:305–306
Knauss JF (1976) A tissue culture method for producing *Dieffenbachia picta* cv. Perfection free of fungi and bacteria. Proc Fla State Hortic Sci 89:293–296
Kromer K, Kukulczanka K (1985) In vitro cultures of meristem tips of *Canna indica* (L.). Acta Hortic 167:279–285
Kusey WE, Hammer PA, Weiler TC (1980) In vitro propagation of *Gypsophila paniculala* L. Bristol Fairy. HortScience 15:600–601
Lane WD (1979) In vitro propagation of *Spirea bumalda* and *Prunus cistena* from shoot apices. Can J Plant Sci 59:1025–1029
Lawson RH, Hearon SD (1973) Symtomatology of *Cattleya* mericlones infected with *Cymbidium* mosaic virus. Am Orchid Soc Bull 42:1071–1074
Leu L (1972) Freeing sugarcane from mosaic virus by apical meristem culture and tissue culture. Rep Taiwan Sugar Exp Stn 57:57–63
Limasset P, Cornuet P (1949) Recherche du virus de la mosaique du tabac dans les meristèmes des plantes infectées. CR Hebd Séances Acad Sci 228:1971–1972
Lloyd G, McCown B (1980) Commercially feasible micropropagation of mountain laurel, *Kalmia latifolia*, by use of shoot-tip culture. Proc Int Plant Prop Soc 30:421–427
Long RD, Cassells AC (1986) In: Withers LA, Alderson (eds) Plant tissue culture and its agricultural applications. Butterworth, London, pp 239–248
Lourens AG, Martin FA (1987) Evaluation of in vitro propagated sugarcane hybrids for somaclonal variation. Crop Sci 27(4):793–796
Lundergan C, Janick J (1979) Low temperature storage of in vitro apple shoots. HortScience 14:514
Manzhulin AV, Butenko RG, Popov AS (1983) Effect of pretreatments of potato apices surviving after deep freezing. Sov Plant Physiol 30:1188–1193
Mater AA (1987) Production and cryogenic freezing of date palm germplasm and regeneration of plantlets from frozen material. Iraqi J Agric Sci 5:35–49
McComb JA, Newton S (1981) Propagation of kangaroo paws using tissue culture. J Hortic Sci 56:181–183
Mellor FC, Stace-Smith R (1969) Development of excised potato buds in nutrient medium. Can J Bot 47:1617–1621
Minocha SC (1980) Cell and tissue culture in the propagation of forest trees. In: Sala F, Parisi B, Cella R, Ciferri O (eds) Plant cell cultures: results and perspectives. Elsevier/North Holland Biomedical, Amsterdam, pp 295–300
Monaco LC, Sondahl MR, Carvalho A, Crocomo OJ, Sharp WR (1977) Application of tissue culture in the improvement of coffee. In: Reinert J, Bajaj YPS (eds) Applied and fundamental aspects of plant cell, tissue, and organ culture. Springer, Berlin Heidelberg New York, pp 109–129
Monteuuis O (1987) In vitro meristem culture of juvenile and mature *Sequoiadendron giganteum*. Tree Physiol 3(3):265–272
Morel G (1960) Producing virus-free cymbidiums. Am Orchid Soc Bull 29:495–497
Morel G (1963) La culture in vitro du meristème apical de certaines orchidées. CR Hebd Séances Acad Sci 256:4955–4957

Morel G, Martin C (1952) Guérison de dahlias atteints d'une maladie à virus. CR Hebd Séances Acad Sci 235:1324–1325
Morel G, Martin C (1955) Guérison de pommes de terre atteintes de maladies à virus. CR Séances Acad Agric Fr 41:472–475
Mori K, Hosokawa D (1977) Localization of viruses in apical meristem and production of virus-free plants by means of meristem and tissue culture. Acta Hortic 78:389–396
Mullins MG, Nair Y, Sampet P (1979) Rejuvenation in vitro: induction of juvenile characters in an adult clone of *Vitis vinifera* L. Ann Bot 44:623–627
Murashige T, Skoog F (1962) A revised medium for rapid growth and bioassays with tobacco tissue cultures. Physiol Plant 15:473–497
Oglevee-O'Donovan W (1986) Production of culture indexed geraniums. In: Zimmerman RH, Griesbach RJ, Hammerschlag FA, Lawson RH (eds) Tissue culture as a plant production system for horticultural crops. Martinus Nijhoff, Boston, pp 119–123
Paludan N (1985) Inactivation of tobacco mosaic virus in *Aeschynanthus hildebrandii* by means of heat treatment, chemotherapy and meristem-tip culture. Tidsskr Planteavl 89:273–278
Papachatzi M, Hammer PA, Hasagawa PM (1981) In vitro propagation of *Hosta decorata* Thomas Hogg using cultured shoot tips. J Am Soc Hortic Sci 106:232–236
Pierik RLM (1987) In vitro culture of higher plants. Martinus Nijhoff, Boston, pp 296–300
Pierik RLM (1988) Handicaps for the large scale commercial application of micropropagation. In Kozai T (ed) Acta Hortic Symp High technology in protected cultivation. Giken, Tokyo, pp 63–71
Pillai SK, Hildebrandt AC (1968) Geranium plants differentiated in vitro from stem tip and callus cultures. Plant Dis Rep 52:600–601
Quak F (1977) Meristem culture and virus-free plants. In: Reinert J, Bajaj YPS (eds) Applied and fundamental aspects of plant cell, tissue, and organ culture. Springer, Berlin Heidelberg New York, pp 598–615
Reed BM (1990) Survival of in vitro-grown apical meristems of *Pyrus* following cryopreservation. HortScience 25(1):111–113
Reed BM, Lagerstedt HB (1987) Freeze preservation of apical meristems of *Rubus* in liquid nitrogen. HortScience 22(2):302–303
Riviers S (1973) Culture in vitro du merestène apical du *Lilium candidum* L. (Liliacées) apres ablation totale des primordiums foliaires. CR Acad Sci Ser D 1989–1992
Robb SM (1957) The culture of excised tissue from bulb scales of *Lilium speciosum* Thun. J Exp Bot 8:348–352
Roggemans J, Claes MC (1979) Rapid clonal propagation of rhubarb by in vitro culture of shoot tips. Sci Hortic 11:241–246
Sakai A, Yamakawa M, Sakata D, Harada T, Yakuwa T (1978) Development of a whole plant from an excised strawberry runner apex frozen to –196 °C. Contrib Inst Low Temp Sci Ser Bull 36:31–38
Schnabdrauch LS, Sink KC (1979) In vitro propagation of *Phlox subulata* and *Phlox paniculata*. HortScience 14:607–608
Seibert M (1976) Shoot initiation from carnation shoot apices frozen to –196 °C. Science 191:1178–1179
Shabde M, Murashige T (1977) Hormonal requirements of excised *Dianthus caryophyllus* L. shoot apical meristem in vitro. Am J Bot 64:443–448
Smith RH, Murashige T (1970) In vitro development of the isolated shoot apical meristem of angiosperms. Am J Bot 57:562–568
Snir I, Erez A (1980) In vitro propagation of Malling Merton apple root stocks. HortScience 15:597–598
Stace-Smith R (1985) Virus-free clones through plant tissue culture. In: Robinson CW, Howel JA (eds) Comprehensive biotechnology Vol 4. The practice of biotechnology: specialty products and service activities. Pergamon Oxford, pp 169–179
Stace-Smith R, Mellor FC (1968) Eradication of potato viruses X and S by thermotherapy and axillary bud culture. Phytopathology 58:199–203
Stevenson JH, Harris RE (1980) In vitro plantlet formation from shoot-tip explants of *Fuchsia hybrida* cv. Swingtime. Can J Bot 58:2190–2199
Stone OM (1963) Factors affecting the growth of carnation plants from shoot apices. Ann Appl Biol 52:199–209
Sutter E, Langhams RW (1979) Epicuticular wax formation on carnation plantlets regenerated from shoot tip culture. J Am Soc Hortic Sci 104:493–496

Takayama S, Misawa M (1980) Differentiation in *Lilium* bulbscales grown in vitro. Effects of activated charcoal, physiological age of bulbs and sucrose concentration on differentiation and scale leaf formation in vitro. Physiol Plant 48:121–125

Ten Houten JG, Quak F, van der Meer FA (1968) Heat treatment and meristem culture for the production of virus-free plant material. Neth J Plant Pathol 74:17–24

Theiler R (1977) In vitro culture of shoot tips of *Pelargonium* species. Acta Hortic 78:403–414

Thimann KV (1977) Hormone action in the whole life of plants. Univ of Massachusetts Press, Amherst

Towill LE (1981) Survival at low temperatures of shoot-tips from cultivars of *Solanum tuberosum* group Tuberosum. Cryo Lett 2:373–382

Towill LE (1988) Survival of shoot-tips from mint species after short-term exposure to cryogenic conditions. HortScience 23(5):839–841

Uemura M, Sakai A (1980) Survival of carnation (*Dianthus caryophyllus* L.) shoot apices frozen to the temperature of liquid nitrogen. Plant Cell Physiol 21:85–94

Van Os H (1964) Production of virus-free carnations by means of meristem culture. Neth J Plant Pathol 70:18–26

Vine SJ, Jones OP (1969) The culture of shoot tips of hop (*Humulus lupulus*) to eliminate viruses. J Hortic Sci 44:281–284

Walkey DGA (1980) Production of virus-free plants by tissue culture. In: Ingram DS, Helgeson JP (eds) Tissue culture methods for plant pathologists. Blackwell Scientific, Oxford, pp 109–117

Walkey DGA, Webb MJW (1968) Virus in plant apical meristems. J Gen Virol 3:311–313

Wang PJ (1977) Regeneration of virus-free potato from tissue culture. In: Barz W, Reinhard E, Zenk MH (eds) Plant tissue culture and its biotechnological application. Springer, Berlin Heidelberg New York, pp 386–391

Wang PJ (1978) Clonal multiplication of *Cryptomeria japonica D. Don* in vitro. In: Studies and essays in commemoration of the Golden Jubilee of Academia Sinica, Taipei, Taiwan, pp 559–566

Wang PJ, Hu CY (1980) Regeneration of virus-free plants through in vitro culture. In: Fiechter A (ed) Advances in biochemical engineering. Vol 18 Plant Cell Cultures II. Springer, Berlin Heidelberg New York, pp 61–99

Wang PJ, Hu CY (1982) In vitro mass tuberization and virus-free seed-potato production in Taiwan. Am Potato J 59:33–37

Wang PJ, Hu CY (1984) In vitro cloning of the deciduous timber tree *Sassafras randaiense*. Z Pflanzenphysiol 113:331–335

Wang PJ, Hu CY (1985) Potato tissue culture and its application. In: Li PH (ed) Potato physiology. Academic Press, New York, pp 503–577

Wang PJ, Huang LC (1974) Studies on the shoot meristem culture of *Allium sativum*. L. Chin Hortic 20:79–87

Wang PJ, Huang LC (1976) Beneficial effects of activated charcoal on plant tissue and organ cultures. In Vitro 12:260–261

White PR (1934) Multiplication of the viruses of tobacco and *Aucuba* mosaics in growing excised tomato root tips. Phytopathology 24:1003–1011

Wochok ZS, Sluis CJ (1980) Gibberellic acid promotes *Atriplex* shoot multiplication and elongation. Plant Sci Lett 17:363–369

Yakuwa H, Oka S (1988) Plant regeneration through meristem culture from vegetative buds of mulberry (*Morus bombycis* Koidz) stored in liquid nitrogen. Ann Bot 62:79–82

Yang YW, Hsing YI, Chang WC (1981) Clonal propagation of *Stevia rebaudiana* Bertoni through axillary shoot proliferation in vitro. Bot Bull Acad Sin 22:57–62

Zatyko JM, Molnar I (1985) Preliminary results on the in vitro mass propagation of grapes from shoot-tip meristem. Fruit Sci Rep 12:83–85

Ziv M (1979) Transplanting *Gladiolus* plants propagated in vitro. Sci Hortic 11:257–260

Ziv M, Halevy AH, Shilo R (1970) Organ and plantlet regeneration of *Gladiolus* through tissue culture. Ann Bot (Lond) 34:671–676

I.4 Micropropagation Through Somatic Embryogenesis in Conifers

S. M. ATREE and L. C. FOWKE[1]

1 General Account

Somatic embryogenesis is the development of embryos from somatic cells. A somatic embryo should ultimately possess both a root and a shoot meristem and be capable of forming a complete plant. Most embryogenic cultures of conifers have been induced from species within the Pinaceae and are similar in appearance. The exception is redwood (*Sequoia sempervirens*) (see sect. 1.4), which is a member of the Taxodiaceae. Embryogenic cultures of conifers within the Pinaceae are translucent to white, and when cultured on solidified medium, glisten due to the production of mucilage. Established embryogenic cultures consist primarily of somatic embryos, or of asexual embryos in the instance of haploid larch (*Larix* sp.) (see sect. 1.5). They are polarized structures possessing an embryonic region subtended by elongate suspensor cells. The somatic embryos initiate further embryos by cleavage, or cellular components of embryos initiate embryos directly, in the absence of true callus proliferation (see Hakman and Fowke 1987b; Hakman and von Arnold 1988). However, embryogenic cultures of *Pinus* species may not always contain recognizable embryos subtended by a suspensor region until after transfer to medium containing ABA (Becwar et al. 1990; Lainé and David 1990).

During zygotic embryo development in conifers (see Fig. 1), nuclear division of the zygote results in two then four free nuclei which migrate to the end of the proembryo opposite the neck cells. The nuclei become orientated into a single tier and each nucleus divides. Cell walls form between the eight nuclei, producing two tiers of four cells; however, the cell walls of the proximal tier remain open to the "egg" cytoplasm. Each cell then divides again, forming a 16-celled proembryo consisting of four tiers of four cells each. The proembryonal stage is completed when the third "suspensor" tier elongates pushing the embryonal cells into the gametophytic tissue (see also Foster and Gifford 1974; Owens and Molder 1984). Embryogenic cultures of conifers contain immature embryos equivalent to the initial developmental stages of zygotic embryos (e.g. Fig. 1E), but have been suggested to also include proembryo stages. For example, Gupta and Durzan (1987a) and Durzan and Gupta (1987) observed free nuclei in cells from loblolly pine (*Pinus taeda*) and Douglas fir (*Pseudotsuga menziesii*) suspension cultures. With larch (von Aderkas and Bonga 1988), a four-nucleate coenocyte was also observed in certain elongate cells believed to be the progenitors of new embryos. All these authors suggest the multinucleate cells approximate the free-nuclear stage initiated during zygotic embryogenesis. The free-nucleate cells observed in

[1]Department of Biology, University of Sasketchewan, Saskatoon, Sasketchewan, Canada, S7H OWO

Biotechnology in Agriculture and Forestry, Vol. 17
High-Tech and Micropropagation I (ed. by Y.P.S. Bajaj)

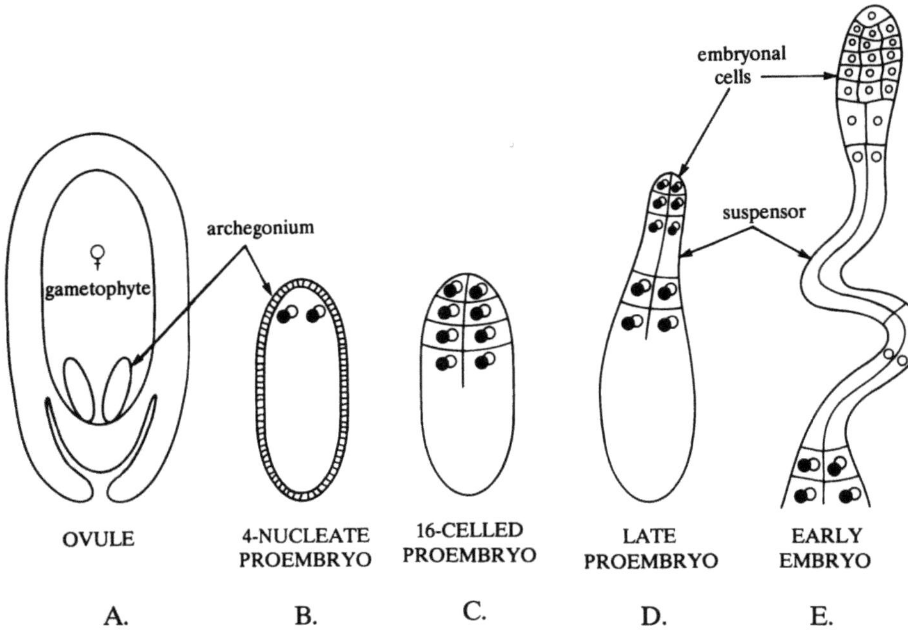

Fig. 1A-E. Zygotic embryo development in conifers. **A** Ovule containing female gametophyte with two archegonia. **B** Proembryo development begins when the fertilized egg nucleus divides into two then four free nuclei which migrate to the end of the proembryo opposite the neck cells, where they are orientated into a single tier. **C** The nuclei have divided twice to form a 16-celled proembryo composed of four tiers of 4 cells each. The proximal tier has remained open to the "egg" cytoplasm. **D** The proembryo stage is completed when the third suspensor tier elongates, pushing the embryonal cells into the gametophyte. **E** The early developing embryo at a stage comparable to somatic embryos in proliferating cultures

loblolly pine, Douglas fir, and larch suspension cultures, were not sequentially followed during nuclear migration and development from proembryo to embryo. To clearly demonstrate that they represent true proembryo stages, these events need to be fully documented. A cytological investigation of cultured white spruce protoplasts (Fowke et al. 1990), however, showed that multinucleate (coenocytic) protoplasts derived from spontaneous fusions underwent synchronous nuclear division and cytokinesis to form multicellular colonies. Each colony then formed a suspensor region, giving rise to an embryo. The formation of embryos from multinucleate protoplasts is unique for plant protoplast cultures, and the resemblance to conifer zygotic embryogenesis suggests that the mechanism for nuclear organization and cell formation is operating in the white spruce multinucleate protoplasts.

1.1 Induction of Somatic Embryogenesis

Only embryonic tissues have so far initiated embryogenic cultures from conifers. These tissues include immature zygotic embryos dissected from seeds collected during the growing season, mature zygotic embryos dissected from stored seed, and excised tissues from seedlings. In addition, megagametophytes of several conifer species have been cultured with varying success. The reports of induction of somatic embryogenesis in conifers are shown in Table 1 and the process of embryogenic induction and subsequent plant regeneration is illustrated in Figs. 2–7. The first report was from immature zygotic embryos of Norway spruce (Hakman et al. 1985), and was followed shortly after with induction of asexual haploid embryos from larch megagametophytes (Nagmani and Bonga 1985) (see sect. 1.5, concerning larch).

1.2 Immature Zygotic Embryos

Immature zygotic embryos dissected from seeds collected during the maturation period were used in many initial studies of conifer somatic embryogenesis. Often just one cone collection date proved superior. This was because the maturity of the developing zygotic embryo influenced embryogenic induction frequencies, which often decreased as the seeds matured, especially with *Pinus* species (e.g., Chandler et al. 1989; Jain et al. 1989; Becwar et al. 1990; Lainé and David 1990), but also with Douglas fir (Durzan and Gupta 1987) and diploid larch (Klimaszewska 1989a). With *Pinus* species precotyledonary zygotic embryos are the most responsive. With *Picea* early cotyledonary embryos appear superior.

The media used to induce somatic embryogenesis from immature zygotic embryos of *Picea* species have usually been modifications of LP medium, after von Arnold and Eriksson (1981), used at full strength. Full strength LP produced significantly higher induction frequencies with black spruce than half strength LP (Tautorus et al. 1990). Half strength LP was used by von Arnold and Hakman (1988) with Norway spruce, and half strength MS (Murashige and Skoog 1962) was used with sitka spruce (Krogstrup et al. 1988). Media used for other genera include modifications of the following: half strength LP with *Pinus caribaea* (Lainé and David 1990); MS at half strength with loblolly pine (Gupta and Durzan 1987a); DCR (after Gupta and Durzan 1985) at full strength with eastern white pine (*Pinus strobus*) (Finer et al. 1989) and loblolly pine (Becwar et al. 1990), and at half strength with Douglas fir (Durzan and Gupta 1987); and SH (Schenk and Hildebrandt 1972) with *Abies alba* (Schuller et al. 1989).

Usually both an auxin and a cytokinin are necessary for induction, and are included at 10 and 5 μM, respectively. However, Durzan and Gupta (1987), and Gupta and Durzan (1987a) included them at 50 and 40 μM, respectively, for both loblolly pine and Douglas fir. *Abies alba* somatic embryos were induced using cytokinin as the sole growth regulator (Schuller et al. 1989). Although growth regulators may be beneficial to embryogenic induction of *Pinus radiata* and *Pinus taeda*, induction in their absence also occurred when embryos were cultured within their megagametophytes (Chandler et al. 1989; Becwar et al. 1990). Continued growth of *Pinus taeda* cultures occurred only in the presence of 2,4-D and BA, therefore the latter authors suggested that the megagametophyte supplied the necessary growth regulators for induction. With both species no induction occurred when isolated immature embryos were cultured either in the presence or absence of applied growth regulators. Cultures of *Pinus radiata*, however, could be maintained in an embryogenic condition in the absence of growth regulators (see also haploid larch, sect. 1.5). Co-cultivation with megagametophytes was also used to establish embryogenic cultures derived from immature embryos of sugar pine (Gupta and Durzan 1986b), loblolly pine (Gupta and

Table 1. Induction of somatic embryogenesis in conifers

Species	Observation	Reference
Abies alba	Induction from immature zygotic embryos	Schuller et al. (1989)
Larix decidua	Induction of haploid embryos from megagametophytes	Nagmani and Bonga (1985), von Aderkas et al. (1987), von Aderkas and Bonga (1988)
L. × eurolepis	Induction of hybrid diploid embryogenic callus from immature zygotic embryos	Klimaszewska (1989a), von Aderkas et al. (1990)
Picea abies	Induction from immature zygotic embryos	Hakman et al. (1985), Hakman and von Arnold (1985)
	Induction from mature zygotic embryos	Gupta and Durzan (1986a), Jain et al. (1988), Verhagen and Wann (1989), von Arnold and Hakman (1986), von Arnold (1987)
	Induction from mature zygotic embryos and 3–7-day-germinated seedlings	Krogstrup (1986), Lelu et al. (1987)
	Quantification of the level of somatic embryogenesis among callus lines	Becwar et al. (1987a)
	Biochemical comparisons of embryogenic and nonembryogenic callus	Wann et al. (1987)
	Induction of haploid embryos from megagametophytes	Simola and Santanen (1990)
Picea mariana	Induction from immature zygotic embryos	Hakman and Fowke (1987a), Tautorus et al. (1990)
	Induction from mature zygotic embryos	Tautorus et al. (1990)
	Induction from 12–21-day seedlings	Attree et al. (1990a)
Picea glauca	Induction from immature zygotic embryos	Attree et al. (1989a), Hakman and Fowke (1987a), Lu and Thorpe (1987)
	Induction from mature zygotic embryos	Tremblay (1990)
	Induction from 12–30-day seedlings	Attree et al. (1990a)
[a]Interior spruce	Induction from immature zygotic embryos	Webb et al. (1989)
Picea sitchensis	Induction from immature zygotic embryos	Krogstrup et al. (1988)
	Induction from mature zygotic embryos	von Arnold and Woodward (1988)
Pinus caribaea	Induction from immature zygotic embryos	Lainé and David (1990)
Pinus elliotti	Induction from immature zygotic embryos	Jain et al. (1989)
Pinus lambertiana	Induction from mature zygotic embryos	Gupta and Durzan (1986b)
Pinus radiata	Induction from immature zygotic embryos	Chandler et al. (1989)
Pinus strobus	Induction from immature zygotic embryos	Finer et al. (1989)
Pinus taeda	Induction from immature zygotic embryos	Gupta and Durzan (1987a) Becwar et al. (1990, 1991)
Pseudotsuga menzesii	Induction from immature zygotic embryos	Durzan and Gupta (1987)
Sequoia sempervirens	Induction from 7-day-germinated seedlings	Bourgkard and Favre (1988)

[a]*P. glauca, P. engelmanni*, and intermediate hybrids.

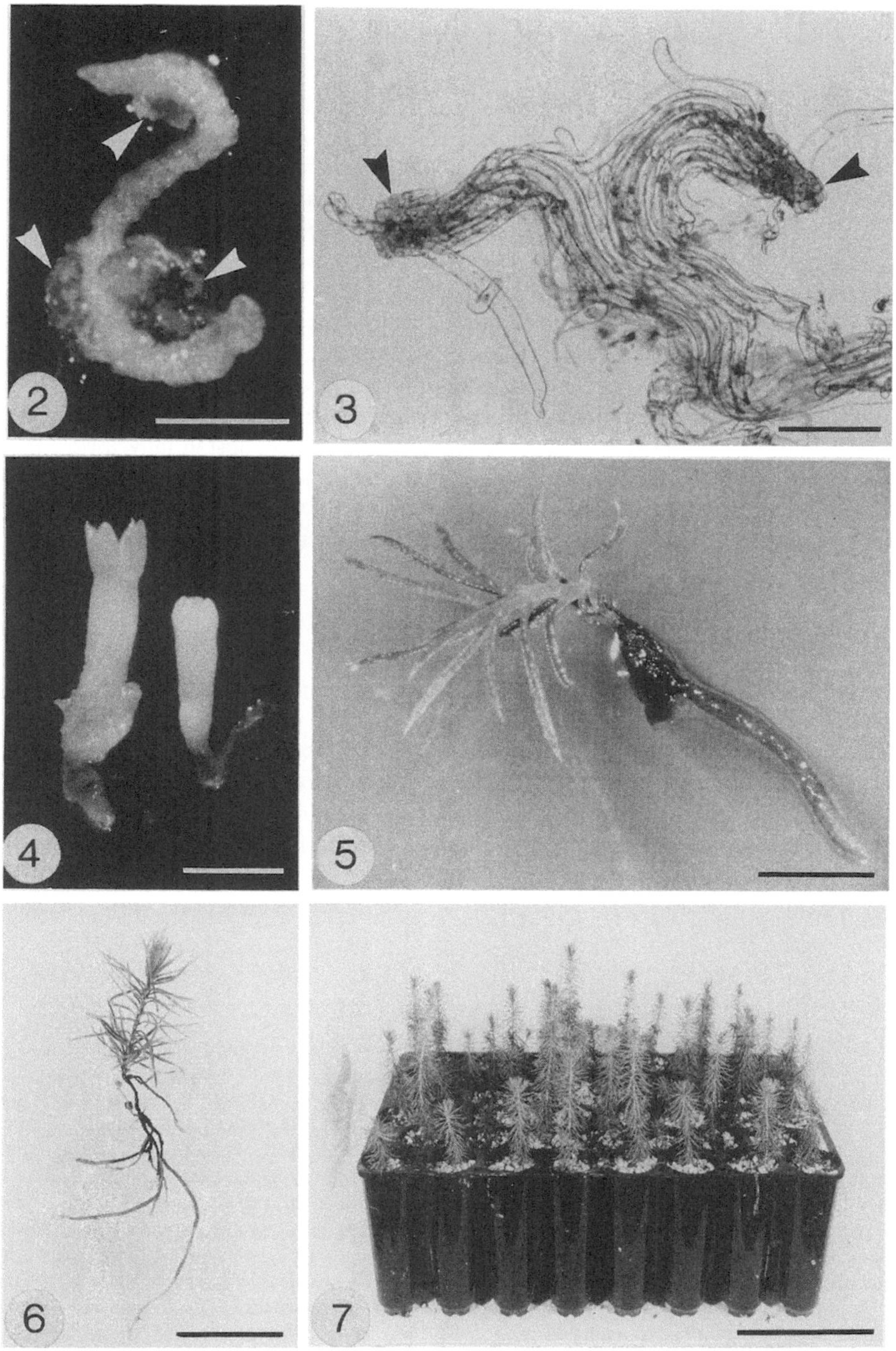

See caption on p. 58

Durzan 1987a), eastern white pine (Finer et al. 1989), *Pinus caribaea* (Lainé and David 1990), Douglas fir (Durzan and Gupta 1987), and *Abies alba* (Schuller et al. 1989).

Sucrose has been included in media usually at 30 mM but up to 100 mM with no evidence that this higher concentration was detrimental. Casein hydrolysate was suggested to improve induction of somatic embryogenesis from immature embryos of Norway spruce (Hakman et al. 1985), but was found to significantly inhibit induction from immature black spruce zygotic embryos (Tautorus et al. 1990).

Observed differences in induction frequencies reflect a variable ability of different species (e.g., Hakman and Fowke 1987a) and seedlots from different localities (e.g., Hakman and von Arnold 1985) or maternal genotype (Webb et al. 1989; Becwar et al. 1990) to undergo induction, in addition to differences in developmental stage of embryos and diverse media used. With *Picea* induction frequencies greater than 65% have been obtained from immature zygotic embryos (e.g., white spruce, Lu and Thorpe 1987; Norway spruce, Hakman and von Arnold 1985; black spruce, Tautorus et al. 1990). However, induction frequencies were more often substantially lower. *Pinus* species have shown induction frequencies in the order of 5%. These lower frequencies coupled with the short annual availability of the suitable explants have impeded the rapid optimizing of media and conditions for these species.

1.3 Mature Zygotic Embryos

Dissecting zygotic embryos from seeds throughout the growing season to provide sufficient embryogenic material for experimentation is very labor-intensive, and material is available for only a short duration annually. The induction of somatic embryogenesis from mature embryos dissected from stored seed can extend the period during which embryogenic material can be generated from just a few months to more than a decade. However, in comparison to immature embryos, the conditions for induction are generally more restrictive and the process of induction takes longer (e.g., Gupta and Durzan 1986b; Tautorus et al. 1990). Other than *Pinus lambertiana*, all reports pertain to mem-

Figs. 2-7. Embryogenic induction, embryo maturation and plant development from black spruce cultures derived from different explant sources. Figures **2–4** show development of embryogenic cultures derived from shoot explants excised from 12-day germinated seedlings. Figures **5** and **6** show plantlets regenerated from embryogenic cultures derived from immature zygotic embryos. Figure **7** shows plants regenerated from embryogenic cultures derived from stored seed. **2** Embryogenic induction from several points of origin (*arrowheads*) of a shoot explant, after 3–4 weeks' culture on medium containing 4.5 μM BA and 9 μM 2,4-D. The explant had previously been given 2 weeks preculture on medium containing 4.5 μM BA. *Bar* 2 mm. **3** Proliferating somatic embryos undergoing cleavage within an embryogenic suspension culture. The somatic embryos are organised into a meristematic embryonal region (*arrowheads*), subtended by elongated suspensor cells. *Bar* 100 μm. **4** Maturation of somatic embryos. Pale embryos with emerging cotyledons developed while cultured for 28 days on medium containing 12 μM ABA. *Bar* 1 mm. **5** Mature embryos turned green and underwent shoot and root formation after 8 weeks' culture on growth regulator-free medium in the light. At this point plantlets can be transferred to pots and grown in the greenhouse. *Bar* 5 mm. **6** An established greenhouse-grown plant with a branching root system after 3 months in vermiculite. *Bar* 2 cm. **7** Established greenhouse-grown plants now in a perlite/vermiculite/soil mix 7 months after ABA treatment. *Bar* 10 cm bers of *Picea*. Since mature seeds contain embryos of similar developmental stage which are available throughout the year, the variability of response observed with immature embryos of differing maturity is eliminated, thus permitting detailed studies of factors influencing induction. Reports have appeared in which consistently high induction frequencies have been obtained from mature embryos. The frequencies of induction compare favorably with those often obtained from immature embryos. Somatic embryogenesis has been initiated from mature embryos of Norway spruce (von Arnold and Hakman 1986; von Arnold 1987; Verhagen and Wann 1989) and Sitka spruce (*Picea sitchensis*) (von Arnold and Woodward 1988), but in these reports the duration of seed storage was not indicated. Somatic embryogenesis has also been initiated from seeds of Norway spruce stored at 4 °C for 1 and 2 years by Jain et al. (1988) and Gupta and Durzan (1986a) respectively, sugar pine (*Pinus lambertiana*) stored for 5 years at 20 °C (Gupta and Durzan 1986b), white spruce stored 11 years at 3–5 °C Tremblay (1990), and from black spruce stored for 13 years at –18 °C (Tautorus et al. 1990).

Embryogenesis has most frequently been initiated in the dark. With Norway spruce, culture under light of a 20-h photoperiod was found to drastically inhibit induction (von Arnold 1987) and darkness was considered essential by Gupta and Durzan (1986a). In a subsequent study with the same species, Verhagen and Wann (1989) suggested that somatic embryogenesis was initiated equally well under light or dark. Lower induction frequencies were obtained in the light-cultured treatments compared to dark, but no significant difference was detected. Treatments were, however, subjected to a shorter photoperiod and substantially lower light intensity than tested by von Arnold (1987).

For induction from mature zygotic embryos, media have been used most frequently at half strength. Basal media tested include the following: LP (Jain et al. 1988; von Arnold and Hakman 1988; von Arnold and Woodward 1988), MS (Gupta and Durzan 1986a; Verhagen and Wann 1989), DCR (Gupta and Durzan 1986b), and Litvay (after Litvay et al. 1981) (e.g., Tautorus et al. 1990; Tremblay 1990). Only von Arnold and Hakman (1986) used LP medium solely at full strength. Comparisons of media strength with Norway and black spruce have shown the reduced salt concentrations to be optimal (Jain et al. 1988; von Arnold 1987; Tautorus et al. 1990). The half strength MS medium of Verhagen and Wann (1989), when modified by omitting NH_4NO_3, decreasing KNO_3 to 0.5 mM, and including 5 mM glutamine, led to induction frequencies up to 32% from Norway spruce, which was significantly higher than controls. However, this is contrary to a previous study of Norway spruce (von Arnold 1987), whereby half strength LP but with NH_4NO_3 increased to 15 mM was found to be optimal with induction frequencies up to 53%. In the latter study, 15 embryos cultured per 10-cm Petri dish and 20 ml medium were best, and a medium of pH 6 or above was superior. In addition, it was important to damage embryos prior to culture as this led to a fourfold increase in induction frequency. A similar induction frequency (55%) was obtained by Jain et al. (1988) using half strength LP medium with all amino acids excluded. With white spruce, an induction frequency of 40% was achieved from 4–11-year-stored seed (Tremblay 1990). The induction frequency was shown to be affected by geographic provenance of the seeds, their storage duration, and their percentage germination.

The growth regulators 2,4-D or NAA were equally effective in promoting induction from mature zygotic embryos (e.g., Verhagen and Wann 1989). Von Arnold (1987) obtained slightly higher induction frequencies with 2,4-D (at 20 µM) compared to

NAA, but NAA was more effective at lower concentrations (5 μM). Jain et al. (1988) found NAA was more effective than either IAA or 2,4,5-T, but did not compare 2,4-D in the same experiment. Sucrose at 30 mM was found superior when using LP medium at full (von Arnold and Hakman 1986) and half (von Arnold 1987) strength.

1.4 Induction from Seedling Explants

For many spruce species more than one fertilized egg in a gametophyte can produce a developing embryo thereby demonstrating simple polyembryony (see Attree et al. 1991). Krogstrup (1986) realized the value of using seedling-derived explants for establishing embryogenic cultures, as seed-derived cultures may consist of more than one genotype due to simple polyembryony. Multiple genotypes were shown to be present in embryogenic tissue derived from individual immature loblolly pine seeds (Becwar et al. 1991). Furthermore, when using stored seed, it may be possible to obtain higher induction frequencies from seedling-derived explants than from zygotic embryos (Attree et al. 1990a). This results from the elimination of nonviable seed embryos, which do not germinate, and permits the selection of the most vigorous seedlings for explant culture. Using seedling-derived explants is also considerably less labor-intensive than dissecting embryos from seeds, especially if many seeds do not contain a developed embryo.

Somatic embryos have been induced from cultured seedling explants of Norway spruce (e.g., Krogstrup 1986; Lelu et al. 1987), redwood (Bourgkard and Favre 1988), and both black and white spruce (Attree et al. 1990a) (Figs. 2–4). Most studies suggested that preculture on medium containing cytokinin as the primary growth regulator was beneficial. In general, pretreatment with 5 μM BA for 1–3 weeks led to improved induction frequencies compared to non-BA-pretreated explants; however, Attree et al. (1990a) found no significant difference between BA pretreatment and controls with black spruce. Reports emphasize a possible relationship between induction of somatic embryogenesis and the ability to form adventitious buds with cytokinin.

With Norway spruce, Lelu et al. (1987) obtained an induction frequency of 23% by preculturing cotyledon explants from 3–5-day-germinated seedlings on the medium of Campbell and Durzan (1975) containing 4.5 μM BA and reduced levels (e.g., 0.054 μM) of NAA for 1 week prior to transfer to a secondary culture medium containing 21.5 μM NAA and 0.45 μM BA. Lower induction frequencies were obtained with cotyledons excised 6–7 days after germination. With no pretreatment, induction frequencies ranged from 11–14%. In the study of Norway spruce by Krogstrup (1986), results from mature embryos and excised cotyledons from 7-day-old seedlings were combined for a total induction frequency of 4% and thus induction frequencies from the cotyledons alone cannot be identified. However, best results were obtained following a preculture on modified MS medium containing 5 μM BA for 3 weeks, prior to transfer to a medium containing 50 μM 2,4-D, and 25 μM each of BA and kinetin for 3–4 weeks.

With redwood (Bourgkard and Favre 1988), 28% of cotyledons excised from 7-day-germinated seedlings gave rise to compact embryogenic material, from each of which on average 2.3 somatic embryos arose. Further multiplication of embryos did not occur. Induction frequencies were higher from cotyledon than hypocotyl or root explants. The embryos were observed after 12 weeks' culture on modified half strength MS medium containing 2 μM each of BA and kinetin, together with 2.5 μM 2,4-D. Higher 2,4-D concentrations led to reduced induction frequencies, but preculture on medium lacking 2,4-D was not tested. Although induction frequencies from redwood were high, the potential of high frequency plant regeneration remains low in comparison to other species, unless secondary embryogenesis can be induced.

With black spruce (Attree et al. 1990a), an induction frequency of 18% was achieved from shoot explants of 12-day-old seedlings germinated from 10-year-stored seed. The explants were given 2 weeks' preculture on full strength LP medium containing 0.45 μM BA, prior to transfer to medium containing 0.45 μM BA and 9 μM 2,4-D. This induction frequency was not significantly different from the induction frequency obtained with explants given no preculture (12%). Seedling age did have a significant effect. Shoot explants from 21-day-old black spruce seedlings showed an induction frequency of 5%. Induction frequencies of 38 and 18% were also obtained from 12-and 30-day-old white spruce shoot explants from 5-year-stored seed given no BA pretreatment.

Conifers are outbreeders, therefore each embryogenic culture is of individual and unproven genotype. The induction of somatic embryos from nonembryonic tissues of mature trees of known phenotype is desirable, but tissues from mature conifers have unfortunately proven extremely difficult to culture in vitro (see Mott 1981; David 1982; Attree et al. 1991). By optimizing the conditions for embryogenic induction with increasingly older seedling tissues, it may be possible to further extend the age of plants from which somatic embryogenesis can be induced, to the point where tissues from mature trees are inducible.

1.5 Induction from Larch

The potential attraction of haploid tree cultures to forestry include unmasking recessive traits, the production of homozygous diploids, genetic mapping, and protoplast somatic hybridization (see Bonga et al. 1988). The larch system is discussed separately here due to novel differences of ploidy, type of explant cultured, and mode of development. Embryogenic cultures of larch generated from cultured megagametophytes are haploid and as these are derived from gametophytic and not somatic tissues, strictly are not somatic embryos but asexual embryos. The haploid asexual embryos of larch exhibit an unusual indirect pattern of initiation which contrasts with the more direct development of diploid somatic and zygotic embryos (see von Aderkas et al. 1991). According to von Aderkas and Bonga (1988), in addition to the free nucleate cells (see sect. 1), which may develop directly into embryos, a more typical pathway is evident, whereby elongated cells first developed from the gametophytes. These elongated cells then divided asymmetrically to yield small densely cytoplasmic terminal cells which subsequently divided, to form loose aggregates of small cells, termed microcalli. A long/short cell interchange was repeated through several subcultures. Eventually, the microcalli became more densely packed and formed embryos with suspensor and terminal meristematic embryonal cells.

In the first report from larch (Nagmani and Bonga 1985), megagametophytes were excised from immature ovules. The gametophytic tissue, when cultured on various media and growth regulator combinations, underwent induction at a frequency of 0.3%. Cultures grew rapidly and continued to proliferate embryos when subcultured on hormone-free medium. Many plantlets developed within the cultures and these were isolated and transferred to soil. Although the authors regarded the embryos and plantlets as haploid, this was not confirmed until a later study by von Aderkas et al.

(1987). In this latter study, a more thorough investigation of embryogenic induction was undertaken. Megagametophytes were amenable to embryogenic induction only for approximately a 4-week period commencing shortly after fertilization. Litvay medium containing 90 mM sucrose, 500 mg/l glutamine, and 1000 mg/l casein hydrolysate was suitable for culture. During initiation 2,4-D at 0.1–10 μM was essential, thereafter embryogenic cultures could be maintained on growth regulator-free medium. Light does not appear to inhibit embryogenic induction from larch, as in all reports, cultures were initiated and maintained under light of 16-h photoperiod.

In a more recent study (Klimaszewska 1989a), diploid somatic embryos were induced from immature zygotic embryos of hybrid *Larix leptolepis* × *L. decidua* (= *L.* × *eurolepis*) and their reciprocal crosses. When immature zygotic embryos were co-cultivated with megagametophytic tissue, diploid somatic embryos developed similarly to diploid somatic embryos of *Picea*, and required auxin and cytokinin for growth as well as induction. The optimal time for induction was 2–3 weeks after fertilization when zygotic embryos were at the precotyledonary stage, and a suitable culture medium was MS supplemented with glutamine. A comparison of both haploid and diploid embryogenic calli of *L. leptolepis* and *L. decidua* and the hybrid *L.* × *eurolepis* (von Aderkas et al. 1990) showed that the mode of callus development and ability to proliferate on growth regulator-free medium displayed by haploid larch were due to differences in ploidy and not species.

Other than larch, haploid embryogenesis has been reported only for Norway spruce (Simola and Santanen 1990). Unfortunately, the report provides no information concerning the developmental maturity of the megagametophyte explants, the nature of the explants (i.e., exclusion of zygotic embryo(s) and their tissues, etc.), or the culture methods used. Additionally, it is not indicated whether the ploidy of the embryogenic cultures was verified, and the induction frequencies were not given. The medium used for induction was similar to that described previously by Simola and Honkanen (1983). This medium is essentially similar to other conifer embryogenic media currently being used. It was supplemented with casein hydrolysate (1000 mg/l), glutamine (500 mg/l), myo-inositol (100 mg/l), and sucrose was included at 2%. The growth regulators were kinetin and 2,4-D, which were included at 0.5 and 2 mg/l, respectively. Simola and Honkanen (1983) previously reported the recovery of haploid plantlets of Norway spruce from nonembryogenic, megagametophyte-derived callus via organogenesis.

2 Plant Regeneration

Conifer somatic embryos maintained on a medium containing both auxin and cytokinin continue to proliferate but do not mature. Somatic embryos must be transferred to different growth regimes to encourage maturation. Mature embryos are then generally cultured on growth regulator-free medium for germination and plantlet development and then finally transferred to soil (see Figs. 4–7). The species of conifers from which plantlets have been recovered and the methods used are shown in Table 2. Limited somatic embryo maturation was first demonstrated for *Picea abies* (Hakman et al. 1985). Shortly afterwards, Hakman and von Arnold (1985) obtained rooted plantlets with the same species. Still greater success was achieved in the same year with haploid larch

(Nagmani and Bonga 1985) and plants were transferred to soil. Only superficial comparisons of maturation and plantlet recovery frequencies can be made. Methods of inducing embryo maturation are variable, and different species, and even genotypes within species, may be expected to display variable competencies to form plantlets using any one method.

ABA promotes the accumulation of storage lipids and proteins in conifer somatic embryos (Hakman and von Arnold 1988; Feirer et al. 1989; Roberts et al. 1990a). In maturation protocols there has been a trend towards increasingly higher concentrations of ABA, to levels above those previously found to be optimal. This trend results from a better understanding of the timing of ABA application and removal, thereby promoting maturation while overcoming the detrimental effects of ABA on subsequent embryo germination. Plantlets from many species studied have been recovered following maturation of somatic embryos on media containing low levels of cytokinin and auxin, cytokinin alone, or no growth regulators (see Table 2). Such treatments, however, often led to poor success in spruce and pines. Studies have also included growth regulator combinations, such as ABA and auxin, or ABA and cytokinin. With diploid larch (Klimaszewska 1989a), 0.38 μM ABA and 1 μM kinetin were better than the same concentrations of kinetin and ABA alone, or no growth regulators, yielding up to 122 cotyledonary embryos per initial 150 mg fresh weight of callus. The following studies achieved maturation of Norway spruce somatic embryos on media containing 1 μM each of ABA and either IBA or NAA. In the study of Verhagen and Wann (1989), embryo maturation led to two to four embryos per sample elongating and developing cotyledons, but only several rooted plantlets were recovered in total. Becwar et al. (1987b) and Jain et al. (1988) obtained plantlets but did not indicate the frequencies obtained. Becwar et al. (1989) obtained an average of 17 mature embryos per gram of material, which corresponded to a maturation frequency of 3%.

With white and Norway spruce, higher frequencies of plantlets were clearly demonstrated following culture using ABA at elevated concentrations, compared to culture in media containing cytokinin alone, or reduced growth regulator concentrations. For example, in the study of von Arnold and Hakman (1988), Norway spruce plantlets developed from about 5% of cultures within 2 months after they had been transferred to medium lacking plant growth regulators. With ABA or zeatin added, the frequency of cultures capable of forming plantlets was doubled after 1 month; however, the average number of plantlets per sample was more than ten times higher after the ABA treatment. When these media formulations were compared using white spruce (Hakman and von Arnold 1988), plantlets developed only on ABA-containing medium. In both studies, 1/2 strength LP, 15 mM ammonium nitrate, 90 mM sucrose, and 7.6 μM ABA gave the best response for plantlet recovery. Somatic embryos of both species took about 28 days to develop cotyledons on the ABA medium. Better germination occurred if cultures were transferred at this stage to phytohormone-free medium in the light. With white spruce, approximately 35 plantlets were recovered per initial 0.5 ml suspension inoculum. The methods differed slightly from those of Dunstan et al. (1988), who found 8–12 μM ABA superior, but exposed white spruce embryos to this treatment for 35 days prior to transfer to growth regulator-free medium. Approximately 29 plantlets were recovered per initial 0.75 ml inoculum, which corresponded to a maturation frequency of 4%. This concentration of ABA also led to the maturation of black spruce embryos derived from 12-day seedling shoot explants (Attree et al. 1990a).

Table 2. Embryo maturation media used for recovery of plantlets from conifer somatic embryos

Species	Basal medium	Growth regulators	Reference
Abies alba	SH (full and 1/2 strength)	None	Schuller et al. (1989)
Larix decidua	MS or L (full and 1/2 strength)	None	Nagmani and Bonga (1985)
	L (full and 1/2 strength)	None	von Aderkas and Bonga (1988)
L. × *eurolepis*	MS	0.38 μM ABA, 1 μM kin	[a]Klimaszewska (1989a)
Picea abies	LP	None	Hakman et al. (1985)
	LP	5 μm BA, 2-iP, kin or Z	Hakman and von Arnold (1985)
	LP	1 μM BA, 1 μM ABA	Becwar et al. (1987b, 1989)
	MS (full and 1/2 strength)	2 μM BA, 1 μM 2,4-D	Gupta and Durzan (1986a)
	LP (1/2 strength)	0.1–1 μM BA, 2-iP or Z	[a]von Arnold (1987)
	MS	3.8–22.7 μM ABA	Boulay et al. (1988)
	LP (1/2 strength)	7.6 μM ABA	[a]von Arnold and Hakman (1988)
	LP (1/2 strength)	1 μM NAA, 2 μM ABA	Jain et al. (1988)
	LP (1/2 strength)	7.6 μM ABA	Mo et al. (1989)
	LP (1/2 strength)	1 μM IBA, 1 μM ABA	Verhagen and Wann (1989)
Picea glauca	LP	5 μM BA, 0.5 μM 2,4-D or none	Hakman and Fowke (1987a)
	SH (1/3 strength)	0.5 μM 2,4-D	Lu and Thorpe (1987)
	LP (1/2 strength)	7.6 μM ABA	Hakman and von Arnold (1988)
	LP	None	Kartha et al. (1988)
	LP	8–12 μM ABA	Dunstan et al. (1988)
	LP (1/2 strength)	12 μM ABA	Attree et al. (1989b)
	LP (1/2 strength)	16 μM ABA	[a]Attree et al. (1990c)
	LP (1/2 strength)	5 μM kin, 1 μM 2,4-D	[a]Tremblay (1990)
Interior spruce	LP	1 μM IBA 1 μM ABA	[a]Webb et al. (1989)
	LP	0.1–10 μM IBA, 30–40 μM ABA	[a]Roberts et al. (1990a,b), Webster et al. (1990)
Picea mariana	LP	5 μM BA, 0.5 μM 2,4-D or none	Hakman and Fowke (1987a)
	LP (1/2 strength)	12 μM ABA	Attree et al. (1990a,c) [a]Attree et al. (1990c)
Picea sitchensis	MS (1/2 strength)	5 μM ABA	Krogstrup et al. (1988, 1990)
Pinus caribaea	LP	15 μM ABA	Lainé and David (1990)
Pinus taeda	MS (1/2 strength)	2 μM kin, 2 μM BA and 5 μM 2,4-D	Gupta and Durzan (1987a)
Pseudotsuga menzesii	DCR (1/2 strength)	0.5 μM ABA	Durzan and Gupta (1987)
Sequoia sempervirens	MS (1/2 strength)	None	Bourgkard and Favre (1988)

[a]Studies showing soil-established plants.
Only best media formulations are shown. DCR, Gupta and Durzan 1985; L, Litvay et al. 1981; LP, von Arnold and Eriksson 1981; MS, Murashige and Skoog 1962; SH, Schenk and Hildebrandt 1972; ABA, abscisic acid; BA, 6-benzyladenine; 2,4-D, 2,4-dichlorophenxyacetic acid; IBA, indole-3-butyric acid; 2-iP, (isopentenyl) adenine; kin, kinetin; Z, zeatin.

In a later study, Attree et al. (1990b) compared a range of ABA concentrations during maturation of white and black spruce somatic embryos. It was observed that shorter exposure times to 32 μM ABA were not as effective as maturation on a medium containing 16 μM ABA for 28 days, and maturation frequencies of 6–8% were recorded. Roberts et al. (1990a) found that with somatic embryos of interior spruce (*Picea engelmannii, Picea glauca* and intermediate hybrids) 30–40 μM ABA with 0.1–

10 µM IBA led to mature embryos that did not undergo precocious germination and contained higher levels of storage proteins compared to lower ABA concentrations.

2.1 Maturation in Liquid Medium

Recovery of plantlets directly from liquid cultures permits easier handling of large quantities of material. This could be advantageous to future large-scale industrial applications. Maturation of Douglas fir and Norway spruce somatic embryos to the globular stage was initiated in suspension cultures with medium containing ABA (Durzan and Gupta 1987; Boulay et al. 1988). Prolonging the suspension culture phase did not yield more mature stages, but led instead to browning of the cultures.

It would be desirable to extend the liquid culture phase to include mature embryos. Methods could also incorporate embryo encapsulation to yield artificial seeds which may better withstand machine handling and prolonged storage. Embryos of Norway spruce and loblolly pine (Gupta and Durzan 1986a, 1987a) have been encapsulated in alginate gel; however, embryo survival and plantlet recovery following the encapsulation process have not yet been documented.

2.2 Plantlet Regeneration from Protoplasts

The recalcitrance of conifer protoplasts to culture has been stressed previously (Attree et al. 1991). To date, protoplasts from six conifer species have demonstrated a morphogenetic capability by regenerating to somatic embryos (white spruce, Attree et al. 1987, 1989ab; Loblolly pine, and Douglas fir, Gupta and Durzan 1987b, Gupta et al. 1988; larch, Klimaszewska 1989b; black spruce, Tautorus et al. 1990; and *Pinus caribaea*, Lainé and David 1990). In a further report (Lang and Kohlenbach 1989), cultured protoplasts derived from embryogenic callus of *Abies alba* differentiated into small meristematic, and suspensor type cells, but somatic embryos were not described.

Protoplasts of white spruce (Attree et al. 1989b), larch (Klimaszewska 1989b), and *Pinus caribaea* (Lainé and David 1990) have regenerated to plantlets. The time from initial protoplast isolation to obtaining rooted plantlets ranged from 15 weeks (white spruce), to approximately 20 weeks (*Pinus caribaea*). Initial choice of genotype was important for success, as suspension lines best capable of regenerating to plantlets showed the lowest protoplast yields and plating efficiencies (Attree et al. 1989b; Lainé and David 1990).

2.3 Soil Establishment

Survival of plantlets derived from somatic embryos in soil is likely to be determined by the quality of the mature embryo. Embryos undergo expansion to form plantlets, during which time storage reserves are mobilized. Thus, best root and hypocotyl elongation would be expected in individuals that have stored adequate reserves (lipids and proteins), and which are of sufficient developmental and physiological maturity. Additionally, however, germination of somatic embryos takes place in the absence of the nutritive megagametophyte. This may impose additional constraints on the vigor of germinated plantlets. With *Picea* species, the absence of the megagametophyte can be largely overcome by the application of sucrose (Becwar et al. 1989), and germinated plantlets are similar to excised zygotic embryos germinated in vitro (Attree et al. 1990b). Such plants may require only a gradual reduction in humidity to acclimatize plants following transfer from in vitro culture to greenhouse conditions. With *Pinus* species, however, somatic and excised zygotic embryos after germination were considerably smaller than seed-germinated plants, consequently survival following transfer to soil was difficult to achieve (Lainé and David pers. commun.).

In initial reports of soil establishment, few surviving plants were described (e.g., Douglas fir, Durzan and Gupta 1987; loblolly pine, Gupta and Durzan 1987a; Norway spruce, von Arnold and Hakman 1988; white spruce, Tremblay 1990). With larch, 35 diploid plantlets were established in soil (Klimaszewska 1989a), but the establishment frequency was not given. The first attempt to quantify the establishment process (Becwar et al. 1989) showed that 56% of Norway spruce somatic embryos germinated and 9 out of 31 plantlets (29%) survived soil establishment. A study of interior spruce (Roberts et al. 1990b) showed that matured somatic embryos were sensitive to desiccation. However, 90% of those subjected instead to a high humidity treatment (termed partial drying) germinated, and of those 47% survived transfer to soil, compared to only 3% of untreated controls. It was suggested that the partial drying treatment switched gene expression from a maturation to a germination program. In a follow-up study 1200 interior spruce plantlets were tested for nursery performance (Webster et al. 1990). Following the first growing season, survival was 80% for most genotypes, and it was observed that growth rates, final height, shoot and root morphology, and frost hardiness were similar to plants derived from zygotic seedlings. In a study of white and black spruce (Attree et al. 1990b), matured embryos were not subjected to a partial drying treatment, but were transferred to ABA-free medium for 14 days' post maturation in the dark, prior to picking out embryos for germination. Over 1800 plantlets were recovered and germination frequencies of 64 and 74% were recorded for white and black spruce, respectively. White spruce plantlets subsequently survived soil transfer at a frequency of only 18% (29/160) after 3 months, and of those, half produced dormant buds. This result was similar to the nonpartial-dried interior spruce plantlets of Roberts et al. (1990b). Black spruce plantlets, on the other hand, survived soil establishment at a frequency of 60% (160/265) with less than 2% dormant bud production. The comparison of black and white spruce showed that there were differences in the ability of closely related species to respond to soil establishment following regeneration under identical conditions, and that a partial drying treatment was not necessary for all conifer species. Regenerated Sitka spruce plantlets similarly did not form dormant buds (Krogstrup 1990). In this study 50% of mature somatic embryos germinated, of which 90% subsequently survived soil establishment.

3 Conclusions

Micropropagation via somatic embryogenesis offers significant advantages over micropropagation via organogenesis for large-scale plant production in forestry, due to the potential for more rapid multiplication rates and lower costs per plantlet. For example, somatic embryos can be proliferated in suspension culture and, when subcultured weekly, bulk up to yield vast quantities of embryos very rapidly. Somatic embryos also mature within a few weeks when transferred to maturation regimes and the total time to obtaining soil-established plants is just a few months. In addition, rooting occurs without the need for complex root-induction treatments. At present, somatic embryos can be induced only from embryonic tissues. This constraint is, however, also present when regenerating plants via adventitious buds induced on seedling explants, which is one of the more successful methods for conifer micropropagation to date. Despite this current limitation, somatic embryogenesis could produce unlimited quantities of improved trees from full-sib elite seed. Such seed derived from controlled crosses is very expensive to produce and limits the technique to providing trees for breeding orchards. Somatic embryogenesis also provides a regeneration system which lends itself to gene transfer studies. Cryopreservation of embryogenic suspensions of conifers permits long-term storage of genotypes while performance trials are undertaken. Currently, our white spruce lines cryopreserved for over 2 years by K.K. Kartha show no observable loss of viability and vigor after thawing. For somatic embryogenesis to directly supply plants to forest nurseries, improvements must be made to the frequency of embryo maturation and plantlet recovery to further reduce costs. Methodologies are, however, improving rapidly. For example, since the first reports of somatic embryogenesis in 1985, there has been a steady increase in the number of conifer species from which somatic embryos have been induced. Methods have progressed from utilizing immature zygotic embryos to mature zygotic embryos and finally seedlings as explant sources. The induction frequencies from the latter have not been unfavorable and, by optimizing conditions, could be further improved, and these methods extended to include additional species or mature tissues. Early reports concerning plantlet regeneration generally described initial embryo maturation and limited shoot production. More recent reports, especially among the *Picea*, have demonstrated higher frequencies of plantlet recovery, more normal root development, and soil-established plants. The availability of large quantities of mature somatic embryos capable of development to plantlets has provided sufficient experimental material to tackle the problems of rooting and soil establishment, and should enable the potential of somatic embryogenesis to forestry to be realized in the very near future. These successes need to be transferred to other genera such as the pines, which currently show narrow windows of induction and low induction frequencies, in addition to poorer maturation, germination, and soil-establishment frequencies.

Acknowledgments. We wish to thank Dennis Dyck and Pat Rennie for assistance preparing the figures. We also acknowledge financial support through the Natural Sciences and Engineering Research Council of Canada.

References

Attree SM, Bekkaoui F, Dunstan D, Fowke LC (1987) Regeneration of protoplasts from an embryogenic suspension culture of white spruce (*Picea glauca*). Plant Cell Rep 6:480–483

Attree SM, Dunstan D, Fowke LC (1989a) Initiation of embryogenic callus and suspension cultures, and improved embryo regeneration from protoplasts of white spruce (*Picea glauca*). Can J Bot 67:1790–1795

Attree SM, Dunstan D, Fowke LC (1989b) Plantlet regeneration from embryogenic protoplasts of white spruce (*Picea glauca*). Bio/Technology 7:1060–1061

Attree SM, Budimir S, Fowke LC (1990a) Somatic embryogenesis and plantlet regeneration from cultured shoots and cotyledons of seedlings from stored seed of black and white spruce (*Picea mariana* and *Picea glauca*). Can J Bot 68:30–34

Attree SM, Tautorus TE, Dunstan DI, Fowke LC (1990b) Somatic embryo maturation, germination, and soil establishment of plants of black and white spruce (*Picea mariana* and *Picea glauca*). Can J Bot 68:2583–2589

Attree SM, Dunstan D, Fowke LC (1991) White Spruce [*Picea glauca* (Moench) Voss] and Black Spruce [*Picea mariana* (Mill) B.S.P.] . In: Bajaj YPS (ed) Biotechnology in agriculture and forestry Vol 16. Trees III. Springer, Berlin Heidelberg New York Tokyo, pp 423–445

Becwar MR, Noland TL, Wann SR (1987a) A method for quantification of the level of somatic embryogenesis among Norway spruce callus lines. Plant Cell Rep 6:35–38

Becwar MR, Noland TL, Wann SR (1987b) Somatic embryo development and plant regeneration from embryogenic Norway spruce callus. Tappi 70:155–160

Becwar MR, Noland TL, Wyckoff JL (1989) Maturation, germination, and conversion of Norway spruce (*Picea abies* L.) somatic embryos to plants. In Vitro Cell Dev Biol 25:575–580

Becwar MR, Nagmani R, Wann SR (1990) Initiation of embryogenic cultures and somatic embryo development in loblolly pine (*Pinus taeda*). Can J For Res 20:810–817

Becwar MR, Blush TD, Brown DW, Chesick EE (1991) Multiple paternal genotypes in embryogenic tissue derived from individual immature loblolly pine seeds. Plant Cell Tissue Organ Cult (in press)

Bonga JM, von Aderkas P, James D (1988) Potential application of haploid cultures of tree species. In: Hanover JW, Keathly DE (eds) Genetic manipulation of woody plants. Plenum, New York, pp 57–77

Boulay MP, Gupta PK, Krogstrup P, Durzan DJ (1988) Development of somatic embryos from cell suspension cultures of Norway spruce (*Picea abies* Karst.). Plant Cell Rep 7:134–137

Bourgkard F, Favre JM (1988) Somatic embryos from callus of *Sequoia sempervirens*. Plant Cell Rep 7:445–448

Campbell RA, Durzan DJ (1975) Induction of multiple buds and needles in tissue cultures of *Picea glauca*. Can J Bot 53:1652–1657

Chandler SF, Bateman C, Blomstedt C, Willyams D, Young R (1989) Forestry Biotechnology at Calgene Pacific. Aust J Biotechnol 3:281–284

David A (1982) In vitro propagation of gymnosperms. In: Bonga JM, Durzan DJ (eds) Tissue culture in forestry. Martinus Nijhoff/Junk, The Hague, pp 72–108

Dunstan DI, Bekkaoui F, Pilon M, Fowke LC, Abrams SR (1988) Effects of ABA and analogues on the maturation of white spruce (*Picea glauca*) somatic embryos. Plant Sci 58:77–84

Durzan DJ, Gupta PK (1987) Somatic embryogenesis and polyembryogenesis in Douglas-fir cell suspension cultures. Plant Sci 52:229–235

Feirer RP, Conkey JH, Verhagen SA (1989) Triglycerides in embryogenic conifer calli: a comparison with zygotic embryos. Plant Cell Rep 8:207–209

Finer JJ, Kriebel HB, Becwar MR (1989) Initiation of embryogenic callus and suspension cultures of eastern white pine (*Pinus strobus* L.). Plant Cell Rep 8:203–206

Foster AS, Gifford EM (1974) Comparative morphology of vascular plants. 2nd edn. WH Freeman, San Francisco

Fowke LC, Attree SM, Wang H, Dunstan DI (1990) Microtubule organization and cell division in embryogenic protoplast cultures of white spruce (*Picea glauca*). Protoplasma 158:86–94

Gupta PK, Durzan DJ (1985) Shoot multiplication from mature trees of Douglas-fir (*Pseudotsuga menziesii*) and sugar pine (*Pinus lambertiana*). Plant Cell Rep 4:177–179

Gupta PK, Durzan DJ (1986a) Plantlet regeneration via somatic embryogenesis from subcultured callus of mature embryos of *Picea abies* (Norway spruce). In Vitro Cell Dev Biol 22:685–688

Gupta PK, Durzan DJ (1986b) Somatic polyembryogenesis from callus of mature sugar pine embryos. Bio/Technology 4:643–645

Gupta PK, Durzan DJ (1987a) Biotechnology of somatic polyembryogenesis and plantlet regeneration in loblolly pine. Bio/Technology 5:147–151

Gupta PK, Durzan DJ (1987b) Somatic embryos from protoplasts of loblolly pine proembryonal cells. Bio/Technology 5:710–712

Gupta PK, Dandekar AM, Durzan DJ (1988) Somatic proembryo formation and transient gene expression of a luciferase gene in Douglas fir and loblolly pine protoplasts. Plant Sci 58:85–92

Hakman I, Fowke LC (1987a) Somatic embryogenesis in *Picea glauca* (white spruce) and *Picea mariana* (black spruce). Can J Bot 65:656–659

Hakman I, Fowke LC (1987b) An embryogenic suspension culture of *Picea glauca* (white spruce). Plant Cell Rep 6:20–22

Hakman I, von Arnold S (1985) Plantlet regeneration through somatic embryogenesis in *Picea abies* (Norway spruce). J Plant Physiol 121:149–158

Hakman I, von Arnold S (1988) Somatic embryogenesis and plant regeneration from suspension cultures of *Picea glauca* (white spruce). Physiol Plant 72:579–587

Hakman I, Fowke LC, von Arnold S, Eriksson T (1985) The development of somatic embryos in tissue cultures initiated from immature embryos of *Picea abies* (Norway spruce). Plant Sci 38:53–59

Jain SM, Newton RJ, Soltes EJ (1988) Enhancement of somatic embryogenesis in Norway spruce (*Picea abies* L.). Theor Appl Genet 76:501–506

Jain SM, Dong N, Newton RJ (1989) Somatic embryogenesis in slash pine (*Pinus elliottii*) from immature embryos cultured in vitro. Plant Sci 65:233–241

Kartha KK, Fowke LC, Leung NL, Caswell KL, Hakman I (1988) Induction of somatic embryos and plantlets from cryopreserved cell cultures of white spruce (*Picea glauca*). J Plant Physiol 132:529–539

Klimaszewska K (1989a) Plantlet development from immature zygotic embryos of hybrid larch through somatic embryogenesis. Plant Sci 63:95–103

Klimaszewska K (1989b) Recovery of somatic embryos and plantlets from protoplast cultures of *Larix* × *eurolepis*. Plant Cell Rep 8:440–444

Krogstrup P (1986) Embryo-like structures from cotyledons and ripe embryos of Norway spruce (*Picea abies*). Can J For Res 16:664–668

Krogstrup P, Eriksen EN, Møller JD, Roulund H (1988) Somatic embryogenesis in Sitka spruce (*Picea sitchensis* (Bong.) Carr.). Plant Cell Rep 7:594–597

Krogstrup P (1990) Effect of culture densities on cell proliferation and regeneration from embryogenic cell suspensions of *Picea sitchensis*. Plant Sci 72:115–123

Lainé E, David A (1990) Somatic embryogenesis in immature embryos and protoplasts of *Pinus caribaea*. Plant Sci 69:215–224

Lang H, Kohlenbach HW (1989) Cell differentiation in protoplast cultures from embryogenic callus of *Abies alba* L. Plant Cell Rep 8:120–123

Lelu M-A, Boulay M, Arnaud Y (1987) Obtention de cals embryogénes à partir de cotylédons de *Picea abies* (L.) Karst. prélevés sur de jeunus plantes âgées de 3 à 7 jours après germination. CR Acad Sci Paris 305:105–109

Litvay JD, Johnson MA, Verma D, Einspahr D, Weyrauch K (1981) Conifer suspension culture medium development using analytical data from developing seeds. Tech Pap Ser Inst Paper Chem 115:1–17

Lu C-Y, Thorpe TA (1987) Somatic embryogenesis and plantlet regeneration in cultured immature embryos of *Picea glauca*. J Plant Physiol 128:297–302

Mo LH, von Arnold S, Lagercrantz U (1989) Morphogenic and genetic stability in longterm embryogenic cultures and somatic embryos of Norway spruce (*Picea abies* {L.} Karst.). Plant Cell Rep 8:375–378

Mott RL (1981) Trees. In: Conger BV (ed) Cloning agricultural plants via in vitro techniques. CRC, Boca Raton, pp 217–254

Murashige T, Skoog F (1962) A revised medium for rapid growth and bioassay with tobacco tissue cultures. Physiol Plant 15:473–497

Nagmani R, Bonga JM (1985) Embryogenesis in subcultured callus of *Larix decidua*. Can J For Res 15:1088–1091

Owens JN, Molder M (1984) The reproductive cycle of interior spruce. Province of British Columbia, Information services Branch, Ministry of Forests. Victoria, BC, Canada

Roberts DR, Flinn BS, Webb DT, Webster FB, Sutton BCS (1990a) Abscisic acid and indole-3-butyric acid regulation of maturation and accumulation of storage proteins in somatic embryos of interior spruce. Physiol Plant 78:355–360

Roberts DR, Sutton BCS, Flinn BS (1990b) Synchronous and high frequency germination of interior spruce somatic embryos following partial drying at high relative humidity. Can J Bot 68:1086–1090

Schenk RU, Hildebrandt AL (1972) Medium and techniques for induction and growth of monocotyledonous and dicotyledonous plant cell cultures. Can J Bot 50:199–204

Schuller A, Reuther G, Geier T (1989) Somatic embryogenesis from seed explants of *Abies alba*. Plant Cell Tissue Organ Cult 17:53–58

Simola LK, Honkanen J (1983) Organogenesis and fine structure in megagametophyte callus lines of *Picea abies*. Physiol Plant 59:551–561

Simola LK, Santanen A (1990) Improvement of nutrient medium for growth and embryogenesis of megagametophyte and embryo callus lines of *Picea abies*. Physiol Plant 80:27–35

Tautorus TE, Attree SM, Fowke LC, Dunstan DI (1990) Somatic embryogenesis from immature and mature zygotic embryos, and embryo regeneration from protoplasts in black spruce (*Picea mariana* Mill.). Plant Sci 67:115–124

Tremblay FM (1990) Somatic embryogenesis and plantlet regeneration from embryos isolated from stored seeds of *Picea glauca*. Can J Bot 68:236–242

Verhagen SA, Wann SR (1989) Norway spruce somatic embryogenesis: high frequency initiation from light-cultured mature embryos. Plant Cell Tissue Organ Cult 16:103–111

von Aderkas P, Bonga JM (1988) Formation of haploid embryoids of *Larix decidua*. Am J Bot 75:690–700

von Aderkas P, Bonga JM, Nagmani R (1987) Promotion of embryogenesis in cultured megagametophytes of *Larix decidua*. Can J For Res 17:1293–1296

von Aderkas P, Klimaszewska K, Bonga JM (1990) Diploid and haploid embryogenesis in *Larix leptolepis, L. decidua*, and their reciprocal hybrids. Can J For Res 20:9–14

von Aderkas P, Bonga J, Klimaszewska K, Owens J (1991) Comparison of larch embryogeny in vivo and in vitro. In: Ahuja MR (ed) Woody plant biotechnology. Plenum, London (in press)

von Arnold S (1987) Improved efficiency of somatic embryogenesis in mature embryos of *Picea abies* (L.) Karst. J Plant Physiol 128:233–234

von Arnold S, Eriksson T (1981) In vitro studies of adventitious shoot formation in *Pinus contorta*. Can J Bot 59:870–874

von Arnold S, Hakman I (1986) Effects of sucrose on initiation of embryogenic callus from mature zygotic embryos of *Picea abies* (L.) Karst. (Norway spruce). J Plant Physiol 122:251–265

von Arnold S, Hakman I (1988) Regulation of somatic embryo development in *Picea abies* by abscisic acid (ABA). J Plant Physiol 132:164–169

von Arnold S, Woodward S (1988) Organogenesis and embryogenesis in mature zygotic embryos of *Picea sitchensis*. Tree Physiol 4:291–300

Wann SR, Johnson MA, Noland TL, Carlson JA (1987) Biochemical differences between embryogenic and non-embryogenic callus of *Picea abies* (L.) Karst. Plant Cell Rep 6:39–42

Webb DT, Webster F, Flinn BS, Roberts DR, Ellis DD (1989) Factors influencing the induction of embryogenic and caulogenic callus from embryos of *Picea glauca* and *P. engelmanni*. Can J For Res 19:1303–1308

Webster FB, Roberts DR, McInnis SM, Sutton BCS (1990) Propagation of interior spruce by somatic embryogenesis. Can J For Res 20:1759–1765

I.5 Graft Compatibilities in Vitro

R. MOORE[1]

1 Introduction

Micrografting is a means of propagation that involves fusing small pieces of tissue, either in vivo or in vitro. Although relatively new and cumbersome, micrografting has already produced startling results. For example, micrografting has been used to produce healthy citrus orchards and rejuvenate trees, and will be a valuable tool for investigating the transition from juvenile to adult growth (see review by Jonard 1986).

In vitro micrografting involves grafting tissues in aseptic conditions. It is less expensive than in vivo grafting, and often enables growers to recognize incompatibility within a few weeks or days rather than the years often required to observe incompatibility in vivo (see reviews by Moore 1984a; Jonard 1986).

The success or failure of a micrograft depends largely on the compatibility responses that occur between the grafting tissues. Similar responses are widespread in plants.

1.1 Importance of Compatibility Response in Plants

Compatibility responses are involved in several developmental processes in plants, including host-pathogen interactions, pollen-stigma interactions, parasitisms, mycorrhizae, and other symbioses (e.g., nodulating bacteria). Plant grafting is a compatibility response which, like the processes mentioned above, is economically important. For example, European grape (*Vitis vinifera*) is susceptible to grape phylloxera, while American grape (*V. lambrusca*) resists phylloxera. Grafting a *V. vinifera* scion onto a *V. lambrusca* rootstock produces a high-yielding plant that resists phylloxera (Mahlsted and Haber 1957; Hartmann and Kester 1975). Grafting is also used to perpetuate and speed production of certain clones of plants, increase the quality and hardiness of plants and fruit, facilitate adaptation to new environments, and obtain special forms of growth (e.g., dwarfs; see discussion in Hartmann and Kester 1975). Despite this economic importance, we know little about the mechanisms underlying graft compatibility and incompatibility in plants.

The objective of this chapter is to summarize our knowledge of graft compatibility-incompatibility in vitro. As an introduction, let us first consider how grafts form in vivo.

[1]Department of Biological Sciences, Wright State University, Dayton, OH 45435, USA

Biotechnology in Agriculture and Forestry, Vol. 17
High-Tech and Micropropagation I (ed. by Y.P.S. Bajaj)

1.2 Development of Compatible Grafts in Vivo

Formation of a compatible graft typically involves three phases: (1) cohesion of stock and scion, (2) proliferation of callus at the graft interface, and (3) redifferentiation of vascular tissue across the graft interface (Moore 1984a).

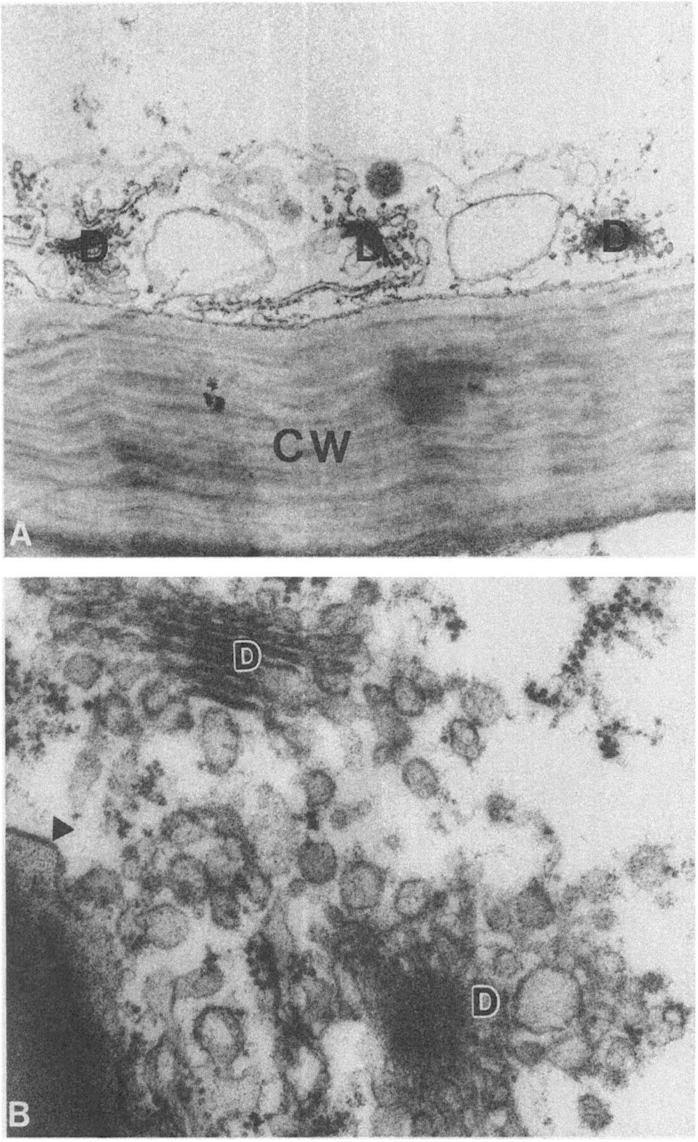

Fig. 1. **A** Accumulation of dictyosomes (*D*) along the graft interface at 6 h after grafting. *CW* Cell wall; × 1700 (Moore and Walker 1981a). *B* Dictyosomes (*D*) and associated vesicles at graft interface 18 h after grafting. Note wavy appearance of plasmalemma (*arrowhead*), suggesting that vesicles are fusing with plasmalemma; × 8200

Cohesion of stock and scion results from deposition of cell wall precursors into the graft interface by dictyosomes (Fig. 1A,B) (Moore and Walker 1981a). Polymerization of these precursors coheres the graft partners. This cohesion does not involve cellular recognition, since cut surfaces will cohere ("graft") to inanimate objects such as porous membranes (Moore 1984b) and wooden sticks (Moore and Walker 1981a).

Callus proliferation at the graft interface is also induced by wounding (Fig. 2A,B). Interdigitation of these proliferating tissues significantly increases the tensile strength of the graft union (Moore 1982, 1984b; Stoddard and McCully 1979). Since callus proliferation occurs in grafted and nongrafted systems, it cannot involve mutual cellular recognition.

Vascular redifferentiation is the final major event that occurs during formation of a successful graft (Fig. 2A,B). However, there are several reports of compatible grafts that lack vascular redifferentiation, indicating that vascular redifferentiation is not necessary for a successful (much less a compatible) graft (Herrero 1951; Muzik 1958; Moore 1984c; see discussion below).

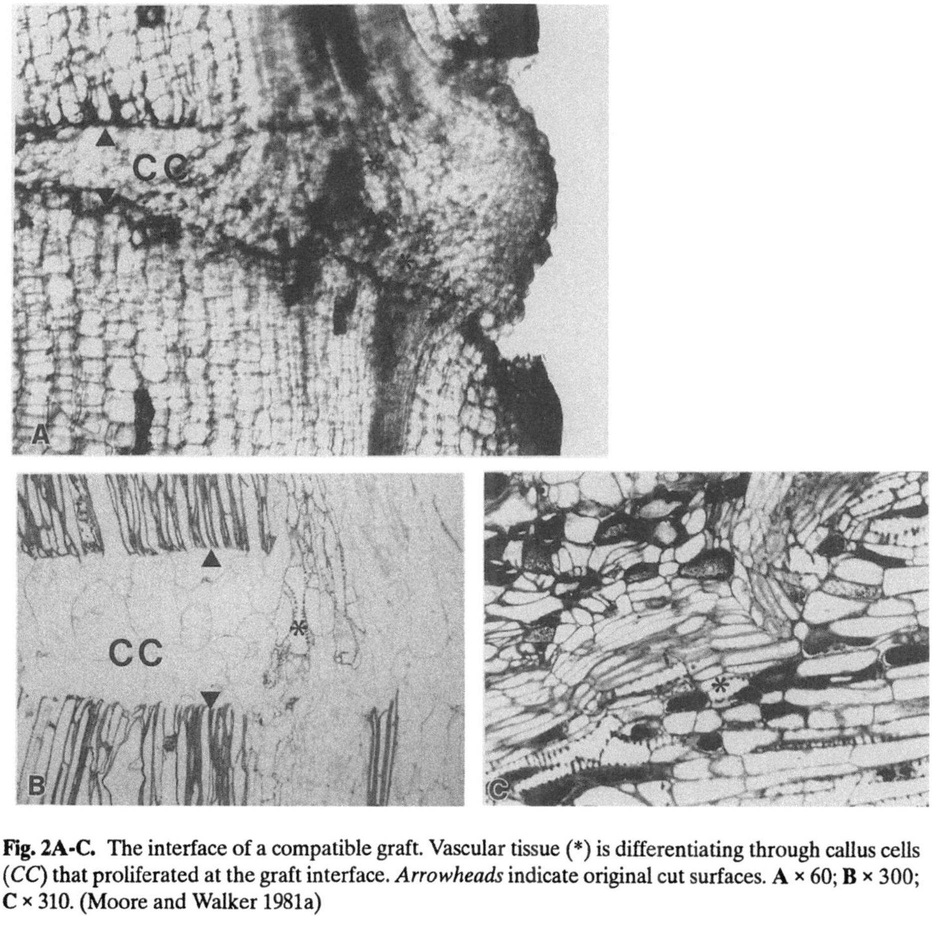

Fig. 2A-C. The interface of a compatible graft. Vascular tissue (*) is differentiating through callus cells (*CC*) that proliferated at the graft interface. *Arrowheads* indicate original cut surfaces. **A** × 60; **B** × 300; **C** × 310. (Moore and Walker 1981a)

1.3 Development of Incompatible Grafts in Vivo

An incompatible graft is not synonymous with an unsuccessful graft. For example, autografts between surfaces of a severed internode are compatible, but may not graft successfully because of problems with environmental conditions (e.g., humidity), infections, or grafting technique. These factors are unrelated to tissue incompatibility, which results from physiological influences between the grafting tissues.

There are three types of graft incompatibility: (1) localized incompatibility, (2) translocated incompatibility, and (3) delayed incompatibility.

Localized incompatibility is overcome by inserting a mutually compatible interstock between the stock and scion. This interstock apparently acts as a sink for substances produced in one partner that elicit incompatibility in the other partner. Examples of localized incompatibility include grafts between pear on quince (Hartmann and Kester 1975), peach on plum (Breen and Muraoka 1975), and apricot on myrobalan (Jonard 1986).

Translocated incompatibility is not overcome by inserting an interstock. In these grafts, toxins from one partner move through the interstock and elicit incompatibility in the other partner. Examples of translocated incompatibility include grafts between plum and almond (Hartmann and Kester 1975), peach and apricot, and peach and myrobalan (Jonard 1986).

Delayed incompatibility develops up to 20 years after grafting. Possible explanations for delayed incompatibility include production and/or accumulation of age-related toxins. Examples of delayed incompatibility include grafts between *Juglans regia* and *J. hindsii* (Hartmann and Kester 1975).

The response that characterizes graft incompatibility is cellular necrosis at the graft interface (Gur et al. 1968; Bevington 1976; Copes 1980; Moore and Walker 1981b; Deloire and Hebant 1982; Jonard 1986). This necrosis is elicited by toxins that move from one graft partner to another (Gur et al. 1968; Bevington 1976; Moore and Walker 1981a; Moore 1984d), where they produce a necrotic layer of collapsed cells that insulate the graft partners from each other (Fig. 3A,B). As a result, the scion typically desiccates and dies. Interestingly, cellular necrosis associated with graft incompatibility is similar to the hypersensitive response (Moore and Walker 1981b; see discussion below), suggesting that cellular necrosis may be a common defense mechanism used by plants in a variety of conditions.

2 In Vitro Studies of Graft Compatibility-Incompatibility

2.1 Studies of Compatible Grafts

Moore and Walker (1983) grafted intact (unwounded) callus masses of *Sedum telephoides* and *Solanum pennellii* to determine which of the early events of graft formation are due to grafting per se as compared to the coincidental wounding necessary to make stem grafts. When placed side by side on culture media and allowed to grow together, callus masses interdigitate and graft successfully. There is no cellular wound response, accumulation of dictyosomes, or dramatic release of cell wall precursors such as occurred in vitro, indicating that these events are due to wounding rather than graft compatibility (see discussion in Moore and Walker 1983). The strong interdigita-

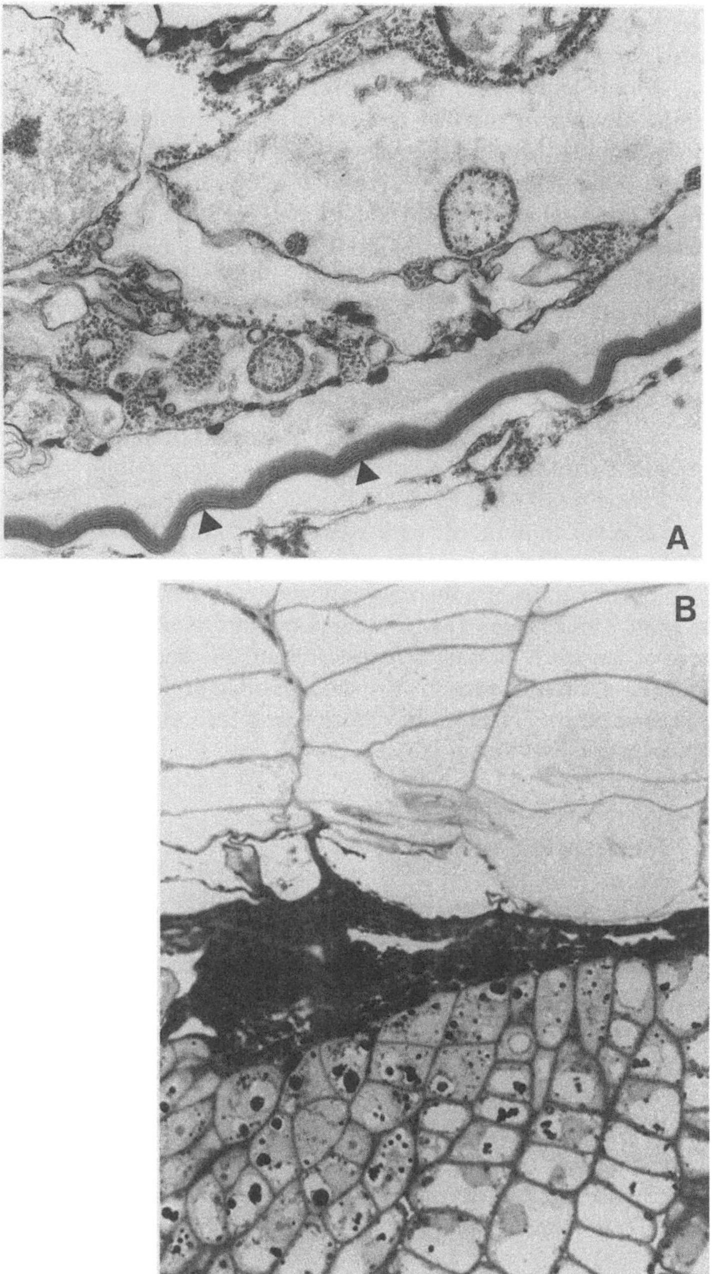

Fig. 3. A Cellular necrosis associated with graft incompatibility between *Sedum telephoides* and *Solanum pennelli*. Necrosis involves aggregation of the cytoplasm and deposition of suberin-like lamellae on walls of *Sedum* cells (*arrowheads*); × 5100 (Moore and Walker 1981b). **B** Cellular necrosis produces a layer of collapsed cells that separate the *Solanum* stock (*bottom*) and *Sedum* scion (*top*); × 500. (Moore and Walker 1981b)

tion of callus cells in vitro is similar to that observed by Ball (1969, 1971), and supports the suggestion of Fujii and Nito (1972) that the responses of callus masses grafted in vitro parallel their responses in vivo.

Moore and Walker (1983) observed that adjacent callus masses graft successfully in the absence of vascular redifferentiation. Similar results were obtained with approach grafts (Moore 1984c), indicating that vascular redifferentiation is unrelated to graft compatibility. However, vascular redifferentiation is typically important for forming a successful graft in field conditions. Parkinson and Yeoman (1982) examined vascular redifferentiation across graft interfaces of *Lycopersicon esculentum, Nicandra physaloides,* and *Datura stramonium* cultured in vitro, and demonstrated that redifferentiation requires apical application of indole-3-acetic acid (IAA) at concentrations of 0.2 to 2.0 mg/l to the apical end of the grafted internode. Kinetin (0.2 mg/l) also stimulated redifferentiation, but gibberellic acid (GA_3 at 0.5 mg/l) was inhibitory. These observations are consistent with other reports of auxin (1) stimulating graft formation (Muller-Stoll 1938; Shimomura and Fuzihara 1977, 1978) by stimulating vascular redifferentiation and production of callus tissue at the graft interface, and (2) being necessary for formation of vascular tissue (Sachs 1981). Auxin for vascular redifferentiation in vivo is supplied primarily by leaves of the scion (Stoddard and McCully 1980; Moore 1985).

In vitro grafts of callus masses also provide information about cellular competence to graft. Unlike stems that contain numerous types of cells, young proliferating callus masses are made of only a single (presumably dedifferentiated) cell type. Successful grafting of callus masses in vitro indicates that graft compatibility does not require the presence of more highly differentiated cells (Moore and Walker 1983). Epidermal cells are incompetent to graft (Moore 1984b).

2.2 Studies of Incompatible Grafts

2.2.1 The Pear/Quince Graft

Since cellular necrosis characterizes graft incompatibility, in vitro studies of incompatibility have focused on understanding what elicits necrosis at the graft interface. The most intensively studied incompatible graft in vitro (and in vivo) is that between pear (*Pyrus communis*) and quince (*Cydonia oblonga*). Interestingly, these grafts are successful in cool weather, but become incompatible when it is hot (Gur et al. 1968). These results suggest that elevated temperatures stimulate production of a toxin that elicits incompatibility. Indeed, quince contains prunasin, a cyanogenic glycoside that ascends into the pear scion, where it is hydrolyzed by β-glycosidase. One of the liberated products is hydrocyanic acid, which includes cellular necrosis at the graft interface (Gur et al. 1968).

In vitro studies supporting this toxin-induced incompatibility between pear and quince have produced the following observations (see Gur et al. 1968).

1. The quality of a graft union (i.e., the "degree" of graft compatibility) is proportional to the amount of β-glycosidase activity in pear at the graft interface. Greater incompatibility correlates with increased activity of β-glycosidase.
2. Varieties of pear containing inhibitors of β-glycosidase are more compatible with quince than varieties lacking the inhibitor.

3. Increased incompatibility between pear and quince at elevated temperatures correlates positively with β-glycosidase of pear having a temperature optimum of 35 °C.
4. Varieties of pear that hydrolyze prunasin rapidly are more incompatible with quince than varieties that hydrolyze prunasin slowly.
5. Toxin-induced necrosis at the graft interface results in less sugar reaching roots of quince, thus stimulating prunasin hydrolysis and cellular necrosis.

Interestingly, graft incompatibility between pear and quince can be overcome experimentally by injecting ferrous polysulfide into the graft interface (Gur 1972). Ferrous polysulfide inactivates β-glycosidase, thus preventing necrosis.

Experiments with callus masses in vitro also support the suggestion that graft incompatibility between pear and quince is due to a toxin(s). For example, callus masses of pear undergo necrosis when grafted to callus masses of quince (Moore 1984d). These and similar results involving in vitro grafts of other plants (Gautheret 1945; Ball 1969, 1971; Moore and Walker 1983) indicate that (1) incompatibility responses in vitro mimic those that occur in vivo, (2) incompatibility does not require a prior wounding of tissues, (3) substances that elicit incompatibility are probably produced by ground tissue rather than by a more highly differentiated tissue (e.g., phloem), and (4) development of incompatibility need not be associated with any particular stage of graft development (see Moore 1984d).

Moore (1986) has used tissue culture to determine the nature of the toxin(s) that elicit incompatibility between pear and quince. As already mentioned, pear cells die after contacting quince cells in vitro. Similar results are obtained when the callus masses of pear and quince are separated by a porous membrane filter (Moore 1984d). These results suggest that graft incompatibility does not require direct cellular contact and probably involves diffusible toxins. This conclusion is supported by results involving suspension cultures of pear and quince (see Moore 1986).

Suspension cultures grow exponentially for approximately 30 days after subculturing, after which growth decreases due to accumulation of metabolites in the medium (Fig. 4). Supplying the cultures with fresh media restimulates growth (Fig. 5).

Cultures of pear and quince are killed by the presence of KCN (as 50 ppm CN^-) in the medium (Fig. 6). Pear is also killed by adding prunasin or amygdalin (another cyanogenic glycoside) as 50 ppm CN^- to the medium (Figs. 7 and 8). Growth of quince in the presence of these cyanogenic glycosides is reduced, but is significantly greater than that of pear (Figs. 7 and 8), indicating that quince may possess a cyanide detoxification system. Indeed, such a system may exist in peach (*Pyrus persica*), another member of the Rosaceae.

Growth of pear cultures slows dramatically when they are added to media in which quince cells grew for 10 days (Fig. 9). This result indicates that quince produces a factor that inhibits growth of pear. This factor is probably prunasin, since (1) prunasin is ubiquitous to quince (Bourquelot and Fichtenholz 1910), (2) prunasin can move from quince into the surrounding medium (Gur et al. 1968), and (3) adding prunasin to the culture medium also inhibits growth of pear cultures (Fig. 9).

Growth of quince cultures decreases slightly when they are placed in media in which pear cells grew for 10 days (Fig. 9). These results suggest that pear cultures produce a factor(s) that slightly inhibits growth of quince. This factor is not a cyanogenic compound, since pear lacks such compounds. The differential responses of pear and quince to each other's metabolites are consistent with ultrastructural analyses of graft incompatibility between pear and quince in vitro (Moore 1984d).

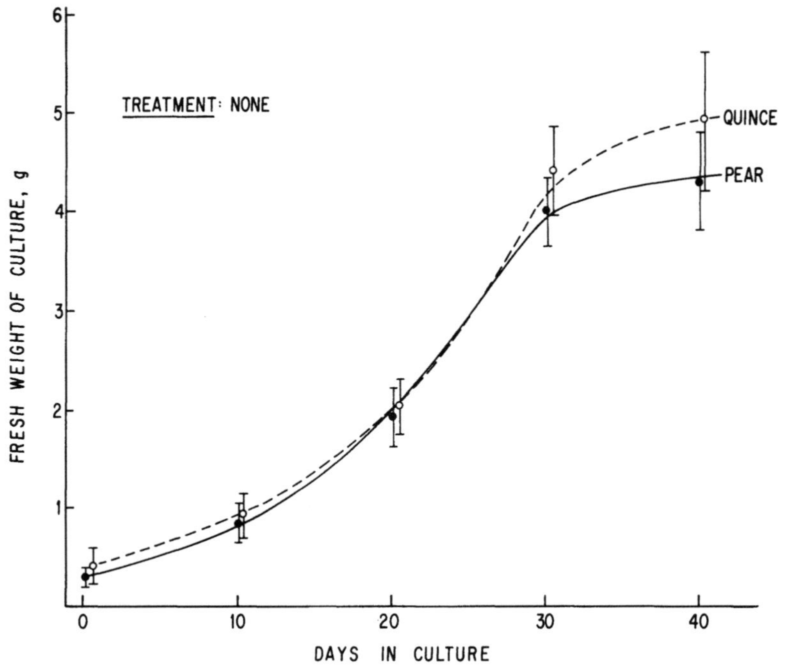

Fig. 4. Growth of suspension cultures of pear and quince

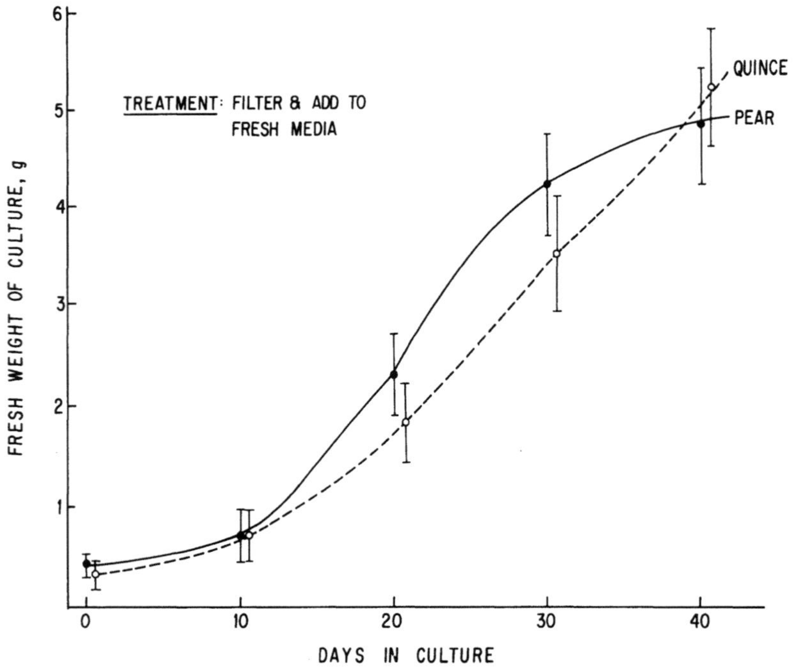

Fig. 5. Growth of suspension cultures of pear and quince when given fresh medium

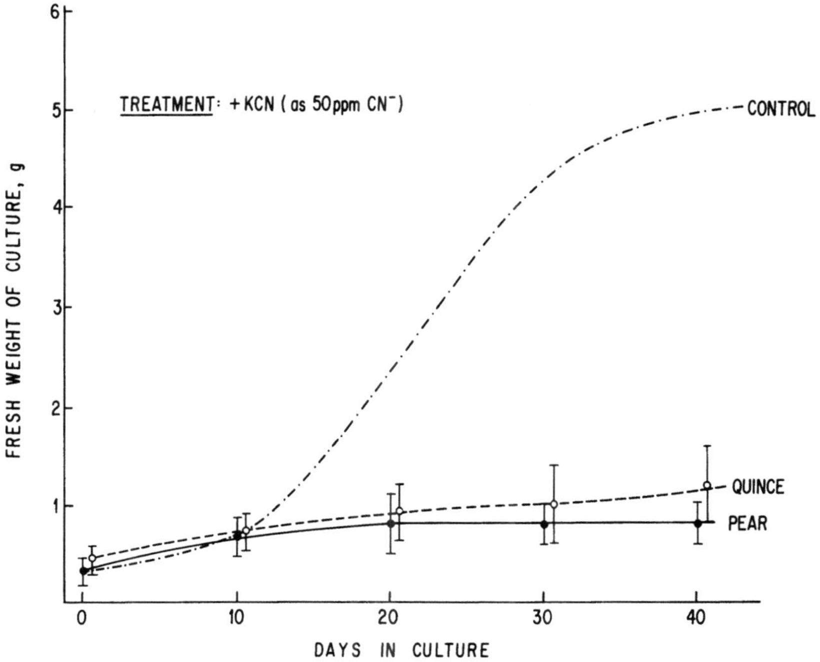

Fig. 6. Influence of KCN (as 50 ppm CN^-) on growth of suspension cultures of pear and quince

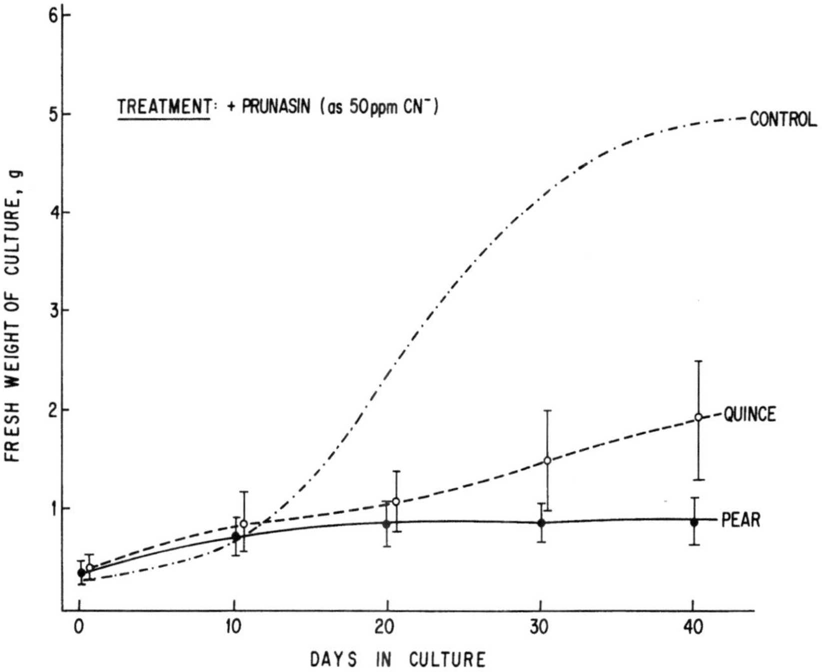

Fig. 7. Influence of prunasin (as 50 ppm CN^-) on growth of suspension cultures of pear and quince

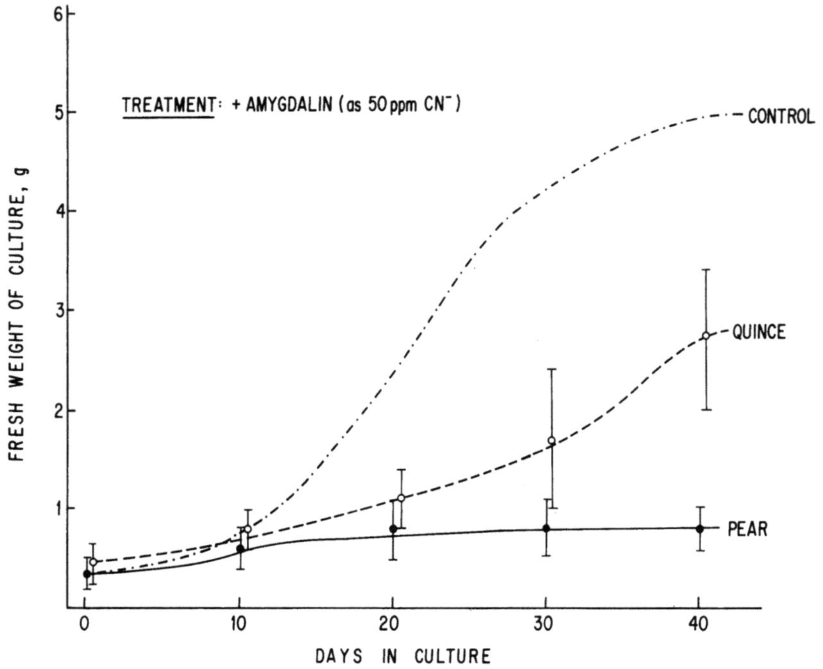

Fig. 8. Influence of amygdalin (as 50 ppm CN^-) on growth of suspension cultures of pear and quince

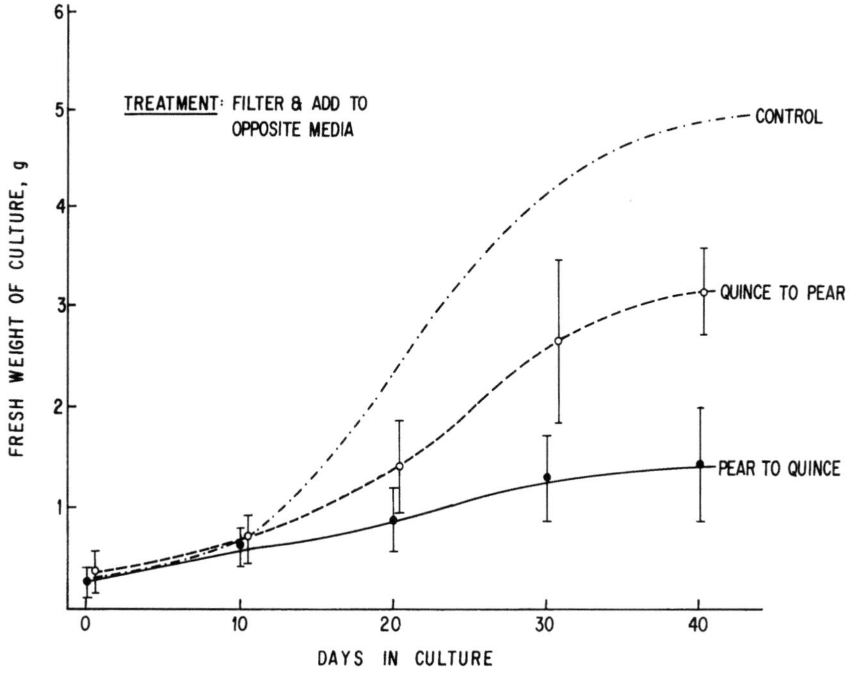

Fig. 9. Effect of adding suspension cultures of pear and quince to media in which the other plant cells had grown for 10 days

2.2.2 Other Incompatible Grafts

The biochemical basis for graft incompatibility between peach (*Prunus persica*) and almond (*Prunus amygdalus*) is similar to that of pear and quince (Gur and Blum 1972). The movement of cyanogenic glycosides from peach into almond stimulates glycosidase activity in almond, thus producing the cellular necrosis that characterizes incompatibility.

Incompatibility between *Sedum telephoides* and *Solanum pennellii* is also similar in vivo and in vitro. Necrosis is restricted to the *Sedum* partner, and correlates positively with the cytoplasmic release and/or synthesis of acid phosphatase (Moore and Walker 1981c), a hydrolytic enzyme. The toxin(s) that move between graft partners and elicit incompatibility in this and most other systems are probably not cyanogenic compounds, and remain to be identified. Furthermore, these toxins need not necessarily induce necrosis to produce an unsuccessful graft. Rather, they need only render callus cells unresponsive to morphogens that promote formation of a successful graft (see below). As a result, vascular redifferentiation would not occur, and the scion would starve.

Several investigators have suggested that various chemical tests can predict the success of grafts. For example, Copes (1978) showed that increased activities of esterase and peroxidase correlate with unsuccessful grafts 90–100% of the time in Douglas fir (*Pseudotsuga manziesii*). Similarly, Deloire and Hebant (1982) reported that incompatible grafts between *Capsicum annuum* and *Lycopersicon esculentum* have greater peroxidase activity than corresponding compatible grafts. Interestingly, the increased levels of peroxidase and esterase characteristic of incompatible grafts also occur in the hypersensitive response (Copes 1978).

3 Models for Graft Compatibility-Incompatibility

Graft compatibility-incompatibility has long been assumed to involve mutual cellular recognition (e.g., involving lectins, glycoprotein receptors, etc.; see Clarke and Knox 1978, 1979; Yeoman et al. 1978; Knox and Clarke 1980; Jeffree and Yeoman 1983; Juniper and Jeffree 1983). However, several observations are inconsistent with such a model. For example,

1. The differential responses of reciprocal grafts suggest cellular recognition does not determine graft compatibility. In these grafts, the A/B (i.e., scion/stock) combination such as *Prunus salicina/P. domestica* is compatible, while the B/A (i.e., *P. domestica/P. salicina*) combination is not (see discussion in Hartmann et al. 1988). It is difficult to imagine how reversing the orientation of the graft would affect interactions between molecules on the surfaces of the contacting cells.

2. Unlike events controlled by cellular recognition, grafts typically exhibit varying degrees of compatibility and incompatibility (Gur et al. 1968).

3. If cellular recognition controls development of a compatible graft, then one would expect all of the events comprising formation of a compatible graft to be linked developmentally. This is not what is observed – rather, the major events of graft formation occur independently of each other (Stoddard and McCully 1980).

4. It is difficult to imagine any selection pressure for a system by which cortical cells of one plant would positively recognize cortical cells in another plant.

In light of these problems, Moore (1984) proposed a novel model for graft compatibility-incompatibility which suggests that grafts will be incompatible only if the naturally occurring morphogens that promote formation of a successful graft (e.g., auxin) are overridden by toxins that elicit graft incompatibility. According to this model, a positive recognition system need not be implicated in explaining graft compatibility-incompatibility. Details regarding this model are presented elsewhere (Moore 1984a,b,c,d).

4 Future Prospects

Plant grafting has been used to study a variety of botanical problems, ranging from nitrogen fixation (Malik 1983) to flowering (Zeevaart 1982). Evidence suggests that it may be applicable to many other problems. For example, grafting an Oldenberg variety of apple onto a Baldwin variety of apple produces a plant that resists cold temperatures (Mahlsted and Haber 1957). Such a system could be important for understanding and inducing cold hardiness in other plants. Other research areas in which plant grafting could play important roles include root and shoot development, abscission, and senescence. However, breakthroughs in these applied areas will probably have to be preceded by improvements in our basic understanding of graft compatibility and incompatibility. For example, what mechanisms are responsible for the different responses of certain clones to grafting? What explanation underlies reciprocal grafts? What kinds of compounds elicit incompatibility? And finally, can graft incompatibility be overcome experimentally? In vitro micrografting involving tissue culture, protoplast fusions, etc. will play an important role in answering these and other questions about graft compatibility and incompatibility. Continued investigations may also lead to in vitro micrografting systems for predicting graft compatibility and incompatibility. Such systems would be economicaly important, since it would drastically reduce the amount of time, space, and effort necessary for determining the compatibility of a particular graft combination.

References

Ball EA (1969) Histology of mixed callus cultures. Bull Torrey Bot Club 96:52–59

Ball EA (1971) Growth of plant tissue upon a substrate of another kind of tissue. I. Qualitative observations. Z Pflanzenphysiol 65:140–158

Bevington KB (1976) Development of union abnormalities in grafts between lemon (*Citrus limon*) and *Poncirus trifoliata*. Aust J Agric Res 27:661–668

Bourquelot EM, Fichtenholz A (1910) Arbutine et methylarbutine, caractères, distinction et recherche dans les vegetaux. J Pharm Chim Paris 1:62–66, 104–109

Breen PJ, Muraoka T (1975) Seasonal nutrient levels and peach/plum graft incompatibility. J Amer Soc Hort Sci 100:339–342

Clarke AE, Knox RB (1978) Cell recognition in flowering plants. Q Rev Biol 53:3–28

Clarke AE, Knox RB (1979) Plant and immunity. Dev Comp Immunol 3:571–589

Copes DL (1978) Anatomical symptoms of graft incompatibility in *Pinus monticola* and *P. ponderosa*. Silvae Genet 29:77–82

Deloire A, Hebánt C (1982) Peroxidase activity and lignification at the interface between stock and scion of compatible and incompatible grafts of *Capsicum* on *Lycopersicon*. Ann Bot 49:887–891
Fujii T, Nito N (1972) Studies on the compatibility of grafting fruit trees. I. Callus fusion between the rootstock and scion. J Jpn Soc Hortic Sci 41:1–10
Gautheret RJ (1945) Une voie nouvelle en biologie végétale. La culture des tissus. 3rd edn, L'Avenir de la Science-21. Gallimard
Gur A (1972) Chemical control of pear-quince graft incompatibility. In: Proc Symp on "pear growing." 1972 September 4–8; Fruit Breeding Station, Angers, France: International Society for Horticultural Science, Fruit Section, Working Group on Pear, 1972, pp 253–264
Gur A, Blum A (1972) The role of cyanogenic glycoside in incompatibility between peach scions and almond rootstocks. Hort Res 13:1–10
Gur A, Samish RM, Lifshitz E (1968) The role of the cyanogenic glycoside of the quince in the incompatibility between pear cultivars and quince rootstocks. Hort Res 8:113–134
Hartmann HT, Kester DE (1975) Plant propagation: principles and practices, 3rd edn. Prentice Hall, Englewood Cliffs, NJ
Hartmann HT, Kofranek AM, Rubatzky VE, Flocker WJ (1988) Plant science: growth, development, and utilization of cultivated plants, 2nd edn. Prentice Hall, Englewood Cliffs, NJ
Herrero J (1951) Studies of compatible and incompatible graft combinations with special reference to hardy fruit trees. J Hortic Sci 26:186–237
Jeffree CE, Yeoman MM (1983) Development of intercellular connections between opposing cells in a graft union. New Phytol 93:491–509
Jonard R (1986) Micrografting and its applications to tree improvement. In: Bajaj YPS (ed) Biotechnology in agriculture and forestry, vol 1. Trees I. Springer, Berlin Heidelberg New York Tokyo, pp 31–48
Juniper BE, Jeffree CE (1983) Plant surfaces. Edward Arnold, London
Knox RB, Clarke AE (1980) Discrimination of self and non-self in plants. In: Marchalonic JJ, Cohen N (eds) Self/non-self discrimination, contemporary topics in immunology, vol 9. Plenum, New York, pp 1–36
Mahlsted JP, Haber ES (1957) Plant propagation. John Wiley, New York
Malik NSA (1983) Grafting experiments on the nature of the decline in N_2 fixation during fruit development in soybean. Physiol Plant 57:561–564
Moore R (1982) Graft development in *Kalanchöe blossfeldiana*. J Exp Bot 33:533–540
Moore R (1984a) A model for graft compatibility-incompatibility in higher plants. Am J Bot 71:752–758
Moore R (1984b) The role of direct cellular contact in the formation of compatible autografts in *Sedum telephoides*. Ann Bot 54:127–133
Moore R (1984c) Cellular interactions during the formation of approach grafts in *Sedum telephoides* (Crassulaceae). Can J Bot 62:2476–2484
Moore R (1984d) Ultrastructural aspects of graft incompatibility between bear and quince in vitro. Ann Bot 53:447–451
Moore R (1985) Studies of vegetative compatibility-incompatibility in higher plants. VII. Influences of individual organs on graft development. Tex J Sci 37:201–211
Moore R (1986) Graft incompatibility between pear and quince: the influence of metabolites of *Cydonia oblonga* on suspension cultures of *Pyrus communis*. Am J Bot 73:1–4
Moore R, Walker DB (1981a) Studies of vegetative compatibility-incompatibility in higher plants. I. A structural study of a compatible autograft in *Sedum telephoides* (Crassulaceae). Am J Bot 68:820–830
Moore R, Walker DB (1981b) Studies of vegetative compatibility-incompatibility in higher plants. II. A structural study of an incompatible heterograft between *Sedum telephoides* (Crassulaceae) and *Solanum pennellii* (Solanaceae). Am J Bot 68:831–842
Moore R, Walker DB (1981c) Studies of vegetative compatibility-incompatibility in higher plants. III. The involvement of acid phosphatase in the lethal cellular senescence associated with an incompatible heterograft. Protoplasma 109:317–334
Moore R, Walker DB (1983) Studies of vegetative compatibility-incompatibility in higher plants. VI. Grafting of *Sedum* and *Solanum* callus tissue in vitro. Protoplasma 115:114–121
Muller-Stoll WR (1938) Versuche über die Verwendbarkeit der B-Indolylessigsäure als verwachsungsörderndes Mittel in der Rebenveredlung. Angew Bot 20:218–238
Muzik IJ (1958) Role of parenchyma cells in graft union in *Vanilla* orchid. Science 127:82
Parkinson M, Yeoman MM (1982) Graft formation in cultured, explanted internodes. New Phytol 91:711–719

Sachs T (1981) The control of the patterned differentiation of vascular tissues. In: Woolhouse HW (ed) Advances in botanical research, vol 9. Academic Press, London, pp 151–262

Shimomura T, Fuzihara K (1977) Physiological study of graft union formation in cactus. II. Role of auxin on vascular connection between stock and scion. J Jpn Soc Hortic Sci 45:397–406

Shimomura T, Fuzihara K (1978) Prevention of auxin-induced vascular differentiation in wound callus by surface-to-surface adhesion between calluses of stock and scion in cactus grafts. Plant Cell Physiol 19:877–886

Stoddard FL, McCully ME (1979) Histology of the development of the graft union in pea roots. Can J Bot 57:1486–1501

Stoddard FL, McCully ME (1980) Effects of excision of stock and scion organs on the formation of the graft union in *Coleus*: a histological study. Bot Gaz 141:401–412

Yeoman MM, Kilpatrick DC, Miedzybrodzka MB, Gould AR (1978) Cellular interactions during graft formation in plants, a recognition phenomenon? Symp Soc Exp Biol 32:139–160

Zeevaart JAD (1982) Transmission of the floral stimulus from a long-short-day plant, *Bryophyllum daigremontianum*, to the short-long-day plant *Echeveria harmsii*. Ann Bot 49:549–552

I.6 In Vitro Induction of Flowering and Its Relevance to Micropropagation

J. VAN STADEN and C.W.S. DICKENS[1]

1 Introduction

After many years of intensive research, which have not solved the mysteries of flowering, there is a need for change and the use of new techniques in order to understand what is probably the most important of all physiological processes. As a novel approach, in vitro techniques are now showing some promise in investigating some of the problems concerned with the flowering process. Such techniques allow for much greater control of the whole or part of the plant under investigation. The whole process can now be linked to micropropagation techniques, more efficient application of exogenous substances can be achieved, effector or affected sites can be isolated or eliminated, and complicatory influences such as bacterial or fungal contamination of wounded surfaces, and of organic substances under test can be avoided. Three main systems are used for in vitro flowering, each with different objectives. (1) Whole plant culture, where plants have all basic organs, even though they may be reduced. (2) The culture of isolated organs or bud explants with meristematic cells which produce flowers directly without the formation of roots or shoots. (3) The culture of non-meristematic tissues such as callus, thin epidermal cell layers, and pith tissue, where direct flower bud differentiation and development without the formation of shoots or roots occur.

The advantage of whole plant investigations is that one can test either a single, or a small number of compounds, and rely on the complete plant in culture to provide all of the other compounds required for growth and/or flowering, in their correct concentrations. Micropropagation is eminently suitable to achieve the desired uniformity. One potential problem here is that in those systems which are maintained in noninductive conditions, and induction is attempted with a single test substance, the physiological conditions occurring in the plant may be directed towards vegetative growth to an extent that the inductive properties of the substances are not realized. Nevertheless, systems could be developed using plants in a marginal flowering condition, which negates some of the problems mentioned above.

The advantage of the second type of system, isolated meristematic regions which do not always form all of the representative organs, is that the confounding influence of certain organs can be eliminated. Micropropagation of shoots, for example, will again ensure a high degree of uniformity, an aspect highly desirable when studying the flowering process. In this way, one can establish the role of each individual organ in

[1]University of Natal/FRD Research Unit for Plant Growth and Development, Department of Botany, University of Natal, P.O. Box 375, Pietermaritzburg 3200, Republic of South Africa

Biotechnology in Agriculture and Forestry, Vol. 17
High-Tech and Micropropagation I (ed. by Y.P.S. Bajaj)

flowering. Conversely, the lack of a certain organ may prevent either the initiation of flowering or the manifestation of the flowering stimulus.

One major factor affects all in vitro flowering systems, and that is the physiological state of the parent plant which provides the explant for culture. This can be reduced or eliminated using micropropagation techniques. If the parent plant is vegetative, then flower induction must be achieved in vitro, whereas if the parent plant is already induced or is flowering, then only the expression of that induction can be investigated. Both systems have their advantages, but the former is likely to be more useful in investigations of the floral stimulus using micropropagation techniques.

In vitro techniques thus provide several ways in which flowering can be examined, most of which cannot be carried out in vivo. The available information is examined in this chapter in the context of the preceding discussion. Investigations supporting a single substance regulating flowering will be dealt with first, followed by investigations relating to multi-component systems.

2 The Existence of Florigen? Evidence Provided by in Vitro Studies

One of the major obstacles to reaching an understanding of the flowering process is the lack of an easily reproducible and sensitive bioassay for flowering.

For some years, in vitro techniques have been used in an attempt to develop a bioassay system where an extracted flower stimulus would bring about flowering under noninductive conditions, or conversely would inhibit flowering in inductive conditions. It is clear that the results of the work that has been done, and which are reviewed here, are not conclusive. Frequently, the results do not give an indication of whether a substance is a single compound (florigen), or extraneous substances which happen to cause the explant to flower, either for pharmacological reasons or because it forms part of the growth requirements of the flower.

There is the possibility that a single substance isolated from extracts could have inductive properties on a wide range of species. Such a substance could be considered to be a florigen even if it is not universally distributed. Once such a compound is identified, its bioassay is possible provided extensive purification of the extract is carried out. This is necessary, unless a highly specific bioassay system can be developed where the explants are responsive only to florigen and not to the wide variety of substances which are known to induce flowering. Such a bioassay system would rely on the existence of a stimulus that is common to at least a large group of plants, or a particular response type. No such substance has yet been identified, or is even known to exist, although some preliminary work has indicated that extracts of *Xanthium* contain flower inducers if applied to *Lemna*, or to *Xanthium* in conjunction with gibberellic acid (Hodson and Hamner 1970). Extracts from *Chrysanthemum* applied to *Xanthium* and *Chrysanthemum* (Biswas et al. 1966) and diffusate from induced scales of Wedgewood *Iris* transmitted to *Iris* apices both induced flowering (Rodriguez-Pereira 1965). All of these examples can be explained in terms of a multi-component model as described by Bernier et al. (1981a,b), which demonstrates the inconclusiveness of these systems as bioassays.

In the determination of the suitability of a particular system as a bioassay, a number of criteria have to be considered. The physiological state of the parent plant is critical. Plants which had been pre-induced, or were already flowering, have often been used as the source of explants, such as in the case of *Streptocarpus* (Simmonds 1982), *Torenia* (Tanimoto and Harada 1981a,b), *Kalanchöe* (Margara and Piollat 1981), *Browallia* (Ganapathy 1969), *Begonia* (Ringe and Nitsch 1968), and also in the much-used *Nicotiana* system, where explants are usually taken directly from the inflorescence tissue (Chouard and Aghion 1961; Aghion-Prat 1965a,b; Tran Thanh Van 1973a). If these systems are used as bioassays, and flowering is brought about by the addition of some substances to the medium, it is probable that this substance simply promotes the manifestation and growth of the buds or flowers, which had already been determined by the stimulus before culture. In vitro flowering in *Nicotiana* has only succeeded in DNP (day neutral plants), which suggests that the stimulus produced by photoperiodic tobacco plants has not been reproduced. Until photoperiodically sensitive tobacco species have been successfully induced in culture, this work is not likely to find acceptance as a bioassay system. Chailakhyan et al. (1974) suggested that in photoperiodic tobacco plants, the stimulus may be the same as in DNP, but that the apex does not develop the ability to synthesize its own requirements for evocation and flowering, and thus relies on a continued input from the leaves. Bridgen and Veilleux (1985) and Dickens (1987) supported this by grafting DNP *Nicotiana* to photoperiodically sensitive tissue in vitro, but could not obtain any flowering in the sensitive explants, or inhibition in the DNP, although this does occur in vivo (Lang 1965).

It is desirable then, that the parent or donor plant used in a bioassay for the flowering stimulus should be strictly vegetative at the time of culture. This can be ensured by using micropropagation techniques.

The nutritional status of the parent plant also affects the ability of the explants to flower, which can again be finely controlled using micropropagation systems. Explants of *Glycine* derived from parent plants grown in vitro on a nutrient-deficient medium fail to flower and fruit to the extent of those from more enriched media (Table 1).

Secondly, the culture medium may play an important role in the effectiveness of a bioassay. A variety of plant growth regulators are usually added to the medium in order to bring about growth (Table 2). Some systems have nevertheless avoided these additions, such as those using *Nicotiana* (Aghion-Prat 1965b), *Torenia* (Tanimoto and

Table 1. Flowering and fruiting response of soybean nodal explants drived from plants grown on two different media, under LD or SD. (Dickens and Van Staden 1985)

Treatment No.	Germination medium	Parent plant day length	Explant plant day length	% Flowering (6 weeks)	% Flowering (10 weeks)	% Fruiting (10 weeks)
1	H_2O	LD	LD	0	0	0
2	H_2O	LD	SD	10	20	20
3	Mod MS	LD	LD	0	0	0
4	Mod MS	LD	SD	50	100	30
5	H_2O	SD	LD	20	20	0
6	H_2O	SD	SD	60	100	20
7	Mod MS	SD	LD	40	50	0
8	Mod MS	SD	LD	100	100	100

Table 2. Species which have been cultured and have successfully produced flowers in vitro

Species	Response group	Parent plant induced	Explant source	Culture daylength	Hormones in medium	Promotion by:	Inhibition by:	Reference
Alium sativum	–	–	Flower	LD	–	GA_3	–	Tizio (1979)
Arabidopsis thaliana	LDP	No	Stem segments	LD	Kinetin (adenine)	–	–	Nitsch (1972)
Baeria chrysostoma	–	No	Seed	SD	–	–	–	Loo (1946a)
Begonia sp.	–	Yes	Leaves and flower stalk	LD and SD	BA, IAA phenylacetic acid	IAA	–	Ringe and Nitsch (1968)
Bouganvillea glabra	–	–	–	–	–	–	–	Chaturvedi and Sharma (1977) Steffen et al. (1986)
Browallia demissa	SDP	Yes	Flower bud	LD	Kinetin or SD 8339	–	IAA, GA_3	Ganapathy (1969)
Carthamnus tinctorious	–	–	Cotyledon	Continuous	IAA, kinetin	–	–	Tejovathi and Anwar (1984)
Cestrum diurnum	LSDP	No	Node	LD to SD	–	Light	–	Caplin and Griesel (1967)
Chenopodium rubrum	SDP	No	Seedling apex	SD	–	–	CCC, ABA, ethrel	DeFossard (1972, 1974)
Cichorium intybus	LDP or CRP	No ?	Raceme root	SD LD	– –	Red light, GA_3, coumaric acid, anti-auxin	Hydration, IAA	Badila et al. (1985) Badila and Paulet (1986) Bouniols (1974) Joseph and Paulet (1975) Paulet and Nitsch (1964) Pierik (1966b, 1970) Margara et al. (1965) Margara et al. (1966)
Chrysanthemum Honeysweet	SDP	No	Apex	SD	–	GA_3	IAA, GA_3	Harada (1967)
Crepis cappilaris	LDP	Yes	Hypocotyl callus	–	IAA, IBA, kinetin	Kinetin	–	Brossard (1979) Jayakar (1970)
Cucumis sativus	–	No	Hypocotyl	Continuous	–	–	–	Rajasekaran et al. (1983)

Cuscuta campestris	–	–	Seedling apex	–	–	–	–	Loo (1946b)
Cuscuta epithymum	–	–	Apex	–	–	Haustoria	–	Bertossi (1956)
Cuscuta reflexa	SDP	No	Apex	Unknown or dark	IAA	–	–	Baldev (1959)
Dianthus caryophyllus	LDP?	Yes	Anthers	LD	IAA, kinetin, coconut milk	–	–	Villalobos (1981)
Dionea muscipula	–	No	Leaf segments	LD	Zeatin	–	–	King (pers. comm.)
Drosera natalensis	–	–	Leaf	LD	BA, NAA	–	–	Crouch (pers. comm.)
Glycine max	SDP	No	Node±leaves	SD or LD to SD	None	–	–	Dickens and van Staden (1985)
Helianthus anuus	DNP or SDP	No	Apex	LD	BA or kinetin	–	–	Henrickson (1954) Paterson (1984)
Iris cv. Wedgewood cv. Ideal	CRP	No	Bulb or apex Primordia	Dark	–	Low temp., GA_3 or induced scale diffusate	IAA	Doss and Christian (1979) Rodriguez-Pereira (1965)
Kalanchöe blossfeldiana	SDP	Yes	Flowers	LD	BA, NAA, 2,4-D	ABA	–	Dickens (1987) Margara and Piollat (1981)
Lemna aequinoctialis (=*paucicostata*)	SDP	No	Whole plant	SD Continuous	–	Zeatin, BA, ABA, CCC, benzoic, tannic, salicylic and nicotinic acids, and vitamin K, dicourmarol	CCC, ABA GA_3, IAA NH_4	Cleland and Tanaka (1979) Fujioka et al. (1986a,b) Gupta and Maheshwari (1970) Kaihara and Takimoto (1985a,b) Kandeler and Hugel (1973) Khurana and Maheshwari (1983a, 1986) Tanaka et al. (1986)
Lemna gibba	LDP	No	Whole plant	12h	–	GA_3, cAMP EDDHA, BA, salicylic, benzoic and nicotinic acids	Chlorogenic glucose, IAA, GA_3, ABA and kinetin	Fujioka et al. (1985) Oota (1963, 1972) Pieterse and Muller 1977 Umemoto (1971)
Lemna minor	LDP	No	Whole plant	–	–	–	Benzoic and salicylic acids	Kaihara et al. (1981)

Table 2. *(Continued)*

Species	Response group	Parent plant induced	Explant source	Culture daylength	Hormones in medium	Promotion by:	Inhibition by:	Reference
Lemna perpusilla (P146)	SDP	Yes/no	Whole plant	SD and LD	–	BA	Allogibb. acid, GA_7	Bennink and de Vries (1975), Pryce (1973)
						ABA		Higman and Smith (1969)
Lunaria annua	CRP, LDP	Yes	Petiole	–	–	Low temps, adenine, GA_3, kinetin, crude extract coconut milk	IAA, NAA, 2,4-D	Pierik (1966a, 1970)
Manihot esculenta	CRP/LDP	–	Apex	LD	IAA, BA, GA_3	–	–	Tang et al. (1983)
Mazus pumilus	–	Yes	Stem internode	–	IAA, kinetin, GA_3	Kinetin	–	Raste and Ganapathy (1970)
Mesembryanthemum floribundum	–	–	–	–	–	–	–	Mehra and Mehra (1972)
Nicotiana	–	–	Leaf epidermis	–	Auxin cytokinin	–	–	Tran Thanh Van (1973b)
Nicotiana rustica	–	–	Leaf protoplasts	–	IAA, kinetin	–	–	Gill et al. (1979)
						–	–	
Nicotiana tabacum cv. Samsun	DNP	Yes	Thin cell layers from inflorescence	Continuous	BA, NAA, IAA, kinetin	BA	–	Cousson and Tran Thanh Van (1983), Croes et al. (1985) Van den Ende et al. (1984c)
Nicotiana tabacum cv. Wisconsin 38	DNP	Yes	Thin cell layers from inflorescence stems	Continuous	IAA, kinetin, or none	RNA base analogs, DNA from induced plants	GA_3	Aghion-Prat (1965a,b) Chouard and Aghion (1961) Hillson and La Motte (1977) Tran Thanh Van et al. (1974) Wardell (1977), Wardell and Skoog (1969a,b)
Nicotiana tabacum cv. Trapezond	DNP	Yes	Stem internodes	LD	IAA, NAA, kinetin	–	–	Chailakhyan et al. (1974, 1975)
Oncidium varicosum	–	Yes	Inflorescence stalk	LD	NAA	–	–	Barbante Kerbauy (1984)

Passiflora suberosa	DNP	Yes	Leaf, tendril, stem segments	LD	BA	BA	–	Scorza and Janick (1978, 1980)
Panax ginseng	–	–	Embryoids and root	LD	BA, GA_4	GA_4	–	Chang and Hsing (1980)
Perilla crispa	SDP	No	Apex and leaves	Continuous	IAA, kinetin	Sucrose, ABA and phloem exudate	–	Wada and Totsuka (1982)
Perilla fructescence	SDP	No	Apex and leaves	SD	BA, NAA, IAA	–	–	Raghavan (1961), Raghaven and Jacobs (1961), Tanimoto and Harada (1979, 1980)
Perilla nankinensis	SDP	No	Apex	–	–	Kinetin, adenine	–	Chailakhyan and Butenko (1959)
Pharbitis nil	SDP	No	Apex seedlings	SD and continuous	–	Low temps, NAA, GA_3, kinetin, BA, GA_3, ABA and benzoic acid	Ethrel, IAA, GA_3	Bhar (1970), Harada (1967) Matsushima et al. (1974) Shinozaki and Takimoto (1982a,b, 1983), Takimoto (1960)
Phlox drummondii	SDP	Yes	Flower bud	–	IAA, coconut milk	IAA, coconut milk	–	Konar and Konar (1966)
Pisum sativum	CRP	–	Apex and axillary bud	–	BA, NAA	–	–	Novak et al. (1985)
Plumbago indica	SDP	No	Internode	SD	–	Ethylene, ABA, adenine	Auxin, GA_3	Nitsch (1972) Nitsch and Nitsch (1967a,b), Nitsch et al. (1967)
Rudbeckia bicolor	LDP	Yes	Stem and leaf	LD	GA_3, BA, NAA	GA_3	–	Tanimoto and Harada (1982a)
Salix tetrasperma	–	–	Buds	–	GA_3, NAA, kinetin	–	–	Angrish and Nanda (1982)
Scrofularia arguta	LDP	–	Node	12 h	Coconut milk, IAA	–	Kinetin	Miginiac (1972)
Silene candida (= *Viscaria*)	LDP	Yes	Apex and leaf primordia	–	GA_3, kinetin	–	*Kalanchöe* extracts	Blake (1966, 1969, 1972)
Silene cardinalis	LDP	Yes	Apex	–	Kinetin	–	–	Blake (1969)
Sinapis alba	LDP	No	Apex	SD	–	Sucrose	Nitrogen	Deltour (1967)
Spinacia oleracea	LDP	No	Apex	–	GA_3, GA_7	–	Sandoz 9789	Culafic and Neskovic (1980)

Table 2. (*Continued*)

Species	Response group	Parent plant induced	Explant source	Culture daylength	Hormones in medium	Promotion by:	Inhibition by:	Reference
Stellaria media	–	–	Apex	–	–	–	–	White (1933)
Spirodela polyrrhiza SP_{20}	–	–	–	–	–	–	–	Khurana and Maheshwari (1980)
Streptocarpus nobilis	SDP	No	–	SD	–	–	GA_3, IAA,	Handro (1977)
		Yes	Leaf		BA	BA	KNO_3, sucrose	Rossini and Nitsch (1966) Simmonds (1982)
Saccharum officinarum	SDP	Yes	Apex	–	–	–	–	Coleman and Nickell (1964)
Thuja plicata	–	–	–	–	GA_3	–	–	Ritchie et al. (1986)
Torenia fournieri	SDP	Yes	Leaves, internode	SD	IAA, kinetin, or none	IAA, ABA, zeatin, sucrose	NAA, BA, kinetin, NH_4NO_3	Bajaj (1972) Chlyah (1973a,b), Tanimoto and Harada (1981a,b,c), Tanimoto et al. (1985)
Vitis vinifera	–	–	Tendril	LD	–	BA, PBA	–	Srinivasan and Mullins (1978)
Wolffia arrhiza	LSDP	No	whole plant	SD	–	BA, kinetin, zeatin	–	Krajncic (1983)
Wolffia microscopica	SDP	No	Whole plant	LD	–	Salicylic acid, BA, kinetin, zeatin, ABA, benzoic acid		Khurana and Maheshwari (1983b) Venkataraman et al. (1970)
Xanthium strumarium	SDP	No	Apex	SD	Kinetin	–	–	Jacobs and Suthers (1971, 1974)

Harada 1981a,b), *Glycine* (Dickens and Van Staden 1985), and *Kalanchöe* (Dickens 1987). It is possible that growth regulators may interfere with the expression of the stimulus, or even simulate its effects, and thus they are not desirable in a bioassay. The composition of the nutrients in the medium may also play an important role (Tanimoto and Harada 1981a; Dickens 1987).

Thirdly, the culture photoperiod is of utmost importance in flowering experiments. The most desirable culture conditions for a qualitative bioassay would be those that keep control plants in a vegetative state, and where the substances under test bring about flowering. One problem associated with this is that the plant may be dominated by vegetative conditions which actively prevent the stimulus from working. For this reason, it would be desirable to develop systems using species which are marginally floristic under certain conditions. Under these conditions, there is less likelihood of strong vegetative determination. DeFossard (1974) suggested that specimens should be grown under inductive conditions in order to avoid natural inhibitor production. Dickens (1987) made use of inductive cycles in a quantitative bioassay, and measured the increased or decreased rate of flowering, and the number of flowers produced by *Kalanchöe* nodal explants grown in vitro. There was no vegetative resistance to flowering, but the issue is whether the stimulus under examination was a flowering one, or simply a growth rate or inflorescence branching stimulus. This is not yet known.

Explant choice is influenced by many factors. DeFossard (1974) claimed that the specimen should be defoliated to reduce its responsiveness to environmental conditions, and should be reduced to the tissues which respond to florigen. This type of bioassay is dependent on a single compound stimulus, and is unlikely to work if flower induction and evocation rely on a more complex stimulus. It is possible that some input from the leaves and other organs may be necessary to support the growth of the flower primordia to a size where they can be evaluated. A lack of these general organs may prevent the expression of the floral stimulus. Conversely to this, isolated apices of *Perilla* would not provide suitable bioassay material, as they flower automatically if stripped of their unfolded leaves (Raghavan and Jacobs 1961).

There is a possibility that a system may be developed where the floral stimulus induces concurrent changes in the plant, which may act as indicators of the presence of the stimulus. *Kalanchöe* is a possibility, as there are several changes associated with the transfer to inductive cycles (Schwabe 1969), such as the onset of leaf succulence, cell sap vicosity, and increased anthocyanin production (Neyland et al. 1963).

The culture of small explants from plants which are already flowering has shown that the determination to flower can be carried through from the flowering parent plant to the explant. This occurs even if the explant grows extensively before flowering takes place, as in the case of *Nicotiana* inflorescence sections. Explants of *Glycine* taken from flowering plants flowered in noninductive long days (LD) (Dickens and Van Staden 1985). Reculturing of *Passiflora* led to the exhaustion of the stimulus (Scorza and Janick 1980). In the DNP *Nicotiana*, the floral stimulus was stable through three subcultures, and is thus thought to be produced in all cells (Chailakhyan et al. 1974).

Results such as these could indicate that some substance which works at a very low concentration is present in the tissue. These results also question the theory that some particular concentration or balance of substances is responsible for flowering, as this is not likely to be transferred and developed during culture. If a substance/florigen is carried over during culture and continues to stimulate flowering, it seems reasonable to assume that its multiplication must occur as the explant enlarges. This suggests some involvement at the gene level, possibly of an epigenetic nature.

The presence of a flower gradient in plants has been shown repeatedly in *Nicotiana* (Chouard and Aghion 1961; Aghion-Prat 1965a) and *Torenia* (Tanimoto and Harada 1979), where only explants taken from the upper regions of the stem will form flowers, while those from the bottom remain vegetative. These upper sections contain either the requisite amount of a stimulus which is produced in the upper regions of the plant and forms a gradient of concentration down the stem, or they contain the correct balance of two or more substances probably derived from different parts of the plant.

This gradient could exist as a result of epigenetic changes which have taken place, and thus no particular substance need be present in the explant to carry over the flowering stimulus besides the altered DNA. Such a situation does not preclude the involvement of florigen, as some signal must be produced by the plant which causes this genetic change, or some substance is produced as a result of this change. This was supported by work on *Helianthus*, where the apex was found to be determinate and various treatments with hormones and other substances had no effect on flowering (Paterson 1984). It is important to remember that much of the "flowering gradient" work is based on DNP cultivars of *Nicotiana tabacum*, and thus there is no production of an environmentally initiated stimulus. Scorza and Janick (1980) did show the existence of some diminishing stimulus in the DNP *Passiflora*. It is not known what events are responsible for the change to the reproductive state in DNP, but it is possible that the same physiological conditions are produced as are found in SDP and LDP. Chailakhyan et al. (1974) claimed that all DNP cells synthesize the stimulus, while in photoperiodically sensitive plants, only the leaves produce the stimulus, and supply it continuously to the apex. This would make their induction in vitro difficult. Ross and Murfet (1985) noted that in *Lathyrus*, the difference between DNP and LDP is under relatively simple genetic control. That there is some similarity in the stimulus is supported by grafting experiments using *Nicotiana* (Lang 1965; Zeevaart et al. 1977), where the transfer of the stimulus was achieved from one response group to another. These grafting results do not provide conclusive evidence, if examined in the light of the Bernier et al. (1981b) multi-component model, where a single substance which forms part of a multi-component system may be transmitted across a graft union and bring about flowering.

The important issue is the separation of florigen from all the other factors involved in flowering. Even though a group of factors may induce flowering, this does not preclude the possible existence of florigen. This was recognized by Sachs (1977) while describing a nutrient diversion hypothesis for the control of flowering. So in reality, it may be difficult to distinguish a florigen theory from the multi-component theory to be discussed later.

3 Studies Concerned with Flower Inhibitors

The presence of inhibitors of flowering is well established. Interpretation is critical here, as it was pointed out by Jacobs and Suthers (1974) that many of the results which show the presence of a flower stimulatory substance can also be interpreted as showing the removal of a flower inhibitor. In support of an inhibitor, Jacobs et al. (1965) had earlier suggested that a florigen may not exist in *Perilla*, and that the apex does not need to be induced to flower, but will flower automatically when the inhibitory effect of the leaves in LD has been removed by SD. Blake (1972) developed a bioassay for flower inhibitors, using *Silene* (=*Viscaria*) apices, and found that extracts of vegetative *Kalanchöe* were significantly more inhibitory than flowering extracts. One problem not dealt with was the fact that flowering leaves of *Kalanchöe* are more succulent than vegetative leaves (Harder 1948), contain more water, and thus the cytoplasm would be more diluted. This may have contributed to the difference in results. Schwabe (1972) did obtain inhibition in *Kalanchöe* by injecting vegetative extract into in vivo plants in SD inductive cycles.

Raghavan and Jacobs (1961) demonstrated that in *Perilla* flowering is controlled by at least two components or events, the first controlling cone formation, and the second controlling flower development on the cone. The former event is controlled by the presence of an inhibitor produced by young folded leaves, while the latter is photoperiodically controlled, by the production of a stimulus by the same leaves. The inhibitor was found to diffuse through the medium, from inhibitory leaves to an isolated apex in inductive conditions.

It is notable from the preceding discussion that the existence or identify of a florigen or specific inhibitor has not been conclusively shown in vitro systems. One major obstacle is the tendency to draw distinction between florigen and multi-component theories, a distinction which may have no justification at all.

4 Flowering as a Multi-Component System: Evidence Provided by in Vitro Studies

Circumstantial evidence from in vitro work suggests that the floral stimulus is composed of several components with different functions. In SDP *Perilla*, isolated apices from vegetative plants cultured in noninductive LD could be made to produce sterile cones, while in SD, fertile flowers were formed (Raghavan and Jacobs 1961), indicating the presence of at least two different components to the stimulus. In *Salix*, cultured dormant buds produced sterile catkins, while nondormant buds produced fertile flowers in the axils of the bracts (Angrish and Nanda 1982). Steffen et al. (1986) found that *Bougainvillea* reproductive meristems could be initiated under any condition, but florets were formed only in inductive conditions. It is now generally recognized that many of the substances involved in flowering may be acting on secondary growth processes after the work of the stimulus has been completed, and are probably produced as a result of the primary stimulus. Kannangara et al. (1986) extracted and separated fractions from flowering *Xanthium* plants which exhibited promotive activity on the rate of development of primordia in induced plants.

Scorza and Janick (1980) suggested that in the DNP *Passiflora*, the flowering stimulus may consist of a flower promoter and/or an inhibitor, and a cytokinin, the critical components of which seem to have a short life and are subject to dissipation as they are translocated through the plant. They concluded that the flower induction component of the stimulus in *Passiflora* is not synthesized in vitro despite the fact that small tissue explants were able to produce flowers. This is in agreement with the situation in DNP *Nicotiana* sp. (Chouard and Aghion 1961; Aghion-Prat 1965b;

Konstantinova et al. 1969). These results show that there is distinction between induction and evocation.

Two classical theories can be interpreted as supporting the multi-component model. Firstly, the antagonism which exists between flowering and rooting. Gaspar (1980) proposed that this antagonism was due to a control mechanism where inverse variations of auxin and peroxidase enzyme are responsible for flowering and rooting. This antagonism was also found in in vitro-grown *Kalanchöe* (Dickens 1987), where reduced root mass was produced in inductive conditions and where hormone-stimulated root growth inhibited flowering. *Helianthus* apices also failed to produce roots once flowers had been initiated (Paterson 1984). Flowering of *Cichorium* apices was inhibited by the presence of root tissue which was in close proximity to the apical bud (Joseph et al. 1985). This antagonism is widely appreciated but not understood, although it suggests some relationships between the physiological control of the roots and the production of flowers. This aspect, which is apparently of some significance, can be studied with a greater chance of success using micropropagation systems, as these allow both the induction and elimination of roots in experimental systems.

The second classical theory that supports multi-component systems, is that a high C/N ratio can stimulate or even induce flowering. This will be examined in detail under the section on nutrients. It has recently been stressed that this theory is worthy of further investigation and that it has been unnecessarily neglected (Trewavas 1983).

Naturally, no investigation of multi-component models would be completed without a detailed look at plant growth regulators. These ubiquitous compounds are known to affect flowering as well as a variety of other growth phenomena, but the disturbing fact in all is their apparent lack of specificity. This implies that either the cell becomes sensitive to them when required (Trewavas 1981), or that they are primarily involved in routine growth and development, but not in the actual initiation of flowering itself. Bernier and Kinet (1985) claim that there is sufficient information available indicating that plant growth regulators are primary controlling agents of flowering. This uncertainty is not likely to be resolved until there is more understanding about the mode of action of plant hormones. In this respect greater use can be made of refined micropropagation techniques as the media and culture conditions are easily manipulated.

5 Hormones and in Vitro Flowering

There are not many reports where auxins stimulate flowering in vitro. *Pharbitis* seedlings could be induced to flower in noninductive conditions in vitro, by the application of NAA, although this was always accompanied by a suppression of root elongation. Flower induction in this case was thought to be a consequence of root suppression (Shinozaki and Takimoto 1983). Auxin stimulation of flowering was also found in *Phlox* callus derived from flower primordia (Konar and Konar 1966) and in *Streptocarpus* if supplied with cytokinins (Rossini and Nitsch 1966). These results are contrary to the findings of Simmonds (1982), who found that IAA was strongly inhibitory of flowering in *Streptocarpus*. NAA was also found to promote bud development in *Perilla* (Tanimoto and Harada 1980). Possibly the most extensive work in this area is that of Tanimoto and Harada (1981b,c), who found that in *Torenia* stem segments,

IAA promoted the initiation and development of flower buds if applied early on in culture. It is important to note that the explants were taken from induced plants, and thus auxin may simply have been supporting the expression of the flower buds. The level of endogenous IAA in the tissue was found to remain constant regardless of the physiological state of the explant, but became undetectable after 2 weeks of culture (Tanimoto et al. 1985).

Van den Ende et al. (1984b,c) found that in thin cell layers of *Nicotiana*, NAA inhibited the development of flowers early on, but became promotive later on in growth. This is also the situation in *Cichorium* (Paulet and Nitsch 1964; Margara and Touraud 1968). Such promotion is likely to be far removed from the flora stimulus.

Auxin is widely recognized as being an inhibitor of flowering in in vitro systems, although its presence in many media may be necessary for growth. Auxin was found to be inhibitory in the SDP *Plumbago* (Nitsch and Nitsch 1967b), *Perilla* (Chailakhyan et al. 1961; Raghavan 1961) *Chrysanthemum* (Harada 1967), *Streptocarpus* (Simmonds 1982), and *Lemna* (Gupta and Maheshwari 1970; Fujioka et al. 1983). In the latter case inhibition is by counteracting the inductive effects of cytokinins, e.g., *Kalanchöe* (Dickens 1987), *Browallia* (Ganapathy 1969), and *Helianthus* (Paterson 1984). Inhibition was also obtained in the LDP *Cichorium* (Paulet and Nitsch 1964; Margara and Touraud 1968), *Begonia* (Ringe and Nitsch 1968), *Lemna* sp. (Fujioka et al. 1983), as well as in DNP *Nicotiana* cultivars (Aghion-Prat 1965a,b; Hillson and LaMotte 1977).

In DNP *Nicotiana* cultures, auxin is usually included in the medium (Tran Thanh Van et al. 1974) and is essential for flowering (Van den Ende et al. 1984b), although according to Croes et al. (1985), increased auxin almost completely eliminates bud formation on older tissues. They also noted a strange situation where NAA strongly suppressed bud formation on internodes, but did not in flower stalk tissue, which is the tissue usually used in this type of culture. Aghion-Prat (1965b) found that this auxin inhibition could be partially overcome by cytokinins and that normal flowers could be produced in the absence of IAA. It was also shown that if the cytokinin level was kept constant, and the IAA level increased, this caused a reversion of buds from the flower to vegetative state (Wardell and Skoog 1969a,b). Yet these authors also found that IAA is required for the development of normal flowers. The apparent contradiction may be a result of using different explant sources or parent plants at different stages of the flowering process.

In *Nicotiana* thin cell layers, IBA was important in bringing about flowering in liquid cultures (Cousson and Tran Thanh Van 1981). Hillson and LaMotte (1977) noted that IAA inhibited both flowering and vegetative bud formation if supplied with low kinetin levels, while at high kinetin levels, IAA inhibited flower bud formation but stimulated vegetative bud formation. This seems to be a more specific inhibition of flower induction, as IAA did not affect flower development.

The inhibition of flowering in *Nicotiana* by IAA was reversed by RNA base analogs, which resulted in an increase in the number of flowers on stem segments (Wardell and Skoog 1969b). These base analogs also caused the production of flowers on stem segments lower down on the plant, thus removing the floral gradient originally identified in vitro by Chouard and Aghion (1961). This gradient was also disturbed by kinetin and auxin (Aghion-Prat 1965b). Yet explants from vegetative plants could not be induced to flower by either of the above treatments, which suggests that they affect only the expression of flowering, but not the induction. Wardell and Skoog (1973) noted that the floral gradient was also reflected by DNA content, and concluded that auxin is not the only substance causing the gradient. This is supported by Noma et al. (1984), who found no correlation between flower-forming ability and endogenous IAA levels in a different cultivar. The significance of this is that the small quantity of IAA in stem segments of *Nicotiana* cannot be used to explain the differing abilities of explants to flower. The only hint of a pattern of distribution was that the concentrations of free IAA and IAA conjugates were highest in the first and second internodes, which normally have the ability to form flower buds, but this trend was not repeated in other tissues which also have this ability.

It was shown by Silberger and Skoog (1953) and Key (1969) that auxin-induced cell enlargement depends on the synthesis of RNA. Vanderhoef and Key (1968) also indicated that kinetin inhibited cell elongation and RNA synthesis. From this information, it is evident that increased cell enlargement and RNA synthesis in response to higher IAA levels are closely related to the inhibition of flowering or the maintenance of vegetative growth. Kinetin counteracts IAA-inhibited flowering possibly by inhibiting cell elongation and RNA synthesis. The situation in *Kalanchöe* seems contrary to this, as flowering is usually accompanied by cell expansion and an increase in leaf succulence (Harder 1948; Gummer 1949; Schwabe 1958), which suggests higher auxin levels, although this has not been investigated. The accumulation and disappearance of starch in cells has been observed in regions of callus where vegetative and floral buds are differentiating (Sadik and Ozburn 1967; Thorpe and Murashige 1970). Auxin may be involved here, as IAA has been shown to reduce starch accumulation in *Nicotiana* callus, while kinetin enhances it (Tetley and Ikuma 1970). The latter authors noted that there is a complex interrelationship between starch metabolism and hormone-induced growth. IAA also inhibited cytokinin-induced flowering in the SDP *Lemna aequinoctialis* (Gupta and Maheshwari 1970). The above information indicates that a possible regulation mechanism of the flowering process may be the balance between auxin and cytokinins.

In their examination of auxin action in flowering, Tanimoto and Harada (1981b) concluded that IAA suppresses flowering if vegetative tissue explants are used, but stimulates flowering if explants are derived from flower inflorescence tissue. Similar conclusions were made by Croes et al. (1985). Again, this supports the idea that auxin promotes differentiation of flower buds, but inhibits the induction and/or evocation, possibly due to its inhibition of DNA synthesis. It thus seems unlikely that auxins form part of the floral stimulus, a conclusion also reached by Gaspar et al. (1985).

In many of the media used for in vitro flowering, cytokinins were necessary inclusions in order to bring about the growth response, as in *Plumbago* (Nitsch and Nitsch 1967a,b), *Streptocarpus* (Rossini and Nitsch 1966), and *Torenia* (Bajaj 1972; Chlyah 1973a,b; Tanimoto and Harada 1982b). In the latter case, Tanimoto and Harada (1981b,c) reported that cytokinins were not necessary if explants were taken from the basal parts of old flowering plants. Cytokinins were also necessary in *Begonia* (Ringe and Nitsch 1968), *Arabidopsis* (Nitsch 1972), and in the DNP *Nicotiana* cultivars (Wardell and Skoog 1969a; Tran Thanh Van et al. 1974; Cousson and Tran Thanh Van 1981; Van den Ende et al. 1984b).

In some situations, cytokinins were able to induce flower bud formation in noninductive conditions, such as in explants of *Perilla* (Chailakhyan and Butenko 1959; Chailakhyan et al. 1961; Tanimoto and Harada 1980), *Plumbago* (Nitsch and Nitsch 1967b) and *Browallia* (Ganapathy 1969). Bud initiation was also achieved in *Begonia* sp. (Ringe and Nitsch 1968), and in the DNP *Passiflora* (Scorza and Janick 1980), where benzyladenine (BA) was the most effective. Inducion was also achieved in *Vitis* tendrils cultured in vitro (Srinivasan and Mullins 1978).

Cytokinins were promotive of flowering in the SDP *Streptocarpus* (Simmonds 1982) and in the DNP *Nicotiana* (Tran Thanh Van et al. 1974). Promotion was also obtained in *Mazus* (Raste and Ganapathy 1970) and in the determinate plant *Helianthus* (Paterson 1984). No effect was found in *Pharbitis, Chrysanthemum* (Harada 1967), and *Phlox* (Konar and Konar 1966).

Inhibition of in vitro flowering by cytokinins has been detected in cultures of *Scrofularia* (Miginiac 1972), *Kalanchöe* (Dickens 1987), although this was accompanied by vegetative inhibition; and in the SDP *Torenia* (Tanimoto and Harada 1981b), although slight promotion was obtained early on in culture.

In the DNP *Nicotiana* cultivars, somewhat contradictory results have been reported, possibly due to the use of different explants taken from plants at different physiological stages. Kinetin inhibited flower bud formation and development of *Nicotiana* irrespective of the accompanying auxin concentration, although vegetative bud formation was stimulated (Hillson and LaMotte 1977). Kinetin was also found to cause IAA inhibition of flower and vegetative bud formation if supplied at low concentrations. If kinetin was supplied at high concentrations, flowering was inhibited, but vegetative bud formation was stimulated. Aghion-Prat (1965a) found that low levels of kinetin could partially overcome IAA inhibition. Wardell and Skoog (1969a,b) found that at constant low kinetin levels, increasing auxin levels caused a change in the buds from a flowering to a vegetative state. Wardell and Skoog (1969a) noted that cytokinins increased the number of vegetative buds, but had no effect on the number of flowers. High concentrations of kinetin caused branching of the flower shoots to form clusters. Also in *Nicotiana*, cytokinin induced a greater number of vegetative buds, but inhibited flowering and root formation (Tran Thanh Van et al. 1974). Hillson and LaMotte (1977) noted that while flower buds increased in number in the presence of high kinetin and high auxin concentrations, vegetative buds increased in number to an even greater extent. This may be due to a stimulation of vegetative growth and consequent inhibition of flower expression, which supports the conclusion of Wittwer and Aung (1969), who found that kinetin inhibited flowering in intact *Lycopersicon* plants, but stimulated vegetative growth. Similarly, in *Scrofularia*, kinetin inhibited flower buds in vitro, but stimulated vegetative buds (Miginiac 1972).

From the above information, it can be concluded that cytokinins have the ability to stimulate flowering in a variety of plants, but tend to be more supportive of vegetative growth. In some cases, cytokinins induce bud formation, and in others are promotive of growth and development, yet they could also be inhibitory of flowering.

Negretskii et al. (1984) have shown that high endogenous cytokinin levels in *Nicotiana* are associated with the upper regions of the plant, a trend which is reversed by stem girdling, which also reverses the flowering gradient. This trend implies that endogenous cytokinins are stimulatory of flowering, if indeed they play any role at all. During cold flower induction of *Cichorium* root explants, the endogenous levels of zeatin riboside increased significantly.

In *Browallia*, cytokinins and adenine, a suggested precursor of cytokinin biosynthesis (Peterson and Miller 1976; Chen et al. 1985), promoted flower bud development but not bud initiation (Ganapathy 1969). In *Begonia* leaf fragments, cytokinins could not induce bud formation alone, while adenine and auxin were necessary inclusions (Ringe and Nitsch 1968). It is possible that adenine plays a dual role, as the precursor needed for the synthesis of the necessary cytokinins, and also as the purine needed by nucleic acid metabolism, and in this way is more beneficial than cytokinin alone. The synergistic relationship between adenine and cytokinin has been extensively studied (Skoog and Miller 1957; Nitsch and Nitsch 1967a,b; Nitsch et al. 1967) but is not fully understood in relation to flowering.

Bernier et al. (1974) found that in the apex of *Sinapis* there is an increase in mitotic activity during flower evocation, with a concomitant rise in protein, RNA, and DNA synthesis. This is supported by other works and takes place possibly as a result of cytokinins (Jacqmard et al. 1972, 1976; Seidlova 1974; Brulfert et al. 1975; Seidlova and Culafic 1982). Adenine may initiate or promote flowering, possibly by promoting RNA synthesis at the apex, or by supporting an increase in adenine-nucleotides (Bodson 1985).

Cytokinins have been shown to stimulate the formation of meristematic zones and bud formation in *Torenia* stem segments (Tanimoto and Harada 1982b), a proportional relationship existing for both phenomena. These results bear a close resemblance to the situation of flower induction in *Nicotiana* epidermal layers (Van den Ende et al. 1984b). It is thus possible that the prime role of cytokinins in flowering is the development of meristematic regions, but how this relates to the different processes of induction, evocation, and differentiation is not clear. There does seem to be a relationship between the accumulation of starch in the regions where buds differentiate in callus (Sadik and Ozburn 1967; Thorpe and Murashige 1970), and the occurrence of cytokinin, which is known to enhance starch accumulation (Tetley and Ikuma 1970).

It has been suggested that although cytokinins do appear to be involved in flower induction and evocation, several other substances are also involved (Scorza and Janick 1980; Bernier et al. 1981b). The former authors concluded that the role of cytokinins in evocation may be to control early mitotic activity and associated cell synchronization, splitting of vacuoles, precocious initiation of axillary meristems, and increased rate of appendage production by the meristem. Sachs (1977) suggested that cytokinins could be involved in nutrient mobilization towards the apex which has been induced to flower. Whatever the role of the cytokinins, it is certain that they do play an important role in flowering, although it is not clear at what physiological stage.

Of all the plant hormones, the gibberellins have been the most successful in the induction of flowering in cold-requiring plants and rosette LDP's grown in vivo. According to Bernier et al. (1981b), they have the distinction of being the most potent florigenic compounds, although these authors do recognize that many plants cannot be induced by these compounds. Zeevaart and Lang (1962) went as far as saying that gibberellins are the physiological precursors of florigen. Chailakhyan and Lozhnikova (1985) recently supported the concept of the flowering stimulus being composed of gibberellins and anthesins. Yet despite their wide acclaim, this group of compounds has been neglected in in vitro investigations. A few workers have utilized them in their media, but seldom with startling results. Indeed, there seems to be little similarity between the situations in vitro and in vivo. No detailed study exists where gibberellin stimulation in vivo has been extended to the in vitro situation.

Rudbeckia stem and leaf explants taken from flowering plants would flower only in vitro in the presence of gibberellin (Tanimoto and Harada 1982a). Explants taken from vegetative plants did not flower under any conditions, suggesting that only the expression of the flower stimulus is being affected here. The development of the calyx and corolla of *Silene* (=*Viscaria*) and the production of pollen were all promoted by gibberellic acid (Blake 1969), which again suggested the involvement in the development of the flowers but not in the induction. Gibberellic acid in combination with BA supported direct flowering and the production of fertile pollen on embryoids of *Panax* (Chang and Hsing 1980). Gibberellic acid also hastened flowering in stem tips of *Chenopodium* that had cotyledons attached, but was ineffectual when no cotyledons were present (DeFossard 1972). This suggests the involvement of the cotyledons, where it is possible that gibberellins catalyze the release of some substance/s necessary for flowering.

According to Tanimoto and Harada (1981b), gibberellins have an indirect action on flowering in the SDP *Torenia*, stimulating at low concentrations but inhibiting at higher concentrations. Gibberellins also promoted the flowering of *Cichorium* explants taken from vernalized plants (Paulet and Nitsch 1964). In the SDP *Pharbitis*, gibberellic acid inhibited the initiation and development of flowers in inductive conditions (Harada 1967). Gibberellic acid had no effect on the flowering of *Kalanchöe* explants, but stimulated both the succulence of the leaves and the elongation of the stem and peduncle (Dickens 1987). This suggests that these growth phenomena are not related to flowering.

Gibberellins have been found to be inhibitory of flowering in several plants including *Plumbago* (Nitsch and Nitsch 1967b), *Streptocarpus* (Rossini and Nitsch 1966), *Torenia* at higher concentrations (Tanimoto and Harada 1981b), and *Mazus*, where flowering was delayed by 2 weeks, and was accompanied by stem elongation (Raste and Ganapathy 1970), *Browallia* (Ganapathy 1969), *Pharbitis and Chrysanthemum* (Harada 1967; Choshi 1979, 1980).

In *Nicotiana*, gibberellic acid inhibited flowering of stem segments completely, although application after initial formation promoted the elongation of the flowers (Aghion-Prat 1965a; Wardell and Skoog 1969a). The latter authors also found that this inhibition could not be reversed by the application of base analogs, as could IAA inhibition of flowering. Gibberellin also reduced starch accumulation in *Nicotiana* in a manner similar to auxin (Tetley and Ikuma 1970). As kinetin caused an increase in starch accumulation and also resulted in bud differentiation, gibberellin may be working to inhibit bud differentiation in a way that is somehow linked to a decrease in starch.

Bernier et al. (1981b) suggested that gibberellin action takes place at the shoot apex, and results in the stimulation of mitotic activity in subapical tissue leading to the bolting of the stem, an action which is not directly related to flowering.

There is an obvious void of information about the role of gibberellins in the flowering of cultured plants, a situation which urgently needs to be rectified.

Despite the fact that inhibitors have been implicated in the regulation of reproductive growth, little work has been done on their action in vitro. Of the known inhibitors, ABA has attracted most attention, but little can be concluded from the information documented. This is unfortunate, since ABA is known to be partially instrumental in the creation of sinks, in particular in developing seeds (Ackerson 1984).

ABA applied to in vitro systems has initiated flowering of vegetative stem segments on young stem explants of the SDP *Torenia* (Tanimoto and Harada 1981b; Tanimoto et al. 1985), but was inhibitory if applied to the apical regions of flowering plants. Stimulation of flowering also occurred in *Pharbitis* (Harada et al. 1971), but the primary effect seemed to be due to the suppression of root elongation (Shinozaki and Takimoto 1983). Flower promotion by ABA, but not initiation, was detected in the

SDP *Plumbago* (Nitsch and Nitsch 1967b) grown under inductive conditions, and in *Perilla* (Purse 1984), possibly as a result of the suppression of vegetative growth, a phenomenon which occurs naturally in many plants at the onset of flower initiation. Promotion by ABA was also obtained in *Chenopodium* (Krekule and Kohli 1981; Seidlova et al. 1981). DeFossard (1972, 1973) noted that cotyledons on excised shoot tips of induced seedlings of *Chenopodium* delayed epicotyl development and flowering, although this was overcome by gibberellic acid. Shoot tips detached from the cotyledons were similarly delayed or inhibited by ABA and CCC, but were not affected by gibberellic acid. It was suggested that this inhibition may be due to the inhibition of leaf development which in turn affects flowering. The number of flowers produced by *Kalanchöe* explants grown under LD conditions were significantly increased by ABA, at a concentration that did not affect vegetative growth (Dickens 1987).

Tanimoto et al. (1985) noted the existence of a flower gradient in *Torenia* plants, with the greatest flowering response occurring at the second internode. The distribution of endogenous ABA exhibited a similar pattern to this, with a decrease in concentration towards the base and a peak at internode two. Explants taken from old plants flowered poorly and were inhibited by applied ABA, while ABA was promotive in younger plants. These old explants contained high levels of endogenous ABA, which was further increased by exogenous application, to become inhibitory, probably due to supra-optimal levels. The optimum endogenous level of ABA required to regulate flower formation was between 16–20 ng/g fresh weight, which could be achieved by exogenous application. ABA was most inductive of flowering if applied to explants from vegetative plants, and in this way acted in the opposite way to auxins, where stimulation was achieved in older reproductive explants (Tanimoto and Harada 1981b). The endogenous levels of IAA did not reflect the flower gradient in any way.

As was discussed earlier, much evidence exists to support the existence of an inhibitor of flowering in plants, although little evidence exists to place ABA in this role. In *Perilla*, an inhibitor of flowering is produced by young leaves (Raghavan and Jacobs 1961), yet applied ABA was slightly stimulatory of flowering, as was the case in *Kalanchöe* (Dickens 1987). Any specific flower inhibitor should ideally be inhibitory of flowering, but not of other growth processes.

Some proof of the existence of inhibitors of flowering has been obtained by in vitro techniques, such as the presence of inhibitors in vegetative *Kalanchöe* extracts applied to partially induced *Silene* (=*Viscaria*) explants (Blake 1972). Dickens (1987) attempted to obtain inhibition with gallic acid using *Kalanchöe* nodal explants, but the results indicated that this substance inhibited both flowering and vegetative growth equally at high concentrations, and had no effect at low concentrations. This was contrary to the suggestions of Pryce (1972).

It is apparent that investigations of ABA and flowering have been neglected. There are indications that ABA may be involved in the regulation of flowering in particular in SDP, in a way that may be related to the inhibition of vegetative growth. Work is urgently needed in this area.

Little work has been done on ethylene and flowering in vitro. According to Nitsch (1972), flowering of *Plumbago* was induced by ethylene and its precursor methionine. In *Chenopodium* (DeFossard 1973) and in *Pharbitis* (Shinozaki and Takimoto 1983), ethrel was totally inhibitory of flowering. In *Kalanchöe*, explants which failed to flower in inductive SD, due to being in air-tight containers, had a slightly higher ethylene content in their atmosphere than did explants flowering in containers which allowed gaseous exchange (Dickens 1987).

The role of ethylene in flowering needs to be examined, especially in view of the interactions which occur between this substance and other hormones, in particular auxins. It is also conceivable that this hormone may frequently accumulate in micropropagation experiments.

6 Nonhormonal Substances and in Vitro Flowering

Numerous substances besides the recognized plant growth regulators have been found to affect flowering in vitro. The most significant group of these substances are the benzoic acid derivatives, which have been fairly extensively researched since it was recognized that salicylic acid may promote flowering. Much of the initial work was done by Cleland (1974) and Cleland and Ajami (1974), on the role of salicylic acid extracted from aphid honeydew obtained from *Xanthium* plants, which had inductive properties if applied to *Lemna gibba* G3. These results were confused by the later findings that the endogenous levels did not fluctuate in response to the photoinductive cycles. Cleland and Kang (1985) also noted that in *Lemna gibba*, salicylic and benzoic acids could reverse the inhibition of flowering caused by a lack of ammonium in the medium. Conversely, gibberellic acid, IAA or ABA inhibited benzoic acid-induced flowering in *Lemna aequinoctialis* 151 and 381, while zeatin supported flowering (Fujioka et al. 1983). BA also enhanced flowering induced by EDDHA or salicylic acid in *Lemna gibba*, but was ineffective on its own (Pieterse and Muller 1977).

Nicotinic acid has been extensively researched in *Lemna*, where it has powerful flower induction properties (Fujioka et al. 1983). This action is strongly daylength-dependent, varying according to the response type of the plant, yet the endogenous levels do not fluctuate according to daylength, which raises the possibility that other substances work in conjunction with them. Similar conclusions were made for salicylic and benzoic acids.

The phenolic β-coumaric acid was stimulatory of both vegetative and flower bud differentiation in in vitro-grown cultures of *Cichorium* (Paulet and Nitsch 1964). Cold was necessary for flowering, during which the endogenous levels of zeatin riboside and dicafeylquinic acid in the roots increased significantly (Mialoundama and Paulet 1975; Paulet 1979). Badilla and Paulet (1986) found that in *Cichorium*, the endogenous levels of hydroxycinnamic acid ester (especially chlorogenic acid) increased during floral induction, reaching a maximum level simultaneously with the maximum level of floral induction.

An increasing variety of substances are being found with florigenic activity, although it is unlikely that any of those noted so far is florigen. These substances could nevertheless be part of a multi-component stimulus, although there is also a distinct possibility that their effect is purely pharmacological.

Most investigations of in vitro flowering have made use of the MS medium (Murashige and Skoog 1962). This medium has a fairly high nutrient concentration, including a high level of NH_4NO_3. It is not known whether this has been detrimental to the general investigation of the flowering process, as intensive investigations of the effects of these nutrients are seldom made. The investigations of Tanimoto and Harada

(1981a,b) have indicated that with the correct nutrient composition, flowering can be brought about in small vegetative explants without the aid of plant hormones. They found that the dilution of the mineral salts of MS medium to one fifth of the recommended concentration enhanced adventitious bud formation and development, and in so doing enhanced flower bud formation, but did not influence the ratio of flowering to vegetative buds. Sucrose was also promotive of flowering, but the addition of the other recommended organic components of MS medium, stimulated callus proliferation and vegetative bud initiation, but inhibited shoot development and flower bud differentiation. Similar work was also done by Wada and Totsuka (1982), who found that low nitrogen, high carbohydrate, and high light intensities would induce flowering in seedlings of *Perilla* cultured on dilute low nutrient White's (1933) medium.

Chlyah (1973a,b) had previously cultured leaf and stem fragments of *Torenia* on Hoagland's nutrient medium, and achieved a high rate of flower bud formation without the use of auxin or cytokinin. Both of the above investigations found that low salts and the absence of NH_4NO_3 were supportive of flowering in vitro. The medium developed by Tanimoto and Harada (1981a,b) was subsequently used successfully for the culture of *Glycine* and *Kalanchöe* nodal explants (Fig. 1), both of which flowered successfully in vitro in response to photoperiodic induction. The MS medium was totally inhibitory of flowering in *Kalanchöe*, probably due to the high salt concentrations. NH_4NO_3 was not in itself inhibitory of flowering (Dickens 1987).

For years, nitrogen in various forms has been recognized as having some influence on flowering. As early as the turn of the century, Kraus and Kraybill (1918) noted that increasing nitrate concentrations caused increasing flowering. In agreement, Loo (1946a) investigated carbohydates and nitrogen in *Baeria*, and concluded that nitrate allows for the most prolific flowering, while all other nitrogen sources depress flowering. This was supported by Tanimoto and Harada (1981a,b), who found it necessary to eliminate NH_4NO_3 from the medium in use, but not KNO_3.

Possibly the most widely researched and controversial of the nutrient effects in relation to flowering is the ratio of carbon and nitrogen. In 1913, Klebs proposed that a high carbon/nitrogen ratio brings about flowering in plants. Many workers have subsequently supported this theory, and some evidence exists in in vitro investigations to support it.

Sucrose supplied to defoliated plants of *Chenopodium* (Lona 1948) and to *Perilla* (Lona 1949; 1950; Wada and Totsuka 1982) promoted flowering in all daylengths, especially if accompanied by high light intensities and low nitrogen levels. Chailakhyan (1968) noted that although sucrose promoted flowering in *Perilla*, as did high light intensities, the supply of nitrogen also stimulated flowering, but only in inductive SD. It is possible that in the latter case, as photoperiodic induction was taking place, the higher nitrogen levels were simply supporting the growth of the flowers. Chailakhyan (1968) stated that the carbon/nitrogen ratio does not condition the start of flowering, but is of some importance, and indeed he recommended the use of up to 7% sucrose for flowering of *Nicotiana* (Chailakhyan et al. 1975). Hinnawy (1956) supported this.

In most cases, the availability of carbohydrate does seem to be essential for flowering, as in *Xanthium*, where a high light intensity is required for induction, but could be partly replaced by sucrose (Liverman and Bonner 1953). In vitro this was supported by Cousson and Tran Thanh Van (1983), where the quantity of light supplied to *Nicotiana* thin cell layers resulted in different organ production, an effect which was mimicked by glucose, but not in the role of an osmoregulator. A ratio of glucose and

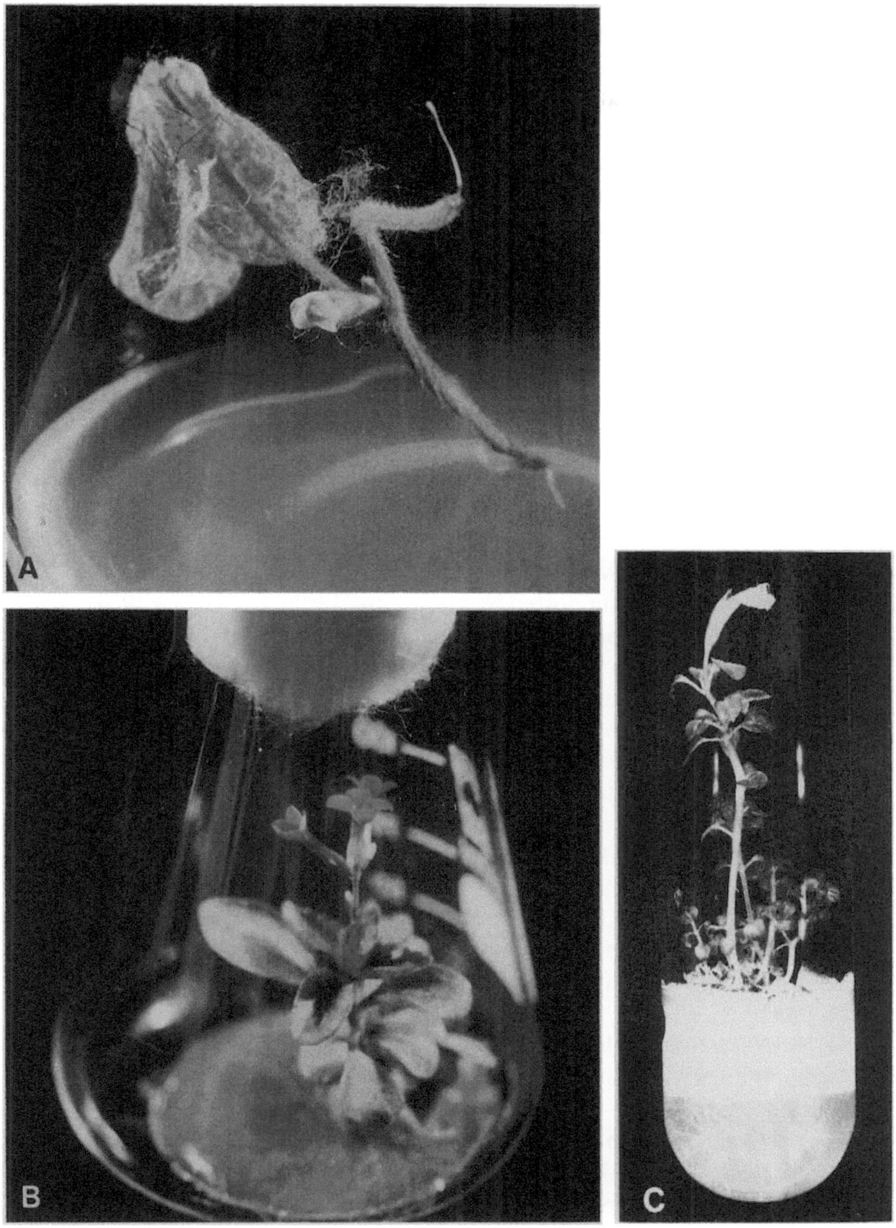

Fig. 1A-C. In vitro flowering of nodal explants of *Glycine* (**A**) and *Kalanchöe* (**B**) under photoinductive conditions. (**C**) In vitro flowering in *Torenia*. (Bajaj 1972)

sucrose was almost able to substitute for the light requirement for the development of anthers and a style. They suggested that during the sequence of events leading to flower differentiation, light acts on energy-dependent sugar uptake and metabolism and on the increase of reducing potential of the chloroplasts.

Sinapis apical buds could be induced to flower by sucrose (Deltour 1967) or by high light intensity (Bodson et al. 1977). Similarly, in *Cuscuta*, for flowering to occur, a high sucrose concentration removed the necessity of having a high light intensity (Baldev 1962). In contrast to the above, light was essential for flowering in *Passiflora*, and could not be replaced by sucrose (Scorza and Janick 1980; Scorza 1982). In *Spinacea*, photosynthesis was not necessary for induction and flowering, provided an organic carbon source was available in the medium (Culafic et al. 1982). High light intensity was able to promote flowering, as was a high sugar concentration, but only during the first 30 days of culture, after which high light intensities inhibited flowering (Wada and Totsuka 1982).

A point raised earlier was the accumulation of starch in the areas of callus where bud differentiation takes place (Thorpe and Murashige 1968, 1970) and in apical buds of cauliflower, which subsequently differentiate flower buds (Sadik and Ozburn 1967). In *Nicotiana* thin cell layers, explants that flower exhibit early starch accumulation and higher mitochondrial activity in the cells (Tran Thanh Van and Chlyah 1976). This accumulation of starch is supported by cytokinins (Tetley and Ikuma 1970), which also support bud differentiation, and is reduced by IAA and GA_3, which tend to inhibit flower bud differentiation (Thorpe and Murashige 1968). It thus seems apparent that there is a complex relationship between carbohydrate and hormone levels in cells and organs, which may in some way be responsible for flowering. This does not detract from the possibility of a flowering hormone, as these events may form part of the evocational process, but not the induction process.

Closer analysis of the role of carbohydrates in flowering has revealed that different responses are obtained with different sugars. Steffen et al. (1986) noted that in *Bougainvillea*, floret development was promoted by fructose (83%), glucose (68%), and least of all by sucrose (24%). The latter buds grew poorly when compared to those grown on fructose, but the leaves showed no difference. In *Plumbago*, sucrose and maltose increased flowering, while lactose, cellobiose, and mannitol were ineffective (Nitsch and Nitsch 1967b).

In *Nicotiana* thin cell layers, glucose was necessary for flowering (Van den Ende et al. 1984a), and even a brief removal from glucose delayed flowering but did not affect initial development. In the absence of sugar, any sign of differentiation is missing. These authors could not confirm the results of Tran Thanh Van (1977) that vegetative buds develop on a mannitol medium, which would imply an osmotic effect. This could be due to differences in explant sources.

In *Plumbago*, mannitol could again not replace the effect of sucrose (Nitsch and Nitsch 1967b), supporting the idea that this is not an osmotic effect. Sucrose also stimulated flowering in intact seedlings of *Nicotiana rustica* (Steinberg 1950) and in *Pharbitis* (Takimoto 1960; Kimura 1963). Concentrations of sucrose in the culture medium of up to 4% stimulated the rate of flowering in *Kalanchöe* nodal explants grown in induction conditions (Fig. 2).

The role of nutrients, and in particular carbohydrates and nitrogen, in the induction and evocation of flowering is far from clear, as many other growth phenomena are also regulated by these substances. Whether the effects of any one of these substances is

Fig. 2. The effect of sucrose concentration (0 to 8%) on the flowering of *Kalanchöe* explants after 11 weeks of culture

specific or is merely coincidental remains to be determined. There is, nevertheless, a role for a floral stimulus which would in turn regulate the activities of these substances. A similar conclusion was made by Sachs (1977) while addressing the nutrient diversion hypothesis.

7 Conclusions

It is clear that micropropagation techniques provide an ideal tool for the investigation of flowering physiology. They allow a degree of experimental control seldom achieved in in vivo studies. Despite the use of in vitro techniques there is still little justification in separating a florigen theory of flowering from a multi-component theory, although the former may be part of the latter. The results reviewed in this chapter show that a wide variety of responses are obtained in different plants, or parts thereof, with the application of an equally wide variety of substances. It remains difficult to draw definite conclusions from all the available results.

Many of the results implicate a role for various substances in the evocation or differentiation and growth of flowers. No unequivocal evidence exists, however, from in vitro work to place any of these known substances into the role of flower inducer. The extensive expansion and perfection of bioassays for flowering will allow such conclusions to be ultimately made. It is disappointing that micropropagation techniques, which lend themselves to fine control of media composition, explant culture, and/or pretreatment and environmental control, are not more extensively used in order to unravel the seemingly complex physiology of flowering. The tighter control which can be obtained with micropropagation techniques should at the very least be able to help pinpoint the role of the endogenous hormones in this process. Their elimination or acceptance as part of the flowering stimulus alone will be a great step forward.

Acknowledgments. The authors gratefully acknowledge the financial support of the University of Natal and the Foundation for Research and Development.

References

Ackerson RC (1984) Regulation of soybean embryogenesis by abscisic acid. J Exp Bot 35:403–413

Aghion-Prat D (1965a) Neoformation de fleurs in vitro chez *Nicotiana tabacum* L. Physiol Veg 3:229–303

Aghion-Prat D (1965b) Floral meristem-organizing gradient in tobacco stems. Nature 207:1211

Angrish R, Nanda KK (1982) Seasonal culture of dormant reproductive buds of *Salix tetrasperma*: Analysis of the flowering process. Plant Cell Tissue Organ Cult 1:181–193

Badila P, Paulet P (1986) Esters hydroxycinnamiques et induction photopériodique florale in vitro chez *Cichorium intybus*. Physiol Plant 66:15–20

Badila P, Lauzac M, Paulet P (1985) The characteristics of light in floral induction in vitro of *Cichorium intybus*. The possible role of phytochrome. Physiol Plant 65:257–262

Bajaj YPS (1972) Effect of some growth regulators on bud formation by excised leaves of *Torenia fournieri*. Z Pflanzenphyiol 66:284–287

Baldev B (1959) In vitro responses of growth and development in *Cuscuta reflexa* Roxb. Phytomorphology 9:316–319

Baldev B (1962) In vitro studies of floral induction on stem apices of *Cuscuta reflexa*, a short-day plant. Ann Bot 26:173–180

Barbante-Kerbauy G (1984) In vitro flowering of *Oncidium varicosum* mericlones (Orchidacea). Plant Sci Lett 35:73–75

Bennink GJH, De Vries JWA (1975) Flower development in *Lemna perpusilla* under long-day conditions with the cytokinin benzyladenine. II. Symp Plant Growth Regulators. Bulg Acad Sci Sofia Abstr 66–67

Bernier G, Kinet JM (1985) The control of flower initiation and development. In: Bopp M (ed) Plant Growth Substances. Springer, Berlin Heidelberg New York Tokyo, pp 293–302

Bernier G, Kinet M, Bodson M, Rouma Y, Jacquard A (1974) Experimental studies on the mitotic activity of the shoot apical meristem and its relation to floral evocation and morphogenesis in *Sinapis alba*. Bot Gaz 135:345–352

Bernier G, Kinet M, Sachs RM (1981a) The physiology of flowering. I. The initiation of flowers. CRC, Boca Raton, Fl

Bernier G, Kinet M, Sachs RM (1981b) The physiology of flowering. II. Transition to reproductive growth. CRC Boca Raton, Fl

Bertossi F (1956) Richerche sulla fisiologie della *Cuscuta epithymun* coltura in vitro. Atti Ist Bot Univ Pavia 14:1–21

Bhar DS (1970) In vitro studies of floral shoot apices of *Pharbitis nil*. Can J Bot 48:1355–1358

Biswas PK, Paul KB, Henderson JHM (1966) Effect of *Chrysanthemum* plant extract on flower initiation in short-day plants. Physiol Plant 19:875–882

Blake J (1966) Flower apices cultured in vitro. Nature 211:990–991

Blake J (1969) The effect of environmental and nutritional factors on the development of flower apices cultured in vitro. J Exp Bot 20:113–123

Blake J (1972) A specific bioassay for the inhibition of flowering. Planta 103:126–128

Bodson M (1985) Changes in adenine-nucleotide content in the apical bud of *Sinapis alba* L. during floral transition. Planta 163:34–37

Bodson M, King RW, Evans LT, Bernier G (1977) The role of photosynthesis in flowering of the long-day plant. *Sinapis alba*. Aust J Plant Physiol 4:467–478

Bouniols A (1974) Néoformation de bourgeons in vitro á partir de fragments de racine d'endive *Cichorium intybus* L.: Influence du degré d'hydration des tissues, et ses consequences sur la composition en acides amines. Plant Sci Lett 2:363–371

Bridgen MP, Veilleux RE (1985) Studies of de novo flower initiation from thin cell layers of tobacco. J Am Soc Hortic Sci 110:233–236

Brossard D (1979) Neoformation of vegetative and reproductive buds from foliar discs of *Crepis capillaris* L. Wallr. cultured in vitro. Z Pflanzenphysiol 93:69–81

Brulfert J, Guerrier D, Queiroz O (1975) Photoperiodism and enzyme rhythms: kinetic characteristics of the photoperiodic induction of crassulacean acid metabolism. Planta 125:33–44
Caplin SM, Griesel WO (1967) Induction of floral primordia in isolated nodal explants of *Cestrum diurnum* in vitro. Am J Bot 54:596–600
Chailakhyan MKh (1968) Internal factors of plant flowering. Annu Rev Plant Physiol 19:1–36
Chailakhyan MKh, Butenko RG (1959) The effect of adenine and kinetin on the differentiation of flower buds in *Perilla* stem tips. Dokl Akad Nauk USSR 129:293–295
Chailakhyan MKh, Lozhnikova VN (1985) The florigen hypothesis and its substantiation by extracting substances which induce flowering in plants. Fiziol Rast 32:1172–1181
Chailakhyan MKh, Butenko RG, Lyubarskaya II (1961) Effects of derivatives of nucleic acid metabolism on the growth and flowering of *Perilla nankinensis*. Fiziol Rast 8:101–113
Chailakhyan MKh, Aksenova N, Konstantinova TN, Bavrina TV (1974) Use of tobacco stem callus for the investigation of some regularities of plant flowering. Phytomorphology 24:86–96
Chailakhyan MKh, Aksenova N, Konstantinova TN, Bavrina TV (1975) The callus model of plant flowering. Proc R Soc Lond Ser B 190:333–340
Chang WC, Hsing Y (1980) In vitro flowering of embryoids derived from mature roots of ginseng (*Panax ginseng*). Nature 284:341–342
Chaturvedi HD, Sharma AK (1977) Induced flowering in excised shoot apices of *Bougainvillea glabra* cv. Magnifica grown in vitro. Indian J Exp Biol 15:402–403
Chen C-M, Ertl SM, Leisner SM, Chang C-C (1985) Localization of cytokinin bionsynthetic sites in pea plants and carrot roots. Plant Physiol 78:510–513
Chlyah H (1973a) Influence de la température sur la mise á fleur de plants de *Torenia fournieri* (Lind) issues de semin ou de boutures de feuilles, ou formées à partir de fragments d'organes cultivés in vitro. CR Acad Sci 277:1451–1454
Chlyah H (1973b) Néoformation dirigée a partir de fragments d'organes d'*Torenia fournieri* (Lind) cultivés in vitro. Biol Plant 15:80–87
Choshi A (1979) Effect of IAA on the flowering of excised plumules in *Pharbitis nil*. Rep Fac Sci, Shizuoka Univ 13:97–104
Choshi A (1980) Effects of gibberellic acid on the flowering of excised plumules in *Pharbitis nil*. Rep Fac Sci, Shizuoka Univ 14:47–54
Chouard P, Aghion D (1961) Modalités de la formation de bourgeons floraux sur des cultures de segments de tige de tabac. CR Acad Sci 252:3864–3867
Cleland CF (1974) Isolation of flower-inducing and flower-inhibitory factors from aphid honeydew. Plant Physiol 54:899–903
Cleland CF, Ajami A (1974) Identification of the flower-inducing factor isolated from aphid honeydew as being salicylic acid. Plant Physiol 54:904–906
Cleland CF, Kang BG (1985) Comparison of salicylic acid, benzoic acid and p-hydroxybenzoic acid for their ability to induce flowering in *Lemna gibba* G3 and for their uptake and metabolism by *Lemna gibba* G3. Abstr 12th Int Conf Plant Growth Substances, Heidelberg
Cleland CF, Tanaka O (1979) Effect of daylength on the ability of salicylic acid to induce flowering in the long-day plant *Lemna gibba* G3 and the short-day plant *Lemna paucicostata* 6746. Plant Cell Physiol 64:421–424
Coleman RE, Nickell LG (1964) Stability of the flowering stimulus in isolated stem tips of sugar cane. Nature 201:941–942
Cousson A, Tran Thanh Van K (1981) In vitro control of de novo flower differentiation from tobacco thin cell layers cultured on a liquid medium. Physiol Plant 51:77–84
Cousson A, Tran Thanh Van K (1983) Light and sugar-mediated control of direct de novo differentiation from tobacco thin cell layers. Plant Physiol 72:33–36
Croes AF, Creemers-Molenaar T, van den Ende G, Kemp A, Barendse GWM (1985) Tissue age as an endogenous factor controlling in vitro bud formation on explants from the inflorescence of *Nicotiana tabacum* L. J Exp Bot 36:1771–1779
Culafic L, Neskovic M (1980) Effect of growth substances on flowering and sex expression in isolated apical buds of *Spinacea oleracea*. Physiol Plant 48:588–591
Culafic L, Konjevic R, Neskovic M (1982) Flowering of in vitro-grown spinach shoots in the presence of the herbicide Sando3 9789. Biol Plant 25:155–157
DeFossard RA (1972) The effect of defoliation, and hypocotyl and root removal on the development and flowering of *Chenopodium rubrum* L. Bot Gaz 133:341–350
DeFossard RA (1973) The effects of CCC, ethrel, abscisic acid, abscisic aldehyde, and abscisic hydrocarbon on the development and flowering of *Chenopodium rubrum* L. Bot Gaz 134:103–117

DeFossard RA (1974) Flower initiation in tissue and organ cultures. In: Street HE (ed) Tissue culture and plant science. Academic Press, New York, pp 193–212

Deltour R (1967) Action du saccharose sur la croissance et la inisea fleurs de plantes issues d'apex de *Sinapsis alba* L. cultivés in vitro. CR Acad Sci 264:2765–2767

Dickens CWS (1987) The physiology of flowering with contributions by in vitro techniques. Ph.D Thesis, University of Natal

Dickens CWS, Van Staden J (1985) In vitro flowering and fruiting of soybean explants. J Plant Physiol 120:83–86

Doss RP, Christian JK (1979) Relationship between bulb size, apex size and flowering in bulbous *Iris* cv. Ideal. Physiol Plant 45:215–218

Fujioka S, Yamaguchi I, Murofushi N, Takahashi N, Kaihara S, Takimoto A (1983) The role of plant hormones and benzoic acid in flowering of *Lemna paucicostata* 151 and 381. Plant Cell Physiol 24:241–246

Fujioka S, Yamaguchi I, Murofushi N, Takahashi N, Kaihara S, Takimoto A, Cleland CF (1985) The role of benzoic acid and plant hormones in flowering of *Lemna gibba* G3. Plant Cell Physiol 26:655–659

Fujioka S, Yamaguchi I, Murofushi N, Takahashi N, Kaihara S, Takimoto A, Cleland CF (1986a) Isolation and identification of nicotinic acid as a flower-inducing factor in *Lemna*. Plant Cell Physiol 27:103–108

Ganapathy PS (1969) Floral morphogenesis and flowering in aseptic cultures of *Browallia demissa* L. Biol Plant 11:165–174

Gaspar TH (1980) Rooting and flowering, two antagonistic phenomena from a hormonal point of view. Aspects and prospects of plant growth regulators. British Plant Growth Regulator Group Symposium Monograph 6

Gaspar TH, Pevel C, Roduit C, Moncousin C, Greppin H (1985) The role of auxin level and sensitivity in floral evocation. Biol Plant 27:325–329

Gill RA, Rashid A, Maheshwari S (1979) Isolation of mesophyll protoplasts of *Nicotiana rustica* and their regeneration into plants flowering in vitro. Physiol Plant 47:7–10

Gummer G (1949) Einfluss der Tageslänge auf den Habitus, vor allem auf die Blattstruktur einiger Langtags- und Kurztagspflanzen (besonders von *Kalanchöe blossfeldiana*). Planta 36:439–465

Gupta S, Maheshwari SC (1970) Growth and flowering of *Lemna paucicostata*. II. Role of growth regulators. Plant Cell Physiol 11:97–106

Handro W (1977) Structural aspects of the neo-formation of floral buds on leaf discs of *Streptocarpus nobilis* cultured in vitro. Ann Bot 41:303–305

Handro W (1984) Effects of some growth regulators on in vitro flowering of *Streptocarpus nobilis*. Plant Cell Rep 2:133–136

Harada H (1967) Flower induction in excised shoot apices of *Pharbitis* and *Chrysanthemum* cultured in vitro. Nature 214:1027–1028

Harada H, Bose TK, Cheruel J (1971) Effects of four growth regulating chemicals on flowering of *Pharbitis nil*. Z Pflanzenphysiol 64:267–269

Harder R (1948) Vegetative and reproductive development of *Kalanchöe blossfeldiana* as influenced by photoperiodism. Symp Soc Exp Biol 11. Growth pp 117

Henrickson CE (1954) The flowering of sunflower explants in aseptic culture. Plant Physiol 29:536–538

Higman BM, Smith H (1969) The induction of flowering by abscisic acid in *Lemna perpusilla* 6746. Life Sci 8:1061–1065

Hillson DT, LaMotte CE (1977) In vitro formation and development of floral buds on tobacco stem explants. Effect of kinetin and other factors. Plant Physiol 60:881–884

Hinnawy EIE (1956) Some aspects of mineral nutrition and flowering. Meded Landbouwhogesch Wageningen 56:1–51

Hodson HK, Hamner KC (1970) Floral inducing extracts from *Xanthium*. Science 167:384–385

Jacobs WP, Suthers HB (1971) The culture of apical buds of *Xanthium* and their use as a bioassay for flowering activity of ecdysterone. Am J Bot 58:836–843

Jacobs WP, Suthers HB (1974) Effects of leaf excision on flowering of *Xanthium* apical buds in culture under inductive and non-inductive photoperiods. Am J Bot 61:1016–1020

Jacobs WP, Raghavan V, Kaushik MP (1965) New applications of tissue culture: floral induction in *Perilla* and petiolar abscission in *Coleus*. In: White PR, Grove AR (eds) Int Conf Plant Tissue Culture. McCutchan, Berkeley, Ca, pp 225–241

Jacqmard A, Miksche JP, Bernier G (1972) Quantitative study of nucleic acids and proteins in the shoot apex of *Sinapis alba* during transition from the vegetative to the reproductive condition. Am J Bot 59:714–721
Jacqmard A, Raju MVS, Kinet JM, Bernier G (1976) The early action of the floral stimulus on mitotic activity and DNA synthesis in the apical meristem of *Xanthium strumarium*. Am J Bot 63:166–174
Jayakar M (1970) In vitro flowering of *Crepis capillaris*. Phytomorphology 20:410–412
Joseph C, Paulet P (1975) Contribution à lé'étude vernalisante du froid sur la racine *Cichorium intybus* L. Physiol Veg 13:517–525
Joseph C, Billot J, Soudain P, Cóme D (1985) The effect of cold anoxia and ethylene on the flowering ability of buds of *Cichorium intybus*. Physiol Plant 65:146–150
Kaihara S, Takimoto A (1985a) Flower-inducing activity of vitamin K in *Lemna paucicostata*. Plant Cell Physiol 26:89–98
Kaihara S, Takimoto A (1985b) Induction of flowering in *Lemna paucicostata* by dicoumarol and 4-hydroxycoumarin. Plant Cell Physiol 26:1465–1472
Kaihara S, Watenabe K, Takimoto A (1981) Flower-inducing effect of benzoic and salicylic acids in various strains of *Lemna paucicostata* and *Lemna minor*. Plant Cell Physiol 22:819–825
Kandeler R (1984) Flowering in the *Lemna* system. Phyton 24:113–124
Kandeler R, Hugel B (1973) Blütenbildung bei *Lemna paucicostata* 6746 durch kombinierte Anwendung von Absisinsäure und CCC. Plant Cell Physiol 14:515–520
Kannangara T, Wightman F, Apsimon J, Abrams SA (1986) Enhanced development of flower primordia in *Xanthium strumarium* supplied with extracts from photoinduced *Xanthium* leaves. Proc Manipulation of Flowering Conf, Univ Nottingham
Key JL (1969) Hormones and nucleic acid metabolism. Annu Rev Plant Physiol 20:449–474
Khurana JP, Maheshwari SC (1980) Some effects of salicylic acid on growth and flowering of *Spirodela polyrrhiza* SP20. Plant Cell Physiol 21:923–927
Khurana JP, Maheshwari SC (1983a) Promotion of flowering in *Lemna paucicostata*, a short day duckweed by cytokinins. Plant Cell Physiol 24:913–918
Khurana JP, Maheshwari SC (1983b) Floral induction in *Wolffia microscopica* by salicylic acid and related compounds under non-inductive long days. Plant Cell Physiol 24:907–912
Khurana JP, Maheshwari SC (1986) Induction of flowering in the duckweed *Lemna paucicostata* under non-inductive photoperiods by tannic acid. Physiol Plant 66:447–450
Khurana JP, Tamot BK, Maheshwari SC (1986) Induction of flowering in a duckweed, *Wolffia microscopica* under noninductive long days by 8-hydroxyquinoline. Plant Cell Physiol 27:373–376
Kimura K (1963) Floral initiation in *Pharbitis nil* subjected to continuous illumination at relatively low temperatures. II. Effect of some factors in culture medium on floral initiation. Bot Mag Tokyo 76:351
Klebs G (1913) Über das Verhältnis der Außenwelt zur Entwinklung der Pflanze. Sitzungsber Akad Wiss Heidelberg Ser B No 5
Konar RN, Konar A (1966) Plantlet and flower formation in callus cultures from *Phlox drummondii* Hook. Phytomorphology 16:379–382
Konstantinova TN, Aksenova NP, Bavrina TV, Chailakhyan MKh (1969) On the ability of tobacco stem calluses to form vegetative and generative buds in culture in vitro. Dokl Acad AUK USSR 187:466–469
Krajncic B (1983) The effect of cytokinins on flowering in the long short-day plant *Wolffia arrhiza* L. Z Pflanzenphysiol 112:281–286
Kraus EJ, Kraybill HR (1918) Vegetation and reproduction with special reference to the tomato. Oreg State Coll Agric Exp Sta Bull 140:1–90
Krekule J, Kohli RK (1981) The condition of the apical meristem of seedlings responsive to a promotive effect of abscisic acid on flowering in the short day plant, *Chenopodium rubrum*. Z Pflanzenphysiol 103:45–51
Lang A (1965) Physiology of flower initiation. In: Ruhland W (ed) Encyclopedia of plant physiology XV/1. Springer, Berlin Heidelberg New York, pp 1380–1536
Liverman JL, Bonner J (1953) Biochemistry of the photoperiodic response. The high intensity light reaction. Bot Gaz 115:121–128
Lona F (1948) La fioritura della brevidiurne *Chenopodium amaranticolor* Coste et Reyn coltivata in solizione nutritiza con saccarosio in assenza di stimulo fotoperiodico euflorigeno. Nuova Giorn Bot Ital, N.S. 46:559–562
Lona F (1949) La fioritura della brevidiurne a notte continua. Nuova G Bot Ital (Nuova Ser) 56:479

Lona F (1950) Il significato dei glucidi e del fattore nutiflorigeno nel meccanismo di fioritura della piante erbecee ed in particolare della brevidiurne. Rend Ist Lombardo Sci Lett, Cl di Sci 83:1

Loo SW (1946a) Preliminary experiments on the cultivation of *Baeria chrysostoma* under sterile conditions. Am J Bot 33:382–389

Loo SW (1946b) Cultivation of excised stem tips of dodder in vitro. Am J Bot 33:295–300

Margara J, Piollat MT (1981) Neoformation de boutons floraux in vitro à partir de fleurs de *Kalanchöe blossfeldiana* Poellin. CR Acad Sci 293:93–96

Margara J, Touraud G (1968) Recherches expérimentales sur la néoformation de bourgeons inflorescentiels ou végétatifs in vitro à partir d'explantats d'endive (*Cichorium intybus* L.) Ann Physiol Vég 10:41–56

Margara J, Rancillac M, Beck D (1965) Recherches expérimentales sur la néoformation de bourgeons inflorescentiels ou végétatifs in vitro à partir d'explantats d'endive (*Cichorium intybus* L.). I. Influence de la polarité. Ann Physiol Vég 7:157–170

Margara J, Rancillac M, Bouniols A (1966) Recherches expérimentales sur la néoformation de bourgeons inflorescentiels ou végétatifs in vitro a partir d'explantats d'endive (*Cichorium intybus* L.) III. Etude critique de la méthode. Ann Physiol Vég 8:285–305

Matsushima T, Itoyuma T, Mashiko Y, Mizukoshi T (1974) Critical time of floral differentiation in *Pharbitis nil* shoot apex. In: Plant growth substances 1973. Hirokawa, Tokyo, pp 967–973

Mehra A, Mehra PN (1972) Differentiation in callus cultures of *Mesembrysanthemum floribundum*. Phytomorphology 22:171–176

Mialoundama F, Paulet P (1975) Variations de la teneur an acides chlorogénique et isochlorogénique au cours du traitement de vernalization des racines de *Cichorium intybus* L., en relation avec leur aptitude à la floraison in vitro. CR Acad Sci 280:1385–1387

Miginiac E (1972) Cinetique d'action comparèe des racines et de la kinétine sur le développment floral de bourgeons cotylédonairs chez le *Scrofularia arguta* Sol. Physiol Vég 10:627–636

Murashige T, Skoog F (1962) A revised medium for rapid growth and bioassay with tobacco tissue culture. Physiol Plant 15:473–495

Negretskii VA, Lozhnikova VN, Chailakhyan MKh (1984) Changes of phytohormone activity with a shift of the flowering gradient in photoperiodically neutral Trapezond tobacco. Sov Plant Physiol 31:896–901

Neyland M, Ng YL, Thimann KV (1963) Formation of anthyocyanin in leaves of *Kalanchöe blossfeldiana* – a photoperiodic response. Plant Physiol 38:447–451

Nitsch C (1972) The role of growth regulators in flowering as demonstrated by in vitro techniques. In: Kaldewey H, Varder Y (eds) Hormonal regulation of plant growth and development. Verlag Chemie, Weinheim, pp 413–421

Nitsch C, Nitsch JP (1967a) The induction of flowering in vitro on stem segments of *Plumbago indica* L. I. The production of vegetative buds. Planta 72:355–370

Nitsch C, Nitsch JP (1967b) The induction of flowering in vitro on stem segments of *Plumbago indica* L. II. The production of reproductive buds. Planta 72:371–384

Nitsch JP, Nitsch C, Rossini LME, Bui Dang Ha D (1967) The role of adenine in bud differentiation. Phytomorphology 17:446–453

Noma M, Koike N, Sano M, Kawashima N (1984) Endogenous IAA in the stem of tobacco in relation to flower neoformation as measured by mass spectroscopic assay. Plant Physiol 75:257–260

Novak FJ, Lucrewtti S, Donini B, Afza R, Hermelin T (1985) In vitro flower induction and pod development in *Pisum sativum*. Genet Agrar 39:333–334

Oota Y (1963) Effects of growth substances on frond and flower production in *Lemna gibba* G3. Plant Cell Physiol 6:547–559

Oota Y (1972) A possible mechanism for sugar inhibition of duckweed flowering. Plant Cell Physiol 13:195–199

Paterson KE (1984) Shoot tip culture of *Helianthus annuus*, flowering and development of adventitious and multiple shoots. Am J Bot 71:925–931

Paulet P (1979) Sur la régulation de la néoformation florale in vitro. Physiol Vég 17:631–641

Paulet P, Nitsch JP (1964) Flower formation on *Cichorium intybus* root tissues cultured in vitro. Ann Physiol Vég 6:333–345

Peterson JB, Miller CO (1976) Cytokinins in *Vinca rosea* L. crown gall tumor tissue as influenced by compounds containing reduced nitrogen. Plant Physiol 576:393–399

Pierik RLM (1966a) The induction and initiation of flower buds in vitro in tissues of *Lunaria annua* L. Naturwissenschaften 53:45

Pierik RLM (1966b) The induction and initiation of flower buds in vitro in root tissues of *Cichorium intybus* L. Naturwissenschaften 53:387

Pierik RLM (1970) Factors affecting the induction and initiation of flower buds in vitro in plant species with a cold requirement for flowering. In: Bernier G (ed) Cellular and molecular aspects of floral induction. Longman, London, pp 408–415

Pieterse AA, Muller LJ (1977) Induction of flowering in *Lemna gibba* G3 under short-day conditions. Plant Cell Physiol 18:45–53

Pryce RS (1972) Gallic acid as a natural inhibitor of flowering in *Kalanchöe blossfeldiana*. Phytochemistry 11:1911–1918

Pryce RS (1973) Allogibberic acid: an inhibitor of flowering in *Lemna perpusilla*. Phytochemistry 12:1745–1754

Purse JG (1984) Phloem exudate of *Perilla crispa* and its effects on flowering of *P. crispa* shoot explants. J Exp Bot 35:227–238

Raghavan V (1961) Studies on the floral histogenesis and physiology of *Perilla*. III. Effect of indoleacetic acid on the flowering of apical buds and explants in culture. Am J Bot 48:870–876

Raghavan V, Jacobs WP (1961) Studies on the floral histogenesis and physiology of *Perilla*. II. Floral induction in cultured apical buds of *P. fructescens*. Am J Bot 48:751–760

Rajasekaran K, Mullins HG, Nair Y (1983) Flower formation in vitro by hypocotyl explants of cucumber (*Cucumis sativus* L.). Ann Bot 52:417–420

Raste AP, Ganapathy PS (1970) In vitro behaviour of inflorescence segments of *Mazus pumilus*. Phytomorphology 20:367–373

Ringe F, Nitsch JP (1968) Conditions leading to flower formation on excised *Begonia* fragments cultured in vitro. Plant Cell Physiol 9:639–652

Ritchie IM, Werkman BR, Longmen KA (1986) Cloning of *Thuja plicata* in aseptic culture. Proc Manipulation of Flowering Conf, Univ Nottingham

Rodrigues-Pereira AS (1965) Physiological analysis of flower formation in Wedgewood Iris. J Exp Bot 16:405–410

Ross JJ, Murfet IC (1985) A comparison of the flowering and branching control systems in *Lathyrus odoratus* and *Pisum sativum*. Ann Bot 56:847–856

Rossini LME, Nitsch JP (1966) Induction de la floraison in vitro chez une plante de jours courts. *Streptocarpus nobilis*. CR Acad Sci 263:1379–1382

Sachs RM (1977) Nutrient diversion: an hypothesis to explain the chemical control of flowering. HortScience 12:220–222

Sadik S, Ozburn JL (1967) Histochemical changes in the shoot tip of cauliflower during floral induction. Can J Bot 45:955–957

Schwabe WW (1958) Effects of photoperiod and hormone treatment on isolated rooted leaves of *Kalanchöe blossfeldiana*. Physiol Plant 11:225–239

Schwabe WW (1969) *Kalanchöe blossfeldiana* Poellniz. In: Evans LT (ed) The induction of flowering. Macmillan, Melbourne, pp 227–246

Schwabe WW (1972) Flower inhibition in *Kalanchöe blossfeldiana*. Bioassay of an endogenous long-day inhibitor and inhibition by ABA and xanthoxin. Planta 113:18–23

Scorza R (1982) In vitro flowering. Hortic Rev 4:106–127

Scorza R, Janick J (1978) Flowering from *Passiflora suberosa* leaf discs in vitro. In: Hughs KW, Henke R, Constantin M (eds) Propagation of higher plants through tissue culture. Proc Intern Symp US Energy Department, Oak Ridge, Tennessee, p 249

Scorza R, Janick J (1980) In vitro flowering of *Passiflora suberosa* L. J Am Soc Hortic Sci 105:892–897

Seidlova F (1974) Changes of growth and RNA synthesis in shoot apical meristems of *Chenopodium rubrum* as a consequence of photoperiodic induction. Z Pflanzenphysiol 73:394–404

Seidlova F, Culafic L (1982) The behaviour of the shoot apical meristem in *Chenopodium rubrum* under conditions noninductive for flowering. Biol Plant 24:471–473

Seidlova F, Kohli RK, Pavlova L (1981) Effects of abscisic acid on the growth pattern of the shoot apical meristem and on flowering in *Chenopodium rubrum* L. Ann Bot 48:777–785

Shinozaki M, Takimoto A (1982a) The role of cotyledons in flower initiation of *Pharbitis nil* at low temperatures. Plant Cell Physiol 23:403–408

Shinozaki M, Takimoto A (1982b) Correlation between long-day flowering and root elongation in *Pharbitis nil*, strain Kidachi. Plant Cell Physiol 23:1055–1062

Shinozaki M, Takimoto A (1983) Effect of some growth regulators and benzoic acid derivatives on flower initiation and root elongation of *Pharbitis nil*, strain Kidachi. Plant Cell Physiol 24:433–439

Silberger J Jr, Skoog F (1953) Changes induced by indoleacetic acid and nucleic contents on growth of tobacco pith tissues. Science 118:443–444

Simmonds J (1982) In vitro flowering of leaf explants of *Streptocarpus nobilis*. The influence of culture medium components on vegetative and reproductive development. Can J Bot 60:1461–1468

Skoog F, Miller CO (1957) Chemical regulation of growth and organ formation in plant tissues cultured in vitro. Symp Soc Exp Biol 11:118–131

Srinivasan C, Mullins MG (1978) Control of flowering in grapevine (*Vitis vinifera* L.). Formation of inflorescenses in vitro by isolated tendrils. Plant Physiol 61:127–130

Steffen JD, Sachs RM, Hackett WP (1986) In vitro growth and development of reproductive meristems on *Bougainvillea* San Diego Red. Proc Manipulation of Flowering Conf, Univ Nottingham

Steinberg RA (1950) Flowering response of a variety of *Nicotiana rustica* to organic compounds in aseptic culture. Am J Bot 37:547–551

Takimoto A (1960) Effect of sucrose on flower initiation of *Pharbitis nil* in aseptic culture. Plant Cell Physiol 1:241–246

Tanaka O, Horikawa W, Nishimura H, Nasu Y (1986) Flower induction by suppression of nitrate assimilation in *Lemna paucicostata* 6746. Plant Cell Physiol 27:127–133

Tang AF, Cappadodia M, Byrne D (1983) In vitro flowering in cassava (*Manihot esculanta* Crantz). Plant Cell Tissue Organ Cult 2:199–206

Tanimoto S, Harada H (1979) Influences of environment and physiological conditions on floral bud formation on *Torenia* stem segments cultured in vitro. Z Pflanzenphysiol 95:33–41

Tanimoto S, Harada H (1980) Hormonal control of morphogenesis in leaf explants of *Perilla fructescence* Britton var. crispa Decaisne f. viride-crispa Makino. Ann Bot 45:321–327

Tanimoto S, Harada H (1981a) Effect of IAA, ammonium nitrate and sucrose on the initiation and development of floral buds in *Torenia* stem segments cultured in vitro. Plant Cell Physiol 22:1553–1560

Tanimoto S, Harada H (1981b) Chemical factors controlling floral bud formation of *Torenia* stem segments in vitro. I. Effects of mineral nutrients and sugars. Plant Cell Physiol 22:533–541

Tanimoto S, Harada H (1981c) Chemical factors controlling floral bud formation of *Torenia* stem segments cultured in vitro. II. Effects of growth regulators. Plant Cell Physiol 22:543–550

Tanimoto S, Harada H (1982a) Physiological and hormonal factors influencing organogenesis in *Rudbeckia bicolor* explants cultured in vitro. Plant Cell Physiol 23:107–113

Tanimoto S, Harada H (1982b) Effects of cytokinin and anticytokinin on the initial stage of adventitious bud differentiation in the epidermis of *Torenia* stem segments. Plant Cell Physiol 23:1371–1376

Tanimoto S, Miyazaki A, Harada H (1985) Regulation by abscisic acid of in vitro flower formation in *Torenia* stem segments. Plant Cell Physiol 26:675–682

Tejovathi C, Anwar SY (1984) In vitro induction of capitula from cotyledons of *Carthamnus tinctorious* (safflower). Plant Sci Lett 36:165–168

Tetley RM, Ikuma H (1970) Effect of plant growth factors upon the starch content in cultured tobacco pith explants. Plant Physiol 45(S):10

Thorpe TA, Murashige T (1968) Starch accumulation in shoot-forming tobacco callus cultures. Science 160:421–422

Thorpe TA, Murashige T (1970) Some histochemical changes underlying shoot initiation in tobacco callus cultures. Can J Bot 48:277–285

Tizio R (1979) Floraison in vitro de l'ail (*Alium sativum* L.). CR Acad Sci 289:401–404

Tran Thanh Van K (1973a) Direct flower neoformation from superficial tissues of small explants of *Nicotiana tabacum*. Planta 115:87–92

Tran Thanh Van K (1973b) In vitro control of de novo flower, bud, root and callus differentiation from excised epidermal tissues. Nature 246:44–45

Tran Than Van K (1977) Regulation of morphogenesis. In: Barz W, Reinhard E, Zenk MH (eds) Plant tissue culture and its biotechnological application. Springer, Berlin Heidelberg New York, pp 367–385

Tran Thanh Van K, Chlyah A (1976) Differentiation du boutons floraux, de bourgeons végétatifs, de racimes et al cal á partir de l'assise sous-épidermique des ramifications florals de *Nicotiana tabacum* Wisc. 38. Etude infrastructurale. Can J Bot 54:1979–1996

Tran Thanh Van K, Dien NT, Chlyah A (1974) Regulation of organogenesis in small explants of superficial tissue of *Nicotiana tabacum* L. Planta 119:149–159

Trewavas AJ (1981) How do plant growth substances work? Plant Cell Environ 4:203–228

Trewavas AJ (1983) Nitrate as a plant hormone. In: Jackson MB (ed) Interactions between nitrogen and growth regulators in the control of plant development. British Plant Growth Regulator Group, Monograph 9, pp 97–100
Umemoto T (1971) Effect of chlorogenic acid on flower production in long-day duckweed, *Lemna gibba* G3. Plant Physiol 12:165–169
Van den Ende G, Barendse GWM, Kemp A, Croes AF (1984a) The role of glucose on flower bud formation in thin-layer cultures of *Nicotiana tabacum*. J Exp Bot 35:1853–1859
Van den Ende G, Croes AF, Kemp A, Barendse GWM (1984b) Development of flower buds in thin-layer cultures of floral stalk tissue from tobacco: Role of hormones in different stages. Physiol Plant 61:114–118
Van den Ende G, Croes AF, Kemp A, Barendse GWM, Kroh M (1984c) Floral morphogenesis in thin-layer tissue cultures of *Nicotiana tabacum*. Physiol Plant 62:83–88
Vanderhoef LN, Key JL (1968) Inhibition by kinetin of cell elongation and RNA synthesis in excised soybean hypocotyl. Plant Cell Physiol 9:343–351
Venkataraman R, Seth PN, Maheshwari SC (1970) Studies on the growth and flowering of short-day plant, *Wolffia microscopica*. I. General aspects and induction of flowering by cytokinins. Z Pflanzenphysiol 62:316–317
Villalobos V (1981) Floral differentiation in carnation (*Dianthus caryophyllus* L.) from anthers cultured in vitro. Phyton 41:71–75
Wada K, Totsuka T (1982) Long-day flowering of *Perilla* plants cultured in nitrogen-poor media. Plant Cell Physiol 23:977–985
Wardell WL (1977) Floral induction of vegetative plants supplied a purified fraction of DNA from stems of flowering plants. Plant Physiol 60:885–891
Wardell WL, Skoog F (1969a) Flower formation in excised tobacco stem segments. I. Methodoloty and effects of plant hormones. Plant Physiol 44:1402–1406
Wardell WL, Skoog F (1969b) Flower formation in excised tobacco stem segments. II. Reversible removal of IAA inhibition by RNA base analogs. Plant Physiol 44:1407–1412
Wardell WL, Skoog F (1973) Flower formation in excised tobacco stem segments. III. DNA content in stem tissue of vegetative and flowering tobacco plants. Plant Physiol 52:215–220
White PR (1933) Plant tissue culture; results of preliminary experiments on the culturing of isolated stem tips of *Stellariamedia*. Protoplasma 19:97–116
Wittwer SH, Aung LH (1969) *Lycopersicon esculentum* Mill. In: Evans LT (ed) The induction of flowering; some case histories. Macmillan, Melbourne, pp 409–429
Zeevaart JAD, Lang A (1962) The relationship between gibberellin and floral stimulus in *Bryophyllum diagremontianum*. Planta 58:531–542
Zeevaart JAD, Brede JM, Cetas CB (1977) Translocation patterns in *Xanthium* in relation to long-day inhibition of flowering. Plant Physiol 60:747–753

I.7 Vitrification in Micropropagation

T. GASPAR[1]

1 Definition

Vitrification is the term generally used to characterize the hyperhydric malformations frequently affecting herbaceous and woody plants during their in vitro vegetative propagation. Glassiness, translucency, glauciness, and vitrescence are terms also used when describing this physiological disorder (Gaspar et al. 1987). The so-called vitrified or vitreous vitroplants appear turgid or hyperhydric (as if the cells were turgescent, at first sight), watery at their surface, and hypolignified. Their organs are somehow translucent, in some cases less green, and easily breakable (Jones 1976; Werner and Boe 1980).

Vitrification is not restricted to organs or whole plants. Some yellowish and hyperhydric callus types may be considered as vitreous (Crèvecoeur et al. 1987). Their high degree of friability therefore has to be questioned in relation to the breakability of vitrified organs.

The process of vitrification is sometimes reversible: new leaves formed in vitrified plants after transfer to a nonvitrifying medium or to the greenhouse often have an anatomy approaching that of normal plants (Von Arnold and Ericksson 1984; John 1986; Böttcher et al. 1988).

To my knowledge, vitrification occurs only during multiplication (on multiplication media) and never during rooting. Plants submitted to intensive multiplication (frequent subculture with high rate of regeneration) are more susceptible to vitrification (John 1986; Boxus and Pâques 1987).

Vitrification was first described in the 1960's (Phillips and Matthews 1964; Hackett and Anderson 1967) and attracted more attention with the increasing use of micropropagation, which has led to several reviews recently (Kevers et al. 1984; Ziv 1986; Gaspar et al. 1987; Pâques and Boxus 1987a).

2 Morphological, Anatomical, Mechanical, and Biochemical Characteristics

In general, stems of vitrified plantlets are broad, and thick in diameter. The lengths of the internodes are shorter than those of plants appearing normal. Leaves are thick, frequently very elongated, wrinked and/or curled, and brittle.

[1]Hormonologie fondamentale et appliquée, Institut de Botanique (B 22), Université de Liège, Sart Tilman, 4000 Liège, Belgium

Biotechnology in Agriculture and Forestry, Vol. 17
High-Tech and Micropropagation I (ed. by Y.P.S. Bajaj)

Anatomically, there is hypertrophy of the cortical and pith stem parenchyma (Leshem 1983a,b; Ziv et al. 1983) with hypertrophy of the cells (Vieitez et al. 1985; Werker and Leshem 1987) and large intercellular spaces. Increased extracellular space and air volume have been shown in vitrifying material (Kevers and Gaspar 1986). The vessels and tracheids in vitreous plants are not properly lignified (Beauchesne 1981; Vieth et al. 1983; Daguin and Letouzé 1985; Vieitez et al. 1987). In vitrified leaves, palisade tissue is nonexistent or drastically reduced (Brainerd et al. 1981; Debergh et al. 1981; Vieitez et al. 1987). The mesophyll looks spongy with large intercellular spaces (Brainerd et al. 1981; Debergh et al. 1981; Vieth et al. 1983; Von Arnold and Ericksson 1984). Severe structural damage is observed in the apical dome (Werker and Leshem 1987). There are no pro-cambium strands descending basipetally from the leaf primordia. Vascular bundles are few and lack the typical arrangement of a normal shoot (Leshem 1983a; Ziv et al. 1983).

The succulent mature leaf shows a thin noncontinuous cuticle, with reduced or absent deposition of epicuticular wax (Grout and Aston 1978; Sutter and Langhans 1979; Ziv et al. 1983, 1987; Sutter 1985; Aitken-Christie et al. 1985; John 1986; Dhawan and Bhojwani 1987) and fewer stomata (often plugged with an unknown material) (Brainerd et al. 1981; Leshem 1983a,b; John 1986) than the mature leaf of normal plants. Most stomata of leaves from in vitro plantlets do not seem to have a closure mechanism and this may be the main cause of rapid water loss during acclimatization under low relative humidity (Brainerd and Fuchigami 1982; Dhawan and Bhojwani 1987; Ziv et al. 1987). Watery callus types are constituted of parenchymatic cells without xylem elements (Crèvecoeur et al. 1987).

Cell wall extensibility properties of vitrifying internodes was kinetically investigated in comparison to those of normal plants using the constant stress method (Kevers et al. 1987). Vitrifying conditions rapidly induced the capability of the walls to be extended. Measurements indicated that the extensibility capabilities involved plastic properties rather than elastic ones. These changes were paralleled by decreased relative levels of cellulose and lignin (see below).

Biochemically, vitrified tissues are characterized by:

- Less dry weight and more water (Kevers et al. 1984; Crèvecoeur et al. 1987). The additional water in vitrifying tissues was shown to be located in the extraprotoplastic space. The amount of intracellular water is less important (Kevers and Gaspar 1986).
- Less lignin (Daguin and Letouzé 1985; Kevers and Gaspar 1985a; Kevers et al. 1987). The lower lignification of vitreous plants was associated to lower activities of enzymes involved in the synthesis of lignin precursors and in their polymerization: phenylalanine ammonia lyase (Kevers and Gaspar 1985a; Hegedus and Phan 1987; Letouzé and Daguin 1987), hydroxycinnamate CoA ligase (Phan and Letouzé 1983; Hegedus and Phan 1987; Letouzé and Daguin 1987), coniferyl alcohol dehydrogenase (Letouzé and Daguin 1987), acidic and wall peroxidases (Kevers et al. 1984; Kevers and Gaspar 1985a; Crèvecoeur et al. 1987; Letouzé and Daguin 1987), flavonone synthase and hydroxyannamate CoA reductase (Hegedus and Phan 1987).
- Less cellulose (Kevers et al. 1987), which was associated to a low C/N ratio favoring the synthesis of amino acids rather than the sugar units for cellulose (see below).

— Low calcium content, low Ca^{2+}/uronic acids ratio, low ratio of uronic acids to neutral sugars due to higher amounts of the latter in isolated cell walls of vitrified tissues (Kevers et al. 1988). All these characteristics may account for the high wall plastic potential previously measured in vitrifying internodes (Kevers et al. 1987).
— Less chlorophyll, which causes their translucency (Hegedus and Phan 1983; Phan and Letouzé 1983; Ziv et al. 1983) and presumably lower photosynthetic capacity (Leshem 1983b). The level of both chlorophyll a and chlorophyll b is also particularly low in "vitreous" calli (Crèvecoeur et al. 1987). It is noteworthy that the level of chlorophyll b is higher than that of chlorophyll a. This is the reverse of what is commonly observed in most other plant tissues, but has already been reported in olive callus (Lavee and Messer 1969). Note that some authors associate a darker green color with the process of vitrification (Navatel 1982; Borkowska and Michalczuk 1987; Boxus and Pâques 1987).
— Less soluble phenols (Kevers et al. 1984).
— Higher activity of basic peroxidases (Kevers et al. 1984; Kevers and Gaspar 1985a), which is thought to be associated to an enhanced auxin catabolism (Kevers et al. 1984; Gaspar 1986; Gaspar et al. 1987).
— Higher activity of glutamate dehydrogenase (Letouzé and Daguin 1983; Daguin and Letouzé 1986), which is thought to bring about the diversion of the carbohydrate pool to amino acid synthesis.
— Less soluble Ca^{2+}, Min^{2+}, and Na^{+} and more K^{+}, at least in carnation (Kevers and Gaspar 1986).
— Reduced ethylene production following a burst initiated with the application of the vitrifying conditions (Kevers and Gaspar 1985b). The levels of ACC and EFE activity were also rapidly increased under vitrifying conditions. The flask atmospheres of vitrifying plants thus accumulate ethylene at a higher rate (Kevers and Gaspar 1985b).
— Reduced polar transport of auxin (Gersani et al. 1986).
— Increase of pH of the culture medium (Pâques and Boxus 1984).

3 External and Internal Factors Inducing Vitrification

All suggested external causes refer exclusively to the culture conditions (Gaspar et al. 1987). As was stated by Debergh et al. (1981), the source and physiological condition of the mother plant at the time of explantation and inoculation do not interfere with vitrification.

3.1 Culture Medium

Most of the reported cases of vitrification have occurred on complete mineral medium of Murashige and Skoog (1962), which has a particularly high ammonium nitrate content (Beauchesne 1981), or on media with high ionic strength, but apparently never on media of low ionic strength (Installé et al. 1985) such as those of Gautheret, White, Heller, or Knop. Letouzé and Daguin (1983) and Daguin and Letouzé (1986) have induced vitrification in *Salix babylonica* plantlets by increasing NH_4^+ in the medium.

The role of large quantities of ammonium ions in the MS medium was also underlined by Vieitez et al. (1985), Daguin and Letouzé (1986), and Phan and Hegedus (1986).

There is also some evidence that the reduction of agar concentration or its avoidance (liquid culture instead of solid ones) causes vitrification (Hauzinska 1974; Davis et al. 1977; Sutter and Langhans 1979; Debergh 1983; Hakkaart and Versluijs 1983; Druart et al. 1984; Kevers et al. 1984; Pâques and Boxus 1984, 1987b; John 1986). However, many data suggest that the control of water potential is not the only way in which agar concentration affects vitrification (Debergh 1983; Bornman and Vogelmann 1984; Vieitez et al. 1985).

The role of cytokinins from the culture medium in vitrification has been mentioned by Debergh et al. (1981), Debergh (1983), Miedema (1984), Bornman and Vogelmann (1984), and Pasqualetto et al. (1986). Leshem et al. (1988) suggest that vitrification is the bud's reaction to an excess of cytokinin. The beneficial effect of raising agar concentration should be explained by decreasing the availability of cytokinins. Some plants, however, vitrify in the absence of cytokinins, which leads to the suggestion that cytokinins are not inducers of the process but may act synergistically with other factors (Gaspar et al. 1987). Vitrification due to growth factor imbalance involving auxin and cytokinin was already hypothesized by Beauchesne (1981). Interacting effects of NAA and cytokinins on vitrification were shown by Leshem and Sachs (1985), Dencso (1987), and Orlikowska (1987).

Chlorine ions (Quoirin and Lepoivre 1977) have also been suspected as external agents of vitrification but their role in the process has never been confirmed.

3.2 Culture Vessel Atmosphere

The size and type of culture vessel, the type of closure, the concentration of agar, and the climate parameters of the culture room all influence vitrification (Hakkaart and Versluijs 1983; Debergh and Maene 1984).

Works from different laboratories (Wardle et al. 1983; Ziv et al. 1983; Debergh and Maene 1984; Böttcher et al. 1988) dealing with the manipulation of the vapor pressure gradient between the leaf and the atmosphere lead to the hypothesis that high relative humidity in the culture atmosphere is a causative agent of vitrification, but as proposed for cytokinins, relative humidity might be an enhancer rather than an inducer of the process.

There has not been sufficient work on the effect of the gases emitted by the plantlets in vitrifying conditions, their accumulation in the culture vessel atmospheres, and their effects as exogenously supplied agents, namely ethylene and carbon dioxide, to assess their role. My feeling, however, is that the excess of ethylene produced by plants placed in vitrifying conditions, which accumulates to some extent in the flask atmosphere (Kevers and Gaspar 1985b) is partly responsible for the morphological changes in these plants: exogenously supplied ethylene induces identical morphological traits (stem thickening, reduction of internode length) in several systems (Gaspar et al. 1989); better gas exchange of the culture atmospheres (Hakkaart and Versluijs 1983) and reduced ethylene emission using inhibitors of ethylene biosynthesis (Gaspar and Kevers 1985) reduce or suppress vitrification.

3.3 Environmental Conditions

No particular effect of light on vitrification has been reported. Temperature may be a stimulating factor (Gaspar et al. 1987).

3.4 Internal Factors

Kevers et al. (1984), Gaspar (1986), and Gaspar et al. (1987) have proposed a hypothetical sequence of reactions leading to vitrification, integrating most of the available biochemical data. These reactions (involving soluble phenols, basic and acidic peroxidases, the metabolisms of auxin and ethylene) lead to a deficiency in cellulose and lignin which may explain the anatomical and hence part of the morphological changes. It was first proposed (Kevers et al. 1984) that these deficiencies might be responsible for the hyperhydric transformations through increased water uptake via reduced wall pressure. It was shown later that surplus of water was located in the intercellular spaces and that protoplasts contained less water (Kevers and Gaspar 1986; Pâques and Boxus 1987b). The first proposal thus no longer holds true. Cells, on the contrary, should be rather plasmolyzed (instead of turgescent) or the initiating mechanism at least might be plasmolysis due to the high osmotic pressure of intercellular fluid constituted by the (liquid) culture medium entered by capillary infiltration (Böttcher and Göring 1987). This could be the foreseen stress factor. The observed ionic variations might be a response to the hyperosmotic shock (Kevers and Gaspar 1986). This might, incidentally, explain why vitrification never occurs on rooted plants where roots constitute a barrier to such an infiltration. Vitrification also does not occur on a rooting medium, which again throws light on, the role of cytokinins. We have recently shown the synergistic effect of cytokinins and auxins in ethylene production during the micropropagation of roses (Gaspar et al. 1989). Such a synergistic effect between cytokinins and the stressing hyperosmotic shock in ethylene production might be suspected to occur and be connected with vitrification.

It is now thought that ethylene acts rather as a promoter of lignification (Roberts and Miller 1982; Fukuda and Komamine 1985). The deficient lignification in vitrification, possibly dependent on the rate of ethylene production, would thus result from the low production following the feedback inhibition of its own synthesis (Kevers and Gaspar 1985b; Gaspar et al. 1987). In parallel, a higher assimilation of NH_4^+ would lead to a higher consumption of sugars and to their being deviated from cell wall polymers biosynthesis (Beauchesne 1981; Gaspar et al. 1987). As remedies, the main following proposals have been made.

4 Remedies

4.1 Removal of Apparent Causes

4.1.1 In the Culture Medium

— Use a solid stationary medium (Hauzinska 1974; Davis et al. 1977; Sutter and Langhans 1979; Monette 1983) or raise agar concentration (Debergh et al. 1981;

Debergh 1983; Hakkaart and Versluijs 1983; Leshem 1983a,b; Bornman and Vogelmann 1984; Von Arnold and Ericksson 1984; Chavdarov 1986). Raising agar concentration, however, often reduces multiplication rate (Leshem 1983a,b; Ziv et al. 1983; Von Arnold and Ericksson 1984).

— Lower cytokinin level or replace one type by another (Werner and Boe 1980; Beauchesne 1981; Debergh et al. 1981; Debergh 1983; Wilkins and Dodds 1983).
— Lower NH_4^+ concentration (Beauchesne 1981; Riffaud and Cornu 1981; Letouzé and Daguin 1983) or dilute Murashige and Skoog's (1962) medium or replace it by Hellers's (1969) formula (Vieitez et al. 1985).

4.1.2 In the Culture Vessel Atmosphere

— Allow a better gas exchange (Hakkaart and Versluijs 1983).
— Lower relative humidity either by raising agar concentration, by using a desiccant (Ziv et al. 1983), or by favoring water condensation (Maene 1985).

4.2 Antidotes

4.2.1 In the Culture Medium

— Add phlorizin or phloroglucinol to the culture medium (Hegedus and Phan 1983, 1987).
— Add $CoCl_2$, 2–3 days before subculture (Gaspar and Kevers 1985). Other inhibitors of ethylene biosynthesis, when added at the beginning of subculture, do not affect the initiation of the vitrification process, except DNP and STS (also $AgNO_3$ as an inhibitor of ethylene action), which slightly reduce the visible symptoms and allow limited growth of axillary buds (Druart et al. 1984; Kevers and Gaspar 1985b).
— Add alimentary pectin or replace part of agar by pectin (Zuccherelli 1979; Druart et al. 1981; Rugini and Verma 1983) or add hydrolyzed agar at the minimum rate of 0.1 g/l of nutritive medium (Pâques and Boxus 1986; Boxus and Pâques 1987). The latter method might allow multiplication in liquid medium.
— Replace sucrose by fructose (Rugini et al. 1987).

4.2.2 In the Environmental Factors

— Apply 3–4 weeks cold (3–4 °C) treatment, possibly without BA and without leaves (Boxus et al. 1978; Rugini and Verma 1983).
— Raise temperature (up to 26 °C) in the 3 days after excision (Hakkaart and Versluijs 1983).
— Cool culture medium during culture (Maene 1985).
— Increase light intensity (up to 17 000 lx) (Maene 1985).

5 Disadvantages

Vitrification counteracts the normal way of micropropagation at the in vitro multiplication and rooting stages and in situ at the acclimatization stage as well. Growth of vitrified shoots is limited, which finally leads to necrosis and death. It is accompanied by low rates of multiplication which finally cannot be pursued. On solid medium, all the vitrified plantlets turn into callus after two or three subcultures (Leshem et al. 1988). Vitrified shoots do not root easily either in vitro or during their attempted weaning in soil. Moreover, they survive with difficulty on transfer to soil due to wilting and high susceptibility to infection (Aitken-Christie et al. 1985; Vieitez et al. 1985). In any case, the growth of new shoots is affected by the presence of vitrified organs.

6 Beneficial Management

Vitrification is a condition that might be of benefit during the multiplication of some conifer species in vitro. Aitken et al. (1981) found that the translucent shoots that formed on vitrified *Pinus radiata* cultures were slow to elongate, but grew into normal shoots if left on the same medium for 2 to 3 months. Aitken-Christie and Thorpe (1985) described three types of shoots that could be formed, i.e., waxy shoots, which were similar to seedling shoots, wet shoots, which were dark green with the needles stuck together and lacked abundant epicuticular waxes, and translucent shoots, which were water-soaked and did not survive transfer from in vitro conditions. Proliferation was much higher in cultures in which wet shoots were present, although survival after transfer from in vitro conditions was poor (Aitken-Christie et al. 1985). Bornman and Vogelman (1984) found that vitrified *Picea abies* cultures could produce ten times as many buds and shoots as normal cultures.

The survival and normal development after subculture of axillary shoots, axillary buds, and induced adventitious buds (whether vitrified or reverted) and shoots formed during some vitrifying manipulations demonstrated that the induction of vitrification is a technique that can be used for the proliferation of Sitka spruce shoots in vitro (John and Pearson 1985; John 1986).

7 Conclusions and Prospects

Vitrification was first described as a teratological morpho- and organogenetic process, and because of its random occurrence in the most commonly used solid media in industrial micropropagation, most efforts at first concentrated on the ways to avoid it. This progressively elucidated the main exogenously inducing or favoring factors which led to design-controlled experimental systems where the biochemical and physiological events leading to the process could begin to be studied more accurately. Vitrification at the moment appears to be the result of a stress phenomenon where several factors (cytokinins, NH_4^+, humidity, etc.) might act synergistically in an orderly sequence. In this system, ethylene appears as a key factor (Kevers and Gaspar 1985b; De Proft et al. 1987; Leonhardt and Kandeler 1987), but a causal relationship with either some steps of its biosynthesis and its final production, or its retention in the tissues versus its emission, or its accumulation in the media or in the direct environment of the plants could not yet be identified. Vitrification in both its morphological and biochemical aspects, shows some analogies with processes occurring in waterlogging, fruit softening, translucency in potato, and abscission, where ethylene is also involved. The vitrification symptoms in calli have been questioned in relation to habituation or auxin autonomy (Crèvecoeur et al. 1987; Gaspar et al. 1988). The more simple question of whether callus formation in fact derives from a process of vitrification has also already been posed (Lamproye et al. 1986); in many instances vitrified plantlets indeed finally turn into callus (Leshem et al. 1988). Vitrification was found later to be accompanied by high potential multiplication rates under particular circumstances (John 1986). The reversion to normal growth of the various shoots formed enabled this new technique of induced vitrification to be used beneficially in the micropropagation scheme, for conifers at least. One may wonder whether vitrification might not be studied and managed further as a method of rejuvenation of woody plants.

References

Aitken J, Horgan KJ, Thorpe TA (1981) Influence of explant selection on the shoot-forming capacity of juvenile tissue of *Pinus radiata*. Can J For Res 11:112–117

Aitken-Christie J, Thorpe TA (1985) Clonal propagation: Gymnosperms. In: Vasil IK (ed) Cell culture and somatic cell genetics of plants. Vol. 1. Academic Press, New York, pp 82–95

Aitken-Christie J, Jones C, Bond S (1985) Wet and waxy shoots in *radiata* pine micropropagation. Acta Hortic 166:93–100

Beauchesne G (1981) Les milieux minéraux utilisés en culture in vitro et leur incidence sur l'apparition de boutures d'aspect pathologique. C R Acad Agric Paris 67:1389–1397

Borkowska B, Michalczuk L (1987) The physiological disorders of sour cherry cultures: necrosis and vitrification. Acta Hortic 212:235–237

Bornman CH, Vogelmann TC (1984) Effect of rigidity of gel medium on benzyladenine-induced adventitious bud formation and vitrification in *Picea abies*. Physiol Plant 61:505–512

Böttcher I, Göring H (1987) Die Vitrifikation der Pflanzen bei der in vitro – Kultur als Infiltration problem. Biol Rundsch 25:191–193

Böttcher I, Zoglauer K, Göring H (1988) Induction and reversion of vitrification of plants cultured in vitro. Physiol Plant 72:560–564

Boxus P, Pâques M (1987) Méthode et composition de lutte contre le phénomène de vitrification au cours de la micropropagation in vitro des plantes. Demande de brevet européen 878770054.1 22.04.87 Publication 0 247 018 European Patent Office

Boxus P, Druart P, Brasseur E (1978) Rapport d'Activités du Centre de Recherches agronomiques de Gembloux, CRA, Gembloux, pp 126–127

Brainerd KE, Fuchigami LH (1982) Stomatal functioning of in vitro and greenhouse apple leaves in darkness, mannitol, ABA and CO_2. J Exp Bot 33:388–392

Brainerd KE, Fuchigami LH, Kwiatkowski S, Clark CS (1981) Leaf anatomy and water stress of aseptically cultured Pixy plum grown under different environments. HortScience 16:173–175

Chavdarov I (1986) Micropropagation and cloning in vitro of decorative plants. Genet Breed 19:373–383

Crèvecoeur M, Kevers C, Gaspar T (1987) A comparative biochemical and cytological characterization of normal and habituated sugarbeet calli. Biol Plant 29:1–6

Daguin F, Letouze R (1985) Relations entre hypolignification et état vitreux chez *Salix babylonica* en culture in vitro. Rôle de la nutrition ammoniacale. Can J Bot 63:324–326

Daguin F, Letouze R (1986) Ammonium-induced vitrification in cultured tissues. Physiol Plant 66:94–98

Davis MJ, Baker R, Hanan JJ (1977) Clonal multiplication of carnation by micropropagation. J Am Soc Hortic Sci 102:48–53

Debergh PC (1983) Effects of agar brand and concentration on the tissue culture medium. Physiol Plant 59:270–276

Debergh PC, Maene LJ (1984) Preparation of tissue cultured plants for rooting and establishment in vivo. In: Novak FJ, Havel T, Dolezel K (eds) Proc Int Symp Plant Tissue Cell Cult Applic Crop Improv, Czechoslovakia Acad Sci, Prague, pp 487–495

Debergh PC, Harbaoui Y, Lemeur R (1981) Mass propagation of globe artichoke (*Cynara scolymus*): evaluation of different hypotheses to overcome vitrification with special reference to water potential. Physiol Plant 53:181–187

Dencso I (1987) Factors influencing vitrification of carnation and conifers. Acta Hortic 212:167–176

De Proft MP, van den Broek G, de Greef JA (1987) Involvement of ethylene on senescence and vitrification of in vitro cultured miniroses. Acta Hortic 212:217–222

Dhawan V, Bhojwani SS (1987) Hardening in vitro and morpho-physiological changes in the leaves during acclimatization of micropropagated plants of *Leucaena leucocephala* (Lam.) de Wit. Plant Sci 53:65–72

Druart P, Boxus P, Liard O, Delaite B (1981) La micropropagation du Merisier à partir de la culture de méristème. In: Proc IUFRO Sect S2 01 5. Int Workshop In Vitro Cultivation for Tree Species, Fontainebleau, France, pp 101–108

Druart Ph, Pâques M, Boxus PH (1984) Morphologie anormale semblable aux symptômes de vitrification rencontrés au cours de la micropropagation de *P. glaudulosa* Thurn. var. *sinensis*. Influence de l'AVG, de l'AIBA et d'$AgNO_3$. Arch Int Physiol Biochim 92(1):V22–V23

Fukuda H, Komamine A (1985) Cytodifferentiation. In: Vasil IK (ed) Cell Culture and somatic cell genetics of plants. Vol. 2. Academic Press, New York, pp 149–212

Gaspar TH (1986) Integrated relationships of biochemical and physiological peroxidase activities. In: Greppin H, Penel C, Gaspar TH (eds) Molecular and physiological aspects of plant peroxidases. Univ Geneva, Switzerland, pp 455–468

Gaspar TH, Kevers C (1985) Cobalt prevention of vitrification process in carnation. Plant Physiol 77 (Suppl):13

Gaspar TH, Kevers C, Debergh P, Maene L, Pâques M, Boxus PH (1987) Vitrification: morphological, physiological, and ecological aspects. In: Bonga JM, Durzan DJ (eds) Cell and tissue culture in forestry. Vol. 1. General principles and biotechnology. Martinus Nijhoff, Dordrecht, pp 152–166

Gaspar TH, Kevers C, Penel C, Crevecoeur M, Greppin H (1988) Biochemical characterization of normal and habituated sugarbeet calli. Relationship with anatomy, habituation and organogenesis. Postdamer Forschungen B57:20–30

Gaspar TH, Kevers C, Bouillenne H, Maziere Y, Barbe JP (1989) Ethylene production in relation to rose micropropagation through axillary budding. In: Clijsters H et al. (eds) Biochemical and physiological aspects of ethylene production in lower and higher plants. Kluwer Acad, Dordrecht, pp 303–312

Gersani M, Leshem B, Sachs T (1986) Impaired polarity in abnormal plant development. J Plant Physiol 123:91–95

Grout BWW, Aston MJ (1978) Modified leaf anatomy of cauliflower plantlets regenerated from meristem culture. Ann Bot 42:993–995

Hackett WP, Anderson JM (1967) Aseptic multiplication and maintenance of differential carnation shoot tissue derived from shoot apices. Proc Am Soc Hortic Sci 90:365–369

Hakkaart FA, Versluijs JM (1983) Some factors affecting glassiness in carnation meristem tip cultures. Neth J Plant Pathol 89:7–53

Hauzinska E (1974) L'organogenèse dans le tissu de cal de l'oeillet (*Dianthus caryophyllus* L.) dans les conditions de culture in vitro. In: Proc 19th Hortic Congr, Warsaw 1A:60

Hegedus P, Phan CT (1983) Actions de phénols sur les malformations observées chez les porte-greffes de pommiers M-26 et 0–3 cultivés in vitro. Rev Can Biol Exp 42:33–38

Hegedus P, Phan CT (1987) Activities of five enzymes of the phenolic metabolism on rooted and acclimated vitreous plants in relation with phenolic treatments. Acta Hortic 212:211–216

Heller R (1969) Biologie Végétale II. Nutrition et metabolisme. Masson et Cie, Paris, pp 415–424

Installe P, Cesar G, Hofinger M, Gaspar TH (1985) Evolution dans la composition en macroéléments des milieux de culture pour tissus végétaux. Meded Fac Landbouwwet Rijksuniv Gent 50/2a:317–325

John A (1986) Vitrification in Sitka spruce cultures. In: Withers L, Alderson PG (eds) Plant tissue culture and its agricultural applications. Butterworth, London, pp 167–174

John A, Pearson DJ (1985) Vitrification in Sitka spruce (*Picea sitchensis*) cultures. News Bull Br Plant Growth Regul Group 7:35

Jones OP (1976) Effects of phoridzin and phloroglucinol on apple shoots. Nature 262:393–393
Kevers C, Gaspar TH (1985a) Soluble, membrane and cell wall peroxidases, phenylalanine ammonia-lyase, and lignin changes in relation to vitrification of carnation tissues cultured in vitro. J Plant Physiol 118:41–48
Kevers C, Gaspar TH (1985b) Vitrification of carnation in vitro: changes in ethylene production, ACC level and capacity to convert ACC to ethylene. Plant Cell Tissue Organ Cult 4:215–223
Kevers C, Gaspar TH (1986) Vitrification of carnation in vitro: changes in water content, extracellular space, air volume, and ion levels. Physiol Vég 24:647–653
Kevers C, Coumans M, Coumans-Gilles MF, Gaspar TH (1984) Physiological and biochemical events leading to vitrification in plants cultured in vitro. Physiol Plant 61:69–74
Kevers C, Prat R, Gaspar TH (1987) Vitrification of carnation in vitro: changes in cell wall mechanical properties, cellulose and lignin content. Plant Growth Regul 5:59–66
Kevers C, Goldberg R, Chu-Ba J, Gaspar TH (1988) Composition of the walls of stem and leaves of vitrifying carnation. Biol Plant 30:219–223
Lamproye A, Kevers C, Gaspar TH (1986) Changes in peroxidase activities, phenol content, and ethylene production in callus-forming basal part of sugarbeet petiole. Bull Groupe Polyphénols (C R Journées Int d'Etude, Montpellier) 13:232–234
Lavee S, Messer G (1969) The effect of growth-regulating substances and light on olive callus growth in vitro. J Exp Bot 20:604–614
Leonhardt W, Kandeler R (1987) Ethylene accumulation in culture vessels, a reason for vitrification? Acta Hortic 212:223–229
Leshem B (1983a) Growth of carnation meristems in vitro: anatomical structure of abnormal plantlets and the effect of agar concentration in the medium on their formation. Ann Bot 52:413–415
Leshem B (1983b) The carnation succulent plantlet, a stable teratological growth. Ann Bot 52:873–876
Leshem B, Sachs T (1985) Vitrified *Dianthus* teratoma in vitro due to growth factor imbalance. Ann Bot 56:613–617
Leshem B, Shaley DP, Izhar S (1988) Cytokinin as an inducer of vitrification in melon. Ann Bot 61:255–260
Letouzé R, Daguin F (1983) Manifestation spontanée et aléatoire d'une croissance anormale en culture in vitro. Recherche de marqueurs métaboliques. Rev Can Biol Exp 42:23–28
Letouzé R, Daguin F (1987) Control of vitrification and hypolignification process in *Salix babylonica* cultured in vitro. Acta Hortic 212:185–191
Maene L (1985) Optimalisering van de commerciële vermeerdering van planten in vitro. Meded. Centr Studie Voortplanting bij Tuinbouwgewassen, IWONL, Gent, 47 p
Miedema P (1984) The effects of growth regulators on vitrification in shoot cultures of *Beta vulgaris*. Acta Bot Neerl 33:375
Monette PL (1983) Influence of size of culture vessel on in vitro proliferation of grape in a liquid medium. Plant Cell Tissue Organ Cult 2:327–332
Murashige T, Skoog F (1962) A revised medium for rapid growth and bio-assays with tobacco tissue cultures. Physiol Plant 15:473–497
Navatel JC (1982) Problèmes liés à la production de porte-greffe d'arbres fruitiers par multiplication in vitro. Fruits 37:331–336
Orlikowska T (1987) Vitrification problem in the in vitro culture of fruit tree rootstocks. Acta Hortic 212:239–244
Pâques M, Boxus P (1984) Comparative study of vitreous and nonvitreous plantlets of apple rootstock M-26 cultivated in vitro Abstr Book, 4th Congr Fed Eur Soc Plant Physiol, Strasbourg, pp 282–283
Pâques M, Boxus PH (1986) Utilization of a hydrolysed agar extract to prevent vitrification of the apple M26 rootstock, cultivated in vitro Arch Int Physiol Biochim 94:pp 33
Pâques M, Boxus PH (1987a) Vitrification: review of literature. Acta Hortic 212:155–166
Pâques M, Boxus PH (1987b) Vitrification: a phenomenon related to tissue water content? Acta Hortic 212:245–252
Pasqualetto PL, Zimmermann RH, Fordham I (1986) Gelling agent and growth regulator effects on shoot vitrification of Gala apple in vitro. J Am Soc Hortic Sci 111:976–980
Phan CT, Hegedus P (1986) Possible metabolic basis for the development anomaly observed in in vitro culture, called "vitreous plants". Plant Cell Tissue Organ Cult 6:83–94
Phan CT, Letouze R (1983) A comparative study of chlorophyll, phenolic and protein contents and of hydroxycinnamate: CoA ligase activity of normal and vitreous plants (*Prunus avium* L.) obtained in vitro. Plant Sci Lett 31:323–327

Phillips DJ, Matthews GJ (1964) Growth and development of carnation shoot tips in vitro. Bot Gaz 125:7–12
Quoirin M, Lepoivre P (1977) Etude de milieux adaptés aux cultures in vitro de *Prunus*. Acta Hortic 78:437–442
Riffaud JL, Cornu D (1981) Utilisation de la culture in vitro pour la multiplication de merisiers adultes (*Prunus avium* L.) sélectionnés en forêt. Agronomie 1:633–640
Roberts LW, Miller AR (1982) Ethylene and xylem differentiation. What's New in Plant Physiol 13:13–16
Rugini E, Verma DC (1983) Micropropagation of difficult to propagate almond (*Prunus amydalus*) cultivar. Plant Sci Lett 28:273–281
Rugini E, Tarini P, Rossodivita ME (1987) Control of shoot "vitrification" of almond and olive grown in vitro. Acta Hortic 212:177–183
Sutter EG (1985) Morphological, physical and the chemical characteristics of epicuticular wax on ornamental plants regenerated in vitro. Ann Bot 55:321–329
Sutter E, Langhans RW (1979) Epicuticular wax formation on carnation plantlets regenerated from shoot tip culture. J Am Soc Hortic Sci 104:493–496
Vieitez AM, Ballester A, San-Jose MC, Vieitez E (1985) Anatomical and chemical studies of vitrified shoots of chestnut regenerated in vitro. Physiol Plant 65:177–184
Vieitez AM, Ballester A, San-Jose MC, Vieitez E (1987) Vitrification in chestnut shoots regenerated in vitro. Acta Hortic 212:231–234
Vieth J, Morisset C, Lamond M (1983) Histologie de plantules vitreuses de *Pyrus malus* cv. M 26 et de *Pelargonium peltatum* cv. Chester Frank, issues de la culture in vitro (Etude préliminaire). Rev Can Biol Exp 42:29–32
Von Arnold S, Ericksson T (1984) Effect of agar concentration on growth and anatomy of adventitious shoots of *Picea abies* L. Plant Cell Tissue Organ Cult 3:257–264
Wardle K, Dobbs EB, Short KC (1983) In vitro acclimatization of aseptically cultured plantlets to humidity. J Am Soc Hortic Sci 108:386–389
Werker E, Leshem B (1987) Structural changes during vitrification of carnation plantlets. Ann Bot 59:377–385
Werner EM, Boe AA (1980) In vitro propagation of Malling-7 apple rootstocks. HortScience 15:509–510
Wilkins CP, Dodds JH (1983) Tissue culture propagation of temperate fruit trees. In: Dodds JH (ed) Tissue culture of trees. Croom Helm, London, pp 56–79
Ziv M (1986) In vitro hardening and acclimatization of tissue culture plants. In: Withers L, Alderson P (eds) Plant tissue culture and its agricultural applications. Butterworth, London, pp 187–196
Ziv M, Meir G, Halevy AH (1983) Factors influencing the production of hardened glaucous carnation plantlets in vitro. Plant Cell Tissue Organ Cult 2:55–65
Ziv M, Schwartz A, Fleminger D (1987) Malfunctioning stomata in vitreous leaves of carnation (*Dianthus caryophyllus*) plants propagated in vitro; implications for hardening. Plant Sci 52:127–134
Zuccherelli G (1979) Moltiplicazione in vitro dei portainnesti clonali del pesco. Frutticoltura 41:15–20

I.8 Acclimatization of Micropropagated Plants

T. Kozai[1]

1 Introduction

In the last 10 years, micropropagation has shown spectacular development. However, at present the widespread use of micropropagation is restricted for several reasons, one of these being that a high percentage of micropropagated plants can be lost or damaged in the acclimatization stage, that is, during transfer from a tissue culture vessel to soil. Due to the special environment in vitro, it is difficult to produce plants which can be acclimatized to the outside environment.

The term acclimatization is defined as the climatic adaptation of an organism, especially a plant, that has been moved to a new environment (Conover and Poole 1984). It takes place under the active guidance of human beings. The term acclimation has a similar meaning but it is a process of nature. This chapter uses the terms acclimatize, acclimatized, and acclimatization in preference to acclimation, conditioned, weaning, habituation, adaptation, etc.

Much research has been conducted to solve the problems related to acclimatization. Brainerd and Fuchigami (1981) examined responses of micropropagated plants to low relative humidity. Wardel et al. (1983) examined the in vitro acclimatization of micropropagated plants to humidity. Sutter and Hutzell (1984) used humidity tents and antitranspirants to increase the survival percentage of micropropagated plants in the acclimatization stage. Read and Fellman (1985) developed a "controlled environment rooting facility" for accelerating the acclimatization of micropropagated woody ornamentals.

In this chapter, features of tissue culture (in vitro) environment and acclimatization (ex vitro) environment are analyzed with special attention to physical environment. Strategies for increasing percentage survival and growth rate of micropropagated plantlets in the acclimatization stage are then summarized. Furthermore, a microcomputer-controlled environmental control system especially designed for acclimatization of micropropagated plants is described with some experimental results.

2 Features of Tissue Culture Environment

In addition to the presence of sugar in the culture medium, the tissue culture environment is different in many respects from the acclimatization environment or ex vitro

[1]Department of Horticultural Engineering, Faculty of Horticulture, Chiba University, Matsudo, Chiba 271, Japan

Biotechnology in Agriculture and Forestry, Vol. 17
High-Tech and Micropropagation I (ed. by Y.P.S. Bajaj)

environment. Table 1 summarizes the features of tissue culture environment in the multiplication and rooting stages and their effects on the growth of explants/plantlets in the multiplication, rooting, and acclimatization stages.

The features of the environment due to low air exchange rate of the vessel are: (1) high relative humidity (Ziv et al. 1983; Vanderscheaghe and Debergh 1987), (2) low CO_2 concentration during the photoperiod and high CO_2 concentration during the dark period (Fujiwara et al. 1987; Kozai et al. 1987; Kozai and Iwanami 1988), and

Table 1. Features of the tissue culture environment in the multiplication and rooting stages and their effects on the growth of explants/plantlets in the multiplication, rooting, and acclimatization stages

Features of the environment	Their effects in vitro	Their effects ex vitro
• Low air exchange-rate		
• High humidity	• No structured epicuticular wax	
	• Malfunction of the stomata low transpiration rate	• Excess transpiration
	• Difficulty in rooting	• Difficulty in water uptake
	• Difficulty in nutrient uptake	• Difficulty in nutrient uptake
• High ethylene concentration		
• High CO_2 during the dark period	• Promote/inhibit differentiation	
• Low CO_2 during the photoperiod	• Low net photosynthetic rate	
	• No or little leaf emergence	• Suppression of photosynthesis
• Low light intensity	• Difficulty in develop in autotrophy	
• Constant air temperatures	• High dark respiration	• Sensitive to environmental stress
• Low dissolved oxygen in the medium	• Difficulty in rooting	• Death of plants
• High inorganic ion concentrations	• High osmotic pressure in the medium, high ethylene production	• Growth retardation due to low net photosynthetic rate
• Sugar-containing medium	• Vitrification or other physiological disorders	
	• Bacterial or fungal contamination	
	• Loss of plantlets	
• Stagnation of the air and the medium composition	• Low transfer coefficients of gas, ion and other substances	
	• Large spatial variations of temp., gas and medium composition, light, etc.	
	• Large variations in size and developmental stage of the plantlets	
• Hormone-containing medium	• High mutation percentage	
• Vitamin-containing medium		

(3) high ethylene concentration (De Proft et al. 1985), in particular when the plantlets are vitrified.

Other features of the environment are: (1) low light intensity (low photosynthetic photon flux density), (2) constant temperatures throughout the day, (3) high inorganic ion concentrations, and hence a high osmotic pressure (low osmotic potential) in the medium (the osmotic pressure is also increased with increasing sugar concentration) (Kimball et al. 1975; Debergh et al. 1981; Kozai et al. 1986), (4) the stagnation of the air in the vessel and the stagnation of nutrients in the gel medium, (5) sugar-, hormone- (plant growth regulator), vitamin- and other organic substance-containing medium.

The responses of plantlets in the multiplication and rooting stages to high humidity are, in many cases, (1) no or ill-structured epicuticular wax (Sutter and Langhans 1982), (2) malfunction of the stomata (Ziv et al. 1987), and (3) a small amount of healthy secondary roots. The plantlets are, therefore, sensitive to environmental stress, in particular, to water stress, in the acclimatization stage.

The high relative humidity, in combination with the stagnation of the inside air and the nutrients in the gel medium, also causes the suppression of water and nutrient uptake by the plantlets in the multiplication and rooting stages, and hence may cause growth retardation, physiological disorders, and a variation in size and developmental stage of the plantlets to a certain extent. The stagnation of the air in the vessel may also cause insufficient gas (CO_2, H_2O, etc.) exchange between the air and the plantlets.

The responses of the plantlets in the multiplication and rooting stages to the combination of low light intensity, low CO_2 concentration during the photoperiod (Fujiwara et al. 1987; Kozai et al. 1988), and the presence of sugar in the medium are: (1) low net photosynthetic rate (Grout and Aston 1978; Grout and Millam 1985) (2) little or no leaf emergence, and hence (3) low growth rate of the plantlets.

Physiological causes of vitrification or glassiness are said to be related to the multiplicity of physical (e.g., gel solidity, osmotic pressure, etc.) and chemical (e.g., ammonium ion, cytokinins, ethylene, etc.) factors involved in the process (Paques and Boxus 1987). Vitrification causes a high percentage of plant losses in the acclimatization stage.

The features of the environment mentioned above are considered to be the main causes of the relatively low percentage of survival and relatively low growth rate of the plantlets in the acclimatization stage.

3 Heterotrophy, Mixotrophy, and Autotrophy

In the conventional micropropagation process, explants or plantlets in the differentiation, multiplication, and rooting stages grow in the presence of sugar in the culture medium. Plantlets in the acclimatization stage, on the other hand, grow in the absence of sugar in the substrate. In other words, the plantlets in the differentiation, multiplication, and rooting stages develop heterotrophy or mixotrophy. The plantlets in the acclimatization stage, on the other hand, are forced to develop autotrophy. The heterotrophic or mixotrophic plantlets need organic nutrients such as sugar as an energy and carbon source for their growth. The autotrophic plantlets, on the other hand, require only an inorganic energy source, primarily light energy, carbon dioxide, water, and minerals.

They are capable, through photosynthesis, of utilizing inorganic substances for producing organic compounds. Mixotrophy means partial heterotrophy or partial autotrophy.

This drastic change at the beginning of the acclimatization stage from heterotrophism or mixotrophism to autotrophism should always be kept in mind when trying to find rational strategies for acclimatization.

4 Increasing Percentage Survival and Growth Rate

In order to increase the percentage survival of the plantlets in the acclimatization stage, the environment at an early stage of acclimatization should be controlled to simulate the environmental conditions under which the plantlets were cultured in the multiplication and rooting stages.

In order to increase the growth rate and/or in order to avoid damage or death of the plantlets after acclimatization (when transplanted in the open or in the greenhouse), on the other hand, the environment at a later stage of the acclimatization should be controlled to simulate the environmental conditions under which the plantlets will be grown in the open or in the greenhouse. Furthermore, the growth and photosynthesis of the plantlets in the acclimatization stage should also be promoted with a minimum loss of the plantlets.

In conventional acclimatization, the main effort of environmental control in the acclimatization stage is to keep the relative humidity high, in particular, at an early stage of acclimatization. The high humidity is generally achieved by covering the plantlets with plastic film under shade, together with frequent mistings. The shading is necessary because firstly the strong solar light itself may directly damage the plantlets and secondly the fluctuating solar light intensity with time leads to the fluctuations of temperature and relative humidity, and hence an excess water loss from the plantlets.

Misting under shade is the easiest way of keeping the humidity high. However, it suppresses photosynthesis, and hence the development of autotrophy and rooting of the plantlets. This vicious circle of environmental control in the acclimatization stage is summarized in Table 2. The degree of shading and the strength of misting are then carefully gradually decreased with time by an expert greenhouse grower, as the plantlets develop autotrophy.

We have to find an efficient way of environmental control in the acclimatization stage, which enables the plantlets to grow vigorously during and after the acclimatization stage, with a nominal loss of the plantlets.

There seem to be two possible approaches to solve the problem. One is to control the acclimatization environment much more precisely, for example, by using a microcomputer-controlled environmental control system. The relative humidity at an early stage of the acclimatization, for example, is precisely controlled at a predetermined high level, with supplying enough light and CO_2 for photosynthesis at the same time.

The other is to control the tissue culture environment in the multiplication and rooting stages similarly to the acclimatization or greenhouse environment. There is then little drastic change in the environment when transferred from culture (in vitro) to the outside (ex vitro).

Table 2. Vicious circle of environmental control in conventional acclimatization

1.	The plantlets are sensitive to water stress and have not developed autotrophy at an early stage of acclimatization
2.	The relative humidity has to be kept high to reduce the damage or death of the plantlets
3.	Heavy shading with misting is necessary during the daytime to keep the humidity high under natural solar light
4.	The photosynthesis of the plantlets is suppressed under heavy shading conditions
5.	Suppression of photosynthesis may suppress the development of autotrophy of the plantlets
6.	Suppression of photosynthesis (autotrophism) may suppress the rooting and leaf emergence of the plantlets
7.	The water and nutrient uptake is suppressed due to the insufficient amount of secondary roots
8.	A small amount of water uptake and excess transpiration leads to damage or death of the plantlets
9.	Part of the leaves should then be removed when acclimatized to reduce the excess transpiration. The photosynthesis is suppressed by the small leaf area
10.	Back to No. 1

The latter approach is called autotrophic (sugarfree) micropropagation, a novel concept, the description of which is given in detail in Chapter II.5 (this Vol.) (Kozai 1991). An example of the former approach with some experimental results is introduced below.

5 Computer-Controlled Acclimatization Unit

5.1 System

A microcomputer-controlled acclimatization unit has been developed (Kozai et al. 1987) and improved (Hayashi and Kozai 1987) to give better environmental control, higher percentage of survival, and more rapid growth of the plantlets. With the acclimatization unit, air temperature, relative humidity, light intensity, CO_2 concentration, air flow rate, and temperature of the nutrient solution can be controlled.

Figure 1 shows the schematic diagram of the improved acclimatization unit. The unit consists of a growth room, an air conditioning room and actuators, and a microcomputer-based control system. The main part of the unit is about 1.5 m wide, 3 m long, and 2 m high. The growth room can accommodate more than 2000 plantlets. The unit has been developed primarily for research purposes, but it can be extended for commercial use without any change in the software system. Figure 2 shows the configuration of the control system. The microcomputer controls the acclimatization unit, and records and analyzes the data.

The environment inside the growth room is controlled on the basis of acclimatization curves. Figure 4 shows a schematic diagram of acclimatization curve for air temperature control. The cooling system starts to operate when the air temperature inside the growth room becomes higher than the upper limit, and stops when it reaches the set point. The heating system is activated when the temperature becomes lower than the

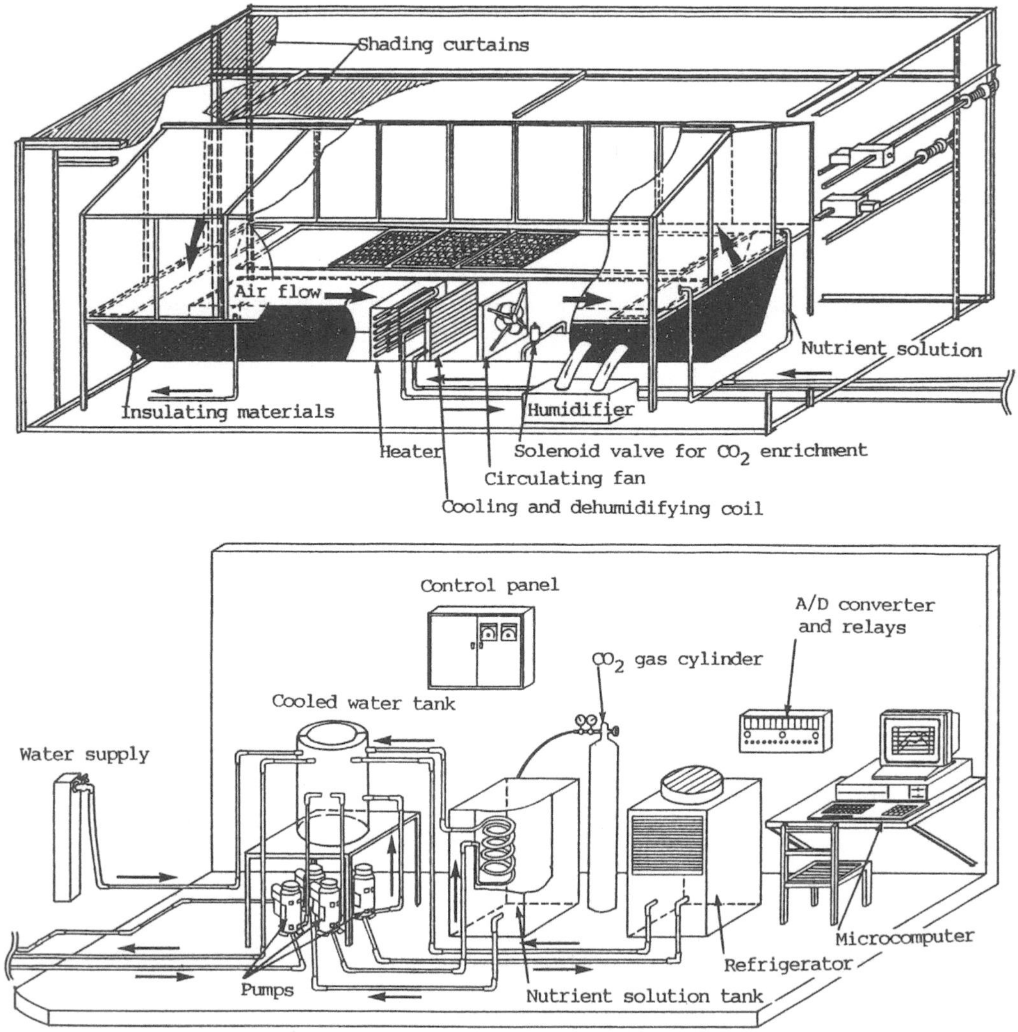

Fig. 1. Schematic diagram of the acclimatization unit. (Hayashi and Kozai 1987)

lower limit and stops when it reaches the set point. Therefore, the air temperature is controlled within the upper and lower limit temperatures shown in Fig. 3.

The set point of the air temperature on the first day of acclimatization is constant, simulating the conditions under which the plantlets were cultured in vitro. The set point of the air temperature on the final day of the acclimatization is similar to that of the greenhouse or the open field where the acclimatized plantlets will be transplanted. Using the acclimatization curve, the diurnal change in the air temperature can be magnified gradually day by day.

The acclimatization curves of relative humidity, light intensity, and temperature of the nutrient solution are also installed in the control system. The curve for each environmental factor can be modified, depending upon the crop and upon the season when the acclimatized plantlets are transplanted to the greenhouse or the open field.

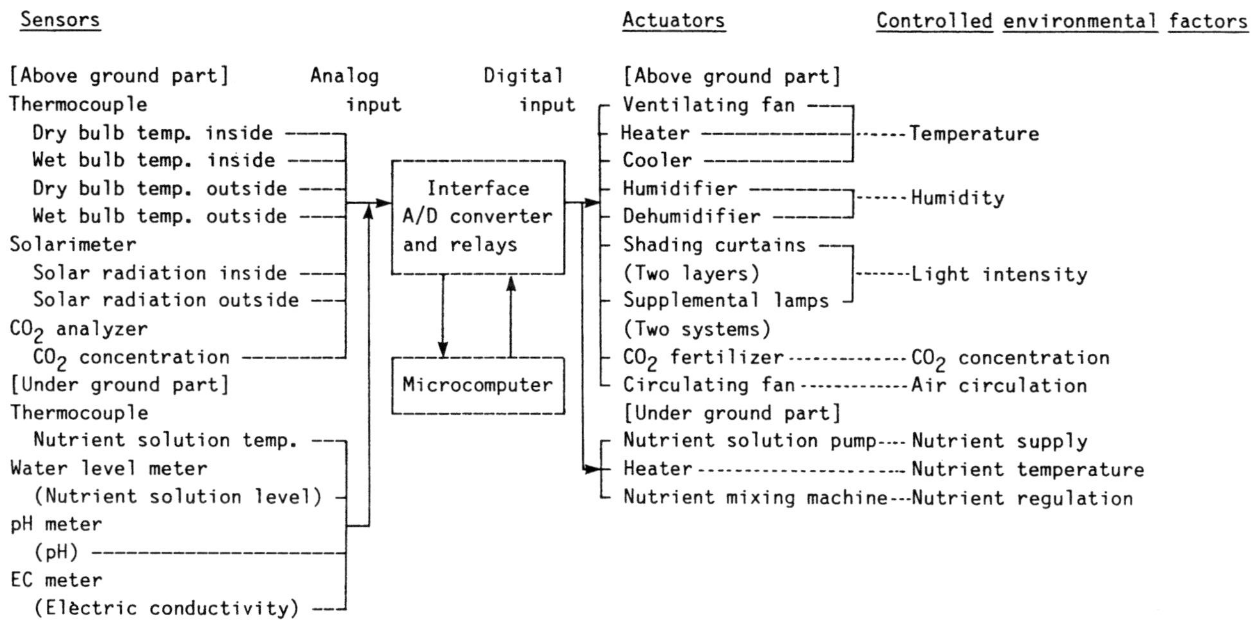

Fig. 2. The configuration of the control system of acclimatization unit. (Kozai et al. 1987)

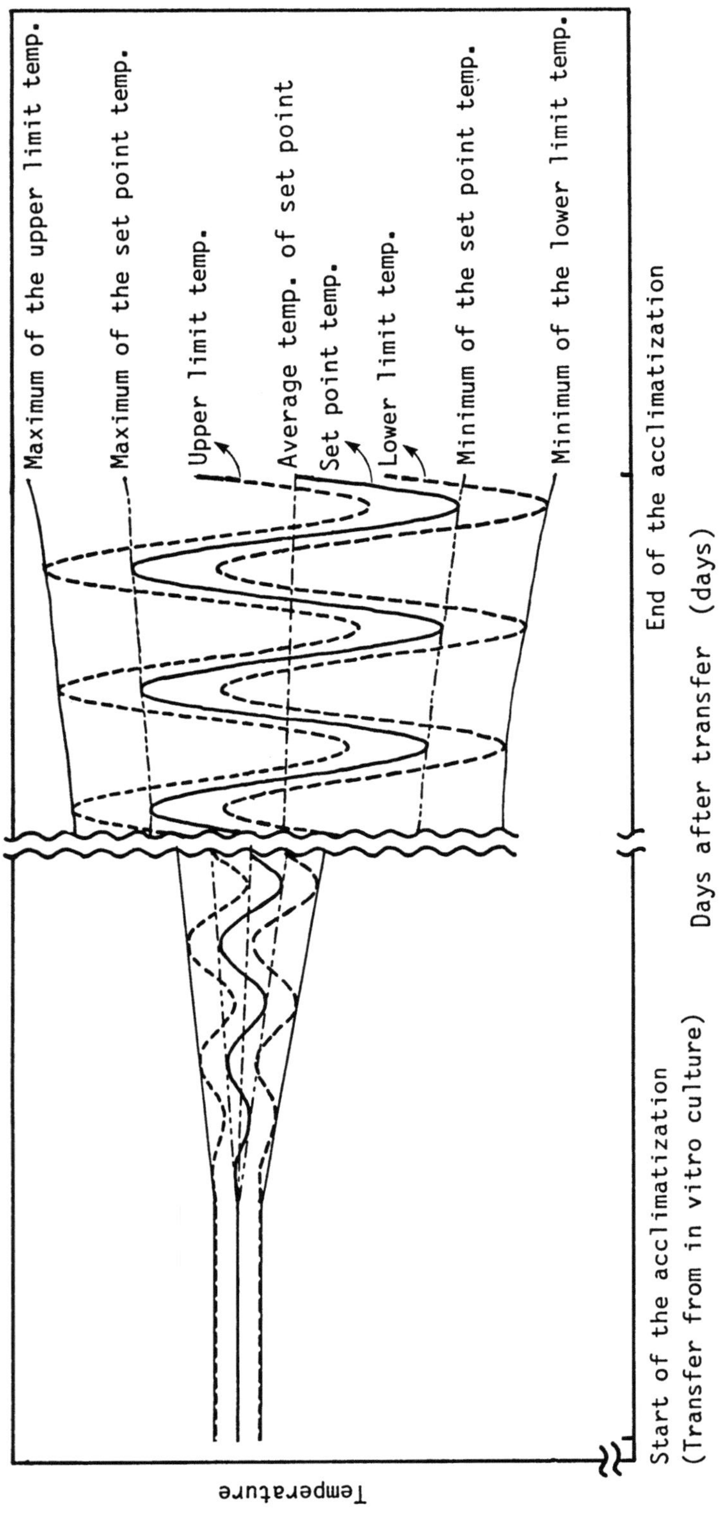

Fig. 3. Schematic diagram of the acclimatization curve for air temperature control. (Kozai et al. 1987)

5.2 Some Experimental Results

5.2.1 Vitrified Plantlets Cultured Submerged in a Liquid Medium

Using the acclimatization unit, a test acclimatization was conducted using strawberry (*Fragaria* × *ananassa* Duch.) plantlets (cv. Houkouwase) cultured submerged in a liquid medium with a bioreactor, most of which were vitrified. The percent survival of the plantlets acclimatized in the unit was 96%, while the percentage was 80% for those carefully acclimatized by an expert greenhouse grower in a greenhouse with the conventional (traditional) method. By 35 days after the start of the acclimatization, dry weight, plant height, and the number of unfolded leaves of the plantlets acclimatized with the unit were 2.8 times, 1.4 times, and 1.3 times those acclimatized in the greenhouse with the conventional method. The changes with time in fresh weight, dry weight, plant height, and the number of unfolded leaves per plantlet are shown in Fig. 4.

The test acclimatization was also conducted for taro (*Colocasia esculenta* Schott) cultured submerged in a liquid medium with the bioreactor. The percent survival of the plantlets acclimatized with the unit was 92% and the percentage for those acclimatized with the conventional method was 77%.

A significant difference in growth rate of the plantlets was also observed between the two methods (Kozai et al. 1987). It is noted that the fresh weight, dry weight, etc. decreased with time after transfer from in vitro to the outside, in particular, in the conventinal method.

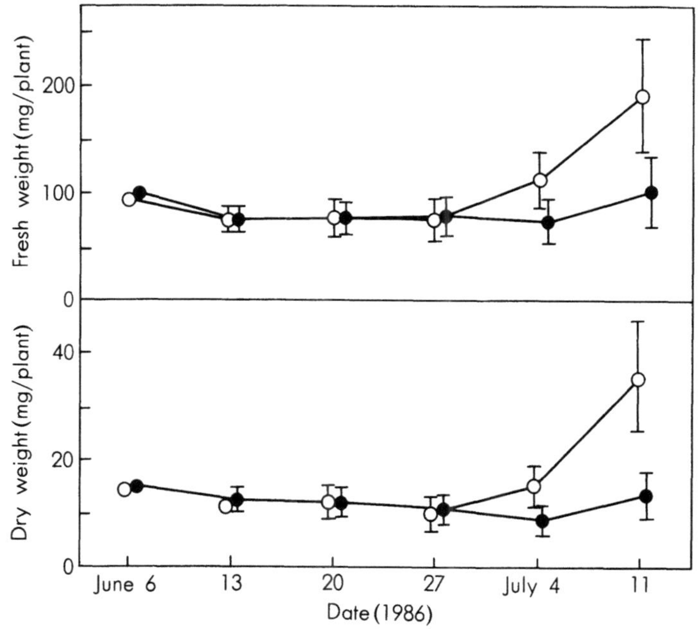

Fig. 4. Changes in fresh weight and dry weight of strawberry (*Fragaria* × *ananassa* Duch.) Houkouwase after transfer from in vitro culture and acclimatized in the acclimatization unit (○) and traditional method (●). *Vertical bars* indicate confidence interval at 5% level. (Kozai et al. 1987)

5.2.2 CO_2 Enrichment in the Acclimatization Stage

Micropropagated strawberries (*Fragaria* × *ananassa* Duch.) cv. Houkouwase cultured on the Murashige and Skoog (1962) medium were washed free of growth medium and transplanted in rockwool cubes (about 3 × 3 × 3 cm) in individual plastic pots and were grown by solution culture in the acclimatization unit. Two identical acclimatization units were used. The CO_2 concentration of one of the units was controlled at about 750 ppm when solar radiation was above 58 W/m^2, starting at the 8th day after transplanting. The CO_2 concentration of the other unit was not controlled.

Figure 5 shows the results of growth of the plantlets in the two units. During the first 14 days, there were no significant differences in the total and root dry weights, plant height, or the number of unfolded leaves per plantlet between the CO_2-enriched and CO_2-nonenriched treatments. On the 21st day, however, the total dry weight and plant height were greater in the CO_2-enriched treatment. On the 28th day, the total dry weight, plant height, leaf area, and root dry weight in the CO_2-enriched treatment were 1.5 times, 1.3 times, 1.6 times, and 2 times those in the CO_2-nonenriched treatment, respectively.

Lakso et al. (1986) showed that CO_2 enrichment was effective for stimulation of growth of in vitro-propagated grapevines after transfer from culture. Desjardins et al.

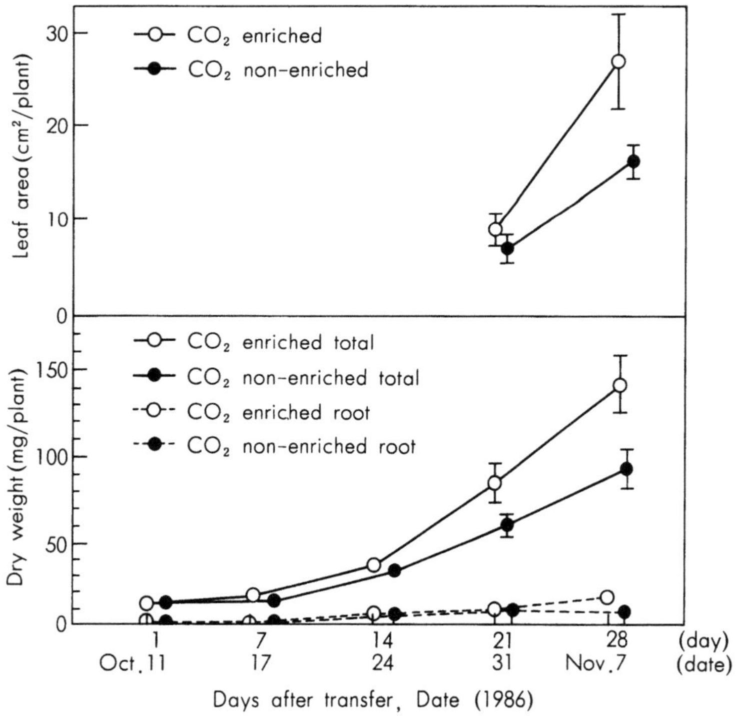

Fig. 5. Changes in leaf area and dry weight of strawberries (*Fragaria* × *ananassa* Duch.) Houkouwase after transfer from in vitro culture to the acclimatization unit under CO_2-enriched (○) and CO_2-nonenriched (●) conditions (Hayashi and Kozai 1987). *Vertical bars* indicate confidence interval at 5% level

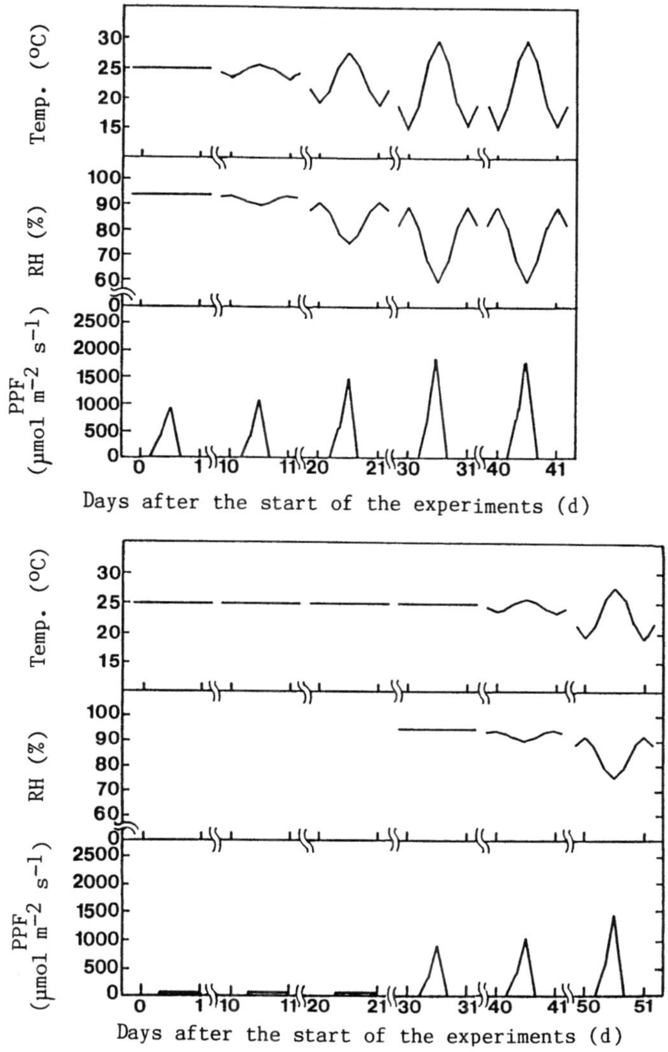

Fig. 6. Changes in daily setpoint patterns of the air temperature (*Temp.*), relative humidity (*RH*), and photosynthetic photon flux (*PPF*) in the acclimatization unit method (*above*) and in the conventional method (*below*). In the conventional method, the setpoint patterns from day 0 to day 30 are for the culture room and those from the 31st to 50th days are for the acclimatization unit. (Hayashi and Kozai 1987)

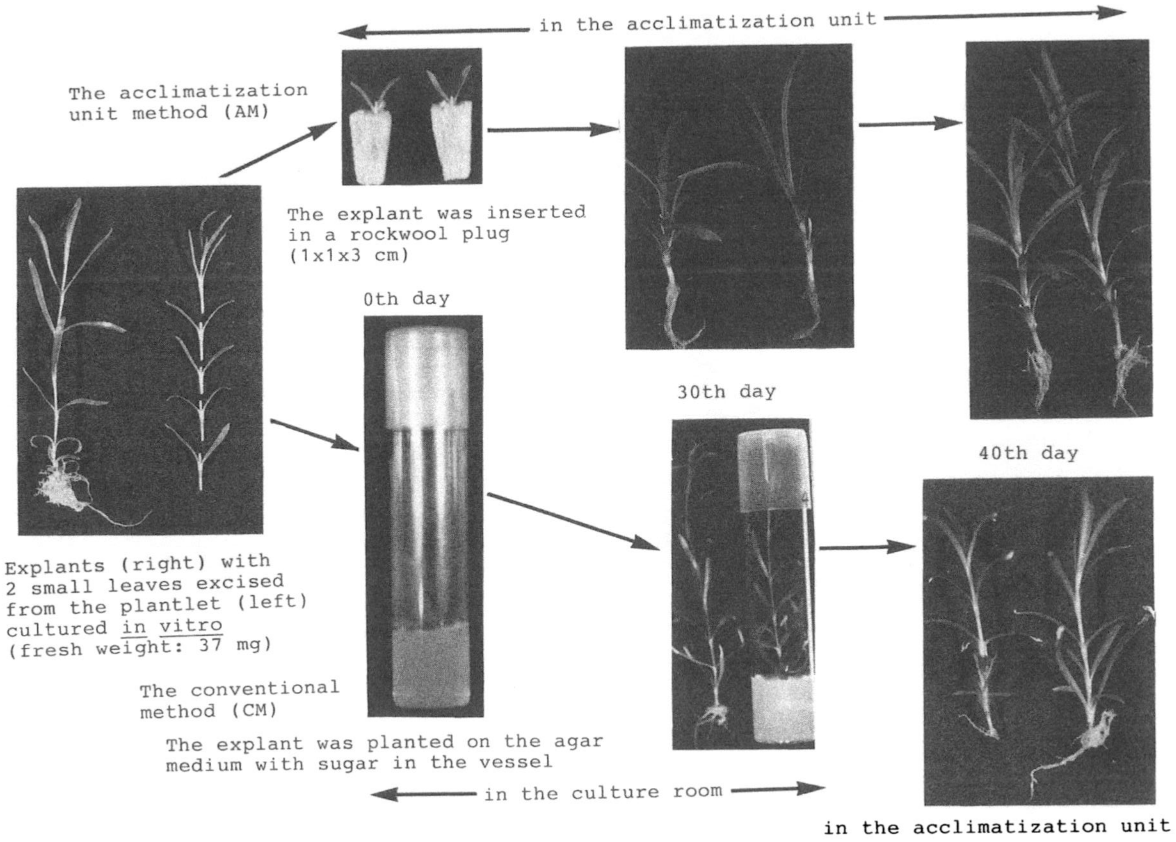

Fig. 7. Carnation explants/plantlets on days 0, 30, and 40 after treatments by the acclimatization unit method and by the conventional method. (Hayashi et al. 1988)

(1987) showed that supplementary lighting and CO_2 enrichment in the acclimatization stage led to a faster outcome of autotrophy, hastened growth, and ultimately improved the productivity of micropropagated strawberry, raspberry, and asparagus plantlets.

5.2.3 Application of the Acclimatization Unit for Growing Explants ex Vitro

Explants, single node cuttings each with two leaves and without roots, of a carnation (*Dianthus caryophyllus* L.), excised from the plantlets cultured in vitro, were inserted in rockwool plugs and were transferred directly to the acclimatization unit. The explants were then cultured with a diluted hydroponic solution (Hayashi et al. 1988).

The growth of the plantlets cultured with the acclimatization unit method was compared with the growth of the plantlets cultured with the conventional method. In the conventional method the explants were cultured and rooted in vitro on half strength MS agar medium containing 20 g/l sucrose for 30 days. The plantlets were then transferred to the acclimatization unit. Daily changes in setpoint patterns of the air temperature (Temp.), relative humidity (RH), and photosynthetic photon flux (PPF) in the acclimatization unit method and the conventional method are shown in Fig. 6.

The plantlets grew faster and were more vigorous with the acclimatization unit method than with the conventional method, as shown in Fig. 7. It can be said that better growth in the acclimatization unit method is due to the better environment for growth during rooting and acclimatization. That is, the rooting and growth of explants can be done ex vitro if the environment is properly controlled.

There are several advantages in the acclimatization unit method over the conventional method in such points such as (1) a shorter culture period, (2) no need for transplanting during rooting and acclimatization, (3) more vigorous plants, and (4) no contamination loss during rooting, etc.

5.2.4 Acclimatization Unit with Lamps

A plant growth chamber especially designed for acclimatizing micropropagated plants has recently been developed by a Japanese manufacturer. The concept and functions of the acclimatization chamber with lamps are similar to the acclimatization unit under solar light but the environment can be more precisely controlled with the acclimatization chamber with lamps.

6 Concluding Remarks

Acclimatization of micropropagated plants is an interesting research subject from both scientific (physiological) and practical (engineering) points of view. Much research has been conducted on acclimatization, most of which, however, was from physiological point of view.

Successful acclimatization can be achieved by careful environmental control, taking physiological changes of micropropagated plants into consideration. In this chapter, some problems related to the acclimatization have been discussed from the point of view of environmental control engineering. It is concluded that many problems on acclimatization remain unsolved and that there are many areas for further study.

References

Brainerd KE, Fuchigami LH (1981) Acclimatization of aseptically cultured apple plants to low relative humidity. J Am Soc Hortic Sci 106(4):515–518

Conover CA, Poole RT (1984) Acclimatization of indoor foliage plants. Hortic Rev 6:120–154

Debergh PS, Harbaoui Y, Lemeur R (1981) Mass propagation of globe artichoke (*Cynara scolymus*): evaluation of different hypotheses to overcome vitrification with special reference to water potential. Physiol Plant 53:181–187

Desjardins Y, Gosselin A, Lamarre M (1987) Utilisation de l'enrichissement carbone et de l'eclairage d'appoint lors de l'acclimatation ex vitro de fraisiers, de framboisiers et d'asperges. Symp Plant micropropagation in horticultural industries. Arlon, Belgium, pp 176–184

De Proft MP, Maene LJ, Debergh PC (1985) Carbon dioxide and ethylene evolution in the culture atmosphere of *Magnolia* cultured in vitro. Physiol Plant 65:375–379

Fujiwara K, Kozai T, Watanabe I (1987) Measurements of CO_2 concentration in stoppered vessels containing tissue cultured plantlets and estimates of net photosynthetic rates of the plantlets. J Agric Meteorol 43(1):21–30

Grout BWW, Aston MJ (1978) Transplanting of cauliflower plants regenerated from meristem culture. 2. Carbon dioxide fixation and the development of photosynthetic ability. Hortic Res 17:65–71

Grout BWW, Millam S (1985) Photosynthetic development of micropropagated strawberry plantlets following transplanting. Ann Bot 55:129–131

Hayashi M, Kozai T (1987) Development of a facility for accelerating the acclimatization of tissue-cultured plantlets and the performance of test cultivations. Symp Plant micropropagation in horticultural industries. Arlon, Belgium, pp 123–134

Hayashi M, Nakayama M, Kozai T (1988) An application of the acclimatization unit for growth of carnation explants, and for rooting and acclimatization of the plantlets. Acta Hortic 230:189–194

Kimball SL, Beversdorf WD, Bingham ET (1975) Influence of osmotic potential on the growth and development of soybean tissue cultures. Crop Sci 15:750–752

Kozai T, Iwanami Y (1988) Effects of CO_2 enrichment and sucrose concentration under high photon flux on the tissue cultured plantlet growth of carnation (*Dianthus caryophyllus* L.) during the preparation stage. J Jpn Soc Hortic Sci 57:255–264

Kozai T, Fujiwara K, Watanabe I (1986) Relation between the culture medium composition and water potential of liquid culture media. J Agric Meteorol 42:1–6 (in Japanese with English summary)

Kozai T, Iwanami Y, Fujiwara K (1987) Effects of CO_2 enrichment on the plantlet growth during the multiplication stage. Plant Tissue Cult Lett 4(1):22–26

Kozai T, Hayashi M, Hirosawa Y, Kodama T, Watanabe I (1988) Environmental control for acclimatization of in vitro cultured plantlet (1) Development of the acclimatization unit for accelerating the plantlet growth and the test cultivation. J Agric Meteorol 42(4):349–358 (in Japanese with English Summary)

Lakso AN, Reisch BI, Mortensen J, Roberts MH (1986) Carbon dioxide enrichment for stimulation of growth of in vitro-propagated grapevines after transfer from culture. J Am Soc Hortic Sci 111(4): 634–638

Lee N, Wetzstein HY, Sommer HE (1985) Effects of quantum flux density on photosynthesis and chloroplast ultra structure in tissue-cultured plantlets and seedlings of *Liquidambar styraciflua* L. towards improved acclimatization and field survival. Plant Physiol 78:637–641

Murashige T, Skoog F (1962) A revised medium for rapid growth and bioassays with tobacco tissue culture. Physiol Plant 15:473–497

Paques M, Boxus Ph (1987) "Vitrification": review of literature. Acta Hortic 230:155–166

Read PE, Fellman CD (1985) Accelerating acclimation of in vitro-propagated woody ornamentals. Acta Hortic 166:15–20

Sutter EG, Hutzell M (1984) Use of humidity tents and antitranspirants in the acclimatization of tissue-cultured plants to the greenhouse. Sci Hortic 23:303–312

Sutter E, Langhans RW (1982) Formation of epicuticular wax and its effect on water loss in cabbage plants regenerated from shoot-tip culture. Can J Bot 60:2986–2902

Vanderschaeghe AM, Debergh PC (1987) Technical aspects of the control of the relative humidity in tissue culture containers. Symp Plant micropropagation in horticultural industries. Arlon, Belgium, pp 68–76

Wardel K, Dobbs EB, Short KC (1983) In vitro acclimatization of aseptically cultured plantlets to humidity. J Am Soc Hortic Sci 108(3):386–389

Ziv M, Meir G, Halevy AH (1983) Factors influencing the production of hardened glaucous carnation plantlets in vitro. Plant Cell Tissue Organ Cult 2:55–65

Ziv M, Schwartz A, Fleminger D (1987) Hardening aspects of micropropagated carnation plants having malfunctioning stomata. Symp Plant micropropagation in horticultural industries. Arlon, Belgium, pp 47–54

I.9 The Effect of the Nature and Origin of Explants on Micropropagation

D. CACHIȚĂ-COSMA[1]

1 Introduction

In order to culture plant explants and to direct the growth and development processes under in vitro conditions, special problems must be solved, as the explants are removed from a complete system, the plant, of multiple interconnections, and structural and functional interrelations (interdependences) that exist within the living organism. Thus, during the in vitro culture phase, it is necessary to create a special ecophysiological frame which allows the cultured cells to take over vital functions as a new system, with a certain degree of autonomy. After being cultured in vitro, the explant constitutes a living unit which, on the one hand, is independent of the mother plant, containing all of its genetic information; and, on the other hand, it expresses all the functions that the explanted tissues or cells performed at their original location. The metabolism of in vitro cultured cells can be manipulated within the limits of their adaptability, which, in turn, depends on their heredity. In vitro techniques generally aim at conserving the genome unchanged, at obtaining copies that are genetically identical to the subject submitted to propagation, i.e., to cloning.

2 What is Meant by the Nature and Origin of Explants?

The success concerning the preservation of cell totipotency and the adaptation of explants to aseptic conditions depends on the *nature* and *origin* of explants, and on their physiological state following sterilization, explantation, and inoculation.

In order to analyze the *reactivity* or the behavior of the various types of explants grown in vitro – as a primary culture or a subculture (both underway for a long or a short time) – we must first define the terms "nature" and "origin" of the plant material. In fact, defining these terms is not a simple task, as they have different meanings, depending on the fields in which they are used. Briefly, these meanings may be systematized as follows:

Nature

— Innate features, natural dispositions, and disposabilities representing the result of these features, etc.

Origin

— Beginning, birth, source, starting point of a being or of a phenomenon
— Cause of things or of phenomena
— Affiliation through genesis to a family, species, cultivar

[1]Biological Research Centre, J.C.B. Str. Republicii no. 48, 3400-Cluj-Napoca, Romania

Biotechnology in Agriculture and Forestry, Vol. 17
High-Tech and Micropropagation I (ed. by Y.P.S. Bajaj)

- Laws of functioning (growth, development, expression of competence, of totipotency, etc.)
- Specific characters, particular properties or features that an individual or a system presents from the beginning, or acquires during its development
- Constitution (mode)
- Characteristic conformation
- The provenance (e.g., a cell, which either lives in vivo, or belongs to an in vitro culture; specifying the origin of an explant implies the specification of the species, cultivar, clone, organ and tissue to which it belongs, as well as specifying that it has derived from in vitro or in vivo conditions, the number of the subculture, etc.)

The nature and origin of explants, i.e., of inocula, constitute "biological filters" which translate, in a certain way, the exciting factors of the medium into living structures. Thus, the nature and origin of explants determine, sooner or later, the reaction of the subject (the organism) at its different levels of organization.

As mentioned previously, the nature and origin, based on the heredity of explants, imply countless aspects, each of them allowing analyses from different points of view and therefore various interpretations. Due to the lack of an explicit vocabulary adopted with scientific rigour by every researcher, and due to the diversity of the plant material used in experiments under varied experimental conditions, an agreement of the experimental data and a convergence of the different viewpoints are difficult to achieve. In this regard, we find valuable, for instance, the various explanations concerning the causes of vitrification in in vitro cultured explants.

Both the nature and the origin of explants, i.e., the phylogeny or the place (emplacement) from which the explants were removed, and regardless of whether they are (mature or immature) embryos or explants derived from plants grown under natural conditions or whether they represent microcuttings of newly formed in vitro plants or organs, or even calluses, cells, and protoplasts, may often play a major role in their in vitro development (evolution) on a certain type of medium and under specific environmental conditions. Detailed theoretical and dynamic studies regarding the kinetics of in vitro and in vivo morphophysiological and biochemical processes, using experimental models, must be undertaken. Once an experiment is completed, it should be repeated using explants of similar origins, taken from individuals of the same species.

The morphogenetic correlations that exist between the cells of a tissue or between organs, as well as the consequences of their presence or absence and the way this absence of correlations or signals can be replaced by environmental, physical, or chemical factors, should be studied comparatively when culturing certain types of explants in vitro, etc. Moreover, the unfolding and evolution of the morphogenetic (embryogenetic or organogenetic) processes should also be studied. The fact that in vitro cultures often use parts of a plant or organ, i.e., fragments removed from a complex organized system which are submitted to certain stress situations with which their cells have to cope due to *endoregulation*, and which then became independent, is forgotten. In such situations, morphogenesis supposes the occurrence of *new* autoregulation phenomena which depend on the conditions imposed from outside. Comparative studies in this direction and the creation of experimental models that allow us to obtain conclusive and repeatable (in terms of quality and quantity) data, under strictly determined conditions, would provide indispensable information to explain the mechanisms of cytodifferentiation and the pathways of in vitro histo- and morphogenesis. These studies would also reveal facts permitting a better understanding of the in vitro and in vivo or in situ organogenesis, embryogenesis, and ontogenesis. Undoubtedly, the technique of in vitro embryo and explant culture has been used for a long time. In order to compare the different data, it is therefore necessary to establish, in detail, the source of the plant material that has been used (in terms of species, organ, developmental stage, subculture, etc.) and the experimental or culture conditions. The intuition and fantasy of each researcher determine the ingeniousness of the experiments and the value of their results; results that may, or may not, solve the problems under study. We emphasize, for example, the fact that even now, the functional (metabolic and biochemical) complex and the molecular bases of the interactions between phytohormones and other morphophysiological systems are not entirely known. The current manipulations concerning the induction, initiation, promotion, and support of morphogenesis are theoretically and practically based on the concept of hormonal balance, especially on the auxin/cytokinin ratio of Skoog and Miller (1957).

Morphogenesis, regardless of whether it concerns embryogenesis or organogenesis, is a complex process that depends on an entire series of biological phenomena which occur before explanting. Until the moment the tissues are explanted, the cells of these future inocula have undergone – together with the system to which they belonged – certain steps of the vital cycle, steps during which the different physiological needs have been more or less satisfied. These needs include the ratio and the endogenous (accumulation) of phytohormones and nutrients, both depending primarily on the light and temperature of the environment, and on the mineral salt supply of tissues and cells, etc. The way in which these needs have been satisfied before explanting determines the reaction of explants, the regenerative and morphogenetic ability of inocula; therefore, the endogenous factor is important regarding whether or not certain reactions are obtained. Thus, the choice of the explant constitutes a crucial factor for its in vitro evolution, i.e., for the morphogenetic processes.

An explanted fragment behaves, on the whole, like a microcutting, which differs from traditional cutting in that the latter is richer in endogenous reserves, and has buds and even leaves which either contain, or synthesize, certain hormones. Therefore, the explant depends entirely on the culture medium. Yet, the size of the explant, the age of the mother plant, the physiological state of the plant and of its organs and cells are all factors that influence the evolution of the in vitro cultured explant.

In cormophytes, only a small part of the cells in the adult organism preserves their embryonic morphophysiological characters and their capacity to divide. As known during the initial growth phases of the embryo, all its cells which are of the "formative" type, divide. As the embryo grows, the areas that present this embryonic character move away and usually remain localized at the extremities of the axis, i.e., at the two poles of the plant, at the tip of roots and shoots. The tissues that have cells which preserve their property to divide, i.e., to give birth to new cells throughout their lifetime, are called meristems.

The growth of most cells is not followed by multiplication, but by a morphofunctional specialization which takes place step by step, passing through stages of structural and functional improvements. These improvements create the biological frame that the plant needs in order to traverse the following developmental stages. The moment the whole plant begins a functional decline, its meristems gradually slow down or stop their activity. The secondary meristems, for example the cambium of the woody plants in temperate zones, arrest their activity during the cold season.

Vegetative multiplication depends on the presence or on de novo formation of meristems in the organs or fragments that are used as initial material. All the phenomena connected with vegetative multiplication are closely dependent on the regenerative and morphogenetic, or on the organogenetic processes. Organogenesis is controlled by hormones and the different phases of the life cycle are characterized by certain organogenetic stages. The passing to an immediately following phenophase is conditioned by progressive accumulations, by the fullfillment of all the morphophysiological phases that constitute the structural and functional basis for the next stages of the biological cycle of the studied organism.

In in vitro cultured inoculi, the regenerative processes play a major role, especially when plant micropropagation is the aim.

Regeneration is the ability of organisms to regenerate any of their parts that has been destroyed or is absent from the very beginning. This is a way of exhibiting the self-preservation property. It was noted that wounding unleashes a complex of normal factors which are still little known in terms of their role and importance for starting regenerative processes. By means of the techniques used in the in vitro culture of plant organs, tissues, and cells, the regenerative ability of the explanted cells is exploited fully. In other words, these techniques use the complete expression of cell *totipotency*.

Theoretically, all the living cells of plants are totipotent; diploid plants have been obtained from somatic cells and haploid plants have been regenerated through androgenesis and gynogenesis starting from generative cells. The regenerative ability may, or may not, be expressed by single cells or by groups of cells, which belong to a certain system (organ, tissue, explant, callus, etc.), depending on many endogenous and exogenous factors. The regenerative aptitude differs from cell to cell, from tissue to tissue, from organ to organ, from species to species and even within the same species from cultivar to cultivar. The age of cells, their degree of differentiation, and their capacity to dedifferentiate also play an essential role in the evolution of inocula. The endogenous hormonal content, its nature, as well as the enzyme content represent factors that regulate the liberation of some metabolic pathways implied in the unblocking and functioning of processes affecting regeneration.

In in vitro cultures, the dedifferentiation and the success in turning the morphofunctional, differentiated cells back to the state of meristematic cells constitute the basic conditions for the beginning of regenerative processes that present the potential of finally recovering a new plant. The phases of cell dedifferentiation are complex. The aptitude of cells belonging to in vitro cultured explants to dedifferentiate is very unevenly distributed within various species, organs, and tissues. Therefore, in order to organize a new way of living, the structurally and functionally differentiated cells of explants cultured on aseptic media must be converted functional differentiated cells into meristematic ones which are capable of multiplication. This transformation implies biochemical and morphostructural changes at the infrastructural level. These changes are similar to physiological ones, i.e., the settlement of the entire functional assembly which is characteristic for the state of the meristematic cell. All these gradual changes at the metabolic and ultrastructural level determine the qualitative transformation of cells. The dedifferentiation process starts when certain cells receive an inductive impulse. In certain tissues, but not all, only a small number of cells are able to dedifferentiate and become the centers that generate new cells, thus making restitution phenomena possible. The wound factor plays an important part, not only because pain causes the wounded organ or tissue to recover its functions, but due to the chemical compounds generated on this occasion as well. It seems that the wounded organs and tissues accumulate certain substances which play an essential role in the induction and stimulation of the dedifferentiation processes and of cell multiplication.

When the meristematic cells, newly formed at the level of different types of explants or inocula, present bipolar growth, the regeneration of somatic embryos takes place, and the meristem is said to be of the embryonic type.

The phenomenon by which a group of cells exerts control over the differentiation and function of another group of cells is called *induction*. Therefore, between neighboring cells there are correlations that provide the basis for tissue formation, i.e., for *histodifferentiation*, while the interrelations between tissues allow their organization in functional formations, i.e., *organogenesis*.

According to Chlyah (1984) there are at least three categories of correlations that can be experimentally demonstrated: organ-cell, tissue-cell, and cell-cell correlations.

Meins, cited by Murashige (1984), considered that a newly initiated tobacco callus contained a meristemoid in 1000 up to 10 000 cells. Moreover, different species have a various number of cells which are capable of turning into meristemoids whose distribution on the surface of plants or explants is not uniform. This difference determines the various reactions of homologous explants removed from different, and even from the same species. Generally, the explants obtained from cotyledons, hypocotyls, inflorescences, roots, stems, leaves, depositing organs, nucellus, and anthers are considered to have a good regenerative capacity. Nevertheless, initial inocula consist most frequently of stem apices or meristems. It is considered that the density of the meristemoid-producing centers decreases with aging, or that the regenerative competence of these centers, already existing within a certain organ, is gradually repressed in the course of time. In maize and sorghum, plantlets can be easily obtained if callus is first generated from immature zygotic embryos. Regeneration takes place mainly in their scutellum.

The number of meristemoids per callus can be increased by periodic selection during its in vitro culture, thus avoiding the decline of the regenerative ability of a continuous culture. No indications regarding objective selection criteria can be given. Generally, the methods are based on the operator's professional intuition to distinguish the color, consistency, friability, etc. of the inoculum, and the morphogenetic competence of its areas. For example, Nabors (1982) considered the "green areas" on the surface of the *Avena sativa* callus to be the markers for its regenerative ability. The expression of totipotency may be the consequence of the transformation of non-meristems into meristemoids under the direct incidence of the conditions under which the cultured cells have been grown, before or during the regenerative processes.

Pretreating the mother plant with growth substances may enhance the regenerative ability of the explants excised from certain organs. The reaction of explants to the culture medium should also be regarded in connection with the relations between the two of them, especially concerning the phytotoxicity of some products that appeared due to autoclaving (for instance, sucrose turns into caramel), or of the substances (e.g., phenols) leached out into the medium by the cultured cells.

According to Vasil (1984a), the in vitro regeneration follows at least three main directions: (1) growth of the shoot meristems already present in the explants; (2) de novo meristem formation; and (3) somatic embryo formation. The first two ways of regeneration allow a "limited" clonal multiplication, while the embryogenic culture represents an unlimited source of clonal multiplication. The degree to which the explants, obtained from different parts of the plant, express their regenerative ability depends on the species, i.e., on its hereditary background. The inoculum should be larger than a certain "critical weight", which is taken into account when the composition of the culture medium is fixed.

Some other points of view concern the genetic predeterminism of the conformation of the different types of meristems. However, according to Margara (1982), the transition from an apex of the vegetative type into one of the floral or inflorescence type, and vice versa, is possible. Nonetheless, flower or inflorescence meristems derive from vegetative ones by some morphophysiological changes. One change consists of an increase in the size of the meristem simultaneously with the special arrangement of the cell layers making up the primordium. Therefore, the prefloral phase, i.e., the time that precedes the development of the reproductive organs, is characterized by structural, physiological, and biochemical changes. Inflorescences contain not only meristems of the vegetative type, but floral meristems as well. The photoperiod plays a decisive part in cell differentiation during the formation of floral meristems.

Young inflorescences and floral meristems show a high regenerative capacity, while also showing an aptitude for becoming vegetative ones. Shoot meristems, regardless of whether apical or axillary, are protected inside the buds. After they are formed, buds generally undergo a period of rest, of functional latency, i.e., they are dormant. Usually, this phenomenon also occurs in depositing organs, lignified branches and zygotic and somatic embryos.

Apical meristems control the development of axillary ones by inhibiting their growth, i.e., they exert apical dominance. The excision of the apical bud stimulates the growth of the axillary meristems, i.e., shoot formation. As long as the apical bud is growing, the development of the lateral buds is inhibited. Horticulturists know this phenomenon well and take it into account when carrying out correction cuttings. Moreover, as the length of the stem increases, the buds at the stem basis escape the influence of apical dominance. In many cases, floral induction is accompanied by the more or less complete suppression of apical dominance that stimulates the development of the vegetative buds on the floriferous stem. The buds exert correlative inhibition if they are located on senescent organs or on organs that have surpassed the mature phase. The growth of meristems and buds presents seasonal variations; their activity diminishes and may even stop during dormancy. The moment when this period begins depends on the species, the season, developmental stage of the organs and hormone ratio that exists within the tissues.

The problem of the functional autonomy of the apical or axillary shoot meristem is still unsolved. It is possible that the meristem, once it is separated from the plant and cultured in vitro, behaves similarly with the meristem that is integrated in the mother plant. The cell layers beneath the meristematic dome might also exert some influence upon the meristem. It also seems that certain "markers" – of the same type with the elements that precede conducting tissue formation which, in turn, precedes primordium formation – participate in stimulating the formation of primordia.

The multiplication, growth, and arrangement of the cell layers in the shoot meristematic dome vary from species to species, from one individual to another, and even within the same plant and branch, depending on the position of the meristem in the crown. The apical bud that originates in the embryo axis, often behaves differently with respect to the terminal meristems of lateral branches, or to the meristems of axillary buds. On the other hand, the cellular organization of the vegetative tips of young shoots is sometimes different from that of the vegetative tip of a mature branch.

Thus, in woody plants, the greater the distance between these meristems and the roots, the less water, nutrients, and phytohormones reach the meristems. Consequently, meristematic cells present a decrease in their regenerative ability due to an "aging" process. When cultured on aseptic media, this type of meristem often becomes necrotic. In order to stimulate the regenerative ability of terminal and axillary meristems, the administration of some nutritional or hormonal treatments to the plants from which the meristems are excised, is often recommended. These treatments aim to stimulate the regenerative power of those meristematic cells that have suffered a functional decline (senescence). Aging consists of a gradual loss of viability and regular metabolism and represents an "erosion", a natural expression of the slow progressive changes as plants grow old. One cause of aging is represented by the alterations that occur in the meristematic cells during the vital cycle of the plant. Such changes have been detected, for example, in the case of grafting between partners of different ages. An old piece grafted upon a young mother plant causes slower growth of the latter and a decrease in the growth rate of the whole plant. The "aging" of the meristematic cells can also be induced, for instance, by means of thermal stress. As a result of this stress, the damage caused to the meristem persists even if it is grafted onto a young mother plant. The newly formed plant presents aging phenomena that resemble the damage observed in the meristematic cells. Therefore, although the changes that occur in the meristems have different causes, their expression is the same in all meristematic cells. Another kind of aging is caused by the changes that occur during traditional vegetative multiplication. These changes cannot be eliminated by rejuvenescent regeneration because they affect the structure and organization of the living matter at the molecular level, as the *cells accumulate genetic errors*. These errors are brought about, for instance, by viral infections of the plant, infections that are transmitted from one individual to another by using cuttings and grafts. Moreover, every cycle of cloning – that uses traditional methods – enhances thenegative effects of viruses which affect more and more the descendants. The techniques of meristem culture on aseptic media, whereby the meristems are excised from plants having many viruses, allows virus elimination. Moreover, this type of plant regeneration not only saves the clone, but recovers the juvenility of the plant material obtained in vitro, thus increasing its quality.

Senescence often does not affect the whole plant, but rather some of its parts or areas. This phenomenon is frequently caused by thermal or light stress, or – in some annual herbaceous plants – by drought and intense heat. Senescence may initiate abrupt changes, thus leading to rapid deterioration of the affected organs. Senescence is induced by the lack of an optimal nutrient supply or of certain chemical elements, by excess ions in the culture medium, by the presence of parasites, or by prolonged darkness. A high CO_2 content of the atmosphere and low, positive temperatures delay the occurrence and progress of senescence is flowers and fruits. It was also noted that auxins, gibberellins and, above all, cytokinins delayed the senescence of organs under certain conditions (depending on the species, the nature of the organ, the time at which these substances were applied and their concentrations), while ethylene and abscisic acid accelerated this process. Free radicals seem to stimulate cell senescence, because they increase cell sensitivity to ethylene.

Nevertheless, it is sometimes possible to recover the juvenescence of plants. This depends on the nature of the biological material, on the factors that have led to the loss of its viability, and on the seriousness and type of disturbances that have occurred. Some methods, such as the cuttings of fruit trees or vines, are currently used in horticulture, as they restore the plants to a younger, more vigorous and productive form. In other cases, e.g., beans, the removal of the plant apex (i.e., the growth tip of the main stem) determines the senescence of old leaves and causes their recovery to the initial state of those leaves that have become partially senescent. The terms "nature" and "origin" imply many aspects. For instance, *origin* may denote (1) the phylogenetic affiliation of the explant, i.e., its genetic information (that may, or may not, be expressed under in vitro conditions); (2) its initial location (emplacement) on the plant or organ; (3) the type of culture from which the inoculum derives; or (4) the type of culture to which it is destined. Regarding the *nature* of explants, it may imply the (morphological) structure, i.e., the constitution and conformation of explants, their consistency and physiological state, or their degree of ontogenetic development. The indications and interpretations regarding different culture conditions for various explants often depend on how different authors understand the notions of the nature and origin of inocula.

Many papers emphasize the fact that the morphogenetic ability of explants depends on their nature or genotype. The decisive, crucial factor for preserving the morphogenetic ability of the inoculum (i.e., of the explant) is its developmental state (Vasil 1984a). The developmental stage determines the regenerative ability and the reaction of the explant: it is believed that a good morphogenetic response can be obtained only if the explant is in a "certain" developmental stage. When explants of an earlier or later stage of development – with respect to the stage that had been proved to be the best for the expression of morphogenetic ability – were used, a nonmorphogenetic callus was often obtained. Nevertheless, in vitro morphogenesis is also greatly influenced by (1) the endogenous phytohormonal content; (2) the exogenous growth regulator supply (these growth regulators either stimulate or inhibit this process); (3) the mineral salt content of the culture medium, e.g., the nitrogen source, especially in somatic embryogenesis; (4) the nature and concentration of the sugar in the culture medium; (5) the presence of other organic compounds, mainly amino acids; (6) the presence, nature, and concentration of some gases (O_2, CO_2, C_2H_4 = ethylene) that are dissolved in the substratum or make up the atmosphere in the culture vessels; (7) the presence of some complex biological extracts or of charcoal in the culture medium; and (8) by physical factors: light, temperature, the osmotic potential of the medium, its physical state, and the type of culture vessels. However, it is not yet known if – and how much – morphogenesis is influenced by the substances that have been introduced in the culture medium in the first place, or by the ionic complex that occurs in the substratum after autoclaving.

There is also another problem, concerning the similarity of the morphogenetic pathways under normal, natural conditions, with those that occur under in vitro conditions.

The nature and origin of explants, the number of previous subcultures, the type of media used, or the incubation conditions, influence, often decisively, the preservation, arrest, or even loss of organogenic or embryogenic competence. While studying the in vitro evolution of explants, or their reaction to the transfer ex vitro, the nature and origin of explants – as reference points – should be analyzed from all points of view, as these experiments are of the multifactorial type. Moreover, we must keep in mind that we work with living material, the reaction of which depends not only on the genetic information of its cells, but on the complex organization of the living system, as well as on various factors that may determine changes in the reaction of the genetic material or even in the variability. Because of these factors, the explants may present a special or paradoxical behavior. The unraveling of the mechanisms of heredity and the factors that are responsible for preservation or change – in a way that can be determined, controlled, and reproduced – in the evolution and development of explants, constitutes a permanent, crucial problem. Research using model plants, tobacco, carrot, rice, etc., allows us to formulate general rules, mainly concerning the reaction (behavior) of explants, i.e., the way certain types of explants develop and evolve in vitro. When we say, types of explants, we mean explants with different phylogenetic origins (species, variety, genotype), provenances (i.e., source or location on the plant), constitutions (morphoanatomical conformation), and features (physiological abilities) that could be, in turn, extended to other species.

The nature and origin of explants should express a certain quality. When we refer to a mother plant whose phylogenetic origin is known, the quality of the explant (or of the inoculum) means a certain competence of reaction which is optimal when: (1) the explant contains a high percentage of meristematic cells or of cells that present the ability of dedifferentiate and generate meristems; (2) the explant contains as many totipotent cells as possible, not only from the regenerative point of view (i.e., cells that generate callus), but from the embryo- and organogenetic one, as well; and (3) the explant has the aptitude to respond in the desired way, depending on the culture conditions and on the aim of the experiment. In other words, our aim represents our "guide" in choosing the explant.

The quality of the explant determines the way in which we promote and maintain its competence to react in a particular, known way or in a desired way, to be obtained by means of different culture conditions. The best conditions for the incubation and growth of the culture should be selected, taking into account that these conditions are specific to the nature of the explants.

The operator directs the culture used for *micropropagation* according to his will, by choosing the type of explant (or inoculum) and the methods that allow the multiplication (cloning) of the subject, to give countless copies without altering their genetic information, i.e., their genotype and phenotype. The chosen type of explant should present great efficiency under the known environmental conditions. While in the micropropagation of different species or genera, various methods and explants (or inocula) of different natures and origins, or even calluses, can be used, cloning imposes the selection of a certain type of explant, represented by fragments having an organized structure, such as apical or axillary shoot meristems (and not by callus). These fragments behave like microcuttings and the incidence of variability, in such cases, is not greater than that occurring during vegetative multiplication using traditional methods. Another method of efficient cloning uses somatic embryos. The use of excess growth regulators (e.g., cytokinins) in the culture medium, or the treatment of the mother plant (or the cultures from which propagules, microcuttings, or cells are obtained) with such substances leads to abundant bud formation per explant and sometimes allows a good micropropagation, but causes mutations. When variability occurs, the industries based on the cloning of plants using micropropagation techniques are damaged. Flower buds (either in one piece or sectioned), flower peduncles, sectioned inflorescences (flower bud or anthodium), etc. can be used as initial explants for cloning, as they present rapid multiplication. Sometimes, cloning may pass through a callus stage, but only in some species or in certain cultivars.

In vitro micropropagation has a double aim: the regeneration of plants that are free of viruses and their propagation in order to obtain crops free of phytopathogens, thus preserving the features of the subject whose origin in terms of health and genetic characters is known. This double aim of in vitro multiplication supposes that the culture is initiated *only* from the meristematic dome with the utmost two leaf primordia, 0.25–0.5 mm in size (depending on the species and cultivar). According to Quak (1977), this type of explant is – in terms of nature, maximal size, and constitution – the only one that is suitable for obtaining virus-free plants; the more leaf primordia excised together with the meristematic dome, the greater the risk of viral infections. During thermotherapy of the mother plant, the viruses cannot migrate from their host cells into the newly formed ones, thus giving us the opportunity to excise explants consisting of apices of about 1 mm and containing the apical dome and some leaf primordia. Generally, in order to best use the qualities of explants, the mother plants should be studied early in their so-called O stage, before the explant is excised. If the subject we want to multiply in vitro presents certain deficiencies or a weak physiological state, the situation can be improved by applying certain treatments to the plants and organs destined to provide the explants for initiating an aseptic culture, before excising them.

Micromultiplication techniques may use explants of various natures and origins to initiate the first micropropagation cycle. Once the aseptic culture is started, the efficiency of micropropagation can be increased using various in vitro methods: (1) one meristem (apex) may regenerate either a new plantlet, or many shoots that can be divided into microcuttings. The microcuttings that have at least one node and one axillary bud are the best, due to their rapid growth and their aptitude to regenerate new shoots; meristems, or other types of explants, can be used for the regeneration of numerous buds; the presence of many buds per inoculum and the possibility of using microcuttings make a subculture necessary. (2) In *Gerbera, Saintpaulia, Exacum,* and *Drosera*, one inoculum, regardless of its kinds, may often develop in vitro masses (bushes) of plantlets that can be separated into propagules which are then subcultured. A propagule represents a unit made up of a basal part with two to three leaves in its upper region, which represents the starting point for a new colony. (3) In many orchids (*Cymbidium* or *Cathleya*), the meristem gives rise to a protocorm which is a morphoanatomical and physiological unit with high autonomy, i.e., it is a green microtuber that can form new protocorms through budding or after it is cut into pieces. In protocorms, organogenetic processes can also be induced and thus whole plants can be regenerated. (4) The explants represented by segments of bulb scales or leaves of bulbiferous plants (e.g., *Hyacinthus*) may form microbulbs that generate plantlets once the dormancy period has been successfully surpassed. (5) In potato, the shoots generated from meristems or other types of explants may, incidentally or not (i.e., in the presence of CCC or coumarin, plus 80 g/l sucrose and in the dark) form microtubers. These tubers may be used in the in vitro multiplication of potatoes, with or without cutting them. Unsectioned microtubers may be successfully transferred to ex vitro conditions and may generate normal plants in a protected area or in the field. (6) Explants may also generate callus, mainly on a medium containing 2,4-D. At the callus level, caulogenetic and somatic embryogenetic processes can be induced, a fact that allows the multiplication of plants by separating the new plantlets or embryos. (7) In vitro cultured cells can generate somatic embryos, which represents one of the most efficient ways of vegetative multiplication.

3 Examples of Explant Reaction

The origin in terms of phylogenetic affiliation or provenance, and the nature of explants influence the speed of callus formation and growth, as well as the consistency, color, and morphogenic aptitude of this callus. However, we cannot determine to which species the callus belongs by just looking at it; we have to wait until shoots differentiate from the callus, as only their morphology gives us the necessary information concerning the species. The organogenetic ability of the callus depends not only on its provenance, but also on the nature and concentration of the growth regulators in the culture medium. For example, carnation meristems cultured on a medium containing 0.1 mg/l K and various IAA concentrations evolve in the way presented in Fig. 1 (A-C); as the IAA concentration decreases from 100 to 0.1 mg/l (Fig. 1A, 1–6), the meristems form normal plantlets; high IAA levels (100, 50, or 20 mg/l, respectively) inhibit meristem development and shoot growth. By increasing K concentration to 1 mg/l, while maintaining the above mentioned IAA levels (Fig. 1B), weaker nor no rhizogenesis was obtained, but more buds and shoots formed on a single meristem; 1 mg/l 2,4-D (Fig. 1C) directs the carnation meristems toward callus formation; in the presence of 0.1 mg/l NAA, the callus tended to form roots; in the presence of 100, 10, and 1 mg/l procaine chlorhydrate, carnation callus presented a larger volume, but no morphogenesis. On semisolid media, CEPA (which gradually produces ethylene) favored a strong multiplication of carnation meristems, but inhibited rhizogenesis; 1 and 0.1 mg/l CEPA (Fig. 1D, 4 and 5) favored the elongation of the new shoots much more than 1 mg/l BA (Fig. 1D, 6) or than the combination of 1 mg/l BA plus 1 mg/l IAA or excess IAA (see Cachiță et al. 1985a, 1987). Those varieties which had a "propensity" toward vitrification also showed a very high multiplication rate in the presence of CEPA; more vitrified plants occurred when the culture vessels were closed with polyethylene foil than with cellophane; we supposed that the polyethylene foil gradually emanated ethylene (on the other hand, the direct contact of potato and chrysanthemum leaves with the polyethylene foil in vitro may determine the vitrification of these plants). The vitrification process may also be closely related to the variety. The following carnation varieties: *Vanya, Sacha, Palma,* and *Anthares* present the highest percentage of vitrified plants (see Cachiță et al. 1989, 1991b). By culturing meristems, a higher percentage of vitrified plants is obtained than by culturing microcuttings. The high multiplication rate and growth in vitrified plants (a wide spread phenomenon occurring in carnation and other plants as well) seem to indicate a high hormone level which may be unleashed or sustained by ethylene. As vitrification greatly increases the multiplication rate (see John 1986), this process could be used in the future, as a method for plant micropropagation, if vitrified plants could be induced to recover their normal functional state. By culturing floral carnation meristems (Fig. 1E) or buds (Fig. 1 F-H) on media with identical composition, extremely varied reactions were obtained, according to the physiological state of the explants, i.e., whether they were in the preanthesis or the anthesis stage. Of course, the variety also had some influence in this case. This was particularly obvious in chrysanthemum buds. *Chrysanthemum* is a species with extremely high regenerative ability. Therefore, chrysanthemum can be used as a test plant. Within 6–10 weeks, according to the variety, chrysanthemum meristems regenerate vigorous plantlets with a rich root system. Various leaf explants (blade, fragments), petiole segments derived from the Super Yellow variety and cultured under the same

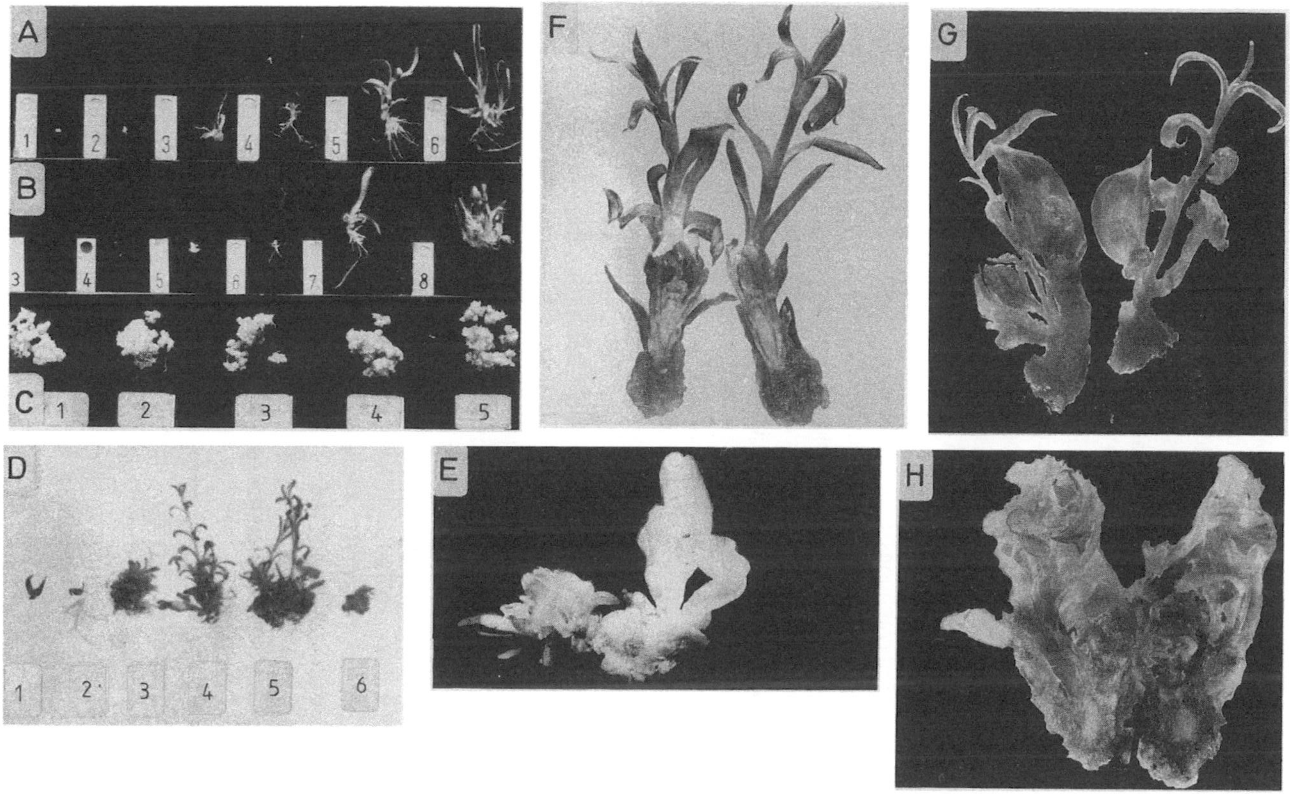

Fig. 1A-H. Carnation (*Dianthus caryophyllus*). **A-E** Meristems cultured in vitro on semisolid medium, basal medium which contained different growth regulators in various concentrations: **A** 0.1 mg/l K plus: *1* – 100 mg/l; *2* – 50 mg/l; *3* – 20 mg/l; *4* – 10 mg/l; *5* – 1 mg/l; *6* – 0.1 mg/l IAA, respectively. **B** 1 mg/l K plus the same range of IAA concentrations as in **A. C** 1 mg/l 2,4-D plus: *1* no other growth regulator, *2* – 0.1 mg/l NAA; *3* – 100 mg/l; *4* – 10 mg/l and *5* – 1 mg/l procaine Hcl. **D** *1* Hormone-free medium; *2* – 10 mg/l IAA; *3* – 10 mg/l; *4* – 1 mg/l; *5* – 0.1 mg/l CEPA; *6* – 1 mg/l BA. **E** Carnation flower meristem; **F-G** flower bud explants; **H** flower buds presenting no organogenesis. (Cachiţă 1980, 1984, 1987)

culture conditions, do not behave in the same way. The morphogenetic ability of chrysanthemum leaf explants depends on whether or not they contain the main vein. Leaf and petiole explants (intact ones or explants cut longitudinally in two) tend to form callus; but when the vein is present, the callus forms many shoots and roots; the explants from the base or center of the leaf blade or sheath have high organogenetic ability. The floral peduncle (Fig. 2A, a and b; and B-D) also presents a high organogenetic capacity. Most frequently, the 0.5 mm peduncle, cut longitudinally in two, generates callus (Fig. 2A, a), which primarily generates shoots and then roots (Fig. 2A, b). In other cases, the flower bud may not open and the shoots may form on the central area of the bud (Fig. 2B) or peduncle (Fig. 2C); sometimes, the flower opens and shoots form on the cut area of the peduncle (Fig. 2D; Cachită 1987). Caulogenesis on explants, consisting of chrysanthemum flower buds which had been cultured on the same type of medium and under identical conditions, depended on the varieity and the anthesis phase. As can be seen (Fig. 2E-G) in Super Yellow and, more rarely, in Westland, bud formation and shoot regeneration occur either in the central part of the bud (Fig. 2B), or in the minute buds from th basis of the bracts that cover the inflorescence (Fig. 2E-F), as well as along the peduncle (Fig. 2C), or on its cut area (Fig. 2D). If the anthesis stage is more advanced, the buds will become flowers in the culture vessels (Fig. 2G). In Aglow (Lazăr and Cachiță 1985b), the process of flowering is accompanied by the transformation of the central flowers into plantlets, i.e., their pistil turns into one or two shoots (Fig. 2H); in this case, a mass of callus showing no organogenesis surrounds the floral peduncle.

The in vitro formed shoots can be divided into microcuttings having one or more nodes: this method increased the efficiency of micropropagation without altering the genotype. Microcuttings containing the apex grow quickly and present a high rhizogenetic ability; microcuttings consisting in the lower two to three nodes also have rapid growth and the ability to form vigorous roots; microcuttings, even single-node ones, have the aptitude to generate many plantlets. The expression of this aptitude depends on the nature and concentration of the growth regulators in the medium (it is important, for instance, whether cytokinins predominated in the medium); usually zeatine determines a higher caulogenetic efficiency, but many genotypic alterations may occur. Chrysanthemum and carnation (or other species) microcuttings grow vigorously and form roots on a well-balanced medium, in terms of its hormonal content, or even on a hormone-free one. CEPA, and especially IAA, induced rhizogenesis, while BA did not (Cachiță 1987; Cachiță et al. 1985a,b). In the latter case, the single-node explant generated callus and two to three smaller shoots with respect to those grown from the axillary buds of the microcuttings cultured on media with CEPA or IAA. It is therefore clear that explants of the same origin, provenance, and nature, or in other words, of the same type, evolve differently on media that are identical in all respects, but contain various growth regulators, with different physiological (morphogenetic) properties. In chrysanthemum, single-leaf microcuttings presented a weak regenerative ability, as they only rarely generated one root from the median vein. In chrysanthemum, the bases of the roots and the neck also have a high caulogenetic ability. Explants of apparently the same provenance and nature, e.g., meristems, apices, nodes, peduncles, buds, inflorescence fragments, and microcuttings, have different internal structures, and metabolism which determine their various reactions.

In *Hortensia* (*Hydrangea opuloides*) it is easy to regenerate vigorous plants from meristems. It is interesting to note that in this species and also in carnation, the in vitro

Fig. 2A-H. Organogenesis in explants consisting of *Chrysanthemum morifolium* (peduncle or flower bud explants). **A** Explants consisting of peduncles which have been cut lengthwise in two (*a* the explant generated a small callus on which the differentiation of buds can be noted; *b* the buds become shoots which root strongly); **B,D** explants consisting of flower buds together with their peduncles; shoots are generated from the bud center (**B** mainly in Super Yellow); **C** caulogenesis along the peduncle (especially in Westland); **D** flower bud that flowered in vitro and presented caulogenesis in the cut area of its peduncle; **E-G** bud formation in explants consisting of flower buds which have been cut in two and have been cultured without their peduncle; **H** flower buds derived from the Aglow chrysanthemum variety, cultured in vitro: in the central area, the flowers become plantlets. (Cachiță 1984, 1987)

formed roots present apogeotropism and prophototropism: this phenomenon is little explained (it may imply the need for oxygen, or a negative chemotropism, or the loss of the correct geotropic aptitude by the root apex; maybe all three explanations are implied in this paradoxical phenomenon). In *Hortensia* the apical microcuttings and those consisting of nodes (intact single-node microcuttings, or microcuttings that have been cut longitudinally in two) regenerate new plants, readily. The explants consisting of internodes form calluses and roots, mainly at the root end of the internode. Flower buds, cultured alone or together with their peduncles, generate a small callus in the area where the peduncle was cut (Lazăr and Cachiță 1984).

In *Exacum afine*, explants consisting of buds, apices, nodes, and leaves (whole leaves or the blade bases) generate plants readily. Flower buds cultured with their peduncles form a small, spherical, green callus on the sectioned end of the peduncle. Further on, this callus generates many shoots and roots which make up a colony (a bush). Rhizogenesis in these explants is very much delayed (Lazăr and Cachiță 1985a; Cachiță 1987).

In *Drosera rotundifolia*, flower buds cultured in vitro with their peduncles give rise to a large colony of rosettes with small leaves. Each rosette can be separated and subcultured, representing a unit (i.e., a *propagule*) that can therefore be used for the micropropagation of this plant, as it generates new colonies of such plantlets. It is important to note that these propagules are remarkably sensitive to the presence (in the culture medium) of some substances which have stimulating or inhibiting effects upon the plants. We therefore used (and recommend) these propagules as very efficient phytotests (Cachiță et al. 1991d in press).

In *Saintpaulia*, leaf explants (represented by blade fragments), petioles, flower buds, and peduncles regenerate plantlets easily and form dense bushes of leaves with petioles (Fig. 3A). Leaf explants of 1 cm^2 (Fig. 3A) and flower buds (Fig. 3B-D) also regenerate such leaf colonies, which may be divided into propagules. In flower buds, organogenesis occurs on the edge of the petals (Fig. 3B) even if the flower bud has its peduncle (Cachiță 1984, 1987); minicuttings consisting of minipetioles (Fig. 3F) or minileaves (Fig. 3G) can be used; minileaves with petioles present even a higher regenerative ability (Fig. 3H).

The calyx of *Saintpaulia* presents no morphogenetic ability, neither in carnation nor in *Exacum*. In *Saintpaulia*, rhizogenesis is slow and the newly formed roots are fragile (Fig. 3E,H).

In *Gerbera* (Fig. 4), the meristems and young inflorescences can regenerate plantlets and then colonies which may also be divided into propagules (Fig. 4B,C); these propagules develop relatively quickly, generate new bushes (Fig. 4D), and present very intense rhizogenesis. The propagules of this species react in a very specific way (Fig. 4C,E) to the presence of growth regulators in the culture medium (1 mg/l CEPA stimulates the growth of *Gerbera* propagules; Cachiță 1987; Cachiță et al. 1987).

In sugar beet (*Beta vulgaris* var. *saccharifera*), as in many other species, mostly in gymnosperms, in vitro cultures are initiated using explants derived from plantlets which were generated from seeds germinated under aseptic conditions. Such explants, consisting of apices with their epicotyls and its cotyledonary nodes, regenerate easily and quickly new plants with a dense root system (Fig. 5A,B). After the inflorescences have been removed, adult plants generate shoots relatively easy at the neck level; when explanted and cultured in vitro, these shoots present a weak rhizogenesis (Fig. 5C). After excising the apices from the main floriferous stem and from its lateral

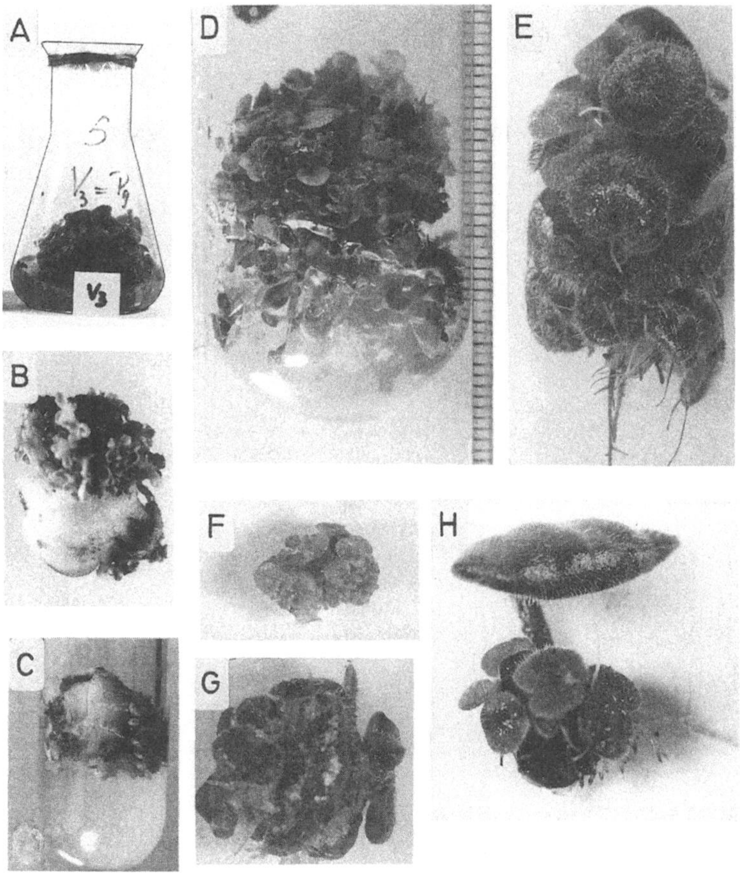

Fig. 3A-H. The in vitro evolution of different types of *Saintpaulia ionantha* explants. **A** A colony (bush) regenerated from a leaf explant of 1 cm^2; **B** bud formation on the edge of the petals; **C-D** bud formation from explants represented by flower buds; **E** colonies of plantlets generated by different explant types; **F-G** organogenesis in explants consisting of mini: petiole (**F**), blade (**G**), or leaf with its petiole (**H**). (Cachiţă 1984)

branches, the growth of the axillary buds located at the ramification points of the first, second, and third order branches, is often stimulated, thus providing the possibility to excise and culture them in vitro. This type of axillary buds, situated in those areas where the branches bearing inflorescences ramify, is very regenerative and forms dense bushes of leaves in vitro (Fig. 5D). These colonies can then be divided into propagules (Fig. 5E-F); however, both these formations and the rachis (Fig. 5G-I), which grows from the apical buds excised from the tip of the branches within the inflorescence, become often vitrified (Fig. 5F). In sugar beet, the changes in the physiology of the in vitro generated vitrified plantlets seem to be very serious; nevertheless, as long as the plant is still in vitro, its leaves grow very well, although they are hypertrophic. This indicates, once more, that habituation may be responsible for the vitrification process. The explants, consisting of young inflorescence fragments (Fig. 5G-I) or flower buds (together with their rachis), may be used for the in vitro cloning of sugar beet in

Fig. 4A-E. The in vitro behavior of explants and propagules of *Gerbera jamesonii*. **A** A colony of plantlets generated by meristems; **B-E** colony of plantlets derived from propagules **B** subcultured propagule; **C** propagule cultured on a medium with 1 mg/l IAA; **D** propagule cultured on a medium with 1 mg/l BA; **E** the evolution of the propagules cultured on media with various CEPA levels with respect to the evolution of the propagules that were cultured for the same period on media with IAA or BA: *1* hormone-free medium, before being cultured, the petioles of the propagule were sectioned; *2–6* basal medium (BM) with: *2* – 10 mg/l CEPA; *3* – 1 mg/l CEPA; *4* – 0.1 mg/l CEPA; *5* – 1 mg/l BA; *6* – 1 mg/l IAA. (Cachiţă 1987; Cachiţă et al. 1987)

Fig. 5A-I. The in vitro evolution of different types of sugar beet (*Beta vulgaris* var. *saccharifera*). **A,B** Regeneration of whole plantlets from explants represented by apices excised from plantlets generated in vitro from seeds germinated under aseptic conditions; **C** the in vitro evolution of explants derived from the shoots that were regenerated at the neck level after the floriferous stem had been removed; **D** colony of plantlets regenerated from explants represented by axillary buds which were excised from the points from which the inflorescence branch out, after the removal of the apices of the floriferous branches; **E-F** propagules detached from the colony of plantlets (**E** normal propagule; **F** vitrified propagule); **G-I** rachis regenerated from explants derived from the apex of an inflorescence. (Cachiţă et al. 1991a, in press)

some cultivars. In many sugar beet cultivars it is difficult to obtain callus, especially organogenic callus. Therefore, the micropropagation of sugar beet may use either propagules, or microcuttings (3 cm long) represented by leaves with petioles (Cachiță et al. 1991a, in press).

In shrubs, such as *Ligustrum, Forsythia,* and *Weigelia,* the regenerative and organogenetic processes occur differently on similar explants (apices or buds) growing on identical media and under identical conditions. Though *Ligustrum* is usually multiplied via cuttings, its in vitro response is surprising: it generates roots readily, but also forms abundant callus. A bud rises in the middle of this callus, and then the bud turns into a shoot with well-developed leaves, the morphology of which is characteristic of the species. On the same medium, stem apices of *Forsythia* develop harmoniously, producing shoots and buds. Nevertheless, in this case, too, the explants present a propensity toward callus formation; as concerns the explants, consisting of flower buds, they become necrotic in *Ligustrum,* but form long, thick roots and no buds in *Forsythia.* In *Spiraea,* the same type of explants (apices and flower buds), cultured on the same media that allowed plant regeneration in *Forsythia* and *Weigelia,* is "refractory".

Potato (*Solanum tuberosum*) could be easily regenerated from various types of explants (Bajaj 1987). For instance, meristems and apices generate shoots, which can be divided afterwards into microcuttings or slips, or calluses, which in turn, generate shoots. These shoots – derived from meristems, or from other explant types – bear simple (not compound) leaves, which is a juvenile character. In vitro cultured apices generated shoots bearing compound leaves in extremely rare cases and mainly on those media containing GA_3. Under insufficient light – in terms of intensity or duration – or in very dense cultures, accompanied by an ethylene or CO_2 excess, the shoots are long and have rudimentary leaves (Fig. 6A-B).

The shoots derived from meristems generate microtubers either above (Fig. 6A) or in the agar medium (Fig. 6B); microtubers may also form at the neck level (Fig. 6C) or they may occur anywhere, when the culture has many shoots (Fig. 6D). Microtubers may also generate shoots from their buds (eyes; Fig. 6E-G). Moreover, these microtubers may be used as such for in vitro plant regeneration, or they may be cut crosswise, i.e., along the breadth of the tuber, in pieces 1–2 mm thick. Further on, these pieces generate one or more shoots and roots, depending on the composition of the culture medium (Fig. 6F-G); CEPA stimulated organogenesis at this level (Fig. 6G). Microtubers may also be transferred directly to ex vitro conditions, as they present a high adaptability to the nonsterile conditions. The method of Wang and Hu (1982) proved to be efficient for obtaining as many microtubers per culture vessel (with a large volume) as possible. A great number of microtubers may be obtained in vitro by controlling the culture conditions (Chandra et al. 1988) as follows: short photoperiod (8 h light/day), liquid media, low sucrose (80 g/l), and high CCC or coumarin levels. It is strange that potato tuberization has been induced by using CCC or coumarin which are not implied in the natural tuberization that occurs in vivo. Good results have also been obtained by us, using the "double layer" culture (Fig. 6H). Here, the culture grows on agar for some time and then a liquid medium containing CCC is injected, so that the culture vessels are not opened. Whether or not tuberization occurs and the intensity of this process depend on the culture conditions and the variety. There are no references concerning the relation between the provenance of explants and their ability to generate shoots and tubers (Cachiță and Zăpîrțan 1991, in press).

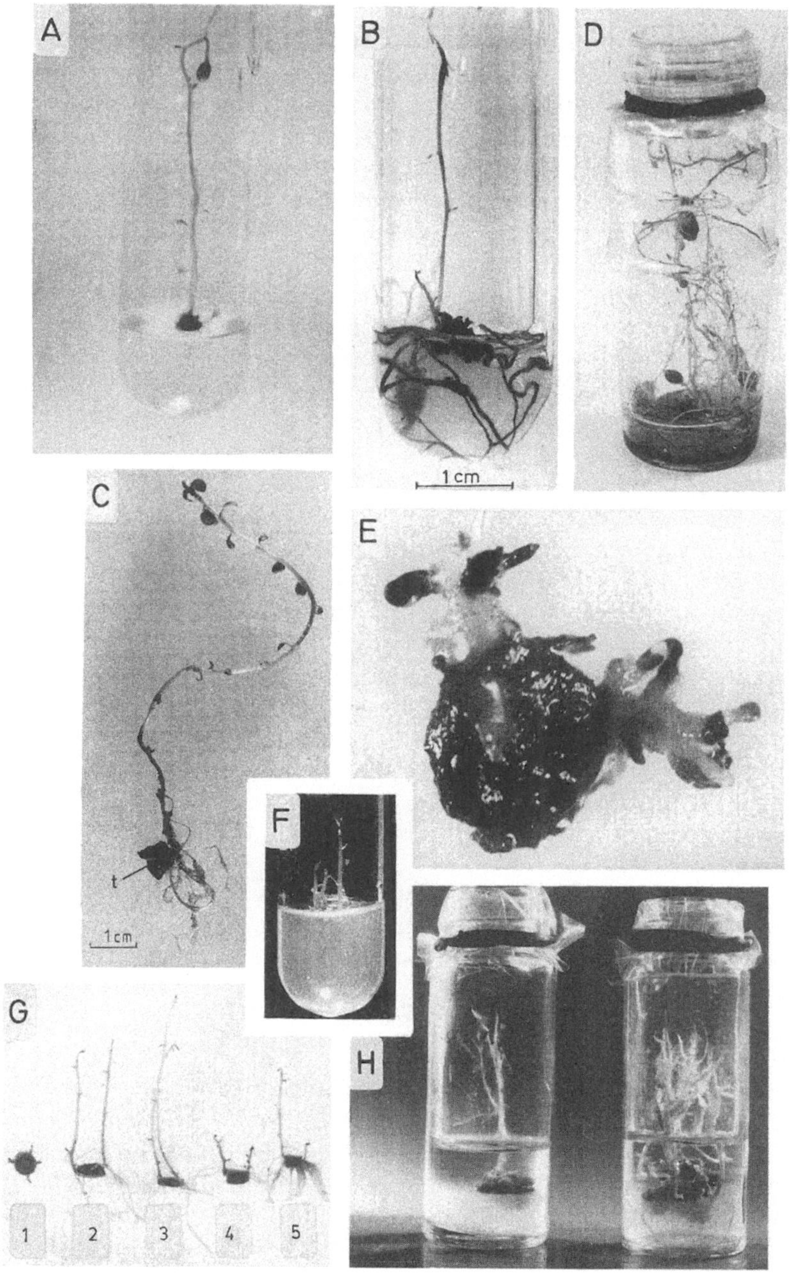

Fig. 6A-H. Microtubers generated in vitro cultured potato shoots. **A-C** Shoots with microtubers; **E-G** shoot regeneration from microtuber cross-sections; **G** *1* hormone-free medium; *2* BM plus 1 mg/l CEPA; *3* basal medium (BM) plus 0.1 mg/l CEPA; *4* BM plus 1 mg/l BA; *5* BM plus 1 mg/l IAA; **H** shoot culture using the double-layer technique (the upper layer is represented by a CCC solution). (Cachiță 1987; Cachiță and Zăpîrțan 1991, in press).

In *Cymbidium* (as well as in other orchids, such as *Cattleya*, *Phalenopsis*, *Pleone*) in vitro cultured meristems generate protocorms; orchids naturally derive from seeds. These protocorms present functional autonomy during their in vitro culture; they are able to multiply without forming any organs (Fig. 7A), especially when cultured deeper, in a stationary, simple (mineral salts and sucrose) medium. The newly formed protocorms remain gathered in blocks (Fig. 7A,B) for a long time. The multiplication of and organogenetic processes in protocorms are influenced by growth regulators; moreover, the presence of BA induces senescence in protocorms, while the culture turns brown (Cachiță 1987) (senescence was perhaps caused by the substances that changed the color of the medium). Certain protocorms may spontaneously present organogenetic processes (Fig. 7B); this happens mainly in spring, under a 16-h light/24 h photoperiod, even on hormone-free media, with no thermic control (see Cachiță et al. 1988). Organogenesis may be favored by separating the protocorms and culturing them on solid media, especially on those containing growth regulators (Fig. 7C), 1 or 0.1 mg/l IAA or CEPA (Fig. 7C, 5, respectively, 1–3); BA promotes protocorm multiplication but strongly inhibits rhizogenesis (Fig. 7C, 4; see Cachiță et al. 1987). In *Cymbidium*, the double-layer technique (which we devised and published in 1983; see Cachiță 1983) gives very good results concerning organogenesis by first stimulating caulogenesis and then rhizogenesis (Fig. 7D). Therefore, in *Cymbidium*, protocorms of the same type (in terms of their origin and nature) behave in a different way, depending, first of all, on the culture conditions and on the growth regulators in the medium (Fig. 7C). Moreover, the response of these inocula also depends on certain *endogenous* factors which are related to some unknown exogenous ones. It is possible that organogenesis in *Cymbidium* protocorms occurs on solid or liquid media (Fig. 7D-F) under some "impulse" received from a "biological clock", which is correlated with natural, seasonal factors, not only with light and temperature, but with other unknown factors related to the earth and the universe, as well as with the interaction of these factors (Cachiță et al. 1988, 1991c, in press).

In monocotyledons, the effect of the nature and origin of explants in micropropagation often varies. The micromultiplication process depends very much on the species, variety and, most of all, on the provenance of the explants with respect to age and culture conditions.

For example, *Freesia* meristems, apices, or nodes (Fig. 8A-C) generate single shoots which root easily. Very young flower buds regenerate countless shoots either at the flower base under the ovary, or from small mass of callus that surrounds the ovary (Fig. 8D-F). In order to regenerate these shoots, the cultured flower buds were maintained 8–10 weeks in the dark and then transferred to continuous light; under 16-h light/day organogenesis occurred slowly. When rachis or rachis fragments are cultured, they produce countless shoots; rhizogenesis may be induced in the areas where bracts are inserted on the rachis fragment (Fig. 8G-H) if these rachis fragments are cultured with their bracts but without flower buds. Therefore, we can conclude that different types of explants, with respect to the nature and provenance, belonging to the same species, express an organogenetic ability, but the intensity of the multiplication processes depends on the explant type. Explants consisting of rachis present a much intenser multiplication (especially at the rachis nodes, where the bracts and flower buds are inserted), than those consisting of flower buds and above all, than the nodes, apices, and meristems, even if all these types of explants are cultured on the same medium type (Cachiță et al. 1985d).

Fig. 7A-F. *Cymbidium* protocorms cultured on various media. **A** Liquid medium; **B** a glomerule made up of several protocorms; some organogenesis, caulo- and rhizogenesis, can also be noted; **C** organogenesis in protocorms cultured on solid media with different growth regulators in various concentrations: *0* hormone-free basal medium (BM); *1* plus 10 mg/l CEPA; *2* BM plus 1 mg/l CEPA; *3* BM plus 0.1 mg/l CEPA; *4* BM plus 1 mg/l BA; *5* BM plus 1 mg/l IAA; **D** protocorms cultured using the double-layer technique; **E-F** organogenesis in protocorms cultured on solid medium (**E**), or submerged in the liquid medium (**F**). (Cachiţă 1983, 1984, 1987)

Fig. 8A-H. The in vitro reaction of different *Freesia hybrida* explants: meristem (**A**), apex (**B**), node (**C**); shoot regeneration from very young flower buds of *Freesia hybrida.* **D-F** (**D** initiation of regenerative processes at the base of the ovary; **E** many shoots were regenerated at the base of the ovary, in time: **F** general aspects concerning the in vitro culture of *Freesia* flower buds); **G** rachis fragments; **H** bracts without flower buds. (Cachiță et al. 1985d)

In fact, inflorescences and flowers generally present a very high ability to regenerate organs, especially in monocots. In *Montbretia crocosmiflora*, a phenomenon similar to that in *Freesia* (see Cachiță 1987; Cachiță et al. 1985) takes place.

Our last example is concerned with the response of explants derived from *Hyacinthus* bulbs to their in vitro culture. Fragments excised from the disk of the bulb (the disk representing the metamorphosed stem and having leaves on its upper side and roots on its lower one) present a high regenerative ability and generate microbulblets. Explants derived from flesh leaves generate microbulblets on their external side, but mainly on their internal one (Fig. 9A). Further on, shoots develop from the middle of these bulblets (Fig. 9B); root regeneration is somewhat delayed (Fig. 9C,D). By merely

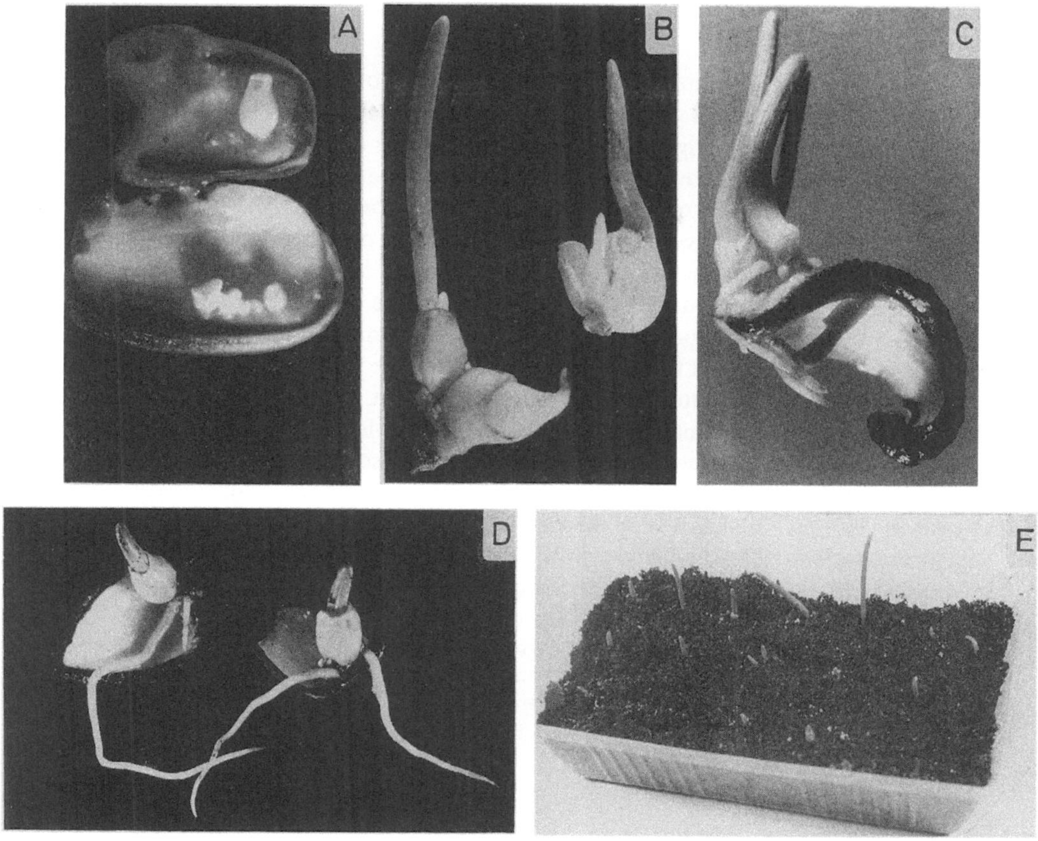

Fig. 9A-E. The in vitro reaction of explants derived from *Hyacinthus* bulbs. **A** The initial phase of microbulblet regeneration; **B** shoots develop from the middle of these microbulblets; **C-D** root regeneration of the base of the microbulblets; **E** shoots growing on soil. (Cachiță 1987)

planting the rooted scales in an aseptic soil (Fig. 9E) they adapt very quickly to nonsterile conditions.

The aspects referred to in the second part of this chapter are derived from the experiments we made to either micropropagate the species we studied, or to analyze their behavior under different culture conditions. The scientific literature describes countless experiments of this type, as well as methods for the industrial micropropagation of plants. These topics are also presented in various books published in the last few years (Evans et al. 1983; Vasil 1984b; Bajaj 1986–1990).

The implications of the nature and origin of explants in micropropagation are very complex. These two factors determine, in almost all cases, the results of our operations. Therefore, we had to limit the area and range of our examples to the most representative aspects encountered during our 15-year work.

4 Summary

An explant constitutes a living unit, independent of the mother plant, but containing all its genetic information and presenting all the functions that the explanted tissues or cells performed at their original location. The response of explants to their in vitro culture depends on their heredity. Yet, this response can be modified within the limits of their adaptability, according to the various culture conditions. This can lead to the preservation of the unchanged genome, to obtaining clones, or to producing variants and creating new genotypes. The potential of the explants to adapt to the culture medium and to the culture conditions depends very much on the nature and on the origin of explants, on their physiological state following sterilization, explantation, inoculation, as well as on the duration of the culture period and on the type of the previous culture.

The problem referred to in the title is extremely vast. However, there is no doubt that future research will continuously increase our knowledge in this direction. It will also solve the countless mysteries concerning the response of the explants cultured on aseptic media. Eventually, we shall be able to "break the resistance" of some explants to their in vitro culture and to keep the explants completely under control, regardless of their organization. This will solve the difficulties related to regeneration, organogenesis, or somatic embryogenesis in certain species or culture types.

Abbreviations: IAA = indole-3-acetic acid; NAA = alpha-naphthalene acetic acid; BA = benzyladenine; K = kinetin; 2,4-D = 2,4-dichlorophenoxyacetic acid; GA_3 = gibberellic acid; CCC = chlorocholine chloride; CEPA = 2 chloro-ethyl-phosphonic acid = Ethrel.

References

Bajaj YPS (ed) (1986–1990) Biotechnology in agriculture and forestry, vols 1–13. Springer, Berlin Heidelberg New York

Cachiță CD (1980) Cresterea si diferentierea meristemelor caulinare de *Dianthus caryophyllus* var. Linda, în cultură aseptică. Șt și Cerc Biol Veget 32, pp 63–68

Cachiță CD (1983) Contributii la tehnica de multiplicare si propagare in vitro a orhideelor. In: II-lea Simp Nat de Culturi de Celule si Tesuturi Vegetale. Acad Șt Agr Silv, Pitesti, pp 191–197

Cachiță CD (1984) Tehnica cultivării "in vitro" a celulelor si tesuturilor vegetale. In: Culturi de celule si tesuturi vegetale – aplicatii în agricultură. Ceres, Bucuresti, pp 7–94

Cachiță CD (1987) Metode "in vitro" la plantele de cultură – baze teoretice si practice. Ceres Bucuresti, 274 pp

Cachiță CD, Achim F, Cristea V, Gergely K (1987) 2-Chloro-ethylphosphonic acid and morphogenesis. Symp "Florizel" 87, Arlon, Belgium. Plant Propagation in Horticultural Industries, pp 234–246

Cachiță CD, Zăpîrțan M (1991) Potato tuberogenesis using in vitro double layer technique. In: IV-lea Simp Nat de Culturi de Tesuturi si Celule Vegetale, Cluj-Napoca (in press)

Cachiță CD, Achim F, Vârban N (1985a) Noutăti în înmultirea "in vitro" a garoafelor. In: III-lea Simp Nat de Culturi de Celule si Tesuturi Vegetale, Bucuresti, pp 335–344

Cachiță CD, Cristea V, Achim F, Gergely K (1985b) Actiunea acidului 2-cloroetilfosfonic asupra inoculilor cultivati "in vitro". In: III-lea Simp, Nat de Culturi de Celule si Tesuturi Vegetale, Bucuresti, pp 463–474

Cachiță CD, Cristea V, Gergely K (1985c) Multiplicarea "in vitro" si micropropagarea pe medii aseptice la *Montbretia crocosmiflora* Lemoine. In: III-lea Simp Nat de Culturi de Celule si Tesuturi Vegetale, Bucuresti, pp 325–330

Cachiță CD, Cristea V, Badiu A (1991a) Researches on the in vitro morphogenesis at the level of different kinds of sugar beet explants. In: IV-lea Simp Nat de Culturi de Tesuturi si Celule Vegetale, Cluj-Napoca (in press)

Cachiţă CD, Lăzar M, Grigoras S (1985d) Morfogeneza la nivelul inoculilor constînd din rahis si boboci de *Freesia hybrida*. In: III-lea Simp Nat de Culturi de Celule si Tesuturi Vegetale, Bucuresti, pp 431–436

Cachiţă CD, Petrescu C, Stănilă M, Cristea V, Gergely M, Pataky L (1988) Organogenesis in vitro-theoretical and practical aspects in ornamentals. In: 2nd Nordic Symp on Cell and tissue culture. EUCARPIA Int Congr on Genetic manipulation in plant breeding, Helsingør (Elsinore), Denmark (Abstr), 16 pp

Cachiţă CD, Zăpîrţan M, Cristea V, Osvath T, Bercea V, Bechis E, Bosica I (1991b) In vitro carnation vitrification – morphophysiological and biochemical study. In: IV-lea Simp Nat de Culturi de Tesuturi si Celule Vegetale, Cluj-Napoca (in press)

Cachiţă CD, Zăpîrţan M, Cristea V, Petrescu C (1991c) Biological clock and in vitro organogenesis. In: IV-lea Simp Nat de Culturi de Tesuturi si Celule Vegetale, Cluj-Napoca (in press)

Cachiţă CD, Zăpîrţan M, Grigoras S (1991d) In vitro cultured *Drosera rotundifolia* – a new biotest. In: IV-lea Simp Nat de Culturi de Tesuturi si Celule Vegetale, Cluj-Napoca (in press)

Cachiţă CD, Zăpîrţan M, Bercea V, Cristea V, Osváth T, Bechis E (1989) Vitrification and some morphological and biochemical modifications of carnation plantlets (14 varieties) formed from meristem cultivated "in vitro". Poster. In: Colture in vitro o micropropagazione in ortoflorofrutticoltura, Cesena, 2. VII. 1989, 37 pp

Chandra R, Dodds JH, Tovar P (1988) "In vitro" tuberization in potato (*Solanum tuberosum*) L. IAPTC Newsletter 55, pp 10–20

Chlyah H (1984) Correlations in organogenesis. In: Novák FJ, Havel L, Doležel J (eds) Plant tissues and cell culture application to crop improvement. Czechoslovak Acad Sci, Prague, pp 57–66

Evans D, Sharp WR, Ammirato PV, Yamada Y (1983) Handbook of plant cell culture, vol 1. Techniques for propagation and breeding. Macmillan, New York

John A (1986) Vitrification in sitka spruce cultures. In: Withers LA, Alderson PG (eds) Plant tissue culture and its agricultural applications. Butterworth, London, pp 167–174

Lazăr M, Cachiţă CD (1984) Propagarea "in vitro" a plantelor de hortensia (*Hydrangea opuloides* Koch). Productia vegetală. *Horticultura* 3, pp 36–39

Lazăr M, Cachiţă CD (1985a) Immultirea "in vitro" a plantei *Exacum afine* Balf. In: III-lea Simp Nat de Culturi Celule si Tesuturi Vegetale, Bucuresti, pp 321–324

Lazăr M, Cachiţă CD (1985b) Morfogeneza la inoculii constînd din boboci de crizanteme, *Chrysanthemum morifolium* Remat., soiurile "Westland" si "Aglow". In: III-lea Simp Nat de Culturi de Celule si Tesuturi Vegetale, Bucuresti, pp 437–442

Margara J (1982) Bases de la multiplication végétative. Les Les méristèmes et l'organogènese. INRA, Versailles

Murashige T (1984) Parameters in regenerating plants "in vitro". In: Novák FJ, Havel L, Doležel J (eds) Plant tissue and cell culture application to crop improvement. Czechoslovak Acad Sci, Prague, pp 23–31

Nabors MW (1982) Tissue culture for crops: stress resistant plants from cell culture. Dep Bot Plant Pathol, Colorado State Univ, I

Quak F (1977) Meristem culture and virus-free plants. In: Reinert J, Bajaj YPS (eds) Applied and fundamental aspects of plant cell, tissue, and organ culture. Springer, Berlin Heidelberg New York, pp 598–615

Skoog F, Miller CO (1957) Chemical regulation of growth and organ formation in plant tissue cultivated "in vitro". Symp Soc Exp Biol 11, pp 118–130

Vasil IK (1984a) Developing biotechnology for the improvement of cereal and grass crops – the consequences of somatic embryogenesis. In: Novák FJ, Havel L, Doležel J (eds) Plant tissue and cell culture application to crop improvement. Czechoslovak Acad Sci, Prague, pp 67–76

Vasil IK (ed) (1984b) Cell culture and somatic cell genetics of plants. Academic Press, Orlando

Wang PJ, Hu CY (1982) "In vitro" mass tuberization and virus-free seed potato production in Taiwan. Am Potato J 59, pp 33–37

I.10 Problems with Explant Exudation in Micropropagation

J.E. PREECE and M.E. COMPTON[1]

1 General Account

During the time of organ establishment for the purpose of micropropagation, discoloration of the medium and browning or blackening of the explant is often observed. This may be a sign of potential problems and may mean that special techniques will be required to obtain a healthy-looking and rapidly growing explant. Similar staining of the medium may be observed throughout the in vitro cycle. Although this is a concern for those involved in micropropagation, it has also been noted by researchers working with protoplasts, callus, somatic embryos, and other forms of plant tissue culture.

Staining of the medium by substances that exude from explants is frequently observed with many (especially woody) species (Compton and Preece 1986) (Fig. 1). The staining varies in both color and intensity. Colors range from various shades of blue, purple, gray, or brown to black. Staining may be confined to the immediate area surrounding the explant, or may result in discoloration of the entire medium. We have noticed visible staining within minutes, hours, or days after explants are placed in vitro. A related problem is frequently referred to as oxidative browning of the original explant or tissues that grow in vitro. This may be either irreversible lethal necrosis, an indication of decline of the cultures, a sign of explant distress, or a signal that additional steps are necessary if one is to produce healthy, growing cultures.

When discoloration of the medium or explant has been noticed, it has usually been interpreted as a sign of problems with the explant. Therefore, a wide variety of techniques have been used to overcome the problem and obtain healthy cultures (Table 1; Hu and Wang 1983; Compton and Preece 1986). These techniques include stock plant treatments, proper selection of explants, special care during explant preparation, use of antioxidants as a presoak or in the medium, explant presoaks in sterile water, choice of another medium, changing the level of sucrose in the medium, use of liquid rather than agar solidified medium, changing or simply pulsing plant growth regulators, use of activated charcoal in the medium, incubation of cultures in lower light or darkness, avoiding high temperatures inside of the transfer chamber, discarding explants that show browning, and either transferring explants to new locations within the same vessel or to a fresh medium within relatively short intervals of time when discoloration is first noticed. These are frequently time-consuming and laborious tasks that may be necessary if explants are to survive. Autotoxicity from exudates is widespread in the plant kingdom across taxonomic classes, orders, families, genera, species, and cultivars.

---

[1]Department of Plant and Soil Science, Southern Illinois University, Carbondale, IL 62901–4415, USA

Biotechnology in Agriculture and Forestry, Vol. 17
High-Tech and Micropropagation I (ed. by Y.P.S. Bajaj)

A  B

Fig. 1A,B. Staining of the agar solidified (**A**) and stationary liquid medium (**B**) is evident on these *Juglans nigra* (black walnut) cultures. **A** The adult shoot tip caused intense black staining of the medium and showed lethal necrosis after only 24 h in vitro on WPM with no plant growth regulators. **B** Etiolated shoots are evident from this culture that originated as a juvenile shoot tip. The medium gradually darkened as the end of the 14-day culture cycle approached

Many associate staining with explant distress; however, this does not appear to be a universally correct assumption. Ashby et al. (1987) reported that some shoot explants from juvenile *Acer saccharinum* (silver maple) caused medium staining; they performed as well as those that did not produce an exudate. Shoot explants from adult silver maple establish more slowly in vitro and stain the agar solidified medium more than those from juvenile stock plants. Comparing rapid transfers to fresh medium with normal fortnightly transfers, there were no advantages gained using the rapid transfer technique for establishing adult silver maple shoot explants (Preece and Brown unpubl.). Ettinger and Preece (1985) observed a dark exudate that diffused into the medium from *Rhododendron* P.J.M. Hybrids shoot tip explants; however, this exudation was not related to explant growth and survival. Although we were unable to adequately quantify the response, it appeared that proliferating shoot cultures of blackberry (*Rubus* sp.) that had the fastest multiplication rates also had the darkest stationary liquid media (Preece and Slivinski unpubl.). Welander (1988) reported that buds from the base of *Betula pendula* (birch) seedlings that had been grown in long days exuded the greatest amount of phenolic substances into the medium and exhibited the best growth and development in vitro.

Explant browning may not always be associated with staining of the medium, in fact Morini and Sciutti (1989) in a review focused on tissue browning and not staining of the medium. We have observed explant necrosis from *Hibiscus rosa-sinensis*, even though there was no visible staining of the medium, and on shoot explants of *Rhododendron* P.J.M. Hybrids; their exudation was shown not to be autoinhibitory (Ettinger and Preece 1985; Compton and Preece 1988b). The length of the necrosis on hibiscus shoot tip explants was increased by the presence of benzyladenine (BA) in the medium. In contrast, explant necrosis of hibiscus and *Rhododendron* P.J.M. Hybrids was reduced when 1 or 10 μM indolebutyric acid (IBA) was added to the medium (Compton

Table 1. Survey of the literature on explant exudation

Plant species	Explant type	Staining of medium	Browning of explant	Treatment to overcome browning	Results	Reference
Acer rubrum	Shoot tip and node from seedlings and adult trees	Yes	Yes	Pre-soak explants in ascorbic acid (150 mg/l), incorporate into medium PVP (5, 10, 20 mg/l), or L-cysteine (50 mg/l), place on preconditioning medium	The chemicals did not control or prevent exudation or injury, the preconditioning treatment was effective	Welsh et al. (1979)
Acer saccharinum	Shoot tip and node from seedlings	Yes		None	The exudate was apparently not autotoxic	Ashby et al. (1987)
Acer saccharum	Shoot		Yes	Collect dormant shoots in February and force new growth to place in vitro	Explants showed a browning response when collected from stock plants in the spring, but not when excised from new growth from Feb. shoots	Kerns and Meyer (1988)
Anigozanthos manglesii, A. pulcherrimus, and three hybrids	Apical shoot, bud, and cortical sections of rhizome	Yes		Ascorbic acid (1 mM), PVP (0.1 mM), dithiothreitol (1 mM), and bovine serum albumin (0.25 mg/l) were added to the medium	The antioxidants did not reduce the dark blue exudate in the medium, but some clones produced less stain	McComb and Newton (1981)
Berberis thunbergii atropurpurea	Shoot from adult plants	Yes		Avoid using Anderson's revised medium, transfer explants to fresh medium weekly	Using these measures, shoots proliferated	Uno and Preece (1987)
Betula pendula	Bud from seedling and adult trees	Yes	Yes	Prior to excising buds, soak stems in 0.1 mM L-cystein for 24 h, or sterile water for 3 h, transfer adult buds to fresh medium after 2 days	Exudate from seedlings was not inhibitory, after the step of transferring adult buds to fresh medium, their growth was satisfactory	Welander (1988)

B. platyphyla var. *szechuanica*	Shoot tip and node from seedlings	Yes		Transfer to fresh medium	Shoots proliferated	McCown and Amos (1979)
Camellia japonica	Shoot tip and nodal from adult trees		Yes	Use Heller medium plus 1 mM $(NH_4)_2SO_4$	Explants on MS and MS-½NO_3 had necrotic buds and leaves became progressively brown	Samartin (1989)
Carica papaya	Shoot tip	Yes		Agitate in sterile water on a shaker at 160 rpm for 2–3 h	This minimized the flow of latex into the medium	Rajeevan and Pandey (1983)
Castanea dentata	Stem segment from adult trees	Yes		Transfer to other locations within same vessel or to fresh medium	The tannin-like exudation appeared to prevent normal growth, but transferring alleviated the problem	Read et al. (1985)
C. mollissima	Nodal from seedlings	Yes		Transfer to other location within the same vessel after 8 days in culture	Explants that were transferred to a new location grew better than those not transferred	Yang et al. (1986)
C. sativa	Embryonic axes	Yes		Transfer away from medium containing IBA	Avoiding IBA resulted in no black exudate	Vieitez and Vieitez (1980a)
C. sativa	Node from seedlings	Yes	Slower shoot development	Soak in sterile distilled water for 2–3 h	Inhibition of shoot development by tannins was partially overcome with the soak	Vieitez and Vieitez (1980b)
C. sativa	Bud from adult trees	Yes	Yes	Use Schenk and Hildebrandt's medium	Polyphenol (tannin) exudation ceased after 2–3 days in vitro; there was less exudation on Schenk and Hildebrandt's medium than MS	Biondi et al. (1981)
C. sativa, *C. crenata* × *C. sativa*	Nodal from juvenile and adult trees	Yes	Autointoxication of explants reported	Treat with caffeine and PVP, soak for 3 h in sterile water, determine influence of bud position, use low salt media	Caffeine and PVP were ineffective, the water soak reduced tannin exudate, tannins increased from the first to fourth bud; there was less diffusion of tannins on low salt media	Chevre et al. (1983)

Table 1. (*Continued*)

Plant species	Explant type	Staining of medium	Browning of explant	Treatment to overcome browning	Results	Reference
C. sativa × *C. crenata*	Shoot apex and node from adult trees	Yes		Sterile distilled water soak for 2–3 h	The soak reduced the amount of tannins that exudated; there were clonal differences in the amount of exudate	Vieitez et al. (1983)
Cephalotus follicularis	Shoot tip	Yes	Yes	Incubate cultures in darkness for the first 6 weeks	This treatment prevented the accumulation of harmful phenolic oxidation products	Adams et al. (1979)
Chamaecyparis nootkatensis	Shoot tip from adult and juvenile trees		Yes	Use antioxidants singly and in combination	Even with the antioxidant treatments, adult explants declined from oxidative browning to the point where they would not respond, juvenile buds established and proliferated	Kurz (1986)
Coffea arabica	Internodes anthers, leaf endosperm, fruit	Yes	Yes	Immerse explant in L-cysteine • HCl, dithiothreitol, Cleland's reagent, glutathione, mercapto-ethanol, ascorbic acid, or diethyldithiocarbamate; avoid high temperatures within transfer chamber; add L-cysteine • HCl to medium, incubate cultures in darkness; use activated charcoal	Various combinations of treatments were necessary because phenols were oxidized to quinones that were inhibitory to plant cellular growth	Monaco et al. (1977)
Corylus avellana	Shoot and cotyledonary segments from seedlings			Soak seeds without testa overnight in running water	Soaking washed out interfering phenolics	Pérez et al. (1985)
Diospyros kaki	Bud from field-grown plants	Yes	Yes	Use 2-iP in the medium	With BA, buds died and medium turned brown, this did not occur with 2-iP	Cooper and Cohen (1985)

Eucalyptus ficifolia	Adult shoot	Yes	Yes	Use medium with only 60 mM sucrose	Exudate problems were eliminated compared to media with higher levels of sucrose	de Fossard et al. (1978)
E. grandis	Node from seedlings	Yes	Yes	Soak explants in sterile distilled water for ≥ 2 h, incubate cultures in darkness for ≥ the first 7 days in vitro	After 2 h, the water used for the soaks was yellowish in color and the pH had changed to 5.6–6.1; if the leaf blade was attached to the node explant, exudation is increased	Cresswell and Nitsch (1975)
E. grandis	Leaf disc, node	Yes	Yes	Use healthy stems with a diameter of < 5 mm, pre-soak explants in distilled water in the light for 3 h, incubate cultures in darkness for the first 8 days	Using these techniques, exudation could be minimized or eliminated	Durand-Cresswell et al. (1982)
E. grandis	Node from 5-year-old trees		Yes	Transfer from medium containing kinetin and NAA to medium with BA and NAA	Cultures were revived	Lakshmi Sita and Shobha Rani (1985)
Euphorbia lathyris	Shoot tip	Yes		Transfer to fresh medium on days 1, 3, and 5 during the first week in vitro, cutting, rather than pulling leaves off during explant preparation	Although the rapid transfer technique was effective, minimizing explant wounding reduced the exudate problem	Ripley and Preece (1986)
Fuchsia hybrida	Shoot tip, petiole, leaf, and root	Yes	Yes	Use B5 medium at 25% normal strength, add 0.01% PVP to medium, use liquid medium	After 4–12 h on agar solidified medium, purple exudate was produced by shoot tips, less was evident with liquid, 25% medium, or with PVP; however, exudate did not seem to inhibit growth	Stevenson and Harris (1980)
C. sativa × *C. crenata*	Shoot apex and node from adult trees	Yes		Sterile distilled water soak for 2–3 h	The soak reduced the amount of tannins that exudated; there were clonal differences in the amount of exudate	Vieitez et al. (1983)

Table 1. (*Continued*)

Plant species	Explant type	Staining of medium	Browning of explant	Treatment to overcome browning	Results	Reference
Juglans hindsii × *J. regia* Paradox	Nodal segment (3–4 cm)	Yes		Tap water rinse and transfer to fresh medium when there was staining	Proliferating cultures were obtained successfully	MacGranaham et al. (1987)
J. hindsii, Pterocarya stenoptera						
Juglans nigra	Apical meristem (0.8–1 mm)		Yes	None reported	Explants died	Lenartowicz and Millikan (1977)
J. nigra	Seedling shoot tip (2 cm)	Yes		Rapid transfer to fresh medium, incorporation of 1% PVP into media	Rapidly transferred shoots established and proliferated, PVP may have been inhibitory	Somers et al. (1982)
J. nigra	Shoot tip		Yes	Overnight pre-soak in 0.01 M Na-DIECA, following sterilization, 5-h soak in water	Axillary shoots proliferated	Stefan and Millikan (1985)
J. nigra	Embryonic axis and seedling shoot tip	Yes		Transfer to fresh medium on days 1,3, and 5 during the first week in vitro, and every 14 days thereafter	Axillary shoots proliferated	Heile-Sudholt et al. (1986)
J. nigra	Shoot tip	Yes		Soak overnight in 0.01 M Na-DIECA solution, following surface sterilization, stir for 5 h, place in vitro on filter paper bridges with medium containing 15 g/l sucrose, 2.0% PVP, and no copper, and incubate in low light for the first 21 days	Shoots proliferated	Stefan and Millikan (1987)
J. regia	Seed			Seeds received a 12-h pre-soak in running water to remove phenolics	Seeds germinated and shoots proliferated	Rodriguez (1982)

Kalmia latifolia	Stem tip from seedlings	Yes	Yes	Transfer to fresh medium	Shoot growth and multiplication occurred, upon subdividing basal shoot masses, tissues deteriorated, exudated, and grew poorly	Lloyd and McCown (1980)
Larix × eurolepis	Immature zygotic embryos		Yes	Use medium solidified with 1.0% agar plus 0.4% gellan gum	When higher levels of agar and gellan gum were used, calli became necrotic	Klimaszewska (1989)
Liquidambar styraciflua	Shoot tip from adult trees		Yes	Transfer frequently to fresh medium during first month in vitro	By transferring every 2–3 days for first 2 weeks, then weekly for the next 2 weeks, problems with explant chlorosis and death were overcome	Sutter and Barker (1985)
Malus sylvestris	Shoot tip	Yes		Add PVP to the medium at 0.5–2.0%	Regardless of the concentration, explants with PVP grew better, without PVP, filter paper bridges were stained with oxidized phenolics	Walkey (1972)
M. sylvestris	Shoot tip		Yes	Use ½ strength MS salts	With this medium there was little or no browning	Werner and Boe (1980)
Malus sp.	Bud and shoot		Yes	Use PVP, transfer to fresh, medium daily, use anti-oxidant	After 4–8 weeks, PVP could be omitted and darkening of the medium by oxidation of polyphenols could be overcome	McComb (1978)
Malus sp.	Shoot tip	Yes		Place explants in liquid medium first in flasks rotated at 1 rpm for 2–4 days, then transfer to solid medium	Growth was better and fewer explants were lost with this protocol	Zimmerman and Broome (1980)
Malus sp.	Shoot tip		Yes	Collect explants during mid-spring or mid-summer	When explants were collected during mid-autumn or mid-winter explant browning was a problem	Hutchinson (1984)
Musa × paradisiaca	Shoot tip from adult plants	Yes	Yes	Dip buds in cysteine (50 mg/l), transfer to fresh medium	Dipping reduced medium and explant discoloration, transfer to fresh medium also overcame the problem	Jarret et al. (1985)

Table 1. (*Continued*)

Plant species	Explant type	Staining of medium	Browning of explant	Treatment to overcome browning	Results	Reference
Musa sp.	Terminal male bud		Yes	None	Although the explant apices became brownish in 4–5 days, it did not appear to be a problem	Swamy and Sahijram (1989)
Picea glauca	Bud from adult trees		Yes	Use Schenk and Hilderbrandt (SH) medium	There was oxidative browning of the tissue on MS, WPM, and B5 media; positive results were obtained with SH medium	Kurz (1986)
Pinus taeda	From seedling and adult trees		Yes	Transfer to fresh medium more frequently, every 2–3 weeks	The decline in the cultures is less pronounced with this transfer schedule	Heyser and Mott (1980)
Pistacia vera	Terminal and axillary bud from in vitro seedlings and 1–2-year-old greenhouse-grown plants	Yes	Yes	Regular transfer to fresh medium	Browning of medium and explant were worse with greenhouse-grown stock plants, shoots that were transferred to fresh medium proliferated shoots	Alderson and Barghchi (1982)
P. vera, *P. khinjuk*	Seedling shoot tip and node	Yes	Yes	Transfer to fresh medium incubate in darkness	Shoots proliferated after transferring to fresh medium, incubation in darkness offered no benefit	Barghchi and Alderson (1983)
P. vera	Seedling shoot tip and node	Yes	Yes	Transfer to fresh medium, incubate in darkness	Browning was only evident in the initial culture, darkness was not beneficial	Barghchi and Alderson (1985)
P. vera	Node from adult trees	Yes	Yes	Transfer within the same jar, pre-soak explants in 50 mg/l malonic acid for up to 20 min, exclude ferrous sulfate from medium	Malonic acid pre-soak only delayed browning somewhat, transfers within the same jar improved establishment, excluding ferrous sulfate nearly eliminated browning	Barghchi (1986a)

P. vera, P. khinjuk, P. atlantica, P. palaestina	Seedling apical and axillary bud		Yes	Pre-wash for 60 min	Prewashing did not have much effect	Barghchi (1986b)
Populus ciliata	Young leaves from adult female trees	Yes	Yes	Grow cell suspensions in darkness	When cell suspensions were grown in the light there was a rapid accumulation of anthocyanin pigment and cell browning	Cheema (1989)
Pseudotsuga menziesii	Bud from adult trees		Yes	Avoid prolonged exposure to cytokinin	After being initially placed onto medium with high levels of BA, explants were transferred to medium without BA, shoots elongated and proliferated	Kurz (1986)
Pyrus communis	Shoot tip	Yes	Yes	Discard explants with tissue or medium discoloration	About 30% of shoot tips showed medium or tissue discoloration	Singha (1980)
Quercus lebani	Embryo from fruit	Yes		Incorporate cystein • HCl (10 mg/l) and PVP (200 mg/l) into medium	This treatment avoided blackening of the medium by phenolics	Srivastava and Steinhauer (1982)
Rhododendron simsii	Shoot and bud			Add 0.25% activated charcoal to the medium	The charcoal was necessary for elongation, perhaps from adsorption of toxic substances released in the medium by the explant	Dabin and Bouharmont (1983)
Rhododendron sp.	Shoot tip from seedlings and adults		Yes	Soak explants in citric acid (150 mg/l) ascorbic acid (100 mg/l), vary concn. NH_4 and NO_3 in MS medium, reduced KNO_3	When KNO_3 was reduced to ½ the MS medium level, browning and death symptoms of explants was eliminated	Anderson (1975)
Rhododendron sp.	Shoot tip from adults	Yes			Exudation was not related to explant growth and survival	Ettinger and Preece (1985)
Ribes grossularia	Shoot apices from adult field-grown plants	Yes	Yes	Use 2.2 μM BA in the medium at all IBA concentrations tested	With this concentration of BA, there were the most axillary shoots and green leaves	Welander (1985)

Table 1. (*Continued*)

Plant species	Explant type	Staining of medium	Browning of explant	Treatment to overcome browning	Results	Reference
R. uva-crispa	Shoot tip		Yes	Add NAA to the medium reduce NH_4NO_3 to 1/10 of the level in MS	Incorporation of 0.1 μM NAA resulted in 100% explant survival and decreased explant browning, changing the NH_4NO_3 level was ineffective	Wainwright and Flegmann (1985)
Rubus sp.	Shoot tip	Yes	Yes	Treat explants with ascorbic acid plus citric acid or cysteine • HCl, transfer to fresh medium after 1–2 days	Treating shoot tips was ineffective and caused further injury, transferring to fresh medium was effective	Broome and Zimmerman (1978)
Strelitzia reginae	Terminal bud, node, and inflorescence stalk	Yes	Yes	Pretreat explant with antioxidant, add 0.04% DTT (dithiothreitol) to liquid media, add 1% activated charcoal to agar solidified media, surface sterilize explants with 0.3% $HgCl_2$, rather than 9% $Ca(ClO)_2$, soak in low pH solution, incubate cultures in darkness	Unless multiple treatments were applied, oxidative browning of explants and staining of the medium led to explant necrosis and total death	Ziv and Halevy (1983)
Tectona grandis	Seedling parts and terminal and axillary buds from 100-year-old trees	Yes	Yes	Agitate for 45 min in 10 ml of various solutions containing 2% sucrose	5% H_2O_2 or 0.005% ascorbic acid reduced exudate, but explants died within a few weeks; multiple shoots only formed with 0.7% Polyclar AT (insoluble PVP) in the solution	Gupta et al. (1980)
T. grandis	Shoot tip from adult trees	Yes		Move explant to other location within same vessel after 48 h		Mascarenhas et al. (1987)

Theobroma cacao	Bud from seedlings		Immerse buds in liquid medium	On agar solidified medium, buds became covered with a viscous exudate that was inhibitory to nutrient uptake; this did not occur in liquid medium	Orchard et al. (1979)
T. cacao	Shoot tip and nonwoody node from seedlings		Wash explant under running tap water for 2.5 h	A mucilaginous exudate covered the newly excised explants, preventing surface sterilization, unless they received the pre-wash	Passey and Jones (1983)
Vaccinium corymbosum	Single-node from adult plants, axillary bud from the cultures established from the single nodes	Yes	Use WPM, rather than other media	Browning of cultures was severe with some media, but with WPM it was the least, and there were the most usable shoots	Wolfe et al. (1983)
Vitis vinifera	Shoot tip from adult plants	Yes	Select weaker shoots for explants; choose axillary, rather than terminal shoots for explants; select shaded shoots, rather than those grown in full sunlight for explants	When these choices were made, explant survival was greater and endogenous phenol content was less	Yu and Meredith (1986)

and Preece 1988b); thus, with these species, IBA may act as a protectant. Wainwright and Flegmann found that incorporation of 0.1 mM naphthaleneacetic acid (NAA) in the medium resulted in 100% survival and decreased explant browning of *Ribes uva-crispa*. However, when IBA was incorporated into the medium, there was a black exudate from *Castanea sativa* and eliminating the IBA resulted in no black exudate (Vieitez and Vieitez 1980a). Most micropropagation systems employ the use of cytokinins to rapidly increase axillary shoot production; the inclusion of auxin may be beneficial with some species if explant necrosis is a problem.

It is, however, well known that staining of the medium by the explant is often a sign of problems, and unless handled properly, lethal browning will overcome the tissue or organ. A dilemma in studying and understanding this phenomenon is that the cause(s) and effect(s) of exudation are confounded. It is not clear whether (1) exudation is a cause of explant distress; (2) if explants that are under a stress or have considerable amounts of wounded tissue stain the medium more than those that are more healthy; or (3) if both happen together, i.e., explants that are more traumatized stain the medium more and this exudation, in turn, is toxic to the plant tissue, thus resulting in faster death. For example, we have seen that staining of the medium from adult shoot tips of silver maple is not autotoxic; when explants were rinsed in scalding rinse water following NaClO sterilization, they all produced copious amounts of stain and died, whereas those that were rinsed with 25 °C water survived and produced little or no stain (Preece and Brown unpubl.). This is an example of the stain being a sign and not a cause of explant distress. Singha (1980) simply discarded the 30% of pear (*Pyrus communis* shoot tip explants that showed varying degrees of tissue or media discoloration. This indicated that, in this case, discoloration was also interpreted as a sign, not cause, of explant distress. Durand-Cresswell et al. (1982) reported two kinds of exudate, one that appeared within 1 h of excision of *Eucalyptus grandis* explants and appeared worse when high levels of sucrose, serine, chlorogenic acid, and cytokinins were added to the medium, or the cultures were incubated in the light. The second exudate appeared to be a product of dying cells and appeared towards the end of the incubation period. With this species, exudation seemed to be both a cause and sign of explant distress. In any case, if medium staining occurs, it may be a sign of trouble, and it may be necessary to take extra steps to avoid excessive explant necrosis.

In order to learn if exudates are toxic, Compton and Preece (1988b) conducted bioassay experiments. They monitored medium staining from shoot tips of blackberry (*Rubus* sp. Dirkson Thornless), black walnut (*J. nigra*), *Euphorbia lathyris*, *Hibiscus rosa-sinensis*, and *Rhododendron* P.J.M. Hybrids, and tested the effects of the exudation on tobacco callus growth and development. They replaced shoot tips of these species that had been on media for 7 days with 1 cm^3 pieces of tobacco callus. They found that tobacco callus growth was inhibited when media had contained *E. lathyris* or black walnut stem tips (532 and 2939 mm^3 of visible stained medium, respectively). Tobacco callus growth was not influenced by blackberry, hibiscus or rhododendron exudate (576, 0, and 1046 mm^3 of visible stained medium, respectively). Although hibiscus explants did not produce a visible brown-black exudate, a mucilaginous exudate was present, as indicated by the stickiness of the explants when they were removed from culture. They concluded that medium darkening varied with each species, and that the response of tobacco callus growth was not related solely to the degree of medium browning. Therefore, the species from which the exudate originated appeared to influence tobacco callus growth more than the degree of visible staining. In a related study, Durand-Cresswell et al. (1982) soaked *Eucalyptus grandis* explants in water and found that the water was inhibitory to cress seed germination and tomato rooting. When the water was treated with PVP, it was no longer inhibitory. Gould and Murashige (1985) found berberine to be among the exudates in the medium from *Nandina domestica* cultures. When added to the medium, berberine stimulated *Nandina* callus growth; at low concentrations it stimulated, but at high concentrations inhibited tobacco callus growth; and rooting of *Sequoia* stem cuttings was not affected by any concentration.

2 Phenolic Compounds

Isolated mesophyll protoplasts of *Brassica napus* and brown exudates in cultures isolated from these protoplasts were tested with two staining procedures for lignin (Heil and Kohlenbach 1989). Both the acidic phloroglucinol test for lignin and toluidine blue 0 showed a positive reaction. These might be useful to determine rather quickly and easily if the staining of a medium is from oxidized polyphenols.

Rarely have the biochemical state of the initial explant and the stain in the medium been elucidated (Durand-Cresswell et al. 1982; Gould and Murashige 1985; Yu and Meredith 1986; Scalbert et al. 1988; Welander 1988). Durand-Cresswell et al. (1982) analyzed the water in which *Eucalyptus grandis* explants had been soaked. They found that leaf discs produced more phenols than nodes, perhaps because the discs had received more wounding. When the phenols were more specifically characterized, they were shown to consist of tannins, flavonoids, and simple phenols. When the leaves were soaked for less than 1 day in the light, there were more total phenols, especially flavonoids, compared to soaking in the dark. Since the flavonoids were inhibitory to seed germination and elongation of tomato roots, it was postulated that they were among the inhibitory factors in the exudation of the eucalyptus explants. Gould and Murashige (1985) analyzed extracts of media that had contained *Nandina domestica* cultures and found that they contained detectable quantities of phenolic compounds, amino acids, proteins, carbohydrates, organic acids, and alkaloids. The yellow component of the stain was identified as the alkaloid berberine. Yu and Meredith (1986) analyzed grape (*Vitis vinifera*) shoot tip explant extracts for total phenol content. They found that there was a highly significant negative correlation ($r = -0.87$) between explant survival and total phenol content. The explants from axillary shoots that were weak and from the portions of the stock plant that were shaded survived better in vitro and had significantly lower levels of total phenols than those that were from vigorous terminal shoots excised from the portions of the stock plant that received full sunlight. Welander (1988) extracted birch buds and analyzed for total phenol content using HPLC. She found that buds that grew well in vitro had similar patterns of extracted phenolic compounds. When extracts from the buds were added to the culture medium, the chloroform extracts promoted growth; the water extracts caused browning of the medium, but did not inhibit growth; and the ethyl acetate extracts also caused a browning of the medium and at a concentration corresponding to five to ten buds, inhibited growth. The lowest concentration of ethyl acetate extract promoted growth in vitro.

Scalbert et al. (1988) analyzed calli and shoots from juvenile and adult trees of pedunculate oak (*Quercus robur*) for polyphenols. The polyphenols produced by the calli were mainly vescalagin and castalagin. Chromatographically, the calli from seedlings gave hexahydroxydiphenoyl esters in the form of a streak, whereas those from young and adult trees gave many other spots. However, only the agar medium from the seedling callus was stained. When the stained agar was extracted, it yielded low amounts of ellagic acid and the same streak of hexahydroxydiphenoyl esters. The authors reported that the agar media from the other calli and the shoot cultures were not stained and contained no polyphenols. The calli from the adult trees contained the most proanthocyanidins of the calli tested. The shoots had a lower phenol and tannin content than the calli that were tested. They identified the following polyphenols in the leaves derived from the shoot cultures: vescalagin, castalagin, and pedunculagin. However, in the stems there were only low amounts of the hexahydroxydiphenoyl esters vescalagin and castalagin. There was a higher degree of polyphenol oxidation in the calli than in the adult trees and in vitro-grown shoots. The authors reported that dark brown staining of agar media could not be related to high amounts of polyphenols in the in vitro-grown plant tissues, but rather to oxidized polyphenols present in the medium.

The frequently cited sources of oxidative browning of the explant and the stain in the medium are phenolics, oxidized phenolics, tannins, or oxidized polyphenols. Phenolic compounds in plants are, in part, synthesized through the shikimic acid, phenylpropanoid, flavonoid, and terpenoid pathways, and are present in abundant quantities in plants. Many important plant growth regulators and amino acids are also produced in, or result from the products of these pathways (Table 2). Complete biosynthetic pathways of these compounds have been outlined by Goodwin and Mercer (1983).

Free phenols exist only in trace amounts in plants; however, their glycosides and esters are present, usually in high amounts, in actively metabolizing tissues (Towers 1964). When phenolics are applied to plants, they detoxify the compounds by converting them to glycosides or other derivatives (Towers 1964).

Phenylalanine ammonia lyase (PAL) is a key enzyme in the production of lignin, phenolics, and some alkaloids (benzylisoquinolines) (Goodwin and Mercer 1983). PAL catalyzes the conversion of phenylalanine to cinnamic acid, which leads to the production of phenylpropanoid compounds or flavonoid glycosides. High PAL activity has been associated with tissues undergoing lignin biosynthesis, and has been proposed by some authors (Hillis and Ishikura 1970) to be controlled by phytochrome. Increases in

Table 2. Biosynthetic pathways of some important endogenous phytochemical

Compound	Pathway	Reference
Abscisic acid	Terpenoid	Walton (1987)
Indoleacetic acid	Shikimic acid	Goodwin and Mercer (1983)
Gibberellins	Terpenoid	Graebe (1983)
Phenylalanine	Shikimic acid	Goodwin and Mercer (1983)
Tyrosine	Shikimic acid	Goodwin and Mercer (1983)
Tryptophan	Shikimic acid	Goodwin and Mercer (1983)
Lignin	Shikimic acid and phenylpropanoid	Goodwin and Mercer (1983)

PAL activity occurred when pea (*Pisum* sp.) seedlings were illuminated with red light; however, a following far-red treatment nullified any red light stimulation. However, PAL is synthesized at relatively high rates in darkness (Smith 1973); therefore the role of light is questionable. Nevertheless, several authors have reported that there was less of a problem with staining of the medium and browning of explants when cultures were incubated in darkness (Table 1).

PAL activity has also been shown to be higher in expanding eucalyptus (*Eucalyptus sideroxylon* and *E. sieberi*) leaves than in fully expanded leaves approaching senescence (Hillis and Ishikura 1970). Margna (1977) stated that there is a balanced relationship between protein metabolism and the production of flavonoids and cinnamic acid derivatives. When protein synthesis is favored (e.g., high nitrogen levels), a decrease in phenylpropanoid accumulation can be observed (low PAL activity), whereas phenylpropanoid biosynthesis is favored under conditions that inhibit protein synthesis. The latter generally occurs under conditions of environmental stress or pathogen attack (e.g., mineral deficiency, water stress, senescence, wounding, and addition of antibiotics). Therefore, it may be possible that explant wounding increases or stimulates the activity of PAL, or other related enzymes in vitro, thus increasing exudation; however, additional research in this area is required.

Phytoalexins are believed to be a major defense component in plant disease resistance. Studies have shown that phytoalexins are produced in response to wounding and pathogen infection (biotic or abiotic elicitors). By definition, phytoalexins are low molecular weight compounds that are produced by the host plant and express broad-spectrum antimicrobial activity, but are not found in healthy tissue (Dixon et al. 1983). Phytoalexins may be derived from the phenylpropanoid pathway (isoflavonoids), the terpenoid pathway, or from acetate-polymalonate and shikimate-polymalonate (Dixon et al. 1983). Members of the Leguminosae (*Pisum sativum, Phaseolus vulgaris*, and *Glycine max*) and the Solanaceae (*Solanum* spp., *Capsicum annuum, Nicotiana* spp., *Lycopersicon esculentum*, and *Datura stramonium*) have been most commonly studied. PAL is one enzyme that has been associated with phytoalexin production. Cramer et al. (1985) observed an increase in PAL synthesis in suspension cultures of *Phaseolus vulgaris* treated with elicitor (*Colletotrichum lindemuthianum*). Increases in chalcone synthase and chalcone isomerase were also observed. The ability to produce phytoalexins may be important when attempting to incorporate disease resistance into important plant species. However, since wounding stimulates the production of these compounds, explant preparation techniques may, in part, account for the production and release of autotoxins in vitro.

In vivo, many plants produce compounds that act as natural herbicides and herbivore deterrents or repellents. These compounds are generally referred to as allelopathic agents (compounds), and may belong to one of the following groups; alkaloids, coumarins, flavonoids, long-chain fatty acids and acetylenes, organic acids, phenolic acids, purines, quinones, steroids, terpenoids, unsaturated lactones, and others (Rice 1984). These agents may regulate plant growth by inhibiting cell division, interacting with hormones, and inhibiting enzyme activity (Einhellig 1986). Allelopathic agents may also influence respiration, photosynthesis, and nutrient and water uptake. These compounds can be transferred to other plants through volatilization, root exudation, leachates from shoots, and residue decomposition.

Perry (1932) demonstrated that actively growing black walnut (*Juglans nigra*) trees can kill white pine (*Pinus strobus*) and black locust (*Robina pseudoacacia*) when they are co-cultivated. However, allelopathic effects of black walnut may depend on soil type. Fisher (1978) reported that black walnut inhibited red pine (*Pinus resinosa*) and white pine on poorly drained sites, but in well-drained areas, white pine appeared to decrease black walnut growth. The following species have also been shown to express allelopathic effects on the growth of surrounding plants: *Ailanthus altissima* (Mergen 1959), *Quercus falcata* Pagodaefolia, *Q. michauxii*, *Q. shumardii* (Hook and Stubbs 1967), and *Festuca arundenacea* Kentucky-31 (Peters 1968). It is possible that such allelopathic compounds are present in explant exudates; however, Compton and Preece (1988a) found that although black walnut exudate was inhibitory to tobacco callus, juglone stimulated tobacco callus growth.

Allelopathic agents have also been shown to act as herbivore deterrents and repellents. Deterrents are compounds that inhibit insect feeding or oviposition, whereas repellents cause an animal to make a directed move away from, or avoid the chemical source (Rice 1983). Many plants in the Solanaceae produce chemicals that deter the potato beetle (Rice 1983) and other insects (Lichtenstein et al. 1962; Lichtenstein and Casida 1963). Apple (*Malus* sp.), cottonwood (*Populus deltoides*), black locust, and bur oak (*Q. macrocarpa*) produce flavonoids that deter the elm bark beetle (Norris 1977). Horse chestnut (*Aesculus octandra*) and white ash (*Fraxinus americana*) produce coumarins that act as deterrents, whereas box elder (*Acer negundo*), silver maple (*A. saccharinum*), and *Magnolia acuminata* produce alkaloid deterrents (Rice 1983). Members of the Juglandaceae produce juglone, which repels the elm bark beetle. Norris (as cited in Rice 1983) reported that in healthy hickory (*Carya* sp.) cells, juglone exists mainly as a glucoside, but also occurs in the atmosphere around the trees. He also stated that hydroxylation of juglone increases its effectiveness as a repellent. Compounds that act as allelopathic agents or herbivore deterrents and repellents in nature may be released in vitro. Because of the limited environment in the culture vessel, these substances may accumulate to autotoxic levels.

3 Oxidation of Phenolics

When tissues are injured, such as when explants are cut off the stock plant and during preparation for the in vitro environment, various compounds are often released that are oxidized. Polyphenols can be oxidized by air (Robertson 1983), peroxidases (Mayer and Harel 1979; Vaugh and Duke 1984), or polyphenoloxidases (Mayer and Harel 1979; Hu and Wang 1983; Vaugh and Duke 1984), and turn brown or black. This may be the reason for the widely reported darkening of both tissue and culture medium. Polyphenoloxidases, including catechol oxidase, are confined to plastids in healthy, nonsenescing tissues (Vaugh and Duke 1984). When plants are injured, as in explant preparation, phenolic compounds that were largely located in the vacuole are mixed with the contents of plastids and other organelles, and then the dark pigmentation associated with oxidized polyphenolic compounds appears. When phenols become oxidized they form compounds called quinones. These are highly reactive compounds that polymerize rapidly and form covalent bonds with proteins (Loomis and Battaile 1966). Oxidized polyphenolic compounds inhibit enzyme activity (Hu and Wang 1983) and may result in the decline or lethal browning of the explants.

Forrest (1969) analyzed callus clumps and cell aggregates of *Camellia sinensis* (tea plant). It was reported that the large cell clumps that grew slowly had greater polyphenol oxidase activity than the faster-growing smaller cell aggregates. It was concluded that the synthesis of anthocyanin, flavonols, and polyphenol oxidase was generally inversely related to growth rate of the cultures.

Many plants contain polyphenolic compounds that may function as inhibitory agents. In addition to polyphenols, plants may contain other inhibitory compounds: coumarins, terpenoids, and steroids (Muhitch and Fletcher 1985). These inhibitory agents have various phytotoxic effects on plants which may result in dwarfing or even death.

4 Conclusions

Micropropagation of woody and other species can be much more difficult and labor-intensive if inhibitory exudates are produced. In some cases, once browning of the explant material begins, it is extremely difficult to rescue the organ and reverse or stop the problem. Although exudation is not always a sign of serious problems with the explant, it is generally a cause for concern because it is frequently a sign of explant distress.

The nature and chemistry associated with staining of the culture medium and browning of tissues in vitro are not yet fully understood. It is well known that juvenile forms of woody plants respond better in vitro than adult forms of many species. Oxidation products may be partially responsible for this. We have observed that shoot explants from adult black walnut produce abundant exudate and are much more difficult than juvenile forms to establish in vitro (Preece, Huetteman, and Heile-Sudholt unpubl.). Welander (1988) reported that exudate from buds excised from birch seedlings was not autoinhibitory; however, she transferred adult buds to fresh medium after 2 days. Kurz (1986) reported that adult shoot tips of *Chamaecyparis nootkatensis* declined from oxidative browning to the point where they would not respond, but juvenile buds could be established and proliferated and no browning was reported. Interestingly, in Table 1, many of the papers that are summarized reported browning with explants from adult stock plants.

Those working with in vitro cultures should be alerted to watch carefully for dark discoloration of the medium and/or explant. This appears quite different from and should not be confused with microbial contamination. If decline of explants seems to be associated with the discoloration, further steps are warranted. If, however, explants appear normal and the medium becomes discolored, perhaps the extra steps with that particular species are a waste of time. Although there are many techniques that have been shown to be successful in overcoming explant establishment and survival problems associated with discoloration, we have had our greatest successes with rapid transfers within the same vessel or to fresh medium, or with explant pre-soaks for times up to 4 h.

References

Adams RM II, Koenigsberg SS, Langhans RW (1979) In vitro propagation of *Cephalotus follicularis* (Australian Pitch Plant). HortScience 14:512–513

Alderson PG, Barghchi M (1982) The potential for *Pistacia* propagation through shoot, leaf and callus culture. In: Fujiwara A (ed) Plant Tissue and Cell Culture 1982, Maruzen, Tokyo, pp 739–740

Anderson WC (1975) Propagation of rhododendrons by tissue culture: Part 1. Development of a culture medium for multiplication of shoots. Comb Proc Int Plant Prop Soc 25:129–133

Ashby WC, Preece JE, Huetteman CA, Bresnan DF, Roth PL (1987) Silver maple tree improvement for biomass production. In: Proc Fifth North Central Tree Improvement Confer, North Central Tree Improvement Assoc, Fargo, North Dakota, pp 6–23

Barghchi M (1986a) In vitro culture of mature commercial *Pistacia vera* L. cultivars. Comb Proc Int Plant Prop Soc 35:331–333

Barghchi M (1986b) In vitro micropropagation of *Pistacia* rootstocks. Comb Proc Int Plant Prop Soc 35:334–337

Barghchi M, Alderson PG (1983) In vitro propagation of *Pistacia* species. Acta Hortic 131:49–60

Barghchi M, Alderson PG (1985) In vitro propagation of *Pistacia vera* L. and the commercial cultivars Ohadi and Kalleghochi. J Hortic Sci 60:423–430

Biondi S, Canciani L, de Paoli G, Bagni N (1981) Shoot formation from bud cultures of mature chesnut. In: Colloque International sur la Culture in vitro des Essences Forestieres. AFOCEL Nangis, France, pp 181–185

Broome OC, Zimmerman RH (1978) In vitro propagation of blackberry. HortScience 13:151–153

Cheema GS (1989) Somatic embryogenesis and plant regeneration from cell suspension and tissue cultures of mature himalayan poplar (*Populus ciliata*) Plant Cell Rep 8:124–127

Chevre A-M, Gill SS, Mouras A, Salesses G (1983) In vitro vegetative multiplication of chestnut. J Hortic Sci 58:23–29

Compton ME, Preece JE (1986) Exudation and explant establishment. Newsl Int Assn Plant Tissue Cult 50:9–18

Compton ME, Preece JE (1988a) Effects of phenolic compounds on tobacco callus and blackberry shoot cultures. J Am Soc Hortic Sci 113:160–163

Compton ME, Preece JE (1988b) Response of tobacco callus to shoot tip exudation from five species. HortScience 23:208–210

Cooper PA, Cohen D (1985) Micropropagation of Japanese persimmon (*Diospyros kaki*). Comb Proc Int Plant Prop Soc 34:119–124

Cramer CL, Bell JN, Ryder TB, Bailey JA, Schuch W, Bolwell GP, Robbins MP, Dixon RA, Lamb CJ (1985) Co-ordinated synthesis of phytoalexin biosynthetic enzymes in biologically stressed cells of bean (*Phaseolus vulgaris* L.) EMBO J 4:285–289

Cresswell R, Nitsch C (1975) Organ culture of *Eucalyptus grandis* L. Planta 125:87–90

Dabin P, Bouharmont J (1983) Application of in vitro cultures in azalea (*Rhododendron simsii*). Acta Hortic 131:89–93

De Fossard RA, Bennett MT, Gorst JR, Bourne RA (1978) Tissue culture propagation of *Eucalyptus ficifolia* F. Muell. Comb Proc Int Plant Prop Soc 28:427–435

Dixon RA, Dey PM, Lamb CJ (1983) Phytoalexins: Enzymology and molecular biology. In: Meister A (ed) Advances in enzymology, Vol 55. John Wiley, New York, pp 1–136

Durand-Cresswell R, Boulay M, Franclet A (1982) Vegetative propagation of eucalyptus. In: Bonga JM, Durzan DJ (eds) Tissue culture in forestry. Martinus Nijhoff, The Hague, pp 150–181

Einhellig FA (1986) Mechanisms and mode of action of allelochemicals. In: Putnam AR, Tang C-S (eds) The science of allelopathy. John Wiley, New York, pp 171–188

Ettinger TL, Preece JE (1985) Aseptic micropropagation of *Rhododendron* PJM hybrids. J Hortic Sci 60:269–274

Fisher RF (1978) Juglone inhibits pine growth under certain moisture regimes. Soil Sci Soc Am J 42:801–803

Forrest GI (1969) Studies on the polyphenol metabolism of tissue cultures derived from the tea plant (*Camellia sinensis* L.). Biochem J 113:765–772

Goodwin TW, Mercer EI (1983) Introduction to plant biochemistry, 2nd edn. Pergamon, Oxford

Gould JH, Murashige T (1985) Morphogenic substances released by plant tissue cultures. Plant Cell Tissue Organ Cult 4:29–42

Graebe JE (1987) Gibberellin biosynthesis and control. Annu Rev Plant Physiol 38:419–465

Gupta PK, Nadgir AL, Mascarenhas AF, Jagannathan V (1980) Tissue culture of forest trees: clonal multiplication of *Tectona grandis* L. (teak) by tissue culture. Plant Sci Lett 17:259–268
Heil BM, Kohlenbach HW (1989) Acidic phloroglucinol test positive reaction in cultures of isolated mesophyll protoplasts of *Brassica napus* L. and with secreted brown exudates. Biol Plant 31: 100–103
Heile-Sudholt C, Huetteman CA, Preece JE, Van Sambeek JW, Gaffney GR (1986) In vitro embryonic axis and seedling shoot tip culture of *Juglans nigra* L. Plant Cell Tissue Organ Cult 6:189–197
Heyser JW, Mott RL (1980) The relationship between the production of phenolic compounds and the decline in growth of loblolly pine cultures. Plant Physiol 65(Suppl):492 (Abstr)
Hillis WE, Ishikura N (1970) The biosynthesis of polyphenols in tissues with low phenylalanine ammonia lyase activity. Phytochemistry 9:1517–1528
Hook DD, Stubbs J (1967) An observation of understory growth retardation under three species of oaks. US For Ser Res Note SE-70
Hu CY, Wang PJ (1983) Meristem, shoot tip, and bud cultures. In: Evans DA, Sharp WR, Ammirato PV, Yamada Y (eds) Handbook of plant cell culture, Vol 1. Macmillan, pp 177–227
Hutchinson JF (1984) Micropropagation of Northern Spy apple rootstock. Comb Proc Int Plant Prop Soc 34:38–48
Jarret RL, Rodriguez W, Fernandez R (1985) Evaluation, tissue culture propagation, and dissemination of Saba and Pelipita plantains in Costa Rica. Sci Hortic 25:137–147
Kerns HR, Meyer MM Jr (1988) Micropropagation of selected *Acer* sp. and cultivars. HortScience 23:452 (Abstr)
Klimaszewska K (1989) Plantlet development from immature zygotic embryos of hybrid larch through somatic embryogenesis. Plant Sci 63:95–103
Kurz ML (1986) Tissue culture of conifers. Comb Proc Int Plant Prop Soc 35:206–216
Lakshm Sita G, Shobha Rani B (1985) In vitro propagation of *Eucalyptus grandis* L. by tissue culture. Plant Cell Rep 4:63–65
Lenartowicz A, Millikan DF (1977) Meristem culture of *Juglans nigra* L. Trans Ill State Acad Sci 70:193
Lichtenstein EP, Casida JE (1963) Myristicin, an insecticide and synergist occurring naturally in the edible parts of parsnips. J Agric Food Chem 11:410–415
Lichtenstein EP, Strong FM, Morgan DG (1962) Identification of 2-phenylethylisothiocyanate as an insecticide occurring naturally in the edible part of turnips. J Agric Food Chem 10:30–33
Lloyd G, McCown B (1980) Commercially-feasible micropropagation of mountain laurel, *Kalmia latifolia*, by use of shoot-tip culture. Comb Proc Int Plant Prop Soc 30:421–427
Loomis WD, Battaile J (1966) Plant phenolic compounds and the isolation of plant enzymes. Phytochemistry 5:423–438
Margna U (1977) Control at the level of substrate supply – an alternative in the regulation of phenylpropenoid accumulation in plant cells. Phytochemistry 16:419–426
Mascarenhas AF, Kendurkar SV, Gupta PK, Khuspe SS, Agrawal DC (1987) Teak. In: Bonga JM, Durzan DJ (eds) Cell and tissue culture in forestry Vol 3. Martinus Nijhoff, Dordrecht, pp 300–315
Mayer AM, Harel E (1979) Polyphenol oxidases in plants. Phytochemistry 18:193–215
McComb JA (1978) Clonal propagation of woody plants using tissue culture, with special reference to apples. Comb Proc Int Plant Prop Soc 28:413–426
McComb JA, Newton S (1981) Propagation of kangaroo paws using tissue culture. J Hortic Sci 56:181–183
McCown B, Amos R (1979) Initial trials with commercial micropropagation of birch selections. Comb Proc Int Plant Prop Soc 29:387–393
McGranahan GH, Driver JA, Tulecke W (1987) Tissue culture of *Juglans*. In: Bonga JM, Durzan DJ (eds) Cell and tissue culture in forestry. Martinus Nijhoff, Dordrecht, pp 261–271
Mergen F (1959) A toxic principle in the leaves of *Ailanthus*. Bot Gaz 121:32–36
Monaco LC, Söndahl MR, Carvalho A, Crocomo OJ, Sharp WR (1977) Applications of tissue culture in the improvement of coffee. In: Reinert J, Bajaj YPS (eds) Applied and Fundamental Aspects of plant cell, tissue, and organ culture. Springer, Berlin Heidelberg New York, pp 109–129
Morini S, Sciutti R (1989) Browning of tissues in in vitro propagation. Plant Propagator 1:10–12
Muhitch MJ, Fletcher JS (1985) Influence of culture age and spermidine treatment on the accumulation of phenolic compounds in suspension cultures. Plant Physiol 78:25–28
Norris DM (1977) Role of repellents and deterrents in feeding of *Scolytus multistriatus*. In: Hedin PA (ed) Host Plant Resistance to Pests. American Chemical Society, Washington, DC, pp 215–230

Orchard JE, Collin HA, Hardwick K (1979) Culture of shoot apices of *Theobroma cacao*. Physiol Plant 47:207–210

Passey AJ, Jones OP (1983) Shoot proliferation and rooting in vitro of *Theobroma cacao* L. type amelonado. J Hortic Sci 58:589–592

Pérez C, Rodríguez R, Tamés RS (1985) In vitro filbert (*Corylus avellana* L.) micropropagation from shoot and cotyledonary node segments. Plant Cell Rep 4:137–139

Perry GS (1932) Some tree antagonisms. Proc Pa Acad Sci 6:136–141

Peters EJ (1968) Toxicity of tall fescue to rape and birdsfoot trefoil seeds and seedlings. Crop Sci 8:650–653

Rajeevan MS, Pandey RM (1983) Propagation of papaya through tissue culture. Acta Hortic 131:131–139

Read PE, Hosier MA, Yang Q (1985) Tissue culture of chestnuts. In: Payne JA (ed) 76th Annu Rep, Northern Nut Growers Assoc, Southwest Missouri State Univ, Springfield

Rice EL (1983) Pest control with nature's chemicals: allelochemicals and pheromones in gardening and agriculture. University of Oklahoma Press, Norman

Rice EL (1984) Allelopathy. 2nd edn. Academic Press, Orlando

Ripley KP, Preece JE (1986) Micropropagation of *Euphorbia lathyris* L. Plant Cell Tissue Organ Cult 5:213–218

Robertson T (1983) The organic constituents of higher plants. 5th edn. Cordus, North Amherst, MA

Rodriguez R (1982) Stimulation of multiple shoot-bud formation in walnut seeds. HortScience 17:592

Samartin A (1989) A comparative study of effects of nutrient media and cultural conditions of shoot multiplication of in vitro cultures of *Camellia japonica* explants. J Hortic Sci 64:73–70

Scalbert A, Monties B, Favre J-M (1988) Polyphenols of *Quercus robur*: adult tree and in vitro grown calli and shoots. Phytochemistry 27:3483–3488

Singha S (1980) In vitro propagation of Seckel pear. In: Proc Conf Nursery Production of Fruit Plants Through Tissue Culture – Applications and Feasibility. USDA, Beltsville, Maryland, pp 59–63

Smith H (1973) Regulatory mechanisms in the photocontrol of flavonoid biosynthesis. In: Milborrow BV (ed) Biosynthesis and its control in plants. Academic Press, New York, pp 303–321

Somers PW, van Sambeek JW, Preece JE, Gaffney G, Myers O Jr (1982) In vitro micropropagation of black walnut (*Juglans nigra* L.). In: Thielges BA (ed) Proc, Seventh North American Forest Biology Workshop. University of Kentucky Press, Lexington, pp 224–230

Srivastava PS, Steinhauer A (1982) In vitro culture of embryo segments of *Quercus lebani*: Organogenesis and callus growth as a differential response to experimental conditions. Z Pflanzenphysiol 106:93–96

Stefan SJ, Millikan DF (1985) Tissue culture of black walnut. In: Payne JA (ed) 76th Annu Report, Northern Nut Growers Association, Southwest Missouri State Univ, Springfield, pp 99–102

Stefan SJ, Millikan DF (1987) Tissue culture of black walnut. Walnut Council Bull 14:4, 7, 15

Stevenson JH, Harris RE (1980) In vitro plantlet formation from shoot-tip explants of *Fuchsia hybrida* cv. Swingtime. Can J Bot 58:2190–2192

Sutter EG, Barker PB (1985) In vitro propagation of mature *Liquidambar styraciflua*. Plant Cell Tissue Organ Cult 5:13–21

Swamy RD, Sahijram L (1989) Micropropagation of banana from male floral apices cultured in vitro. Sci Hortic 40:181–188

Towers GHN (1964) Metabolism of phenolics in higher plants and micro-organisms. In: Harborne JB (ed) Biochemistry of phenolic compounds. Academic Press, London, pp 249–294

Uno S, Preece JE (1987) Micro- and cutting propagation of 'Crimson Pygmy' barberry. HortScience 22:488–491

Vaugh KC, Duke SO (1984) Function of polyphenol oxidase in higher plants. Physiol Plant 60:106–112

Vieitez AM, Vieitez E (1980a) Plantlet formation from embryonic tissue of chestnut grown in vitro. Physiol Plant 50:127–130

Vieitez AM, Vieitez ML (1980b) Culture of chestnut shoots from buds in vitro. J Hortic Sci 55:83–84

Vieitez AM, Ballester A, Vieitez E (1983) In vitro plantlet regeneration of mature chestnut. J Hortic Sci 58:457–463

Wainwright H, Flegmann AW (1985) The micropropagation of goose berry (*Ribes uva-crispa* L.): I. Establishment in vitro. J Hortic Sci 60:215–221

Walkey DG (1972) Production of apple plantlets from axillary-bud meristems. Can J Plant Sci 52:1085–1087

Walton DC (1987) Abscisic acid biosynthesis and metabolism. In: Davies PI (ed) Hormones and their role in plant growth and development. Martinus Nijhoff, Dordrecht, pp 113–131
Welander M (1985) Micropropagation of gooseberry. Sci Hortic 26:267–272
Welander M (1988) Biochemical and anatomical studies of birch (*Betula pendula* Roth.) buds exposed to different climatic conditions in relation to growth in vitro. In: Hanover JW, Keathley DE (eds) Genetic manipulation of woody plants. Plenum, New York, pp 79–99
Welsh K, Sink KC, Davidson H (1979) Progress on in vitro propagation of red maple. Comb Proc Int Plant Prop Soc 29:383–386
Werner EM, Boe AA (1980) In vitro propagation of Malling 7 apple rootstock. HortScience 15:509–510
Wolfe DE, Eck P, Chin C-K (1983) Evaluation of seven media for micropropagation of highbush blueberry. HortScience 18:703–705
Yang Q-G, Read PE, Fellman CD, Hosier MA (1986) Effect of cytokinin, IBA, and rooting regime on Chinese chestnut cultured in vitro. HortScience 21:133–134
Yu D-H, Meredith CP (1986) The influence of explant origin on tissue browning and shoot production in shoot tip cultures of grapevine. J Am Soc Hortic Sci 111:972–975
Zimmerman RH, Broome OC (1980) Apple cultivar micropropagation. In: Proc Conf Nursery Production of Fruit Plants Through Tissue Culture – Applications and Feasibility. USDA, Beltsville, Maryland, pp 54–58
Ziv M, Halevy AH (1983) Control of oxidative browning and in vitro propagation of *Strelitzia reginae*. HortScience 18:434–436

I.11 Woody Plant Micropropagation with Cytokinins

J.W. Einset[1]

1 Introduction

Micropropagation is an important technology for multiplying several species of woody plants. In the US alone, about 30 million woody plant propagules for the nursery trade, with a market value of approximately US $15 million, are produced annually by micropropagation. When compared with other potential markets in plant biotechnology (e.g., hybrid seed corn is a $1.5 billion-per-annum market in the US), the commercial value of woody plant micropropagation is small. Nevertheless, micropropagation's impact on woody plant technology during the last 15 years has been substantial (Torrey 1985; Zimmerman et al. 1986). If the technology could be extended to hard-to-propagate species that are not currently amenable to micropropagation, its commercial use as a tool for rapid clonal multiplication and for producing disease-free propagules could be expanded significantly (Kyte 1983; Dirr and Heuser 1987; Einset 1987b).

Two basic strategies have been used to micropropagate woody species; namely, callus regeneration and shoot multiplication (Einset 1986b). Unfortunately, callus regeneration often results in a high percent of somaclonal variants, making this strategy of questionable value for large-scale clonal multiplication of superior woody plant cultivars. Shoot multiplication, on the other hand, is a reliable method that can be used when the production of large numbers of true-to-type clones is required. In fact, it is estimated that shoot multiplication has been used to produce well over 5 billion plants (primarily nonwoody species, i.e., foliage plants such as syngoniums and ferns, as well as plants for flower production such as pyrethrums and *Gerbera*) during the last 15 years (Zimmerman et al. 1986).

In principle, micropropagation by shoot multiplication exploits the effect of cytokinin in regulating shoot growth under controlled in vitro conditions (Skoog and Miller 1957). By artificially elevating the endogenous cytokinin concentration of shoots, continued growth is sustained and apical dominance is overcome. The result is the production of highly branched axes which are subdivided at the end of each passage in vitro. Individual shoots can then be used for a subsequent round of shoot multiplication or, alternatively, for plantlet production after a rooting treatment (Murashige 1974; Lloyd and McCown 1980).

Because of cytokinins' crucial role in micropropagation by shoot multiplication, research to improve and expand the scope of this technology with woody plants has focused on this important class of phytohormones. This review summarizes recent work dealing with the effects of cytokinins on a broad range of woody species as well as

[1]EniChem Americas, 2000 Princeton Park, Corporate Center, Monmouth Junction, NJ 08852, USA

Biotechnology in Agriculture and Forestry, Vol. 17
High-Tech and Micropropagation I (ed. by Y.P.S. Bajaj)

with endogenous cytokinin production in woody plants using *Actinidia* (kiwifruit) as a model. Two fundamental questions are addressed:

1. To what extent is the principle of cytokinin-controlled shoot growth (i.e., micropropagation by shoot multiplication) generally applicable to woody dicotyledonous species?
2. How and where do woody plants make cytokinins? What is the cytokinin requirement of shoots grown in vitro?

2 Comparative Physiology

2.1 Survey of Taxa

Although the precise physiological role of cytokinins is still unknown, early studies focused on their possible involvement in controlling shoot (stem and leaf) growth. The classic demonstration that shoot-produced auxin is responsible for the inhibition of lateral bud growth (Thimann and Skoog 1934) led to the development by Skoog and Tsui (1948) of an in vitro system, using tobacco stem segments, to study stimulatory as well as inhibitory factors regulating shoot growth. This line of investigation, in turn, resulted in the discovery of kinetin (Miller et al. 1956), the first cytokinin, and the demonstration that growth and organogenesis in tobacco tissue cultures could be regulated by the balance of cytokinin and auxin in the culture medium (Skoog and Miller 1957).

When cytokinins are incorporated into media for micropropagation by shoot multiplication, the artificially high cytokinin concentration produced within shoots stimulates continued growth and it also inhibits apical dominance resulting in a branched axis. To determine whether this same strategy could be applied to woody species generally, we conducted a broadly based survey examining the in vitro response of shoots from several different taxa (Einset 1984a, 1985b). The tests used explants from seedlings (<1 year old) incubated on a basal nutrient medium containing inorganic (Murashige and Skoog 1962) and organic (Linsmaier and Skoog 1965) constituents as specified plus 5 mg/l pyridoxine-HCl and 5 mg/l nicotinic acid with or without a cytokinin supplement. Species were screened by transferring growing shoot tips (0.5–1 cm) from seedlings to four different media: (1) the basal medium without a cytokinin supplement, (2) basal medium with 30 μM i^6 Ade, (3) basal medium with 2.2 μM thidiazuron (Einset and Alexander 1984) and (4) basal medium with 33 μM N^6-benzyladenine (bzl^6 Ade) and 0.6 μM indole-3-acetic acid (IAA). Each species was tested on at least three different occasions, with four replications per treatment in each experiment. The cultures (20 ml per tube) were incubated at 27 °C with a 16-h light (40–65 $\mu E/m^2/s$) and 8-h dark photoperiod. Continued shoot growth by a majority of the shoots on at least one or more of the three different cytokinin media was considered a positive (+) response, while no growth or only sporadic growth during the 6-week incubation period was considered a negative (–) response. In most cases of positive response, shoots were subcultured for at least one passage to demonstrate sustained growth in vitro.

As Table 1 shows, 126 different woody species, representing 5 subclasses, 33 families, and 67 genera, were tested for in vitro response (Einset 1985b). As is apparent, shoot responses vary dramatically between taxa. Growth is obtained consistently, for

Table 1. Taxa of woody plants tested for cytokinin response in tissue cultures

Subclass	Order	Family	No. of species +	–	Total
Magnoliidae	Magnoliales	Magnoliaceae	0	6	6
		Annonaceae	0	1	1
	Laurales	Calycanthaceae	0	3	3
		Lauraceae	1	2	3
	Ranunculales	Ranunculaceae	1	1	2
Hamamelidae	Fagales	Betulaceae	6	0	6
Dilleniidae	Theales	Dipterocarpaceae	0	1	1
		Theaceae	1	3	4
		Guttiferae	1	0	1
	Malvales	Malvaceae	1	0	1
		Tiliaceae	0	6	6
		Sterculiaceae	0	3	3
	Urticales	Ulmaceae	0	6	6
		Moraceae	2	1	3
	Ericales	Actinidiaceae	5	0	5
		Ericaceae	3	0	3
		Clethraceae	1	0	1
		Cyrillaceae	1	0	1
Rosidae	Rosales	Rosaceae	5	0	5
		Saxifragaceae	1	0	1
	Fabales	Leguminosae	13	0	13
	Cornales	Cornaceae	3	3	6
		Nyssaceae	0	1	1
	Sapindales	Aceraceae	0	10	10
		Staphylaceae	0	2	2
		Sapindaceae	0	1	1
		Rutaceae	7	1	8
		Anacardiaceae	2	3	5
Asteridae	Gentianales	Oleaceae	8	0	8
	Lamiales	Verbenaceae	2	0	2
	Scrophulariales	Scorphulariaceae	1	0	1
		Bignoniaceae	5	0	5
	Rubiales	Rubiaceae	2	0	2

example, among species of subclass Asteridae but it is rare with species in subclass Magnoliidae. With species in subclasses Dilleniidae and Rosidae, shoot responsiveness is scattered. In Dilleniidae, for instance, positive species were concentrated in the order Ericales while in the Rosidae all of the species tested in the orders Rosales and Fabales were positive and species in other orders were usually negative.

Taxa designated as negative (–) can be subdivided further into two categories: those showing a sporadic response versus those exhibiting a zero response in all replicates. Approximately 5% of the shoot tips of *Magnolia soulangiana* tested, for example, gave a significant growth response on 30 μM i^6 Ade. When this occurred, it was possible to subculture the resulting shoots by monthly passages on fresh medium for at least 1 year. A similar sporadic response is observed with other Magnoliaceae as well as with *Stewartia pseudocamellia* (Theaceae), *Ulmus parvifolia* (Ulmaceae), and ten

Acer (Aceraceae) species. On the other hand, shoots from six different *Tilia* species were consistently unresponsive, in spite of repeated attempts to obtain viable cultures.

Thus, as far as micropropagation is concerned, woody species can be classified into three different groups (Einset 1985b). The first group contains those species that exhibit an inherent response as explants, i.e., positive (+) species in our screen. The second group contains species that respond sporadically as explants or only after repeated subculturing, a phenomenon that has been referred to as acclimation, phase transition, or habituation (Lloyd and McCown 1980). Finally, the third group consists of species that are unresponsive to cytokinin treatments.

2.2 Evolutionary Considerations

Figure 1 (Einset 1985b) illustrates the shoot response data in relation to the systematic placement of taxa according to Cronquist (1981). Clearly, the distribution of responsive versus unresponsive species is nonrandom.

It is also of interest that several of the taxa responding positively as seedlings give a negative response when shoot tips from older (i.e., >1 year) plants are used (Murashige 1974). This phenomenon, which is often cited as an example of the "juvenile" to "adult" transition in woody plants, obviously restricts the range of species amenable to micropropagation. It also probably reflects an important aspect of the

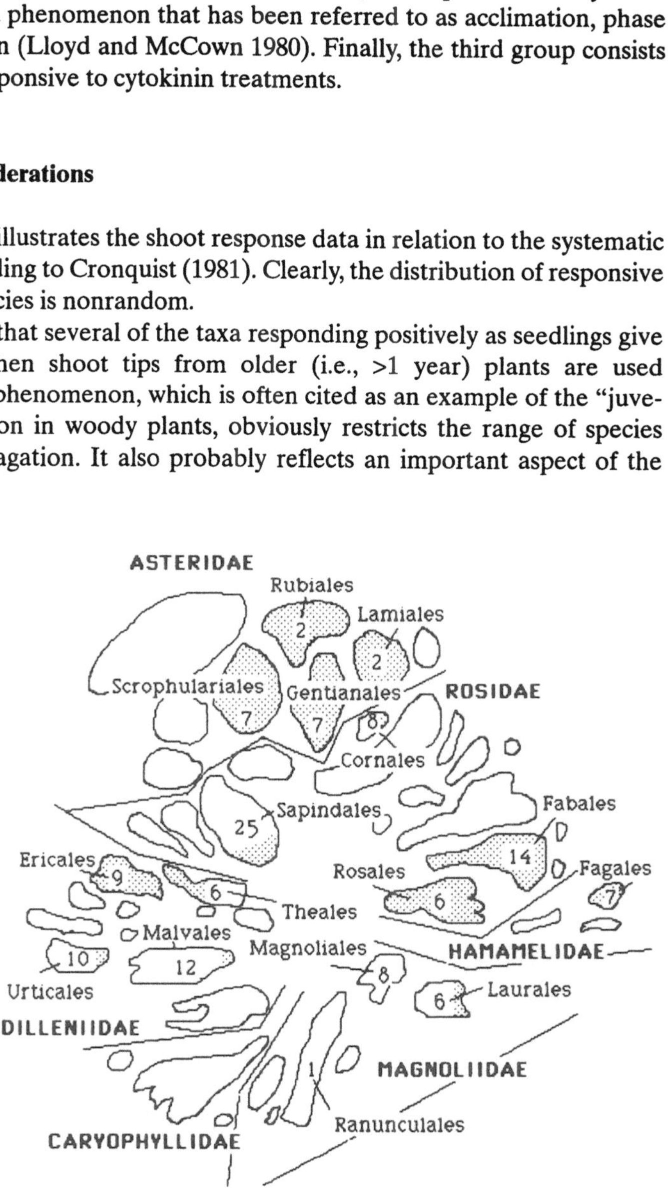

Fig. 1. Framework relating the cytokinin responses of woody shoots in vitro to dicot systematics. *Numbers* indicate species within each order that have been tested, while *stippling* shows the percent giving a positive growth response. Subclasses: 2+, 3+/–; orders: 7+, 7+/–, 2–; families: 15+, 5+/–, 11–; species: 70+, 56–. (Einset 1985b)

developmental processes occurring during shoot maturation, especially when one considers it together with the fact that no species are known in which shoot tips from older plants respond in vitro while tips from seedlings fail to grow.

If one assumes, as a hypothesis, that shoot maturation in plants normally proceeds through a responsive to unresponsive developmental sequence, then differences between taxa can be thought of in terms of changes in the relative timing of events in ontogeny (heterochrony) as a result of evolution (Gould 1977; Einset 1987a). Consider the six *Cornus* species tested, for instance, using shoot tips from both seedlings and older (>5 years) individuals. Three of the species respond as seedlings while only a single one of these (*C. canadensis*) responds as an adult as well. Theoretically, *C. canadensis* could be considered as retarded in shoot maturation compared to its ancestors and also, presumably, compared to related species. After all, it retains shoot responsiveness, a "juvenile" character, even as an older plant. Similarly, the other two *Cornus* species responding as seedlings (*C. kousa* and *C. officinalis*) could be viewed as retarded in their shoot development relative to *C. drummondii*, *C. florida* and *C. macrophylla*, all of which fail to grow in culture when seedlings are used as explant sources.

Extending the theory to other systematic groups, species in subclass Asteridae, for instance, might be thought of as retarded in shoot ontogeny relative to Magnoliidae. Subclass Dilleniidae, order Ericales, also would be ontogenetically retarded compared to most species in the order Theales. Likewise, the families Rosaceae and Leguminosae in subclass Rosidae are retarded in relation to the families Aceraceae and Sapindaceae.

Although the theory, as any evolutionary hypothesis (Stebbins 1974), is probably impossible to prove experimentally, it does place differences in shoot physiology into a potentially useful conceptual framework. Theory predicts, for example, that even nonresponding species must proceed through a responsive stage at some time during their development. Based on this, one might expect that very young *Magnolia* seedlings would respond in vitro provided shoot tips are taken at an early enough stage. Moreover, one would expect that any plant responding as a seedling should also respond when tips from older plants are used as long as the appropriate "rejuvenation" procedures are employed.

3 Cytokinin Biology

The failure of a shoot tip to respond positively to cytokinin treatments in vitro can have several alternative explanations.

1. The requirement for cytokinin by the explant is greater than the quantity supplied. Perhaps the explant is especially active in destroying cytokinin or, alternatively, its uptake capacity for cytokinin is low (Letham and Palni 1983).
2. The explant is not being supplied with the appropriate chemical form of cytokinin (Einset 1986a).
3. It needs another phytohormone(s) for growth in tissue culture (Einset 1985a).
4. It needs additional vitamins, amino acids, micronutrient salts, etc.
5. A component(s) of the medium is toxic.

To distinguish between some of these different explanations, we have taken a close look at cytokinin metabolism in micropropagated *Actinidia* shoots and in growing plants. Our rationale is that basic information on the biology of cytokinin, generally, and on the in vitro shoot response, in particular, will enable one to understand factors involved when shoots fail to grow in culture.

3.1 Why Use *Actinidia*?

On the assumption that the choice of experimental material in biological research (e.g., *E. coli, Drosophila, Xenopus*) can be just as important as methodology, a survey was conducted of woody species at the Arnold Arboretum of Harvard University (Boston, MA) for a species or genus that combines especially desirable characteristics for cytokinin studies. Some of the critical features of *Actinidia* species that led to their choice as experimental subjects include the ability of shoots to be micropropagated using conventional methodology, the ease of cytokinin analysis using HPLC, and the availability of cytokinin-autonomous tissue lines (Einset 1986a).

3.2 How Does *Actinidia* Make Cytokinin?

Figure 2 shows the proposed pathway of cytokinin (zeatin, io^6 Ade) biosynthesis in *Actinidia*. Based on metabolic-feeding experiments (Einset 1984b, 1986a,b), the pathway involves three critical steps leading up to io^6 Ade: (1) synthesis of i^6 AMP (N^6-Δ^2-isopentenyladenine mononucleotide) by dimethylallylpyrophosphate transferase (DMAAP transferase), (2) conversion of i^6 AMP to i^6 Ade via i^6 AMP hydrolase/phosphorylase and (3) O_2-dependent, i^6 Ade hydroxylase.

The simplicity of the *Actinidia* pathway contrasts sharply with alternative mechanisms of cytokinin biosynthesis proposed by Letham, Chen, and Morris (Letham and Palni 1983; Morris 1986). In all of these other schemes, O_2-dependent hydroxylation of the isopentenyl chain is postulated to occur on several alternative substrates. Thus,

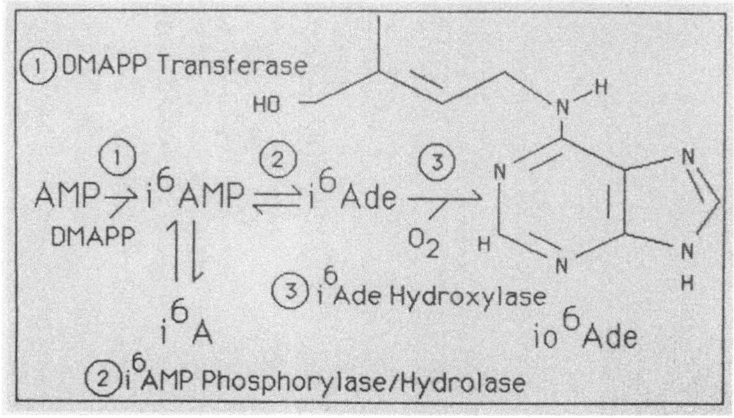

Fig. 2. Cytokinin biosynthesis in *Actinidia*

hydroxylation of i^6 AMP would produce io^6 AMP, hydroxylation of i^6 A produces io^6 A, and i^6 Ade hydroxylation produces io^6 Ade.

While no evidence has been presented for plant cytokinin biosynthesis involving multiple sites of hydroxylation, this is clearly not the mechanism in *Actinidia.* Figure 3, for example, illustrates the kinetic parameters for hydroxylation of i^6 Ade and related compounds when fed to *Actinidia* tissue cultures (Silverstone and Einset unpubl.). As indicated, i^6 Ade (the natural substrate) has the lowest K_m of the compounds tested; i.e., 12 μM compared to values of 44 μM for N^6-(Δ^3-isopentenyl)adenine and 470 μM for 7-(3-methylbutylamino) pyrazolopyrimidine. Neither N^6-(Δ^2-isopentenyl)amino-8-azapurine nor N^6-(Δ^2-isopentenyl)adenosine are hydroxylated, even in feeds with concentrations as high as 200 μM.

Interestingly, all three of the hydroxylated substrates (Fig. 3) have similar V_{max} values. This suggests that the differences in rates of hydroxylation can be attributed to differences in substrate affinity for the hydroxylase, indicated by widely divergent K_m values, rather than to variations in catalysis after the substrate binds to the enzyme's active site.

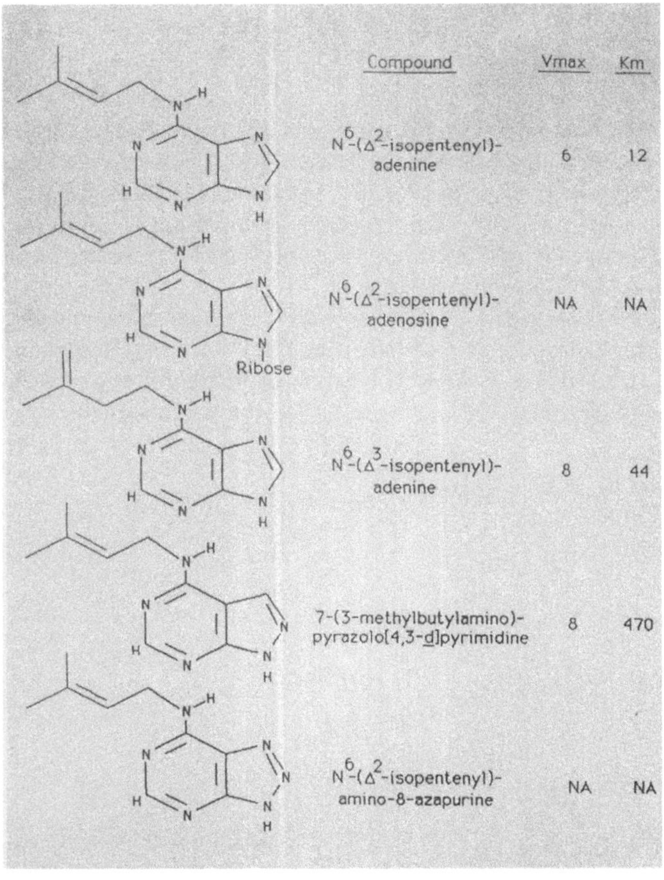

Fig. 3. Summary of kinetic parameters for hydroxylation of i^6 Ade and analogs by *Actinidia* tissue. K_m is expressed as μM concentration while the units for V_{max} are nmol/g/h

Alterations on the pyrimidyl ring of i^6 Ade apparently have profound effects on the ability of compounds to be hydroxylated by *Actinidia*. Substitution of a ribose at N9, for example, completely abolishes substrate activity, as does replacement of N for C at the 8-position (i.e., azapurine). Thus, given the strict substrate specificity of i^6 Ade hydroxylase, it is apparent that io^6 Ade is made in *Actinidia* by hydroxylating only i^6 Ade. A logical extension of this finding is that metabolites of io^6 Ade such as io^6A (ribosylzeatin) and io^6 AMP, as well as glucosylated derivatives, most likely are all synthesized subsequent to the i^6 Ade-to-io^6 Ade reaction.

3.3 When and Where Are Cytokinins Produced?

Understanding cytokinin biosynthesis in *Actinidia* enables one to ask the question of the sites of phytohormone production within intact plants. In fact, when growing plants are examined for their distribution of i^6 Ade hydroxylase, a critical step in the biosynthetic pathway, enzyme activity is found to be localized in two major regions, one centered in the root and the other in the stem (Einset 1984b, 1986a,c). Leaves and fruits, on the other hand, have very low i^6 Ade hydroxylase activity.

The high biosynthetic potential of *Actinidia* roots is not surprising, especially in view of several lines of evidence implicating the root as a major organ of cytokinin production in plants generally (Kulaeva 1962; Loomis and Torrey 1964; Torrey 1976; Wareing et al. 1977): (1) cytokinins are found in high concentrations in roots, primarily root tips, and indirect data implicates the quiescent center as a site of very active production (Feldman 1979), (2) bleeding sap of decapitated plants contains abundant quantities of the cytokinins io^6 Ade, io^6 A and io^6 AMP, and (3) water stress and the process of senescence, both of which cause characteristic cytokinin-deficient symptoms in shoots, are correlated with decreases in the cytokinin levels of bleeding sap.

Actually, analyses of thin (1 mm) tissue sections from growing *Actinidia* roots show that the maximum i^6 Ade hydroxylase activity exists not at the tip but rather a few millimeters behind the tip in a region of cellular differentiation corresponding to the tissues where secondary cambial activity is initiated (Einset 1986c). A similar gradient of activity exists in stems with the highest enzyme levels occurring somewhat behind the tip and progressively lower activities in older tissues. A growing stem of *Actinidia arguta*, for example, has relatively low i^6 Ade hydroxylase in the terminal 0.5 cm, increasing levels 0.5–15 cm from the tip and then a gradual decline in activity 15–50 cm.

Interestingly, the rapidly growing leaves associated with the most cytokinin-active regions of the stem have relatively low biosynthetic potential. This is consistent with the long-standing theory that leaves are cytokinin-requiring organs (Richmond and Lang 1957). Fruits, also, appear to be low in i^6 Ade hydroxylase, raising the interesting possibility that large quantities of cytokinin might be imported into fruits during development (Nitsch 1971).

On an organ-by-organ basis, it is estimated that approximately 68% of an *Actinidia* plant's cytokinin production is localized in roots, with stems and leaves accounting for about 26 and 6%, respectively, of the total activity (Einset and Silverstone 1987; Silverstone 1987). A critical question concerns the identity of tissues supplying the growing shoot tip, apparently a cytokinin-requiring structure, with cytokinin. The two possibilities, of course, are the root system and the stem. Although roots contain three fold more i^6 Ade hydroxylase than stems, the proximity of the active stem tissues to the shoot tip makes these tissues a likely source of a significant proportion of the tip's cytokinin supply.

3.4 Uptake During Micropropagation

The picture that emerges from studies on *Actinidia*, then, is of a cytokinin-requiring shoot meristem generating cells that proceed through a developmental program involving cellular elongation and differentiation, and resulting in the production of mature stem tissue. At some time during the developmental program, presumably at the stage of cellular differentiation, cells begin to produce cytokinin, and this cytokinin is then used by cytokinin-requiring tissues such as leaves, mature stem tissue and, of course, the shoot tip itself.

Figure 4 illustrates the effect of different medium concentrations of i^6 Ade on endogenous cytokinin contents of growing *Actinidia* shoots from 20-day cultures (Einset 1985b). By applying the concept of "critical concentration" from mineral nutrition research (Epstein 1972), the data indicate that actively growing shoots require a very high internal cytokinin concentration, i.e., 100–200 nmol zeatin per g fresh weight. Presumably this elevated cytokinin level is generated as a result of i^6 Ade uptake and metabolism, followed by io^6 Ade accumulation in tissues at levels far in excess (greater than 30-fold at 15 days after initiating the cultures) of the medium's cytokinin concentration.

In fact, as Table 2 (Einset 1986d) shows, *Actinidia* shoots consume i^6 Ade at >100 nmol i^6 Ade per g fresh weight per day. Shoots of other woody species also consume cytokinin at high rates, ranging from 18–160 nmol i^6 Ade per g fresh weight per day, depending on the species. Given an uptake rate of 160 nmol/g per day (*Actinidia arguta*) and an increase in fresh weight corresponding to one doubling per week, a typical micropropagated shoot (0.1 g) would deplete the cytokinin in 20 ml of 30 μM i^6 Ade within 18 days. Under the same conditions, shoots with consumption rates corresponding to 150, 100, 50, and 25 nmol/g fresh weight per day would deplete the cytokinin in a standard medium in 20, 28, 56, and 112 days, respectively. It is therefore probable, in most micropropagation protocols (e.g., Kyte 1983), that shoots are trans-

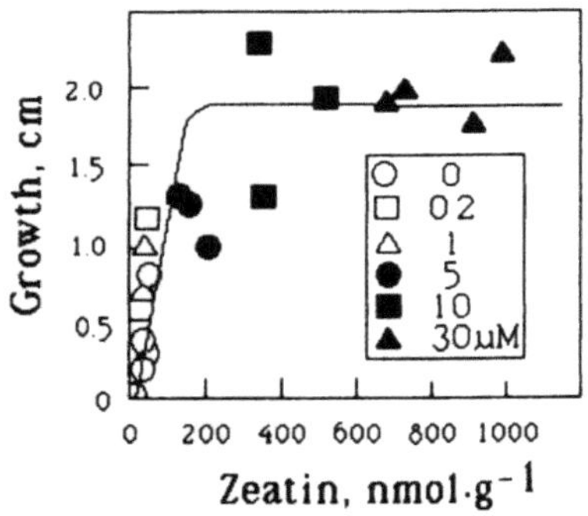

Fig. 4. Growth of *Actinidia arguta* shoots in tissue cultures in relation to their internal cytokinin content. Shoots (1 cm) from established cultures were transferred to media containing the designated i^6 Ade concentrations (*symbols*) and incubated for 20 days. After measuring the growth increment, stems were extracted and analyzed for cytokinin contents by HPLC. (Einset 1985b)

Table 2. Cytokinin consumption rates by micropropagated shoots growing in a nutrient medium supplemented with 30 μM i^6 Ade. Stock cultures of micropropagating shoots were subdivided and individual shoots (0.1 g per shoot) were transferred to fresh medium; in each case values are means of at least four independent determinations +/– SE

Family	Species	Rate of consumption (nmol/g FW/day)
Taxodiaceae	*Metasequoia glyptostroboides*	120+/–43
Magnoliaceae	*Magnolia soulangiana*	18+/–3
Lauraceae	*Sassafras albidum*	53+/–11
Actinidiaceae	*Actinidia arguta*	160+/–50
	Actinidia kolomikta	100+/–23
Oleaceae	*Syringa × hyacinthiflora*	62+/–10
Solanaceae	*Nicotiana tabacum*	33+/–8
Scrophulariaceae	*Paulownia coreana*	130+/–30

ferred to fresh medium long after the cytokinin in the medium has become exhausted. The so-called "maturation" phase of tissue culture growth, for example, conceivably could involve a period of cytokininless development. On the other hand, problems with rooting of micropropagated cuttings may conceivably be traceable to high internal cytokinin levels.

4 Discussion

The principle of micropropagation via cytokinin-controlled shoot growth, based on findings made by Skoog and coworkers nearly 45 years ago, does not appear to apply to all woody plants. According to the extensive comparative survey reviewed here, the ability of shoots to be manipulated by cytokinin treatments in vitro appears to be restricted to taxa such as subclass Asteridae (including tobacco) or orders Rosales and Fabales in subclass Rosidae that are usually considered to be relatively specialized (Stebbins 1974; Cronquist 1981). Less specialized taxa are either unresponsive or only sporadically responsive. Making the reasonable assumption that the physiological diversity among plants must be a product of evolution, it is suggested that specialized (i.e., cytokinin-responsive) taxa have abbreviated a normal pattern of shoot maturation (cytokinin-responsive juvenile shoots become cytokinin-unresponsive adult shoots) in a classical heterochronic fashion (cf. Gould 1977). Perhaps by pursuing the obvious implications of this concept, progress can be made in multiplying hard-to-propagate species in vitro.

A complementary approach would involve research to identify the particular defect involved when a species fails to grow in tissue culture. The limited data so far suggests that species that grow slowly in vitro consume very little cytokinin from the medium. Perhaps more effective means (mode of application and/or chemical form) for supplying the phytohormone can be devised. Much more research will be required in this area, especially in view of the fact that the cytokinin content of explants can be markedly different from the medium's phytohormone concentration.

References

Cronquist A (1981) An integrated system of classification of flowering plants. Columbia University Press, New York

Dirr MA, Heuser CW Jr (1987) The reference manual of woody plant propagation: from seed to tissue culture. Varsity Press, Atlanta

Einset JW (1984a) Biotechnology at the Arnold Arboretum. Arnoldia 44:27–33

Einset JW (1984b) Conversion of N^6-isopentenyladenine to zeatin by *Actinidia* tissues. Biochem Biophys Res Commun 124:470–474

Einset JW (1985a) Chemicals that regulate plants. Arnoldia 45:28–34

Einset JW (1985b) Role of cytokinin in woody plant micropropagation. Proc Int Plant Prop Soc 35:608–615

Einset JW (1986a) Zeatin biosynthesis from N^6-(Δ^2-isopentenyl)adenine in *Actinidia* and other woody plants. Proc Natl Acad Sci (USA) 83:972–975

Einset JW (1986b) A practical guide to woody plant micropropagation. Arnoldia 46:36–44

Einset JW (1986c) Biosynthesis of zeatin from N^6-(Δ^2-isopentenyl)adenine in *Actinidia*: sites and seasonal changes in activity. Proc Natl Acad Sci (USA) 83:7751–7754

Einset JW (1986d) Cytokinin consumption by micropropagated shoots. Proc Int Plant Prop Soc 36:635–640

Einset JW (1987a) How development's clock guides evolution. Arnoldia 47:20–25

Einset JW (1987b) Expanding the horizons of woody plant micropropagation. Am Nurseryman 166:187–194

Einset JW, Alexander JH Jr (1984) Multiplication of *Syringa* varieties and species in tissue cultures. Proc Int Plant Prop Soc 34:628–636

Einset JW, Silverstone A (1987) Hydroxylation of N^6-(Δ^2-isopentenyl)adenine to zeatin; relative activities of organ systems from *Actinidia arguta*. Plant Physiol 84:208–209

Epstein E (1972) Mineral nutrition of plants: principles and perspectives. John Wiley, New York

Feldman LJ (1979) Cytokinin biosynthesis in roots of corn. Planta 145:315–321

Gould SJ (1977) Ontogeny and phylogeny. Belknap, Harvard Press, Cambridge

Kulaeva ON (1962) The effect of roots on leaf metabolism in relation to the action of kinetin on leaves. Sov Plant Physiol 9:182–189

Kyte L (1983) Plants from test tubes: an introduction to micropropagation. Timber, Portland

Letham DS, Palni LMS (1983) The biosynthesis and metabolism of cytokinins. Annu Rev Plant Physiol 34:163–197

Linsmaier EM, Skoog F (1965) Organic growth factor requirements of tobacco tissue cultures. Physiol Plant 18:100–127

Lloyd G, McCown BH (1980) Commercially-feasible micropropagation of mountain laurel, *Kalmia latifolia*, by use of shoot-tip culture. Proc Int Plant Prop Soc 30:421–427

Loomis RS, Torrey JG (1964) Chemical control of vascular cambium initiation in isolated radish roots. Proc Natl Acad Sci (USA) 52:3–11

Miller CO, Skoog F, Okumura FS, Von Saltza MH, Strong FM (1956) Isolation, structure and synthesis of kinetin, a substance promoting cell division. J Am Chem Soc 78:1375–1380

Morris RO (1986) Genes specifying auxin and cytokinin biosynthesis in phytopathogens. Annu Rev Plant Physiol 37:509–538

Murashige T (1974) Plant propagation through tissue cultures. Annu Rev Plant Physiol 25:135–166

Murashige T, Skoog F (1962) A revised medium for rapid growth and bioassays with tobacco tissue cultures. Physiol Plant 15:473–497

Nitsch JP (1971) Perennation through seeds and other structures: fruit development. In: Steward FC (ed) Plant physiology, a treatise. Academic Press, New York, pp 413–501

Richmond AE, Lang A (1957) Effect of kinetin on protein content and survival of detached *Xanthium* leaves. Science 125:650–651

Silverstone A (1987) Analysis of N^6-(Δ^2-isopentenyl)adenine hydroxylase in *Actinidia* through precursor and analog feeding experiments. Thesis, Bachelor of Arts with Honors (Biology), Harvard College, Cambridge

Skoog F, Miller CO (1957) Chemical regulation of growth and organ formation in plant tissues cultured in vitro. Symp Soc Exp Biol 11:118–131

Skoog F, Tsui C (1948) Chemical control of growth and bud formation in tobacco stem segments and callus cultured in vitro. Am J Bot 35:782–787

Stebbins GL (1974) Flowering plants: evolution above the species level. Harvard University Press, Cambridge
Thimann KV, Skoog F (1934) On the inhibition of bud development and other functions of growth substance in *Vicia faba*. Proc R Soc, Ser B 114:317–339
Torrey JG (1976) Root hormones and plant growth. Annu Rev Plant Physiol 27:435–459
Torrey JG (1985) The development of plant biotechnology. Am Sci 73:354–363
Wareing PF, Horgan R, Henson IE, Davis W (1977) Cytokinin relations in the whole plant. In: Pilet P (ed) Proceedings of the 9th International Conference on Plant Growth Substances. Springer, Berlin Heidelberg New York, pp 147–153
Zimmerman RH, Griesbach RJ, Hammerschlag FA, Lawson RH (eds) (1986) Tissue culture as a plant production system for horticultural crops. Martinus Nijhoff, Dordrecht

I.12 The Effect of the Culture Vessel on Micropropagation

K. KAVANAGH[1], A.P. DREW[2], and C. MAYNARD[2]

1 Introduction

The natural conditions under which plant growth occurs allow for the dissipation of secondary products into the soil or atmosphere. Tissue culture systems lack an effective method of secondary product disposal. The use of enclosed containers to prevent microbial contamination and to retard desiccation interrupts the flow of secondary products away from the developing plant. This interruption can have a variety of consequences, both harmful and beneficial, depending on the type of tissue being cultured, the species, and the culture vessel type. The type of vessel used in micropropagation and the type of closure employed have been shown to have a variety of effects on culture growth and development.

Closure types and vessels vary widely depending on species requirements, efficiency in washing and filling, historic preferences of a lab, cost, and space requirements. Most closures fall into one of the following categories.

1. Screw lids, opaque or solid.
2. Tight-or loose-fitting snap-on lids.
3. Plastic film fixed to the vessel top, Parafilm (American Can Company, Greenwich CT) or polypropylene film.
4. Petri dish with cover.
5. Cotton plug.
6. Lids which have been altered by small holes or slits to promote aseptic ventilation.
7. Fluorocarbon polymer film.

The vessel closure controls the exchange of gases with the surrounding environment, releasing volatile compounds which build up during plantlet development. Volatile compounds released by a variety of cultures were identified by Thomas and Murashige (1979). They found that carbon dioxide, ethylene, ethane acetaldehyde, and ethanol were released from actively growing plant tissue cultures. No detectable levels of the above gases were detected in vessels containing only nutrient medium. With callus production, ethylene levels were found to be the highest, although this varied with the species being cultured. Fully developed shoot cultures had lower levels of carbon dioxide with a further decrease in the presence of light.

Hussey and Stacey (1984) showed that ethylene accumulation in tightly sealed vessels inhibited the formation of potato (*Solanum tuberosum* L.) tubers unless ethyl-

[1]Department of Forest Science, Oregon State University, Corvallis, OR 97330, USA
[2]SUNY College of Environmental Science and Forestry, Syracuse, NY 13210, USA

Biotechnology in Agriculture and Forestry, Vol. 17
High-Tech and Micropropagation I (ed. by Y.P.S. Bajaj)

ene-absorbing agents were used. Accumulation of ethanol by tissues in culture was found to inhibit somatic embryogenesis in citrus (Tisserat and Murashige 1977). Zobel (1987) found that under reduced aeration conditions, volatile compounds may accumulate, modifying differentiation and growth in soybean hypocotyl segments during callus induction. The ability of a culture vessel to release volatile compounds may therefore have a profound effect on the success of a tissue culture system.

Sha et al. (1985) reported that the use of Parafilm to seal culture vessels of actively growing cultivars of potato resulted in shoot-tip necrosis, reduced weight, and increased branching. The saturated conditions within the vessel reduced transpiration rates, thereby decreasing translocation of mineral nutrients. These abnormalities were overcome by the addition of calcium to the media and/or increased aeration of the vessel. While culturing daylilies (*Hemerocallis* spp.), Fitter and Krikorian (1985) observed a change in growth form associated with vessel closure. Poor aeration of the vessel resulted in a more mature growth form which was readily reversed following aeration.

Hussey and Stacey (1984) reported a change in morphology in potato cultures which was dependent on container closure. Poor aeration resulted in stoloniferous shoots with small leaves. This effect was shown to be a direct result of ethylene accumulation in the vessel. De Proft et al. (1985) found that ethylene can have both beneficial and negative physiological effects on the developing culture, depending on the species being cultured. Bromeliad (*Achema dactilyna*) benefited from a suppression of apical dominance and the production of numerous side shoots. Mini-roses reacted negatively with the induction of leaf abscission in the presence of ethylene.

Drew et al. (1988) proved that culture vessel affected stomatal density of developing black cherry (*Prunus serotina*) plantlets. Plantlets grown in culture were found to have stomatal densities 230% higher than seedlings grown under field conditions. Plantlets grown in larger vessels had a higher stomatal density than plantlets grown in 25-mm tubes. Black cherry seedlings grown under standard culture conditions in an enclosed vessel had elevated stomatal densities compared to field-grown seedlings.

Tricoli (1983) discovered that the type of cap used on 25-mm tubes affected the growth of black cherry plantlets regardless of the medium. Caps with poor gas exchange produced epinasty of the leaves, reduced elongation, and resulted in a lower multiplication rate than plantlets cultured in tubes with a more permeable cap. The effect of the culture vessel on plantlet physiology and morphology may affect the subsequent ability of the plantlet to acclimatize to ambient conditions following transfer to soil.

Vessel size and dimension also affect the culture environment. While culturing grapevines (*Vitis vinifera*) in liquid culture, Monette (1983) reported that larger vessels yielded a higher number of shoots following 6 weeks in culture. Shoot height was affected by vessel size, with taller shoots occurring in larger vessels. Adams (1972) showed that the rooting of strawberry (*Fragaria*) shoots in liquid medium was greatly accelerated following the use of larger culture vessels. Debergh et al. (1981) found no effect of vessel size on the level of vitrification of artichoke (*Cynara scolymus*). Vessel volume was found to affect the induction and growth of pollen embryos from anthers of *Nicotiana tabacum*. The optimum number of plantlets was obtained from a vessel size allowing for 5.5 ml per anther. Small-volume vessels were found to severely inhibit plantlet formation (Dunwell 1978).

Tanaka et al. (1988a) developed an inexpensive disposable culture envelope/vessel utilizing a 25-μm-thick fluorocarbon polymer film which had been heat sealed. The fluorocarbon polymer film was shown to have desirable characteristics in thermal stability, chemical inertness, light transmittance, and gaseous permeability, making it a suitable vessel material. In experimental trials with *Cymbidium* protocorm-like body production and shoot development, they reported no growth differences between culture envelopes and standard glass vessels. Further studies were undertaken to examine the effects of polymer thickness, vessel size, plantlet density, and support substrate on *Sparthiphyllum* plantlet growth and development (Tanaka et al. 1988b). Vessels made out of 25-μm film, with a low plantlet density, were found to be superior in promoting plantlet development. Of the three media substrates examined; agar, polyester fiber blocks/liquid medium, and rock wool/liquid medium, the latter was found to be the most promising in terms of plantlet growth and handling. Polymer vessels in combination with liquid medium were found to be very efficient during the acclimatization phase since the medium can be drained off and the vessel gradually perforated as the plantlet becomes acclimated.

The conditions under which plantlets develop in vitro are the result of a complex interaction of the microenvironment, including volatiles that affect the differentiation, morphology, and physiology of the final product. There is a need to better understand the effects of accumulating volatile compounds and the effect of elevated relative humidity levels. Because the type of vessel used in tissue culture work affects results, an experiment was set up to measure the moisture conductance of different vessel-closure combinations.

2 Moisture Conductance

2.1 Materials and Methods

The effect of container shape and closure on moisture conductance was explored using two vessel types and three container closures. The two vessel types used were polycarbonate Magenta cubes (GA-7, Magenta Corp., Chicago IL) 5.8 cm per side, 9.6 cm in height, and cylindrical glass tubes with a radius of 12.5 mm and a height of 10.3 cm. The cylindrical glass tubes had two different polypropylene caps, Magenta 2-way (Magenta Corp., Chicago IL) and the Kimble PM (Grand Island Biological Company, Grand Island NY) (Fig. 1). The cubes were covered with a Magenta polypropylene lid designed to allow for gas exchange, although air has to flow down and up through five turns or bends before exiting the vessel.

The Kimble PM cap locks down tightly, sealing around the lip of the tube with a thin gasket. The Magenta 2-way has no gasket and has a narrow gap between the cap and the lip of the culture tube to allow for gas exchange while maintaining aseptic conditions.

Forty tubes were filled with 10 ml of modified Murashige and Skoog medium used for the propagation of *Prunus serotina* shoot tips (Tricoli 1983), and randomly assigned to receive either a Magenta or Kimble cap. Ten Magenta cubes were filled with 88 ml of the same medium and capped. All vessels were then autoclaved at 121 °C and 15 mPa for 15 min.

Following autoclaving, the vessels were transferred to a growth room with a 24 °C constant temperature and a relative humidity of 48%. Initial weight was taken following solidification of the medium and drying of moisture on the vessel exterior. Half of the vessels from each cap type were then randomly placed on a shelf under cool white

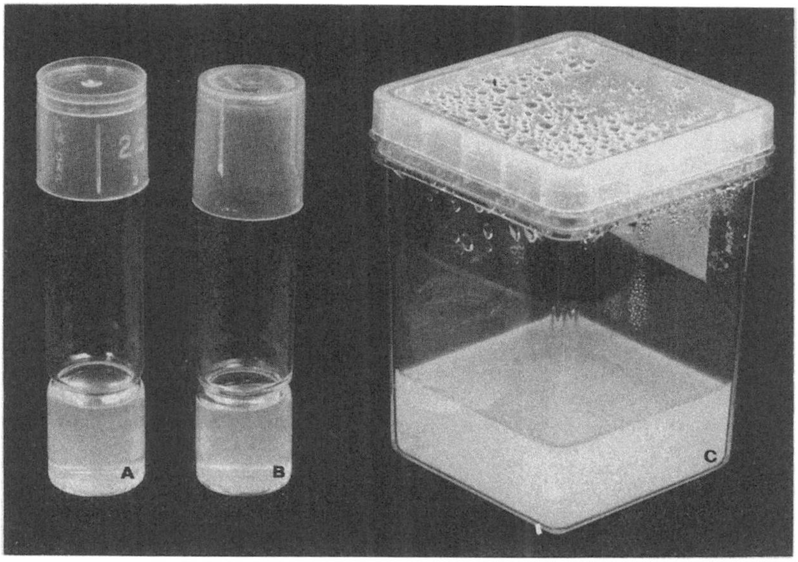

Fig. 1A-C. Three vessel-cap combinations used in experiment. **A** Kimble PM cap on a glass tube. **B** Magenta 2-way cap on a glass tube. **C** Magenta cap on a Magenta polycarbonate cube

Table 1. Dimensions of vessels used in moisture loss experiment

	Tube	Cube
Total volume	40.0 cm^3	380.0 cm^3
Surface area	3.8 cm^2	33.6 cm^2
Medium volume	10 ml	88 ml

fluorescent tubes (40 $\mu mol/m^2/s$ in the 400–700 nm range of photosynthetically active radiation) with the light source 18 cm from the vessel lid, or on a shelf with no direct lighting source (10 $\mu mol/m^2/s$).

The vessels were weighed weekly for the next 6 weeks on a pan balance to the nearest 10 mg to determine moisture loss. The data were analyzed using analysis of variance.

To determine the effects of vessel configuration on moisture diffusion, five Magenta cubes and five cylindrical tubes were filled with 88 and 10 ml of black cherry stage II medium (Tricoli et al. 1985), respectively, left without closures, and placed on a shelf in the growth room. Initial weight was taken following solidification of medium and drying of moisture on vessel surfaces. Weight was then recorded for each vessel after a 24-h period. Fick's law was used to calculate moisture diffusion coefficients for the different vessel types based on dimensions (Table 1) and rate of moisture loss.

2.2 Results

No significant difference ($\alpha = 0.05$) was found in moisture loss according to shelf lighting level. The vessels placed under direct fluorescent lights had a slightly higher loss, but it was not statistically significant (Fig. 2). No interaction was found between light level and vessel type.

Container closure affected moisture loss (Fig. 3). Weight loss per cm^2 of medium surface area was significantly higher in the Magenta 2-way capped tube, with a mean total weight loss of 0.45 g over 6 weeks. The Magenta capped GA-7 cube showed less weight loss, 0.21 g per cm^2 medium surface area, although total weight loss was much greater over the 6-week period. The Kimble capped tube had a significantly lower mean total weight loss per cm^2 at 0.14 g/cm^2 medium surface area.

In addition to having the lowest rate of diffusion, the Kimble capped tubes had the highest within treatment variation, eight times greater than the Magenta capped tubes or cubes (Table 2). This is due to the tendency of the cap to pop up, releasing the gasket from the tube lip. Following re-use, some of the gaskets are torn, allowing for increased gas exchange.

Moisture loss per week followed a very similar pattern for all treatments (Fig. 3). A decrease in weight loss in week 2 is followed by a general increase over the next 4 weeks for all treatments.

Relative coefficients of diffusion for the Magenta cube and cylindrical tube were calculated using Fick's Law: the rate of diffusion of a gas is directly proportional to the cross-sectional area of the path and to the difference in density of the gas along the path, and is inversely proportional to the length of the path. The Magenta cube had a coefficient of diffusion 4.7 times greater than the glass tube.

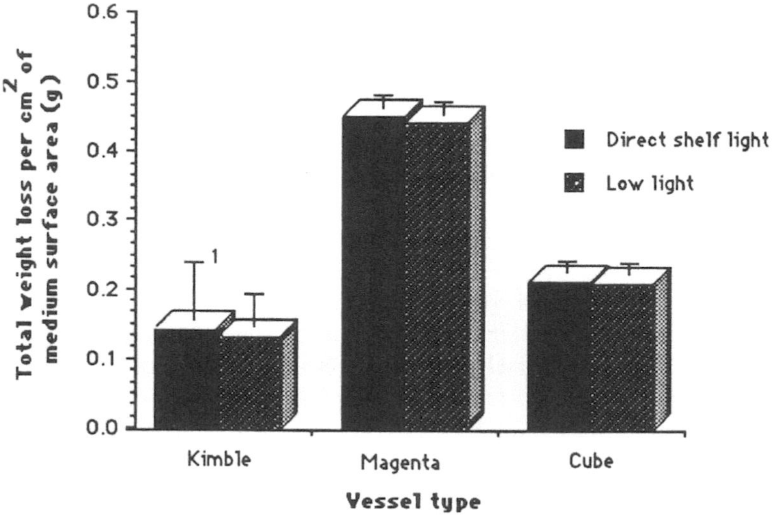

Fig. 2. Total weight loss (*g*) by treatment over a 6-week period. *1:bars* represent standard error

Table 2. Moisture loss per cm^2 of medium surface area by culture vessel-cap type over a 6-week period

Vessel	Moisture loss per cm^2 medium[a] (g)	n	Coefficient of Variation (%)
Kimble	0.141±0.95A[b]	20	67.24
Magenta	0.450±.029B	20	6.55
Cube	0.213±.028C	10	13.36

[a]±One standard deviation.
[b]Means followed by the same letter are not significantly different at $\alpha = 0.05$.

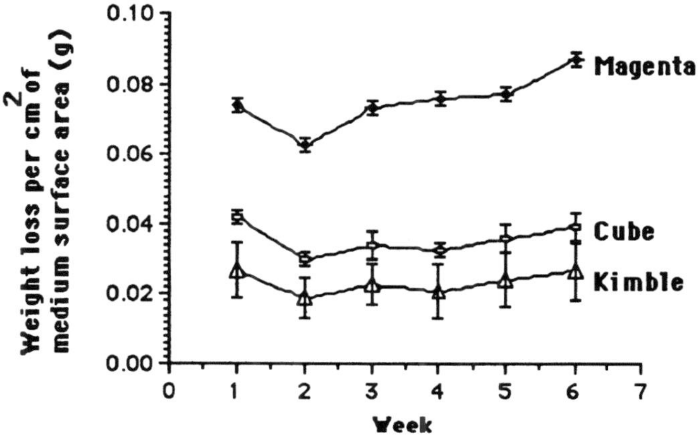

Fig. 3. Weekly weight loss per cm^2 medium surface by vessel-cap type for a total of 6 weeks. *Bars* represent ± standard error

3 Discussion

The culture environment is affected by the type of culture vessel used, its configuration, and the type of closure on the vessel. Vessel shape affects the amount of medium exposed to the atmosphere inside the vessel (Table 1), which can affect the rate of moisture loss and diffusion of gaseous compounds according to Fick's Law of diffusion. Vessel closures are important due to the control of the release of gaseous compounds built up during plantlet growth and the control of moisture loss.

Without caps, diffusion of moisture from the Magenta cubes is much greater than from the tubes, due to increased surface area of medium exposed to the atmosphere. Following capping, however, the tubes exhibit a higher rate of moisture loss than the cubes (Fig. 3) on an area of medium basis, indicating the Magenta cube cap is more effective in retarding moisture loss than the Magenta 2-way capped tube, in spite of the cube dimensions.

Caps or vessels that permit gas exchange also encourage moisture loss from the medium. Variability in the rates of moisture loss and/or increased levels of gas exchange may affect experimental results and plantlet production. The high variability in moisture loss with Kimble PM caps (Table 2) could decrease precision in an experiment. Sha et al. (1985) indicate that poor aeration can affect translocation of mineral nutrients through decreased transpiration rates. This may affect shoot condition and performance. If moisture retention is important, Kimble PM caps may be desired.

Short-term changes in weekly moisture loss could also alter culture performance. Lower loss of moisture in week 2 indicates lower gas exchange rates which may affect developing cultures. The effects of moisture loss and/or gas exchange following the opening of a vessel during an experiment, to add liquid medium and/or to subsample, may affect the culture atmosphere enough to change experimental results. The culture vessel is a critical element in any experiment, especially one empirical in nature, and other treatment effects can be masked by subtle changes in culture vessel atmosphere.

The conditions under which media-filled vessels are stored can have an effect on moisture content. Storage under low humidity conditions or in a frost-free refrigerator can deplete moisture levels dependent on the type of closure used and the length of storage. The time of the year can also affect ambient humidity conditions in the growth room, depending on whether or not the air is humidified and on the use of cooling or heating systems. In our growth room, humidity levels vary from 43% in March to 62% in September.

Environmental conditions under which tissue culture plantlets develop in vitro can be characterized by high relative humidity (100%), no air movement, low light levels (20–60 $\mu mol/m^2/s$), and a heterotrophic mode of nutrition (Conner and Thomas 1981). Very different environmental conditions are found in the field or greenhouse, with variable rates of humidity and wind speed, increased light levels, and an autotrophic mode of nutrition.

Tissue culture plantlets reflect the environmental conditions under which they are grown, and their biological state can affect subsequent acclimatization to greenhouse or field conditions. Culture conditions reduce the formation of epicuticular wax and thus reduce the ability of plantlets to retain water following transfer to soil. Sutter and Langhans (1982) studied the formation of epicuticular wax on cabbage plantlets (*Brassica oleracea* var. *capitata*), pointing out that environmental conditions such as high humidity and low light intensity appear to be responsible for the decreased amounts of epicuticular wax formed on plantlets grown in culture. The relationship between reduced epicuticular wax and high humidity was found to be true for developing seedlings as well as plantlets (Grout and Aston 1976). Wardle et al. (1983) found that cauliflower (*Brassica oleracea* L. Botrytis group) plantlets grown under reduced humidity conditions, within the culture vessel, developed large quantities of surface wax and had a lower rate of water loss, thereby establishing a direct cause and effect relationship between high humidity conditions and formation of epicuticular wax.

Stomatal frequency has also been found to be altered by the tissue culture environment. Brainerd et al. (1981) examined cultured Pixy plum (*Prunus insititia* cv. Pixy) leaves and found stomatal frequency to be lower (150 per mm^2) than for noncultured leaves (300 per mm^2). Although they did not quantify their results. Wetzstein and

Fig. 4A-C. Acetate leaf impressions taken from *Prunus serotina* plantlet leaves cultured in different vessel/cap combinations. **A** Kimble PM cap on a glass tube. **B** Magenta 2-way cap on a glass tube. **C** Magenta cap on a Magenta polycarbonate cube

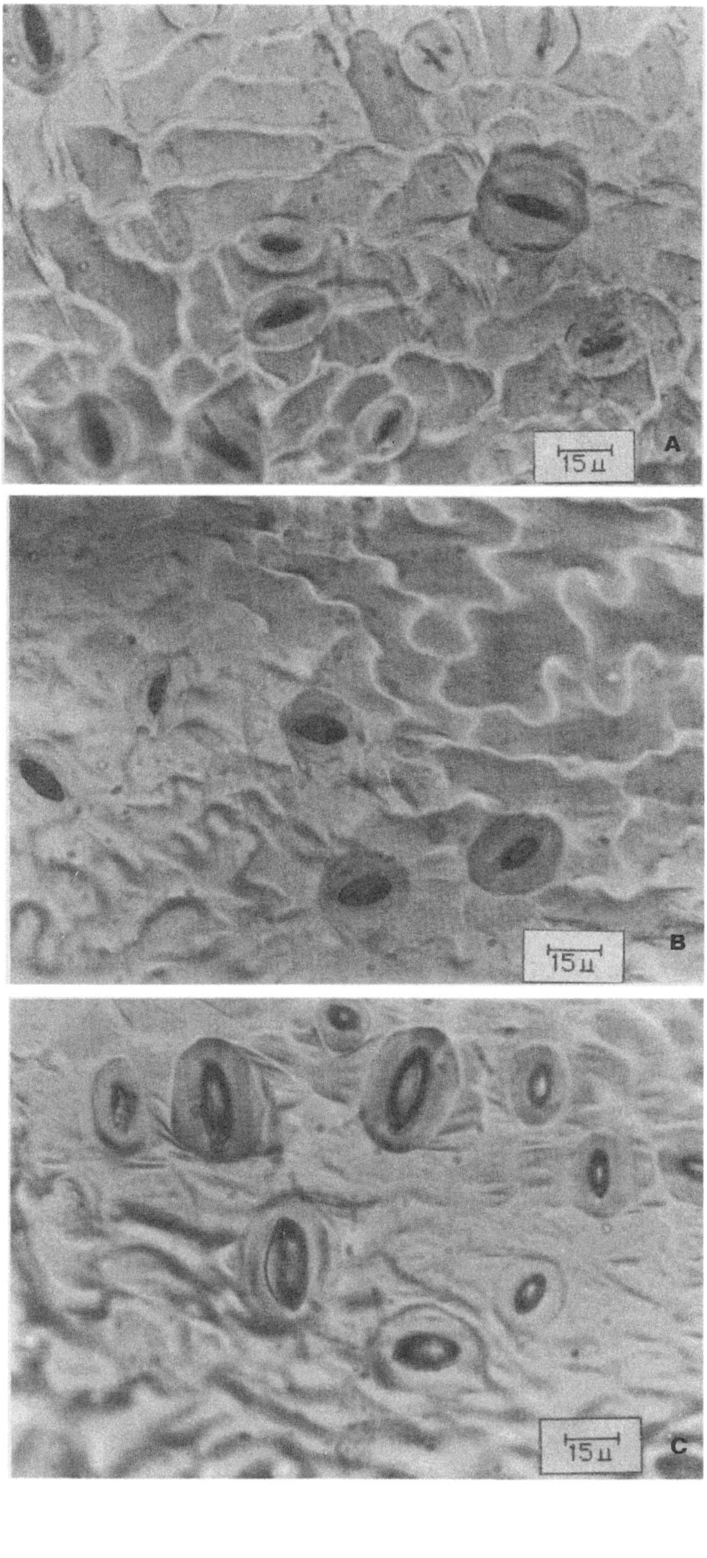

Fig. 4A-C

Sommer (1982) observed stomata to be more numerous in micropropagated sweetgum than in noncultured sweetgum seedlings. The effect of culture environment on stomata density is illustrated in research done by Drew et al. (1988), in which black cherry plantlet leaves were sampled and stomatal density assessed by means of acetate impressions. Stomata density was found to be significantly higher in plantlet leaves grown in a Magenta cube than in plantlets grown in Kimble capped or Magenta 2-way capped tubes (Fig. 4). These stomatal changes were not found to be long-lasting, but they can influence the plantlet survival during the very critical time of adjusting to ambient greenhouse or field conditions.

It is desirable to adapt the vessel system to better fit the needs of the plantlet rather than laboratory conditions. High relative humidity levels have been shown to affect plantlet development. Ziv et al. (1983) found that by adjusting the culture environment through the decrease of humidity and by increasing the amount of sucrose and agar in the medium they were able to produce carnation (*Dianthus caryophyllus*) plantlets that were better able to adjust to low humidity conditions. Work has also been done by Short and Roberts (1986) to lower the humidity within the growth vessel using a glycerol solution while culturing chrysanthemum plantlets in vitro. They found no reduction in plantlet growth at vessel humidity levels of 80% with the resulting foliage exhibiting normal water relations and stomatal physiology. Plantlets were directly transferable to soil.

4 Summary and Conclusion

Culture vessel environment is an important component of the tissue culture system. Additional studies are necessary to better understand how vessel size, shape, and cap affect culture performance. Ultimately, it may be necessary to redesign the entire vessel system to encourage the removal of harmful by-products and to decrease humidity levels, promoting a growth form capable of more ready acclimatization to the outside environment.

Vessel size and cap type have an effect on gaseous diffusion during culture. Diffusion of moisture from the media is a function of exposed surface area and cap closure. Control of the diffusion rate is important in terms of moisture loss and gaseous exchange during plantlet growth and it is necessary to find a proper balance between retarding moisture loss and promoting gaseous exchange. Of the three culture vessel/cap combinations tested, no single one could be said to be preferred since the environment created by each is unique and may be suitable for individual cultural needs.

References

Adams AN (1972) An improved medium for strawberry meristem culture. J Hortic Sci 47:263–264

Brainerd KE, Fuchigami LH, Kwartkowski S, Clark CS (1981) Leaf anatomy and water stress of aseptically cultured Pixy plum grown under different environments. Hortscience 16(2):173–175

Conner AJ, Thomas MB (1981) Re-establishing plantlets from tissue culture: a review. Int Plant Propagators Soc Combined Proc 28:342–357

Debergh P, Harbaoui Y, Lemeur R (1981) Mass propagation of globe artichoke (*Cynara scolymus*): evaluation of hypotheses to overcome vitrification with special reference to water potential. Physiol Plant 53:181–187

De Proft MP, Van Den Broek, Van Dijck R (1985) Implications of container atmosphere during micropropagation of plants. Med Fac Landbouw Rijksuniv Gent 50(1):129–132

Drew AP, Kavanagh KL, Maynard CA (1988) Comparative growth of micropropagated *Prunus serotina* plantlets and seedlings in different containers. In: Valentine FA (ed) Progress and prospects in forest and crop biotechnology. Springer, Berlin Heidelberg New York Tokyo, pp 99–111

Dunwell JM (1978) Anther culture in *Nicotiana tabacum*; The role of the culture vessel atmosphere in pollen embryo induction and growth. J Exp Bot 30(116):419–428

Fitter MS, Krikorian AD (1985) Mature phenotype in *Hemerocallis* plantlets fortuitously generated in vitro. J Plant Physiol 121:97–101

Grout BW, Aston MJ (1976) Transplanting of cauliflower plants regenerated from meristem culture I. Water loss and water transfer related to changes in leaf wax and xylem regeneration. Hortic Res 17:1–7

Hussey G, Stacey NJ (1984) Factors affecting the formation of in vitro tubers of potato (*Solanum tuberosum* L.) Ann Bot 53:565–578

Monette PL (1983) Influence of culture vessel on in vitro proliferation of grape in liquid medium. Plant Cell Tissue Organ Cult 2:327–332

Sha Liu, McCown BH, Peterson LA (1985) Occurrence and cause of shoot-tip necrosis in shoot cultures. J Am Soc Hortic Sci 110(5):631–634

Short KC, Roberts AV (1986) Growth and stomatal physiology of chrysanthemum plantlets cultured in vitro and at transplanting into soil. In: Sommers DA, Gengenbach BG, Biesboer DD, Hackett WP, Green CE (eds) VI Intern Congr Plant Tissue and Cell Culture, Aug 3–8, Minneapolis, MN. Alan Liss, New York, pp 432

Sutter E, Langhans W (1982) Formation of epicuticular wax and its effect on water loss in cabbage plants regenerated from shoot tip culture. Can J Bot 60:2869–2902

Tanaka M, Goi M, Higashiura T (1988a) A novel disposable culture vessel made of fluorocarbon polymer films for micropropagation. Acta Hortic 226:663–670

Tanaka M, Jinno K, Goi M, Higashiura T (1988b) The use of disposable fluorocarbon polymer film culture vessel in micropropagation. Acta Hortic 230:73–80

Thomas D, Murashige T (1979) Volatile emissions of plant tissue cultures: I Identification of the major components. In Vitro 15(9):654–658

Tisserat B, Murashige T (1977) Probable identity of substances in citrus that repress asexual embryogenesis. In Vitro 13(11):785–789

Tricoli DM (1983) In vitro propagation of *Prunus serotina*. MS Thesis, SUNY College of Env Sci and Forestry, Syracuse, NY, pp 84

Tricoli DM, Maynard CA, Drew AP (1985) Tissue culture propagation of mature trees of *Prunus serotina* Ehrh. I. Establishment, multiplication and rooting in vitro. Forest Sci 31(1):201–208

Wardle K, Dobbs EB, Short KC (1983) In vitro acclimatization of aseptically cultured plantlets to humidity. J Am Soc Hortic Sci 108(3):386–389

Wetzstein HY, Sommer HE (1982) Leaf anatomy of tissue cultured *Liquidambar styraciflua* (Hamamelidaceae) during acclimatization. Am J Bot 69(10):1579–1586

Ziv M, Meir G, Halevy AH (1983) Factors influencing the production of hardened glaucous carnation plantlets in vitro. Plant Cell Organ Tissue Cult 2:55–65

Zobel RW (1987) Gaseous compounds of soybean tissue cultures: carbon dioxide and ethylene evolution. Environ Exp Bot 27(2):223–226

I.13 Disposable Film Culture Vessels

M. TANAKA[1]

1 Introduction

Although numerous papers have been published on micropropagation of plants, only a few workers have studied the importance of culture vessels. Monette (1983) reported the influence of the size of culture vessel on micropropagation of grape. Some new culture systems, such as the airlift system (Koch 1973), the fermentor system (Takayama and Misawa 1981), the tilting and rocker system (Harris and Mason 1983), and the automated system (Tisserat and Vandercook 1985; Aitken-Christie and Jones 1987) have been established, but these systems are costly when used in micropropagation. Hence, in most commercial tissue culture laboratories, conventional glass or plastic culture vessels are normally used (see also Kavanagh et al. Chap. I.12, this Vol.).

What are the properties required for these culture vessels? Generally, it should have high light transmittance, gas exchangeability, low water vapor transmittance, and chemical inertness for normal growth and development of plantlets. A glass or plastic vessel satisfies these properties, hence their wide usage. These culture vessels, however, may have the following disadvantages: they are relatively expensive, less durable, less transportable, inflexible in shape and size, and the chance of contamination with microorganisms through their opening is relatively high due to gaseous exchange caused by temperature changes when using conventional closures.

To overcome these disadvantages, we have developed disposable culture vessels by using fluorocarbon polymer films (Tanaka et al. 1988a,b). This chapter reviews the various aspects of these novel film vessels and discusses their possible practical application in micropropagation or in other fields of plant tissue culture.

2 Properties of Fluorocarbon Polymer Films

Fluorocarbon polymers or fluorine-containing resin are being used in a wide range of fields such as chemistry, electronics, and medicine. They have superior properties such as thermal stability, chemical inertness, and electrical reliability. The films of fluorocarbon polymers are sold under trade names such as: Neoflon PFA (tetrafluoroethylene perfluoroalkyl vinyl ether copolymer) and Neoflon FEP (tetrafluoroethylene hexafluoropropylene copolymer) produced by Daikin Industries, Ltd. (Chemical division,

[1]Department of Horticulture, Faculty of Agriculture, Kagawa University, Miki-cho, Kagawa 761–07, Japan

Biotechnology in Agriculture and Forestry, Vo. 17
High-Tech and Micropropagation I (ed. by Y.P.S. Bajaj)

1–1 Nishihitotsuya, Settsu-shi, Osaka 566, Japan). These films also possess superior properties (Table 1) in comparison with some other polymer films. Neoflon films have a melting point of 265–310 °C, and hence can be autoclaved safely. Oxygen, nitrogen, or carbon dioxide gases are more permeable to Neoflon films than other polymer films. Therefore they make a more suitable vessel material for cultures which require gas permeability compared to their counterparts. They have low water vapor transmittance as polyvinylene chloride (wrapping film), but a high light transmittance like glass (Fig. 1). These films can be easily made to any desired size and shape by heat-sealing.

Neoflon films possess all the above properties required for culture vessels. Furthermore, to prevent microbial contamination during culture, such film vessels may have to be sealed completely after inoculation of explants. Two types of culture vessels for micropropagation have been developed for trial by using these films.

Table 1. Properties of Neoflon films. (Tanaka et al. 1988a)

	Melting point (°C)	Gas-permeability × 10^{-8} ($cm^3 \cdot cm/cm^2$ s • atm • 25 °C) O_2	N_2	CO_2	Moisture-permeability ($g/m^2 \cdot 24$ h; 45 μm)
Neoflon PFA[a]	302–310	4.8	1.4	10.3	3.1
Neoflon FEP[b]	265–275	3.7	1.2	9.7	2.0
Polyethylene (low density)	106–124	2.2	0.74	9.6	5.8
Polyvinylene chloride	170–180	0.00035	0.00009	0.0011	1.8
Polycarbonate	140–150	1.1	0.2	3.8	>20

[a]Tetrafluoroethylene perfluoroalkyl vinyl ether copolymer.
[b]Tetrafluoroethylene hexafluoropropylene copolymer.

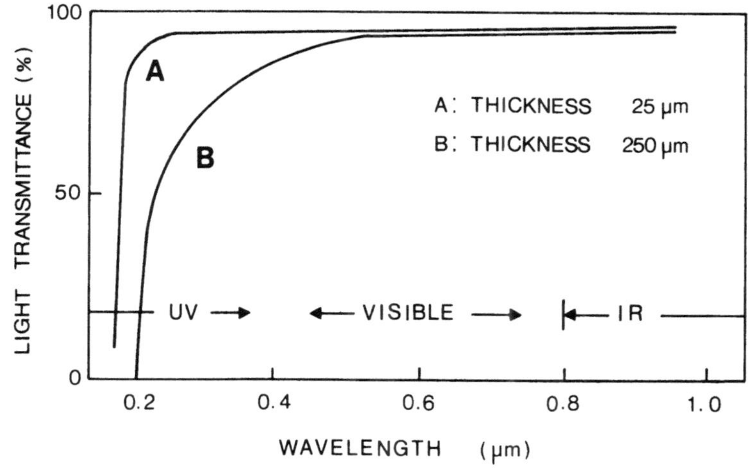

Fig. 1. Light transmittance of Neoflon FEP film. (Tanaka et al. 1988a)

3 Box-Shaped Vessel, the "Culture Pack"

In commercial tissue culture laboratories, a number of large-sized culture vessels are required in the later stages of micropropagation, especially for rooting and shoot development. For this stage of micropropagation, a culture vessel in the shape of a box was made using Neoflon films. This vessel is referred to as the "Culture Pack".

3.1 Procedure

A sheet of Neoflon film (PFA and FEP; 22 cm long by 32 cm wide) is folded in the manner shown in Fig. 2A. The overlapped vertical and bottom horizontal edges (the shadowed portion) are then heat-sealed by impulse sealer (model K-250EW, Shigahosoki Co., Ltd., Nagoya, Japan). To form into a vessel, air is blown into the Culture Pack (Fig. 2B). A frame (7.5 × 7.5 × 10.5 cm, outer diameter) made of stainless steel wire of diameter 2 mm is then placed into the Culture Pack as support. After pouring 140 ml of agar medium into the Culture Pack (Fig. 2C), the opening is folded and covered with an aluminium foil. The Culture pack is then sterilized by autoclaving (121 °C, 17 min). Tissues can then be planted into each Culture Pack. The opening of the Culture Pack is heat-sealed (Fig. 2D). Microbial contamination after sealing is completely prevented.

3.2 Shoot Development in the Culture Pack

As shown in Fig. 2, the Culture Pack provides a completely closed system. The first question to ask is whether the shoots planted in such a Culture Pack could develop normally to plantlets or not. To prove this, 16 shoots of *Spathiphyllum* cv. Merry having three leaves were cultured on agar medium in the Culture Pack made of 25 μm Neoflon PFA or FEP films. The medium used for the shoot development was that of Murashige and Skoog (1962) supplemented with 2 mg/l NAA, 0.5 mg/l BA, 1 g/l activated charcoal, 25 g/l sucrose, and 10 g/l agar. For comparison, shoots were also cultured on the medium (140 ml) in a 500-ml Erlenmeyer flask covered with aluminum foil.

The plantlets grown in the Culture Pack were as normal as those grown in flasks. The number of roots and fresh weight of plantlets in the Culture Pack made of PFA film were in fact higher when compared with those in the Erlenmeyer flask (Fig. 3). Further work showed that larger plantlets were obtained in the Culture Pack made of PFA film compared to those of FEP film. This could be due to the better gas permeability property of PFA film than FEP film (Table 1). From these data, there is no doubt that the Culture Pack can be successfully used for the micropropagation of *Spathiphyllum*.

The Culture Pack can also be used in the same way as a conventional glass vessel in the micropropagation of *Cymbidium*. Sixteen shoots of *Cymbidium* cultivars, each having three leaves and no root, were excised from a mass of protorcorn-like bodies (PLBs) 3 months after subculture. They were planted on modified Vacin and Went (1949) medium supplemented with Nitsch et al. microelements (1967), 2 g/l tryptone, 0.1 mg/l NAA. 0.1 mg/l kinetin, 20 g/l sucrose, and 10 g/l agar. The plantlets grown on solid medium in the Culture Pack made of 12.5 μm Neoflon PFA, or FEP film were normal (Fig. 4), and the size and fresh weight were almost the same as those of the plantlets grown in flasks (Fig. 5).

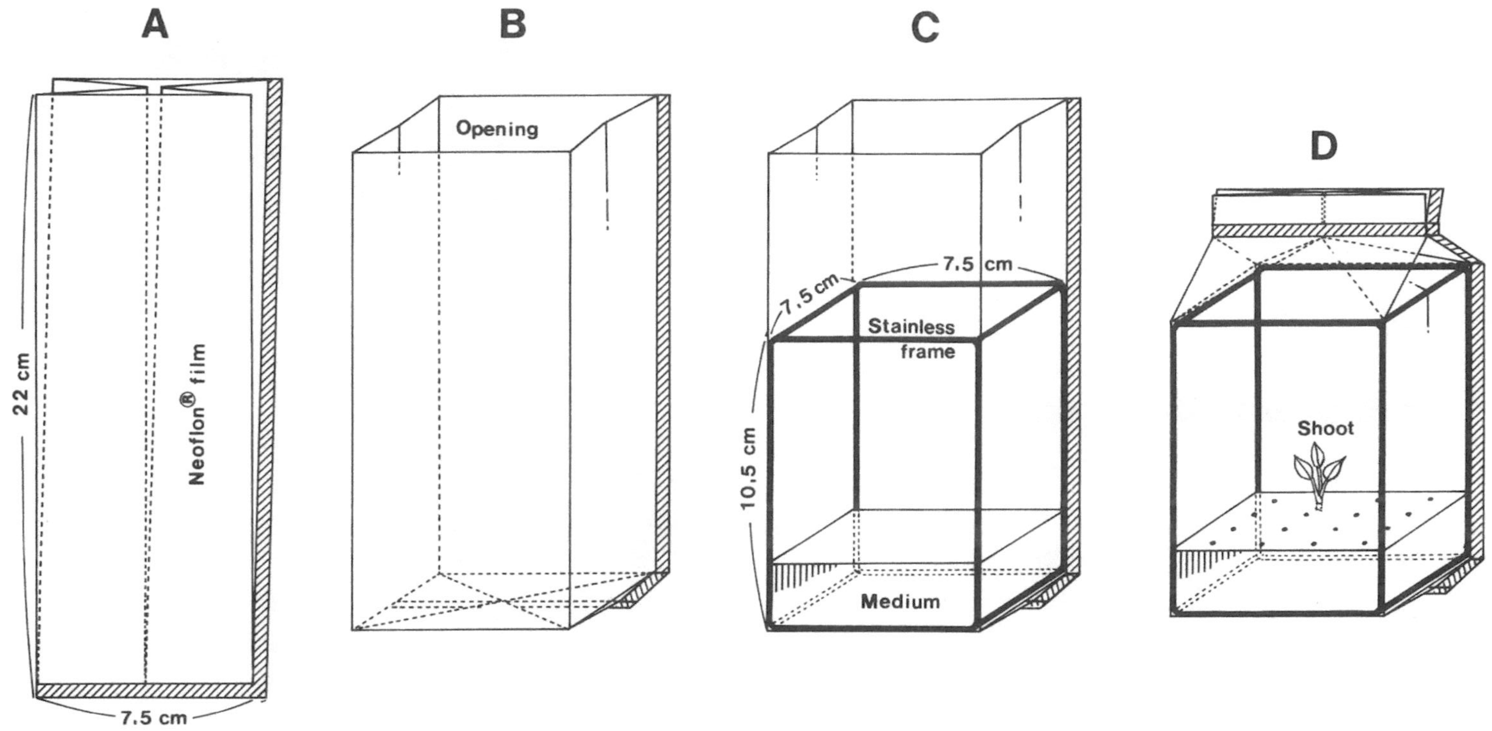

Fig. 2A-D. Diagram illustrating the procedure to make and use the Culture Pack of standard size. The shoots planted on agar medium are omitted in this diagram except one in the Culture Pack (**D**). (Tanaka et al. 1988a)

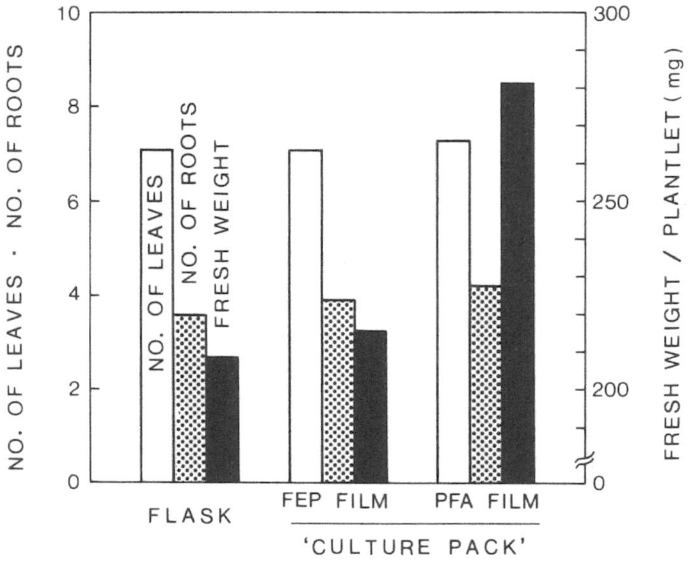

Fig. 3. Effect of culture vessels on shoot development of *Spathiphyllum* cv. Merry 75 days after planting. The Culture Pack of standard size used was made of Neoflon PFA or FEP 25 μm film. (After data of Tanaka et al. 1988b)

Fig. 4. A comparison of the development of *Cymbidium* Lois Kelly Cherry and Cream shoots grown in different culture vessels 90 days after planting. *Left* 500 ml flask; *right* the Culture Pack made of 12.5 μm Neoflon FEP film

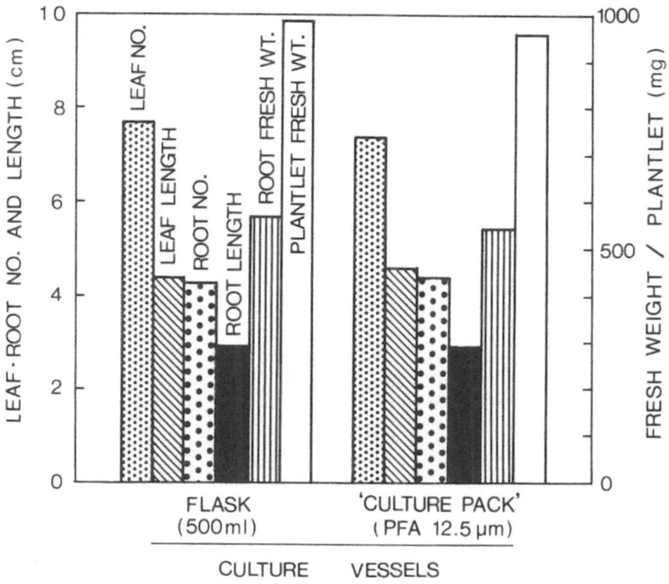

Fig. 5. Effect of culture vessels on shoot development of *Cymbidium* Gawain Empress 90 days after planting. (Tanaka et al. 1988a)

In further investigation, 16 shoots of *Gerbera* cv. Beatrix with three leaves were cultured on agar medium in the Culture Pack made of 25 μm PFA film and in the various culture vessels. Plantlet development was enhanced in the Culture Pack compared with those in the various types of conventional culture vessels (Table 2).

We have been carrying out trials to use the Culture Pack for the production of tissue-cultured plants of many other horticultural plants. Our results showed that the Culture Pack is suitable for use and has no adverse effect on shoot development (unpubl.).

Leaves and roots of tissue-cultured plantlets are often damaged or broken when taking them out from a conventional culture vessel by using forceps. Such damage, however, can be avoided when using the Culture Pack, from which the plantlets could be easily removed by cutting the film with a blade (Fig. 6).

3.3 Effects of Thickness of Neoflon Film

In general, the various properties of polymer films, such as gas permeability or light transmittance, are affected by the thickness of the film. Therefore, it is necessary to examine the effect of thickness of film used for the Culture Pack on shoot development.

Shoots of *Spathiphyllum* and *Cymbidium* were cultured in the Culture Pack made of PFA film of 12.5, 25, 50, and 100 μm thickness. For both plant materials, enhanced normal growth of plantlets was obtained in the Culture Pack made of thinner film. This could be due to better gas permeability. In *Cymbidium*, the plantlets in the Culture Pack made of 50 and 100 μm films were abnormal, forming elongated stems with very short leaves (Fig. 7). With increased thickness of PFA film from 12.5 to 100 μm, the

Table 2. Effect of culture vessels on shoot development of *Gerbera* cv. Beatrix 50 days after planting. (Tanaka et al. unpubl.)

Culture vessels	No. of leaves	Plant height (cm)	Top fresh weight (mg)	No. of roots	Root fresh weight (mg)
500 ml Erlenmeyer[a]	6.6b[f]	8.4a	499.0b	3.4ab	109.1b
500 ml Mason jar[b]	7.3a	7.4b	535.9b	3.5ab	110.8b
500 ml culture bottle[c]	7.2a	7.6b	536.1b	3.5ab	106.9b
Plant box[d]	6.6b	8.4a	530.4b	3.3b	115.0ab
Culture Pack[e]	7.1a	8.6a	617.9a	3.8a	134.0a

[a]Covered with aluminum foil.
[b]Covered with polycarbonate screw cap.
[c]Mason jar-shaped plastic culture vessel made of polycarbonate with the same polymer screw cap.
[d]Magenta GA7-shaped plastic culture vessel made of polycarbonate with the same polymer cap.
[e]Culture Pack made of 25-μm Neoflon PFA film.
[f]Numbers in a column followed by the same letter are not significantly different at 5% level (t-test).
Each culture vessel contained 100 ml of Modified Kyoto medium (Tsukamoto et al. 1963) [2 g/l Hyponex ($N:P_2O_5:K_2O$ = 6.5:6:19), 100 mg/l NH_4NO_3, 100 mg/l $MgSO_4$, 30 mg/l Fe-EDTA, 1 mg/l thiamine • HCl, 20 g/l sucrose, 8 g/l agar].

Fig. 6. *Cymbidium* plantlets grown in the Culture Pack. They can be easily removed from the Culture Pack by cutting the film with a blade

Fig. 7. Effect of thickness of Neoflon PFA film used for the Culture Pack on shoot development of *Cymbidium* Katydid Yasuko Oji 90 days after planting

number of leaves and roots per plantlet increased, but the average length of leaves and roots decreased with increased thickness of the film. Likewise, in *Spathiphyllum*, with decreased thickness of film, enhanced growth of plantlets was obtained (Fig. 8).

From these results, the Culture Pack made of 25 μm Neoflon PFA film is recommended for use in shoot development in micropropagation. The 12.5-μm thickness can also be used, but it is too thin and inconvenient to handle.

3.4 The Larger-Sized Culture Pack

To improve efficiency and easy of work in transferring shoots onto appropriate established medium and to save space in the environmentally controlled culture room, we tried to increase the volume of a culture vessel. Since the Culture Pack is made of fluorocarbon polymer film, it is easy to make it into any desired size. Therefore, the Culture Pack in a larger size (22.5 × 15 × 10.5 cm, the outer size of a frame made of 30 mm stainless steel wire), which is six times as large as that of the standard size (Fig. 9A), has been developed. A sheet of 25-μm Neoflon film (PFA, 39 cm long by 52 cm wide) is folded and heat-sealed in the manner described earlier, except that the opening is at the side (Fig. 9C). To prevent the vessel from collapsing, a frame is placed in the Culture Pack. The Culture Pack of medium size (15 × 15 × 10.5 cm), which is four times as large as the standard size, is also made of the same film (31 cm long by 52 cm wide) (Fig. 9B). After 140 ml, 560 ml, and 840 ml of the agar medium are poured into the Culture Pack of standard size (A), medium size (B) and larger size (C), respectively, the opening is folded and covered with an aluminum foil and then sterilized by autoclaving. To examine whether these large Culture Packs can be used for culture, 16, 64, and 96 shoots of *Spathiphyllum* were planted in the medium in the Culture Packs A, B, and C, respectively. Their openings were then heat-sealed.

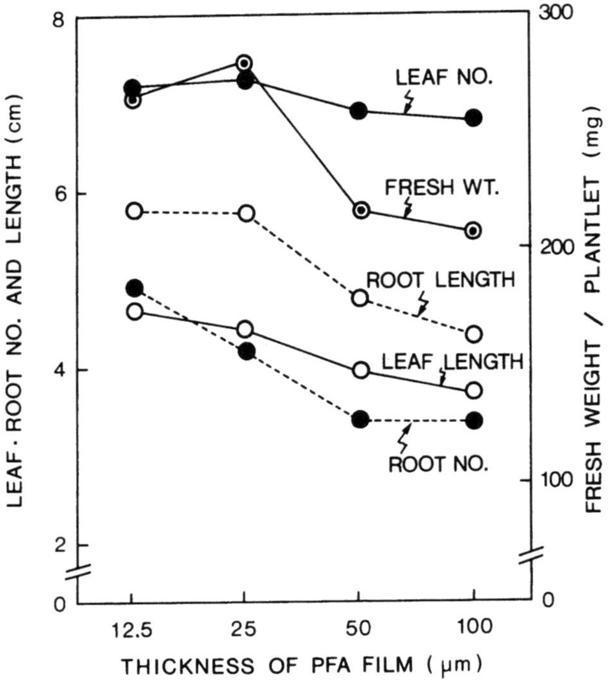

Fig. 8. Effect of thickness of Neoflon PFA film used for the Culture Pack on shoot development of *Spathiphyllum* cv. Merry 75 days after planting. (After data of Tanaka et al. 1988b)

The plantlets grown on agar medium in the Culture Pack of larger and medium size were normal. Their size and fresh weight were almost the same as those of the plantlets grown in the standard-sized Culture Pack (Fig. 10). The results showed that there is no problem in using the larger-sized Culture Pack for in vitro shoot development. Hundreds of plantlets could be cultured in each larger Culture Pack and the use of such a vessel is suggested for commercial tissue culture laboratories.

Enhanced growth of plantlets was also obtained in the larger Culture Pack made of 12.5 μm PFA film when compared to 25 or 50 μm film (Fig. 11). However, the use of 12.5 μm film for the larger Culture Pack is risky, due to possible damage before and after autoclaving, since a large quantity of medium (840 ml) is in the Culture Pack. On the other hand, if the larger Culture Pack is made of 50 μm film, the cost of the film is four times that of the 12.5 μm. For these reasons, from the practical point of view the larger Culture Pack made of 25 μm Neoflon film is recommended.

3.5 Use of Culture Pack with Rockwood

One of the major costs in micropropagation in commercial laboratories is manual labor, especially when carrying out manipulation at later stages (Maene and Debergh 1985). Before transferring the plantlets to soil, they should first be acclimatized to enable them to adjust to the external growing conditions. During this acclimatization, especially after transferring, some plantlets fail to establish or die off because of excessive water loss or poorly developed root system. For these reasons we inves-

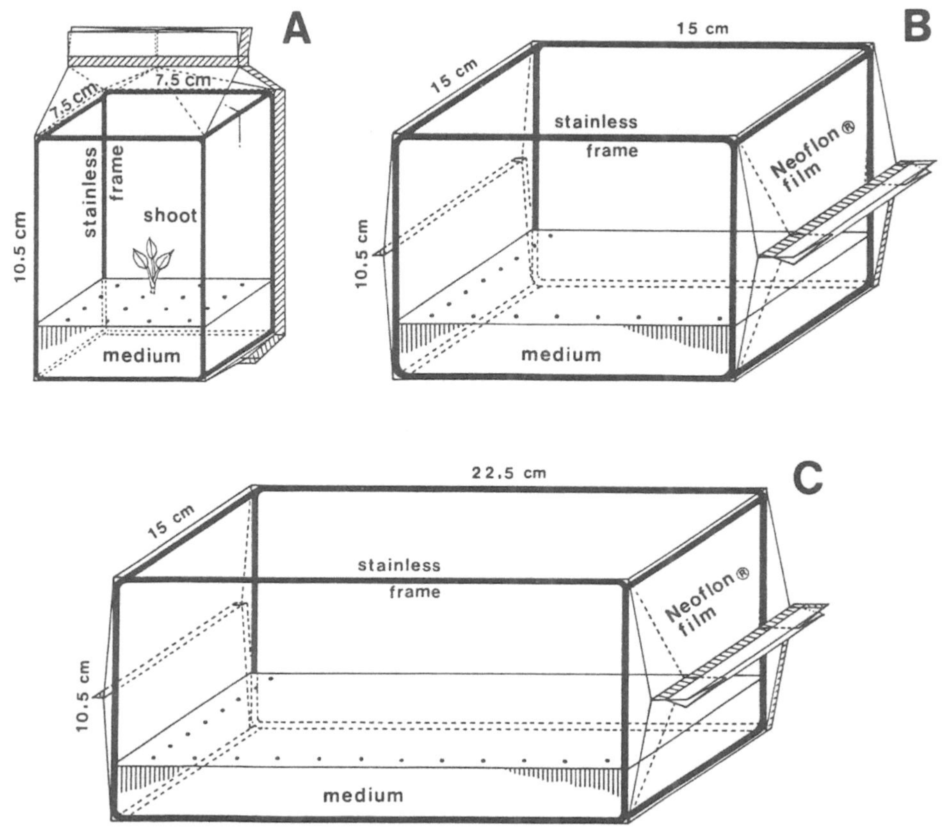

Fig. 9A-C. Diagram illustrating the various sizes of the Culture Pack which were filled with agar medium. The opening of these Culture Packs has been heat-sealed on either the *upper shadowed portion* (**A**) or the *side shadowed portion* (**B** and **C**). The shoots planted on agar medium are omitted in this diagram except one in the Culture Pack of standard size (**A**). **A** Standard size. **B** Medium size. **C** Larger size. (After Tanaka et al. 1988b)

tigated the possibilities of saving manual labor and of minimizing losses of tissue-cultured plants by using an artificial substrate, for example, rockwool, in the Culture Pack.

Sixteen joined blocks (four by four blocks) of Grodan Rockwool Multiblock AO 18/30 (Grodania A/S, Denmark) or a polyester-fiber block (7.5 × 7.5 × 3 cm, Toyobo Co., Ltd., Osaka, Japan) are placed into the standard sized Culture Pack made of 25 µm PFA film. This is autoclaved (25 min at 121 °C) and 140 ml of the sterile liquid medium is poured into the Culture Pack. In the case of the Multiblock, the liquid medium attains a depth of about 1 cm in the Culture Pack. On the other hand, a polyester-fiber block absorbs the liquid medium completely. The shoots of *Spathiphyllum* are inserted into the small hole made on the surface of the Multiblock or polyester-fiber block. The opening of the Culture Pack is then heat-sealed (Fig. 12). For comparison, shoots are also cultured on agar medium in the Culture Pack.

The plantlets grown on both the artificial substrates were normal, and were more vigorous than those on agar medium (Fig. 13). The number of leaves, leaf length, and number of roots increased when cultured on polyester-fiber block compared to those

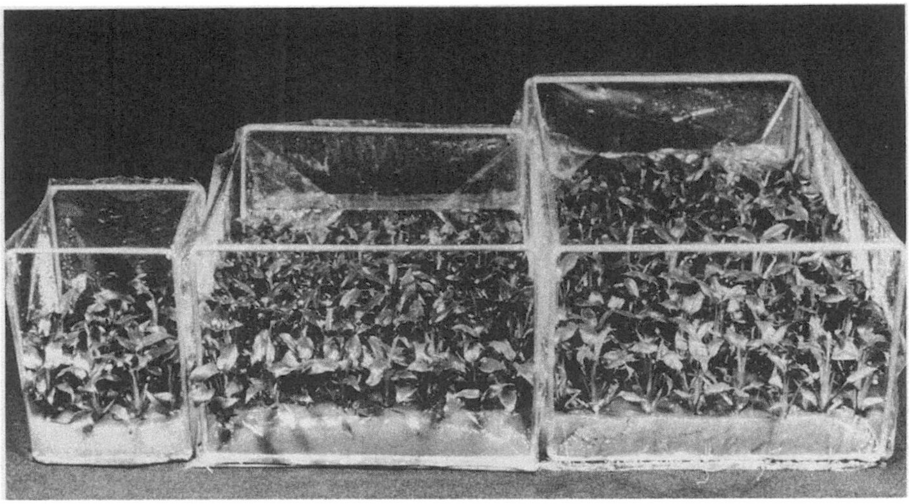

Fig. 10. A comparison of shoot development of *Spathiphyllum* cv. Merry cultured in the Culture Pack of different sizes. *Left* standard size; *middle* medium size; *right* larger size (see Fig. 9)

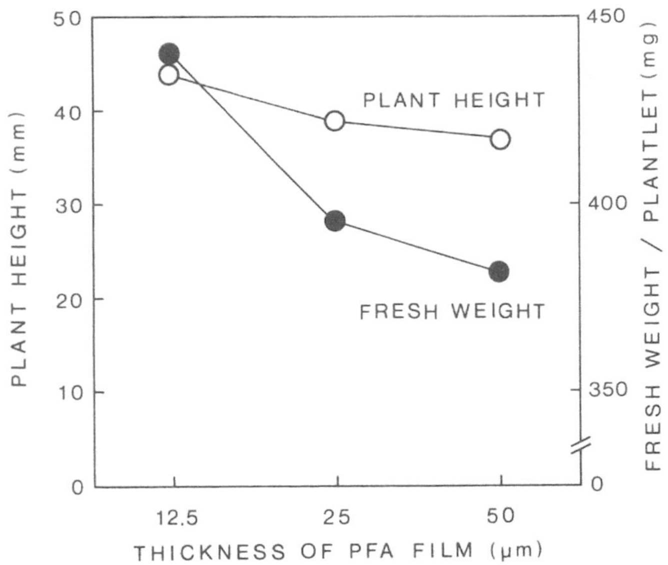

Fig. 11. Effect of thickness of Neoflon PFA film used for the Culture Pack of larger size on shoot development of *Spathiphyllum* cv. Merry 75 days after planting. (After data of Tanaka et al. 1988b)

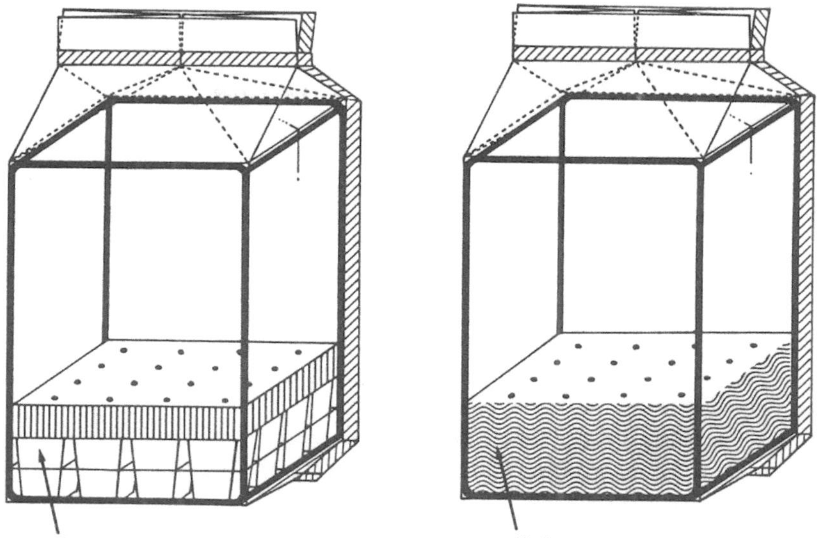

Fig. 12. Diagram illustrating the Culture Pack of standard size in which rockwool Multiblock (*left*) and polyester-fiber block (*right*) are placed with liquid medium instead of agar medium. These Culture Packs have been heat-sealed on the *upper shadowed portion*. The shoots planted on artificial substrates are omitted in this diagram. (Tanaka et al. 1988b)

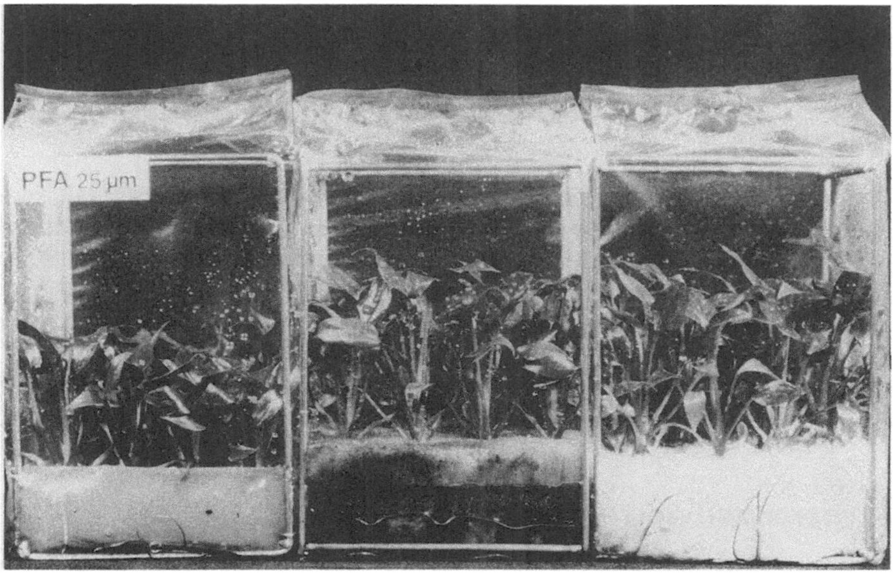

Fig. 13. A comparison of the development of *Spathiphyllum* cv. Merry shoots grown on differenι substrates in the Cjlture Pack of standard size 75 days after planting. *Left* agar medium; *middle* rockwool Multiblok; *right* polyester-fiber block

grown on the Multiblock and agar medium. However, there was no significant difference in the plant height and fresh weight of plantlets between the various substrates. From these data, the Culture Pack in combination with an artificial substrate was found to be suitable for the shoot development in micropropagation of *Spathiphyllum*. The plantlets grown on polyester-fiber block, however, could not be separated from each other, since their roots grew and remained within the block. With the Multiblock, on the other hand, each block could be separated easily without causing any damage to the roots. Therefore, it appears that the Multiblock in the Culture Pack is preferable.

The plantlets grown in the Culture Pack can be acclimatized by gradually making a larger slit on the top of the film. Liquid medium is drained off by making several holes at the bottom. The plantlets on the Multiblock could easily be taken out from the Culture Pack by cutting the film, without damage to the leaves and roots. Furthermore, it is very easy to transfer each plantlet block by block into larger blocks of rockwool (e.g., Grodan Rockwool Growingblock DM4).

The cost for plantlet production is expected to be reduced by this system, since manual labor could be saved and losses due to manual transfer to soil and to failure of acclimatization could be minimized.

4 Envelope-Shaped Vessel, the "Culture Bag"

For the proliferation of protocorm-like bodies (PLBs) in the micropropagation of orchids, the cultures are maintained routinely on agar medium or in liquid medium on a shaker. Using Neoflon film, another vessel in the shape of an envelope has also been developed for the proliferation of *Cymbidium* PLBs. This vessel is referred to as the "Culture Bag".

4.1 Procedure

The procedure to make and use the Culture Bag is shown in Fig. 14. A sheet of Neoflon film (15 cm long by 17 cm wide) is shaped into an envelope and sealed as indicated by the shadowed side. The upper edge of the envelope is cut on a slant (Fig. 14A). After autoclaving the Culture Bag, 80 ml of sterile modified Vacin and Went's liquid medium is poured into the bag through the opening. Explants are then inoculated into the liquid medium through the opening (Fig. 14B). After purging the air in the upper space in the Culture Bag, the opening is sealed with a closing element the Closure (Spectrum Medical Industries, USA), and then heat-sealed (Fig. 14C).

4.2 PLB Proliferation in the Culture Bag

The explants used are segments of PLBs derived from shoot-tip culture of *Cymbidium* Showgirl Husky Honey. Each PLB, 3 mm in diameter, is longitudinally divided into four. Sixty of these segments were cultured on agar medium, stationary liquid medium, and in shaken liquid medium in Erlenmeyer flasks, and in the Culture Bag made of Neoflon FEP film of 100 μm thickness.

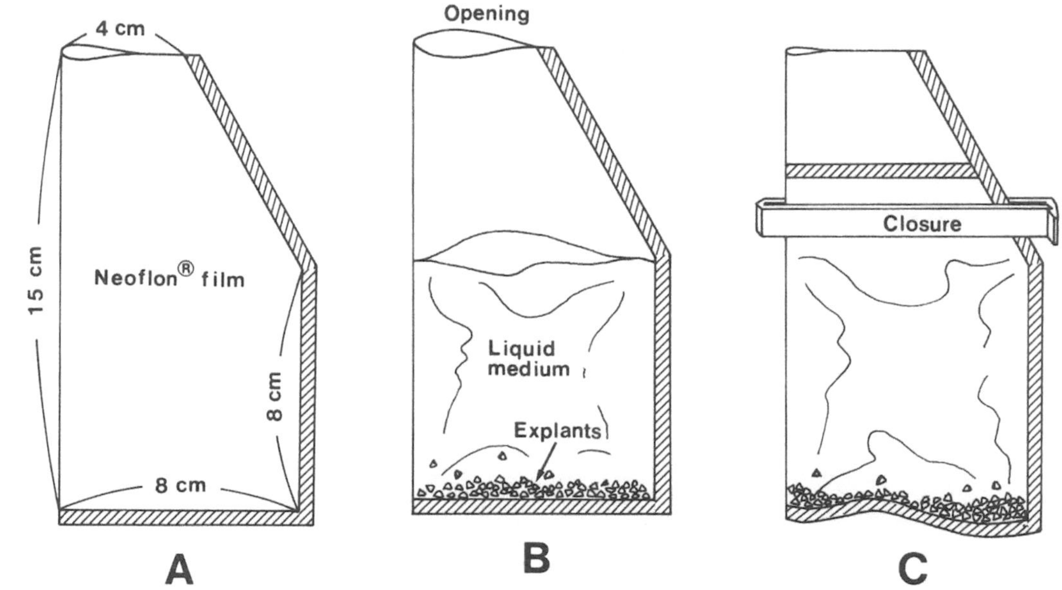

Fig. 14A-C. Diagram illustrating the procedure to make and use the Culture Bag. (Tanaka et al. 1988a)

PLB proliferation on the explants was observed in the Culture Bag. This also occurred with tissue in the conventional culture vessels (Fig. 15). Culture survival in the Culture Bag or on solid medium was about 95%, higher than in stationary or shaken liquid medium in flask. All surviving segments produced PLBs. The average number of PLBs per segment in the Culture Bag was about 12. However, on solid medium, the average was 4.3, in stationary liquid medium 9.3, and in shaken liquid medium 4.3. No shoot formation was observed in the Culture Bag, while in the flasks explants produced shoots (Fig. 16). Only PLBs were produced in the Culture Bag. The fresh weight of cultures grown on solid medium and in shaken liquid medium was high, but the number of PLBs was less when compared with those grown in the Culture Bag. This could be due to the many shoots that developed.

From these data, the Culture Bag is found to be a suitable vessel for the proliferation of *Cymbidium* PLBs. During culture, the Culture Bag can be maintained by hanging without shaking. This saves space in the environmentally controlled chamber (Fig. 17). The proliferated PLBs can also be easily taken out from the Culture Bag by cutting the upper part of the bag with scissors.

4.3 Other Possible Uses of the Culture Bag

In other experiments, the use of the smaller-sized Culture Bag (9 cm × 4 cm) is also found suitable for the culture of tobacco mesophyll protoplasts, in which the division frequency 5 days after plating and the total cell volume 30 days after plating are higher than those cultured in 60 × 15 mm plastic Petri dishes (Tanaka and Ikeda 1987). This

Fig. 15. Proliferation of protocorm-like bodies (PLBs) of *Cymbidium* in the Culture Bag. Note that no shoot formation is observed. (75 days after inoculation)

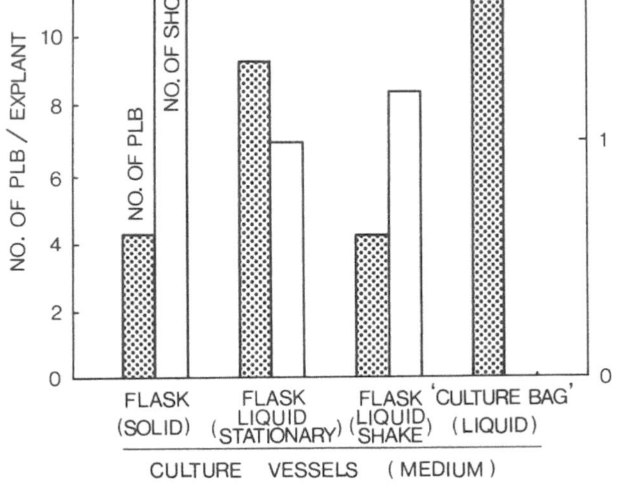

Fig. 16. Effect of culture vessels on the proliferation of PLBs and shoot formation of *Cymbidium* Showgirl Husky Honey 75 days after inoculation. (Tanaka et al. 1988a)

Fig. 17. The Culture Bag maintained by hanging without shaking in environmentally controlled chamber. This saves space

suggests that the Culture Pack could be used not only for organ culture but also for cell culture.

Valuable culture materials, clones or cell lines, placed in the Culture Bag can be sent via air mail. For the gene bank, we have already succeeded in preserving PLBs in the Culture Bag under low temperature for several months (unpubl.).

5 Conclusions

Culture vessels discussed above: the Culture Pack (CP) and the Culture Bag (CB), should provide several advantages when used in plant tissue culture: they are (1) cheap [CP, CB], (2) disposable [CP, CB], (3) transportable [CP, CB], (4) preservable [CB], (5) convenient for handling of cultures [CP, CB], (6) flexible in shape and size [CP, CB], (7) they provide a completely closed sterile system [CP, CB], (8) can save space in an environmentally controlled culture room [CB], and (9) also be useful for acclimatizing tissue-cultured plants [CP].

A prototype of the Culture Pack and Culture Bag were developed to overcome the many disadvantages of conventional culture vessels. These film vessels are at the present at the stage of practical application, and are not yet available on the market for commercial use. The price of these disposable vessels is expected to be cheaper than that of glass vessels, since they are composed of fluorocarbon polymer film and frame only. Recently, Daikin Industries Ltd. has made a trial manufacture of the Culture Pack to produce samples for further trials.

References

Aitken-Christie J, Jones C (1987) Towards automation: radiata pine shoot hedges in vitro. Plant Cell Tissue Organ Cult 8:185–196

Harris RE, Mason EB (1983) Two machines for in vitro propagation of plants in liquid media. Can J Plant Sci 63:311–316

Koch L (1973) Vergleich zweier Verfahren zur Vermehrung von *Cymbidium*-Protokormen. Gartenbauwissenschaft 38:419–426

Maene L, Debergh P (1985) Liquid medium additions to established tissue cultures to improve elongation and rooting in vitro. Plant Cell Tissue Organ Cult 5:23–33

Monette PL (1983) Influence of size of culture vessel on in vitro proliferation of grape in a liquid medium. Plant Cell Tissue Organ Cult 2:327–332

Murashige T, Skoog F (1962) A revised medium for rapid growth and bioassays with tobacco tissue cultures. Physiol Plant 15:473–497

Nitsch JP, Nitsch C, Rossini LME, Bui Dang Ha D (1967) The role of adenine in bud differentiation. Phytomorphology 17:446–453

Takayama S, Misawa M (1981) Mass propagation of *Begonia* × *hiemalis* plantlets by shake culture. Plant Cell Physiol 22:461–467

Tanaka M, Ikeda S (1987) Application of fluorocarbon polymer films to plant tissue culture. 3. Culture of tobacco mesophyll protoplasts in a Culture Bag made of fluorocarbon polymer film. Jpn J Breed 37 (Suppl 1):126–127

Tanaka M, Goi M, Higashiura T (1988a) A novel disposable culture vessel made of fluorocarbon polymer films for micropropagation. Acta Hortic 226:663–670

Tanaka M, Jinno K, Goi M, Higashiura T (1988b) The use of disposable fluorocarbon polymer film culture vessel in micropropagation. Acta Hortic 230:73–80

Tisserat B, Vandercook C (1985) Development of an automated plant culture system. Plant Cell Tissue Organ Cult 5:107–117

Tsukamoto Y, Kano K, Katsuura T (1963) Instant media for orchid seed germination. Am Orchid Soc Bull 32:354–355

Vacin EF, Went FW (1949) Some pH changes in nutrient solutions. Bot Gaz 110:605–613

Section II
Rooting, Artificial Substrates, Autotrophic Micropropagation, Nursery Planting

II.1 Rooting of in Vitro Cuttings

CH. MONCOUSIN[1]

1 Introduction

Most perennial plants, herbaceous or woody, have a genetical structure that does not allow them to be consistently multiplied other than by vegetative multiplication. The horticultural industry is thus largely dependent on adventitious rooting, whether by classical multiplication methods or by in vitro micropropagation. It is thus surprising to see that interest in work on adventitious rhizogenesis has declined over the past 10 years, when it could be of great economic importance due to the development of in vitro microcutting techniques.

In addition, from the scientific point of view, the study of adventitious rooting is likely to provide essential information for understanding the mechanisms involved in organogenesis in plants. A cutting is in fact nothing more than plant separated from its source of minerals and water supplied by the roots. The re-establishment of the rooting system requires such deep modifications in the metabolism of the cutting that the majority of cuttings without pre-formed roots will not be able to form a new rooting system.

To understand, adventitious rhizogenesis needs a description of the biochemical modifications that take place in the cutting, especially in the rooting zone. As precise analysis cannot be made at the initial cell level, it has to be based on the cytohistological changes from the time of cutting to the emergence of the first root. Bouillenne and Went (1933) and Libbert (1956) have shown that the presence of a root inhibits the setting up of other roots. With in vitro cuttings of *Vitis*, the newly formed root meristems progressively induce secondary formations of the stem and thus hide its real potential (Favre 1973a). The emergence of the first root is therefore the most direct expression of the rooting capacity of cuttings.

In many publications, even recent ones, the authors do not emphasize that rhizogenesis is divided into precise stages: induction, cellular activation, orientation, organization, and rooting expression. For each of these, one can distinguish precise cytohistological and biochemical states, but also varying sensitivity to any external treatment. Often the study of rhizogenesis is limited to comparing the effects of various external treatments on the percentage of rooted cuttings and on the number of roots formed per cutting, instead of studying the effects on the evolution of the internal contents and the metabolism of the plant.

In this chapter, we use data collected from the literature as well as from our laboratories in the attempt to answer questions on the evolution of rooting ontogenesis,

[1]Centre horticole de Lullier, Ecole d'ingénieurs ETS, 1254 Lullier, Switzerland

Biotechnology in Agriculture and Forestry, Vol. 17
High-Tech and Micropropagation I (ed. by Y.P.S. Bajaj)

correlations between organs, the influence of the origin of the plant material and its in vitro history, the role of mineral and carbohydrate nutrition, the influence of environment, the role of rooting substances, the characteristics of adventitious rooting, and finally to deduce practical consequences.

2 Stages of Root Ontogenesis

The formation of adventitious roots has been described in detail for in vitro microcuttings of *Vitis* by Favre (1973a,b,c), who distinguishes four stages. The first is marked by cytological modifications and consists essentially of a general activation at the base of the cutting. The second stage is characterized by the appearance of cellular divisions, which end with the formation of a primary mass of meristematic cells: the morphogenetical rooting zone. The third stage covers the organization of the morphogenetical zone in a root meristem. The last stage corresponds to the initial growth stages of the root meristem and the emergence of the first root.

The general activation at the base of the cutting takes place in the context of a return to the meristematic state of cytologically differentiated cells. This is shown by a considerable enlargement of the nuclei, which move from a parietal location to a central position within the cells, as well as by the reduction and fragmentation of the vacuole system (Mahlstede and Watson 1952). This reactivation affects not only the cells which will form the rooting morphogenetic area, but also all the cells at the base of the cutting which will form the healed callus. The development of mitotic activity in the tissues at the base of the cutting in fact results on the one hand in setting up the initial cells of the morphogenetic area and on the other in forming the healing tissue.

The formation of the healed callus is shown by cellular growth localized first at the level of the primary phloem cells, then reaching all the secondary nonlignified cells, and sometimes even the cortical parenchyma and the marrow (Favre and Medard 1969). The formation of morphogenetical areas is described by the same authors in the following manner: "while the process of cell differentiation begins, (at the same time) the secondary tissues progressively lose their characteristic way of functioning, and evolve into a new organization in which the most active cells, whilst still retaining a tangential orientation in their divisions, tend to be localized on the external face of the central cylinder. Then a few of the cells in this area lose all their polarity (of division) and recover, cytologically and histologically, their primary meristematic state." The meristematic cells set in place by this process then multiply without any polarity and form a small globulous mass, rapidly increasing in size, which represents the morphogenetical rooting area (Favre and Medard 1969).

The appearance of lengthening and division polarities after the previous multiplication stage leads to an initial organization in the morphogenetical field. In the basal and internal region of the field, the cells grow longer in the radial direction, break up crosswise and constitute the central cylinder. In the apical and superficial region, the septations of the cells lead to the installation of cellular depositions at the origin of the cortex and the epidermis of the future root.

With microcuttings of *Vitis* the adventitious rhizogenesis process is sequential and can be divided into six distinct stages: no visible modifications, nuclei puffing, periclinal divisions of the cambium, return to the meristematic stage, presence of the morphogenetical field, and appearance of the radical meristem. The sequential character of adventitious rhizogenesis was confirmed with chestnut microcuttings by Vieitez et al. (1981). Transverse sections of the base of in vitro chestnut tree shoots show a discontinuous ring of sclerenchyma surrounding the primary phloem. This sclerenchyma ring is composed of fibers separated into groups by regions of parenchymatic tissue. Analysis of transverse sections taken between the day of cutting and the day of root emergence revealed a sequence of four anatomically distinguishable structures: meristemoid, root primordium, root primordium with its own vascular system, and adventitious root. Significant cellular changes occurred as early as 2 days after cutting; a certain number of cells in the phloem parenchyma and vascular rays can be observed during division, giving rise to cells of smaller size with pre-eminent nuclei and dense cytoplasm (reversion to meristematic state). Four days later, the number of dividing cells increases so that meristematic cell masses form in the phloem region near the cambium; this increase of cells in division leads to meristem-like aggregations of smaller isodiametric cells which are nonvacuolated and could be identified as meristemoids which start to separate on the 6th day; polarization of the division induces the typical appearance of root primordia (8th day). The primordia, at this stage, are generally located at the level of the sclerenchyma ring, which they cross between day 9 and 10. Between day 11 and 12, most of the primordia develop their own vascular system. Similar evolutions, but out of phase, are noted with *Picea* (Patel et al. 1988) and *Pinus* microcuttings (Mott and Amerson 1981).

With *Vitis* and *Malus* microcuttings, we found, as described by Favre (1973a), identical stages in root ontogenesis but we noted important temporal overlappings, (even with homogeneous clones) (Table 1) probably originating from a lack of precision in definition of the more forward ontogenetic stages, but also certainly heterogeneity of behavior, as we found the same phenomenon at later stages.

The principal conclusion deduced from this kind of study is the sequential character of adventitious rhizogenesis. The course of the stages is very similar to the three phases described by Smith and Thore (1975a) with *Pinus radiata* hypocotyl cuttings. In fact, the pre-initiative phase is characterized by a lack of visible histocytological indications, the initiative phase corresponds to the first cytological modifications and to cellular cutting, which leads to the formation of morphogenetical fields, and the postinitiative phase characterizes the ultimate stages of the rooting process. Kinetin and gibberellic acid inhibited rooting, but indole-butyric acid had a stimulating effect during the pre-initiative phase (Smith and Thorpe 1975b). Smith and Thorpe (1977)

Table 1. Average time of appearance of various ontogenesis rooting stages with *Vitis* microcuttings

Ontogenetic stages (h)	40%	60%	80%[a]
Nuclei puffing	33	39	45
Beginning of periclinal division of the cambium	45	54	68
Constitution of morphogenetic radical fields	78	81	92
Radical primordium appearance	106	115	126
First root emergence	128	136	142

[a]Ratio of cuttings presenting ontogenetic stage described.

demonstrated that the effectiveness of the auxin is dependent on the nature of the substance, its concentration, and time of application, but also on the initial rhizogenic capacity of the plants. This underlines the relationship between the rhizogenic capacity of the cuttings and an optimal content of various factors.

3 Rooting Correlations: Influence of Lateral Organs

The presence of young leaves or active buds generally stimulates adventitious rhizogenesis. The stimulating influence of the buds on rooting in cuttings has been widely demonstrated (Bouillenne and Went 1933; Molisch 1935; Nysterakis 1954; Harada and Nakayama 1959; Julliard 1963). More precise analysis performed on stable, homogenous in vitro material has emphasized that the positive effect of the buds on adventitious rooting can be divided into activating, organizing, and inhibiting action (Favre 1973c). However, it seems to be more a matter of how far the buds are away from the base of the cutting than their theoretical role (Favre 1973a).

The leaf (like the bud) generally has a stimulating effect on adventitious rooting. Its presence is sometimes found to be compulsory at cutting (Zimmerman and Hitchcock 1929; Halma 1931; Evans 1952). The similarity of the effects of leaves and buds can be illustrated by the behavior of in vitro cuttings of *Vitis* with only one bud or leaf and with the axillary bud removed (Favre 1973a,b). The elements responsible for the activation of rooting seem to originate in the leaf during photosynthesis and are then stored in the buds. However, according to Julliard (1963), the meristematic zone rather than the developing foliar organs is at the origin of the rooting stimuli. The comparative study of the rooting ontogenesis of mono-nodal *Vitis* cuttings, with or without leaves, shows that the presence of the leaf has no influence on the speed of appearance of the earliest stages (dilation of the nuclei to regain the primary meristematic state), but that the cuttings with leaf more quickly attain the stage of formation of rooting of morphogenetic areas. The time advantage of these cuttings with leaf increases as they come closer to the stage of emergence of the roots (Moncousin 1987).

4 Influence of Juvenility

The genetic basis of the phenomenon of rhizogenesis explains why certain genotypes can be propagated by cuttings more easily than others, but generally little attention is paid to the heritability of this character in plant breeding programs.

However, apart from their genetic quality, the ontogenetic age of the mother plants clearly influences the possibilities of adventitious rooting, especially in woody plants. Cuttings taken from seedlings or young plants generally exhibit rooting potential which disappears when the plants have reached the adult stage, and even before they acquire the ability to flower. A reversal of this state, corresponding to rejuvenation, can be achieved by various manipulations (Franclet 1974, 1977, 1979).

The juvenile stage of a plant seems to depend on the size of the meristems, their diapause, and environmental conditions. Roux (1968) has shown that during development, the size and cellular amount of the meristems increase when the juvenile stage is over. Microsurgical reduction of the meristem in adult *Phyllanthus distichus* orthotropic axis (Bancilhon 1969) permits the return to a juvenile type of morphogenesis. The utilization of in vitro methods also allows a miniaturization of the plant material.

Morphogenesis is of a juvenile type and can be maintained by cloning (Morel 1963; Champagnat 1977; Nozeran et al. 1977; Favre and Grenan 1979; Mullins et al. 1979). Comparison of apex of *Dahlia* plants cultivated in vitro or in greenhouse conditions showed that the terminal meristem was reduced in size due to a reduction in the number of cells (Watelet-Gonod and Favre 1981). The reacquisition of the ability for adventitious rhizogenesis following the use of in vitro methods has been shown in many cases: *Citrus* (Bouzid and Lasram 1971), *Vitis* (Favre 1970, 1973a), *Pinus* (Franclet et al. 1980), *Cynara* (Moncousin 1982; Moncousin and Ducreux 1984), etc.

The better rhizogen behavior of juvenile material has been attributed to various causes. Vasquez and Gesto (1982) propose an auxin content, which is more important in juvenile shoots of *Castanea sativa*. Heuser and Hess (1972) consider that the greater ability for rooting of juvenile forms of *Hedera* cuttings was due to the presence of rooting lipid-like co-factors. However, differences exist in the juvenile form of *Hedera*; the basal internodes root more easily than those in the apical part of the twig (Poulsen and Andersen 1980). On the same material, Rogler and Hackett (1982) have shown that the plant can pass from an adult state to a juvenile one by the application of gibberellic acid, and that the reverse was possible by the use of abscisic acid. These various transformations correspond to an increase or decrease of the ability for adventitious rooting.

From the complete study of rejuvenation by Franclet et al. (1987) the following are the most important points:

- the principal cause of the rejuvenating effect is the duration of exposure of the plant to cytokinin (BAP),
- the number of subcultures of old clones of a BAP-rich medium increases the rejuvenation effect, in particular, the shorter they are
- the juvenile clones cannot stand exposure to BAP for a long period; the cytokinin loses its organogenetic capacity and vitrification of the cuttings,
- the rejuvenized clones preserve their "youth" by alternating subcultures with and without BAP.

These points explain the contrary effects noticed when increasing the number of subcultures with woody plants (Pierik 1987).

5 Basal Medium and Carbohydrate Nutrition

The mineral nutrition before and after the cutting process influences the rooting behavior of the cuttings.

Before the cutting process, the need for a medium with a high salt concentration is recognized. In the micropropagation of *Cynara scolymus*, the dilution of the mineral fraction in the proliferation medium caused a sharp drop in rooting percentage (Fig. 1) (Moncousin 1982). Maene and Debergh (1985) came to similar conclusions for microcuttings of various plants in the elongation phase, i.e., before rooting treatment.

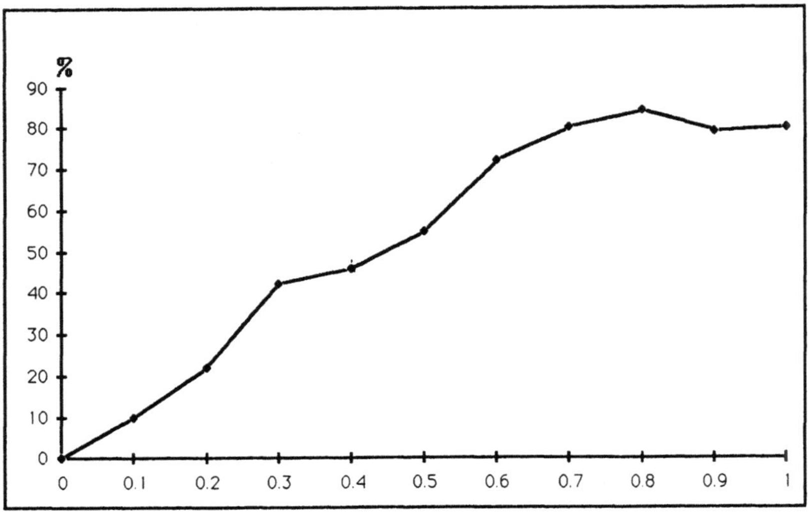

Fig. 1. Development of the rooting percentage of microcuttings of *Cynara* in relation to the degree of dilution of the mineral function during the three subcultures preceding adventitious rooting. (Moncousin 1982)

Replacing the macromineral solution (MCX-87.3 mM) by a more diluted one (QL/2–33.8 mM) during the four subcultures preceding microcutting results in an important reduction in the rooting percentage of *Malus* (Table 2) (Collet and Le 1988).

Comparison of the MCX and QL/2 media shows that the difference is not limited to the molarity but also lies in the "richness" in NH_4^+ (20.6 mM instead of 2.5 mM), NO_3^- (39.4 mM instead of 16.4 mM), and K^+ (20.05 mM instead of 9.9 mM).

During rhizogenesis, a more diluted mineral fraction is used than during micropropagation. This practice is applied for herbaceous plants (Welander 1979; Simmonds 1984; Maene and Debergh 1985), as well as for woody ornamentals (Hyndman et al. 1982; Maene and Debergh 1985), fruit trees (Sriskandarajah et al. 1982; James 1983; Simmonds 1983; Broome and Zimmerman 1984; Zimmerman 1984) or forestry trees (Boulay and Franclet 1977; Chalupa 1977; Coleman and Thorpe 1977; Boulay 1979; David 1982; Hammerschlag 1982; Cornu 1984; Patel et al. 1986).

The basic mineral fraction is very often MS diluted by 50 to 75%, sometimes Knop (1959) or Quoirin and Lepoivre (1977) diluted by half. In some cases a medium is especially made for adventitious rooting (Collet 1985; Collet and Le 1987).

Eliasson (1978) considers that the addition of mineral elements is not essential to achieve adventitious rooting, as most of the ions are transferred basipetally in the stem, but according to Appelgren and Heide (1972), too great a reduction in ionic concentration can have negative effects on the formation of roots. During the adventitious rooting of *Cynara*, after reduction of the ionic concentration of the mineral fraction of the MS medium, the rooting percentage went from 25 to 62% (dilution factor 50%) then to 40% (70% dilution), and finally to 15% (90% dilution) (Moncousin unpubl.).

The favorable effect of a diluted mineral solution on rooting can be explained by the reduction in nitrogen (Hyndman et al. 1982; Driver and Suttle 1987). However, a strong reduction in nitrate disturbs the rooting of microcuttings of *Eucalyptus* (Depommier 1981).

Table 2. Evolution of rooting percentage on an identical medium after four subcultures on more or less mineral-rich media. (Collet and Le 1988)

Rooting percentage	M25		M26		M27		MM106	
After	MCX	QL/2	MCX	QL/2	MCX	QL/2	MCX	QL/2
14 days	83	7	43	15	67	20	33	–
21 days	88	22	68	19	70	24	49	–

Nitrogen must be added by using both NO_3^- and NH_4^+ (Hyndman et al. 1982): the optimal ratio between these ions is 3 of rose cuttings (Hyndman et al. 1982) and 4 for *Cynara scolymus* (Moncousin 1982). Trewavas (1983) suggested that the nitrate ion, at a low concentration, could mimic the action of IAA and in any case supports its role.

Other ions are more or less implicated in the rooting process. Boron has a beneficial effect on cuttings (Middleton et al. 1980; Jarvis and Booth 1981; Jarvis et al. 1983, 1984) or microcuttings (Moncousin and Gaspar 1983) but its role is very complex, as it is implicated in the transport of carbohydrates and the metabolism of auxins and phenols. Knowing that high auxin concentrations are necessary for the first stages of rhizogenesis and that they can then inhibit the evolution of the morphogenetical areas (Middleton 1977), a regulation of endogenous auxin by boron, either by destruction through oxidation or through auxinic protectors, would explain the lack of effects shortly after cutting and the later effectiveness of boric treatment. Manganese intervenes in rhizogenesis: a low concentration increases the rooting percentage but disturbs the development of roots of *Vitis* (Chee 1986). These effects presume a direct or indirect action on the auxinic activity (Gaspar et al. 1982). The effect of potassium iodine on the growth of roots can also be explained by the action of iodine on the auxin metabolism (Chee 1986).

The application of systematic variants (Homes 1961) to the definition of a medium adapted to adventitious rhizogenesis in various plants has shown that the optimal ionic concentration varied between 35 and 45 mM, the best nitrate level was 80 to 85% of the total mineral nitrogen, the major anion was NO_3^-, the main cation K^+, and the presence of chlorine and sodium was detrimental (Moncousin 1982). The rooting medium developed by Collet (1985) shows the same trend in concentration, major ions, NO_3^- ratio, and the role of Cl^- and Na^+ ions. In comparison with that of MS, these media are characterized by an ionic concentration that is halved, by a lower amount of total nitrogen – especially ammonium – more as proposed by Hyndman et al. (1982) and Fletcher (1982). However, calcium and phosphate are more abundant, in agreement with McComb and Bennett (1982).

The study of mineral nutrition on root formation is still very fragmentary; the explant reserves certainly influence the needs during this delicate stage. Detailed information on this subject was given by Nemeth (1986) and Gaspar and Coumans (1987).

The successful weaning of plants from in vitro cultivation is the ultimate aim of rhizogenesis and the carbon nutrition not only contributes most to this success, but favors the setting up of a good vascular system (Sommer and Caldas 1981) and accelerates lignification (Thorpe 1978; Thorpe and Biondi 1981; Driver and Suttle 1987). Furthermore, the rooting of microcuttings demands continuous availability of free sugars (Haissig 1974; Thorpe 1978, 1982b, 1984; Gaspar and Coumans 1987).

Even though enrichment in sucrose undeniably improves rooting in certain plants by biochemical intervention, energetic support, or osmotic effect (Greenwood and Berlyn 1973; Leach 1979; Amerson and Mott 1982; Hyndman et al. 1982; Aldrufeu and Mele 1984; Thorpe 1985), the greatest improvements come from the preparation of the cuttings ex vitro (Maene and Deberg 1985, 1987; Zimmerman et al. 1987; Driver and Suttle 1987).

By pre-treating cuttings of *Juglans* with a solution of 68.5 mM (29 g/l) of sucrose, the rooting percentage increased from 55.5% (with a pre-treatment of 7 days) to 86.7% (pre-treatment 21 days); but the best results were obtained after 7 days of pre-treatment and an increase in sucrose (Fig. 2) (Driver and Suttle 1987).

In rhizogenesis, high additions of sucrose have negative effects, which can be explained by the accumulation of rooting inhibitors, the reduction of promotors (Eliasson and Brunes 1980), and the transformation of the added sugars into soluble and storage forms (Haissig 1984), but also by the fact that the ratios between nitrogen and sucrose (Hyndman et al. 1982; Driver and Suttle 1987) and auxin and sucrose (Welander 1976) influence the rooting process. An interaction between carbohydrates and light has been pointed out by Lovell et al. (1972): the sucrose which inhibits the rooting of *Sinapis alba* in the light has a promoting effect in darkness or when a photosynthesis inhibitor is added to cuttings growing under light. This is why the majority of rooting media contain little sucrose: in the literature we find media with 5 g/l of sucrose (Chalupa 1977; Cheng and Voqui 1977; Boulay 1979; Timmis and Ritchie 1984), but in 70% of the cases doses between 10 and 20 g/l were mentioned for both herbaceous and woody plants.

Trials systematically carried out in our laboratories have tested these doses for various plants such as globe artichoke, vines, fuchsias, platycerium, or apple trees (Fig. 3) as well as rooting in the absence of sucrose.

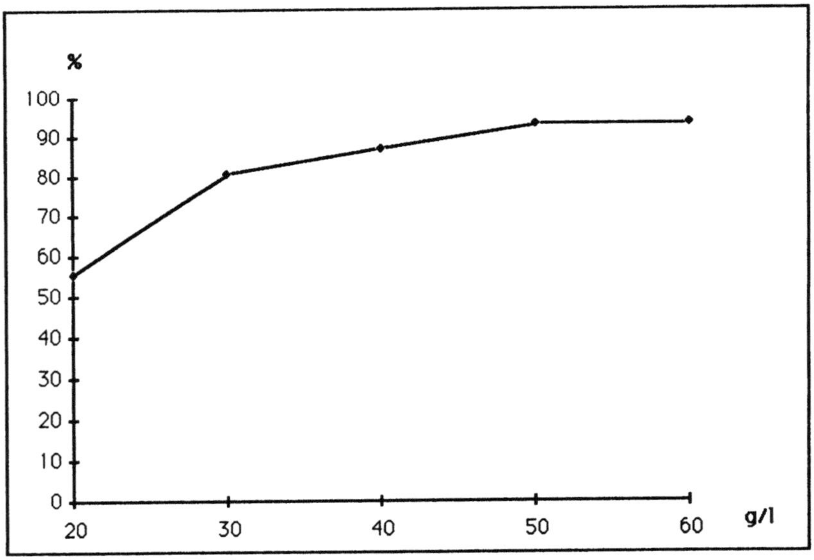

Fig. 2. Development of the rooting percentage in *Juglans regia* × *Juglans hindsii* in relation to sucrose after 7 days of pretreatment. (Driver and Suttle 1987)

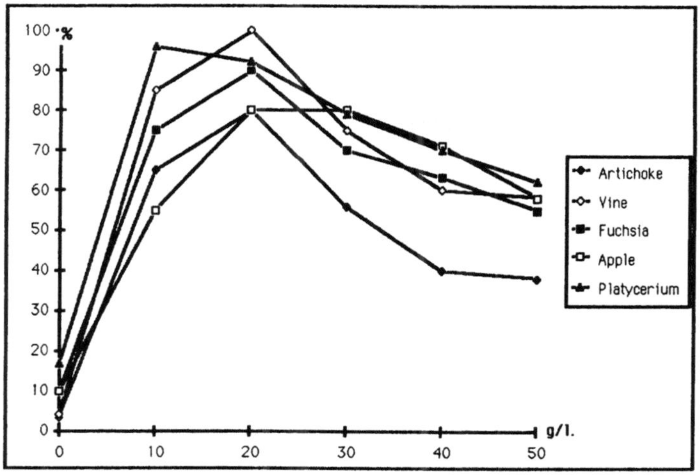

Fig. 3. Rooting percentage of various plants according to the amount of sucrose. (Moncousin unpubl.)

All carbon sources are not able to sustain rooting. Mannitol and sorbitol allow the formation of stems in potatoes, but only sucrose allows rooting (Kikuta and Okazawa 1984). The intensive ramification of *Malus* EMLA 9 is possible on a medium containing sucrose or sorbitol, but the cuttings originating from the sorbitol medium will only root correctly on a rooting medium containing sucrose (Moncousin unpubl.).

The influence of the carbon/nitrogen ratio is clearly illustrated by Hyndman et al. (1982) and Driver and Suttle (1987). With MS medium, the greatest number of roots was obtained with 204.5 mM (70 g/l) sucrose, whereas the normal dose is 87.64 mM (30 g/l). At 87.64 mM sucrose, the best results were obtained by reducing nitrogen to between 7.5 and 15 mM (Hyndman et al. 1982).

The best proportion between the molarity of sucrose and total mineral nitrogen has to be higher than 10 to obtain the greatest number of roots in the rose bush (Hyndman et al. 1982), but for microcuttings of *Juglans*, a rooting percentage greater than 90% is only obtained with a ratio of 30 during preparation for ex vitro rooting (Driver and Suttle 1987).

Amino acids can intervene, besides the addition of NH_4^+ and NO_3^-, by reacting with the sugars (Gaspar and Coumans 1987).

The intervention of oligopeptides in neoformation has been illustrated by Klampt (1983), who suggests that tryptophane and phenylalanine mimic the central structure of phenyl acetic acids.

Vitamins are generally considered to be useless, even detrimental, in adventitious rooting (Skirvin et al. 1981a). However, Novak and Juvova (1983) consider that they are essential for the rooting of vine. Riboflavin is particularly inhibitory (Miller et al. 1982), but by taking advantage of its photolability, it is possible to use it as a regulator to the auxin activity (Druart 1982). Vitamin D and its analogs (ergosterol, ergocalciferol, and cholecalciferol) are used as auxin cofactors in some root-inducing media. These components modify peroxidase activity during root induction (Moncousin and Gaspar 1983; Berthon et al. 1987), probably by affecting peroxidase secretion through the calcium (Kevers et al. 1983).

The pH of the culture medium modulates the rooting process; in general a pH between 4.5 and 5.5 seems to be the most suitable, but contradictory results allow no definite conclusions (Barlass et al. 1980; Sharma et al. 1981; Williams et al. 1984; Nemeth 1986).

6 Environmental Influences

As relative humidity is never a limiting factor in in vitro conditions, the environment of the microcuttings can be divided as influence of temperature, light, and rooting support.

High temperatures, in increasing cellular metabolism (Veierskov and Andersen 1982; Ooishi et al. 1978), favor the initiation of primordium (Favre 1973a) or callus (Cheng and Voqui 1977), depending on the auxin state of the cutting. According to its organogenetic state, it can be useful to place it either at high temperatures (30–35 °C) or lower ones (20–25 °C), varying the choice depending on whether the cutting is in light or darkness. Druart et al. (1981) have illustrated this phenomenon, using wild cherry microcuttings: the rooting percentage is 26% at a permanent 28 °C; with the same temperature but in darkness during the first week, it increases to 79%. Zimmerman (1984) has noted that the optimal temperature for a 1-week treatment in darkness depended on the *Malus* cultivar and on the presence or absence of phloroglucinol. Without the latter, the best temperature is 25 °C for Golden Delicious, 30 °C for Delicious and 35 °C for Royal Red Delicious; with phoroglucinol temperatures of 25 or 30 °C give the same results for Golden Delicious. For conifers, the best results were obtained with relatively low temperatures (Rancillac 1979; Mohammed and Vidaver 1988).

Light is often considered to inhibit rhizogenesis especially at the setting of the primordia. Quite a few works note the beneficial effect of darkness during the first phase of the rooting process. Hammerschlag et al. (1987), working with apple trees, combine a temperature of 26 °C and a dark period of 14 days. In comparing the rooting of two clones of wild cherry, Druart et al. (1981) proved the efficiency of dark treatment for 10 days: the rooting percentage of the least rhizogen clone being increased from 53.6 to 84.3%. The positive effect of darkness is sometimes emphasized by raising the temperature (Zimmerman 1984; Zimmerman and Fordham 1985) or adding phloroglucinol in the medium (James 1983; Zimmerman 1984). The optimal length of the dark period varies between 3 and 10 days depending on the species and cultivar (Depommier 1981; Druart et al. 1982; James 1983; Moncousin and Gaspar 1983; Welander 1983; Zimmerman 1984; Al Maari et al. 1986).

Darkness during the whole of the rooting phase has a stimulating effect on *Prunus* (Tricioli et al. 1985) and *Rhododendron* cuttings (Pierik 1987); but the absence of light sometimes has an inhibiting effect: Williams et al. (1985) pointed out a reduction of rooting percentage from 70 to 20% with *Grevillea.*

Variation of peroxidase activity and the amount of phenolic compounds with darkness is probable and has undeniable consequences on rooting (Druart et al. 1982; Moncousin and Gaspar 1983; Moncousin 1986a,b,c). In the light, a clone of *Vitis* which is difficult to root can be distinguished from one that is easy to root by the absence of a decrease in peroxidase activity, an increase in the amount of phenols during the first 24 h, a net tapering of the peak of the specific peroxidase activity, and reduction of the amount of total phenols. In the dark, the differences in rooting disappear and the evolution curves of the two markers are similar (Moncousin 1986a). Treatment on the proliferation medium gives lower results than when performed after wounding or auxin addition (Patience and Anderson 1987).

For treatments in the light, the results are difficult to compare, as photoperiods (12 to 24 h) and light intensities (20 to 280 $\mu E/m^2$ s) differ greatly. Walker et al. (1987) increased the rooting percentage of *Sequoia sempervirens* microcuttings from 52 to 78% by increasing light from 70 to 280 $\mu E/m^2$ s, whereas Hansen et al. (1978) on *Pinus sylvestris* and Stromquist and Eliasson (1979) on *Picea abies* have noted an opposite effect. By darkening the culture medium, the optimal lighting can be multiplied by 5 (Stromquist and Eliasson 1979), which revives the debate as to the influence of light on the medium components. Activated charcoal, whose effects were quoted by Nemeth (1986), could play a role simply by its blackening effect. The quality of the light can also be of influence; even though replacing fluorescent day light tubes by Grolux tubes does not improve rooting in wild cherry cutting (Druart et al. 1981), the combination of fluorescence and incandescence has positive effects (Leach 1979; John and Mason 1987). The promoting effect of white hot light can be explained by the phytochrome; red light favors rhizogenesis (Beauchesne 1976; Economou and Read 1986), whereas far red has an inhibiting effect (Fletcher et al. 1965; Economou and Read 1986). Comparing red and blue radiations, Chee (1986) has demonstrated that blue light, which increases the destruction of IAA by a strengthening of PAL and by mono-phenol production, had no influence on the number of roots, but that their growth was improved.

Support of the culture plays a great role in the rooting process and in successful weaning. The effects of agar have been analyzed by Nemeth (1986); oxygen reduction, due to the colloid, can diminish rooting (Gebhardt 1985; Gebhardt and Friedrich 1987). Even though media containing less agar are more favorable to rhizogenesis, they are not a good preparation for a return to normal growing conditions, as they favor hyperhydricity (Von Arnold and Eriksson 1984; Marin and Gella 1987). Replacing agar with various more or less inert supports, Aldrufeu (1987) suggested the use of cellulose, perlite, and vermiculite for the in vitro rooting of *Pelargonium* in either the presence or absence of sucrose. Hutchinson (1984) with apple tree microcuttings, grouping results of rooting and weaning, retained coarse sand, perlite, and agar, clearly excluding cellulose and vermiculite.

7 Relations Between Rhizogenesis, Auxins, Peroxidases, Phenols, and Ehtylene

The amount of endogenous auxin determines the relationship between auxins and adventitious rooting. It has been noted that endogenous auxin components accumulate in areas such as nodes, where root development is frequent (Kulescha 1952), and that the auxin content of the stems can increase or decrease during the neoformation of roots.

Exogenous application of indole-3-acetic acid and of various auxinic substances also confirms the importance of auxin activity for adventitious rooting (Nemeth 1986), even though in some cases this was either detrimental or did not work.

As the auxin is transported basipetally, a relationship with the polarity of adventitious rooting might be considered. However, the addition of substances inhibiting auxin transport leads mainly to the inhibition of adventitious rooting at the basal pole without any appearance of adventitious roots at the other pole (Jarvis et al. 1983, 1984; Moncousin unpubl.).

Hess (1962), analyzing cultivars both easy and difficult to root, noted the presence of phenolic type co-factors, specific to rhizogenesis, which acted synergetically with the auxins. Various other findings also seem to support the hypothesis of the intervention of phenolic compounds in the induction process of adventitious rhizogenesis (Mosella-Chancel et al. 1980; Poessel et al. 1982). It should also not be forgotten that phenols can also vary the auxinic metabolism (Stonier et al. 1979; Sembdner et al. 1980).

The formation of adventitious roots is generally inhibited by the addition of auxins (Nemeth 1986; Gaspar and Coumans 1987). The initiation of adventitious roots seems to require a minimal concentration of cytokinin and a favorable auxin/cytokinin ratio (Skoog and Miller 1957).

The inhibiting effect of gibberellic acid on rooting was shown for several species (Brian and Radley 1955; Kato 1958; Hartung et al. 1980; Fabidjan et al. 1981a,b). The application of substances inhibiting the formation of gibberellic acid or its antagonists favors the adventitious rooting of cuttings (Fabidjan et al. 1981a,b). The time of application of gibberellic acid and the light conditions of the cuttings can play an important role in the response of the plant. In the micropropagation of *Asparagus*, inhibitors of gibberellic acid are often used to stimulate rooting (Khunachak et al. 1987).

7.1 Auxins

The major role played by auxins in adventitious rooting is universally admitted even if this role has not yet been proved. Auxins are in fact the only substances that noticeably increase adventitious rooting by external addition; as long as the time of application and the dose are suitable!

7.1.1 Influence of External Auxinic Additions

Stem cuttings generally respond favorably to external additions but, in every case, the reaction is dependent on the plant material, its ontogenetic age, the nature of the auxin, the contact duration and the time of application (Nemeth 1986; Gaspar and Coumans 1987).

The most often used auxinic compounds to stimulate adventitious rhizogenesis are indole-3-acetic acid, indole-3-butyric acid, and naphthalene-acetic acid, but the former two seem to be more active (George and Sherington 1984).

2,4-dichlorophenoxyacetic acid, which is not metabolized by plants, is less active except for monocotyledons (Robert et al. 1987). Auxinic stimulation is generally at its maximum when high concentrations are applied immediately or just after cutting (Jarvis et al. 1983; Collet 1985). When the auxins are added at too low doses, rhizogenesis is not stimulated and sometimes even retarded (Moncousin 1982; Jarvis et al. 1983). In the opinion of some authors, constant auxin addition during the whole adventitious rooting phase is necessary (Gaspar and Coumans 1987; Pierik 1987).

A considerable addition of auxin in the first stages following cutting should be sufficient to initiate the first cellular divisions and ensure the development of the primordia (Kantharaj et al. 1979; Moncousin and Gaspar 1983; Moncousin 1986c; Collet and Le 1987, 1988).

7.1.2 Internal Auxins

7.1.2.1 Free Auxins

Some studies show that the concentration of auxin-like substances increases in the basal zone of the cuttings before the appearance of roots (Collet 1985; Moncousin 1986a,b,c; Berthon et al. 1989) and that the number of roots formed per cutting is related to the total amount of these substances present in the neoformation zone during cutting (Bastin 1966).

The absence of adventitious rooting has been attributed to an internal auxin level that is too low (Smith and Wareing 1972a,b), but in other cases the auxin level does not appear to be a limiting factor (Greenwood et al. 1976). Nakano et al. (1980) have demonstrated that auxin activity at the base of a *Vitis* cutting decreases during cutting, and that it was lower in rooted cuttings than in nonrooted ones. Trefois and Brunner (1982) found a clear correlation between rooting percentage and richness in internal auxin-like substances in *Prunus* cuttings. In the cuttings of *Cotinus coggygria*, the existence of a correlation between the rooting percentage and the initial amount of IAA has been proposed (Blakesley et al. 1985a,b).

Treatment of *Pinus sylvestris* cuttings with IBA provokes a high increase in the amount of IAA in the base of the cuttings (Dunberg et al. 1981). Having found no difference in metabolism or auxin transport between treated or untreated cuttings, these authors concluded that IBA was converted into IAA. This hypothesis was confirmed by Epstein and Lavee (1984).

However, comparison between these various studies cannot be made without making some reservations, as the frequency of treatment is very variable, from doses either daily or at cutting and root emergence, the physiological state of the starting material being rarely taken into consideration. There is no systematic comparison of zones suitable or unsuitable for rhizogenesis in the same plant material.

7.1.2.2 Conjugated Auxins

Conjugated auxins are frequently present in plant tissues, and their quantity is often clearly more important than that of free IAA (Collet and Le 1987; Moncousin et al. 1987; Moncousin 1988). The physiological role of conjugated IAA has been considered in translocation, storage, biosynthesis, and auxin action. Auxin conjugation is generally considered as a means of detoxification or regulation in the internal amount of 'active' auxin, especially after external auxin treatment. In their works on the study of rooting by a short inductive pre-treatment, Collet and Le (1987, 1988) have shown that there was first an accumulation of free IAA in the base of the cutting and then a rapid (24-h) decrease in the amount, with the formation of conjugated IAA.

Several examples show important seasonal variations in the amount of conjugated IAA. In *Rhododendron* cuttings, both easy and difficult to root, the amount of free IAA does not differ significantly and changes very little during the year. However, the amount of conjugated IAA undergoes important fluctuations, with a maximal increase in summer. Between summer and autumn, the amount of conjugated IAA decreases more sharply in *Rhododendron britannia* (difficult to root) than in *R. ponticum* (easy to root), whereas the reverse is true for free IAA (Wu and Barnes 1981).

In the same way, the amount of IAA esters present in *Cotinus* cuttings is 10 to 20 times greater in autumn than in spring. However, in the autumn cuttings, which root less easily than spring ones, the ratio between the free IAA and the IAA total is close to 1, whereas it is 0.02 in the spring (Blakesley et al. 1985a,b).

7.2 Peroxidases

The general involvement of peroxidases, and of certain isoforms in particular, in the control of internal auxin level and in the physiological processes that depend on is largely recognized (Gaspar et al. 1982, 1985). The most important relation between peroxidases and auxins is based on indirect statements made following the addition of either positive or negative effectors of IAA oxidase action of peroxidases or deduced from variations in the amount of these effectors during particular phases of adventitious rooting. Peroxidases also interfere in the lignification mechanism (Lewis 1980a,b; Lobarzewski 1986).

Various authors have elaborated theoretical schemes on the evolution of peroxidase activity during adventitious rhizogenesis (Gaspar et al. 1982; Moncousin and Gaspar 1983; Moncousin and Ducreux 1984; Moncousin 1987, 1988; Mato et al. 1988; Berthon et al. 1989).

The addition of substances capable of stimulating or inhibiting peroxidase activity during one of the two phases has direct impacts on rooting (Smith and Thorpe 1977; Moncousin 1982; Moncousin and Gaspar 1983). The modulation of peroxidase activity by environmental factors such as darkness and temperature is also possible, with undeniable results for adventitious root formation (Druart et al. 1982; Moncousin 1982, 1986a).

Significant fluctuations in peroxidase activity have been noted during the rooting of *Prunus* (Quoirin et al. 1974), *Asparagus* (Van Hoof and Gaspar 1976), *Nicotiana* (Gaspar et al. 1977b; Thorpe et al. 1978), *Pinus* (Gaspar et al. 1977a), *Malus* (Druart et al. 1982), *Cynara* (Moncousin 1982), *Sequoiadendron* (Berthon et al. 1987), *Vitis* (Moncousin 1986a,b, 1988; Mato et al. 1988), etc.

7.3 Phenolic Compounds

Bouillenne and Bouillenne-Walrand (1955) had suggested that the "rhizocalin" complex was formed from auxin, orthodiphenol, and an enzyme, probably a polyphenol oxidase. The role attributed to auxin – acceleration of cellular divisions – was not specific to rhizogenesis, whereas the phenolic compound and the enzyme localized in cells susceptible to producing root morphogenetic areas were considered as being specific to rhizogenesis. Even if the polarized transport of the phenolic compound has never been clarified, this theory is the origin of a school of thought which attempts to explain the behavioral difference between cuttings that are easy or difficult to root by the difference in their phenolic compound content (Foong and Barns 1981; Nemeth 1986).

Let us recall that monophenols do not behave as cofactors favoring the destruction of IAA, while o-diphenols inhibit the action of IAA-oxidase (Marigo 1979). Stonier et al. (1979) have incidentally shown the protective role exerted by chlorogenic acid towards the auxin; in this case the phenol would in preference be degraded by peroxidases and the auxin would be destroyed only later. The p-coumaric and ferrulic acids would also interfere in the regulation of the establishment of conjugated and linked forms of the auxin IAA (Lee 1980).

7.3.1 Exogenous Phenolic Compounds

The stimulating action of certain phenolic compounds on rhizogenesis has been described by several authors (Nemeth 1986; Gaspar and Coumans 1987), but was best realized by Mosella-Chancel et al. (1980) and Hammerschlag et al. (1987), who studied the influence of various phenolic compounds on rhizogenesis, keeping in mind the application time of the so-called substances.

Van Raalte (1954) emphasized a synergy of action between IAA and the phenyl acetic or phenyl butyric acids in the adventitious rooting of various cuttings. Gorter (1958, 1962) confirmed this phenomenon and also showed that the application of indol could stimulate rhizogenesis even without external auxin additions. The very nature of the substances used allows us to presume that they would not act directly, but by modulation of the IAA oxidase activity. It is interesting to note that, according to Gorter (1969), the indols would act in synergy with IAA, NAA, and 2,4-D, whereas the naphthols would act with IAA or NAA, and the phenols only with IAA. The substances used would have a preparatory role in sensitizing the cells to auxin (Gorter 1969). The positive influence of phloroglucinol and of phloridzin has been pointed out by many authors (Jones 1976; Zimmerman 1983, 1984; Hegedus and Phan 1987).

However, these results must be weighed against the fact that the additions of phenolic compounds are influenced by the natural phenolic content of the cuttings, by the phenol metabolism adapted to such plant, and by the specific conditions of each experiment.

7.3.2 Endogenous Phenolic Compounds

The extraction and identification of natural substances capable of stimulating rhizogenesis have in many cases led to emphasizing phenol-like substances (Bassuk and Howard 1981; Bassuk et al. 1981).

The role of anthocyans is without doubt important, given their great solubility in water and therefore their easy circulation (Lee 1971). Bachelard and Stowe (1962) have shown a clear relation between the formation of anthocyans and adventitious rooting in *Acer rubrum*. Identical statements have been made by Mullins (pers. commun.) and by ourselves on microcuttings as varied as *Betula, Cynara, Malus,* or *Prunus.*

An increase in total phenolic content in the root regeneration zone was shown in various species (Druart et al. 1982; Moncousin et al. 1988). On the other hand, Montain et al. (1983a,b) and Moncousin et al. (1988, 1989) have shown an accumulation of soluble phenols in the area producing roots, which leads us to suppose an implication in the lignification process. It was demonstrated that PAL activity achieved a maximum when the phloem and xylem nodules were formed (Haddon and Northcote 1975), and the substances capable of inhibiting PAL activity inhibit rooting (Tripathi and Schlosser 1979). Inadequate lignification has a negative influence on the vascularization of adventitious roots and has dramatic consequences on the development of cuttings (Lewis 1980a,b). However, PAL activity registered during adventitious rooting can be linked to the synthesis of compounds other than lignin. Transcinnamic acid can produce benzoic acid, phenolic esters, flavonoids, and anthocyans, which can play a role in adventitious rooting.

The metabolism of simple phenols during rooting has been emphasized by the study of polyphenol oxidase and peroxidase activity. Polyphenol oxidases extracted from root callus can, after purification, oxidize the o-diphenols but not the monophenols (Habaguchi 1977). The latter author has also shown an increase in polyphenol oxidase activity before the appearance of roots in untreated carrot callus, and a close relationship between the rooting of callus treated with adenosin monophosphate and polyphenol oxidase activity. Given the localization of polyphenol oxidase, it is possible that their action on phenols is the consequence of wounds induced during cutting (Vaughn and Duke 1984). Al Barazi and Schwabe (1984) have shown that rooting ability was directly related to polyphenol oxidase activity at the time of cutting. Given the role of polyphenol oxidase in the evolution of the phenol pool, one can suppose that these enzymes contribute to ensuring the protection of free IAA during the first stages of adventitious rhizogenesis.

An inverse variation of peroxidase activity and total phenolic content has, however, been described during rooting (Druart et al. 1982); Balasimha and Subramonian 1983; Moncousin 1986a,b, 1987, 1988).

The stimulating effect of rutin on adventitious rhizogenesis (Bhattacharya and Kumar 1980; Berthon et al. 1987; Hammerschlag et al. 1987) can without doubt find an explanation in the action of rutin on total peroxidase activity (Moncousin and Gaspar 1983).

7.4 Ethylene

The involvement of ethylene in adventitious rhizogenesis has been illustrated through external additions (Kawase 1971) or by inhibiting its production (Le Guyader 1987). Krishnamoorthy (1970) and Robbins et al. (1983) have shown a positive influence of ethylene additions on adventitious rooting, whereas Geneve and Heuser (1983), Batten and Mullins (1978) and Mudge and Swanson (1978) conclude an inhibiting effect. The great diversity of results obtained can without doubt be explained by the variety of plant material used, the cutting type, the application doses, techniques, duration, and time.

Research aimed at studying the effects of auxin application on ethylene production have also shown surprising results. For Geneve and Heuser (1983), there is no correlation between the total amount of ethylene, ethylene production, and the number of roots formed after auxin treatment of the cuttings. For Batten and Mullins (1978), on the contrary, exogenous auxin does induce ethylene production, but it does not play a role in rhizogenesis to the extent that cuttings untreated with auxin can root without being able to increase ethylene production. According to Linkens et al. (1973), the application of auxin can induce the first cellular divisions but in the absence of ethylene the morphogenetical areas, and naturally the roots, would not form. On vine cuttings, two peaks of ethylene production appear; if the first peak is not linked to rooting, the second one is more marked in the rhizogen zone (Moncousin 1987; Moncousin et al. 1989; Gaspar et al. 1990). Mullins (1972) has stated that the addition of high doses of exogenous auxins and relatively low doses of ethylene favor adventitious rhizogenesis; this could mean that the time of ethylene addition would condition the rooting response. The use of inhibitors of ethylene synthesis during the first stages of adventitious rhizogenesis would have a stimulating effect on the rooting of *Helianthus* cuttings (Fabidjan et al. 1981a). On the other hand, the long application of amino ethoxyvinyl glycine (AVG), a well-known ethylene inhibitor, would decrease the number of roots formed (Robbins et al. 1983, 1985).

To conclude, it is interesting to note that Epstein (1982) showed that ethylene favored the formation of conjugated IAA in olive tree leaves; it must be recalled that conjugated IAA, as we have seen before, can play an important role in rhizogenesis either by interfering at the level of the regulation of free IAA content, or by direct action.

8 Relations Between Ontogenetic States and Rooting Markers

A recent study makes it possible to link histocytological stages to the evolution of several markers (Fig. 4) (Moncousin 1987; Moncousin et al. 1989; Gaspar et al. 1990). Adventitious rhizogenesis can be divided into well-defined histocytological stages: absence of visible cytological phenomena, dilation of the nucleus of the target cells, cambium periclinal division, presence of radical morphogenetical areas, root meristems included in the cutting itself, and first root emergence.

The metabolic markers studied were more or less implicated in each of these stages. In the absence of visible cytological phenomena, an increase is recorded in the amounts of four markers: ethylene production, free IAA, o-dihydroxyphenols, and phenols. Ethylene production increases spectacularly during the first hours. The increase of the last three markers was more pronounced at the base than at the top of the cuttings, especially for free IAA. The o-dihydroxyphenols have an important role to play in the increase of the content of total soluble phenols. At the same time, peroxidase activity falls, so that this marker appears to be an antagonist to the previous ones. At nucleus dilation, the situation changes. The development of the amount of free IAA, o-dihydroxyphenols, and total soluble phenols is reversed. Free IAA falls to levels lower than at the beginning. The antagonism between these markers and the specific activity of peroxidases, which increases appreciably, is confirmed. Ethylene, IAA conjugates, and monophenols remain minor markers.

At the beginning of the periclinal divisions of the cambium, peroxidase activity appears to be the main marker. Its increase, initiated during the previous phase, continues, and reaches maximum activity at the base of the cuttings. Free IAA, o-dihydroxyphenols, and phenols are always antagonists to specific peroxidasic activity.

On the appearance of radical morphogenetic areas, the cuttings start a new phase, within which none of the chosen markers is particularly noticeable.

The appearance of root meristems in the cutting itself: this stage corresponds for the markers to a return to the starting levels. The o-dihydroxyphenols, and to a lesser

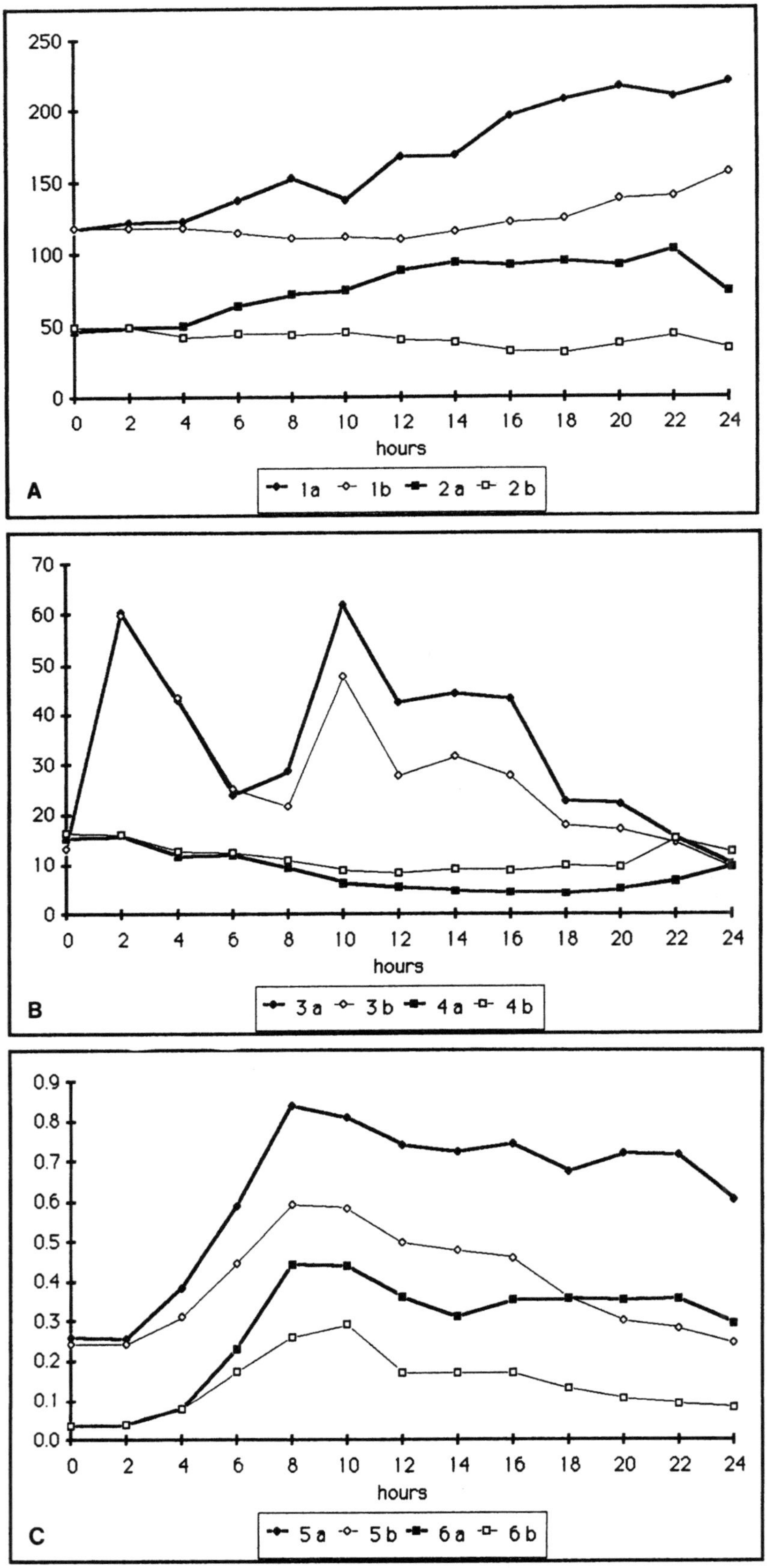

250
200
150
100
50
0
0 2 4 6 8 10 12 14 16 18 20 22 24
hours
A
1a 1b 2a 2b
70
60
50
40
30
20
10
0
0 2 4 6 8 10 12 14 16 18 20 22 24
hours
B
3a 3b 4a 4b
0.9
0.8
0.7
0.6
0.5
0.4
0.3
0.2
0.1
0.0
0 2 4 6 8 10 12 14 16 18 20 22 24
hours
C
5a 5b 6a 6b

Fig. 4A-C

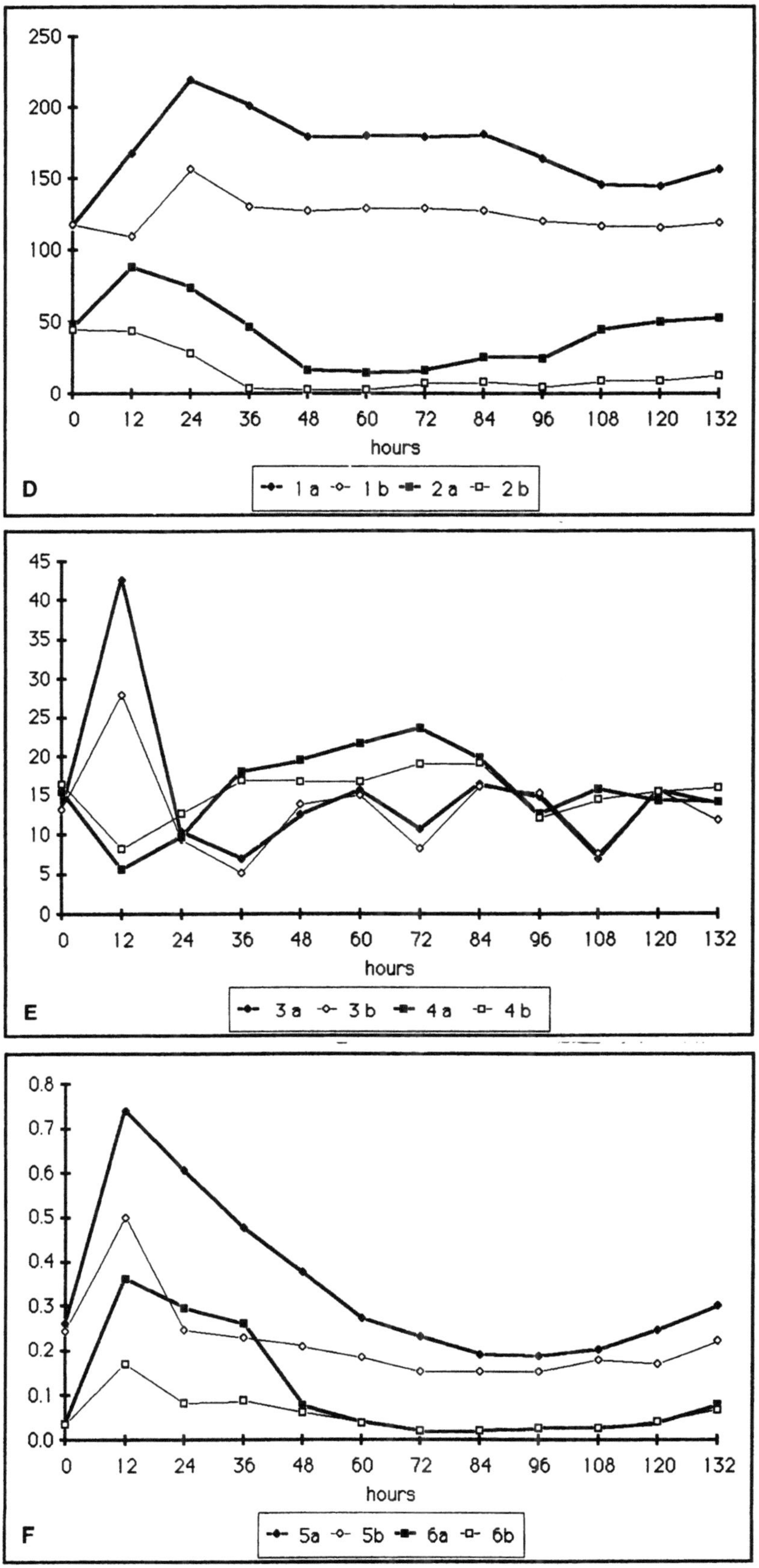

250
200
150
100
50
0
0 12 24 36 48 60 72 84 96 108 120 132
hours
D
1a 1b 2a 2b
45
40
35
30
25
20
15
10
5
0
0 12 24 36 48 60 72 84 96 108 120 132
hours
E
3a 3b 4a 4b
0.8
0.7
0.6
0.5
0.4
0.3
0.2
0.1
0.0
0 12 24 36 48 60 72 84 96 108 120 132
hours
F
5a 5b 6a 6b

Fig. 4D-F

extent the total soluble phenols, continue to increase progressively, especially in the bases of the cuttings. The increase in free IAA, initiated in the previous stage, is enhanced, and finds itself at its initial level. The increase is more distinct at the bases and is probably due more to neo-synthesis in relation with bud activation than to deconjugation; the amount of conjugated IAA remains significantly higher than the starting level. Specific peroxidase activity integrates values close to its initial levels.

The reaction to cutting can be divided into two stages: the first, directly related to the wound, lasts a few hours, the second extends over several days. Different mechanisms based on the same markers also develop successively. The hours that follow cutting are characterized by a set of phenomena which are more or less closely linked.

Ethylene production after wounding has been reported by several authors (Abeles 1973; Hyodo et al. 1983). Hanson and Kende (1976) explain it by the presence of a link between methionine and an enzymatic system. Hyodo et al. (1983) propose an additional argument by showing that the wound provokes an increase in ACC content by increasing the activity of ACC synthase. The reverse control of ACC synthase by ethylene (Riov and Yang 1982; Imaseki et al. 1982) can explain the successive wave pattern of development. Confirmation of this feedback was obtained by Yoshii and Imaseki (1982) and Yang (1983). Etylene production could also be due to the direct or indirect intervention of membrane peroxidase (Rohwer and Mader 1981; Boyer and De Jaegher 1986).

Accumulation of phenolic compounds, after wounding, was reported by Tanaka and Uritani (1977). This accumulation is accompanied by an increase in the activity of enzymes related to polyphenol synthesis (PAL and transcinnamic hydroxylase acid). In so far as small doses of ethylene favor an increase of PAL and the accumulation of chlorogenic acid (Imaseki et al. 1968; Imaseki 1970), one can suppose that ethylene is a key element in the modification of the phenolic pool.

The increase in the content of o-dihydroxyphenols and total soluble phenols, prominent during the first 24 h, confirms the results obtained by Fernqvist (1966). The accumulation, which is more important at the base of the cuttings, leads us to think that there is very rapid migration of the phenols (Montain et al. 1983a,b).

The reduction of specific peroxidase activity a few hours after cutting could be the consequence of a strengthening in the amount of negative effectors of IAA oxydase. This hypothesis rests not only on a reinforced production of diphenols, but also on the migration of phenols towards the base of the cutting, since it is here that the production is strongest and the reduction of peroxidase activity greatest. Soekarjo and Janssen (1969) have shown an accumulation of auxinic protectors after wounding; our results add a further element to this remark.

A rapid increase in free IAA content during the cutting process has already been reported (Brunner 1978). This is confirmed and the amounts contained in the bases or the tops of the cuttings do not vary in the same way. The increase in the basal part could

Fig. 4A-F. Evolution of various markers in the base (*a*) or in the top (*b*) of *Vitis* microcuttings during the first 24 h (**A-C**) or the first 132 h (**D-F**). **1** IAA conjugated content (ng/g F.W.). **2** Free IAA content (ng g F.W.). **3** C_2H_4 production (nl/h g F.W.). **4** Specific peroxidasic activity (10 × optical density/min g F.W.). **5** o-diOH phenol content (mg/g F.W.). **6** Total soluble phenols (mg/g F.W.)

be the consequence of an auxin migration (perhaps in a conjugated form) and of an added protection of the auxin in the basal zone.

The second increase of ethylene (10th hour) could be the consequence of the increased free IAA. Several authors (Sakai and Imaseki 1971; Imaseki et al. 1975) reported an increase in ethylene production following the addition of auxins. The amount of ACC increases after addition of auxin (Yoshii et al. 1980) following activation of ACC synthase (Imaseki et al. 1982).

The rise and then the fall of the specific peroxidase activity which we have observed after the second stage of rhizogenesis confirms the scheme proposed by Gaspar (1986), based on the works of Gaspar et al. (1977a,b), Thorpe et al. (1978), Druart et al. (1982), and Moncousin and Gaspar (1983). As in these works, we notice that the appearance of morphogenetic areas, and therefore of root primordia, takes place when specific peroxidase activity diminishes, after having reached a peak.

The increase in the amount of total phenols during the first hours following cutting confirms the results of several authors (Fernqvist 1966; Bhattaracharya and Kumar 1980; Balasimha and Subramonian 1983; Moncousin 1986a, 1987, 1988).

The antagonism peroxidase-phenol is not sufficient to reveal whether there is a true relationship between phenols and peroxidase activity or whether the dosage is only influened by the presence, in varying quantities, of phenolic compounds in the extracts (Mato and Vieitez 1985; Mato et al. 1985). Taking into account the role played by the o-dihydroxyphenols in the protection of IAA by inhibiting the activity of IAA oxidase (Stonier et al. 1970, 1979), and because their evolution is similar to that of the total soluble phenols, our results confirm the implication of these compounds only in the mechanisms leading to adventitious rhizogenesis.

Doses of free IAA, using reliable modern techniques, were given during rooting. Norcini et al. (1985) observed an increase of free IAA during the first 12 h, then a fall until the 24th h, followed by a slight increase. Maldiney et al. (1986) have shown an increase during the first 2 days and, after 4–5 days, a return to the initial level. Our findings are identical, but we were able to prove that increase in amounts of free IAA during the first 24 h takes place only in the lower part of the cutting, under the bud. Above it, the amount decreases considerably from the 12th h. This difference between the base and the top remains through the whole rooting process. Norcini et al. (1985) showed that the content of conjugated IAA increased from the 10th h, to reach its maximum value at the 30th h, and then fall progressively, always remaining above its initial value. Our results move in exactly the same direction.

9 Conclusion

The success in rooting microcuttings is possible only if material from the elongation stage is carefully prepared. Any technique that sets plants under conditions of heterotrophy, at the same time avoiding hyperhydricity, will be positive. The double-phase technique developed by Maene and Debergh (1985), and improved by Aitken-Christie and Jones (1987) makes it possible to place the plant under the best possible conditions for rooting. Kozai (1988a,b) has shown that adapted culture vessels can cause autotrophy in place of hetero- or mixotrophy, thus avoiding drastic change at weaning. The system developed by Fujiwara et al. (1988) anticipates the micropropagation of the future.

The best rooting response will be obtained with plant material which, at the time of cutting, shows the strongest activity, the lowest content of o-dihydroxyphenols, and poor ethylene production, meaning that only sufficiently rejuvenated material placed under optimal hormonal conditions will fulfill these characteristics (Moncousin 1988).

Long auxin treatment will be replaced by short periods, from a few hours to a few days. Auxin concentration, being inversely related to the length of the treatment, will vary from micro- to milli-molar. Natural auxin is preferable in short-length treatments because the subsequent possibilities for detoxification are more important with IAA than with IBA or NAA.

During the first week of in vitro or ex vitro rooting, darkness and higher temperatures are combined with more successful results than culture followed by maximal lighting, lower temperatures, and high humidity to favor best root expression. Progressively going back to ranges of temperatures, light and hydricity under natural or greenhouses condition will have a positive effect on future growth (Kozai 1988a,b).

References

Abeles FB (1973) Ethylene in plant biology. Academic Press, Lond New York, pp 87–108

Aitken-Christie J, Jones C (1987) Towards automation: Radiata pine shoot hedges in vitro. Plant Cell Tissue Organ Cult 8:185–196

AL Barazi Z, Schwabe WW (1984) The possible involvement of polyphenol-oxidase and the auxin-oxidase system in root formation and development in cuttings of *Pistacia vera*. J Hort Sci 59(3):453–461

Aldrufeu A (1987) Rooting and acclimatization of *Pelargonium* zonale plantlets. Acta Hortic 212:361–366

Aldrufeu A, Mele E (1984) Saccharose and substrate effects in the rooting of in vitro *Rosa* plantlets. In: Plant Tissue Culture and its Agricultural Applicaitons Symp Nottingham Univ, (Abstr) 68

Al Maari K, Duron M, Arnaud Y, Miginiac E (1986) Etude comparative de l'aptitude à la micropropagation, par culture in vitro, du poirier "Passe-Crassane". C R Acad Agric Fr 72:413–421

Amerson HV, Mott RL (1982) Improved rooting of western white pine shoots from tissue cultures. Forest Sci 28(4):822–825

Appelgren M, Heide OM (1972) Regeneration in *Streptocarpus* leaf and its regulation by temperature and growth substances. Physiol Plant 27:417–423

Bachelard EP, Stowe BB (1962) A possible link between root initiation and anthocyanin formation. Nature (Lond) 194:209–210

Balasimha D, Subramonian N (1983) Role of phenolics in auxin induced rhizogenesis and isoperoxidases in cacao (*Theobroma cacao* L.) stem cuttings. Indian J Exp Biol 21:65–68

Bancilhon L (1969) Etude expèrimentale de la morphogenése, et plus spécialement de la floraison d'un groupe de *Phyllanthus* à rameaux dimorphes. Ann Sci Nat Bot 10:127–224

Barlass M, Grant WJR, Skene KGM (1980) Shoot regeneration in vitro from native Australian fruit-bearing trees – quandong and plum bush. Aust J Bot 28:405–411

Bassuk NL, Howard BH (1981) A positive correlation between endogenous root inducing cofactor activity in vacuum-extracted sop and seasonal changes in M26 winter apple cuttings. J Hortic Sci 56:301–312

Bassuk NL, Hunter LD, Howard BH (1981) The apparent involvement of polyphenol oxidase and phloridzin in the production of apple rooting cofactor. J Hortic Sci 56:313–322

Bastin M (1966) Root initiation, auxin level and biosynthesis of phenolic compounds. Photochem Photobiol 5:423–429

Batten DJ, Mullins MG (1978) Ethylene and adventitious root formation in hypocotyl segments of etiolated mung bean [*Vigna radiata* (L.)] Willczek seedlings. Planta 138:193–197

Beauchesne G (1976) Multiplication végétative et culture in vitro. In: Jacques R (ed) Etudes de biologie végétale. Hommage au Prof Chouard, Masson, Paris, pp 165–176

Berthon JY, Boyer N, Gaspar T (1987) Sequential rooting media and rooting capacity of *Sequoiadendron giganteum* in vitro. Peroxidase activity as a marker. Plant Cell Rep 6:341–344
Berthon JY, Maldiney R, Sotta B, Gaspar T, Boyer N (1989) Endogenous levels of growth regulators during the course of adventitious rooting in cuttings of *Sequoiadendron giganteum* in vitro. Biochem Physiol Pflanz 184:405–412
Bhattaracharya NC, Kumar A (1980) Physiological and biochemical studies associated with adventitious root formation in *Phasoleus mungo* L. inrelation to auxin-phenol synergism. Biochem Physiol Pflanz 175:421–435
Blakesley D, Alderson PJ, Weston GD (1985a) Rooting studies on *Cotinus*. Abstr Book 12th Int Conf Plant Growth Substances, Heidelberg, p 86
Blakesley D, Weston G, Alderson P, Hall J, Elliott MC (1985b) The effect of time of year and the induction of etiolation on rooting and the concentration of IAA and cytokinin in cuttings of *Cotinus coggygria* royal purple. News Bull Br Plant Growth Regulator Group 7:51–52
Bouillenne R, Bouillenne-Walrand M (1955) Auxines et bouturage. 14th Hortic Congr Schweninger 1:231–238
Bouillenne R, Went F (1933) Recherches expérimentales sur la néoformation des racines dans les plantules et les boutures des plantes supérieures. Ann Jardin Bot Buitenzorg 43:25–202
Boulay M (1979) Propagation in vitro du douglas [*Pseudotsuga menziesii* (Mirb.)] par micropropagation de germination aseptique et culture de bourgeons dormants. Micropropagation d'Arbres Forestiers, Ann AFOCEL 12(6):67–75
Boulay M, Franclet A (1977) Recherches sur la propagation vegetative du Douglas: *Pseudotsuga menziesii* (Mirb.). Franco, CR Acad Sci Paris 284:1405–1407
Bouzid S, Lasram M (1971) Utilisation de cultures in vitro pour l'obtention de clones de *Citrus* homogènes et de bon état sanitaire. 8th Congr Int Agrum Méditer Holland Pearl CLAM (Madrid) V 2:1–6
Boyer N, De Jaegher G (1986) Direct or indirect role of peroxidase in ethylene biosynthesis? In: Greppin H, Penel C, Gaspar T (eds) Molecular and physiological aspects of plant peroxidases. Univ Geneva, pp 47–60
Brian PW, Radley M (1955) A physiological comparison of gibberellic acid with soma auxins. Physiol Plant 8:899–912
Broome OC, Zimmerman RH (1984) Culture of shoot meristems: fruit plants. In: Vasil IK (ed) Cell culture and somatic cell genetics of plants. Academic Press, Lond New York, pp 111–122
Brunner H (1978) Einfluss verschiedener Wuchsstoffe und Stoffwechselgifte auf wurzelregenerierendes Gewebe von *Phaseolus vulgaris* L. Veränderungen des Wuchsstoffgehaltes sowie der Peroxidase und der IAA-Oxydase Activität. Z Pflanzenphysiol 88:13–23
Chalupa V (1977) Organogenesis in Norway spruce and Douglas fir tissue cultures. Commun Inst For Tech 10:79–87
Champagnat M (1977) Culture de bourgeons et multiplication végétative des Orchidées. La culture des tissus végétaux (Travaux dédiès à la mémoire de Georges Morel). Masson Paris, pp 238–253
Chee R (1986) In vitro culture of *Vitis*, the effects of light spectrum, manganese sulfate and potassium iodide on morphogenesis. Plant Cell Tissue and Organ Culture 7:121–134
Cheng TY, Voqui T (1977) Regeneration of Douglas fir plantlets through tissue culture. Science 198:306–307
Coleman WK, Thorpe TA (1977) In vitro culture of Western red cedar (*Thuja plicata* Don.). I. Plantlet formation. Bot Gaz 138(3):298–304
Collet GF (1985) Enracinement amélioré lors de la production in vitro de rosiers. Rev Suisse Vitic Arboric Hortic 17(4):259–263
Collet GF, Le CL (1987) Role of auxin during in vitro rhizogenesis of rose and apple-trees. Act Hortic 212:273–280
Collet GF, Le LC (1988) Micropropagation de porte-greffes de pommier et de poirier. II Enracinement in vitro de *Pyrus malus* L. (M25, 26, 27, MM106, M9 type Jork) et de *Cydonia oblonga* Mill. (A). Rev Suisse Vitic Arboric Hortic 20(2):131–138
Cornu D (1984) Micropropation at the forest tree improvement laboratory of the INRA-Orléans, France. Proc Int Symp In vitro propagation of Forest Tree Species, Bologna, pp 85–91
David A (1982) In vitro Propagation of gymnosperms. In: Bonga JM, Durzan DJ (eds) Tissue Culture in Forestry. Martinus Nijhoff, Dordrecht, pp 72–108
Depommier D (1981) Micropropagation d'*Eucalyptus* résistant au froid: influence de quelques facteurs sur l'allongement et l'enracinement des plantules. Colloque international sur la culture in vitro des essences forestières. IUFRO Fontainebleau, pp 127–132

Driver JA, Suttle GRL (1987) Nursery handling of propagules. In: Bonga JM, Durzan DJ (eds) Cell and tissue culture in forestry. Martinus Nijhoff, Dordrecht, pp 320–335
Druart P (1987) Contribution à l'élaboration de techniques de production en masse in vitro d'espèces ligneuses utilisables en culture fruitière. Thèse Docteur en Sciences Agronomiques, Faculté des Sciences Agronomiques, Gembloux (Belgique), 329 p
Druart P, Boxus P, Liard O, Delaite B (1981) La micropropagation du merisier à partir de la culture de méristème. Colloque international sur la culture in vitro des essences forestières. IUFRO Fontainebleau, pp 101–108
Druart P, Kevers C, Boxus P, Gaspar T (1982) In vitro promotion of root formation by apple shoots through darkness effect on endogenous phenols and peroxidases. Z Pflanzenphysiol 108:429–436
Dunberg A, Hsihan S, Sandberg G (1981) Auxin dynamics and the rooting of cuttings of *Pinus sylvestris*. Plant Physiol 67:5–11
Economou AS, Read PE (1986) Light quality control of shoot and root formation in vitro *Petunia* cultures. Hortscience 21:761
Eliasson L (1978) Effects of nutrients and light on growth and root formation in *Pisum sativum* cuttings. Physiol Plant 43:13–16
Eliasson L, Brunes L (1980) Light effects on root formation in aspen and willow cuttings. Physiol Plant 48:261–265
Epstein E (1982) Levels of free and conjugated indole-3-acetic acid in ethylene-treated leaves and callus of olive. Physiol Plant 56:371–373
Epstein E, Lavee S (1984) Conversion of IBA to IAA by cuttings of grapevine (*Vitis vinifera*) and olive (*Olea europea*). Plant Cell Physiol 25:697–703
Evans H (1952) Physiological aspects of the propagation of *Cacao* from cuttings. XIIIth Int Hort Congr Lond. Sect VII, pp 1169–1190
Fabidjan D, Yeung E, Mukherjee I, Reid DM (1981a) Adventitious rooting in hypocotyls of sunflower seedlings (*Helianthus annuus*). I. Correlative influence and developmental sequence. Physiol Plant 53:578–588
Fabidjan D, Taylor JS, Reid DM (1981b) Adventitious rooting in hypocotyls of sunflower (*Helianthus annuus*) seedlings. II. Action of gibberellins, cytokinins, auxins and ethylene. Physiol Plant 53:589–597
Favre JM (1970) Influence de facteurs internes sur la cinétique d'enracinement de la Vigne cultivée in vitro. Rev Gen Bot 77:519–562
Favre JM (1973a) Effets corrélatifs de facteurs internes et externes sur la rhizogenèse de la Vigne cultivée in vitro. Thèse d'Etat. Cent Sci d'Orsay, Univ Paris-Sud, 89 p
Favre JM (1973b) Effets corrélatifs de facteurs internes et externes sur la rhizogenèse d'un clone de vigne (*Vitis riparia* × *Vitis rupestris*) cultivée in vitro. Rev Gén Bot 80:279–361
Favre JM (1973c) Divers aspects du rôle du bourgeon et des noeuds sur la rhizogenèse de la vigne cultivée in vitro. Rev Cytol Biol Végét. Paris 37:393–406
Favre JM, Grenan S (1979) Sur la production de vrilles, de fleurs et de baies chez la vigne cultivée in vitro. Ann Amélior Plantes 29:247–253
Favre JM, Medard R (1969) Ontogénie des racines adventives chez la vigne cultivée in vitro. Rev Gén Bot 76:455–467
Fernqvist I (1966) Studies on factors in adventitious root formation. Lantbrukshogsk Ann 32:109–244
Fletcher JS (1982) Control of nitrogen assimilation in suspension cultures of Paul's scarlet rose. In: Fujiwara A (ed) Plant tissue culture 1982. Maruzen, Tokyo, pp 229–230
Fletcher RA, Peterson R, Zalik S (1965) Effect of light quality on elongation, adventitious root production and the relation of cell number and cell size to bean seedling elongation. Plant Physiol 40(3):541–552
Foong TW, Barnes MF (1981) Rooting "cofactors" in *Rhododendron*: the fractionation and activity of components from an easy-to-root and a difficult-to-root variety. Biochem Physiol Pflanz 176:507–523
Franclet A (1974) Recherches préliminaires pour l'installation d'ensouchements de taillis de peuplier par "semis" de courtes boutures. AFOCEL 73:11–18
Franclet A (1977) Manipulation de pieds-mères et amélioration de la qualité des boutures. AFOCEL 20 pp
Franclet A (1979) Rajeunissement des arbres adultes en vue de leur propagation végétative. In Micropropagation des Arbres Forestiers. AFOCEL 12:3–18

Franclet A, David A, David H, Boulay M (1980) Première mise en évidence morphologique d'un rajeunissement de méristèmes caulinaires de pin maritime âgé (*Pinus pinaster* Sol.). CR Acad Sci Paris 290:927

Franclet A, Boulay M, Bekkaoui F, Fouret Y, Verschoore-Martouzet B, Walker N (1987) Rejuvenation. In: Bonga JM, Durzan DJ (eds) Cell and Tissue culture in forestry. Nijhoff, Dordrecht, pp 232–248

Fujiwara K, Kozai T, Watanabe I (1988) Development of a photo-autotrophic tissue culture system for shoots and/or plantlets at rooting and acclimatization stages. Acta Hortic 230:153–158

Gaspar T (1986) Integrated relationships of biochemical and physiological peroxidase activities. In: Greppin H, Penel C, Gaspar T (eds) Molecular and physiological aspects of plant peroxidases. Univ Geneva, pp 455–453

Gaspar T, Coumans M (1987) Root formation. In: Bonga JM, Durzan DJ (eds) Cell and tissue culture in forestry, Nijhoff, Dordrecht, pp 202–217

Gaspar T, Smith D, Thorpe TA (1977a) Arguments supplémentaires en faveur d'une variation inverse du niveau auxinique endogène au cours des deux premières phases de la rhizogenèse. Comptes Rendus Acad Sci Paris, 185:327–330

Gaspar T, Thorpe TA, Tran Thanh Van M (1977b) Changes in isoperoxidases during differentiation of cultured tobacco epidermal layers. Acta Hortic 78:61–73

Gaspar T, Penel C, Thorpe T, Greppin H (1982) Peroxidases 1970–1980. A survey of their biochemical and physiological roles in higher plants. Univ Genève-Cent Bot 324 p

Gaspar T, Penel C, Roduit C, Moncousin C, Greppin H (1985) The role of auxin level and sensitivity in floral induction. Biol Plant 27:325–329

Gaspar Th, Moncousin Ch, Greppin H (1990) The place and role of exogenous and endogenous auxin in adventitious root formation. In: Millet B, Greppin H (eds) Intra- and intercellular communications in plants. INRA, Paris, pp 125–139

Gebhardt K (1985) Development of a sterile cultivation system for rooting of shoot tip cultures in duroplast foam. Plant Sci Lett 39:141–148

Gebhardt K, Friedrich M (1987) Micropropagation of *Calluna vulgaris* cv HE BEALE. Plant Cell Tissue Organ Cult 9:137–145

Geneve RL, Heuser CW (1983) The relationship between ethephon and auxin on adventitious root initiation in cuttings of *Vigna radiata* (L.) R. Wilcz. J Am Soc Hortic Sci 108(2):330–333

George EF, Sherington PD (1984) Plant propagation by tissue culture. Exegetics Ltd, Eversley, England, 709 pp

Gorter CJ (1958) Synergism of indole and indole-3-acetic acid in the root production of *Phaseolus* cuttings. Physiol Plant 11:1–9

Gorter CJ (1962) Further experiments on auxin synergists. Physiol Plant 15:88–95

Gorter CJ (1969) Auxin synergists in the rooting of cuttings. Physiol Plant 22:497–502

Greenwood MS, Berlyn GP (1973) Sucrose-indole-3-acetic acid interactions on root regeneration by *Pinus lambertiana* embryo cutting. Am J Bot 60:42–50

Greenwood MS, Atkinson OR, Yawney HW (1976) Studies of hard and easy-to-root ortets of sugar maples: difference not due to endogenous auxin content. Plant Propagators 22:3–6

Habaguchi K (1977) A possible function of cyclic AMP in the induction of polyphenol oxidase preceding root formation in cultured carrot-root callus. Plant Cell Physiol 18:191–198

Haddon LE, Northcote DH (1975) Quantitative measurement of the course of bean callus differentiation. J Cell Sci 17:11–26

Haissig BE (1974) Metabolism during adventitious root primordium initiation and development. NZ J For Sci 4:324–337

Haissig BE (1984) Carbohydrate accumulation. Physiol Plant 61:13–19

Halma HH (1931) The propagation of *Citrus* by cuttings. Hilgardia 6:131–157

Hammerschlag F (1982) Factors influencing in vitro multiplication and rooting of the plum rootstock myrobalan (*Prunus cerasifera* Ehrh.). J Am Soc Hortic Sci 107:44–47

Hammerschlag FA, Bauchan GR, Scorza R (1987) Factors affecting in vitro multiplication and rooting of peach cultivars. Plant Cell Tissue Organ Cult 8:235–242

Hansen J, Stromquist LH, Ericsson A (1978) Influence of the irradience on carbohydrate content and rooting of cuttings of pine seedling (*Pinus sylvestris*). Plant Physiol 61:975–979

Hanson AE, Kende H (1976) Methionine metabolism and ethylene biosynthesis in senescent flower tissue on morning glory. Plant Physiol 57:528–537

Harada S, Nakayama H (1959) The influence of the bud and leaf on root formation in tea cuttings. Proc Jpn Soc Crop Sci 3:105–113

Hartung W, Ohl B, Kummer V (1980) Abscisic acid and the rooting of runner bean cuttings. Z Pflanzenphysiol 98:95–103
Hegedus P, Phan CT (1987) Activities of five enzymes of the phenolic metabolism on rooted and acclimated vitreous plants in relation with phenolic treatments. Acta Hortic 212:211–216
Hess CE (1962) A physiological analysis of root initiation in easy and difficult-to root cuttings. 16th Int Hortic Congr Brussells 4:375–381
Heuser CW, Hess CE (1972) Isolation of three lipid root-initiating substances from juvenile *Hedera helix* shoot tissue. J Am Soc Hortic Sci 97(5):571–574
Homes M (1961) Alimentation minérale équilibrée des végétaux. Univ Wetteren, 287 p
Hutchinson JF (1984) Factors affecting shoot proliferation and root initiation in organ cultures of the apple Northern spy. Sci Hortic 22:347–358
Hyndman SE, Hasegawa PM, Bressan RA (1982) Stimulation of root initiation from cultured rose shoots through the use of reduced concentrations of mineral salts. HortScience 17:82–83
Hyodo H, Tanaka K, Watanabe K (1983) Wound-induced ethylene production and 1-aminocyclopropane-1-carboxylic acid synthase in winter squash fruit. Plant Cell Physiol 46:172–174
Imaseki H (1970) Induction of peroxidase activity by ethylene in sweet potato. Plant Physiol 46:172– 174
Imaseki H, Uchiyama M, Uritani I (1968) Effect of ethylene on the inductive increase in metabolic activities in sliced sweet potato roots. Agric Biol Chem 32:387–389
Imaseki H, Kondo K, Watanabe A (1975) Mechanism of cytokinin action on auxin-induced ethylene production. Plant Cell Physiol 16:777–785
Imaseki H, Yoshii H, Todaka I (1982) Regulation of auxin-induced ethylene biosynthesis in plants. In: Wareing PF (ed) Plant growth substances. Academic Press, London, pp 259–268
James DJ (1983) Adventitious root formation in vitro in apple rootstock (*Malus pumila*). Physiol Plant 57:149–153
Jarvis BC, Booth AB (1981) Influence of indole-butyric acid, boron, myo-inositol, vitamin D_2 and seedling age on adventitious root development in cuttings of *Phaseolus aureaus*. Physiol Plant 53:213–218
Jarvis BC, Ali AHN, Shaheed AI (1983) Auxin and boron in relation to the rooting response and ageing of mung bean cuttings. New Phytol 95:509–518
Jarvis BC, Yasmin S, Ali AHN, Hunt R (1984) The interaction between auxin and boron in adventitious root development. New Phytol 97:197–204
John A, Mason B (1987) Vegetative propagation of Sitka spruce. Proc R Soc Edinb 93B:197–203
Jones OP (1976) Effect of phloridzin and phloroglucinol on apple shoots. Nature (Lond) 262:392–393
Julliard B (1963) Influence du bourgeon sur la rhizogenèse des boutures de vigne. CR Acad Sci Paris 257:3200–3203
Kantharaj GR, Mahadevan S, Padmanaban G (1979) Early biochemical events during adventitious root initiation in the hypocotyl of *Phaseolus vulgaris*. Phytochemistry 18:383–387
Kato J (1958) Studies on the physiological effect of gibberellin. II. On the interaction of gibberellin with auxin and growth inhibitors. Physiol Plant 11:10–15
Kawase M (1971) Causes of centrifugal root formation. Physiol Plant 25:64–70
Kevers C, Sticher L, Penel C, Greppin H, Gaspar T (1983) The effect of ergosterol, ergocalciferol and cholecalciferol on calcium-controlled peroxidase secretion by sugarbeet cells. Physiol Plant 57:17–20
Khunachak A, Chin CK, Le T, Gianfagna T (1987) Promotion of *Asparagus* shoot and root growth by growth retardants. Plant Cell Tissue Organ Cult 11:97–110
Kikuta Y, Okazawa Y (1984) Control of root and shoot bud formation from potato tuber tissue cultured in vitro. Physiol Plant 61:8–12
Klampt D (1983) Oligopeptides and plant morphogenesis: a working hypothesis. J Theor Biol 100:411–435
Knop W (1959) In: Gautheret RJ (ed) La culture des tissus végétaux. Masson, Paris, p 12
Kozai T (1988a) Symposium on high technology in protected cultivation. Acta Hortic 230:121–128, 153–158, 159–166
Kozai T (1988b) High technology in protected cultivation in horticulture in high technology era. Society of Agricultural Meteorology of Japan, Japan Greenhouse Association, Japan Management Association, Tokyo, pp 1–44
Krishnamoorthy HN (1970) Promotion of rooting in mung bean hypocotyl cuttings with ethrel, an ethylene releasing compound. Plant Cell Physiol 11:979–982

Kulescha Z (1952) Recherches sur l'élaboration de substances de croissance par les tissus végétaux. Rev Gén Bot 59:92–111, 127–157, 195–208, 241–264
Leach GN (1979) Growth in soil of plantlets produced by tissue culture, loblolly pine. Tappi 62:59–61
Lee CI (1971) Influence of intermittent mist on the development of anthocyanins and root inducing substances in *Euonymus alatus* Sieb. 'Compactus'. PhD Diss, Cornell Univ, Ithaca
Lee TT (1980) Effects of phenolic substances on metabolism of exogenous indole-3-acetic acid in maize stems. Physiol Plant 50:107–112
Le Guyader H (1987) Ethylène et différenciation de racines par des cultures de couches cellulaires minces de tabac. CR Acad Sci Paris 305(III):591–594
Lewis DH (1980a) Boron, lignification and the origin of vascular plants – a unified hypothesis. New Phytol 84:209–229
Lewis DH (1980b) Are there inter-relations between the metabolic role of boron, synthesis of phenolic phytoalexins and the germination of pollen? New Phytool 84:261–270
Libbert E (1956) Untersuchungen über die Physiologie der Adventivwurzelbildung. I. Die Wirkungsweise einiger Komponenten der "Rhizokalinkomplexes". Flora, 144:121–150
Linkens AE, Lewis LN, Palmer RL (1973) Hormonally induced changes in the stem and petiole anatomy and cellules enzymes patterns in *Phaseolus vulgaris* L. Plant Physiol 52:554–560
Lobarzewski J (1986) Peroxidase role in lignin biotransformation. In: Greppin H, Penel C, Gaspar T (eds) Molecular and physiological aspects of plant peroxidases. Univ Genevè, pp 231–246
Lovell P, Illsley A, Moore KG (1972) The effects of light intensity and sucrose on root formation, photosynthetic ability and senescence in the target cotyledons of *Sinapis alba* L. and *Raphanus sativus*. L. Ann Bot 36:123–134
Maene L, Debergh P (1985) Liquid medium additions to established tissue cultures to improve elongation and rooting. Plant Cell Tissue Organ Cult 5:23–33
Maene L, Debergh P (1987) Optimalisation of the transfer of tissue cultured shoots to in vivo conditions. Acta Hortic 212:335–348
Mahlstede JP, Watson DP (1952) An anatomical study of adventitious root development in stems of *Vaccinium corymbosum*. Bot Gaz, 113:279–285
Maldiney R, Pelese F, Pilate G, Sotta B, Sossountzov L, Miginiac E (1986) Endogenous levels of abscissic acid, indole-3-acetic acid, zeatine and zeatine-riboside during the course of adventitious root formation in cuttings of Craigella and Craigella lateral suppressor tomatoes. Physiol Plant 68(3):426–430
Marigo G (1979) Polyphénols et croissance végétale. Essai d'évaluation du rôle joué in vivo par les composés phénoliques chez *Lycopersicum esculentum*, Thèse Doct Etat, Physiol végétale, Univ Toulouse, 161 p
Marin JA, Gella R (1987) Acclimatization of the micropropagated cherry rootstock Masto de Montanaña (*Prunus cerasus* L.). Acta Hortic 212:603–606
Mato MC, Vieitez AM (1985) "Protecteurs" auxiniques et activité peroxydasique chez des plantules de châtaignier in vitro. Bull Liaison – Groupe Polyphénols, Journées Int d'études et Ass Gen, Plovdiv, Bulgarie, 12:91–94
Mato MC, Vasquez A, Gesto MDV (1985) Treatment of bean cuttings by chestnut extracts during rooting modifies their IAA-oxidase activity. Physiol Plant 65:63–66
Mato MC, Rua ML, Ferro E (1988) Change in levels of peroxidases and phenolics during root formation in *Vitis* cultured in vitro. Physiol Plant 72:84–88
McComb JA, Bennett IJ (1982) Vegetative propagation of *Eucalyptus* using tissue culture and its application to forest improvement in Western Australia. In: Fujiwara A (ed) Plant Tissue Culture 1982. Maruzen, Tokyo, pp 721–722
Middleton W (1977) Root development in cuttings of *Phaseolus aureus* Roxb. PhD Thesis, Univ Sheffield, UK
Middleton W, Jarvis BC, Booth A (1980) The role of leaves in auxin and boron dependent rooting of stem cuttings of *Phaseolus aureus* Roxb. New Phytol 84:251–259
Miller GA, Coston DC, Denny EG, Romeo ME (1982) In vitro propagation of Nemaguard peach rootstock. Hortscience 17:194
Mohammed GH, Vidaver WE (1988) Root production and plantlet development in tissue cultured conifers. Plant Cell Tissue Organ Cult 14:137–160
Molisch H (1935) Das Knospenlose Internodien als Stockling behandelt. Ber Dtsch Bot Ges 53:575–586

Moncousin C (1982) Contribution à la caractérisation biochimique et physiologique de la phase juvénile de l'artichaut (*Cynara scolymus* L.) au cours de sa multiplication végétative conforme et accélérée en culture in vitro. Thèse Doct Ingénieur, Univ Paris-Sud, Cent Sci d'Orsay, 237 p

Moncousin C (1986a) Peroxidase as a marker for rooting improvement of clones of *Vitis* cultured in vitro. In: Greppin H, Penel C, Gaspar T (eds) Molecular and Physiological Aspects of Plant Peroxidases. Univ Genève, pp 379–385

Moncousin C (1986b) Biochemical events in mono-nodal stem cuttings of *Vitis* hybrid during in vitro adventitious rooting. IAPTC Congr, Minneapolis, p 148

Moncousin C (1986c) Evolution du contenu auxinique endogène au cours de l'enracinement in vitro de boutures de *Malus*. Colloq arbres fruitiers et Biotechnol, Moët-Hennessy, Paris, p 61

Moncousin C (1987) Evolution de l'AIA, des composés phénoloques, de l'activité spécifique des peroxydases et de la production d'éthylène, au cours de la rhizogenèse, sur des microboutures de Vigne (*Vitis riparia* Michx. × *V. rupestris* Scheele). Thèse Doct d'Etat, Biol Physiol végétale. Univ Nancy, 164 p

Moncousin C (1988) Adventitious rhizogenesis control: new developments. Acta Hortic 230:97–104

Moncousin C, Ducreux G (1984) Activité peroxydasique et rhizogenèse dans le cas de *Cynara scolymus* L.: évolution au cours de repiquages successifs de boutures cultivées in vitro. Comparaison avec des jeunes plantes issues de graines. Agronomie 4(2):105–111

Moncousin C, Gaspar T (1983) Peroxidase as a marker for rooting improvement of *Cynara scolymus* L. cultured in vitro. Biochem Physiol Pflanz 178:263–271

Moncousin C, Favre JM, Gaspar T (1988) Changes in peroxidase activity and endogenous IAA levels during adventitious root formation in vine cutting. In: Kutacek M, Bandurski RS, Krekule J (eds) Physiology and biochemistry of auxins in plants. Akademia, Praha, pp 331–337

Moncousin C, Favre JM, Gaspar T (1989) Early changes in ethylene production in vine cuttings before adventitious rooting. Plant Cell Tissue Organ Cult 19:235–242

Montain CR, Haissig BE, Curtis JD (1983a) Initiation of adventitious root promordia in very young *Pinus banksiana* seedling cuttings. Can J For Res 13:191–195

Montain CR, Haissig BE, Curtis JD (1983b) Differentiation of adventitious root primordia in callus of *Pinus banksiana* seedling cuttings. Can J For Res 13:195–200

Morel G (1963) La culture in vitro du méristème apical de certaines Orchidées. CR Acad Sci Paris 256:4955–4957

Mosella-Chancel L, Macheix JJ, Jonard R (1980) Les conditions du microbouturage in vitro du pêcher (*Prunis persica* Batsch.): influences combinées de certaines substances de croissance et de divers composés phénoloques. Physiol Vég 18:597–608

Mott RL, Amerson HV (1981) A tissue culture process for the clonal production of loblolly pine plantlets. NC Agric Res Serv Tech Bull 271:14

Mudge KW, Swanson BT (1978) Effect of ethephon, indole-butyric acid and treatment solution pH on rooting and ethylene levels with mung bean cuttings. Plant Physiol 61:271–273

Mullins MG (1972) Auxin and ethylene in adventitious root formation in *Phaseolus aureus* (Roxb.). In: Carr DJ (ed) Plant Growth Substances. Springer, Berlin Heidelberg New York, pp 526–533

Mullins MG, Nair Y, Sampet P (1979) Rejuvenation in vitro: induction of juvenile characters in an adult clone of *Vitis vinifera* L. Ann Bot 44:623–629

Murashige T, Skoog P (1962) A revised medium for rapid growth and bioassays with tobacco tissue cultures. Physiol Plant 15:473–497

Nakano M, Yuda E, Nakagawa S (1980) Studies on rooting of hardwood cuttings of grapevine, cv. Delaware. J Soc Hortic Sci 48:385–394

Nemeth G (1986) Induction of rooting. In: Bajaj YPS (ed) Biotechnology in Agriculture and Forestry, Vol 1: Trees I. Springer, Berlin Heidelberg New York Tokyo, pp 49–64

Norcini JG, Heuser CW, Hamilton RH (1985) Changes in free and conjugated indole-3-acetic acid during initiation and early development of adventitious roots in mung bean. J Am Soc Hortic Sci 110(4):528–533

Novak FJ, Juvova Z (1983) Clonal propagation of grapevine through in vitro axillary bud culture. Sci Hortic 18:231–240

Nozeran R, Bancilhon-Rossignol L, Grenan S (1977) Nouvelles possibilités d'obtention et de multiplication rapide de clones sains de pomme de terre (*Solanum tuberosum* L.). CR Acad Sci Paris 285:37–49

Nysterakis F (1954) Contribution à l'étude du rôle de l'auxine sur la rhizogenèse. VIIIe Congr Int Bot 11e Sect: 134–135

Ooishi A, Machida H, Hosoi T, Komatsu H (1978) Root formation and respiration of the cuttings under different temperatures. J Jpn Soc Hortic Sci 47:243–247
Patel KR, Kim HR, Thorpe TA (1986) Plantlet formation in pitch pine by tissue culture methods. For Ecol Manage 15:147–160
Patel KR, Rumary C, Thorpe TA (1988) Plantlet formation in black and white spruce. III. Histological analysis of in vitro root formation and the root-shoot union. NZJ Sci 16:289–296
Patience PA, Anderson PG (1987) Development of a system to study peroxidases during the rooting of lilac shoots produced in vitro. Acta Hortic 212:267–272
Pierik RLM (1987) In vitro culture of higher plants. Nijhoff, Dordrecht, 344 p
Poessel JL, Macheix JJ, Mosella-Chancel L (1982) Importance physiologique des composés phénoliques chez les arbres fruitiers. Applications au microbouturage in vitro. 2ème Colloq sur les recherches fruitières, Bordeaux 1982, pp 61–71
Poulsen A, Andersen AS (1980) Propagation of *Hedera helix*: Influence of irradiance to stock plants, length of internode and topophysis of cutting. Physiol Plant 49:359–365
Quoirin M, Lepoivre P (1977) Etude de milieux adaptés aux cultures in vitro de *Prunus*. Acata Hortic 78:437–442
Quoirin M, Boxus P, Gaspar T (1974) Root initiation and isoperoxidases of stem tipcuttings from mature *Prunus* Plants. Physiol Vég 12:165–174
Rancillac M (1979) Mise au point d'une méthode de multiplication végétative in vitro du pin maritime (*Pinus pinaster* Sol.) pour la constitution de clones à partir de semences. Micropropagation d'Arbres Forestiers, Ann AFOCEL, 12/6/79:41–48
Reuveni O, Raviv M (1980) Importance of leaf retention to rooting of avocado cuttings. J Am Soc Hortic Sci 106:127–130
Rohwer F, Mäder M (1981) The role of peroxidase in ethylene formation from 1-aminoxycyclopropane-1-carboxylic acid. Z Pflanzephysiol 104:363–369
Riov J, yang SF (1982) Autoinhibition of ethylene production in *Citrus* peel discs. Plant Physiol 69:687–690
Robbins JA, Kays SJ, Dirr MA (1983) Enhanced rooting of wounded mung bean cuttings by wounding and ethephon. J Am Soc Hortic Sci 108(2):239–325
Robbins JA, Reid MS, Paul JL (1985) The effect of ethylene on adventitious root formation in mung bean (*Vigna radiata*) cuttings. J Plant Growth Regul 4:147–157
Robert ML, Herrera JL, Contreras F, Cooper KN (1987) In vitro propagation of *Agave fourcroydes* Lem (Henequen). Plant Cell Tissue Organ Cult 8:37–48
Rogler CE, Hackett WP (1982) Phase change in *Hedera helix*. Induction of the nature to juvenile phase change by gibberellin A_3. Physiol Plant 34:141–147
Roux J (1968) Sur le comportement des axes aériens chez quelques plantes à rameaux végétatifs polymorphe: le concept du rameau plagiotrope. Ann Sci Nat Bot (12)9:109–131
Sakai S, Imaseki H (1971) Auxin-induced ethylene production by mung bean hypocotyl segments. Plant Cell Physiol 12:349–354
Sembdner G, Gross D, Liebisch HW, Schneider G (1980) Biosynthesis and metabolism of plant hormones. Encycl Plant Physiol NS, Springer, Berlin Heidelberg New York 9:281–444
Sharma AK, Prasad RN, Chaturvedi HC (1981) Clonal propagation of *Boungainvillea glabra* Magnifica through shoot apex culture. Plant Cell Tissue Organ Cult 1:33–38
Simmonds J (1983) Direct rooting of micropropagated M26 apple rootstocks. Sci Hortic 21:223–241
Simmonds J (1984) Induction, growth and direct rooting of adventitious shoots of *Begonia* × *hiemalis*. Plant Cell Tissue Organ Cult 3:282–289
Skirvin RM, Chu MC, Gomez E (1981) In vitro propagation of thornless trailing blackberries. Hortscience 16:310–312
Skirvin RM, Chu MC, Kearnes H (1981b) In vitro proliferation and rooting of Harbrite peach and Montmorency sour cherry. Hortscience 16:459
Skoog F, Miller CO (1957) Chemical regulation of growth and organ formation in plant tissues cultured in vitro. Symp Soc Exp Biol 11:118–131
Smith DR, Thorpe TA (1975a) Root initiation in cuttings of *Pinus radiata* seedlings. I. Developmental sequence. J Exp Bot 26(91):184–192
Smith DR, Thorpe TA (1975b) Root initiation in cuttings of *Pinus radiata* seedlings. II. Growth regulators interaction. J Exp Bot 26(91):193–202
Smith DR, Thorpe TA (1977) Root initiation in cuttings of *Pinus radiata* seedlings: effects of aromatic amino acids and simple phenylpropanoids. Bot Gaz 138:434–437

Smith DR, Wareing PF (1972a) Rooting of hardwood-cuttings in relation to bud dormancy and the auxin content of the excised stems. New Phytol 71:63–80
Smith DR, Wareing PF (1972b) The rooting of actively growing and dormant leafy cuttings in relation to the endogenous hormone levels and photoperiod. New Phytol 71:483–500
Soekarjo R, Janssen MGM (1969) The liberation of inhibitors of indole acetic oxidase activity out of *Coleus* internodes treated with potassium hydroxide and sulphuric acid. Acta Bot Neerlandica 18:651–653
Sommer HE, Caldas LS (1981) In vitro methods applied to forest trees. In: Thorpe TA (ed) Plant Tissue Culture: Methods and Applications in Agriculture. Academic Press, Lond New York, pp 349–358
Sriskadarajah S, Mullins MG, Nair Y (1982) Induction of adventitious rooting in vitro in difficult-to-propagate cultivars of apple. Plant Sci Lett 24:1–9
Stonier T, Singer RW, Yang HM (1970) Studies on auxin protectors. IX Inactivation of certain protectors by polyphenol oxidase. Plant Cell Environ 2:79–82
Stonier T, McGladrie K, Shaw G (1979) Studies on auxin protectors. XIV Chlorogenic Acid, a low molecular weight auxin protector in sunflower. Plant Cell Environ 2:79–82
Stromquist LH, Eliasson L (1979) Light inhibition of rooting in Norway spruce (*Picea abies*) cuttings. Can J Bot 57:1314–1316
Tanaka Y, Uritani I (1977) Polarity of production of polyphenos and development of various enzyme activities in cut-injured sweet potato root tissue. Plant Physiol 60:563–566
Thorpe TA (1978) Physiological and biochemical aspects of organogenesis in vitro. In: Thorpe TA (ed) Frontiers of Plant Tissue Culture. Proc 4th Int Cong Plant Tissue Cell Cult Calgary, 49–58
Thorpe TA (1982a) Callus organization and de novo formation of shoots, roots and embryos in vitro. In: Tomes DT, Ellis BE, Harvey PM, Kasha KJ, Peterson RL (eds) Application of plant cell and tissue culture to agriculture and industry. Univ Guelph, Guelph, pp 115–138
Thorpe TA (1982b) Carbohydrate utilization and metabolism. In: Bonga JM, Durzan DJ (eds) Tissue culture in forestry. Martinus Nijhoff, Dordrecht, pp 325–368
Thorpe TA (1984) Clonal propagation of conifers. Proc Int Symp In vitro Propagation of Forest Tree Species Bologna, pp 35–50
Thorpe TA (1985) Carbohydrate utilization and metabolism. In: Bong JM, Durzan DJ (eds) Tissue culture in forestry. Martinus Nijhoff, Dordrecht, pp 325–368
Thorpe TA, Biondi S (1981) Regulation of plant organogenesis. In: Maramorosch K (ed) Advances in cell culture. Academic Press, Lond New York, 1:213–219
Thorpe TA, Tran Than Van M, Gaspar T (1978) Isoperoxidases in erpidermal layers of tobacco and changes during organ formation in vitro. Physiol Plant 44:388–394
Timmis R, Ritchie GA (1984) Progress in Douglas fir tissue culture. Proc Int Symp Recent Adv For Biotechnol Traverse City, pp 37–46
Trefois R, Brunner T (1982) Influence du contenu auxinique endogène sur la réponse au bouturage et sur l'effet nanifiant de quelques *Prunus*. Bot Kozl 69:197–204
Trewavas AJ (1983) Nitrate as a plant hormone. In: Jackson MB (ed) Interactions between nitrogen and growth regulators in the control of plant development. Monogr 9, British Plant Growth Regulator Group ARC Letcombe, Wantage, pp 97–110
Tricoli DM, Mayard CD, Drew AP (1985) Tissue culture of propagation of mature trees of *Prunus serotina*. Establishment, multiplication and rooting in vitro. For Sci 31:201–215
Tripathi RK, Schlosser E (1979) Effects of fungicides on the physiology of plants. II. Inhibition of adventitious root formation by carbendazim and kinetin. Z Pflanzenkr Pflanzenschutz 86:12–17
Van Hoof P, Gaspar T (1976) Peroxidase and isoperoxidase changes in relation to root initiation of *Asparagus* cultured in vitro. Sci Hortic 4:27–31
Van Raalte MH (1954) On the synergism of indole and indole acetic acid in root production. Ann Bogarl 1:167
Vasquez A, Gesto DV (1982) Juvenility and endogenous rooting substances in *Castanea sativa* Mill. Biol Plant 24:48–52
Vaughn KG, Duke SO (1984) Function of polyphenol oxidase in higher plants. Physiol Plant 60:106–112
Veierskov B, Andersen AS (1982) Dynamics of extractable carbohydrates in *Pisum sativum*. III. The effect of IAA on content and translocation of carbohydrates in pea cuttings during rooting. Physiol Plant 55:179–182
Vieitez AM, Vieitez L, Ballester A (1981) In vitro chestnut regeneration anatomical and chemical changes during the rooting process. Colloq Int sur la culture in vitro des assences forestières. IUFRO, Fountainebleau, pp 149–152

Von Arnold S, Eriksson T (1984) Effect of agar concentration on growth and anatomy of adventitious shoots of *Picea abies* (L.) Karst. Plant Cell Tissue Organ Cult 3:257–264

Walker N, Jaques R, Miginiac E (1987) Action of light on rooting in vitro and acclimatization of *Sequoia sempervirens* to soil. Acta Hortic 212:289–301

Watelet-Gonod MC, Favre JM (1981) Miniaturisation et rajeunissement chez *Dahlia variabilis* cv. Télévision, cultivé in vitro. Ann Soc Nat Bot 13(2–3):51–68

Welander M (1979) Influence of medium composition on organ formation in explants of *Begonia* × *hiemalis* in vitro. Swed J Agric Res 9:163–168

Welander M (1983) In vitro rooting of the apple rootstock M26 in adult and juvenile growth phases and acclimatization of the plantlets. Physiol Plant 58:231–238

Welander T (1976) Effects of nitrogen, sucrose, IAA and kinetin on explants of *Beta vulgaris* grown in vitro. Physiol Plant 36:7–10

Williams RR, Taji AM, Bolton JA (1984) In vitro propagation of *Dampiera diversifolia* and *Prostanthera rotundifolia*. Plant Cell Tissue Organ Cult 3:273–281

Williams RR, Taji AM, Bolton JA (1985) Specificity and interaction among auxins, light and pH in rooting of Australian woody species in vitro. Hortscience 20:1052–1053

Wu FT, Barnes MF (1981) The hormone levels in stem cuttings of difficult-to-root and easy-to-root rhododendrons. Biochem Physiol Pflanz 176:13–22

Yang SF (1983) Mechanism and regulation of ethylene biosynthesis. In: Akazawa T, Asahi T, Imaseki H (eds) The new frontiers in plant biochemistry. Jpn Sci Soc Press, pp 133–151

Yang SF, Hoffman NE (1984) Ethylene biosynthesis and its regulation in higher plants. Annu Rev Plant Physiol 35:155–189

Yoshii H, Imaseki H (1982) Regulation of auxin-induced ethylene biosynthesis. Repression of inductive formation of 1-amino-cyclopropane-1-carboxylate synthase by ethylene. Plant Cell Physiol 23:639–649

Yoshii H, Watanabe A, Imaseki H (1980) Biosynthesis of auxin-induced ethylene in mung bean hypocotyls. Plant Cell Physiol 21:279–291

Zimmerman PW, Hitchcock AE (1929) Vegetative propagation of Holly. Am J Bot 15:556–570

Zimmerman RH (1983) Factors affecting in vitro propagation of apple cultivars. Acta Hortic 131:171–178

Zimmerman RH (1984) Rooting apple cultivars in vitro: interactions among light, temperature, phloroglucinol and auxin. Plant Cell Tissue Organ Cult 3:301–311

Zimmerman RH, Fordham I (1985) Simplified method for rooting apple cultivars in vitro. J Am Soc Hortic Sci 110:34–38

Zimmerman RH, Yae BW, Fordham I (1987) Comparison of rooting methods for apple cultivars in vitro. Acta Hort 212:303–310

II.2 Rooting of in Vitro-Produced Shoots in Nonsterile Sand – an Inexpensive and Efficient Technique for en masse Micropropagation

S.B. RAJBHANDARY[1] and Y.P.S. BAJAJ[2]

1 Introduction

Due to losses by rapid deforestation and the depletion of genetic stocks, concerted efforts are being made to evolve innovative methods for mass clonal propagation, production of short-duration trees with a rapid turnover of biomass, induction of genetic variability for the production of novel fruit and forest trees, high-yielding medicinal and aromatic plants for the pharmaceutical industry, and other rare and elite species of plants.

Although micropropagation has been routinely employed for clonal multiplication of a large number of ornamentals, lately much literature has also accumulated on fruit and forest trees (see Bajaj 1986, 1989), and medicinal and aromatic plants (Bajaj et al. 1988). However, the in vitro-produced shoots often fail to root, or die due to contamination, resulting in total loss of genetic stocks.

The generally used in vitro cloning procedure involves four steps, (1) culture of explants, (2) shoot multiplication, (3) rooting of shoots in culture medium, and (4) transfer of plants to the field. One of the main problems to be solved in en masse production of plants through micropropagation is the induction of rooting on the in vitro-produced shoots (see Németh 1986; Moncousin 1988). In the conventional method such shoots are usually rooted in the auxin-enriched medium followed by transfer of the rooted plants to the hardening process, and then to the field. Such a procedure limits the scope of using micropropagated clones for large-scale plantations because a skilled person can handle only a few shoots or shoot clumps per hour for in vitro rooting, which results in high costs.

In this chapter we describe a simple, efficient, and inexpensive method by which rooting can be induced by inserting the in vitro shoots directly into nonsterile sand, thus skipping step (3) above.

2 Method and Observations

Rooting has been induced in nonsterile sand, and plants of many diverse species have been established in soil (see Table 1) (Rajbhandary et al. 1989). The technique and results with some selected species are summarized here. For establishment of cultures,

[1]National Herbarium and Plant Laboratories, Godawari, Kathmandu, Nepal
[2]A-137 New Friends Colony, New Delhi 110065, India

Biotechnology in Agriculture and Forestry, Vol. 17
High-Tech and Micropropagation I (ed. by Y.P.S. Bajaj)

the shoot tips from field-grown plants of *Pyrethrum* and *Ficus lacor*, tuber sprouts of *Solanum tuberosum*, *Ullucus tuberosus*, *Oxalis tuberosa*, and cotyledonary nodes of *Dalbergia sissoo*, *Ficus auriculata*, *Ficus nemoralis*, and *Poncirus trifoliata* were cultured. Murashige and Skoog medium (1962), supplemented with various concentrations and combinations of kinetin, BAP, and NAA, was used in all the cultures. The pH of the media was adjusted to 5.6 before autoclaving. The cultures were incubated under 16-h photoperiod with light provided by fluorescent tubes at 3000 lx.

For rooting, the elongated shoots (3–5 cm long) were transferred to ordinary nonsterile sand under a high humidity polythene cover (see sect. 3, Protocol).

Table 1. Various species of plants in which rooting has been induced in nonsterile sand

Trees	
Fodder trees	Explant[a]
1. *Ficus auriculata*	CN
2. *F. glaberrima*	CN
3. *F. lacor*	ST
4. *F. nemoralis*	CN
5. *F. semicordata*	CN
6. *Leucaena leucocephala*	CN
Fuel wood tree	
7. *Eucalyptus camaldulensis*	ST
Timber trees	
8. *Bischofia javanica*	ST
9. *Dalbergia sissoo*	CN
Ornamental tree	
10. *Ficus elastica*	ST
Fruits	
11. *Poncirus trifoliata*	CN
Vegetables	
12. Brassica *oleracea* var. *botrytis*	CN
13. *Brassica oleracea* var. *capitata*	CN
14. *Ipomoea batata*	TS
15. *Solanum tuberosum*	TS
16. *Oxalis tuberosa*	TS
17. *Ullucus tuberosus*	TS
18. *Allium fistulosum* (anther-derived dihaploid plant)	
Ornamental plants	
19. *Dianthus caryophyllus*	ST
20. *Rosa* sp.	ST
21. *Chrysanthemum* sp.	ST
Medicinal plants	
22. *Atropa belladonna*	ST
23. *Chrysanthemum cinerariaefolium*	ST
24. *Solanum laciniatum*	L
Experimental plant	
25. *Nicotiana tabacum* (anther-derived dihaploid plant)	

[a]*Abbreviations:* CN = Cotyledonary nodes. ST = Shoot tips. TS = Tuber sprouts. L = Leaves.

2.1 *Dalbergia sissoo* (Sissoo)

This tree is popular for plantation in subtropical areas in Nepal and India and is a source of fuel wood, fodder, and expensive timber. Three-year-old field-grown trees regenerated via the protocol of Suwal et al. (1988) exhibited a uniform growth habit. Although high rooting (over 40%) could be achieved when the microshoots were set as cuttings in the ordinary sand under suitable day and night temperature regime (Table 2), a humid atmosphere resulting from continuous rain caused severe damage to the cuttings by damping-off.

2.2 *Eucalyptus camaldulensis*

Eucalyptus camaldulensis has been planted in an area of over 3000 ha in the fuel wood plantation scheme in Nepal Terai. A marked degree of variation occurred amongst the individual trees, whereby the oil contents varied from 0.3 to 2% and the cineole content from 5 to 86%. Superior trees had a wood volume three times greater than average. Proliferating cultures were induced from shoot tips (5–7 mm) excised from 6-month-old selected plants. Plants regenerated following the method described by Gurung and Rajbhandary (1989) exhibited a high growth rate within 2 years. Although very encouraging results were obtained from the field performance of the regenerated plants, at present we face the problem of promoting shoot elongation in the multiplied shoots which had undergone regular subculture passages for more than 2 years. We overcame the problem of nonelongation of shoot buds by reducing the BAP level to 0.25 mg/l and supplementing the medium with 10% coconut milk, after which both shoot multiplication and elongation were observed in subsequent subcultures.

2.3 *Ficus auriculata, F. lacor* and *F. nemoralis* (Fig. 1)

These are important fodder trees; however, their seed collection is difficult and the percentage of germination low, creating the need for micropropagation.

Shoot tips from standing tree of *F. lacor* (2 m high) were used for culture establishment. A high shoot multiplication rate, 20–30 times in 10–12 weeks, was obtained on MS medium supplemented with a combination of BAP (1 mg/l), kinetin (1.5 mg/l), and casein hydrolysate (1000 mg/l). More than 80% rooting was observed from April to October, except in July when the percentage dropped to 40 (Table 2). Saplings of plantable size could be obtained 2–3 months after transfer of the microcuttings from flask to sand. More than 10 000 plants have been established in the soil and are now undergoing field trials.

In *Ficus auriculata* and *F. nemoralis* cultures were initiated from the cotyledonary nodes. BAP at 1 mg/l and NAA at 0.1 mg/l induced the shoots of *F. auriculata* to multiply at the rate of 20–25 times in 8 weeks, as was observed in the MS containing casein hydrolysate (1000 mg/l). Field-established plants from the sand rooting technique are being studied for their field performance. A few *F. auriculata* plants have grown to a height of over 2.5 m in 3 years in the field in Kathmandu. The growth of *F. auriculata* and *F. nemoralis* inside the greenhouse was observed to be very fast; four plants each of the above species planted in September 1989 had grown to a height of 2.5 m by September 1990.

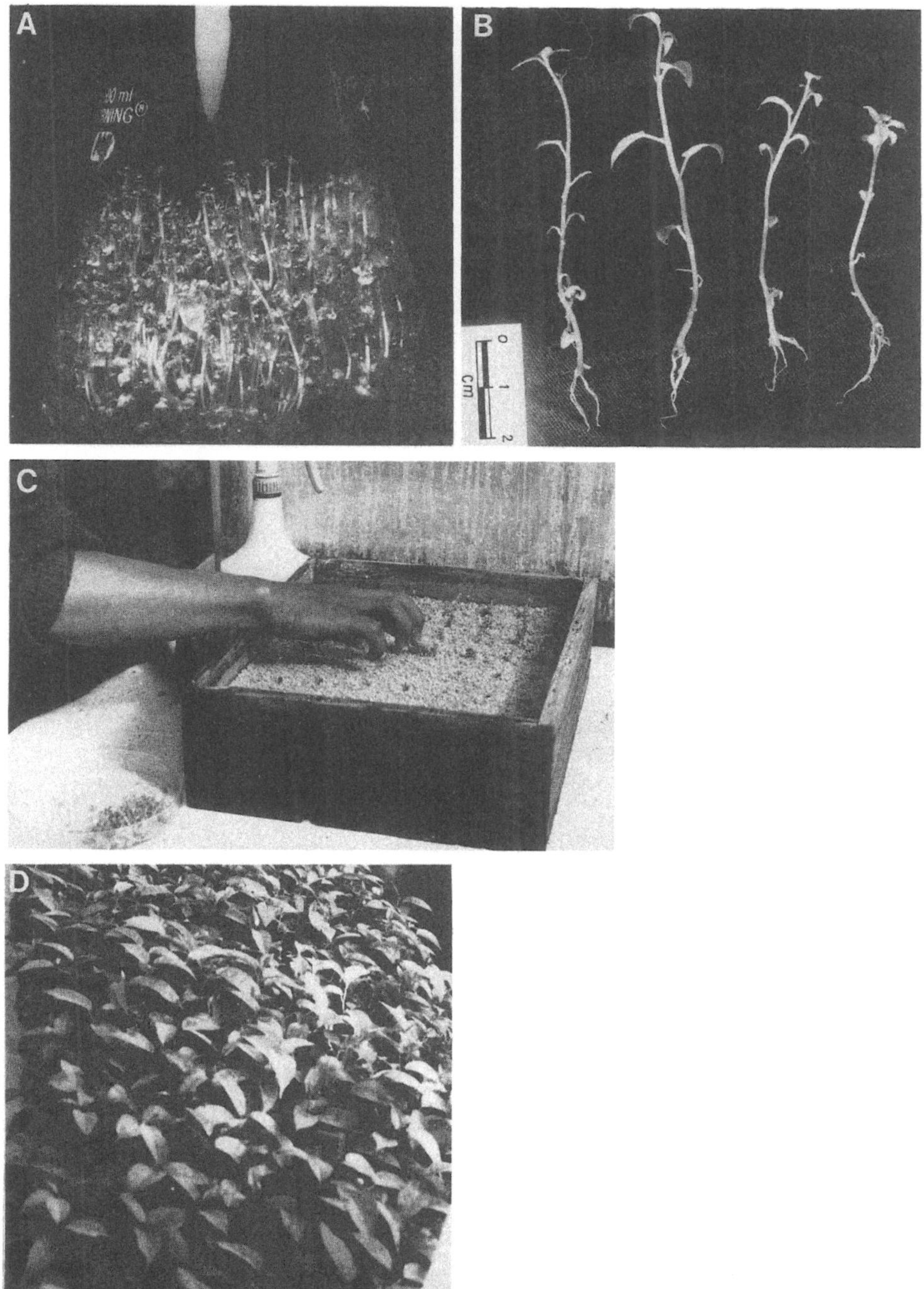

Fig. 1A-D. Micropropagation and induction of rooting in sand in *Ficus lacor.* **A** Multiple shoot formation after 12 weeks in culture. **C** Microshoots transferred to sand bed. **B** Rooted plantlets, 12 days after microshoots transferred to sand bed. **D** Twelve-week-old soil-established plants

Table 2. Effect of monthly variation of day/night temperatures on rooting of microshoots. One hundred and twenty shoots, 2–4 cm, were employed in a month, using 30 shoots each week

Month	Max temp.	Min temp.	*Dalbergia sissoo*	Eucalyptus camaldulensis	*Ficus auriculata*	*Ficus lacor*	*Ficus nemoralis*	*Poncirus trifoliata*
Nov.	20	11	–	20	31	40	0	12
Dec.	24	6	–	5	10	25	0	0
Jan.	24	4	20	0	0	10	0	0
Feb.	26	6	0	0	0	15	0	36
Mar.	26	8	0	5	10	40	10	58
Apr.	28	9	40	35	35	80	15	56
May	34	13	40	49	80	100	50	56
June	34	17	65	59	80	100	50	59
July	31	18	43	52	65	40	30	54
Aug.	27	18	45	43	50	95	25	–
Sep.	29	19	47	40	40	100	20	85
Oct.	29	15	34	40	38	100	5	75

2.4 *Poncirus trifoliata*

The rootstock of this species is often used for grafting the scions of orange. Proliferating cultures were established by using cotyledonary nodes as explant. A shoot proliferation rate of 15–20 times in 8–10 weeks was achieved in the BAP level of 0.25 mg/l. In ordinary sand a high percentage (over 50%) of rooting of shoots was observed from March to September except in August (Table 2).

2.5 *Chrysanthemum cinerariaefolium* (Pyrethrum)

Pyrethrum is an insecticidal plant showing a wide range of variation in growth and pyrethrin content. Its cultures were initiated and plants obtained following the protocol described by Karki and Rajbhandary (1984). The shoot multiplication rate remained the same over 10 years of 8–12 weekly subcultures. Plants established from 5–6-year-old cultures showed no morphological variation. Over 80% rooting of microcuttings could be achieved at a day/night temperature regime of 29/17 °C in nonsterile sand.

2.6 *Solanum tuberosum* (Potato, Fig. 2)

Protocols for the micropropagation of potato are now well established (see Bajaj 1987). Of the 25 cultivars planted in the field, 16 cultivars were initiated from tuber sprouts and nine from the test tube plants which were received from CIP, Peru. Only one CIP cultivar showed no response in culture and died. From January to May, 50–75% rooting of multiple shoots in the ordinary sand was observed. Rooted plants could be easily established in the field with no loss of plants during transplantation (Manandhar and Rajbhandary 1986).

Fig. 2A-D. Micropropagation and induction of rooting in potato. **A** Multiple shoot formation after 10 weeks in culture. **B** Plantlets 3 weeks after microshoots transferred to sand bedding. **C** Rooted plantlet 3 weeks after microshoot transferred to sand bed. **D** Plants in polythene greenhouse, 10 weeks after transfer to field

2.7 *Ullucus tuberosus* and *Oxalis tuberosa*

These are proteinaceous tuber crops grown from Venezuela to Argentina at elevations between 2500 and 3000 m. Their introduction in high altitude terrains in the Himalayas may help to alleviate protein deficiency (King 1987). Cultures of these two species were raised from material obtained from the New York Botanical Garden. A sustained rate of multiplication, 15–20 shoots in 3-weekly subcultures, was observed after the fourth subculture in both the species. For *U. tuberosus*, elongated shoots after 3 weeks of growth period were set as individual microcuttings in the sand bed under high humidity polythene cover to obtain rooted plants. Except for *Oxalis tuberosa*, the multiplied shoot buds had to be transferred to the basic medium containing 1 mg/l GA to induce elongation. The elongated shoots, however, rooted readily in the nonsterile sand. Plants transplanted in September 1988 and harvested in February 1989 had a yield of 200 g per plant with *U. tuberosus* and 70 g per plant with *O. tuberosa.*

3 Protocol for Sand Rooting

For rooting of the in vitro-derived shoots, the sand is first passed through a sieve of 2 × 2 mm pore size, washed several times with tap water, and then dried in bright sunlight for 4–6 h. To the dry sand 10% (by weight) water (boiled and cooled) is added. The sand box, 36 × 30 × 7 cm in size (Fig. 1C), holds 8 kg of dry sand.

The microshoots, 3–5 cm in length, are transferred into the sand so that about 1 cm remains below the surface of sand. Misting is carried out by a handspray whenever the surface of the sand becomes dry.

4 Summary and Conclusions

One of the reasons that limits the use of micropropagation, especially those of the forest tree species, is the high cost of production of in vitro plants (Chu 1989). In our protocol, multiple shoots from in vitro cultures are directly transferred to nonsterile sand beds to induce rooting, thus skipping the rooting of shoots in the culture medium. Over 25 species including forest trees have been produced by this sand-rooting technique. More than 10,000 trees regenerated by this procedure are undergoing field trials. It is believed that this technique can substantially reduce the cost of producing tissue-cultured plants since in vitro rooting has been reported to account for 30–70% of micropropagation production costs. This non-sterile sand method has the potential of enlarging the current world market of micropropagation of plants.

References

Bajaj YPS (1986) Biotechnology of tree improvement for rapid propagation and biomass energy production. In: Bajaj YPS (ed) Biotechnology in agriculture and forestry, Vol 1, Trees I. Springer, Berlin Heidelberg New York Tokyo, pp 1–23

Bajaj YPS (ed) (1987) Biotechnology in agriculture and forestry, Vol 3. Potato. Springer, Berlin Heidelberg New York Tokyo

Bajaj YPS (ed) (1989) Biotechnology in agriculture and forestry, Vol 7. Trees II. Springer, Berlin Heidelberg New York Tokyo

Bajaj YPS, Furmanowa M, Olszowska O (1988) Biotechnology of the micropropagation of medicinal and aromatic plants. In: Bajaj YPS (ed) Biotechnology in agriculture and forestry, Vol 4, Medicinal and aromatic plants I. Springer, Berlin Heidelberg New York Tokyo, pp 60–103

Chu IYE (1989) The role of micropropagation in commercialization of plant biotechnology. In: Proc Conf Int Plant Biotechn Net Work. Tissue culture for crops project, Colorado, pp 35–43

Gurung S, Rajbhandary SB (1989) Micropropagation of *Eucalyptus camaldulensis* – a fast-growing essential oil-bearing tree. In: Proc Int workshop tissue culture Biotechn Med aromatic plants. CIMAP, Lucknow, India, pp 17–21

Karki A, Rajbhandary SB (1984) Clonal propagation of *Chrysanthemum cinerariaefolium* Vis. (pyrethrum) through tissue culture. Pyrethrum Post 15:118–121
King SR (1987) Four endemic Andean tuber crops: promising food resources for agricultural diversification. Mountain Res Dev 7:43–52
Manandhar A, Rajbhandary SB (1986) Rooting in non-sterile potting mix of in vitro potato and its field establishment. Indian J Hortic 43:235–239
Moncousin C (1988) Adventitious rhizogenesis control: new developments. Acta Hortic 230:97–104
Murashige T, Skoog F (1962) A revised medium for rapid growth and bio-assays with tobacco tissue cultures. Physiol Plant 15:473–497
Németh G (1986) Induction of rooting. In: Bajaj YPS (ed) Biotechnology in agriculture and forestry, Vol 1, Trees I. Springer, Berlin Heidelberg New York Tokyo, pp 49–64
Rajbhandary SB (1989) Rooting of in vitro-produced shoots of agricultural, horticultural and forestry species in nonsterile sand. Proc 6th Int Congr SABRAO, pp 627–630
Suwal B, Karki A, Rajbhandary SB (1988) The in vitro proliferation of forest trees. 1. *Dalbergia sissoo* Roxb ex DC. Silvae Genet 37:26–28

II.3 Use of Artificial Substrates for in Vitro Propagation

P. BROCHARD[1]

1 Introduction

Apart from cells or isolated root cultures, tissues or vegetative organs generally react poorly to continual immersion in a liquid culture medium. Agar or various other gels still constitute the ideal basis for most in vitro cultures. However, there are some cases where agar is not suitable, namely, rhizogenesis, micrograftings, and studies of dual culture plants; microorganisms: pathology, symbiotic interaction studies (rhizobacteria and mycorhizal fungi). The use in these cases of substrates other than agar proves not only useful but even essential.

In this chapter the aspects listed above are elucidated and for each of them the use of artificial substrates is proposed.

2 Rhizogenesis and Acclimatization

2.1 Problems

Planning an in vitro propagation program for a woody or herbaceous plant must always include the plant rooting and acclimatization phases, which are decisive for the future of the young plant.

Final success depends on a succession of good results in the following operations:

- induction and expression of rhizogenesis;
- obtaining a rooting system of good quality;
- transfer and acclimatization;
- technical success of these operations at the lowest cost.

Each of these different components is considered here in order to briefly mention the different factors influencing each of them.

2.1.1 Induction and Expression of Rhizogenesis

These two phases, although often confused, are in fact two separate phenomena (Christianson 1987; Maene and Debergh 1987; Nougarède et al. 1987) with very different mechanisms.

[1]Laboratoire de Génétique Appliquée, Challet-Herault Créations, BP Route de Cholet, 49340 Nuaille, France

Biotechnology in Agriculture and Forestry, Vol. 17
High-Tech and Micropropagation I (ed. by Y.P.S. Bajaj)

Induction itself, affecting only a few cells (Nougarède et al. 1987), in fact, consists of two distinct subphases: activation of the cells, then the response of these cells to the hormonal balance induced by the medium (Christianson 1987) in order to make root meristems. The growth of these meristems is visibly expressed in the root.

These very complicated phenomena are controlled, on the one hand, by the endogenous hormone levels and nutrition (Nougarède and Rondet 1983), and on the other, by external factors.

Among these can be mentioned the following: temperature (Gautheret 1971), light (Eliasson and Brunes 1980; Maene and Debergh 1987), pH (Zatyko and Molnar 1987), and finally the aeration of the medium, which in some cases seems to have an important, sometimes decisive, effect.

It is well known that in traditional propagation, nursery gardeners use only light and well-aerated substrates. In contrast, it is also well known that the majority of in vitro rooting trials are made on agar, a rather asphyxiating medium.

However, some studies on this subject have proved that rooting can be much better on an aerated medium: e.g., *Glycerhiza glabra* roots grow more rapidly on filter paper than on an agar medium (Shaw and Dalal 1982). In the same way, the quantity and length of the roots of *Malus* are superior in an agitated liquid medium or in coarse sand moistened with medium than in agar; Hutchinson (1982) ascribes this difference to the aeration of the medium.

It is also quite frequently observed that in many plants, which seem to root well on agar, the rooting is often very superficial and the roots have a tendency to come to the surface of the gel instead of growing into it (e.g., *Hydrangea quercifolia*, Fig. 1).

This depressive reaction of agar on rooting can lead to total inhibition: in our laboratory we never managed to obtain rooting of *Mimosa dealbata* on agar. On the con-

Fig. 1. *Hydrangea quercifolia* in jar showing many roots above agar

Fig. 2. *Mimosa dealbata* rooted on polyurethane foam Stratair

trary, on an aerated substrate (polyurethane foam moistened with nutritive medium) rooting rates of about 40% were obtained (Fig. 2).

2.1.2 Obtaining a Good Quality Rooting System

Fortunately, for in vitro propagation, agar, although not ideal in most cases results in acceptable induction and rooting. It has even been proved that on certain plants, such as *Pinus radiata*, rhizogenesis is apparently independent of the substrates used (Reilly and Brown 1976).

Unfortunately, the quality of roots grown and developed on agar is often poor; most of the time, roots are obtained without absorbant hair, which is typical of hydroponic cultures. Aldrufeu (1987), comparing the use of different substrates such as cellulose, sand, perlite, etc. with agar for the rooting of *Pelargonium*, noted that the roots show root hairs in all substrates except agar. Deformed, short, thick roots (e.g., *Primula obconica*; Fig. 3) can also result.

2.1.3 Problems in Removal from the Culture Vessel and Acclimatization After Rooting in Agar

This is for many plants often the most delicate phase of the program. As Bonga (1985) writes: "in many tissue culture cloning programs, one of the most difficult problems is

Fig. 3. *Primula obconica* rooted in agar showing short and thick roots

the transfer of propagules from culture to soil". These difficulties occur at different levels, e.g., when transferring the rooted plants from in vitro culture, very fragile thin roots are obtained, which often need to be washed, thus causing injuries.

In any case, the rooting system is deeply disturbed. The roots, sometimes few in number, are often limited in function, and in the case of *Acacia* even nonfunctional (Skolmen and Mapes 1978). The plants must readapt or make new roots and must combat not only the stress due to the change from a saturated to a drier atmosphere, but also attack by bacteria or fungi in a new, nonsterile medium.

All these different technical problems may provoke heavy losses, increasing the already high costs of in vitro micropropagation.

2.1.4 Manpower

The transfer from the culture vessel is often difficult and slow. Washing, sometimes necessary for the roots, increases manpower time. Finally, the planting out with bare roots is time-consuming and difficult to mechanize.

2.2 Proposed Solutions

2.2.1 Use of Artificial Substrate in Vitro

Different researchers have often successfully used substrates other than agar for in vitro culture. To facilitate transfer and acclimatization, these substrates are either solid, more or less porous and friable, or with fibers sufficiently close to give them a good structure. A few materials are given in the literature: peat plugs for the rooting of apple trees (Zimmerman et al. 1987), cotton plugs or textiles (Zatyko and Molnar 1987), plastic foam cubes (Hussey 1980), or plugs in polypropylene fibers (Brochard 1987). Others, without being explicitly named, are: cigarette filters, rockwool cubes, glass wool, etc. Therefore, in a few particular cases, the idea of using artificial substrates has already been described, and, especially in the case of commercial propagation, the idea of an individual substrate.

This idea is supported by the present tendency to shorten the phases by inducing all or part of the rhizogenesis ex vitro, and so combining it with acclimatization (Maene and Debergh 1987).

2.2.2 Use of Artificial Substrate ex Vitro

Transfer and acclimatization are traditionally made on substrates derived from peat, sometimes sterilized (Mridula et al. 1983), vermiculite, perlite, sand, in fact, those used frequently by nursery gardeners using classical propagation methods. Although these products can be satisfactory in a research and development program, where a few dozen plants are directly acclimatized in their final culture pot, it is not the same in commercial propagation, where special conditions must be faced, all more or less interdependent:

— strong replanting density
— little thickness of culture substrate
— damping off problems due to parasitic attacks
— easy handling of young plants
— possibility of mechanization for this handling
— delivery facilities
— phytosanitary constraints for exporting to certain countries.

These constraints necessitate certain characteristics in the substrates used:

1. Acclimatization in trays on bulk substrates, although on the whole advantageous, makes the mechanized grasping of the young plant difficult. The mixing with neighbor roots can cause injuries upon uprooting, and stress during replanting.
2. The use of individual alveoles (in trays) filled with bulk substrates based on compost, peat, or sand is an improvement on the above scheme and can allow some mechanization to grasp the plant. But as the alveoles are often, for reasons of economy, very small, the moisture and aeration control of such substrates becomes very difficult.
3. The incorporation of fiber materials in the substrate (wood fibers, textile, rock wool in bulk) can improve the air/water ratio, but the regular filling of small culture pots becomes very difficult.

4. Those different solutions which may, in spite of their defects, be judged satisfactory by many commercial laboratories, cannot help in the export of young plants to countries where phytosanitary reasons prohibit the import of compost.

Thus, just as for in vitro rooting, we come to the use of artificial and individual substrates. Different products have been tested and recommended: rock wool (Debergh 1987), plastic foam (McCown 1985; Constantine 1986), pellets of pressed peat (Zimmerman et al. 1987), small bags of peat rolled around the plantlets (Franclet 1981), etc. However, at the moment, none of the systems available has penetrated the market of the majority of laboratories.

Our work in a commercial research laboratory induced us to try some of the artificial substrates on the market, but we did not find the qualities in which we were interested. Finally, we decided to develop our own products. Our aim was to obtain a sterilizable product, chemically neutral, and easy to introduce into a test tube.

Our second objective was facility in inserting a brittle propagule into the substrate without damaging the plantlet.

With these objectives in mind, we have developed two very different kinds of products: the Milcap plug and the Stratair foam.

1. The Milcap plug is made of artificial fur folded and sewn to form a cylinder of about 2.5 cm in diameter, where the hairs are radially oriented to the inside. The cylinder is then cut to the right length. Its composition: 65% polypropylene fibers and 35% viscose fibers give a water retention of about 40% after drainage. It can be used in vitro after sterilization or directly outdoors for simultaneous rooting and acclimatization. The Milcap plug is presented ready to use in a heat-formed polystyrene tray. We have tested it successfully on many herbaceous as well as woody plants. According to the production of our laboratory, we use it at the moment mostly for aquatic herbaceous plants (Fig. 4).

2. Stratair foam. Despite its qualities, the Milcap plug cannot be a universal substrate. Its main disadvantage is the risk of "strangling", as woody plants, e.g., the rose, grow very quickly.

We worked on foams that would avoid the disadvantages of the products already on the market: too high water retention, difficulty in planting because of their hardness, and sometimes traces of toxicity.

We have thus developed a polyurethane foam that is half hydrophobic, half absorbant, complemented by active charcoal and varied nutrients, with a special, very friable composition.

This product is slightly more difficult to use in vitro: it does not easily tolerate more than 110 °C and it may be preferable to carry out two sterilizations at 105–110 °C at intervals of 24 h rather than one at 115 °C. Its use in rooting and acclimatization has also proved very satisfactory (Figs. 5 and 6).

3 Use of Artificial Substrates for in Vitro Micrografting

The in vitro micrografting technique, relatively seldom employed because it is expensive in manpower, can be very useful in two cases:

1. Recovery of pathogen-free clones from fruit trees infected by virus (Murashige et al. 1972).

Fig. 4. *Cryptocoryne* sp. in acclimatization phase on Milcap plug

Fig. 5. *Daphne mezereum* well-rooted on Stratair

2. Rejuvenization before in vitro propagation from adult trees (Monteuuis 1986).

The method consists of grafting, under sterile conditions, an apex or a meristem of dormant or actively growing buds on a rootstock plantlet derived from seed or by micropropagation.

The use of this technique, theoretically simple, is in fact not easy. One of the keys to its success seems to be the necessity of having a well-rooted rootstock plantlet, and of disturbing the roots as little as possible.

Fig. 6. *Ficus benjamina* in rooting and acclimatization phase on polyurethane plug Stratair, in trays with water tank. The watering of the substrate is done automatically by drip irrigation (*arrows*)

Mosella et al. (1980) used germinations on a bulk substrate, vermiculite. To avoid disturbing the roots, manipulations must take place inside the culture tube, the success rate being correspondingly low.

More recently, Monteuuis (1986), in order to rejuvenize some adult *Sequoïadendron* before their in vitro propagation, improved the technique by using the Milcap plugs in polypropylene fibers for his rootstocks as a help in germination.

This method allowed him, without disturbing the roots that were well established in the substrate, to extract the rootstock plantlets from the culture tube, put them under the stereomicroscope, and to micrograft under optimal conditions.

In order to recover virus-free *Prunus avium*, Deogratias et al. (1985) also used this technique of micrografting the apex and compared two kinds of rootstocks: seedlings germinated in a liquid medium or seedlings on Milcap plugs moistened with the same medium. They noted success rates from 0.5% for cultivated seedlings in liquid medium to 60% for seedlings in synthetic substrate. They attribute this difference, on the one hand, to the transfer facilities in tubes, on the other, to the harmonious development of the rootstock whose vigor is important for the growth of the graft.

These two examples show the advantages of using artificial substrates in the micrografting technique, which until now has been rarely used due to lack of success and often deceptive results. With these improvements the method can be of great help in micropropagation laboratories.

4 Use of Artificial Substrates in a Mixed Culture: Plant and Microorganisms

Under normal ex vitro conditions, the roots of the plants are in contact and in constant contact with numerous microorganisms. These relations can be profitable for both partners in the case of symbiotic microorganisms (mycorrhizal fungi, rhizobacteria) or can be injurious for one or the other (pathogenic effect of these microorganisms, or in contrast, can show repressive and selective action from the root exudates at the rhizosphere level).

The difficult study of these interactions between microorganisms and plants at the root level can be facilitated by the in vitro production of mixed cultures (Ingram 1980).

4.1 Host-Pathogen Interactions

Our experience in plant pathology has shown the difficulties in obtaining cohabitation between soil microorganisms and whole plants: the more or less anaerobic conditions of agar make it difficult for the fungi, bacteria, and nematodes to colonize their roots, and their growth often remains very superficial on the surface of the agar. The use of synthetic substrates for the study of these host-pathogen interactions could certainly be very useful to pathologists.

4.2 Rhizobacteria and Mycorrhiza

Several authors (Kloepfer et al. 1980; Suslov 1982; Digat 1983) have pointed out the power of rhizobacteria to attack pathogenic microorganisms, and have demonstrated the effects in plant growth.

Using these observations, Digat et al. (1987) tried to inoculate in in vitro and on synthetic substrate in polypropylene fibers, *Hydrangea quercifolia* plantlets, *Rosa* X, and *Primula obconica*, with bacteria such as *Pseudomonas fluorescens* and *P. putida.* They noted good development of the bacteria at the root level and, particularly for *Primula obconica*, promotion of in vitro growth.

Concerning the mycorrhizal fungi, a scientific team from INRA in Dijon (France) has recently presented (Gianinazzi 1988) a review on the results of in vitro inoculation of in vitro plants by mycorrhizal fungi: this inoculation, most of the time very beneficial, can stimulate the acceleration of rooting and better root ramification, and has a beneficial effect on growth.

It seems, however, that this inoculation is difficult on agar (Mason 1980) and the latter author recommends the use of a substrate like vermiculite, which is more aerated than agar.

Despite the lack of results published on this subject, it is clear from the preceding paragraphs that the use of synthetic substrate could greatly facilitate mycorrhization and its control in vitro as well as ex vitro during acclimatization; the method still has to be defined and we are planning experiments in our laboratory.

5 Qualities Needed in an Ideal Artificial Substrate

The differing applications of artificial substrates for in vitro culture make products with several qualities necessary.

McCown (1985) and Constantine (1986) have examined the qualities needed for an "ideal substrate". To their propositions we have added some of our own reflections, and we propose a checklist of each single characteristic, which in our opinion, would be required for such a substrate.

5.1 Qualities with Regards to the Plant

1. Steadiness, reproducibility.
2. Sterility or at least cleanliness.
3. No toxicity.
4. No mechanical gene in the short or long term (e.g., strangling of the roots).
5. Facility of penetration by the roots, even thin ones.
6. pH easily adaptable to every plant.
7. "Buffering capacity" of the pH.
8. Nutrition reserves or capacity in ion exchange.
9. Air/water ratio close to 50% for any irrigation method used.
10. Absorption or neutralization of toxic exudates possibly produced during rhizogenesis.
11. Capacity for possible storage of rooted plants without further growth. This would be especially interesting for outdoor woody plants which have to be delivered at a precise time, to plan production for the whole year.
12. Possibility of inoculation by mycorhizal fungi or by rhizobacteria.

5.2 Qualities with Regards to the User

13. Storage facilities before use.
14. Handling facilities and possibility of mechanization.
15. Presentation "ready to use".
16. Planting facilities, even with soft plantlets, which are often necessary when coming from in vitro culture.
17. High density per surface unit.
18. Presentation in different volumes, adaptable to different plants.
19. Possibility of easy regulation for the necessary moisture.
20. Light weight for air transport.
21. Possible autoclave sterilization for use under sterile conditions in vitro.
22. Possibility of exporting to countries that prohibit the import of compost.
23. Attractive presentation in order to appeal to horticulturists, who are sensitive to the appearance of the product.
24. Competitive ratio quality/price.

It is certain that presently this "ideal substrate" does not exist and will probably not exist in the future, at least in one single form: acclimatization and culture conditions can be very different from one laboratory to another, from both technical and economical aspects, and the plants have neither the same needs nor the same sensitivities. In regard to their high pH, glass wool plugs are not suitable for Ericacea; synthetic fiber plugs, being indestructible, may provoke strangling in quickly growing woody plants; the agglomerated latex from the peat plugs seems to be toxic to a few clones of *Kalmia* (McCown 1985).

Table 1 lists the best-known artificial substrates on the market, describing the positive or negative qualities of each. In this table we have omitted points 22 and 23. Point 22, "esthetic judgement", is too subjective, and point 23, "price", varies too greatly between countries, depending on production location, transport costs, and customs duties.

6 Summary and Conclusion

Artificial substrates can be very useful in many aspects of in vitro culture: rhizogenesis, acclimatization, micrografting, mixed culture of plant microorganisms. Their use is still limited, but it will certainly increase with the increasing practice of rooting ex vitro.

The inoculation of in vitro plants with rhizobacteria or mycorrhiza, proposed at present nearly exclusively by research laboratories, will certainly become increasingly common in laboratories, and the use of synthetic substrates will facilitate this operation.

After the first amateur attempts, from cigarette filters or personal compositions, newly developed products are now appearing on the market, whereby each of these products presents some of characteristics required for the "ideal substrate", but none possesses all of them.

For a commercial laboratory, however, even if all these characteristics are useful, only some of them are essential: cleanliness, quick and easy handling, and the possibility of mechanization. In general, all the substrates listed have at least these qualities.

On the other hand, we made this list without mentioning two important characteristics, which are:

— disposability on the market,
— cost of the product,

which depends, on the one hand, on technical choices such as the raw materials or the production facilities by which the producers must aim at the best product at the lowest cost, and on the other, on economic data such as transport, distribution network, and customs duties, which are less easy to influence.

In conclusion, as none of the products listed is perfect, the most important quality needed for an artificial substrate is its adaptability to the plant which is produced in the laboratory, and while the list in Table 1 can help the user, only his own experience and trials can decide his final choice.

Table 1. List of some artificial substrates available on the market

	General composition	Most important trademark	Production company	Characteristics[a] Present	Lacking
1	Products made from peat				
	Peat conglomerated by latex	Techniculture	Castle and Cooke Techniculture Inc. Salinas, CA USA	1–2–4–7–8? 10–12?–13– 14–15–16– 17–18–20	3–5–6–9–11 19–21–22
	Peat or compost surrounded by a cylindrical envelope	Fertiss	Fertil 8 rue Dulac 75015 Paris France	1–2–3–4–5– 7–8–10–12? 13–14–15– 17–20	6–9–11–16 18–19–21 22
	Peat or compost in flat sacks to be wrapped around the plantlet	Melfert	Fertil 8 rue Dulac 75015 Paris France	1–2–3–4–5– 6–7–8–9?– 10–12–16	9–11–13–14 15–17–18– 19–20–21 22
2	Products made from textile fibers				
	Fur cylinders in polypropylene fibers	Milcap	Milcap 49340 Trémentines France	1–2–3–5–6– 13–14–15– 16–17–19– 20–21–22	4–7–8–9–10 11–18
3	Rock wool	Cultilene	Isover Cultilène Saint Gobain Les Miroirs 92096 Paris La Defense France	1–2–3–4–5 12?–13–14 15–17–18– 20–21–22	6–7–8–9–10 11–16–19
		Grodan	Grodania Hovedgaden 483 2640 Hedehusene Denmark	Same	Same

Table 1. *(Continued)*

	General composition	Most important trademark	Production company	Characteristics[a] Present	Lacking
4	Plastic foam				
	Polyurethane foam colored and hydrophilic	Oasis Horticubes	Smithers Oasis Kent OH USA	1–2–3?–4–6 12?–13–14– 15–17–18?– 20–21–22	5–7–8–9–10 11–16–19
		Green Cube	Resma 31018 Gaiarine via Resteinzza 3 Trevizo Italy	1–2–4–6–12? 13–14–15–17 18?–20–21– 22	3–5–7–8–9 10–11–16 19
	Polyurethane foam half hydrophilic half hydrophobic improved with adsorbants and fertilizers	Stratair	Milcap 49340 Trémentines France	1–2–3–4–5– 6–9–10–12? 13–14–15– 17–18–19– 20–21–22	7–8–11–16
5	Cellulose fibers surrounded by a cylindrical envelope	Sorbarod	Baumgartner Papiers SA Case postale 195 R. de la Vernie 12 CH-1001 Lausanne- Crissier Switzerland	1–2–3–4–5–6 11?13–14–15 17–20–21–22	7–8–9–10 16–18–19

[a]The numbers in these two columns refer to the list of the qualities needed for an ideal artificial substrate (see sects. 5.1 and 5.2).

References

Aldrufeu A (1987) Rooting and acclimatization of *Pelargonium zonale* plantlets. Acta Hortic 212:361–366

Bonga JM (1985) Tissue culture techniques. In: Bonga JM, Durzan DL (eds) Tissue culture in forestry. Martinus Nijhoff Dordrecht, pp 4–35

Brochard P (1987) Use of Milcap substrates for in vitro rooting and coming out of plants obtained by micropropagation. Acta Hortic 212:322

Christianson ML (1987) Causal events in morphogenesis. Proc 6th Int Congr Plant Tissue and Cell Culture. Plant Tissue and Cell Culture. Alan R. Liss, New York, pp 45–55

Constantine DR (1986) Micropropagation in the commercial environment. In: Withers LA, Alderson PG (eds) Plant tissue culture and its agricultural application. Butterworth, London, pp 175–186

Debergh P (1987) Recent trends in the application of tissue culture to ornamentals. Proc 6th Int Congr Plant Tissue and Cell Culture. Plant Tissue and Cell Culture. Alan R. Liss, New York, pp 383–393

Deogratias JM, Lutz A, Dosba F (1985) Micro-greffage d'apex de cerisiers (*Prunus avium* L.) multiplies in vitro en vue de l'élimination de trois types de particules virales (CLSV, PDV et NRSV). Proc 5e Coll sur les recherches fruitières. Bordeaux, 1985, 13–14 Nov, pp 85–96

Digat B (1983) Modes d'action et effects des rhizobactéries promotrices de la croissance et du développement des plantes. In: INRA (ed) Les antagonismes microbiens (Coll INRA No 18), Versailles, pp 239–253

Digat B, Brochard P, Hermelin V, Touzet M (1987) Interest of bacterized synthetics substrates Milcap for in vitro culture. Acta Hortic 212:375–378

Eliasson L, Brunes L (1980) Light effects on root formation in aspen and willow cuttings. Physiol Plant 48:261–265

Franclet A (1981) La motte de culture Melfert. AFOCEL-ARMEF: Informations Forêt 165:1–15

Gautheret RJ (1971) Actions de variation de températures sur la rhizogenèse des tissus de rhizomes de topinambour cultivés in- vitro. In: CNRS (ed) Les cultures de tissus de plantes. Paris 1971:187–199

Gianinazzi S (1988) La mycorrhization des vitro plants. IX Coll Sect Fr. IAPTC, Dijon, 25 Nov 1988

Hussey G (1980) In vitro propagation. In: Ingram DS, Helgeson JP (eds) Tissue culture methods for plant pathologists. Blackwell Scientific, Oxford, pp 51–61

Hutchinson JF (1982) In vitro propagation of apple using organ culture. Plant Tissue Culture 1982. Maruzen, Tokyo, pp 729–730

Ingram DS (1980) Tissue culture methods in plant pathology. In: Ingram DS, Helgeson JP (eds) Tissue culture methods for plant pathologists. Blackwell Scientific, Oxford, pp 3–9

Kloepfer JW, Leong J, Teintze M, Schroth MN (1980) Enhanced plant growth by siderophores produced by plant growth promoting rhizobacteria. Nature 286:885–886

Maene LJ, Debergh PC (1987) Optimalization of the transfer of tissue cultured shoots to in vitro conditions. Acta Hortic 212:935–348

Mason PA (1980) Aseptic synthesis of sheathing ecto-mycorrhizas. In: Ingram DS, Helgeson JP (eds) Tissue culture methods for plant pathologist. Blackwell Scientific, Oxford, pp 173–175

McCown DD (1985) Plug systems for micropropagules. In: Zimmerman RH, Griesbach RJ, Hammerschlag FA, Lawson RH (eds) Tissue culture as a plant production system for horticultural crops. Martinus Nijhoff, Dordrecht, pp 53–60

Monteuuis O (1986) Microgreffage de points végétatifs de *Sequoiadendron giganteum* Buchholz séculaires sur des jeunes semis cultivés in vitro. CR Acad Sci Paris, t 302; III, 6:223–225

Mosella C, Signoret PA, Jonard R (1980) Sur la mise au point de technique de microgreffage d'apex en vue de l'élimination de deux types de particules virales chez le pêcher. (*Prunus Persica*, Batsch.). CR Acad Sci 290, D:287–290

Mridula MK, Gupta PK, Mascarenhas AF (1983) Rapid multiplication of *Sapium sebiferum* by tissue culture. Plant Cell Tissue Organ Cult 2:133–139

Murashige T, Bitters WP, Rangan TS, Nauer EM, Roistacher CN, Holliday PB (1972) A technique of shoot apex grafting and its utilization towards recovering virus-free *Citrus* clones. Hortscience 7(2):118–119

Nougarede A, Rondet P (1983) Base cytophysiologique de l'induction rhizogène en réponse à un traitement auxinique dans l'épicotyle du pois nain. Ann Sci Nat Bot Paris 13, 5:121–149

Nougarede A, Chriqui D, Bercetche J (1987) Natural and experimental expression of rhizogenesis: behaviour of target and non-target cells. Acta Hortic 212:349–359

Reilly K, Brown CL (1976) In vitro studies of bud and shoot formation in *Pinus radiata* and *Pseudotsuga menziesii*. Ga For Res Rap 86:1–9

Shaw RR, Dalal KC (1982) *Glycyrhiza glabra* from test tube to field. Plant Tissue and Cell Culture 1982. Maruzen, Tokyo, pp 685–686

Skolmen RG, Mapes MO (1978) After care procedures required for field survival of tissue propagated *Acacia koa*. Proc Int Plant Prop Soc 28:156–164

Suslow TV (1982) Role of root colonizing bacteria in plant growth. Phytopathogenic prokaryotes, vol 1. Academic Press, New York, pp 187–223

Zatyko JM, Molnar I (1987) Influence du pH du milieu de culture sur la formation des racines adventives chez diverses espèces fruitiéres. Proc Congr Moët Hennessy Electronique et pilotage des plantes. Monaco, Sept 87, p 59

Zimmerman RH, Yae BW, Fordham I (1987) Comparison of rooting methods for apple cultivars in vitro. Acta Hortic 212:303–310

II.4 Rockwool as a Substrate for Greenhouse Crops

C. SONNEVELD[1]

1 Introduction

In the Scandinavian countries, the first crops were grown on rockwool in the beginning of the 1970's. From 1975, rockwool was used by Dutch growers, and soon a growing system was developed for fruit vegetable crops, like tomato, cucumber, sweet pepper, and eggplant. Later, systems were developed also for cut flowers, like gerbera, rose, and carnation. In the meantime systems came into being for plant propagation, rooting of cuttings, and growing of container plants. For these purposes, propagating cubes, plugs, and granulated rockwool were produced.

At present over 3000 ha of greenhouse crops are grown on rockwool slabs in the Netherlands. The reason that within a few years Dutch growers changed from soil growing to growing on rockwool lies in the very good results obtained with the latter growing system. Growing on rockwool has proved to be a safe, practicable, economic, and high-yielding system.

In this chapter, the system will be described as developed in the Dutch horticulture industry. The contents of this chapter are based on a lecture delivered at the Symposium on High Technology in Protected Cultivation, Tokyo, 1988 (Sonneveld 1988).

2 Rockwool

Rockwool is mostly made of diabase and limestone. These raw materials are melted with coke at a temperature of 1600 °C. The liquid material is poured onto a fast-running disc in a spinning chamber to form fibers. The fibers are supplied with wetting agent, pressed to a certain density, and stabilized with a special binder to slabs. The slabs have a stable structure, are very porous, and have a high water retention capacity. The water retention capacity is determined by the thickness of the fibers, the density of the fibers in the slabs, and the amount of the binder and wetting agent added. Slabs of different qualities are used for horticultural purposes. In Fig. 1 the water retention capacities of two different types of rockwool slabs are shown at pressure head values between 0 and –25 cm. The water retentivity curves show that a great difference in water retention capacity may exist between different rockwool qualities. The wetting curves differ strongly from the drying curves. Under growing conditions the water retention capacity will mostly fluctuate between the two curves.

[1]Glasshouse Crops Research Station, POB 8, 2670 AA Naaldwijk, The Netherlands

Biotechnology in Agriculture and Forestry, Vol. 17
High-Tech and Micropropagation I (ed. by Y.P.S. Bajaj)

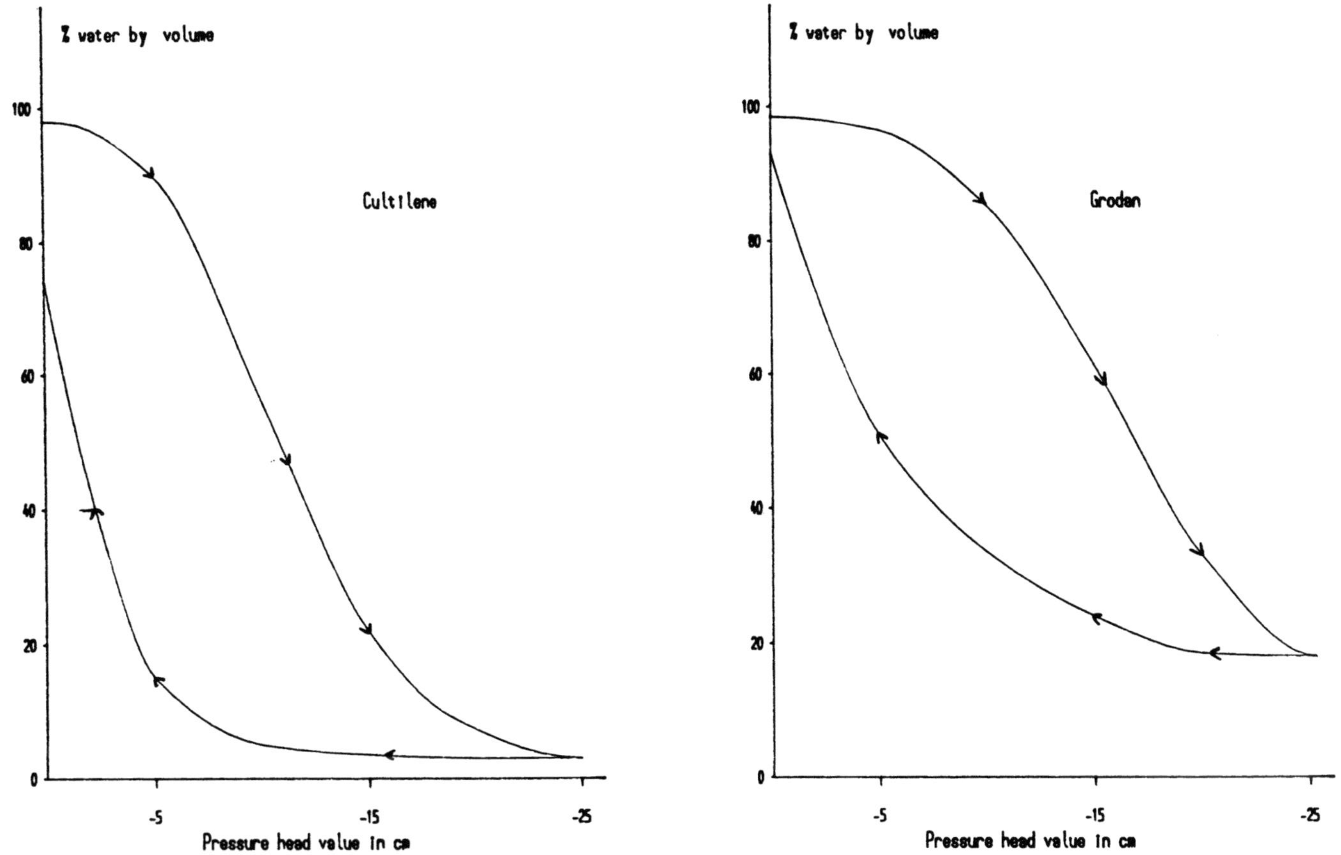

Fig. 1. The water retention curves of two types of rockwool slabs of 7.5 cm thickness. For both trade marks the type for more than 1 year use is shown

3 Growing Systems

The Dutch standard growing system rests on rockwool slabs in bags of plastic film. On top, the bags are provided with planting holes and at the bottom with slits to drain out the surplus water.

The thickness of the slabs varies between 7.5 and 10 cm. Generally, a thickness of 7.5 cm is sufficient. Some crops, however, require a dry surface on top of the slabs to prevent diseases such as phytophthora footrot in gerbera. In such cases a thickness of 10 cm is preferred.

For fruit vegetable crops, often slabs of 30 or 15 cm width are used. In a Dutch greenhouse span of 3.2 m width, such slabs are used for two- or four-plant rows, respectively. In Fig. 2 a four-plant row system is shown. The quantity of rockwool used in these cases is 14 l per m^2 greenhouse.

Flower crops are often grown on beds of 100 cm width (Fig. 3). In such a bed, two strips of rockwool slabs of 30 cm width or four strips of 15 cm width are employed. Two of such beds are laid out in a 3.2 m span. For gerbera, however, three raised beds with two plant rows each are set out in a span (Fig. 2). The rockwool slabs used in this

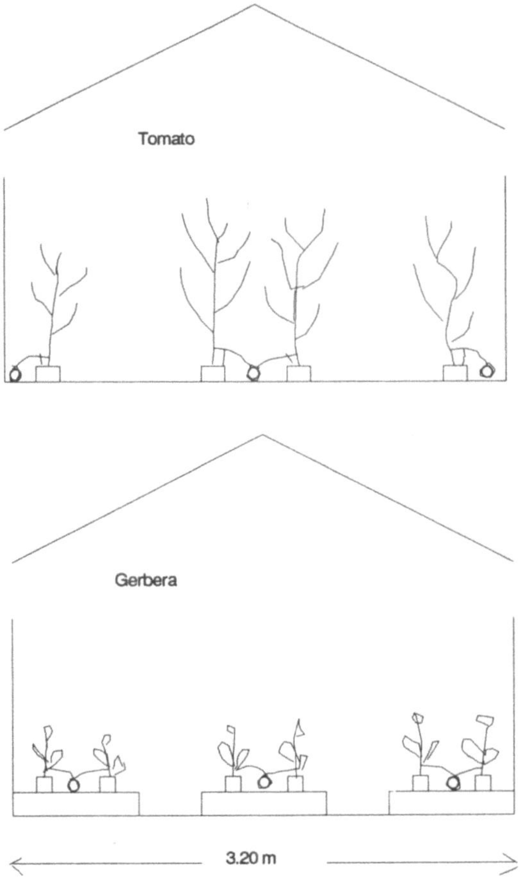

Fig. 2. Rockwool systems as laid out for tomato and gerbera in a Dutch standard greenhouse

Fig. 3. A rose crop grown on rockwool. Two strips of rockwool slabs cover a planting bed of 100 cm width

case are 12.5 cm wide. It will be clear that the quantity of rockwool used per m^2 greenhouse is generally higher for flower crops than for vegetable crops; it amounts to approximately 25 l per m^2. These high quantities are a result of the higher plant densities.

The young plants are propagated in rockwool cubes or plugs, and placed on the top of the slabs (Fig. 4) or in planting holes in the slabs, respectively. The water is given by means of a trickle irrigation equipment. Preferably, each plant is provided with a dripper. At most, eight drippers per m^2 are installed. If more than eight plants are present, one dripper for two or four plants is used.

In the years past, especially with fruit vegetable crops, many experiments were carried out in which comparisons were made between different growing systems (Sonneveld and Welles 1984). Results of these experiments are summarized in Table 1.

It has been found that it is possible to reduce the quantity of rockwool to 10 l per m^2. A further reduction tends to result in a lower yield. The gutter systems mentioned in

Fig. 4. Tomato plants propagated in rockwool cubes placed on top of rockwool slabs

Table 1. Yields of fruit vegetable crops in experiments with different growing systems. The yields are expressed in percentages of the yield of the standard growing system: drained slabs providing 14 l rockwool/m^2 greenhouse

Systems compared with standard system[a]	Number of experiments	Yields in percentages
Standard system		
14 l rockwool/m^2	–	100
10.5 l rockwool/m^2	11	100
7 l rockwool/m^2	10	96
Gutter system		
7–10 l rockwool/m^2	13	98
2.7–4.5 l rockwool/m^2	8	94
Standard system		
Steam-sterilized		
Re-used slabs	16	102

[a]For description of the systems see text.

Table 1 employed a water layer of 1 to 2 cm thickness in which rockwool strips (7–10 l/m^2) or cubes (2.7–4.5 l/m^2) were placed. A strong reduction of the quantity of rockwool reduces yield. Re-use of rockwool slabs after steam sterilization gave good results. Generally, the slabs are not used for more than 3 years. By then the thickness has been reduced too much. A minimum of 6.5 cm is required.

4 Root Zone Temperature

Experiments with root zone warming with rockwool-grown tomatoes mostly did not affect the yield of this crop (Sonneveld and Welles 1984). Therefore the majority of the growers work with ambient temperatures. Results of experiments with root zone warming reported by other workers often show highest yields at temperatures in the root zone of 24–25 °C (Gosselin and Trudel 1982, 1983; Hurd and Graves 1985). However, sometimes the results show discrepancies (Moorby and Graves 1980; Chong and Ito 1982). Probably the different reaction can be explained by differences in air temperature in the various experiments, as was found by Maher (1978), favorable effects of root zone warming can be expected especially at suboptimal air temperatures (Went 1944). Sometimes root zone warming lessened the occurrence of magnesium deficiency, which may be explained by a better uptake of magnesium (Kabu and Toop 1970; Sonneveld 1987).

For cucumbers and sweet peppers root zone warming is normal practice in the Netherlands. Temperatures of 22–24 °C in the root zone seem to be optimal. High temperatures revealed negative effects on yield (Sonneveld and Welles 1984).

The root zone warming is realized by an alkathene tube under the rockwool slabs.

5 Water Quality

The requirements for the quality of the irrigation water depend strongly on the leaching fraction accepted. If this fraction is zero, as in recirculation systems, salt contents of the irrigation water should not exceed the contents absorbed by the crop. Guide values for such contents are listed in the fourth column of Table 2. In the third column, values are given for a leaching fraction of 25%. For bicarbonate, high values can be accepted, because the presence of this ion can be controlled through addition of acid. The guide values for iron rest on the susceptibility of the trickle irrigation equipment for blocking.

Table 2. Guide values for salt contents of irrigation water for rockwool systems. For the drainage systems a leaching fraction of 25% is taken into account

Determination	Units	Drainage systems	Recirculation systems
Na	mmol/l	< 2.5	< 0.5
Cl	mmol/l	< 3.0	< 1.0
Ca	mmol/l	< 4.0	< 2.0
Mg	mmol/l	< 2.0	< 1.0
SO_4	mmol/l	< 2.0	< 1.0
HCO_3	mmol/l	< 10.0	< 10.0
EC	dS/m (25°C)	< 1.5	< 1.0
Fe	µmol/l	< 10.0	< 10.0
Mn	µmol/l	< 10.0	< 10.0
B	µmol/l	< 40.0	< 20.0

Iron easily precipitates in the small pores of the emitters. Manganese and boron may be toxic at high concentrations.

Besides the soluble solids, dust and suspended and colloidal material, such as clay, humic acids and silicon compounds, may also cause problems by precipitation in the emitters. Therefore filtration of the irrigation water is very important. In the Netherlands sand filtration is popular. However, it cannot prevent all precipitations in the emitters. Soluble humic substances pass the filters but may precipitate later gradually after a change of the pH or the salt content. Another problem is the formation of bacterial slimes in the transport lines.

6 Water Supply

In rockwool systems, pure water is never supplied to the substrate, but it is always mixed with nutrients. After placement of the slabs in the greenhouse, they have to be thoroughly soaked with nutrient solutions, as they are a little hydrophobic. If the slabs are wrapped in bags of plastic film, at the start only the planting holes on top of them are made and not yet the drainage holes. Then the slabs are totally soaked with nutrient solution by means of the trickle irrigation system or by hose. The drainage slits are made a couple of days after planting. If the slabs are not wrapped in plastic film, soaking with nutrient solution is only possible by hose.

The quantity of nutrient solution to be given during the growing period depends on the transpiration of the crop grown and on the leaching fraction needed. The transpiration in greenhouses is mainly determined by the global radiation, the energy introduced into the greenhouse by the heating system, and the plant size (De Graaf 1988). For tomato the relationship between the transpiration calculated with the parameters mentioned and the water consumption measured is shown in Fig. 5. On average, the transpiration was highest in the afternoon, as is shown in the figure. In another investigation, it was found that transpiration of individual plants may differ strongly (Van der Burg and Hamaker 1987), as shown for a casual day (Fig. 6). The water absorption per m^2 greenhouse, covered by two tomato plants, varied in this case between 3.2 and 4.6 mm. This means that the uptake of individual plants varies between 1.5 and 2.4 l.

Another factor that causes variability is the water supply of the emitters of the trickle irrigation system. This is especially the case if the system is polluted. In Fig. 7 the variability of the water supply is shown for the same situation as the transpiration variability in Fig. 6. The water supply for 1 m^2 (two emitters) varied between 5.2 and 6.8 m^2, which means that the supply per emitter varied between 2.4 and 3.6 l. One thing and another leads to an excessive supply of water, which excess will drain out. In practice the amount of drainage water soon reaches 20–30% of the supply.

With the drainage water, a substantial part of the plant nutrients supplied is washed out, which pollutes the environment. A good solution for this problem would be recirculation of the nutrient solution. However, recirculation has the risk of an outbreak of plant diseases in the whole system, because bacteria, fungi, and viruses in the nutrient solution are also recirculated; for this reason recirculation is considered too risky.

Research is being carried out to avoid the risk of such outbreak of pathogens. To this purpose, the rockwool slabs are placed in gutters with a slope of 0.3% and covered

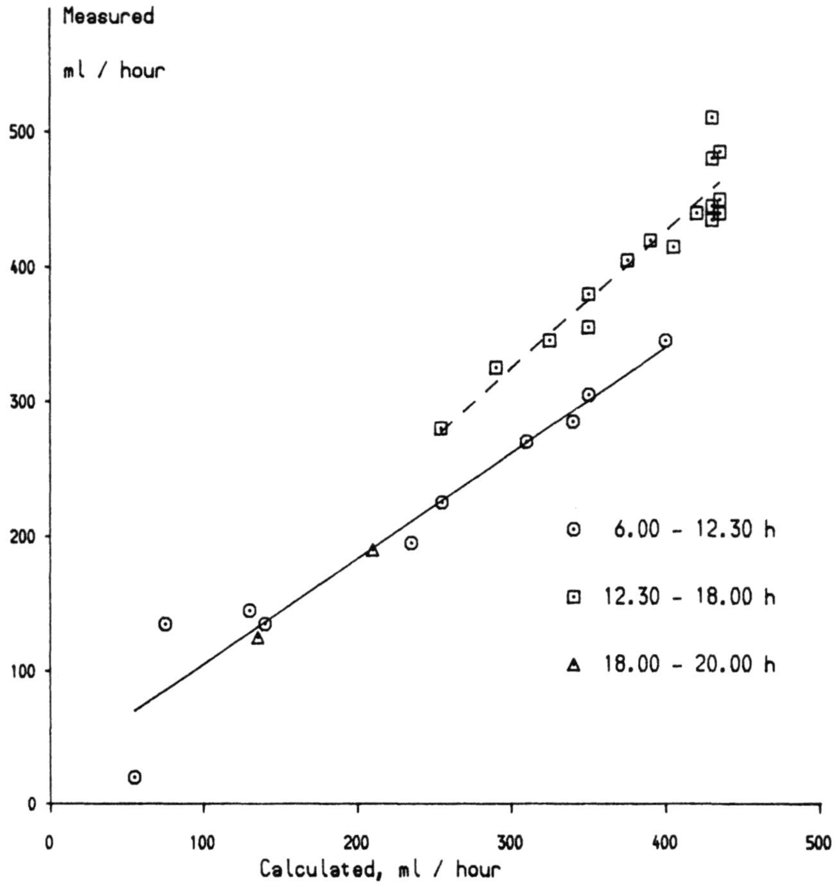

Fig. 5. Relationships between water consumption calculated by the method of De Graaf (in press) for a tomato crop on a summer day and the water absorption measured (De Graaf 1988)

with "black and white" plastic film. Each 2–4 m, a space is left between the slabs to prevent water transport throughout the strips of rockwool slabs. At the lower end of the gutter, the drainage water is gathered. Heating of the drainage water proved to be a safe and practicable sterilization method. A temperature of 95 °C or more during 30 s seems to be sufficient to kill all plant pathogens (Runia et al. 1988).

The difference between re-use of drainage water in a rockwool system and a true recirculation system like the nutrient film technique (Graves 1983) is the availability of a substrate in the root environment. Rockwool has sufficient air space in the root environment to supply the roots with oxygen. Such is not the case in systems without a substrate, in which all the oxygen necessary for the plant has to be supplied by aeration and recirculation of the nutrient solution. Thus, in a nutrient film technique system the quantity of water recirculated may reach 2000 m^3/ha/day, while the quantity of drainage water will mostly not exceed 20 m^3/ha/day. It will be clear that sterilization is economically impossible for systems with a high recirculation rate.

The low recirculation rate of the nutrient solution in a rockwool system requires the use of trickle irrigation systems. A central water supply at the high end of the gutter

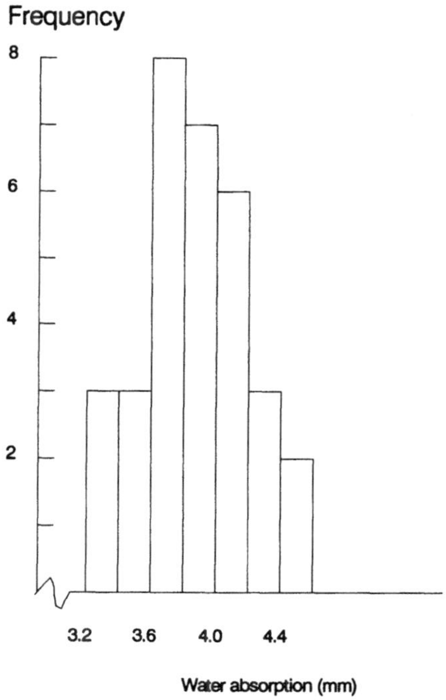

Fig. 6. Frequency diagram of the water absorption of tomato plants on 32 locations in a greenhouse during a summer day. A location consisted of 1 m^2 grown with two plants each supplied with a dripper

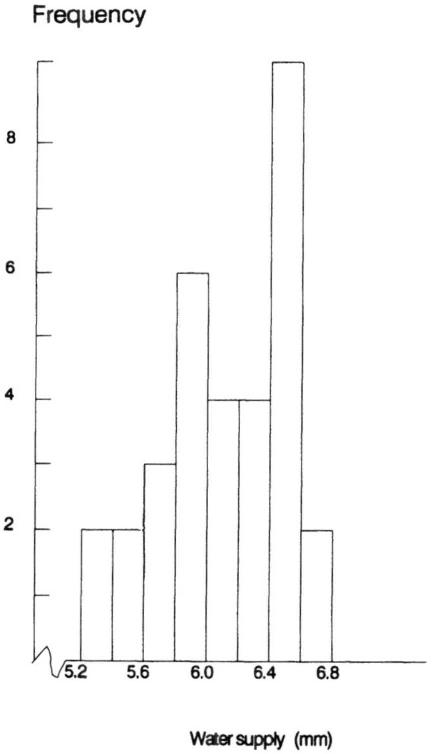

Fig. 7. Frequency diagram of the water supply on 32 locations in a greenhouse. For other conditions see Fig. 6

as in a nutrient film technique system is insufficient for an even distribution of the plant nutrients over the length of the gutters.

Two types of systems developed to automatize the water supply can be distinguished. One is based on measurement of the water absorption by plants and the other on assessment of the transpiration. This assessment is based on data of global radiation, energy supply by the heating system, and plant size. For both systems it is advisable to measure the quantity of drainage water, which quantity can be used as a feedback for readjustment of the water supply. The quantity of drainage water, expressed as a percentage of the water supply, is used as a setpoint in such systems.

A strong drying-up of the rockwool slabs during the growing period has to be prevented, as it is quite difficult to wet such slabs again. This is shown in Fig. 8. After a drying-up, the water content does not reach the original level. The recovery of the water retention capacity depends on the amount of drying up, as is shown in Fig. 9. A complete recovery of the water absorption after a drying-up can only be attained after soaking.

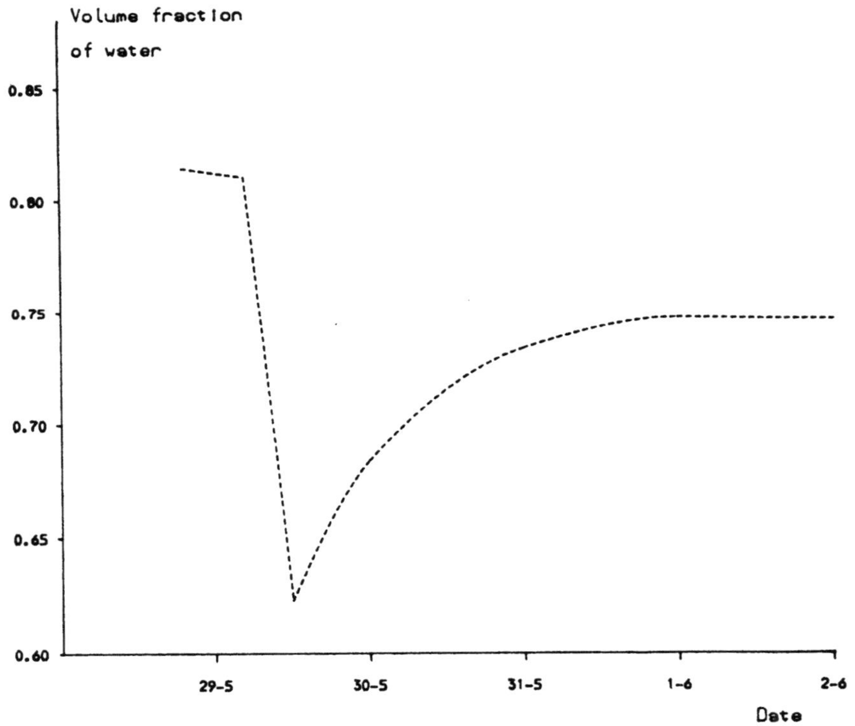

Fig. 8. The volume fraction of water in a rockwool slab after a drying-up and the recovery of the water retention capacity after irrigation in a cucumber crop

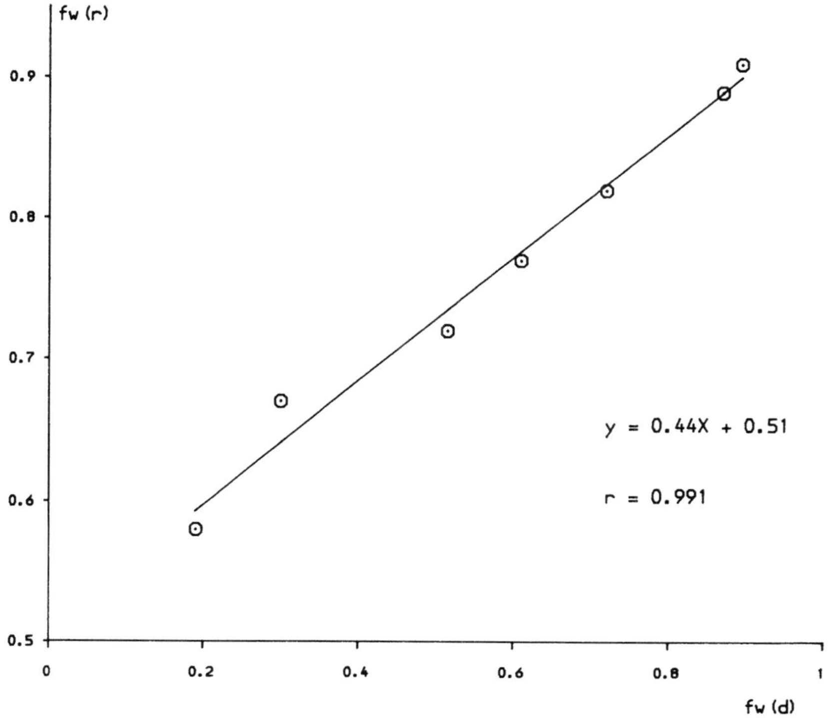

Fig. 9. The relationship between the volume fraction water of rockwool after different levels of drying-up [$f_w(d)$] and after re-wetting with trickle irrigation [$f_w(r)$]

7 Other Rockwool Systems

In the propagation of vegetable plants the seed is sometimes sown in rockwool plugs (Fig. 10) which fit well in holes of the propagating cubes (Fig. 11). When the plants in the plugs reach a manageable size, they are transplanted into the cubes. In other cases the seed is sown in a loose mineral substrate and transplanted with bare roots in the hole of the cube. Then the planting hole is filled with vermiculite to support the seedling. Before planting, the cubes have to be soaked thoroughly with nutrient solution.

The cubes are mostly placed on a concrete floor covered with white plastic film (Fig. 12). Depending on the need, the plants are supplied with nutrient solution, generally by means of a sprinkler system.

During propagation it is very important to maintain a sufficiently high temperature in the root environment, so the concrete floor is equipped with a heating system. In the cubes temperatures between 21 and 24 °C are desirable. Too low temperatures easily induce phythium in young plants.

Cuttings of roses and carnations are stuck in cubes or plugs for rooting and after obtaining the required size transplanted on slabs.

Anthurium and *Cymbidium* are often grown in containers of 5 to 15 l filled with granulated rockwool. The container is supplied with one or two drippers. Dependent on the requirements of the crops, mixtures of hydrophobic and hydrophilic materials are used.

Fig. 10. Young tomato plants grown from seed sown in rock-wool plugs

Fig. 11. A gerbera plant grown in a plug fitting well in the hole of a propagating cube

Fig. 12. Young eggplants placed on a concrete floor covered with white plastic film

8 pH

The pH in the rockwool slabs is maintained at a level between 5.5 and 6.0 for most crops. However, some crops show special needs. Gerbera for example reacts favorably to somewhat lower pH values.

The pH in the rockwool slabs is controlled by adjustment of the pH of the nutrient solution added. The pH values in the nutrient solution are never below 5, otherwise the fibers of the slabs will be affected. pH values in the nutrient solution higher than 6.2 are not advisable because of the precipitation of phosphate, which will easily block the trickle irrigation system. High pH values can also be controlled by addition of ammonium to the nutrient solution. Low pH values in the rockwool slabs can be adjusted by addition of potassium bicarbonate. This salt is not added to the irrigation water together with the nutrient solution, because then the pH of the solution becomes too high and precipitation of phosphates and carbonates will occur. Therefore the potassium bicarbonate is added separately to a restricted amount of water.

9 Electrical Conductivity

Electrical conductivity (EC) values in the root environment between 1.5 and 2 are sufficient to supply rockwool-grown crops with nutrients. However, higher values improve the quality of the produce of fruit vegetable crops. For tomatoes, for example, it

Table 3. The effect of two EC values in the root environment on yield and quality of tomatoes. (Sonneveld and Welles 1988)

Characteristics	EC 2.6	EC 3.5
Number of fruits/m^2	224	222
Fruit yield kg/m^2	12.7	11.9
Average fruit weight g	56	54
Fruit shape index	6.4	6.6
Coloring in days	4.4	4.1
Shelflife in days	17.5	19.2
EC fruit sap dS/m	5.8	6.2
Acid in fruit sap mmol/l	75	84
Refraction fruit sap % Brix	4.8	5.0

was shown that a higher EC value prolongs the shelf life of the fruits, and increases their acid content (see Table 3). Too high EC values reduce yields.

The schematic relationship between the EC in the root environment and the yield of crops is shown in Fig. 13. The curve shown in this figure is determined by the values of n, t, and b. EC values in the root environment in the domain O-n are too low for a sufficient supply of the crop with nutrients, and no maximal yields will be acquired. In the domain n-t, the nutrient supply is sufficient and maximal yields may be obtained. The value t is equivalent to the salinity threshold value (Maas and Hoffman 1977). So values in the root environment above t cause a more or less linear decrease of the yield with increasing EC value. b is the percentage yield decrease per dS/m (25 °C). Values for n, t, and b depend on the crop and the growing conditions. For rockwool-grown tomatoes values of n = 1.5, t = 2.5, and b = 7% were found (Sonneveld and Welles 1988).

In the Netherlands, for fruit vegetable crops EC values around 3 are maintained. However, young plants under poor light conditions are grown at values between 5 and 8, with a view to prevent too lush a growth and to improve fruit setting. For flower crops values around 2 are recommended. In experiments higher values did not improve the quality of the flowers (Van Os and De Kreij 1987; Verdegaal 1987).

10 Nutrients

The relationship between element contents in the root environment and those in plant tissues is often linear. However, such a linear model is only correct in a certain domain, for the functions adjusted generally show an intercept of considerable value. Negative as well as positive intercepts occur. Negative values indicate that the true model for the relationships is concave and positive values indicate convex relationships. This is made clear in Fig. 14. For potassium, phosphate, nitrate, and microelements convex and for calcium and magnesium concave relationships have often been found. Examples of such relationships are shown in Fig. 15.

The values of the parameters of functions adjusted may differ strongly for crops, as is shown for magnesium in Fig. 16 (Sonneveld and Voogt 1985b). Furthermore, an-

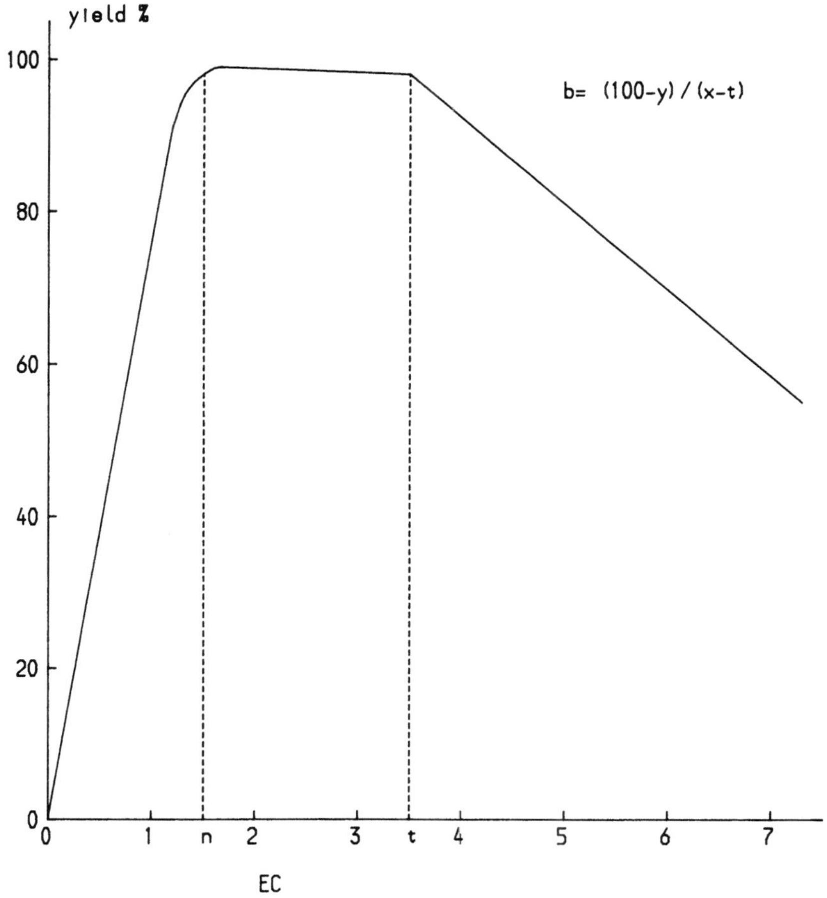

Fig. 13. The relationship between the EC value in the root environment and the yield of crops. For explanation see text

tagonism between ions may affect the parameter values, as is apparent from Fig. 16 for calcium, which reduces the magnesium uptake. The cultivar type and the climatic conditions may also affect the nutrient uptake. This is shown for phosphate uptake of cucumbers in Table 4. Under low light conditions (February) the uptake is higher for the cultivar Millio than for the cultivar Lucinde.

With respect to phosphate, the supply of this element has a special aspect, as the phosphate concentration in the root environment in soilless cultures is high in comparison with the concentration in soil solutions. This high concentration results from the fact that in soilless cultures phosphate is not buffered as is the case in soils. Therefore, the phosphate contents of rockwool-grown crops are often higher than those of soil-grown crops (Table 5). Excessive uptake of phosphate in hydroponic systems has been found for powdery mildew-resistant cucumber cultivars grown under poor light conditions (Zijlstra et al. 1987). The high phosphate uptake induced chlorosis and necrosis. Symptoms of phosphate toxicity are known for other crops also (Howell and Bernard 1961). They are mostly restricted to some cultivars.

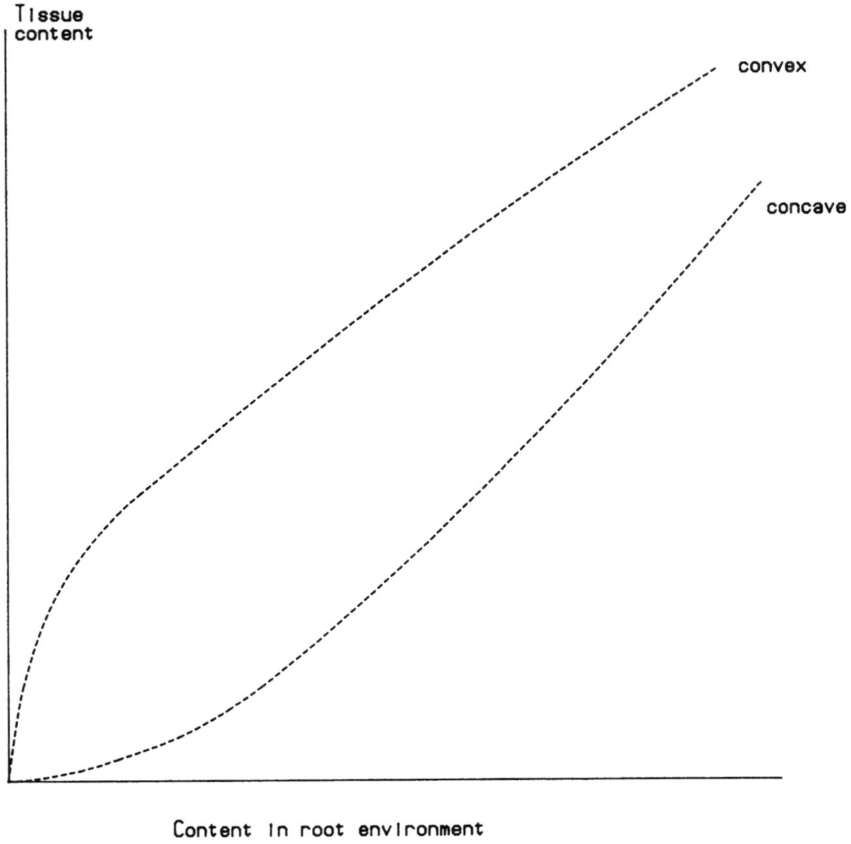

Fig. 14. Models for relationships between element contents of nutrient solutions in rockwool slabs and those of plant tissues. *Convex* if a positive and *concave* if a negative intercept is found with linear regression

In Fig. 17 the relationship is shown between phosphate supply and phosphate content in the rockwool slabs, fruit yield of cucumbers and phosphate content of the leaves. Maximum yield is already obtained with a supply of 1 mmol/l, a content of 0.4 mmol/l in the rockwool slabs and a content in the leaves of 200 mmol per kg dry matter.

High phosphate supplies can strongly increase the phosphate content of plants, which besides the toxicity mentioned, may also hinder the uptake of some other elements. Such hindrance is found for magnesium (Hipp and Gerard 1969), zinc (Royle 1985), and iron (De Kock et al. 1979).

Ammonium is supplied to control the pH value in the root environment. High concentrations of this cation in the nutrient solution may be toxic, especially at high pH values (Bennett 1974), and may disturb the uptake of other cations. For most crops, especially calcium and magnesium uptake is hindered by ammonium application (Barker and Mills 1980).

Small quantities of ammonium nitrogen may be favorable for crops, which is especially the case for crops sensitive to manganese or iron chlorosis. In experiments,

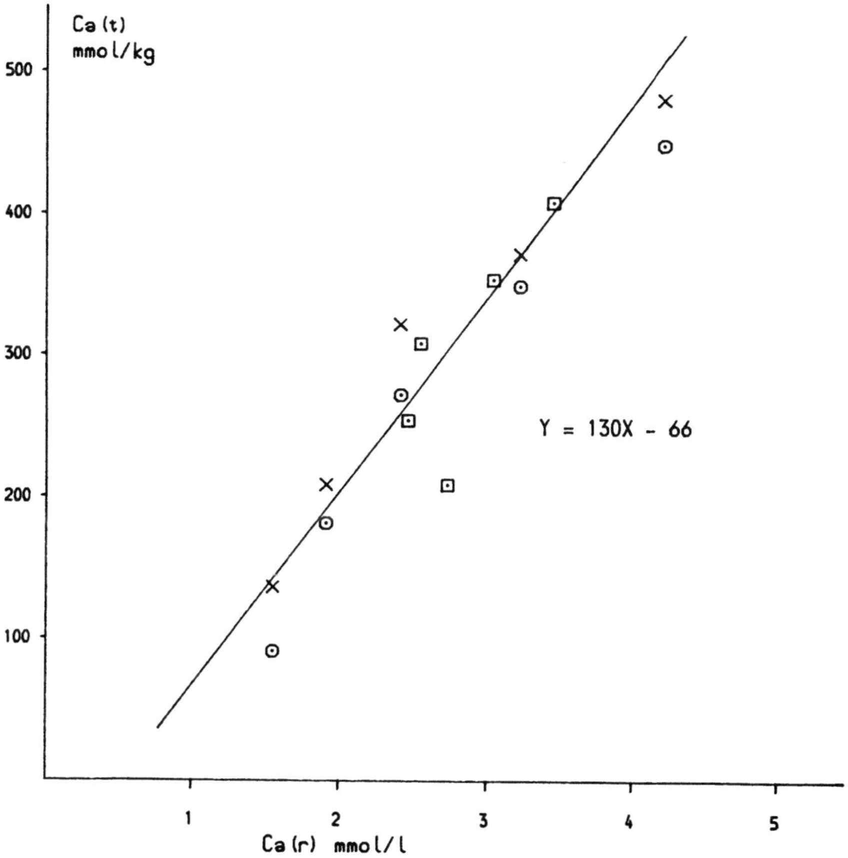

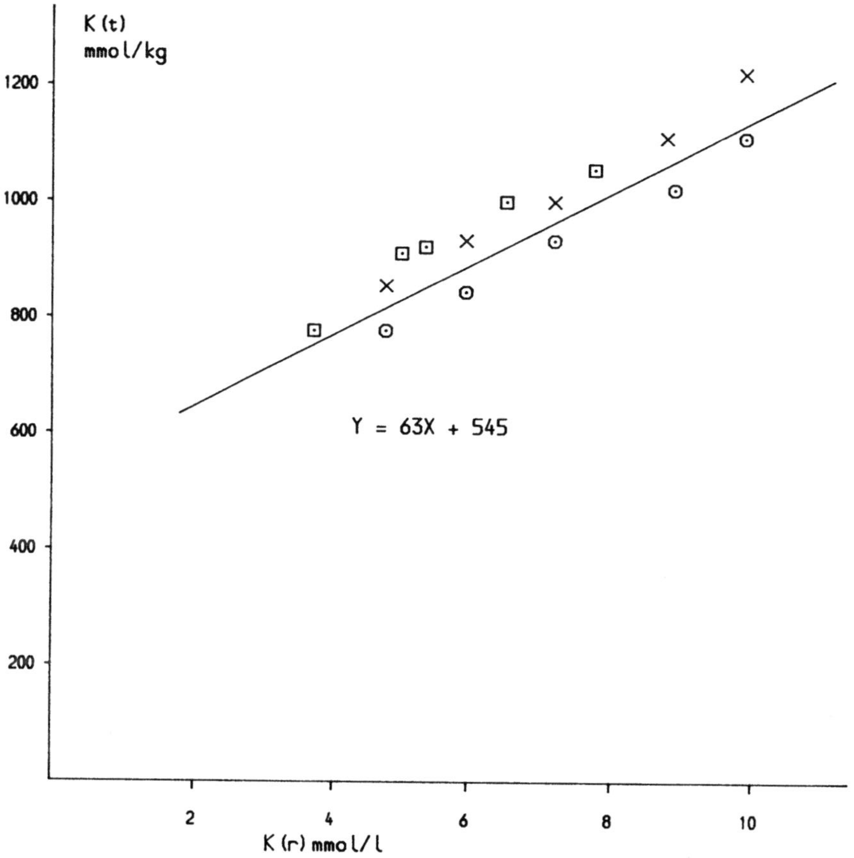

Fig. 15. The relationships between potassium (*left*) and calcium (*right*) contents in the root environment (*r*) on the one hand, and in the leaves of carnations (*t*) on the other hand. Legend ⊡ cv. Westpink, grown 1982; × cv. Westpink, grown 1983; ⊙ cv. Ministar, grown 1983. (Sonneveld and Voogt 1986)

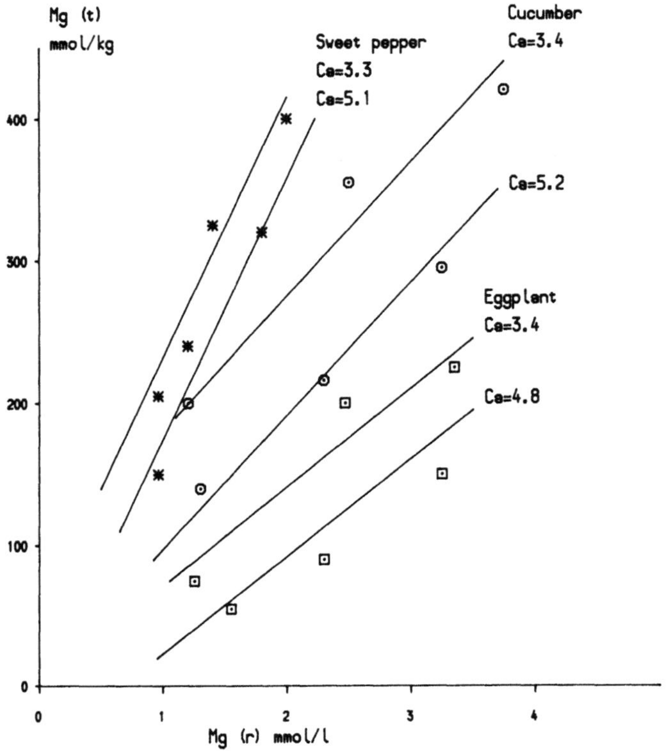

Fig. 16. The relationship of magnesium content in the root environment (*r*) and magnesium content of leaves of different crops (t) at two calcium levels in the root environment. (Sonneveld and Voogt 1985b)

Table 4. Phosphate contents of young fully grown leaves of two cucumber cultivars as affected by phosphate supply and growing period. Contents expressed as mmol/kg dry matter

Cultivar	Low P supply		High P supply	
	Sampling date		Sampling date	
	Feb.	Jun.	Feb.	Jun.
Lucinde	166	135	312	296
Millio	196	140	382	300

control of chlorosis and sometimes also higher yields were found. In Table 6 effects of ammonium on chlorosis and on yield are shown for tomatoes and cucumbers. Effects of ammonium on the course of the pH in the root environment is shown in Fig. 18 for the cucumber crop. For gerbera it was found that adjustment of the pH in the rockwool slabs through ammonium additions may increase the flower production with 15% (Voogt 1988b).

Studies on uptake of microelements from nutrient solutions in rockwool slabs showed great differences for elements and crops. Data published for cucumbers show

Table 5. Phosphate contents of young fully grown leaves of tomatoes and cucumbers grown in rockwool or soil. The contents are expressed as mmol/kg dry material

Crop	Rockwool-grown	Soil-grown
Tomato	174	145
Cucumber	210	161

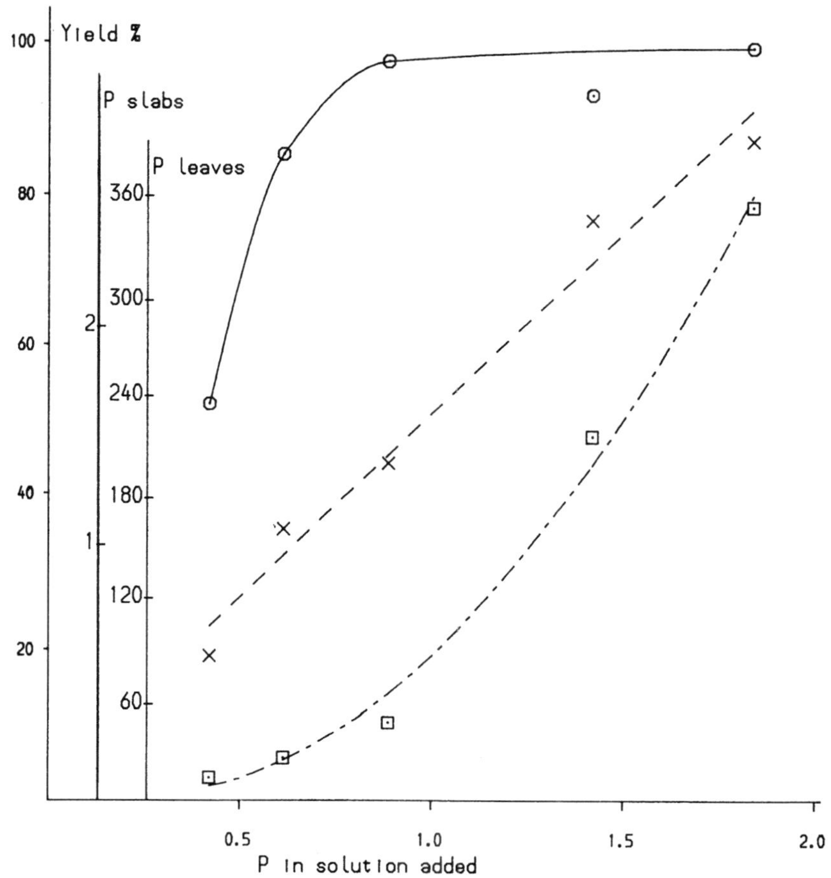

Fig. 17. Effects of phosphate supply (mmol/l) in the nutrient solution for rockwool-grown cucumbers in a recirculation system. The effects on yield (⊙), phosphate content (mmol per kg dry matter) of young fully grown leaves (×) and phosphate contents (mmol/l) of the nutrient solution in the slabs (⊡) are shown

close relationships between contents available in the root environment and contents in plant tissues. An exponential function of the model $y = a\ x^b$ fitted very well for all microelements, except iron (Sonneveld and De Bes 1984). In this function model, x represents the element content in the root environment and y the content in the plant tissue. If b approximates 1, the function becomes linear. For cucumber, however, all b-values were below 1, which means that the relationships found were convex.

Table 6. The effect of different $NH_4/NH_4 + NO_3$ ratios on yield and chlorosis of tomato and cucumber

Cucumber			Tomato		
$\frac{NH_4}{NH_4 + NO_3}$	Chlorosis index[a]	Yield kg/m^2	$\frac{NH_4}{NH_4 + NO_3}$	Chlorosis index	Yield kg/m^2
0	1.7	13.1	0	2.3	17.7
0.08	0.6	14.2	0.06	1.4	17.3
0.15	0.4	16.5	0.12	1.1	18.5

[a]Index of chlorosis: 0 = no symptoms, and 3 = severe symptoms.

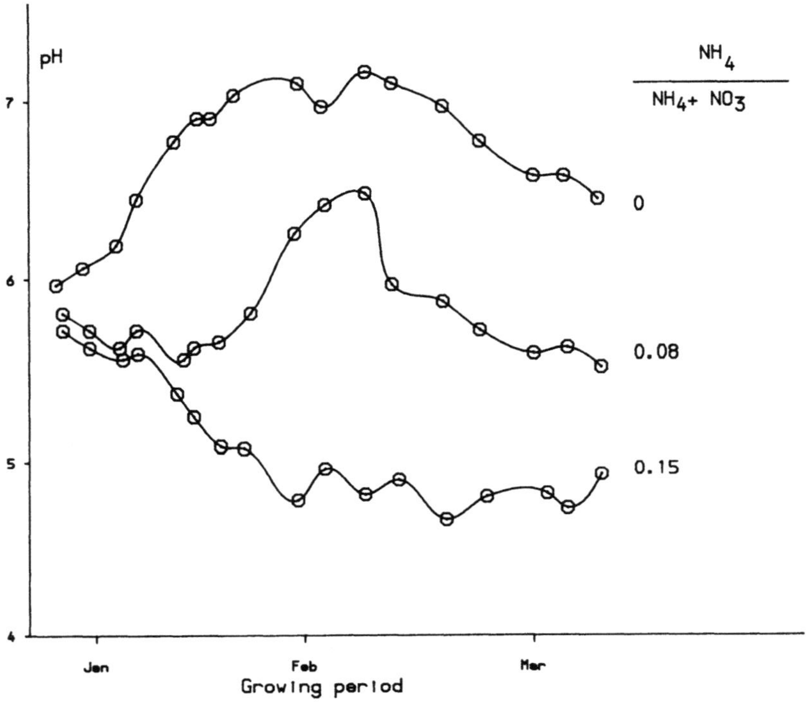

Fig. 18. The effect of the addition of ammonium to the nutrient solution on the pH in the root environment of a rockwool-grown cucumber crop

Differences in uptake between crops are shown in Fig. 19 for the element boron. In the same figure, a great difference in translocation of boron in the plant is shown. In comparison with contents in the leaves, fruit contents are mostly very low. The translocation of minor elements in the plants depends strongly on the element and on the level of supply. An example is given in Table 7 for boron and zinc. With a low supply, the boron content is highest in the petioles and with a high supply in the laminae. For zinc it is the other way round.

Manganese can soon achieve a level toxic to sensitive crops. For such crops a great difference in sensitivity to manganese is mostly found between cultivars (Broer and

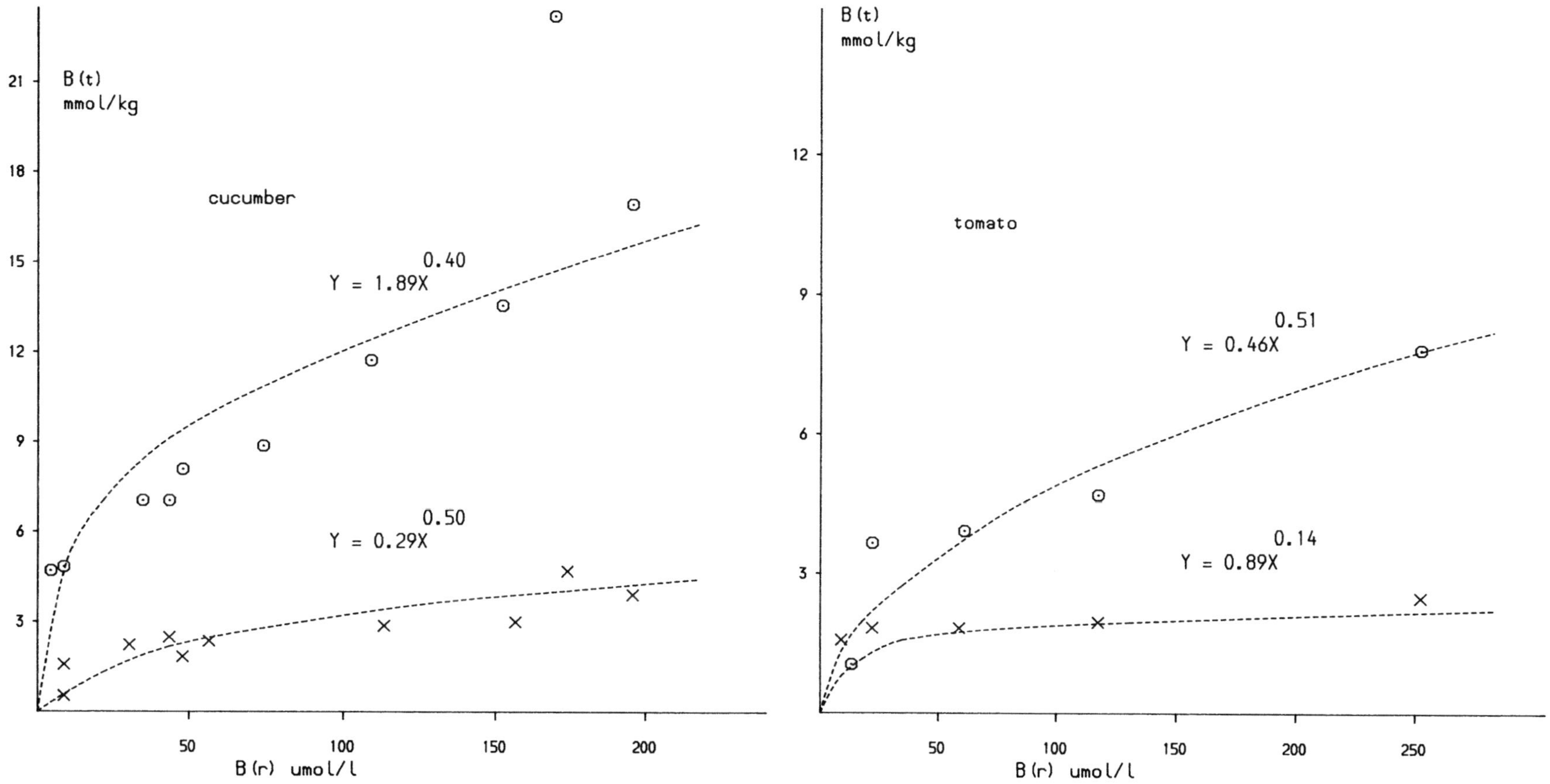

Fig. 19. The relationships between boron contents in the root environment (*r*) and in plant tissues (*t*) of cucumber (*left*) and tomato (*right*). ⊙ – young leaves; × – fruits

Table 7. Translocation of boron and zinc in tomato leaves at a low and at a high supply. The contents are expressed as mmol/kg dry matter

Element	Supply	Laminae	Petioles
B	Low	1.03	2.00
B	High	7.70	2.84
Zn	Low	0.22	0.16
Zn	High	0.38	2.29

Sonneveld 1973; Sonneveld and Voogt 1975). Then the standard supply of 10 μmol manganese per 1 nutrient solution must be reduced to 5 or sometimes even to 0 to prevent toxicity. When the supply must be 0, the manganese liberated from the rockwool fibers and that originating from pollution appear to be sufficient for the crop (Sonneveld and De Bes 1984).

In rockwool systems, for most crops the need for iron addition is not as great as in full hydroponic systems like NFT (Sonneveld and Voogt 1985a). At low pH values some crops can absorb sufficient iron from the slabs themselves. However, a supply of 10–15 μmol per 1 nutrient solution seems to be safer, especially at moderate and higher pH values. For crops sensitive to iron chlorosis, a supply of 20–30 μmol per 1 is advisable.

For rockwool systems, the iron is mostly supplied as Fe-DTPA. Fe-EDTA is also usable (Sonneveld and Voogt 1985a). However, iron chelated as DTPA is stable at higher pH values than that chelated as EDTA (Lindsay et al. 1967), and in rockwool slabs the pH value may at times go up without the possibility of a rapid control. For this reason iron chelate stable at a somewhat higher pH value is preferred.

11 Handling of Plant Nutrition

On the basis of experimental results as discussed in the preceding section, nutrient solutions for different crops have been developed together with guide values for the solutions in the rockwool slabs (Sonneveld and Straver 1989). The nutrient solutions are given as ionic compositions to be supplied to the irrigation water if no ions are available in the raw water. Guide values for the solution in the rockwool slabs generally differ from the composition of the nutrient solution supplied. This difference is necessary as the rate of uptake between different ions varies strongly. In Table 8 an example is given of the compositions of both solutions, that supplied, and that in the rockwool slabs.

The compositions of the nutrient solution developed reflect only the average quantities of ions to be added. Adjustments are necessary depending on growing conditions and plant stage. For example, with a high plant transpiration a lower concentration (EC) is desirable to realize a given concentration in the root environment than with a low plant transpiration. For fruit vegetable crops, the supply of potassium and calcium must be adjusted to the growth stage, because of the strong fluctuations in the ratio in which these elements are absorbed. This is clearly shown by Voogt (1988a). He calcu-

Table 8. The composition of the nutrient solution supplied to a rockwool grown tomato crop and the guide values for the solution in the rockwool slabs

Parameter	Supply	Root environment
EC dS/m (25 °C)	2.3	3.0
NO_3 mmol/l	13.75	17.0
H_2PO_4	1.25	0.7
SO_4	3.75	5.0
NH_4	1.25	< 0.5
K	8.75	7.0
Ca	4.25	7.0
Mg	2.0	3.5
Fe µmol/l	15	15
Mn	10	7
Zn	5	7
B	30	50
Cu	0.75	0.7
Mo	0.5	–

lated that the ratio between the uptake of potassium to that of calcium for a tomato crop varied with time between 2.2 and 7.5. On average the ratio was 3.7. In Fig. 20 the course of the ratio during the growing season is shown

A comparison of the composition of the nutrient solution supplied with the guide values of the solution in the rockwool slabs shows that the ratios between macroelements differ for both solutions. Generally the concentrations of the divalent ions are relatively high in the root environment, while those of the monovalent ions are relatively low. This difference has to do with a difference in rate of uptake between mono- and divalent ions. For the microelements also great differences exist between both solutions. Manganese, for example, is absorbed quite easily by crops, while for boron high contents in the root environment are necessary to serve the crops sufficiently with this element. For rose crops, however, high boron contents in the root environment soon lead to toxicity symptoms (De Kreij et al. 1985).

If the raw water contains sodium and chloride, special attention has to be paid to the accumulation of these ions in the root environment. Generally, sodium accumulates faster than chloride. This is caused by the fact that most crops absorb more chloride than sodium. Thus, the sodium content of the nutrient solution is an indicator for the requirement of leaching.

It was found by Adams (1987), Van der Burg and Sonneveld (1988), and Van der Burg et al. (1987) that with an adequate management of the nutrient supply some accumulation of sodium and chloride can be accepted without appearance of yield reduction. However the EC value may not exceed the salinity threshold value. Thus, for fruit vegetable crops it has been found that sodium and chloride contents between 6 and 10 mmol/l can be accepted. With such high contents the interpretation of the macronutrient contents has to be adjusted. This adjustment is realized by a proportional reduction of all macronutrient contents except phosphate, as is shown in Table 9.

The composition of the solution in the rockwool slabs is controlled by the grower with the aid of analytical data obtained from different samplings. Two or three times a

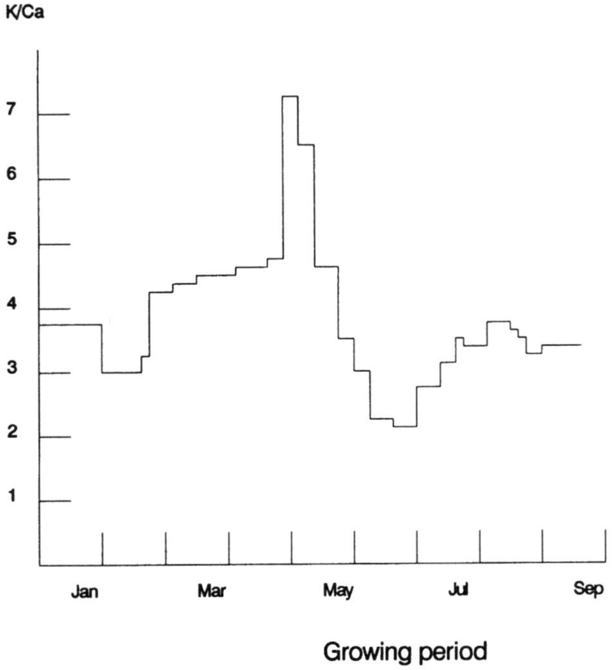

Fig. 20. The course of the ratio between the uptake of potassium and calcium during the growing period of a tomato crop. (Voogt 1988a)

Table 9. Guide values for the solution in the rockwool slabs for a tomato crop for low and high sodium chloride salinity

Parameter	Sodium chloride salinity	
	Low	High
EC dS/m (25 °C)	3.0	3.0
Na mmol/l	2.0	10.0
Cl	2.0	10.0
NO_3	17.0	12.0
H_2PO_4	0.7	0.7
SO_4	5.0	3.5
K	7.0	5.0
Ca	7.0	5.0
Mg	3.5	2.5

week samples are gathered from the slabs in which the grower himself measures the EC- and pH values with portable instruments. Moreover, every fortnight a sample is sent to the laboratory to determine the macroelements. Once a month, in these samples the micro-elements are also determined.

With the aid of the EC- and pH values measured in the rockwool slabs the grower adjusts the EC- and pH values and the ammonium content of the nutrient solution to be added. The analytical data of the laboratory procure the possibility to adjust the ratios between the nutrient elements. For this adjustment a computer program has been developed with which adjusted nutrient solutions are calculated for the growers (see Table 10). Such solutions are valid for 2 weeks only. The computer program is based on the standard nutrient solution for the crop grown, the recommendation for adjustment of the solution, the quality of the water used, the guide values for the nutrient solution in the rockwool slabs, and the data found by the laboratory.

Table 10. An example of computerized advice for tomatoes grown in rockwool, based on analytical data of the nutrient solution in the rockwool slabs

pH 6.3
EC 3.5

Cations (mmol/l)						Anions (mmol/l)				
	NH_4	K	Na	Ca	Mg	NO_3	Cl	SO_4	HCO_3	P
	0.1	11.3	1.6	7.1	2.5	24.3	1.1	2.7	0.1	1.2
		high			low			low		
Micronutrients (μmol/l)										
	Fe	Mn	Zn	B	Cu					
	4.4	7.7	6.6	45	0.9					
	low			low						

Figures not indicated with 'low' or 'high' are normal for a tomato crop. The advice is based on the use of rainwater. The quantities of fertilizers are expressed per m^3 stock solution (100 times concentrated)

Solution A	Standard	Adjustment	Advice	Unit
Calcium nitrate	92.5	0.0	92.5	kg
Ammonium nitrate	4.0	0.0	4.0	kg
Potassium nitrate	13.4	– 1.4	12.0	kg
Iron chelate (6%)	1400	700	2100	g
Solution B				
Potassium nitrate	24.5	–13.7	10.8	kg
Mono potassium phosphate	20.4	0.0	20.4	kg
Epsom salt	43.1	12.3	55.4	kg
Potassium sulfate	34.9	0.0	34.9	kg
Manganese sulfate	170	0	170	g
Zinc sulfate	145	0	145	g
Borax	240	60	300	g
Copper sulfate	19	0	19	g
Sodium molybdate	12	0	12	g

Adjustments must not be maintained longer than 2 weeks. Dependent on the growing stage of the crop, extra adjustments may be necessary. Those are not incorporated in the advice. For such adjustments see Volume 3 of the series *Nutrient Solutions for Glasshouse Horticulture.*

Fig. 21. A dilution unit (Vocom, Brinkman)

The fertilizer mixture calculated is distributed over two stock solutions mentioned A and B. To solution A no sulfates and phosphates and to solution B no calcium salts are added to prevent precipitation of calcium sulfate and calcium phosphate in these concentrated solutions. The stock solutions are prepared in such a way that addition of equal quantities of the two solutions to the irrigation water results in the nutrient ratios intended. The stock solutions are added to the water by a dilution unit on basis of the EC value adjusted. Separately, nitric acid or potassium bicarbonate is added to correct too high or too low pH values in the nutrient solution respectively. A picture of a dilution unit is shown in Fig. 21.

12 Conclusions

The success of the Dutch rockwool system in protected cultivation resulted also in a wide acceptance of this growing system in the countries of Western and Eastern Europe. It has also found acceptance on the American and Australian continents and in Japan.

It may be expected that the growing system will extend and that it will be developed further. At the moment in the Netherlands only 4300 ha greenhouse crops are grown on substrates, namely 3300 ha vegetables and cut flowers on rockwool and 1000 ha potplants on peat mixtures. This is 45% of the total greenhouse area, which covers 9500 ha. Thus, in the Netherlands alone, another 5200 ha still may switch to substrate systems. In the next years a great change from soil to rockwool may be expected for some cut flowers, for which usable systems have recently been developed. The efforts to develop an economic system for crops with a high plant density and a short growing period, however, have up till now not been successful. In this field more research has to be done.

One advantage of rockwool culture is that the free drainage system mainly used nowadays can easily be changed into a system in which the drainage water is re-used. Such an adjustment involves a great increase of fertilizer efficiency and a clean production process. The latter is especially important with respect to environmental pollution in densely populated countries like the Netherlands.

Growing on rockwool has good prospects in protected cultivation all over the world. At present, Dutch growers harvest 60 kg of cucumbers, 50 kg of tomatoes, and 20 kg of sweet peppers per m^2 annually, which yields are higher than those obtained with any other growing system. Currently rockwool costs 0.25–0.30 guilder (0.12–0.15 US$) per liter. The costs of the substrate are thus low in comparison with total costs in protected cultivation. The substrate costs can be lowered if the quantity of rockwool used per m^2 greenhouse is reduced, which is possible, as has already been found in research.

References

Adams P (1987) The test of raised salinity. Grower 107; 2:(supplement) 23–27

Barker AV, Mills HA (1980) Ammonium and nitrate nutrition of horticultural crops. Hortic Rev 2:395–423

Bennett AC (1974) Toxic effects of aqueous ammonia, copper, zinc, lead, boron and manganese on root growth. In: Carson EW (ed) The plant root and its environment. University Press of Virginia, Charlottesville, pp 669–683

Broer S, Sonneveld C (1973) Mangaanvergiftiging bij roos. Vakbl Bloemisterij 28; 48:10–11

Chong PC, Ito T (1982) Growth, fruit yield and nutrient absorption of tomato plant as influenced by solution temperature in nutrient film technique. J Jpn Soc Hortic Sci 51(1):44–50

de Graaf R (1988) Automation of the water supply of glasshouse crops by means of calculating the transpiration and measuring the amount of drainage water. Acta Hortic 229:219–231

De Kock PC, Hall A, Inkson RHE (1979) Active iron in plant leaves. Ann Bot 43:737–740

de Kreij C, van den Berg Th, Warmenhoven M (1985) Optimaal barium en zinkgehalte bij "Motrea" in steenwol. Vakbl Bloemisterij 40, 27:34

Gosselin A, Trudel MJ (1982) Influence de la temperature du substrat sur la croissance, le développement et le contenu d'eléments minéraux de plants de tomate (cv. Vendor). Can J Plant Sci 62:751–757

Gosselin A, Trudel MJ (1983) Interactions between air and root temperature on greenhouse tomatoes. I Growth, development, and yield. J Am Soc Hortic Sci 108:901–905

Graves CJ (1983) The nutrient film technique. Hortic Rev 5:1–44

Hipp BW, Gerard CJ (1969) Magnesium-phosphorus interrelationship in tomatoes. Agron J 61:403–405

Howell W, Bernard RL (1961) Phosphorus response of soybean varieties. Crop Sci 1:311–313

Hurd RG, Graves CJ (1985) Some effects of air and root temperature on the yield and quality of glasshouse tomatoes. J Hortic Sci 60:359–371

Kabu KL, Toop EW (1970) Influence of soil temperature and potassium fertilization on magnesium content of tomato plants. Can J Plant Sci 50:740–742

Lindsay WL, Hodgson JF, Norvell WA (1967) The physico-chemical equilibrium of metal chelates in soils and their influence on the availability of micro nutrient cations. Trans Meet Commns II and IV, Int Soc Soil Sci 1966, pp 305–316

Maas EV, Hoffman GJ (1977) Crop salt tolerance – current assessment. J Irrigation and Drainage Div, Proc Amer Soc of Civil Engineers, 103, (IR 2): 115–134

Maher MJ (1978) The effect of root-zone warming on tomatoes grown in nutrient solution at two air temperatures. Acta Hortic 82:113–120

Moorby L, Graves CJ (1980) Root and air temperature effects on growth and yield of tomatoes and lettuce. Acta Hortic 98:29–43

Proefstation voor Tuinbouw onder Glas (1988) Kwantitatieve informatie voor de Glastuinbouw, 1988–1989, Glasshouse Crops Research Station, Naaldwijk, The Netherlands, p 22

Royle D (1985) Zinc deficiency causes problems in tomatoes on rockwool. Grower 104; 13:11

Runia WTh, van Os EA, Bollen GJ (1988) Disinfection of drainwater from soilless cultures by heat treatment. Neth J Agric Sci 36:231–238

Sonneveld C (1987) Magnesium deficiency in rockwool-grown tomatoes as affected by climatic conditions and plant nutrition. J Plant Nutr 10:1591–1604

Sonneveld C (1988) Rockwool as a substrate in protected cultivation. In: Organizing Committee of Intern. Symp High Technology in Protected Cultivation. Horticulture in High Technology Era, Tokyo, pp 173–191

Sonneveld C, de Bes SS (1984) Micro nutrient uptake of glasshouse cucumbers grown on rockwool. Commun Soil Sci Plant Anal 15:519–535

Sonneveld C, Straver N (1989) Nutrient solutions for vegetables and flowers grown in water or substrates. Proefstation voor Tuinbouw onder Glas, Naaldwijk, The Netherlands (Voedingsoplossingen glastuinbouw, nr. 8)

Sonneveld C, Voogt SJ (1975) Studies on the manganese uptake of lettuce on steam-sterilised glasshouse soils. Plant Soil 42:49–64

Sonneveld C, Voogt W (1985a) Studies on the application of iron to some glasshouse vegetables grown in soilless culture. Plant Soil 85:55–64

Sonneveld C, Voogt W (1985b) Growth and cation absorption of some fruit-vegetable crops grown on rockwool as affected by different cation ratios in the nutrient solution. J Plant Nutr 8:585–602

Sonneveld C, Voogt W (1986) Supply and uptake of potassium, calcium and magnesium of spray carnations (*Dianthus caryophyllus*) grown in rockwool. Plant Soil 93:259–268

Sonneveld C, Welles GWH (1984) Growing vegetables in substrates in the Netherlands. Proc Sixth Intern Congr Soilless Culture, Lunteren, ISOSC, Wageningen, The Netherlands, pp 613–632

Sonneveld C, Welles GWH (1988) Yield and quality of rockwool-grown tomatoes as affected by variations in EC-value and climatic conditions. Plant Soil 111:37–42

van der Burg AMM, Sonneveld C (1988) Specific effects of sodium chloride on sweet pepper grown on rockwool in a recirculation system. Glasshouse Crops Res Station, Naaldwijk, The Netherlands. Annu Rep, 1987, 21

van der Burg AMM, Hamaker Ph (1987) Variatie in waterafgifte druppelaars en wateropname. Groenten en Fruit 42; 19:30–33

van der Burg AMM, Theune D, Sonneveld C (1987) Specific effects of sodium chloride on tomatoes grown in nutrient film. Glasshouse Crops Res Station, Naaldwijk, The Netherlands. Annu Rep 1986, 16

van Os PC, de Kreij C (1987) EC van 2 het beste voor produktie en kwaliteit. Vakbl Bloemisterij 42; 43:56–57

Verdegaal J (1987) Hoge EC beïnvloedt kwaliteit negatief. Vakbl Bloemisterij 42; 42:65

Voogt W (1988a) K and Ca ratios in the nutrient solution with beefsteak tomatoes. Acta Hortic 222:155–165

Voogt W (1988b) pH and Mn levels with gerbera. Glasshouse Crops Res Station, Naaldwijk, The Netherlands, Annu Rep 1987, 16

Went FW (1944) Plant growth under controlled conditions. II Thermoperiodicity in growth and fruiting of tomato. Am J Bot 31:135–150

Zijlstra S, den Nijs APM, Sonneveld C, Vos G (1987) Een nieuwe aanpak van het necroseprobleem bij meeldauwresistente komkommers. Prophyta 41:138–140

II.5 Autotrophic Micropropagation

T. Kozai[1]

1 Introduction

Use of in vitro micropropagated plantlets is currently expanding worldwide; however, its widespread commercial use is still mainly restricted to horticultural crops, because its production cost is relatively high. In order to expand the use of micropropagation beyond horticulture to include agriculture and forestry, a novel mass-propagation system for reducing the production cost by 80–90% is required. The use of autotrophic or sugar-free micropropagation methods described in this chapter may provide a basis for such a system.

2 Micropropagation Process and Reasons for High Production Cost in the Conventional Micropropagation Method

Plantlets cultured in vitro grow relatively slowly in the multiplication stage, taking, in general, several weeks for most plants when tissue culture vessels are used. During multiplication, a loss of plantlets due to fungal or bacterial contamination may occur.

After multiplication and/or rooting in vitro, it takes a few weeks or more to acclimatize the plantlets ex vitro (outside the vessel, under protected environmental conditions) before transfer to the field conditions. During acclimatization, about 20 to 50% of the plantlets are often lost due to their low tolerance to environmental stresses and a negative carbon balance of the plantlets (see Table 1).

Table 1. Possible reasons for low survival and low growth rates of plantlets at the acclimatization stage

1.	Low photosynthetic rate of the plantlets in vitro
2.	Incomplete autotrophy
3.	High transpiration rate due to a thin cuticular layer (a low cuticular resistance) and an abnormal stomatal functioning
4.	Incomplete rooting
5.	Physiological disorders such as vitrification, etc.

[1]Department of Horticultural Engineering, Faculty of Horticulture, Chiba University, Matsudo, Chiba 271, Japan

Biotechnology in Agriculture and Forestry, Vol. 17
High-Tech and Micropropagation I (ed. by Y.P.S. Bajaj)

Table 2. Main reasons which give rise to the high production cost of in vitro plantlets

1.	A few weeks or even a few months are required for multiplication, preparation, and acclimatization of the plantlets
2.	Contamination loss during the multiplication and preparation stages cannot be neglected
3.	Survival rate during acclimatization stage is not high
4.	Labor costs account for 60–70% of the total production cost
5.	Overproduction or underproduction are likely to occur due to the insufficient prediction of the demands
6.	Energy cost for lighting, cooling, sterilization etc. cannot be neglected
7.	Costs of basic components, sugar, agar, vessels, caps, etc. are significant
8.	Mutations are likely to occur partly due to exogenous hormones and abnormal environments

Table 3. Strategies for reducing the production cost of in vitro plantlets

1.	Control of environmental factors for promoting the autotrophic growth of the plantlets in vitro and for reducing the physiological disorders during the multiplication and preparation stages
2.	Labor saving by introducing a tissue culture robot for grasping, cutting, and transplanting during the multiplication stage and for transplanting during the preparation and the acclimatization stages
3.	Energy saving, resource saving, resource recycling
4.	Improving the design and operation of the culture room, culture vessel, and the acclimatization unit
5.	Computerization of the mass propagation process
6.	Introduction of new materials for artificial substrate, vessels, lamps, etc.

The main reasons which give rise to the high production costs for in vitro micropropagated plantlets are summarized in Table 2. Possible strategies for reducing production costs of the plantlets are given in Table 3. Reasons why the strategies are considered potentially effective in reducing production costs will be discussed in the following sections.

3 Carbon Sources for Hetero- and Autotrophic Growth

When a piece of a plant, an explant, such as a shoot or a single node cutting with or without a (part of) small leaf is placed in vitro, i.e., in a tissue culture vessel, the explant will be differentiated and will grow to a plantlet in the presence of sugar (sucrose, glucose, fructose, etc.), a balance of nutrient salts, and possibly plant growth regulators in the culture medium. In a conventional plant tissue culture, sugar is said to be indispensable as a carbon or energy source for the heterotrophic growth of the explant/plantlet in vitro (e.g., George and Sherrington 1984; Thompson and Thorpe 1987).

What prevents the explants in vitro from growing into plantlets in the absence of sugar in the medium? What prevents the use of atmospheric CO_2 as the carbon source for photosynthesis (autotrophy)? It is believed that the explant does not have sufficient photosynthetic abilities to develop autotrophy, even when it has a small leaf or any tissue containing chlorophyll. If the growth of plantlets is dependent on both atmospheric CO_2 (photosynthesis) and sugar in the medium, the growth is said to be mixotrophic.

In conventional vegetative propagation, leaf cuttings and stem cuttings are grown ex vitro on an artificial substrate such as vermiculite in the absence of sugar in a misted greenhouse (Hartmann and Kester 1983). Why do they grow in the absence of sugar? Is the photosynthetic ability of plantlets cultured in vitro so inferior to that of the plantlets grown ex vitro? If so, what is the explanation? This question will be answered in the following sections of this chapter.

4 Fundamental Photosynthetic Characteristics of C_3 plants

Grout and Aston (1978) found that regenerating cauliflower plants in tissue culture showed no net carbon dioxide uptake and had very low levels of photosynthesis during development. They concluded from the results that growth of chlorophyll-containing cultures was totally dependent on an exogenous carbon source in the growth medium and that full photosynthetic competence did not develop until the plants had been removed from the culture medium and transplanted to soil.

This is in contrast to our recent results (Kozai et al. 1987b, 1988a,b; Kozai and Iwanami 1988) showing that most chlorophyll-containing explants and plantlets in vitro have a photosynthetic ability similar to that of plants grown ex vitro, and that they can develop autotrophy, provided that physical environmental factors such as CO_2 and light in vitro are properly controlled for promoting photosynthesis; under these conditions no sugar is required for growth.

This apparent contradiction between the two research groups on the photosynthetic ability of in vitro plantlets can be explained by measuring the diurnal changes in CO_2 concentration in the tissue culture vessel and measuring the photosynthetic characteristics of in vitro plantlets, as will be demonstrated in the following sections.

It is known that most C_3 plants grown ex vitro can never develop a positive carbon balance when the CO_2 concentration at a leaf surface is lower than the CO_2 compensation point (ca. 30–80 ppm) and the photosynthetic photon flux (called PPF hereafter) or visible light intensity at a leaf surface is lower than the light compensation point (ca. 20 $\mu mol/m^2/s$ or ca. 1.5 klx) (Jones 1983).

It is also known that the net photosynthetic rate increases with increasing CO_2 concentration and PPF until the rate is nearly saturated at a CO_2 concentration of 500–1000 ppm or more and a PPF of 100 (for shade-grown leaves)-2000 (for sun-grown leaves) $\mu mol/m^2/s$ (Jones 1983).

Thus, PPF and CO_2 may also be important environmental factors for photosynthesis and growth of the plantlets in vitro. However, little research has so far been conducted on the measurements and control of PPF and CO_2 in the culture vessel and on the effects of the physical environment in vitro on the photosynthesis and growth of the intact plantlets in vitro.

Another important photosynthetic characteristic of C_3 plants is the presence of photorespiration in light (Jones 1983). The photorespiration rate is roughly proportional to the gross photosynthetic rate and accounts for about 50% of the gross photosynthesis at 21% O_2. Photorespiration is, however, reduced with decreasing O_2 concentration and is almost completely repressed in 2% O_2. The net photosynthetic rate of C_3 plantlets in vitro should then increase with decreasing O_2 concentration, provided that the plantlets have photosynthetic characteristics similar to those of the C_3 plants grown ex vitro.

5 Number of Gas Changes of the Vessel per Hour

The gas exchange rate between the room air and the air inside the vessel varies with different culture vessels depending on shape and closure mechanism. Air-tightness of the vessel is expressed as the number of gas changes of the vessel per hour, abbreviated N hereafter. N is defined as the hourly gas exchange rate divided by the air volume of the vessel (Kozai et al. 1986b).

N, for example, is 0.18/h for glass test tubes (air volume: 45 ml) with an aluminum foil cap, 1.5/h for test tubes with a formed plastic cap and 0.6/h for test tubes with a silicon foam rubber plug (Kozai et al. 1986b).

Net CO_2 exchange rate per vessel under steady-state conditions, G (g/h), can then be estimated by the following equation as a function of N:

$$G = a\,N\,V\,D, \tag{1}$$

where a is a conversion factor of CO_2 from volume to weight (1.8×10^{-3} g/cm^3 at 25 °C), V is the inside air volume of the vessel and D is the difference in CO_2 concentration between the inside and outside of the vessel (ppm or μmol/mol) under steady-state conditions. The G for a lighted condition is considered as the net photosynthetic rate per vessel, while the G for darkness is the dark respiration rate per vessel. The mathematical relationships among these factors and some experimental results are given by Kozai et al. (1986b) and Fujiwara et al. (1987a).

6 Diurnal Changes in CO_2 Concentration Inside the Tissue Culture Vessel

Fujiwara et al. (1987a) found that the CO_2 concentration in the relatively air-tight tissue culture vessel containing ornamental plantlets (N = 0.01–0.04/h) decreased sharply to 70–80 ppm or μmol/mol) within 2 h after the start of the photoperiod under a PPF of 65 μmol/m^2/s. The concentration of CO_2 is as low as the CO_2 compensation point of C_3 plants and is much lower than that for outside the vessel (the ambient air being about 350 ppm).

Even in the vessels loosely closed with a formed plastic cap (N = 1.5/h) or with a gas permeable filter (N = 2.7/h), the inside CO_2 concentration decreased to about 100–200 ppm during the photoperiod at a later stage of multiplication for strawberry plantlets (Kozai and Sekimoto 1988). Similar results have been obtained by Kozai et al. (1987b, 1988a,b; Kozai and Iwanami 1988) and Desjardins et al. (1988).

On the other hand, the CO_2 concentration in the vessel during the dark period increased with time and often reached as high as 5000–10,000 ppm (Fujiwara et al. 1987a) or more (De Proft et al. 1985). It can then be considered that the concentration in the vessel containing nongreen callus or explants with little chlorophyll will be much higher than the atmospheric CO_2 concentration, due to the heterotrophic or mixotrophic growth, even during the photoperiod.

7 Interpretation of the Results of Low CO_2 Concentration in the Culture Vessel During the Photoperiod

The following conclusions can be made based on the experimental results given above.

1. The plantlets growing in vitro have photosynthetic ability and may develop autotrophy.
2. However, the photosynthesis of the plantlets in vitro is inhibited mainly by the low CO_2 concentration in the vessel during most of the photoperiod.
3. The low CO_2 concentration in the vessel is mainly due to the net CO_2 uptake by the plantlets in vitro, to the small quantity of CO_2 in the vessel, and to the small number of gas changes of the vessel per hour.
4. Due to the low CO_2 concentration in the vessel, the plantlets in vitro are forced to develop hetero- or mixotrophy for growth, by absorbing sugar from the medium as their main carbon source.
5. In normal practice, a minimum PPF or a minimum visible light intensity is given to the plantlets in vitro. This is because a higher PPF is ineffective toward significantly increasing the net photosynthetic rate of the plantlets when the CO_2 concentration is kept close to the CO_2 compensation point.

These assumptions can be tested experimentally by culturing the plantlets in vitro on the sugar-free medium under conditions of higher CO_2 concentrations and higher PPF in the vessel.

8 Growth of Plantlets in Vitro Under Different Sugar Concentration, CO_2 Concentration, and PPF Conditions

8.1 Statice

Figure 1 shows the effects of CO_2 enrichment and sucrose concentration in the agar (0.8%) medium on the time courses of fresh and dry weight increases of statice (*Limonium* hybrid) plantlets (Kozai et al. 1987b). The explants were tissue-cultured on Murashige and Skoog (1962) medium with either 0 to 3% sucrose in test tubes with plastic-formed caps ($N = 1.5/h$) under a high PPF (150 $\mu mol/m^2/s$). During CO_2-enriched treatment, CO_2 concentration outside the vessel during the photoperiod was maintained at about 950 ppm to increase the concentration inside the vessel.

The growth of plantlets on the medium with 0 or 3% sucrose was promoted better in CO_2 enrichment treatments than in CO_2 nonenrichment treatments under the high PPF. There was greater growth on the medium with 3% sucrose than on the medium with 0% sucrose, in both CO_2-enriched and CO_2-nonenriched treatments. It should be noted, however, that the plantlets cultured on the sucrose-free medium slightly increased their weight, especially when enriched with CO_2, i.e., they seemed to have an ability to develop autotrophy.

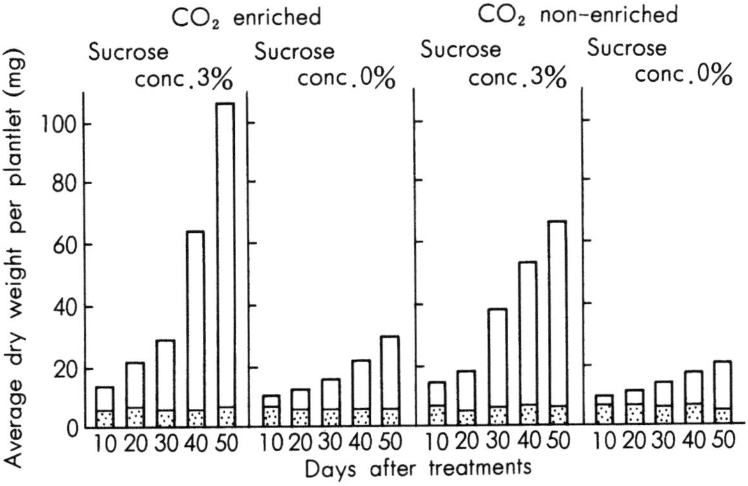

Fig. 1. Effects of CO_2 enrichment and sucrose concentration on the dry weight increases of statice (*Limonium* hybrid) plantlets in vitro. (Kozai et al. 1987b)

Kozai et al. (1987b) showed that the steady-state CO_2 concentration inside the vessel during the photoperiod decreased daily both in CO_2-enriched and CO_2-non-enriched treatments. The decrease was more remarkable in cultures treated with 3% sucrose than with 0% sucrose.

It follows that the growth could be promoted further in all the treatments if the CO_2 concentration inside the vessel was kept at a higher level during the photoperiod throughout the experiments by gradually increasing the concentration outside the vessel with time.

8.2 Cymbidium

Kozai et al. (1988a) measured the changes in dry weight with time of *Cymbidium* cv. Reposa plantlet in vitro during the preparation or rooting stage in CO_2-enriched and CO_2-nonenriched treatments with 0, 1, and 2% sucrose concentrations in the medium. CO_2 was provided for CO_2-enriched cultures as described for experiments with statice. The CO_2 concentration outside the vessel during the photoperiod was maintained at about 950 ppm throughout the experiment.

Growth was considerably promoted by CO_2 enrichment under high (230 $\mu mol/m^2/s$) and low (50 $\mu mol/m^2/s$) PPF. Dry weight increases with time were 1.7 to 2.0 times greater in high PPF and CO_2-enriched treatments than normal culture (namely, low PPF and CO_2-nonenriched treatment), regardless of the sucrose concentrations.

Again, autotrophy developed in the plantlets cultured in vitro on the medium with 0% sucrose, in particular, under high PPF and CO_2-enriched treatments.

The difference in CO_2 concentration between the inside and outside of the vessel during the photoperiod increased with time from about 100 ppm to 800 ppm in CO_2-enriched treatments. Consequently, the CO_2 concentration in the vessel decreased with time from about 800 ppm to 100 ppm.

As given in Eq. (1), the net photosynthetic rate per plantlet is nearly proportional to the difference in CO_2 concentration [D in Eq. (1)], provided that N, V, and the number of plantlets per vessel are the same for all treatments, as is the case in the above experiments. The differences in net photosynthetic rate per plantlet were also remarkable among the treatments.

Figure 2 shows the CO_2-PPF-net photosynthesis curves for *Cymbidium* cv. Reposa plantlet grown in vitro at 25 °C (Kozai et al. 1987d). The net photosynthetic rate per unit dry weight of the plantlet seems to be nearly saturated at a CO_2 concentration of around 2000 ppm and a PPF of around 200 μmol/m^2/s.

It should be noted again that even in the CO_2-enriched treatments the CO_2 concentration in the vessel during the photoperiod is far below saturation CO_2 concentration especially at a later stage of growth.

8.3 Carnation

Figure 3 shows the increases in dry weight with time of carnation plantlets grown in vitro during multiplication in CO_2-enriched and CO_2-nonenriched treatments with 0, 1, and 2% sucrose in the medium. A photograph of the plantlets 30 days after treatment is shown in Fig. 4.

Weight increases with time were greater for CO_2-enriched treatments than for CO_2-nonenriched treatments, regardless of the sucrose concentrations. The effect of

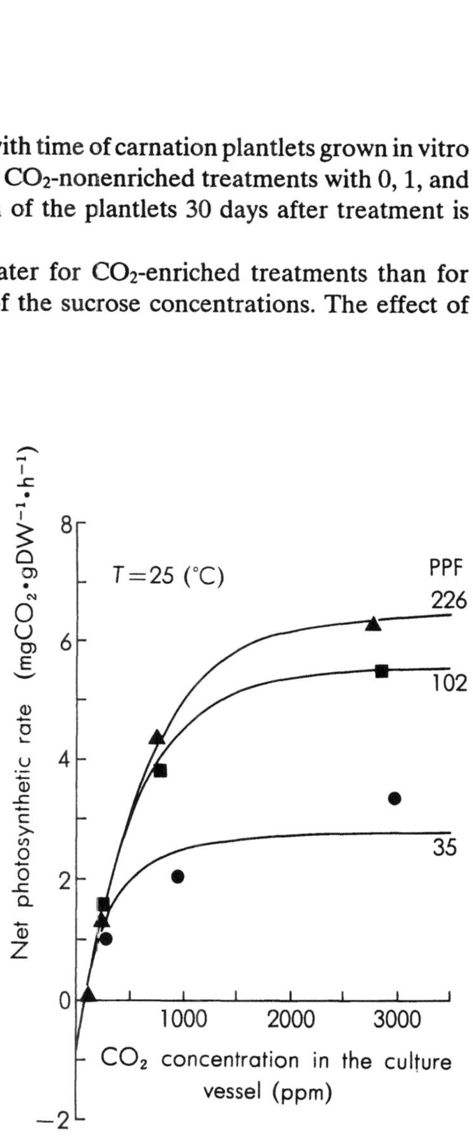

Fig. 2. CO_2-PPF-net photosynthetic rate curves for *Cymbidium* plantlet in vitro. T and PPF denote air temperature and photosynthetic photon flux, respectively (Kozai et al. 1987d)

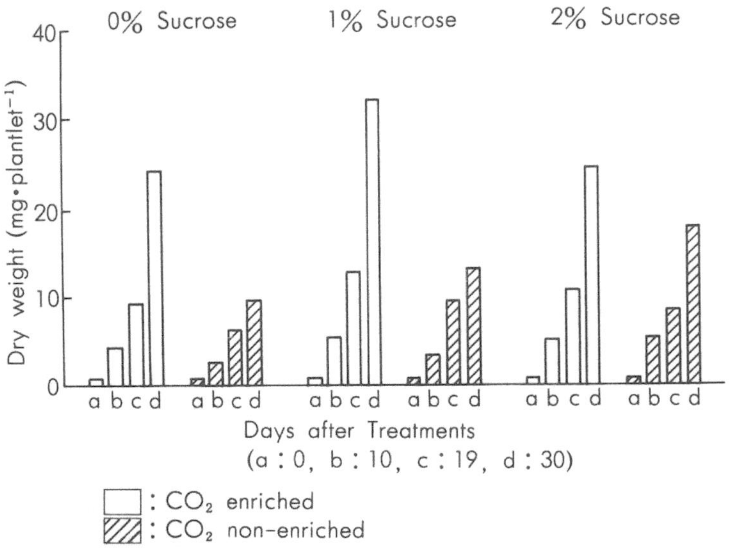

Fig. 3. Time courses of the dry weight of the carnation (*Dianthus caryophyllus* L.) plantlets cultured in vitro under CO_2-enriched and CO_2-nonenriched conditions. (Kozai and Iwanami 1988)

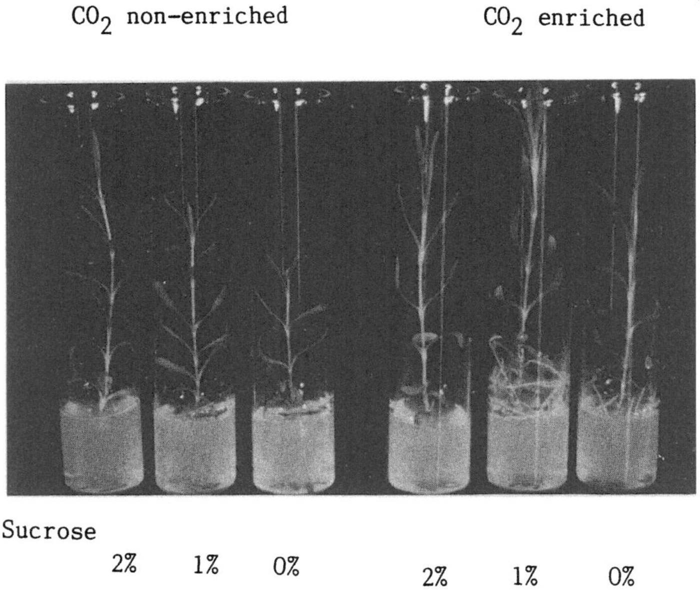

Fig. 4. Carnation plantlets 30 days after CO_2-enriched or CO_2-nonenriched treatments with 0, 1, or 2% sucrose. (Kozai and Iwanami 1988)

CO_2 enrichment on the weight increase was the largest for the medium with 1% sucrose. However, it should be emphasized that the dry weight of the plantlets cultured on the medium without sucrose increased in both CO_2-enriched and CO_2-nonenriched treatments; the increase was greater in the CO_2-enriched treatments than in CO_2-nonenriched treatments.

Figure 5 shows the increases in the difference in CO_2 concentration inside and outside the vessel under steady-state conditions during the photoperiod and dark period. The differences in the photoperiod increased steadily with time for CO_2-enriched treatments. On the other hand, the differences did not significantly increase with time for CO_2-nonenriched treatments.

The trends of the differences in CO_2 concentration during the dark period were similar to those during the photoperiod. However, the differences in the CO_2 concentration difference between CO_2-enriched and CO_2-nonenriched treatments were smaller in the dark period than in the photoperiod.

Figure 6 shows the dependence of the net photosynthetic rate of the plantlets upon the CO_2 concentration in the vessels. The unit of the vertical axis is mg CO_2/g DW/h. The rates at 1, 10, 19, and 30 days after treatment increased with an increase in the CO_2 concentration. It was noted that the rates at 1 day after treatment are several times higher than those at 10, 19, and 30 days after the treatments.

The development of full photosynthetic potential of in vitro plantlets is important for successful acclimatization. However, the results given in Fig. 6 suggest that the

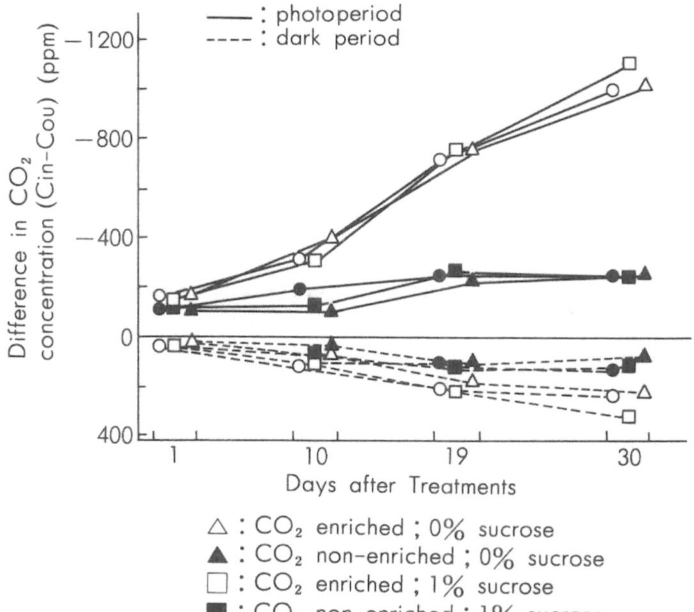

Fig. 5. Time courses of the difference in CO_2 concentration between inside and outside the vessel, during the photoperiod and dark period. (Kozai and Iwanami 1988)

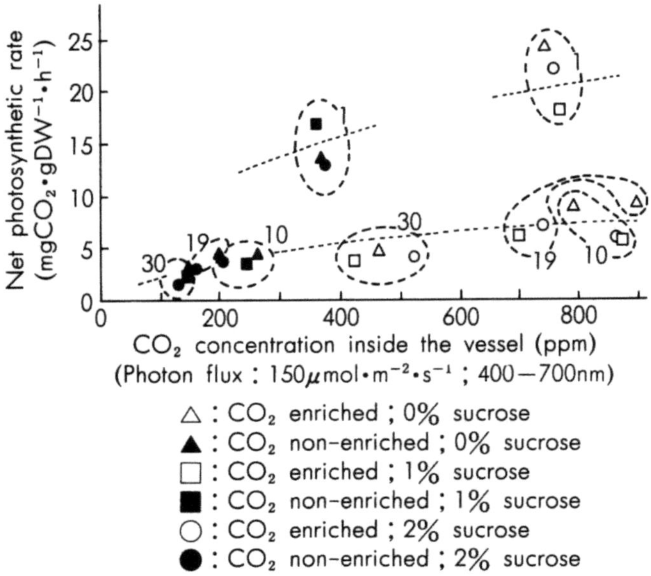

Fig. 6. Dependence of the net photosynthetic rate of carnation plantlets in vitro upon CO_2 concentration. *Numbers* in the figure indicate the days after treatments. (Kozai and Iwanami 1988)

photosynthetic ability is the highest at the start of the culture and decreases with time afterwards. Photosynthetic characteristics of in vitro plantlets and its changes with time during the culture of in vitro plantlets is an area for further study.

Table 4 shows the changes with time of the sucrose, glucose, and fructose concentrations in the medium during the growth in vitro. As the sucrose concentration decreased, the glucose and fructose concentrations increased in the medium.

The total sugar decreases in CO_2-enriched treatments after 30 days were 8 and 4 mg per vessel for the medium with 1 and 2% sucrose, respectively. Those in CO_2-non-enriched treatments were 4 and 3 mg per vessel for the medium with 1 and 2% sucrose, respectively. That is, the amount of sugar absorbed by the plantlets as either sucrose, glucose, or fructose was only about 2–8% in weight of the initial sucrose content. This is partially in agreement with the results of Pieters et al. (1987), who noticed that the total sugar concentration in the medium decreased by only about 20% for the first 28 days of artichoke (*Cynara scolymus*) and *Chrysanthemum indicum* cultures.

It should be noted that the osmotic pressure or absolute value of osmotic potential of the medium, caused by sugar, should have increased with time as the hydrolysis of sucrose to monosaccharides proceeded in the medium, even though the total weight of sugar in the medium decreased by 10%. The effects of osmotic potential, or more precisely, the effects of water potential of the medium on the growth and development of the plantlet, or generally, on the water relations of the air, plantlets, and the medium, are also an area that warrants further study. Changes in water potential of the medium with time in relation to the hydrolysis of sugar and its effect on the growth of cultures are also emphasized by Pieters et al. (1987).

Table 4. Time courses of sucrose (S), glucose (G), and fructose (F) contents in the medium with initial sucrose concentrations of 1 and 2%, during the multiplication of carnation plantlets in vitro[a]. (Kozai and Iwanami 1988)

Treatments	CO_2 enrichment	Sugar	Days after treatment		
Sucrose			0	19	30
1%	Yes	S	107	53	–
		G	0	24	–
		F	0	24	–
		S	101	–	12
		G	0	–	40
		F	0	–	41
	No	S	105	59	–
		G	0	21	–
		F	0	21	–
		S	105	–	11
		G	0	–	45
		F	0	–	45
2%	Yes	S	207	107	–
		G	0	44	–
		F	0	35	–
		S	221	–	48
		G	0	–	75
		F	0	–	74
	No	S	216	140	–
		G	0	35	–
		F	0	35	–
		S	214	–	53
		G	0	–	79
		F	0	–	79

[a]Unit = mg per vessel.
Note : Samples taken for the measurements on the 30th day were different from those taken for the measurements on the 19th day. Therefore, the initial sucrose content was slightly different for the medium of the same treatment. Sucrose content at the start of the experiments was measured before autoclaving.

8.4 Potato

Figure 7 shows the changes with time of the fresh weight of whole potato plantlets in Treatments 1–9 (Kozai et al. 1988b). The description of the treatments is given in Table 5. The C_{out} in Treatments 4–9 was increased with time from 500 to 2000 ppm so as to keep the CO_2 concentration in the vessels during the photoperiod at around 400 ppm throughout the experiments. The increase in fresh weight was two times greater in Treatments 4–9 than in Treatments 1–3. The weight increase was greater with increasing sucrose concentration in Treatments 1–3. On the other hand, the weight increase with increasing sucrose concentration was not significant in Treatments 4–9.

A photograph of the plantlets taken 29 days after the start of the experiments is given in Fig. 8. The plantlet height increased with increasing sucrose concentration in

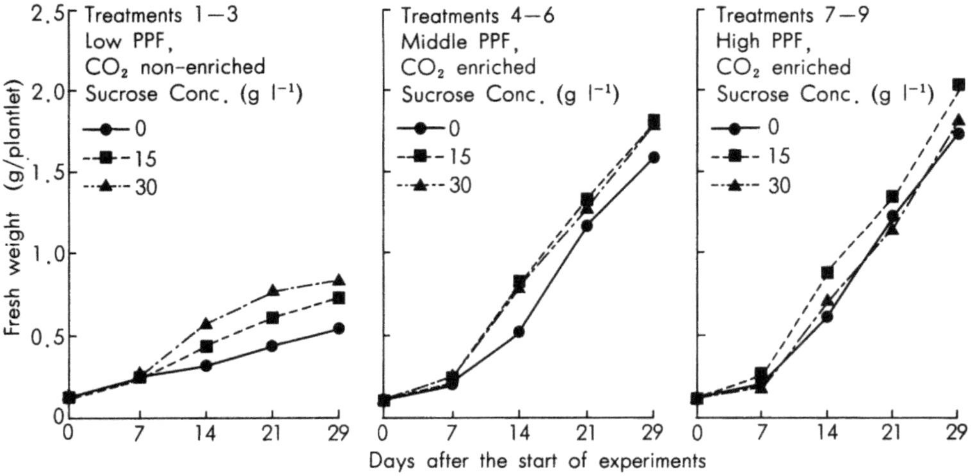

Fig. 7. Changes in fresh weight of whole potato plantlet during growth in Treatments 1–9. For description of treatments, see Table 5. (Kozai et al. 1988b)

Table 5. Photosynthetic photon flux (PPF), CO_2 concentration in culture room (C_{out}), number of air changes per hour (N), and sucrose concentration in the medium (S) for Treatments 1–9

Treatment No.	Type (g/l)	PPF ($\mu mol/m^2/s$)	C_{out} (ppm)	N (per h)	S (g/l)
1	Low PPF, CO_2-nonenriched Suc. conc. 0	65	450	1.3	0
2	Low PPF, CO_2-nonenriched Suc. conc. 15	65	450	1.3	15
3	Low PPF, CO_2-nonenriched Suc. conc. 30	65	450	1.3	30
4	Middle PPF, CO_2-enriched Suc. conc. 0	210	500–2000	5.6	0
5	Middle PPF, CO_2-enriched Suc. conc. 15	210	500–2000	5.6	15
6	Middle PPF, CO_2-enriched Suc. conc. 30	210	500–2000	5.6	30
7	High PPF, CO_2-enriched Suc. conc. 0	400	500–3000	5.6	0
8	High PPF, CO_2-enriched Suc. conc. 15	400	500–3000	5.6	15
9	High PPF, CO_2-enriched Suc. conc. 30	400	500–3000	5.6	30

Treatments 1–3. On the contrary, it decreased with increasing sucrose concentration in Treatments 4–9. The plantlets cultured on sugar-free medium under high PPF conditions developed enough healthy roots so that no particular rooting and acclimatization processes were required for transfer to ex vitro conditions. The plantlets formed 2–3 cm diameter mini-tubers within 2 months of transfer from tissue culture conditions.

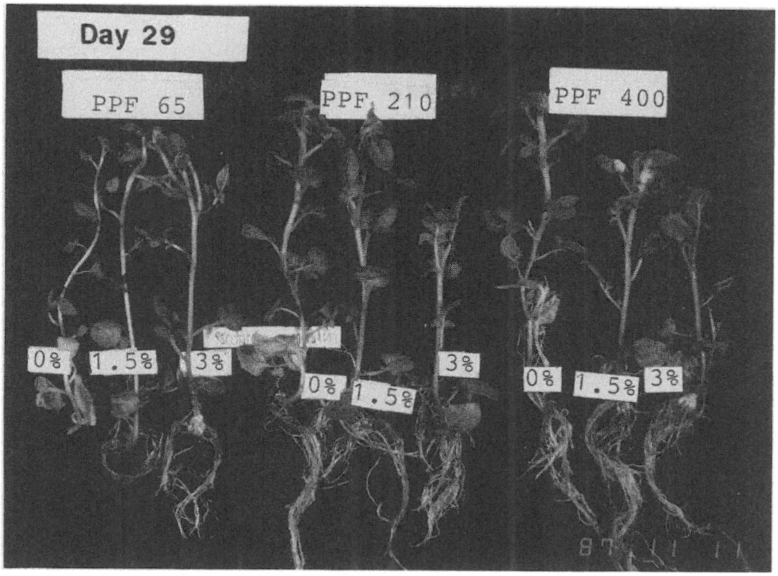

Fig. 8. Potato plantlets cultured in vitro, 29 days after the start of experiments. *From left to right* Treatments 1–9. For description of treatments, see Table 5. (Kozai et al. 1988b)

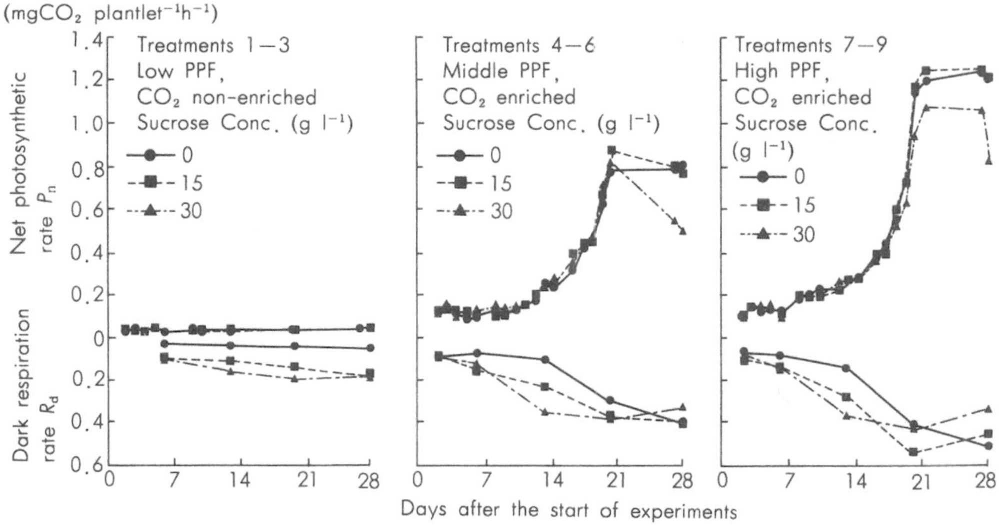

Fig. 9. Time course of net photosynthetic rate (P_n) and dark respiration rate (R_d) per plantlet in Treatments 1–9. For description of treatments, see Table 5. (Kozai et al. 1988b)

Table 6. Growth of potato plantlets in Treatments 1–9, 29 days after the start of experiments. (Kozai et al. 1988b)

Treatment No.	Leaf area[a] (cm^2)	Number of[a] leaves	Fresh weight (g)[a] shoot	root	Dry/fresh[a] weight ratio
1	7	11	0.53	0.03	0.04
2	8	11	0.65	0.11	0.07
3	10	10	0.72	0.12	0.08
4	22	13	1.21	0.39	0.08
5	19	12	1.24	0.57	0.10
6	14	10	1.09	0.72	0.12
7	20	13	1.30	0.48	0.08
8	20	11	1.26	0.81	0.11
9	14	10	1.10	0.75	0.13

[a]Per plantlet.
Note : For the description of treatments, see Table 5.

A time study of the net photosynthetic rate, P_n, and dark respiration, R_d, per plantlet are shown in Fig. 9. P_n does not increase with time in Treatments 1–3. On the other hand, for Treatments 4–6, P_n increased sharply with time until 20 days after the start of the experiments. Thereafter, it decreased slightly with time or remained the same in Treatments 4–9. There seemed to be a tendency in Treatments 1–6 that the dark respiration rate increased with increasing sucrose concentration in the medium. This is perhaps due to the hetero- or mixotrophic growth of plantlets in Treatments 2, 3, 5, and 6 during the dark period.

The number of leaves and the leaf area per plantlet were much greater under CO_2-enriched and high PPF conditions than under CO_2-nonenrichment and low PPF conditions (see Table 6). An explant used for autotrophic micropropagation contained chlorophyll in its leafleat at the beginning of the experiments.

8.5 Lettuce (*Lactuca sativa* L.)

Kaneko et al. (1988) found that green-colored adventitious shoots with only small leaf primordia originated from lettuce callus. These shoots, only several millimeters in length, developed autotrophy on the agar medium with one-tenth the strength of Murashige and Skoog (1962) basal composition. Roots were produced under conditions of 10,000 ppm CO_2, 25 °C, 90% relative humidity, about 100 μmol/m^2/s and a 12-h photoperiod.

8.6 *Bidens* and *Tagetes* (Marigold)

Flores et al. (1988) recently found that cultured (transformed) roots of *Bidens* and *Tagetes* turn green under light and have the ability to photosynthesize. *Bidens* roots have been cultured in light under 1–2% CO_2, with as little as 0.1–0.3% sugar. These cultures grew as fast as heterotrophic cultures did on medium containing 3% sucrose.

Roots grown with little sugar more closely resemble normal roots than do roots grown on conventional medium, which contains higher levels of sucrose.

9 Other Physical Environmental Factors Affecting the Auto- or Mixotrophic Growth of Plantlets in Vitro

9.1 O_2 Concentration in the Vessel

Figure 10 shows the effect of O_2 concentration on the net photosynthetic rate of *Primula malacoides* plantlets (C_3 plant) in vitro at different CO_2 concentrations (Shimada et al. 1988). The measurements were made at a PPF of 110 μmol/m²/s and 25 °C. It is shown that at a CO_2 concentration of 200 ppm, for example, the net photosynthetic rate in 1% O_2 is about three times greater than that in 21% O_2 and is 1.5 times greater than that in 10% O_2.

These results indicate that the photorespiration of the plantlets cultured in vitro was repressed by lowering the O_2 concentrations, as is well known for C_3 plants grown ex vitro. The repression of photorespiration by lowering O_2 concentrations should be observed for all C_3 plantlets in vitro if the photosynthetic characteristics of the in vitro and ex vitro plantlets are similar to each other.

If this is the case, this would have a practical value for promoting the auto- or mixotrophic growth of the plantlets in vitro, because it is relatively easy to keep the O_2

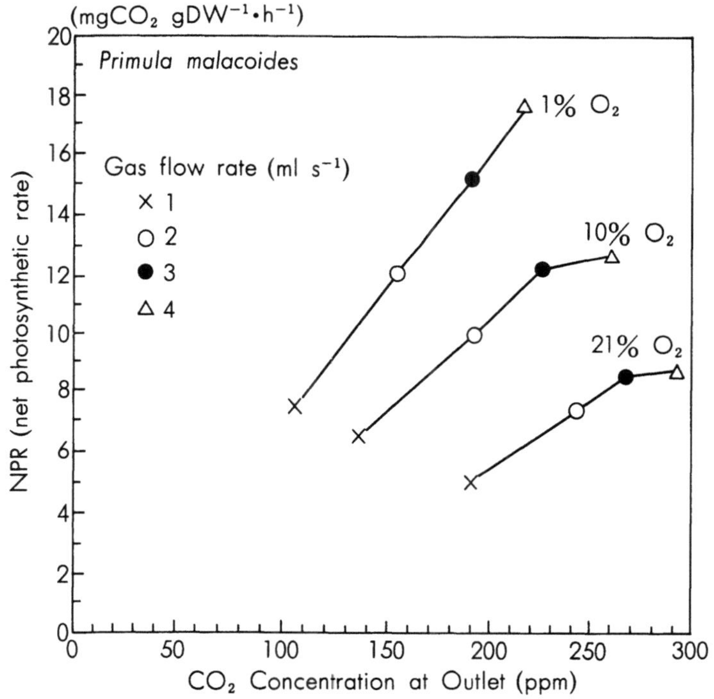

Fig. 10. Net photosynthetic rate of *Primula malacoides* plantlets in vitro during the photoperiod at 1, 10, and 21% O_2 concentrations. (Shimada et al. 1988)

concentration in the vessel at a low level. Oxygen gas absorbent (iron powder) (with CO_2-producing material) is widely used for preventing the bacteria and fungi from growing in food. The absorbent is placed in an air-sealed bag with food. The oxygen concentration in the bag is then kept at about 2–10%, depending on the amount of the absorbent and air volume of the bag. The effect of long-term exposure to low O_2 concentrations during the photoperiod on the growth of C_3 plantlets in vitro is an area for further study.

Another advantage of lowering O_2 concentration in the vessel is that low O_2 concentrations will inhibit the growth of most fungi and bacteria in the vessel. A loss of plantlets due to fungal and bacterial contamination can then be reduced.

9.2 Basal Medium Composition

Table 7 gives the fresh and dry weights of carnation plantlets cultured in loosely capped vessels (N = 6.2/h) under high PPF (210 $\mu mol/m^2/s$) on the sugar-free and sugar-containing (2% sucrose) media with half-strength MS salts (Murashige and Skoog 1962) and full-strength Enshi (Hori 1966) salts used for hydroponics. No growth regulators were added to the media.

The growth of carnation plantlets was best on the sugar-free Enshi medium. The NH_4^+ and Mn^{2+} ion concentrations of the 1/2 MS salts are, respectively, 6.3 and 2.4 times those of the Enshi salts.

The MS salts have been developed so as to promote the heterotrophic growth of plantlets or callus. Therefore, it is not surprising that the Enshi salts developed for hydroponics are more favorable than the 1/2 MS for autotrophic or mixotrophic culture in vitro. Optimal basal composition for auto- and mixotrophic micropropagation is also an area for further study.

Table 7. Fresh and dry weights of the carnation plantlets in vitro on the 15th and 30th days after the start of the experiment. (Kozai et al. 1988c)

Treatment No.	Name	No. of days after treatment	Fresh weight (mg per plantlet)				Dry weight (mg per plantlet)			
			Leaf	Stem	Root	Total	Leaf	Stem	Root	Total
1	1/2 MS	15	66	30	6	102a	11	4	0.3	15a
	no sugar	30	184	67	32	283 x	24	8	2	34 x
2	1/2 MS	15	63	24	0.8	88b	14	4	0.1	18b
	2% sucrose	30	139	68	21	228 x	24	11	2	37 x
3	Hydroponics	15	81	28	7	116c	13	5	0.3	18b
	no sugar	30	259	67	84	410 y	31	8	6	45 y
4	Hydroponics	15	80	27	3	110ac	16	5	0.2	21b
	2% sucrose	30	228	73	58	359 y	34	11	5	50 z

Note: The fresh and dry weights were, respectively, 42 and 5 mg per plantlet at the start of the experiment. Means within a column followed by different letters are significantly different at the 5% level.

9.3 Temperature During the Dark Period

Air temperature in the tissue culture room is kept more or less constant throughout the day at around 25 °C in most cases. This is reasonable for conventional heterotrophic tissue culture, because in heterotrophic tissue culture, a high respiration rate can be considered, in many cases, as an indicator of active plantlet growth. In autotrophic or mixotrophic tissue culture, however, growth may be inhibited if the dark respiration rate is too high. A high, dark respiration rate may indicate a loss of carbon which was fixed by photosynthesis during the preceding photoperiod.

For many plants grown ex vitro, the optimum day temperature for photosynthesis is lower than the optimum temperature for translocation of carbohydrates and growth during the dark period. Kozai et al. (1988) showed that the low air temperature (ca. 15 °C) during the dark period promoted the growth of strawberry plantlets in vitro in autotrophic tissue culture. The temperature regime in autotrophic micropropagation would be an important area for further study because it is not only related to plantlet growth, but also to the consumption rate of artificial energy (oil, electricity, etc.) for the temperature control in the culture room.

9.4 Humidity

Relative humidity in the vessel is not always 100%, even when the vessel is almost air-tight and/or there is water condensation inside the vessel. Kozai and Tanaka (unpubl.) found that the relative humidity in the air-tight vessel (N = 0.05/h) was 90–95% during the photoperiod when the air temperature, the relative humidity, and PPF outside the vessel are kept at about 23 °C, 55%, and 150 $\mu mol/m^2/s$, respectively. The air temperature inside the vessel was about 3 °C higher than that outside, under previously mentioned outside conditions. The relative humidity inside the vessel was almost 100% during the dark period, when the relative humidity and the air temperature in the room were 90% and 23 °C, respectively. They also found that the relative humidity inside the air-leaky vessel (N = 3.8/h) was about 80% during the photoperiod and about 95% during the dark period under previously mentioned conditions.

As shown above, the relative humidity inside the vessel varies with the relative humidity, PPF, and the air temperature outside the vessel, the number of air changes per hour of the vessel, evaporation, transpiration, and condensation rates in the vessel, etc. The morphology and physiological functions of in vitro plantlets are interrelated with the relative humidity in the vessel. Water relations in the vessel is an area for further research.

10 Use of a Gas-Permeable, Clear Plastic Film as a Vessel Closure

The use of a gas-permeable, clear plastic film offers a simple method for closure of the vessel in a normal tissue culture room. Figure 11 is an example of such a closure. A detailed description of the closure is given by Kozai et al. (1988c). It should be noted that explants to be transplanted onto the sugar-free medium in the loosely capped vessel

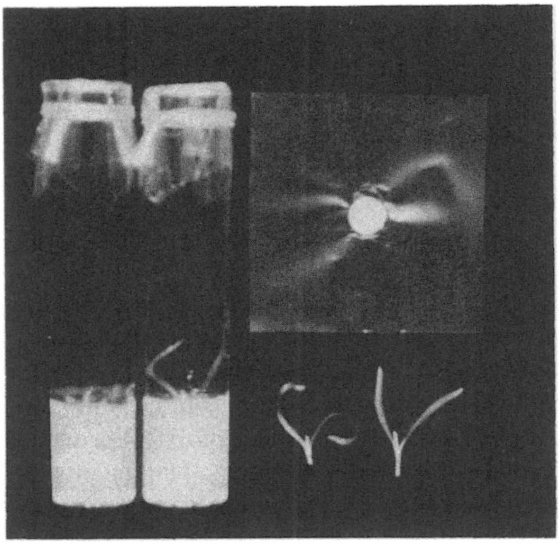

Fig. 11. Gas-permeable, clear plastic film used as a vessel cap for the sugar-free (autotrophic) tissue culture (Kozai et al. 1988c). *Left* two test tubes each containing a carnation explant, capped with the film; *right (above)* a plain view of the film; *right (bottom)* two single node cuttings each with two small leaves, used as explants

should have a node or a shoot tip (with a piece of leaf) containing chlorophyll, as shown in Fig. 11.

With this closure, (1) the CO_2 concentration in the vessel during the photoperiod is kept at a higher level than that in the airtight vessel, (2) the relative humidity in the vessel is kept at a lower level than that in the airtight vessel, and (3) PPF (light intensity) in the vessel is kept at a higher level than in the vessel with aluminum foil.

Modifications of the microenvironment previously mentioned will promote photosynthesis of plantlets and make the plantlets more tolerable to water stress ex vitro. The humidity inside the vessel and a loss of water from the medium and plantlets by evapotranspiration to the outside can partially be controlled to a certain extent by controlling humidity in the culture room.

10.1 Potato

Figure 12 shows potato (*Solanum tuberosum* cv. Benimaru) plantlets cultured for 21 days under a PPF of 160 μmol/m^2/s (12.6 klx) on the sugar-free medium in vessels with (a) a formed plastic cap, (b) an aluminum foil cap, and (c) a gas-permeable, clear film. The number of air changes per hour of the vessel with caps a, b, and c were 1.5, 0.4, and 7.6/h, respectively.

The average fresh weights of whole plantlets cultured in the vessels for 21 days with the caps a, b, or c were 390, 280, and 730 mg per plantlet, respectively. The average fresh weights of roots cultured in those vessels were 50, 10, and 220 mg per plantlet, respectively. The average number of unfolded leaves per plantlet cultured in those vessels were 7, 6, and 8, respectively. The average leaf area per plantlet in those vessels were 7, 6, and 12 cm^2, respectively.

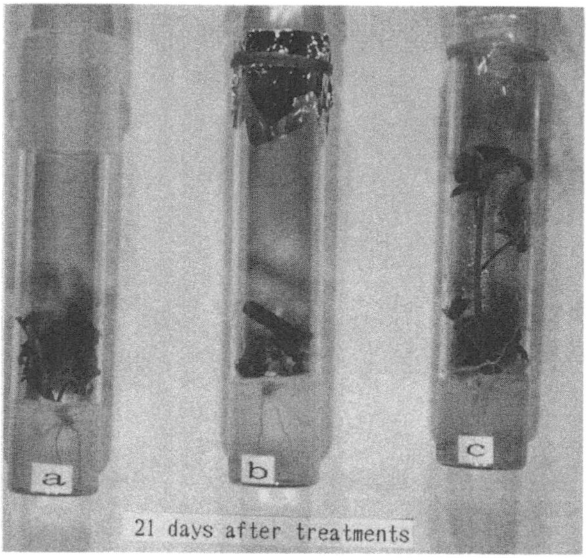

Fig. 12. Potato (*Solanum tuberosum*) plantlets cultured under high photosynthetic photon flux for 21 days on the sugar-free medium in the vessel with three different caps. The explant was a single node cutting with one leaf (Kozai and Lee unpubl.). Basal medium composition: half strength Murashige and Skoog (1962). Closures *a* formed plastic cap; *b* aluminum foil; *c* gas-permeable, clear plastic film

10.2 Chrysanthemum

Similar results were obtained for chrysanthemum (*Chrysanthemum morifolium* Ramat. cv. Syuho-no-chikara) plantlets cultured for 27 days under the same conditions as for the potato plantlets previously mentioned (Kozai and Lee unpubl.). The average fresh weights of the whole plantlet cultured in the vessels with either caps a, b, or c were 248, 139, and 532 mg per plantlet, respectively. The average fresh weights of roots were 27, 17, and 64 mg per plantlet, respectively. The average number of unfolded leaves in those vessels were 9, 5, and 10 per plantlet, respectively.

10.3 Carnation

Figure 13A-C show carnation (*Dianthus caryophyllus* L.) plantlets cultured for 37 days on agar (0.8%) medium in the vessel capped with the gas-permeable, clear film under a PPF of 210 μmol/m^2/s (15 klx). The number of air changes per hour of the vessel was 6.2/h. Favorable growth was observed using sugar-free medium with Enshi (Hori 1966) salts. The fresh weights for the total and roots were, respectively, 610 and 110 mg per plantlet for the sugar-free medium with the Enshi salts; those for the sugar-containing medium with MS salts are 490 and 34 mg per plantlet, respectively.

Figure 13B shows carnation plantlets cultured for 37 days on the sugar-containing (2% sucrose) medium with MS salts in the vessel capped with an aluminum foil and cellophane tape (N = 0.2/h) under a PPF of 70 μmol/m^2/s (5 klx). The total and root fresh weights were 220 and 5 mg per plantlet, respectively.

Comparing the plantlets in Fig. 13A and B, it can be said that the use of gas-permeable, clear plastic film under high photosynthetic photon flux effectively promotes the growth of shoots and roots and makes the plantlets tolerable to water stress.

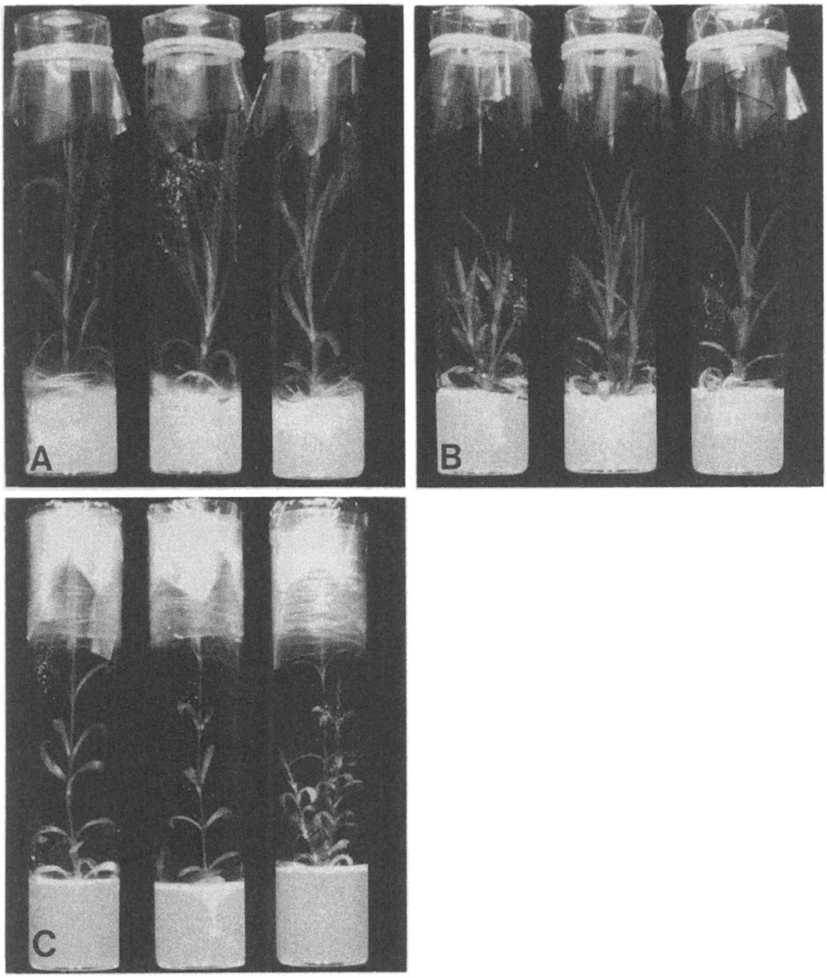

Fig. 13A-C. Carnation plantlets cultured in vitro for 37 days under two different culture conditions. (Kozai and Kubota 1988). **A** Medium: Hydroponic solution (Full strength); Sucrose: 0%; PPF: 210 μmol/m^2/s; Closure: Transparent film with air filter **B** Medium: Murashige and Skoog (1962) (Half strength); Sucrose: 2%; PPF: 210 μmol/m^2/s; Closure: Transparent film with air filter **C** Medium: Murashige and Skoog (1962) (Half strength); Sucrose: 2%, PPF: 70 μmol/m^2/s; Closure: Aluminum foil and Parafilm

11 An Autotrophic Micropropagation System

Based on the ideas of autotrophic micropropagation previously mentioned, a prototype of a novel autotrophic micropropagation system consisting of a culture box assembly, a gas flow assembly, and a culture solution assembly was developed by Fujiwara et al. (1988). A schematic diagram of the configuration of the system is given in Fig. 14.

Micropropagated strawberry plantlets, cultured autotrophically in the photoautotrophic tissue culture system (PTCS) for 28 days at the rooting and acclimatization stages, were about 1.7 times as heavy and more than four times as tall as plantlets cultured using traditional tissue culture conditions (see Fig. 15).

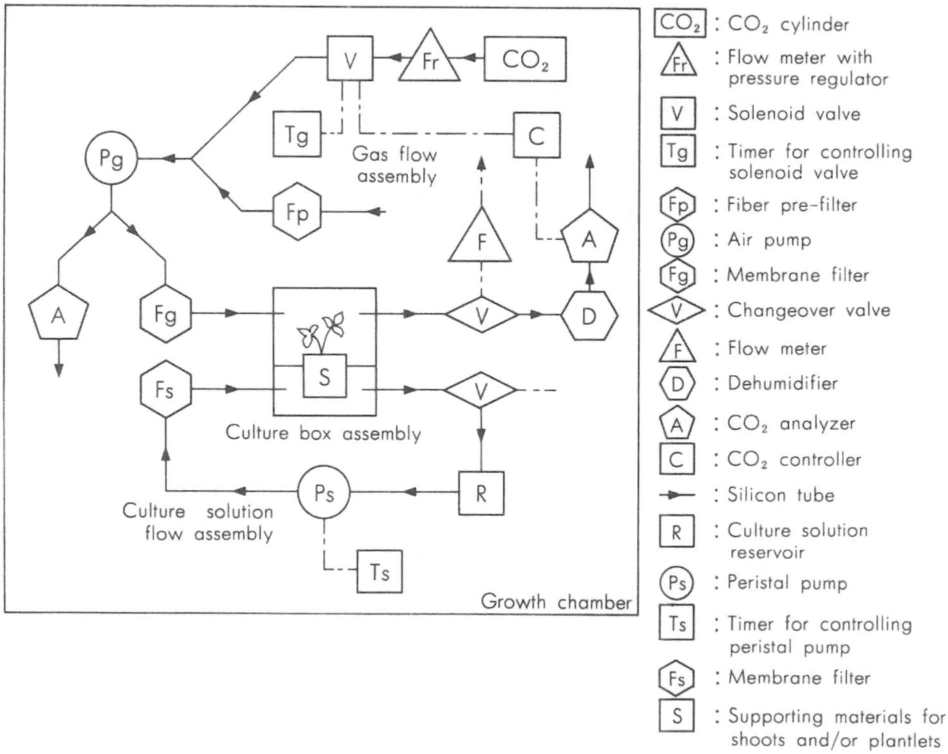

Fig. 14. Configuration of the PTCS (Photoautotrophic Tissue Culture System) and its components. (Fujiwara et al. 1988). Gas and culture solution flow lines are represented by *solid lines*, while electrical lines are represented by the *dash-dotted lines*

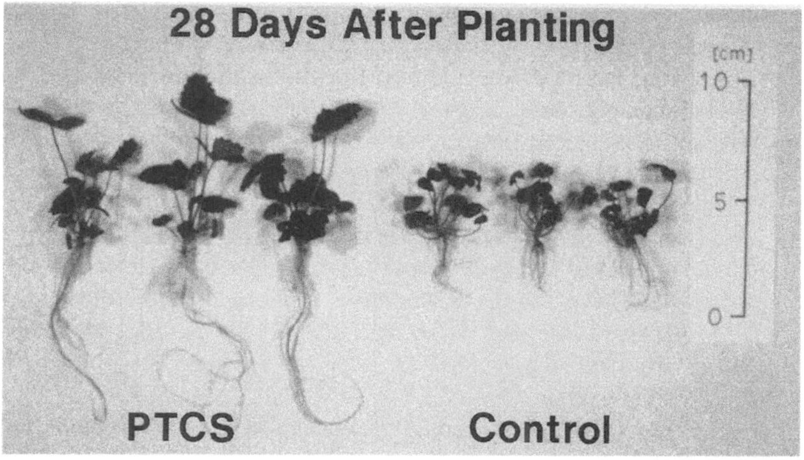

Fig. 15. Strawberry plantlets at rooting stage cultured with the PTCS (Photoautotrophic Tissue Culture System) and the traditional (heterotrophic) tissue culture method. (Fujiwara et al. 1988)

12 A Novel Approach to Auto- or Mixotrophic Micropropagation

A novel approach to the micropropagation of in vitro (disease-free) plantlets can be summarized based on our experimental results of the photosynthetic ability and the acclimatization of in vitro plantlets as follows:

1. The growth of in vitro plantlets can be promoted by increasing the CO_2 concentration in the vessel to 350 ppm or more during the photoperiod, with a PPF of around 200 μmol/m^2/s.
2. The growth rate of plantlets cultured in vitro on the sugar-free medium with CO_2 enrichment under high PPF may be higher than that of plantlets cultured in vitro on the medium with sugar under low PPF with CO_2 nonenrichment.
3. Exogenous growth regulators, vitamins, and other organic substances may not be necessary for plantlets in vitro which are developing autotrophy. Under these circumstances, the plantlets may endogenously produce most of the necessary organic substances.
4. Mutation rates of in vitro plantlets may be reduced in hormone-free, sugar-free tissue culture.
5. By promoting the development of autotrophy of plantlets in vitro, the growth rate and survival rate of plantlets after transfer from in vitro conditions may be increased.
6. With no or minimum use of sugar in the medium, a loss of in vitro plantlets due to bacterial or fungal contamination medium will be remarkably reduced.
7. A much larger vessel than the one traditionally used in tissue culture can be used, if there is little or no fear of loss of plantlets due to contamination.
8. A robot or a micromanipulator with color image sensors can be introduced to automate the processes of picking, cutting, and transplanting the in vitro plantlets, if a larger vessel is used. Several prototypes of robot with electric eyes for grasping, cutting, and transplanting the explants/plantlets in vitro have been developed by Japanese manufacturers, in cooperation with the author and others.
9. Humidity control in the vessel to increase the water stress resistance can be achieved comparatively easily with a larger vessel.
10. Nutrient salts currently used in the growth medium for traditional tissue culture should remain suitable for the in vitro plantlets developing hetero- or mixotrophy. There may be a better nutrient salt composition for autotrophic growth of in vitro plantlets, perhaps similar to those used for hydroponic culture (Kozai et al. 1988a).
11. At present, the air temperature setpoint is constant throughout the day at around 20–25 °C for most plants. Perhaps the setpoint should be lowered during the dark period as compared to the photoperiod, to promote the autotrophic growth of in vitro plantlets. Lowering the temperature at night was effective in promoting the growth of strawberry in vitro on the sugar-free medium (Kozai, Iwabuchi and Watanabe unpubl.).
12. Lowering the O_2 concentration of the air during the photoperiod should increase the net photosynthetic rate of the plantlets in the autotrophic culture and may increase the growth of the plantlets.
13. Dynamic control of physical environmental factors other than temperature, light, and CO_2 in the vessel will become essential.

14. Considerable reduction of the total production cost of the in vitro plantlets should be attained by means of: (a) promoting the growth, (b) deleting sugar, (c) preventing microbial contamination, (d) using automated equipment, (e) reducing mutation, (f) deleting exogenous growth regulators, (g) reducing salt concentration of the medium, (h) energy-efficient control of the in vitro environments, and (i) increasing the growth rate and survival rate during the acclimatization stage.

13 Increase of Production Costs for Lighting, Cooling, and CO_2 Enrichment

A disadvantage of autotrophic micropropagation may be the increase in costs for lighting, cooling, and CO_2 enrichment. The PPF in autotrophic micropropagation should be about five times higher than that in the heterotrophic or conventional micropropagation. Electricity consumption should then also be about five times that of conventional micropropagation, if no means for reducing electricity consumption is introduced.

The electricity consumption for lighting and cooling accounts for approximately 60 and 30% of the total electricity consumption, respectively, assuming that the C.O.P. (Coefficient of Performance) of the cooling system is about 2, in the culture room for autotrophic micropropagation. C.O.P. is the ratio of the heat energy absorbed by the system for cooling to the electric energy required to operate the system. The remainder (about 10%) of the electricity is used for operating miscellaneous equipment.

When the walls of the culture room are thermally well insulated, as is often the case, the cooling load or heating load due to heat transmission through the walls and air infiltration through air leaks in the walls and doors is relatively small.

When the set point of air temperature in the culture room is lower than the outside air temperature, the cooling load due to heat transmission through the walls and air infiltration through the leaks would account for approximately 10% of the total cooling load. The cooling load due to lighting accounts for the rest (about 90%) of the total cooling load.

The electricity consumption per unit cooling load generally increases with increasing the outside air temperature when an air-to-air cooling system is employed, because the C.O.P. of an air-to-air cooling system generally decreases with increase in the outside temperature.

In colder regions or in winter, on the other hand, the culture room must be heated during the dark period. However, the heating load can be minimized if the photoperiod is set during the night and the dark period set during the day. Furthermore, heating can be achieved using less expensive fuels such as oil and gas in order to reduce total heating cost to about 10% of the cooling cost for even in the coldest regions.

The cooling cost in summer or in warmer regions can be reduced in the same way. That is, the photoperiod is set during the night and the dark period during the day.

Therefore, the most important strategy for minimizing the electricity consumption is to find an efficient way of lighting the plantlets. The second important strategy is to find an efficient way for removing the heat generated by the lamps.

13.1 Reducing Lighting Costs

When trying to find efficient ways to reduce lighting costs, the most important factor is that a considerable portion of photosynthetically active radiation (PAR) emitted from lamps is absorbed by the walls, shelves, vessels, floor, etc. and that only a small portion of it is absorbed by the plantlets in vitro.

We must find ways to increase the ratio of PAR received by the plantlets to PAR emitted from the lamps. The ratio can be relatively easily increased, for example, by use of reflective materials for the inner walls and culture shelves (racks) and by use of transparent materials for vessels closures.

Fujiwara et al. (1987b) showed the effects of vessel type, closure type, and color of the culture shelf on the light environment in vessels placed on the culture shelf.

Figure 16a shows the distribution of PPF on an empty black-colored shelf which has one tube-type cool white fluorescent lamp in its upper part. Figure 16b shows the PPF distribution on an empty white-colored shelf with the same lamp. The average PPF on the white shelf is about three times that on the black shelf. Use of reflective material for the inner walls of the culture shelves and the culture room also increases the PPF on the shelf.

Figure 17a shows the distribution of PPF in the test tubes with aluminum foil on the white shelf with one tube-type cool white fluorescent lamp. Figure 17b shows the PPF distribution in the test tubes with formed semi-transparent plastic caps. The average PPF in the test tubes with plastic caps is more than two times that in test tubes with aluminum foil. That is, use of transparent caps and transparent vessels is important for obtaining a higher PPF in the vessels.

Development of a novel lighting system using normal (transparent) and diffuse-type optical fibers to reduce the lighting cost and to improve the quality of the light environment for the plantlets in vitro is now underway (Kozai unpubl.). With this system, a high-pressure metal halide lamp is used as a light source. The lamp itself is placed outside the culture room. Using a thermal filter, infrared radiation emitted from the lamp is removed outside the culture room and only PAR is sent to the culture room through normal optical fibers. The heat removed by the thermal filter can be used as heat source to cool the culture room with an absorption refrigerating machine (see Fig. 18).

The diffuse-type fibers which emit mostly PAR along the fibers (not from the end of the fibers) are connected with the normal fibers and are placed on the culture shelf adjacent to the plantlets, not above the plantlet, but at the sides of the plantlets. The bottom of the upper shelf is covered with reflective film. The ratio of PAR received by the plantlets to that emitted from the fibers can then be increased to a considerable degree.

In addition, when the plantlets become relatively taller, the lower leaves of the plantlets will receive equivalent levels of PAR to the upper leaves of the plantlets, because light is emitted from the sides. With this lighting system, the lighting costs can be reduced and the light environment improved.

Furthermore, the number of plantlets which are placed in a culture room can be increased with this system, because several vessels can be stacked on the shelf or the distance between the shelves can be made shorter, without decreasing the PPF at the plantlet level.

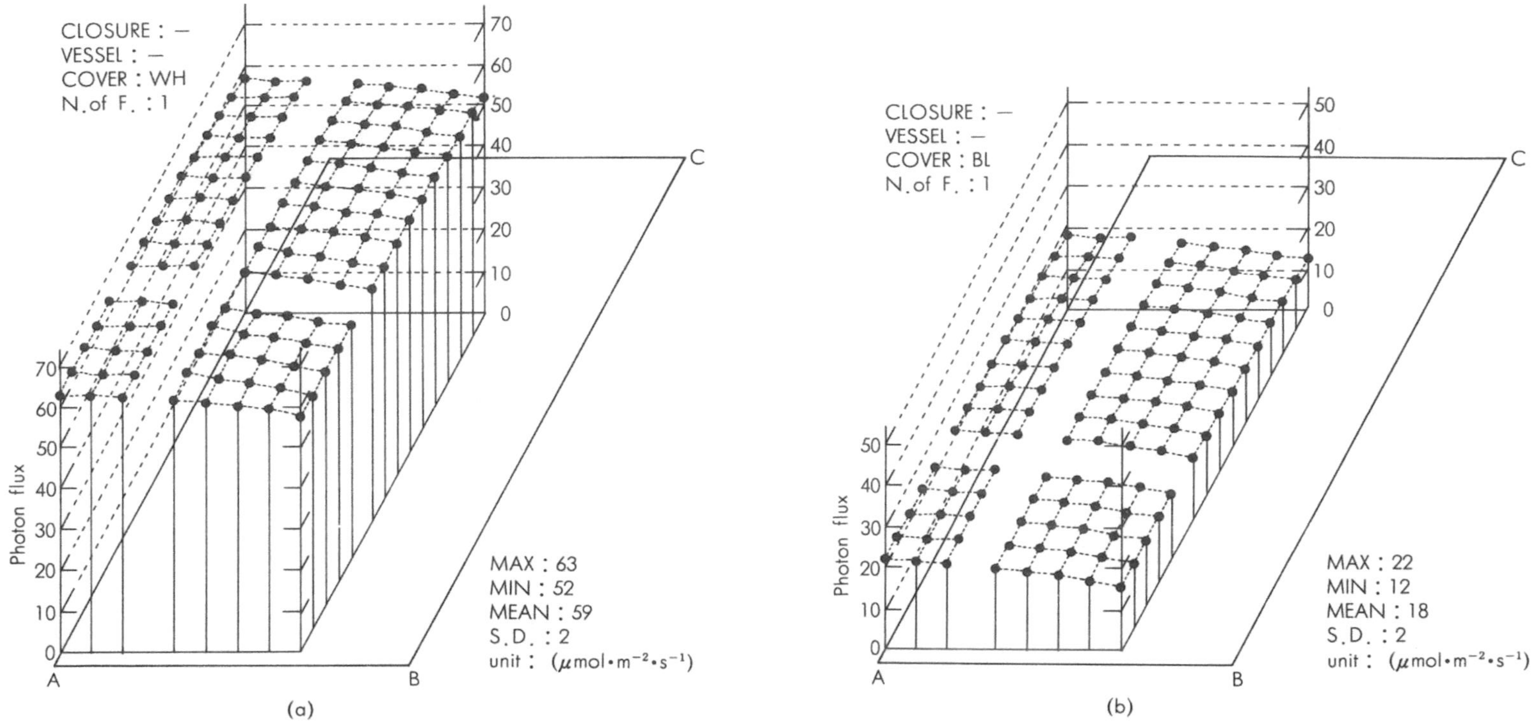

Fig. 16a,b. Distribution of PPF (photosynthetic photon flux) on **a** an empty white-colored culture shelf or **b** an empty black-colored culture shelf. Number of fluorescent lamps = 1. (Fujiwara et al. 1987b)

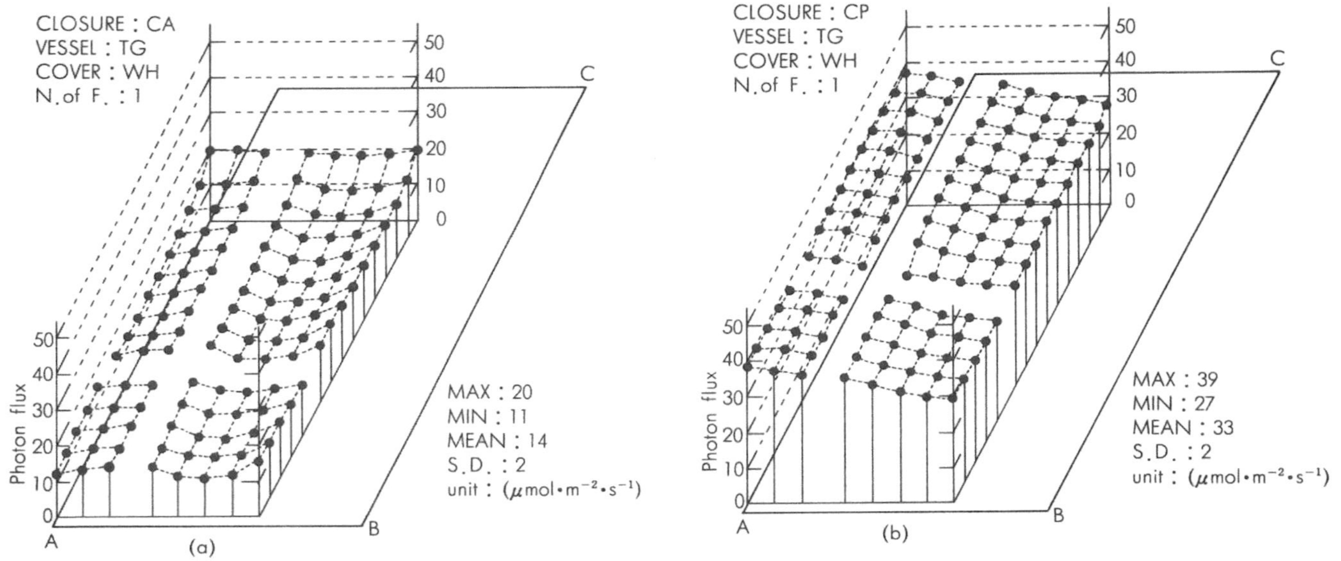

Fig. 17a,b. Distribution of PPF on the white-colored culture shelf. PPF was measured in glass test tubes (*TG*) closed with **a** aluminum foil (*CA*), or **b** semi-transparent formed plastic cap (*CP*). Number of fluorescent lamps = 1. (Fujiwara et al. 1987b)

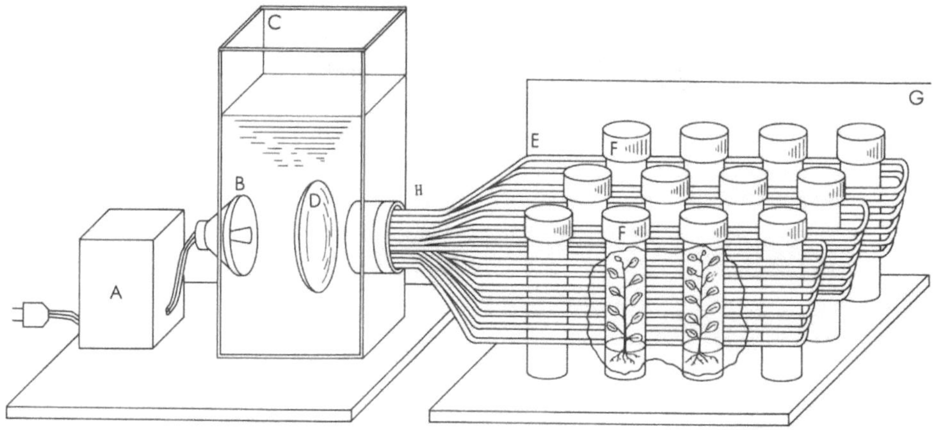

Fig. 18. Schematic diagram of a novel lighting system with diffusive optical fibers for autotrophic micropropagation (Kozai et al. unpubl.). *A* Electric power source; *B* metal halide lamp; *C* water tank for heat absorption; *D* condenser lens; *E* diffusive optical fibers; *F* culture vessels; *G* reflective mirror; *H* transparent (normal) optical fibers

13.2 Reducing the Cooling Costs

The energy balance of a high pressure sodium lamp with and without a reflector is given in Fig. 19. The energy balance of a lamp varies with the thermal and optical characteristics of the reflector. The photosynthetic radiation efficiency of lamps, defined as input power divided by photosynthetically active radiation (PAR, 400–700 nm) emitted from the lamp ranges between 0.17 and 0.26. The rest of the energy input is wasted as longwave (3000 nm and over) and non-PAR shortwave radiation (from 700 nm to 3000 nm), convective, and conductive heat from the lamp, reflector, and ballast, and simply for increasing the cooling load.

Furthermore, some portion of the PAR is uselessly absorbed by the inner walls, shelves, vessels, etc. and some portion of the PAR received by the plantlets is converted into sensible and latent heat. More than 98% of the energy input is probably converted into heat, with only less than 2% of the energy input being converted into chemical energy as carbohydrates by photosynthesis. Therefore, it is essential to reduce the generation of heat in the culture room in order to reduce the cooling load.

Kozai et al. (1987c) showed that by using a water-cooled reflector, schematically shown in Fig. 20, the generation of convective and conductive heat and longwave radiation from the reflector can be reduced (see Fig. 19), without reduction of PAR.

Under conditions of the water temperatures at the inlet being 13 °C and the water flow rate being 3.2 g/s, 50% of the energy input was removed by the water, resulting in a water temperature at the outlet of 25 °C. The temperature distribution of the lamps with different reflectors is given in Table 8.

The warmed water coming out of the reflector can be used as low-temperature heat source and for washing, because the water will not be polluted in the closed-water distribution system. The cooling load can be also reduced using the optical fiber shown in Fig. 18, because in this case, the lamps are placed outside the culture room and only PAR is transmitted through the fiber.

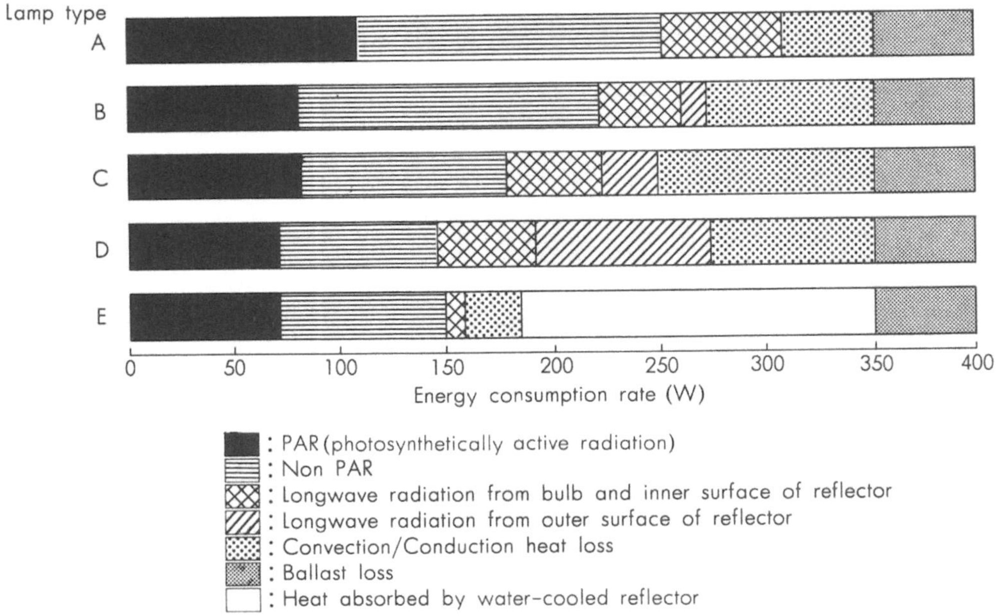

Fig. 19. Energy distribution of a high pressure sodium lamp bulb with or without a reflector (Kozai et al. 1987c). Lamp *A* (without reflector), Lamps *B, C, D,* and *E* (with reflector). Lamp type: *A* Bare lamp bulb (without reflector); *B* with polished aluminum reflector; *C* with white-colored aluminum reflector; *D* with white-colored enameled iron reflector; *E* with water-cooled white-colored enameled iron reflector

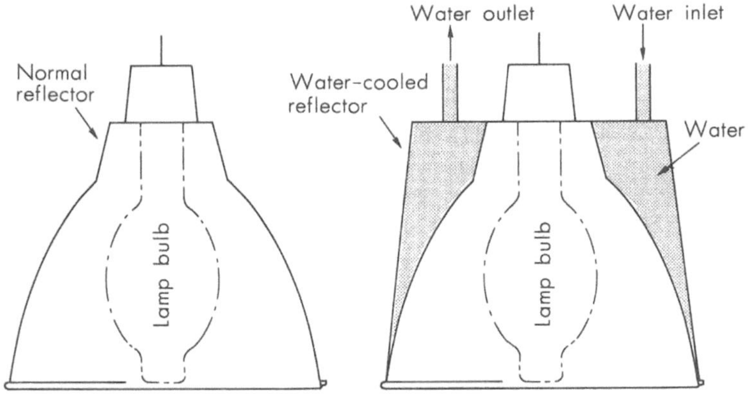

Fig. 20. Schematic diagram of a lamp bulb with normal and water-cooled reflectors. (Kozai et al. 1987c)

Table 8. Temperatures for different positions of the lamp and its surroundings[a]

Lamp type	A	B	C	D	E
Lamp bulb	160	177	175	205	180
Inner surface of reflector	–	46.6	58.0	92.3	30.4
Outer surface of reflector	–	46.6	57.0	78.8	24.6
Ballast	62.6	62.6	62.6	62.6	62.6
Room air	25.1	25.0	25.1	25.0	25.0
Floor	24.7	25.5	25.2	25.2	25.2
Wall	25.7	25.0	25.2	25.0	24.7
Ceiling	25.1	25.0	24.7	24.7	24.7

[a]Unit = °C
For lamp type, see legend to Fig. 19.

13.3 Cost of CO_2 Enrichment

A CO_2 controller widely used for CO_2 enrichment in commercial greenhouses can be used for CO_2 enrichment in the culture room at a cost of about $1500 US. About half of the supplied CO_2 is fixed as carbohydrate by photosynthesis of plantlets and the rest is released to the outside due to air infiltration between the inside and outside of the culture room.

Assume an explant increased its fresh weight from 0.2 g to 1.7 g at the multiplication stage and the dry/fresh weight ratio is 0.13. The dry weight increase is then about 0.2 g [= 0.13 × (1.7–0.2)] and the net CO_2 uptake to increase the dry weight by 0.2 g is roughly 0.3 g [= 0.2 × (44/30)] per plantlet. Assume also that the number of explants micropropagated in a culture room is 200 000 and the multiplication cycle is 1 month. Then, the monthly consumption of CO_2 is 60 kg (= 60 000 g = 0.3 g × 200 000).

Assume the CO_2 concentration in the culture room is kept at 1000 ppm during the photoperiod (16 h/day) and the number of air changes of the culture room per hour is 0.5/h. Assume also that the air volume of the culture room is 250 m^3 and the CO_2 concentration outside the culture room is 350 ppm. The daily loss of CO_2 to the outside is then about 70 kg (= about 40 m^3 = 16 h/day × 0.5/h × 250 m^3 × (1000–350) × 10^{-6} × 30 days). Therefore, the total monthly CO_2 consumption can be estimated to be 130 kg (60 kg + 70 kg). The price of liquid CO_2 is about 1 US$ per kg. Liquid CO_2 in a container will be the best CO_2 source.

Of course, no sugar is required in autotrophic micropropagation. Assume the sucrose concentration in conventional micropropagation is 3% and 10 ml of medium is required for one plantlet. The total sucrose consumption to multiply 200 000 plantlets in the conventional method is then 60 kg (= 60 000 g = 0.3 g per plantlet × 200 000 plantlets).

Therefore, if the cost of 130 kg CO_2 is lower than that of 60 kg sucrose, there is no net increase in cost for carbon source. Even if the former were to be higher than the latter, the net increase in cost should be minimum.

14 Concluding Remarks

In plant micropropagation research much attention has been given to revealing the effects of basal medium composition, plant growth regulators, and other organic and inorganic materials on the differentiation, dedifferentiation, redifferentiation, and growth of explants in vitro under heterotrophic conditions. Much more research should be conducted in the future on the problems related to in vitro environmental physiology, in vitro ecology, and in vitro environmental control engineering under autotrophic or mixotrophic conditions, in order to develop a large-scale automated/robotized mass-propagation system for producing low-cost micropropagated plantlets. The concept of autotrophic micropropagation may contribute toward developing such a system.

Acknowledgments. Some portions of this chapter were previously published as a review paper entitled *High Technology in Protected Cultivation* in Horticulture in the High Technology Era, held in Japan (1988) and a paper presented at the International Symposium on Application of Biotechnology for Small Industries Development in Developing Countries held in Thailand (1988).

References

Charles-Edwards DA (1981) The mathematics of photosynthesis and productivity. Academic Press, New York, pp 22–29

De Proft MP, Maene LJ, Debergh PC (1985) Carbon dioxide and ethylene evolution in the culture atmosphere of *Magnolia* cultured in vitro. Physiol Plant 65:375–379

Desjardins Y, Laforge F, Lussier C, Gosselin A (1988) Effect of CO_2 enrichment and high photosynthetic photon flux on the development of autotrophy and growth of tissue-cultured strawberry, raspberry and asparagus plants. Acta Hortic 230:47–57

Flores H, Pickard J, Hoy M (1988) Production of polyacetylenes and thiophenes in heterotrophic and photosynthetic root cultures of Asteraceae. In: Bioactive molecules Vol 7. Elsevier, Amsterdam, pp 233–254

Fujiwara K, Kozai T, Watanabe I (1987a) Fundamental studies on environments in plant tissue culture vessels (3) Measurements of carbon dioxide gas concentration in stoppered vessels containing tissue-cultured plantlets and estimates of net photosynthetic rates of the plantlets. J Agric Meteorol 43(1):21–30 (in Japanese with English summary)

Fujiwara K, Kozai T, Nakajo Y, Watanabe I (1987b) Measurements and control of physical environment in plant tissue culture vessels (3). Abstr Annu Meet Agric Meteorol Jap, pp 196–197 (in Japanese)

Fujiwara K, Kozai T, Watanabe I (1988) Development of a photoautotrophic tissue-culture system for plantlets at rooting and acclimatization stages. Acta Hortic 230:153–158

George EF, Sherrington PD (1984) Plant propagation by tissue culture. Exegetics, Hants, England, pp 223–227

Goto E, Takakura T (1987) Relationship between light intensity and light energy requirement in the growth of leaf vegetables under artificial light. J Agric Meteorol 43(3):229–232

Grout BWW (1988) Photosynthesis of regenerated plantlets in vitro, and the stresses of transplanting. Acta Hortic 230:129–135

Grout BWW, Aston MJ (1978) Transplanting of cauliflower plants regenerated from meristem culture. II. Carbon dioxide fixation and the development of photosynthetic ability. Hortic Res 17:65–71

Grout BWW, Price F (1987) The establishment of photosynthetic independence in strawberry cultures prior to transplanting. Proc Symp Plant micropropagation in horticultural industries. Arlon, Belgium, pp 55–60

Hayashi M, Kozai T (1987) Development of a facility for accelerating the acclimatization of tissue-cultured plantlets and the performance of test cultivations. Proc Micropropagation in Horticultural Industries (Belgium), pp 123–134

Hayashi M, Nakayama M, Kozai T (1988) An application of the acclimatization unit for growth of carnation explants, and rooting and acclimatization of the plantlets. Acta Hortic 230:189–194

Hartmann HT, Kester DE (1983) Plant propagation. Prentice-Hall, Englewood Cliffs, pp 316–326
Hirosawa Y, Kozai T (1987) Environmental control system for an acclimatization unit. Nogyo Oyobi Engei 62(6):776–781
Honjo T, Takakura T (1987) Effects of CO_2 concentration, light intensity and liquid medium composition for the growth of *Cymbidium* PLB. J Agric Meteorol 43(3):223–227 (in Japanese with English summary)
Hori H (1966) Gravel culture of vegetable and ornamental crops. Agri Hort Yoken-do, Tokyo, pp 210 (in Japanese)
Jones HG (1983) Plants and microclimate. Cambridge University Press, Cambridge, pp 130–169
Kaneko T, Kato R, Takatsuji M (1988) Effects of carbon source on adventitious root formation of butter-head type lettuce. Abstr Annu Meet Environ Control Biol, Japan, pp 44–45 (in Japanese)
Kinoshita T, Kozai T, Fujiwara K (1988) Changes with time in water potential of the culture medium during the multiplication of in vitro plantlets. Abstr Japan Soc Hort Sci Spring Meet, pp 234–235 (in Japanese)
Kozai T (1988) High technology in protected cultivation. In: Horticulture in high technology era. A lecture note of a special lecture held in Tokyo, May 1988, pp 1–43
Kozai T, Iwanami Y (1988) Effects of CO_2 enrichments and sucrose concentration under high photon flux on the tissue cultured plantlets growth of carnation (*Dianthus caryophyllus* L.) during the preparation stage. J Jap Soc for Hort Sci 57:255–264
Kozai T, Kubota C (1988) Basic components for promoting autotrophic growth of in vitro plantlets at multiplication and preparation stages. Abstr Japan Soc Hort Sci Spring Meet, pp 402–403 (in Japanese)
Kozai T, Sekimoto K (1988) Effects of the number of air changes per hour of the stoppered vessel and the photosynthetic photon flux on the carbon dioxide concentration inside the vessel and the growth of strawberry plantlets in vitro. Environ Control Biol 26(1):21–29 (in Japanese with English summary)
Kozai T, Fujiwara K, Watanabe I (1986a) Fundamental studies on environments in plant tissue culture vessels. (1) Relation between the culture medium composition and water potential of liquid culture media. J Agric Meteorol 42(1):1–6 (in Japanese with English summary)
Kozai T, Fujiwara K, Watanabe I (1986b) Fundamental studies on environments in plant tissue culture vessels. (2) Effects of stoppers and vessels on gas exchange rates between inside and outside of vessels closed with stoppers. J Agric Meteorol 42(2):119–127 (in Japanese with English summary)
Kozai T, Hayashi M, Hirosawa Y, Kodama T, Watanabe I (1987a) Environmental control for acclimatization of in vitro cultured plantlets. (1) Development of the acclimatization unit for accelerating the plantlet growth and the test cultivations. J Agric Meteorol 42(4):349–358 (in Japanese with English summary)
Kozai T, Iwanami Y, Fujiwara K (1987b) Effects of CO_2 enrichment on the plantlet growth during the multiplication stage. Plant Tissue Cult Lett 4(1):22–26 (in Japanese with English summary)
Kozai T, Koyama Y, Watanabe I (1987c) Energy balance of a plant growth lamp with water cooled reflector for reducing cooling load, Abstr Annu Meet Agric Met of Japan, pp 186–187 (in Japanese)
Kozai T, Oki H, Fujiwara K (1987d) Environmental control for mass propagation of tissue-cultured plantlets (2). Abstr Japan Soc Hort Sci Spring Meet, pp 366–367 (in Japanese)
Kozai T, Oki H, Fujiwara K (1988a) Effects of CO_2 enrichment and sucrose concentration under high photosynthetic photon flux on growth of tissue-cultured cymbidium plantlets during the preparation stage. Proc Micropropagation in Horticultural Industries, Arlon, Belgium, pp 135–141
Kozai T, Koyama Y, Watanabe I (1988b) Multiplication and rooting of potato plantlets in vitro with sugar-free medium under high photosynthetic photon flux. Acta Hortic 230:121–127
Kozai T, Kubota C, Watanabe I (1988c) Effects of basal medium component on the growth of carnation plantlets in auto- and mixotrophic tissue culture. Acta Hortic 230:159–166
Kozai T, Iwabuchi K, Watanabe I (1988d) Effects of low dark period temperatures and CO_2 enrichment during the photoperiod on the growth of strawberry plantlets in vitro (1), (2). Abstr Annu Autumn Meet Jpn Soc Hort Sci: 270–271, 272–273
Murashige T, Skoog F (1962) A revised medium for rapid growth and bioassay with tobacco tissue cultures. Physiol Plant 15:473–497
Pierik RLM (1987) In vitro culture of higher plants. Martinus Nijhoff, Dordrecht, pp 344
Pieters B, Maene L, Buggenhout P (1987) Evolution of some growth factors during tissue culture. Meded Fac Landbouwwet Rijksuniv Gent 52(4):1409–1416
Shimada N, Tanaka F, Kozai T (1988) Effects of low O_2 concentration on net photosynthesis of C_3 plantlets in vitro. Acta Hortic 230:171–175
Thompson MR, Thorpe TA (1987) Metabolic and non-metabolic roles of carbohydrates. In: Bonga JM, Durzan DJ (eds) Cell and Tissue Culture in Forestry. Martinus Nijhoff, Dordrecht, pp 89–112

II.6 In Vitro Nursery System for Vegetable Crops – Tomato and Melon

S. OHKI, K. NASUDA, Y. MORI, and H. KATSUTA[1]

1 Introduction

1.1 Necessity of Tissue Culture Cloning for Seed-Propagated Vegetables

In the last few decades, the tissue culture technique of higher plants has been well developed and used as a vegetative multiplication method in many species, at least on an experimental scale. However, application of this technique in horticulture is restricted to perennial plants. It is considered that annual vegetable crops are not suitable for in vitro propagation, because the cost is high and hundreds of thousands of young plants are required at one time.

In many cases, it takes many years to create one new cultivar of annual vegetable crops, and it is almost impossible to obtain a cultivar accumulating several recessive characteristics even employing the F_1 breeding system, while we always need new good-quality cultivars. On the other hand, during cross-breeding, one can find several plants expressing numerous recessive characteristics. So we regarded tissue culture propagation as a useful method for these plants obtained in the F_2 or F_3 generations (Ohki et al. 1988); studies on tomato and melon are detailed here.

1.2 Tomato

Tomato is an important fruit vegetable. Bacterial wilt caused by *Pseudomonas solanacearum*, a soil-borne and vascular pathogen, is one of the most important diseases of tomato in warm or temperate areas with high humidity. This pathogen is found in countries that fall within latitudes 45° N and S (Thurston 1976). Especially due to successive cropping, the density of this pathogen increases and it becomes impossible to culture tomato even using the resistant cultivars at present on the market. In Japan, to avoid this soil-transmitted disease, grafted nursery plants are used in 9.8% of total tomato cropped areas, and this limits the tomato production. These resistance genes are introduced by wild varieties of tomato, such as *L. esculentum* var. *pyriforme*, var. *cerasiforme*, or var. *pimpinellifolium*. However, use of these wild varieties causes excessive vegetative growth, resulting in insufficient levels of fruit setting or poor soluble solids content. McGuire (1960) encountered considerable difficulty in combining com-

[1]Laboratory of Horticulture and College Farm, Fukui Prefectural College, 97–21–3 Obatake-cho, Fukui 910, Japan

Biotechnology in Agriculture and Forestry, Vol. 17
High-Tech and Micropropagation I (ed. by Y.P.S. Bajaj)

mercial fruit size and good horticultural characters with optimum levels of resistance in this material. Furthermore, one type of resistance to bacterial wilt in tomatoes was interpreted as polygenic by Singh (1961). Resistance appeared to be conditioned largely by recessive genes and it is not easy to fix this character for seed propagation.

1.3 Melon

The origin of *Cucumis melo* is not clear, but it appears to be indigenous to Africa (Whitaker and Bemis 1976). Once domesticated, it exploded into numerous cultivars that rapidly dispersed throughout Europe and Asia. European cultivars, which are classified as *C. melo* var. *reticulatus* Naudin or var. *cantalupensis* Naudin, have high commercial value but the cultivation is difficult in rainy and humid zones. On the other hand, Asiatic cultivars belonging to *C. melo* var. *makuwa* Makino or var. *conomon* Makino have resistance to diseases such as gummy stem blight. However, the soluble solids content of these Asiatic cultivars is low and the dessert quality of the fruit is poor. Many efforts have been made to improve disease resistance and fruit quality by the hybridization between different varieties. As shown in Table 1, many desirable characteristics of fruit are recessive and the resistance to diseases, although some of them are dominant, is difficult to fix. Tamai et al. (1962) reported that it took nine generations to create one cultivar with powdery mildew resistance which is considered as incomplete dominant (Bohn and Whitaker 1964). Using tissue culture cloning techniques for the mother plants with desirable characters which are selected in F_2 or F_3 generations, it is possible to distribute nursery plants to growers.

Table 1. Inheritance in the muskmelon. (Summarized from Hayes and Garber 1921; Shoda and Sadaoka 1936a,b; Hagiwara and Kamimura 1937)

Characters	Dominant or incomplete dominant	Recessive
Color of skin	Yellow	Green
Form of fruit	Round	Elliptical
Ribbing	Deep	None
Netting[a]	Netted	Smooth
Size of seed	Large	Small
Size of fruit	Large	Small
Taste	Sour	Sweet

[a]Netted was recessive in the cross var. *reticulatus* × *makuwa*. (Shoda and Sadaoka 1936b)

2 Breeding Programs and Their Results

2.1 Tomato

2.1.1 Selection Method

The selection was realized in 1986–1987. The bacterial wilt-resistant female used in this study was line RP. This line was selected from a cross between cultivar Kyoryoku-TVR and TBa, obtained from crosses involving NC1953–64N. Good-quality commercial cultivar Momotaro was used as the pollen parent to improve fruit quality. From the F_2 generation of this cross 216 plants were cultured from April in the College Farm's plastic greenhouse highly infected by *P. solanacearum*. The selection objective was bacterial wilt resistance, high yield, precocity, and high soluble solids content of fruits.

2.1.2 Selection Results in the F_2 Generation

As shown in Table 2, among 216 plants cultured, we obtained 86 plants (39.8%) without any symptoms of bacterial wilt. Acosta et al. (1964) reported heritability of bacterial wilt resistance in tomato by crossing between susceptible variety and *L. pimpinellifolium* HES 5808–2. The percentage of surviving plants after 17 weeks was similar to our result, i.e., about 25 to 30%. Among these surviving plants, 21 plants were selected which bore marketable fruits. We propagated these 21 lines with cuttings and carried out the secondary selection using ten plants per line. As a result, four plants were selected from a view point of overall characteristics. Thus, the selection rate was 1.9%. For the tissue culture propagation, we chose only one plant, line No. 31, which showed the characteristics closest to the breeding objectives. Acosta et al. (1964) suggested that bacterial wilt resistance is controlled by a comparatively small number of major genes. However, bacterial wilt resistance is the only one in which this unfavorable association of higher resistance levels and smaller fruit sizes has continued to persist. In the course of our selection, many plants showed excessive vegetative growth, which may be caused by the introduction of wild species. Therefore it will be quite difficult to fix bacterial wilt resistance and good fruit quality characteristics, such as high solids soluble content and large fruits, for seed propagation.

Table 2. Appearance of the desirable characters in F_2 generation of the tomato. A total of 216 plants were investigated. (Ohki et al. 1988)

Characters	Rate of appearance
Disease resistance[a]	86 (39.8%)
Disease resistance + Marketable fruits	21 (9.7%)
Disease resistance + Good quality	4 (1.9%)
Totally good quality	1 (0.5%)

[a]Without any symptoms of bacterial wilt.

2.2 Melon

2.2.1 Selection Method

The hybridization between cultivars of muskmelon (*Cucumis melo* var. *reticulatus*) and the selection of these filial generations were carried out from 1984 to 1987. Young seedlings were submitted to disease-resistance selection. For the powdery mildew resistance selection, aqueous suspension of the spores of *Sphaerotheca fuliginea* were sprayed on 10-day-old seedlings (Ko et al. 1981). For the selection of gummy stem blight resistance, homogenized mycelium of *Mycosphaerella melonis* (strain No. 79704) cultured 1 week on the PDA medium was inoculated into young seedlings, using needles. These selections were carried out in the incubator with artificial lighting. The surviving seedlings were planted in the plastic greenhouse highly infected by *Fusarium oxysporum* f.sp. *melonis* for the selection of fusarium wilt resistance. The selection objectives were disease resistance and good fruit-bearing quality, i.e., high soluble solids content, good appearance, thick and smooth mesocarp tissue. The soluble solids content expressed by degrees Brix were measured with a reflactometer sampling the innermost part of the mesocarp.

2.2.2 Field Selection Results in the F_2 Generation

The percentage of plants expressing desirable characteristics in the filial generations of the hybrid between cv. Melody No. 2 and cv. Earl's Favourite is shown in Table 3. In the F_2 generation, 84.0% of plants set fruits up to the 17th node, but in the F_5 generation this rate declined to 52%. The plants not diseased by gummy stem blight appeared at a low rate and this rate declined as the generation advanced, although Norton and Prasad (1966) reported that this disease is conditioned by two pairs of dominant genes. We observed the same tendency in the fruit weight, fruit appearance, and high soluble solids content. Hence these three characters are considered to be recessive. However, according to Lumsden, heavy fruit weight and netting are dominant (Hayes and Garber 1921), but the latter is the main factor for good fruit appearance. Sagared described the acidity of fruit as dominant and sweetness as recessive (in Shoda and Sadaoka 1936a). In our

Table 3. Appearance rate of desirable characters in F_2 to F_5 generations of melon. Fifty plants were investigated for F_2 and F_3 generations, 48 plants for F_4 generation, and 25 plants for F_5 generation. (Ohki et al. 1988)

Filial generations	Set fruit up to 17th node	Powdery mildew resistance	Gummy stem blight resistance	Fruits heavier than 1.4 kg	Fruits with good Netting[a]	Fruits with high Brix[b]
F_2	84.0%	64.0%	38.0%	70.0%	58.0%	42.0%
F_3	78.0%	82.0%	34.0%	64.0%	42.0%	48.0%
F_4	64.6%	79.2%	27.1%	47.9%	31.3%	29.2%
F_5	52.0%	84.0%	20.0%	44.0%	32.0%	36.0%

[a]Netting was marked on the basis of 5 points. Good means higher than 4 points.
[b]Higher than 13.5% of soluble solids content.

breeding process, the soluble solids content declined after the F_4 generations, which indicates recessiveness of this gene. On the other hand, the plants showing resistance to powdery mildew appeared at a high rate and did not change up to the F_5 generation. Three plants and four plants, respectively, showing desirable characteristics on the whole, were selected from among 50 plants in each of the F_2 and F_3 generations. Therefore the selection efficiency was 6% for the second generation and 8% for the third generation.

Several authors have reported on the heredity of the melon plant. However, there is disagreement in their results and also in ours. For example, round fruit form is estimated as dominant by Lumsden (1914) but as incomplete dominance by Hagiwara and Kamimura (1937). Netting is reported as dominant by Sagared (in Shoda and Sadaoka 1936a) but as recessive by Shoda and Sadaoka (1936b) in the crossing between var. *reticulatus* and the non-netting variety *makuwa*. These disagreements may be caused by the difference in varieties used. Thus, the heritability of these characters is not clear. Our results clearly showed inbreeding depression in F_4 and F_5 generations, so the selection should be carried out in F_2 and F_3 generations.

3 Tissue Culture Propagation of Selected Plants

3.1 Media and Culturing Conditions

The basal MS culture medium consisted of macroelements and Fe-EDTA (Murashige and Skoog 1962), micro- and organic elements of Ringe and Nitsch (1968), sucrose 30 g/l, and jelling agent (agar: 7 g/l or Gellun gum: 2 or 3 g/l). As growth regulators, indole-3-acetic acid (IAA), indole-3-butylic acid (IBA), α-naphthaleneacetic acid (NAA), 6-benzylaminopurine (BAP), kinetin (KIN), and 6-(γ,γ-dimethylallylamino)-purine (2-iP) were used. The pH of the medium was adjusted to 5.8 with 0.1 N NaOH prior to the addition of the gelling agent. A test tube of 25 mm diameter and 150 mm length with a metal cap, polycarbonate cylinder of 100 mm diameter and 110 mm height, and polycarbonate box of 70 × 70 × 95(H) mm were used as culture vessels. For better aeration of polycarbonate culture vessels, two holes of 10 or 15 mm diameter sealed with gas-permeable film were made. The media were autoclaved for 15 min at 120 °C.

All plant materials were cultured in the greenhouse. The upper three nodes of the lateral branches were sterilized for 10 min in 1% NaClO containing 0.1% of polyoxyethylenesorbitan monolaurate, and rinsed three times in sterile tap water. Cultures were placed in the culture room (12-h daylength, light intensity 6000 or 12 000 lx at culture level, 23 ± 2 °C) or in the growth chamber (12-h daylength, light intensity 5000 lx at culture level, 23 ± 0.5 °C, 65 ± 5% RH).

3.2 Tomato

Many authors have reported on in vitro multiplication of the tomato using various organs (see Sink and Reynolds 1986). Whole plant regeneration from callus is also possible, but abnormality was often reported. Ohki et al. (1978) found that 12% of morphologically abnormal plants regenerated from callus formed slightly on the cut end of hypocotyl segments. They also noted that this abnormality was not always accompanied by a change in the root chromosome number, which was found to be tetra- or mixoploid in about 49% of plants. Sibi (1982) also reported that local morphological abnormality in the plants regenerated from cotyledon-derived callus was not clearly related to the chromosome number. Therefore, we used node culture techniques for the multiplication of selected line No. 31. For the shoot regeneration from leaf or cotyledon explants, many authors described that IAA was the most effective auxin. However, the kind of cytokinin was different according to the authors: 2-iP (Ohki et al. 1978), BAP (Kartha et al. 1976), zeatin (Montagno 1986), or kin (Uddin et al. 1988). This may be caused by different sensitivity to growth regulators in different cultivars. We used 10 μM of 2-iP combined with 2 μM of IAA for the initial culture. 2.3 ± 0.95 shoots per explant were obtained in the initial culture. Only 61% of these shoots obtained in the initial culture rooted on the rooting medium with 5 μM of IBA. When these shoots were subcultured on the proliferating media supplemented with 5 μM of 2-iP and 1 μM of IAA every 25 days, the number of shoots multiplied about five times and 100% of these shoots rooted on the rooting medium.

3.3 Melon

Little is known about in vitro propagation of melon. Moreno et al. (1985), Oridate and Oosawa (1986) and Niedz et al. (1989) reported plant regeneration from callus obtained from the juvenile organs, i.e., hypocotyls and cotyledons. We are not interested in such juvenile organs, because our culture must be started after the fruit quality is estimated. We tried to regenerate shoots from leaf callus, but were unsuccessful. Therefore, we applied node culture for the multiplication (Ohki et al. 1986). Table 4 shows the effect of three cytokinins (Ohki et al. 1990). Only BAP was effective for shoot formation. 2-iP promoted callus proliferation. A screening test was carried out with 16 combinations of BAP and NAA concentrations. As shown in Figs. 1 and 2, the following results were obtained:

- BAP was essential for shoot regeneration. With NAA alone, very few shoots were formed.
- BAP concentration higher than 5 μM caused vitrification and malformation of shoots.
- Abundant callus was produced when both BAP and NAA were added. Abundant callus inhibited shoot differentation.
- We could obtain two to three shoots per explant on the medium supplemented with 0.5 to 1 μM of BAP alone.

We subcultured the nodal sections of differentiated shoots in the initial culture. (Fig. 3). As in the initial culture, BAP at a concentration of 0.5 μM was suitable for shoot proliferation (2.8 ± 0.4 shoots per explant). With this culturing system, we could obtain at least 200 plants from a single selected plant within 3 months, and 3×10^5 plants can be produced per year by subculturing continuously.

However, the propagation rate greatly depended on lines and cultivars (Ohki et al. 1990). The difference of shoot-forming capacity was clear between varieties of *C. melo: reticulatus, makuwa, cantalupensis*, and their hybrids (Table 5). The shoot-forming capacity estimated by percentage of explants differentiated shoots was highest in

Table 4. Effect of different cytokinins on percentage of shoot formed explants (SP), mean number of shoots per explant (SN), and callus dry weight (DW) in the initial node culture of melon cv. Melody No. 2. Data were recorded 30 days after the initiation of culture. (Ohki et al. 1990)

Cytokinins		Concentration of cytokinins (μM)			
		0	0.5	1	5
BAP	SP(%)	18.2	30.0	25.0	63.3
	SN	0.2 ± 0.12[a]	0.4 ± 0.22	0.3 ± 0.19	1.0 ± 0.30
	DW(mg)	26.6 ± 1.8	38.0 ± 14.0	57.0 ± 11.2	119.6 ± 13.0
Kin	SP(%)	0	0	0	0
	DW(mg)	37.4 ± 6.4	30.3 ± 1.7	43.3 ± 8.8	74.7 ± 13.6
2-iP	SP(%)	0	0	0	0
	DW(mg)	40.3 ± 4.2	64.5 ± 10.0	96.2 ± 15.9	174.2 ± 10.5

[a]Each value represents the mean ± SE of 12 replications.

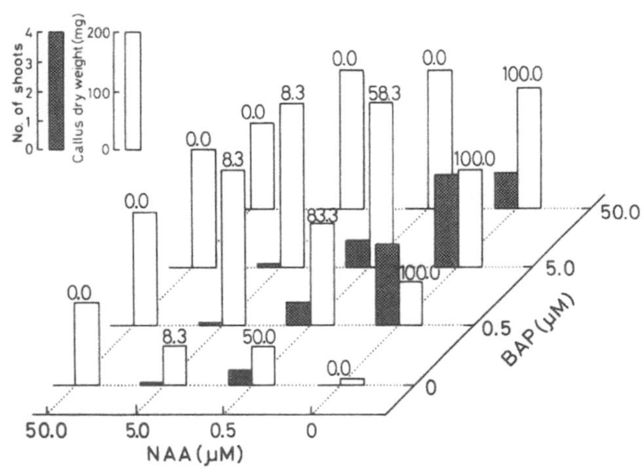

Fig. 1. Effects of BAP and NAA on shoot regeneration and callus formation from nodal sections of melon (hybrid F1: progeny of cv. Ams × cv. Ochiuri) 35 days in culture. *Figures above each bar* indicate percentage of explant-regenerated shoots. Twelve replications per treatment. (Ohki et al. 1988)

var. *makuwa* cv. Ginsen and was lowest in var. *cantalupensis* cv. Cantor. The hybrid showed the intermediate capacity on the medium with 0.5 μM of BAP. Bouabdallah and Branchard (1986) also reported the difference of shoot formation rate from cotyledon-derived callus between var. *cantalupensis* cv. Cantaloup Charentais T and var. *makuwa* cv. Ogon No. 9, in which the former was higher than the latter. Among the lines selected, line 15–6 had the highest capacity for shoot formation. Ohki et al. (1978) reported the maternal effect and heterosis in shoot-forming capacity in hypocotyl culture of the tomato. When raising new lines for tissue culture propagation, these findings should be taken into consideration.

Fig. 2. Shoot regeneration from nodal sections of melon (the same hybrid as Fig. 1) cultured on media supplemented with BAP for 35 days. *Left to right* 0, 0.5, 5, 50 μM of BAP. (Ohki et al. 1988)

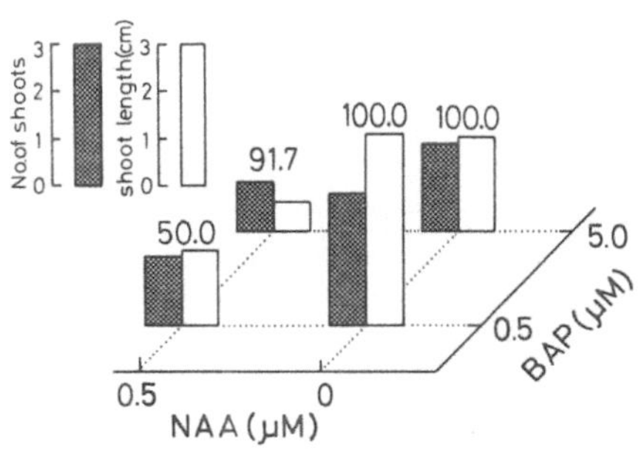

Fig. 3. Effects of BAP and NAA on shoot regeneration from nodal sections of melon (the same hybrid as Fig. 1) 30 days in subculture. *Figures above each bar* indicate percentage of explant-regenerated shoots. Twelve replications per treatment. (Ohki et al. 1988)

Rooting from the shoots obtained was achieved on the medium with NAA or IBA at a concentration of 0.1 to 0.5 μM. However, some lines tested were quite difficult to root. To improve root formation and development, we tested the solid half strength basal media containing 1 μM of IBA solidified with agar (0.65%) or Gellun gum (0.2%), and the same liquid media with support means (rock wool or vermiculite). As shown in Table 6, Gellun gum was the best in the incidence of root formation and their development (Ohki et al. 1990). On vermiculite, although the number of rooted shoots was small, root hair development was promoted. This root hair development makes acclimatization easy. Melon roots formed in vitro are thin and are detached easily from shoots when they are transferred from test tubes. Gellun gum is less sticky and it is comparatively easy to pull out the root system. Therefore, we chose Gellun gum solidified media for the rooting.

Table 5. Percentage of shoot-formed explants (SP), mean shoots number per explant (SN), and mean maximum shoot length (SL) in three melon cultivars and their hybrid. Data were recorded 30 days after the initiation of culture. (Ohki et al. 1990)

Cultivars	Concentration of BAP					
	0.1 μM			0.5 μM		
	SP(%)	SN	SL(cm)	SP(%)	SN	SL(cm)
Ginsen (GIN)	76.9	1.0 ± 0.0[a]a[b]	12.0 ± 4.2 a	100.0	1.7 ± 0.2 ab	14.8 ± 2.4 ab
Honey Dew (HD)	52.6	1.0 ± 0.0 a	1.5 ± 0.3 a	85.7	1.3 ± 0.1 a	8.4 ± 2.1 a
HD × GIN	17.4	1.0 ± 0.0 a	1.0 ± 0.0 a	70.3	2.1 ± 0.3 b	22.4 ± 3.9 b
Cantor	28.0	1.1 ± 0.1 a	6.4 ± 3.2 a	64.0	1.9 ± 0.2 b	19.7 ± 3.9 b

[a]Each value represents the mean ± S.E. of 12 replications.
[b]Mean separation within columns by Duncan's multiple range test, 5% level.

Table 6. Effect of media type on root formation from subcultured shoots of melon (F_3 of cv. Melody No. 2 × cv. Fukamidori). Half strength basal media supplemented with 5 μM of IBA were used. Data were recorded 40 days after the initiation of culture. (Ohki et al. 1990)

Media type	Gelling agents or supporting means	Percentage of explants rooted (%)	Mean number of roots per explant	Mean maximum shoot length (cm)
Solid	Agar	75.0	2.2 ± 0.7[a]a[b]	2.8 ± 0.6 a
	Gellun gum	91.7	2.9 ± 0.5 a	3.4 ± 0.6 a
Liquid	Rock wool	58.3	1.5 ± 0.4 ab	2.3 ± 0.9 a
	Vermiculite	41.7	0.6 ± 0.2 b	1.3 ± 0.7 a

[a]Each value represents the mean ± S.E. of 12 replications.
[b]Mean separation within columns by Duncan's multiple range test, 5% level.

Another difficulty in tissue culture of melon was vitrification. Pierik (1987) summarized the cause of this physiological condition. In our case, the high cytokinin level and excessive water uptake may be the main factors. Debergh et al. (1981) concluded that they succeeded in overcoming vitrification in mass propagation of glove artichoke only by raising the agar concentration to 1.1% instead of 0.6%. Vitrified shoots were almost incapable of forming roots, so we used polycarbonate culture vessels with two holes sealed with fluorocarbon membrane filters (pore size 0.5 μM, Milliseal by Nihon Millipore Kogyo K.K.) to reduce air humidity in the vessel (Fig. 4), and we increased the concentration of Gellun gum to 0.3%.

To increase the percentage of successfully acclimatized plantlets, a well-developed root system and two or three leaves of at least 10 mm diameter are needed. Kozai and Iwanami (1988) pointed out the importance of autotrophic growth in vitro under a high photosynthetic photon flux for healthy plantlet production. We obtained good results for this purpose by using high light intensity of 12 000 lx by fluorescent lamps.

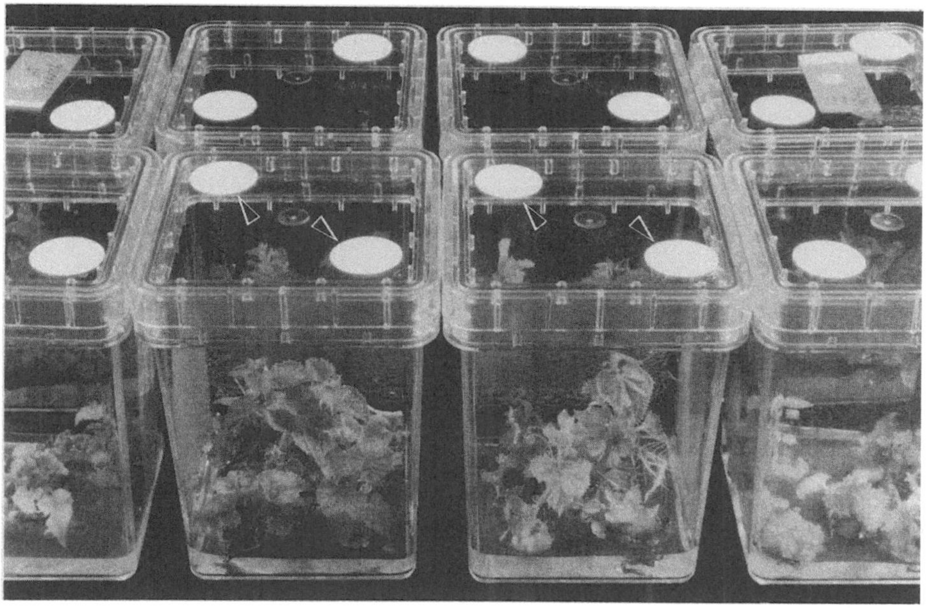

Fig. 4. Polycarbonate culture vessels with two holes (15 mm diameter) sealed by fluorocarbon membrane filters (*arrow heads*). (Unpubl.)

4 Culture of in Vitro-Propagated Plants

4.1 Tomato

Tissue culture-propagated plants showed no morphological malformation or bacterial wilt symptoms. Compared with plants propagated by cuttings, they showed less vegetative growth, resulting in good fruit set. However, the internode number of the first floral cluster was varied from 4 to 11 (Table 7). The floral clusters lower than the seventh internode should be removed by hand. The yield of the tissue culture-propagated plants was 7352.5 kg/ha in the fall cropping without heating. This yield was about 1000 kg higher than normal cultivars.

4.2 Melon

To ascertain the variation of growth rate among the in vitro-propagated plants, 16 in vitro-propagated plants of line 15–6 were cultured in hydroponics and their stem elongation was measured every week (Fig. 5). As compared with F_1 cultivar Melody No. 2, the coefficient of variation was not very high, except 2 to 4 weeks after planting.

Tissue culture-propagated line 965–8 was cultured to compare with three F_1 cultivars in the same plastic greenhouse. One of the F_1 cultivars was diseased by gummy stem blight before setting fruits. As shown in Fig. 6, line 965–8 was superior to F_1 cultivars in all characteristics except soluble solids content. Compared with the original mother plant, fruits of tissue-cultured plants showed the same appearance but were

Table 7. Fruiting internode order of in vitro-propagated tomato. A total of 72 plants were investigated. (Ohki et al. 1988)

Internode order	Number of plants
4	2(2.8%)
5	8(11.1%)
6	12(16.6%)
7	10(13.9%)
8	9(12.5%)
9	19(26.4%)
10	11(15.3%)
11	1(1.4%)

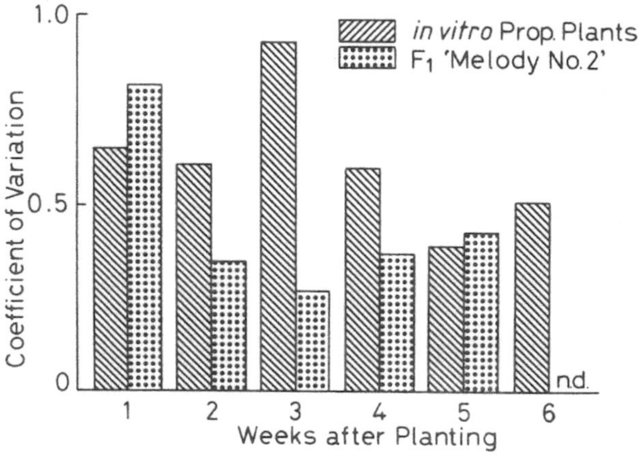

Fig. 5. Coefficient of variation in stem elongation of in vitro-propagated line 15–6 and F_1 cultivar Melody No. 2 of melon cultured in hydroponics. Sixteen plants of each were observed. *n.d.* no data. (Ohki et al. 1988)

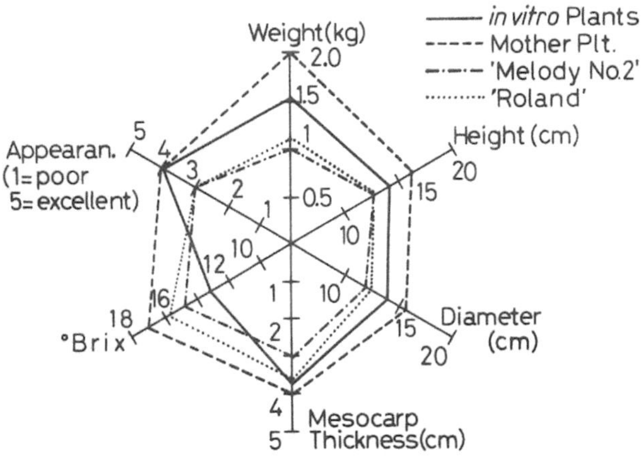

Fig. 6. Fruit characters of in vitro-propagated line 965–8 of melon compared with F_1 cultivars and the original mother plant. (Ohki et al. 1988)

smaller. Almost all lines of the melon selected, except line 15–6, were difficult to root, and development of the root system at the beginning of the culture was limited. This might cause reduction in leaf area and fruit size, although the fruit appearance was excellent.

In order to improve initial growth after acclimatization, shoots of line 15–6 formed in vitro were directly grafted upon pumpkin or melon (Fig. 7). These grafted and nongrafted plants were cultured in the plastic greenhouses of the College Farm and of melon growers. The quality of the fruits of these plants is shown in Fig. 8. The non-grafted plants cultured in the College Farm, where the soil condition is quite poor, produced the smaller fruits. On the other hand, in growers' greenhouses, nongrafted plants grew vigorously (Fig. 9) and the fruits were largest. For this line 15–6, the average soluble solids content was higher than 15%, regardless of being grafted or nongrafted and of soil conditions. However, for the stable production it is better to distribute grafted plants. The acclimatization process can be substituted by this procedure.

5 Conclusions

In our experiment, for both tomato and melon, no morphological disorders were noticed. The morphological characteristic was the same as those of the mother plants

Fig. 7. Grafted melon plants ready to transfer to the greenhouse after 10 days' acclimatization in the growth chamber. In vitro-propagated shoots were directly grafted onto 10-day-old seedlings of melon cv. Barnet. (Unpubl.)

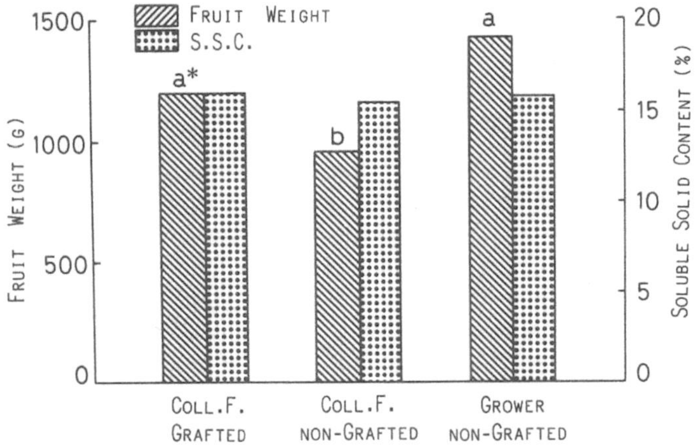

Fig. 8. Fruit weight and soluble solids content (S.S.C.) of in vitro-propagated line 15–6 of melon. Grafted and nongrafted plants were cultured in the College Farm and nongrafted plants were cultured by the melon grower. Means of ten plants each. (Ohki et al. 1988). *Mean separation by Duncan's multiple range test, 5% level

Fig. 9. In vitro-propagated melon line 15–6 nongrafted cultured by the melon grower. Spring cropping in 1987. (Ohki et al. 1988)

initially selected. The resistance to diseases was stable during the tissue culture process. It may be concluded at this point that lines obtained with good quality during the hybridization program and lines which are easy to propagate by tissue culture (such as No. 31 of tomato or 15–6 of melon) are suitable for our in vitro nursery system. For this system, the classical breeding process is indispensable, but it is possible to raise new, better cultivars more easily within shorter periods than is the case with seed production. To distribute these tissue culture-propagated plants commercially, an effort should be made to reduce the cost per plant. For easy-to-root species like tomato, rooting and acclimatization can be carried out simultaneously, using the system proposed by the research group of Kozai (Kozai et al. 1987; Fujiwara et al. 1988). This nursery system can be applied on a comparatively small but intensive scale of vegetable crop production, such as protected horticulture.

References

Acosta JC, Gilbert JG, Quinon VL (1964) Heritability of bacterial wilt resistance in tomato. Proc Am Soc Hortic Sci 84:455–462

Bohn GW, Whitaker TW (1964) Genetics of resistance to powdery mildew race 2 in muskmelon. Phytopathology 54:587–592

Bouabdallah L, Branchard M (1986) Regeneration of plants from callus cultures of *Cucumis melo* L. Z Pflanzenzücht 96:82–85

Debergh P, Harbaoui Y, Lemeur R (1981) Mass propagation of globe artichoke (*Cynara scolymus*): evaluation of different hypotheses to overcome vitrification with special reference to water potential. Physiol Plant 53:181–187

Fujiwara K, Kozai T, Watanabe I (1988) Development of a photoautotrophic tissue culture system for shoots and/or plantlets at rooting and acclimatization stages. Acta Hortic 230:153–158

Hagiwara T, Kamimura K (1937) Genetic studies in *Cucumis*. Jpn J Breed 13:71–79

Hayes HK, Garber RJ (1921) Muskmelon. In: Hayes HK, Garber RJ (eds) Breeding crop plants. McGraw-Hill New York, pp 257–258

Kartha KK, Gamborg OL, Shyluk JP, Constabel F (1976) Morphogenetic investigation on in vitro leaf culture of tomato (*Lycopersicon esculentum* Mill. cv. Starfire) and high frequency plant regeneration. Z Pflanzenphysiol 77:292–301

Ko K, Matuzawa Y, Akutsu F, Kajikawa A, Watanabe T (1981) Effect of moisture by artificial inoculation of some powdery mildew pathogens. Ann Phytopathol Soc Jpn 47:368–369

Kozai T, Iwanami Y (1988) Effects of CO_2 enrichment and sucrose concentration under high photon fluxes on plantlets growth of carnation (*Dianthus caryophyllus* L.) in tissue culture during the preparation stage. J Jpn Soc Hortic Sci 57:279–288

Kozai T, Hayashi M, Hirosawa Y, Kodama T, Watanabe I (1987) Environmental control for acclimatization of in vitro cultured plantlets. (1) Development of the acclimatization unit for accelerating the plantlet growth and the test cultivation. J Agric Meteorol 43:349–358

Lumsden D (1914) Mendelism in melons. Bull New Hampshire Agric Exp Sta 172:1–25

McGuire DC (1960) Bacterial wilt – resistant tomatoes. Hawaii Agric Exp Sta Bienn Re 1958–1960:74

Montagno TJ (1986) Somaclonal and radiation induced variation in *Lycopersicon esculentum*. MS Thesis, The Ohio State University, Columbus

Moreno V, Garcia-Sogo, Granell I, Garcia-Sogo B, Roig LA (1985) Plant regeneration from calli of melon (*Cucumis melo* L., cv. 'Amarillo Oro'). Plant Cell Tissue Organ Cult 5:139–146

Murashige T, Skoog F (1962) A revised medium for rapid growth and bioassays with tobacco tissue cultures. Physiol Plant 15:473–497

Niedz RP, Smith SS, Dunbar KB, Stephens CT, Murakishi HH (1989) Factors influencing shoot regeneration from cotyledonary explants of *Cucumis melo*. Plant Cell Tissue Organ Cult 18:313–319

Norton JD, Prasad K (1966) Inheritance of resistance to *Mycosphaerella melonis* (Pass.) Chiu & Walker (gummy stem blight) in *Cucumis melo* L. (muskmelon). Proc. XVII Int Hortic Congr, Maryland, 1:444

Ohki S, Bigot C, Mousseau J (1978) Analysis of shoot-forming capacity in vitro in two lines of tomato (*Lycopersicon esculentum* Mill.) and their hybrids. Plant Cell Physiol 19:27–42

Ohki S, Nasuda K, Mori Y, Katsuta H, Miyajima D (1986) In vitro propagation of disease-resistant melon. Abstr VI Int Congr Plant Tissue Cell Cult, University of Minnesota, p 225

Ohki S, Nasuda K, Mori Y, Katsuta H (1988) Establishment of breeding = in vitro nursery system for vegetable crops. Acta Hortic 230:89–96

Ohki S, Nasuda K, Mori Y, Katsuta H (1990) Application of in vitro multiplication for the annual vegetable crops associated with classical breeding. In: Sangwan RS, Sangwan-Norreel BS (eds) The impact of biotechnology in agriculture. Kluwer Academic, Dordrecht, pp 67–84

Oridate T, Oosawa K (1986) Somatic embryogenesis and plant regeneration from suspension callus culture in melon (*Cucumis melo* L.). Jpn J Breed 36:424–428

Pierik RLM (1987) In vitro culture of higher plants. Martinus Nijhoff Dordrecht, pp 59–60

Ringe F, Nitsch JP (1968) Conditions leading to flower formation on excised *Begonia* fragments cultured in vitro. Plant Cell Physiol 9:639–652

Shoda S, Sadaoka S (1936a) Genetic studies of occidental and oriental varieties of melon (1). Agric Hortic 11:2622–2632

Shoda S, Sadaoka S (1936b) Genetic studies of occidental and oriental varieties of melon (2). Agric Hortic 11:2846–2852

Sibi M (1982) Heritable epigenic variations from in vitro tissue culture of *Lycopersicon esculentum* (var. Monalbo). In: Earle ED, Demarly Y (eds) Variability in plants regenerated from tissue culture. Praeger, New York, pp 228–244

Singh K (1961) Inheritance of North Carolina type of bacterial wilt resistance in tomato (*Lycopersicon esculentum* L.). Master Thesis, University of Hawaii

Sink KC, Reynolds JF (1986) Tomato (*Lycopersicon esculentum* L.). In: Bajaj YPS (ed) Biotechnology in agriculture and forestry, Vol. 2. Crops I. Springer, Berlin Heidelberg New York Tokyo, pp 319–344

Tamai T, Kamido S, Tomari I, Shinohara K (1962) Raising disease resistant melon cultivar. Agric Hortic 37:557–558

Thurston HD (1976) Breeding for resistance to bacterial wilt. In: Sequeira L, Kelman A (eds) Proc 1st Int Planning Conf Workshop Ecol Control Bacterial Wilt North Carolina State Univ, Raleigh, North Carolina, pp 150–155

Uddin MR, Berry SZ, Bisges AD (1988) An improved shoot regeneration system for somaclone production in tomatoes. HortScience 23:1062–1064

Whitaker TW, Bemis WP (1976) Cucurbits. In: Simmonds NW (eds) Evolution of crop plants. Longman, London, pp 64–69

Section III
Storage/Cryopreservation of Cultures, Quiescent Somatic Embryos, Artificial Seed

III.1 Storage and Cryopreservation of in Vitro Cultures

Y.P.S. Bajaj[1]

1 General Account

Cells and tissue cultures are normally stored in a well-lighted room at 25 °C (Fig. 1). However, this requires periodic transfer to fresh media, involving not only man-power and high costs, but also the hazards of contamination, and sometimes the loss of the entire material/germplasm. Moreover, the cell cultures retained on periodic subcultures undergo genetic erosion, such as change in ploidy level, mutations, endomitosis, translocation, gene amplification, etc. Thus, genetic stability of some of the useful somaclonal cell lines generated in vitro cannot be ensured. Another important aspect is the conservation of germplasm. The existing germplasm banks for plants are customarily supplied by seed collections. Although seeds offer an effective storage system for genetic material, they are usually variable and the plants obtained from them do not show the actual traits of the mother plants. At present, there are no reliable methods for the long-term conservation of germplasm from vegetatively propagated crops and those of plants with recalcitrant seed. Moreover, the germplasm of rare, elite, and endangered plant species must be conserved. To circumvent these problems, in vitro methods are being developed/refined, both for short- and long-term storage, and for the conservation of germplasm (see Bajaj 1986a). Cryopreservation of in vitro cultures in liquid nitrogen has enabled plant recovery in a number of species and has been recommended for the long-term storage of germplasm, especially that of the vegetatively propagated crops (Bajaj 1990a).

2 Nonfrozen Storage (Table 1, Figs. 1, 2)

Various methods have been shown to reduce the growth rate of cultures and thus to delay/reduce the subculture frequency. Some of these methods are:

1. *Minimal Medium and Growth Retardants*: The use of minimal medium and growth retardants such as abscisic acid (Henshaw et al. 1978) or growing cultures in a sucrose-free medium (Jones 1974) have helped to delay the subculture/transfer period.

---

[1]Former Professor of Tissue Culture, Punjab Agricultural University, Ludhiana, India.
(Present address: A-137 New Friends Colony, New Delhi 110065, India)

Biotechnology in Agriculture and Forestry, Vol. 17
High-Tech and Micropropagation I (ed. by Y.P.S. Bajaj)

2. *Mineral Oil Overlay*: Caplin reported in 1959 that the mineral oil overlay of carrot callus tissue considerably reduced the rate of growth and subsequently delayed the transfer frequency. Recently, Augereau et al. (1986) stored tissue cultures of various medicinal plants using mineral oil overlay for 4–6 months without subculture.
3. *Desiccation*: Nitzsche (1980) reported the growth of dried carrot callus after 1 year of storage. Recently, Gray (1987; see also Chap. III.2, this Vol.) observed that somatic embryos of carrot, grape, and orchardgrass showed normal germination after desiccation.
4. *Low Pressure/Low Oxygen*: Bridgen and Staby (1981) stored tissue cultures under low atmospheric pressure and low oxygen.
5. *Low Temperature Storage*: The storage of cultures at nonfreezing temperatures (2–8 °C) has been successfully applied to a large number of species (Table 1; see also Aitken-Christie and Singh 1987).

Of all these methods, low temperature storage has been extensively used. The storage of medicinal plant cell cultures at nonfreezing temperatures and its effect on secondary metabolites has been reviewed (Hiraoka 1988). The in vitro plantlets of *Chrysanthemum* (Fig. 1) and *Petunia*, mass propagated and stored at 4–5 °C for up to 6 years (1978–1984), with occasional exposure to light in the culture room, flowered after transfer to pots, and no abnormalities were observed (Fig. 2).

Table 1. Various methods employed for the storage of tissue cultures (Bajaj 1986a)

Method of storage	Plant species	Culture	References
1. Mineral oil overlay	*Daucus carota*	Callus	Caplin (1959)
2. Desiccation	*Daucus carota*	Callus	Nitzsche (1980)
3. Growth retardants	*Solanum* spp.	Shoot tip	Henshaw et al. (1980)
4. Low atmospheric pressure	*Chrysanthemum morifolium*	Callus/ plantlet	Bridgen and Staby (1981)
5. Low temperature			
9 °C	*Vitis vinifera*	Shoot tip	Galzy (1969)
1–4 °C	*Fragaria* sp.	Meristem/ plantlet	Mullin and Schlegel (1976)
1–4 °C	*Malus domestica*	Buds/ shoots	Lundergen and Janick (1979)
2–4 °C	*Lolium multiflorum*	Tiller buds	Dale (1980)
2–6 °C	*Medicago sativa*	Shoot tip	Cheyne and Dale (1980)
2–5 °C	*Beta vulgaris*	Plantlets	Miedema (1982)
2–4 °C	*Pinus radiata*	Shoots	Smith (1985)
4–5 °C	*Chrysanthemum morifolium*	Explants/ plantlets	Bajaj (1986a)
4–5 °C	*Petunia hybrida*	Explants/ plantlets	Bajaj (1986a)
8 °C	*Actinidia chinensis*	Shoot tip	Monette (1986)

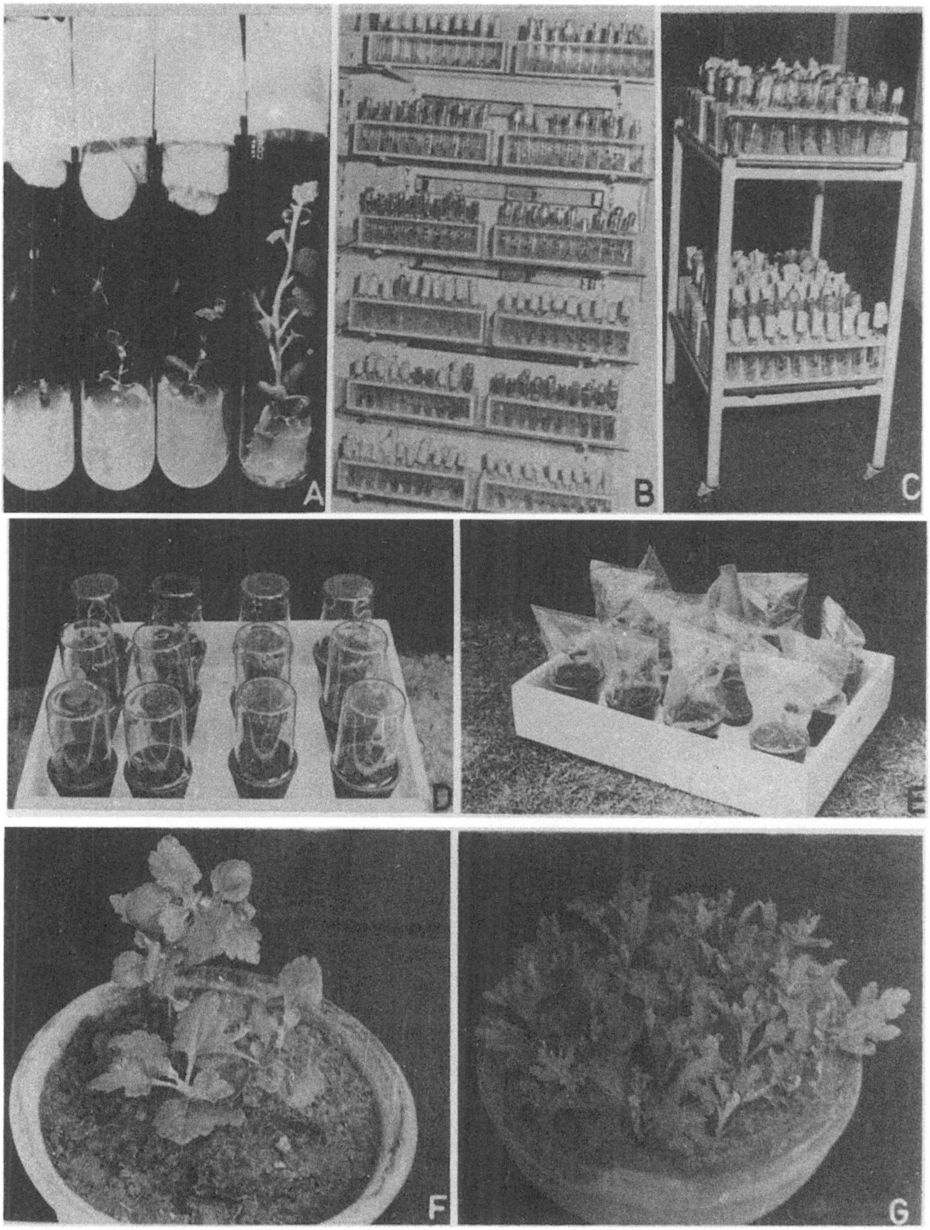

Fig. 1A-G. Mass multiplication and storage of in vitro cultures of *Chrysanthemum morifolium*. **A** Various stages in the formation of plants from shoot tips cultured for 8, 20, 30, and 40 days. **B,C** Large-scale incubation and storage of cultures. **D,E** In vitro obtained plantlets immediately after transfer to pots and covered with glass beakers and polyethylene bags to avoid desiccation. **F,G** Propagation of plants through terminal cuttings of an in vitro obtained plant. **F** A plant suitable for taking terminal cuttings. **G** Terminal cuttings planted in sand for the induction of roots. (Bajaj 1986a)

Fig. 2A-D. Regeneration of plants from cultures of *Chrysanthemum morifolium* and *Petunia hybrida* stored up to 6 years at 4 °C. **A** Cultures of *Chrysanthemum* stored for 3 years. **B** Stocks of germplasm from **A** cold-stored for 6 years (1978–1984). **C** A *Chrysanthemum* plant obtained from a stored culture capable of flowering. **D** Various stages in the development of plants from cultures of stem segments of *Petunia* stored for 3 years at 4 °C

3 **Cryopreservation (–196 °C) of Cultures** (Tables 2–8, Figs. 3–7)

The underlying principle of cryopreservation is to cease metabolism, which can be attained by freezing the cultures in liquid nitrogen (–196 °C). The general protocol involves treating the cultures with an appropriate cryoprotectant such as dimethylsulfoxide (DMSO), glycerol, sucrose, proline, etc., and then subjecting them to liquid nitrogen (LN). The freezing method varies, i.e., prefreezing, a slow and regulated rate of cooling, and sudden immersion in LN. After freeze-storage the cultures are generally thawed at 35–40 °C, washed, and recultured.

Complete plants have been regenerated from frozen cultures of a variety of species (see Tables 2–8). This subject has been recently reviewed by the author on vegetatively propagated crops (Bajaj 1990a), cereals (Bajaj 1983e, 1989a, 1990b), legumes and oilseed crops (Bajaj 1990c), and medicinal and aromatic plants (Bajaj 1988a). The general aspects of cryopreservation, and its importance for the international exchange of genetic stocks and the establishment of germplasm banks have been discussed earlier (Bajaj and Reinert 1977; Bajaj 1979a,b, 1983a, 1984a, 1985a,b, 1986a, 1987). The following general guidelines have emerged from a survey of the literature.

Table 2. Summary of the author's work on the cryopreservation (–196 °C) of plant cell, tissue, and organ culture

Species	Explant	Remarks	Reference
Arachis hypogaea	Meristems	Plants	(Bajaj 1983b)
	Pollen embryos	Callus, plants	(Bajaj 1983c)
Atropa belladonna	Pollen embryos	Plants	(Bajaj 1977a)
	Protoplasts	Plants	(Bajaj 1988b)
Brassica sp.	Pollen embryos	Callus, plants	(Bajaj 1983c)
Cicer arietinum	Meristem	Plants	(Bajaj 1983b)
Citrus sp.	Ovules	Plants	(Bajaj 1984b)
Cocos nucifera	Embryo segment	Callus	(Bajaj 1984b)
Datura innoxia	Callus	40% Cell survival	(Bajaj 1976a)
	Protoplasts	Shoots	(Bajaj 1988b)
Daucus carota	Cell suspension, somatic embryos	Plants	(Bajaj 1976a)
Glycine max	Callus	50% Cell survival	(Bajaj 1976a)
Nicotiana tabacum	Cell suspension	Plants	(Bajaj 1976a)
	Pollen embryos	Plants	(Bajaj 1977a)
	Protoplasts	Plants	(Bajaj 1988b)
Manihot esculenta	Meristems	Plants	(Bajaj 1985c)
Oryza sativa	Endosperm	Callus, plants	(Bajaj 1981a)
	Pollen embryos	Plants	(Bajaj 1984c)
	Rice × pea Protoplasts	Survival	(Bajaj 1983d)
Primula obconica	Anthers	Callus, plants	(Bajaj 1981b)
Saccharum officinarum	Meristems, plantlets	Shoots, plants	(Bajaj et al. 1987)
Solanum tuberosum	Meristems	Plants, tubers	(Bajaj 1985c)
	Somaclones	Plants	(Bajaj 1986b)
Triticum aestivum	Pollen embryos	Plants	(Bajaj 1984c)
	Zygotic embryos		

Table 3. Cryopreservation of in vitro cultures of some vegetatively propagated crops (Bajaj 1990a)

Species	Explant	Response	Reference
Potato	Meristems from sprouts and axillary buds	Callus and Shoots	Bajaj (1977b)
	Meristems from sprouts and axillary buds	Plants with normal tubers	Bajaj (1978a)
	Root tips	Resumed growth	Bajaj (1978a)
	Shoot meristems	Plants	Grout and Henshaw (1978)
	Meristems from three cultivars	2 years of storage, normal plants	Bajaj (1981c)
	Meristems from shoot tips	Multiple shoots	Towill (1981)
	Meristems from axillary buds	4 years of storage, normal tubers	Bajaj (1985c)
	Somaclones (protoplast-derived callus colony)	Regeneration of plants	Bajaj (1986b)
Cassava	Meristems	Callus, plants	Bajaj (1977c, 1982a)
	Meristems	Plants	Kartha et al. (1982)
	Meristems	Survival improved, plants after 3 years of storage	Bajaj (1983f)
	Meristems	Plants after 4 years of storage, no loss in viability	Bajaj (1985c)
Sugarcane	Cell suspensions	Plants	Chen et al. (1979), Finkle and Ulrich (1979)
	Cell suspensions	60% Survival	Bajaj et al. (1987)
	Meristems, plantlets	Plants	Bajaj et al. (1987)

Table 4. Periodical studies on viability of excised meristems of cassava and potato freeze-preserved for various lengths of time (Bajaj 1985c)

Preservation period	Cassava cv. GS-1				Potato cv. Alankar			
	No. of meristems cultured	No. of cultures resuming growth	Survival (%)	Survival % of control	No. of meristems cultured	No. of cultures resuming growth	Survival (%)	Survival % of control
Control (unfrozen)	38	31	81.5	–	43	37	86.0	–
3 months	23	8	34.7	42.5	35	12	34.2	39.7
6 months	26	9	34.6	42.4	27	10	37.0	43.0
1 yr	30	10	33.3	40.8	24	9	37.5	43.6
2 yr	31	11	35.4	43.4	32	11	34.3	39.8
3 yr	25	9	36.0	44.1	25	9	36.0	41.8
4 yr	29	10	34.4	42.2	27	9	33.3	38.7

Table 5. Cryopreservation of in vitro cultures of some medicinal and alkaloid-producing plants (Bajaj 1988a)

Plant species	Culture	Cryoprotectant	Freezing	Thawing (°C)	Response	Reference
Anisodus acutangulus	Cell suspension	DMSO or lactalbumin hydrolysate	−196 °C	37	Hyoscyamine and scopolamine synthesis	Zheng Guang-zhi et al. (1983)
Atropa belladonna	Cell suspension	DMSO (5%)	2 °C/min, −100 °C, −196 °C	37	Callus	Nag and Street (1975)
A. belladonna	Pollen embryos	DMSO + sucrose + glycerol	1–3 °C/min, −196 °C	35–40	Haploid plants	Bajaj (1976b, 1977a, 1978b)
A. belladonna	Protoplasts	DMSO + mannitol	Exposed to vapors immersed in LN	35	Plants	Bajaj (1988b)
Catharanthus roseus	Cell suspension	DMSO (5%) + 1M sorbitol	0.5–3 °C/min to −35 °C, −196 °C	40	Callus	Chen et al. (1984)
Cinchona ledgeriana	Callus	Mannitol + DMSO	Slow freezing	Rapid thawing	Callus	(Hunter 1986)
Citrus sp.	Ovule/nucellar embryo	DMSO 7% + sucrose 7%	−196 °C	35–40	Callus plants	Bajaj (1984b)
Datura stramonium	Cell suspension	DMSO 7%	1 °C/min, −196 °C	36–38	40% Cell survival	Bajaj (1976a)
D. innoxia	Cell suspension	1M sorbitol	1 °C/min, −196 °C	33	Callus	Weber et al. (1983)
D. innoxia	Protoplasts	DMSO + mannitol	Exposed to vapors, immersed in LN	35	Callus	Bajaj (1988b)
Digitalis lanate	Cell suspension	DMSO + glycerol + sucrose	1 °C/min, −35 °C, −196 °C	40	Callus, biotransformation of β-methyldigitoxin to β-methyldigoxin	Diettrich et al. (1982), Seitz et al. (1983)
Dioscorea deltoidea	Cell suspension	7% DMSO	Hardening at 2–10, −30, −70, then −196 °C	20 or 40	Diosgenine synthesis	Butenko et al. (1984)
Lavandula vera	Cell suspension	DMSO (10%) + gluc (20%)	1 °C/min, −40, −196 °C	Rapid	Metabolic potential retained	Watanabe et al. (1983)
Nicotiana tabacum	Cell suspension	DMSO, sucrose	1–2 °C/min, −20, −70, −196 °C		Plants, callus	Bajaj (1976a) Hauptmann and Widholm (1982) Maddox et al. (1983)
N. tabacum	Pollen embryos	DMSO + sucrose + glycerol	1–3 °C/min, −196 °C	35–40	Plants	Bajaj (1976b, 1977a)
N. tabacum	Protoplasts	DMSO + mannitol	Exposed to vapors, immersed in LN	35	Plants	Bajaj (1988b)
N. plumbaginifolia *N. sylvestris*	Cell suspension	DMSO + glycerol	−40 to −70 °C for 4 h, then −196 °C	45	Callus	Maddox et al. (1983)
Panax ginseng	Cell suspension	Sucrose/glycerol	Hardening, −30, −70, then −196 °C	20 or 40	Retention of biosynthetic capacity	Butenko et al. (1984)

Table 6. Summary of cryopreservation studies on wheat (Bajaj 1990b)

Material	Response	Reference
Seed	94.5% Germination, normal plants	Bajaj (1980a)
Zygotic Embryo	41.8–70% Callused/developed plants	Bajaj (1980a, 1984c)
	Plants matured, set seed capable of normal germination	Bajaj (1986a,c)
Pollen-embryos/androgenic anther segments	5–19% Survival, callus, shoots	Bajaj (1983c, 1984c)
Callus cells	11% Survival (26% of control) Plant regeneration	Bajaj (1980a)
Cell suspension	2.2–59.4% Survival, plants	Chen et al. (1985)
Somatic hybrid protoplasts (wheat × pea)	12.2–17.9% Survival	Bajaj (1983d)

Table 7. Summary of the state of the art on the cryopreservation of rice germplasm (see also Bajaj and Sala 1991)

Material	Response	Reference
Seed	98% Germination, normal plants	Bajaj (1981a)
Dehusked seed	87% Germination, normal plants	Bajaj (1981a)
Zygotic embryos	52.5–83% Callused, plant regeneration	Bajaj (1981a, 1984c)
Endosperm	11% Callused, plant regeneration	Bajaj (1981a)
Pollen-embryo/androgenic anther	Anthers callused, plant regeneration	Bajaj (1980b, 1984c)
Callus/cell suspension	60–65% Survival	Sala et al. (1979)
	70–100% Survival	Cella et al. (1982)
	21–42% Survival	Bajaj (1982b)
Somatic hybrid protoplasts		
Rice × pea	10.8–13.9% Survival	Bajaj (1983d)
Rice × sorghum	10.3–13.8% Survival	Bajaj (1983e)

Experience has shown that age, nature, and physiological state of the cells are of utmost importance. The actively growing and periodically transferred, highly cytoplasmic, thin-walled, nonvacuolated, small cells and aggregates of cells are able to withstand freezing, and give high viability. This also avoids selection pressure.

Before freezing, the cultures may be pregrown for a couple of days on a medium containing low concentrations of the cryoprotectant, e.g., DMSO 5%.

A mixture of various cryoprotectants at low concentrations, rather than a single cryoprotectant at a high concentration, yields better results.

The results on the various freezing methods are rather contradictory. Cell survival has been obtained in cultures subjected to sudden freezing, prefrozen material, as well those frozen at a low and regulated rate of cooling. However, efforts should be made to concentrate studies on the regulated rate of cooling in order to reproduce results.

Sudden and brief thawing at 35–40 °C has proved better than thawing at room temperature. The cultures stored at –20 or –70 °C have been observed to undergo deterioration with the passage of time. Thus, for long-term preservation, material should be maintained in liquid nitrogen.

Any results on cell survival based on the staining methods alone can be misleading; thus, resumption of growth should be the criterion for determining viability.

3.1 Cryopreservation of Meristems (Tables 3, 4, Figs. 3, 4)

For the storage of germplasm, meristems are preferred over cell and callus cultures because they are genetically more stable. Meristems would be ideally suited especially for vegetatively propagated crops (Table 3). The meristems of a number of plant species have survived at –196 °C, i.e., carnation (Seibert 1976; Dereuddre et al. 1988), potato (Bajaj 1977b, 1981c, 1985c; Grout and Henshaw 1978), cassava (Bajaj 1977c, 1983f, 1985c; Kartha et al. 1982), lettuce (Seibert and Wetherbee 1977), strawberry (Sakai et al. 1978), pea (Haskins and Kartha 1980), chickpea (Bajaj 1979c, 1983b), peanut (Bajaj 1979c, 1982a, 1983b), apple (Katano et al. 1983), sugarcane (Bajaj et al. 1987), *Rubus* (Reed and Lagerstedt 1987), sugarbeet (Braun 1988), *Pyrus* (Reed 1990), and mulberry (Yakuwa and Oka 1988).

The regeneration of complete plants from frozen meristems of potato and peanut is shown in Figs. 3 and 4. The meristems of potato and cassava stored in LN for up to 4 years regenerated normal plants capable of forming tubers and thick roots, respectively (Table 4; Bajaj 1985c).

3.2 Cryopreservation of Cell Suspensions (Fig. 5)

Callus and cell suspensions, maintained on periodic subculture and storaged, undergo genetic erosion resulting in changes in ploidy, mutation, translocations, gene amplification, etc. To avoid this, freeze-preservation in liquid nitrogen is an alternative method. Earlier work in the 1970's showed that cell suspensions frozen and stored in LN for various lengths of time retained their morphogenetic potential and regenerated complete plants (Fig. 5; Bajaj 1976a). Since then, callus and cell suspensions of a variety of plant species, such as legumes and oilseed crops (Bajaj 1990c), vegetatively propagated crops (Table 3; Bajaj 1990a), and cereals (Tables 6, 7; Bajaj 1983e, 1990b) have yielded positive results. Likewise, cell suspensions of alkaloid-producing and medicinal plants (Table 5; Bajaj 1988a) retained their biosynthetic potential after freezing. Recently, cultured asparagus (Uragami et al. 1989) and navel orange (Kobayashi et al. 1990) cells have been successfully cryopreserved.

3.3 Cryopreservation of Embryos

3.3.1 Cryopreservation of Somatic Embryos (Fig. 5)

Since somatic embryos can now be produced in large quantities in bioreactors, with cryopreservation they can be stored at the appropriate stage, until they are needed. The somatic embryos, especially those of carrot (Bajaj 1976a; Withers 1979), oil palm (Engelmann et al. 1985), sweet orange (Marin and Duran-Vila 1988), and asparagus (Uragami et al. 1989) frozen in LN yielded high viability and regenerated complete plants.

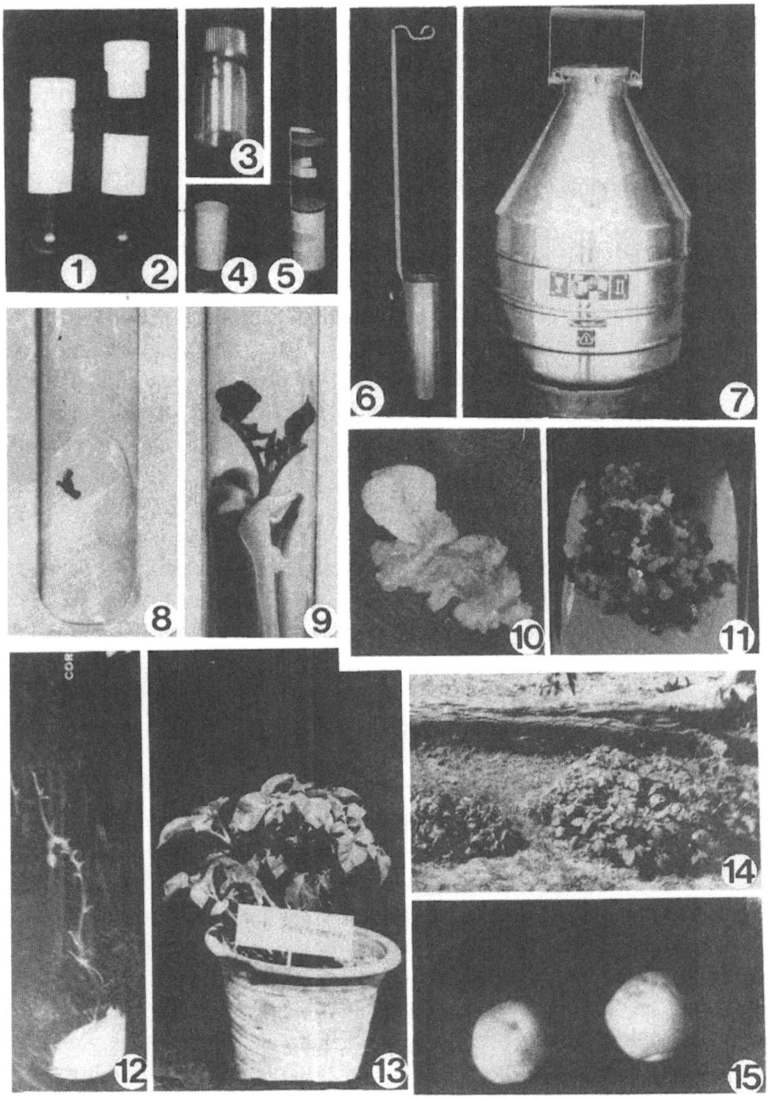

Fig. 3(1–15). Tuberization of potato plants regenerated from frozen meristems. **1–3** Various types of ampoules and vials used for freezing. **4–6** Containers for freezing and storing ampoules. **7** A liquid nitrogen cylinder. **8,9** Frozen-thawed meristems 4 and 10 weeks after culture on agar medium, and on filter paper (liquid medium), respectively. **10** 11-week-old culture of retrieved meristem undergoing proliferation to form callus. **11** Same, after next 4 weeks on MS + IAA + BAP showing initiation of shoot primordium from callus. **12** A plantlet obtained from meristem freeze-preserved for 15 days at –196 °C; note the formation of a tiny tuber. **13–15** Transfer of plants from test tube to the pot and field, and the production of tubers (Bajaj 1978a)

Fig. 4A-H. Various stages in the regeneration of plants of *Arachis hypogaea* from frozen (–196 °C) meristems. **A-C** Frozen-thawed meristems after 6, 9, and 11 weeks of culture, respectively; note the initial alround callusing (perhaps due to cryoinjury) of the meristem in **A. D,E** Normal growth of the retrieved meristem (preserved for 20 months), and regeneration of a complete plant 4 and 12 weeks, respectively, after culture on MS + IAA + BAP. **F,G** Test-tube plants 3 and 7 weeks, respectively, after transferring to soil. **H** Normal seeds obtained from such plants (total time 24 weeks from the culture of excised meristems to seed setting). (Bajaj 1983b)

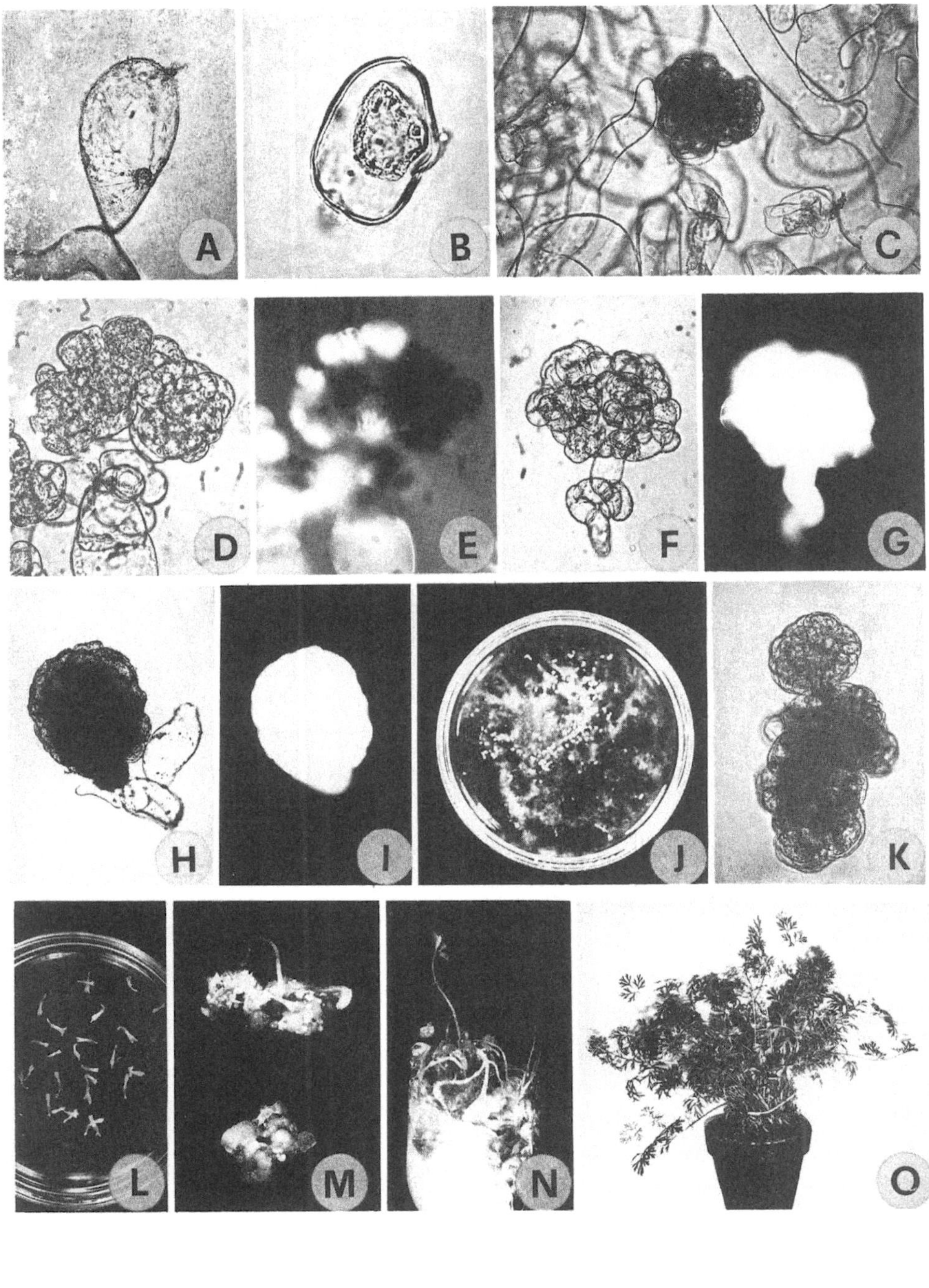
A
B
C
D
E
F
G
H
I
J
K
L
M
N
O

Another method of considerable importance for storage of somatic embryos is to coat them with sodium alginate beads to form a synthetic seed (see Chap. III.3, this Vol. for details).

3.3.2 Cryopreservation of Zygotic Embryos (Fig. 6)

The complete organs and large embryos are particularly vulnerable to cryodamage, and the main problem is the prevention of recrystallization of the ice, and also intracellular freezing which is invariably lethal. Thus, the water content of large embryos must be controlled (by desiccation) to bring it to a level at which intracellular freezing is minimal. The cryodamage can be further prevented by adopting a thawing method which discourages growth of ice crystals.

The potential of the cryopreservation of zygotic embryos is manifold, especially in the conservation of germplasm of plants with "recalcitrant" seed. In a number of fruit and timber trees, and also in plantation crops, the seeds are large and recalcitrant and thus cannot be preserved under ordinary conditions. In such cases, the cryopreservation of zygotic embryos would be a logical alternative. Moreover, in wide hybridization programs, especially those dealing with the intergeneric crosses, which are incompatible due to degeneration or abortion of the embryos, they can be possibly dissected out at immature stages and cryopreserved.

There are a few reports on the cryopreservation of zygotic embryos, i.e., rice (Bajaj 1981a), wheat (Bajaj 1984c), barley and mustard (Withers 1982), and coconut (Bajaj 1984b). In all these cases quick freezing, followed by thawing at 35–40 °C, was employed: however, the viability varied considerably.

The immature embryos of wheat and rice frozen in LN in the presence of DMSO and sucrose were retrieved (Bajaj 1984c). Their survival varied from 42–73%. The retrieved cultures underwent proliferation, or formed plants (Fig. 6) that matured after transfer to the field and set normal seed.

The partially desiccated, immature coconut (recalcitrant) embryos and their segments subjected to sudden freezing showed a lag period of up to 4 months without any sign of growth. However, subsequently, in some of the cultures the embryos elongated and proliferation of the segments at the cut end was observed (Bajaj 1984b). Likewise, embryonal axes of rubber desiccated for 1–5 h and frozen by direct immersion in LN formed seedlings (Normah et al. 1986).

Fig. 5A-O. Regeneration of plants from cell suspensions of *Daucus carota* subjected to super-low temperatures. **A,B** Free cells from a suspension just before (**A**) and after freezing (**B**) in liquid nitrogen; note the plasmolysis of cell in **B**. **C** 12-day-old suspension containing vacuolated free cells and an actively growing clump of cells. **D** A clump of cells (in tungsten light) taken from a culture frozen at –196 °C for 3 months. **E** Same, in ultraviolet light; note the partial survival of the clump (cryoprotectant DMSO 7%, cooling rate 2 °C/min, thawing at 37 °C). **F** A clump from cultures frozen at 20 °C for 1 week. **G** Same, in ultraviolet light; note the complete survival of cells. **H,I** A proembryo along with free cells subjected to 20 °C (1 h) and 70 °C (30 min), and then immersed in liquid nitrogen (cryoprotectant DMSO 5%, thawing at 37 °C); note the complete survival of the proembryo, and the death of single cells. **J,L** Various stages in the differentiation of embryos obtained from cell suspensions frozen in liquid nitrogen; the cultures were raised in a liquid medium. **M,N** Differentiation of plantlets in a 2,4-D-free agar solidified medium, 3 (**M**) and 6 weeks (**N**), after transfer. **O** Plants of carrot obtained from cell suspensions stored in liquid nitrogen for 3 months (6 weeks after transfer to pot from cultures in **N**). (After Bajaj 1976a)

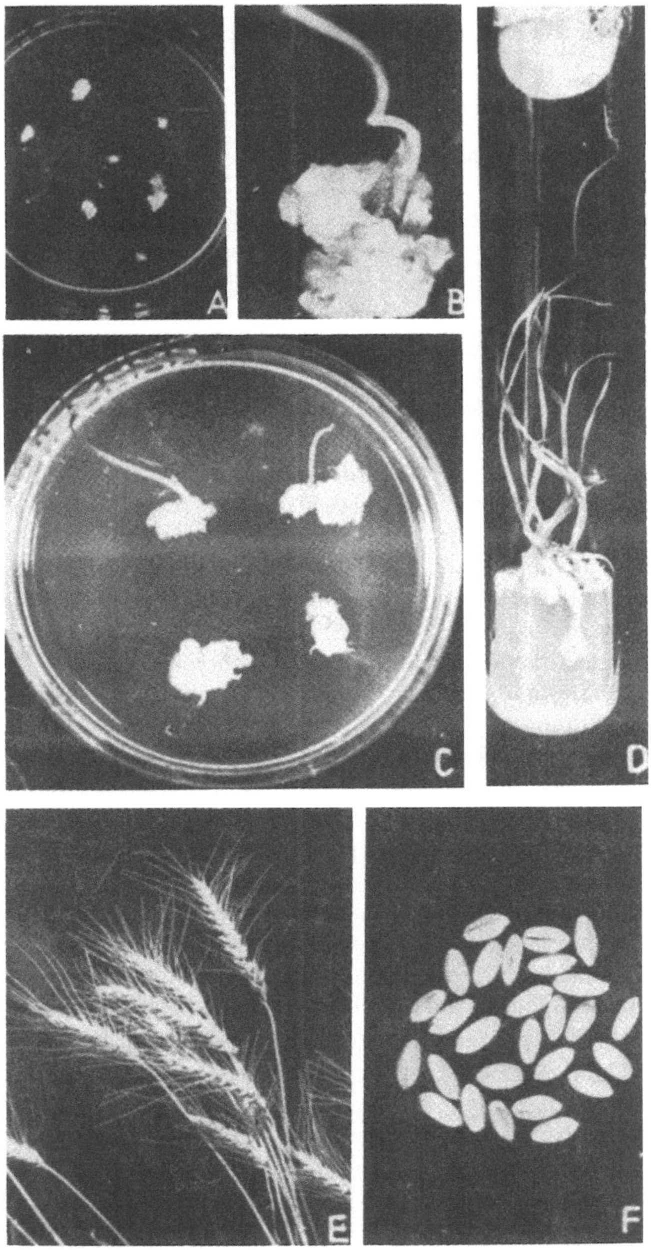

Fig. 6A-F. Regeneration of plants from frozen anthers and immature zygotic embryos of wheat. **A** Callus mass obtained from frozen segments of the androgenic anthers. **B** Same, showing the regeneration of a shoot. **C,D** Survival of immature zygotic embryos in liquid nitrogen; note the profuse callus and the emergence of shoots and plants (Bajaj 1984c). **E,F** Normal spikes and seeds obtained from plant in **D** (Bajaj 1986a)

3.3.3 Cryopreservation of Nucellar Embryos

The nucellar embryos in polyembryonic *Citrus* species are ovular in origin, and thus are somatic in nature. They are located around the zygotic embryo in the upper part of the ovule (micropylar half). Entire young ovules, and the cut micropylar half of the ovules obtained from unripe young fruits, frozen in the presence of DMSO and sucrose, showed a survival of 28.8 and 24.3%, respectively (Bajaj 1984b). The retrieved material, like the controls, when cultured on a medium containing casein hydrolysate (500 mg/l), proliferated to form pseudobulbils. In some cases polyembryonic masses and shoots were formed. Recently (Kobayashi et al. 1990), plants were regenerated from cryopreserved nucellar cells of navel orange.

3.3.4 Cryopreservation of Pollen Embryos (Fig. 6)

The storage of pollen has been of prime interest to plant breeders and horticulturists involved in fruit tree breeding. Cryopreservation of pollen has been successful with a large number of species (see Bajaj 1987), however, only limited work has been done on pollen embryos/androgenic anthers. The cryopreservation of pollen embryos enables: (1) the production en masse of haploid plants; (2) the maintenance of stability of haploid cultures; (3) and the conservation of genetic resources.

The cryopreservation of pollen embryos involved the following steps:

The pollen embryos are removed from the androgenic anthers and suspended in an appropriate liquid medium in 2 ml ampoules. To 0.5 ml of the suspended pollen embryos, a few drops of cryoprotectant solution are gradually added at 10-min intervals (sudden addition may cause damage) three times, and left for about 30–40 min. During this treatment, the ampoules are kept in ice. The ampoules are then frozen by different methods, thawed in warm water (35–40 °C), washed with fresh medium, and recultured.

Segments of the androgenic anthers and pollen embryos of *Arachis hypogaea, Atropa belladonna, Brassica campestris, Nicotiana tabacum, Oryza sativa, Primula obconica,* and *Triticum aestivum* have been successfully frozen, and in some cases entire plants were regenerated after 1 year of storage (Bajaj 1983c).

The suspension of pollen embryos of *Brassica campestris* and *B. napus* showed survival rates of 31 and 44%, respectively, after a lag period of about 4 months. The embryos resumed growth and either developed directly into plantlets or were accompanied by callus. Likewise, frozen embryos of *Arachis villosa* underwent callusing and thus formed shoots.

In wheat, however, the segments taken from 5- to 6-week-old androgenic anthers and frozen gave poor responses (Bajaj 1984c) and only 5% of the anther segments resumed growth after 2–4 weeks and callused. The callus, after transfer to a hormone-free medium or to a medium supplemented with IAA + BA + casein hydrolysate, differentiated shoots/plantlets (Fig. 6).

From the studies conducted so far, it seems that the embryos can withstand a wide range of freezing regimes. For instance, the young pollen embryos of *Atropa* and *Nicotiana*, frozen at a rate of 1–5 °C/min, underwent a lag phase of 2–6 weeks and then resumed growth (Bajaj 1977a). The pollen embryos of *Arachis, Brassica,* and *Triticum* (Bajaj 1983c), after sudden immersion in LN, showed 38, 44, and 19% survival rates, respectively. It is thus obvious that, if other factors are well manipulated, the embryos in general can withstand different freezing methods.

3.4 Cryopreservation of Protoplasts (Table 8, Fig. 7)

Freshly isolated protoplasts of a number of plant species (Table 8) subjected to freezing in liquid nitrogen for various lengths of time have survived and retained their morphogenetic potential (Fig. 7; Bajaj 1988b). Although for all practical purposes freezing of protoplasts is similar to that of cell suspensions, protoplasts differ from cells in their lack of a cell wall. Thus, they require a critical concentration of an appropriate osmoticum and cryoprotectant to prevent them from bursting or causing undue damage to the plasma membrane. Three types of cryoinjury may damage the protoplasts, i.e., intracellular ice formation, expansion-induced lysis, and loss of osmotic responsiveness. Thorough understanding of various cryobiological aspects of protoplasts will help to prevent damage and to further higher survival. For details regarding the literature on various basic, fundamental, and technical aspects of cryopreservation of plant protoplasts, see Bajaj (1989b).

Table 8. Summary of the work done on the cryopreservation of isolated plant protoplasts (Bajaj 1989b)

Plant species	Viability and growth response	Reference
Atropa belladonna	44% Cultures underwent differentiation compared to 71% controls; plant regeneration	Bajaj (1988b)
Bromis inermis	Some protoplasts were viable, as shown by trypan blue dye	Mazur and Hartmann (1979)
Daucus carota	30% Survival, plant regeneration	Takeuchi et al. (1980), Withers (1980), Hauptmann and Widholm (1982)
Datura innoxia	Colony formation in 5–6 weeks, occasional regeneration	Bajaj (1988b)
Glycine max	60% Viability, callus	Takeuchi et al. (1982)
	Cell suspension	Weber et al. (1983)
Hordeum vulgare	68% Viability, callus	Takeuchi et al. (1982)
Marchantia polymorpha	40% Viability, normal thalli and rhizoids	Takeuchi et al. (1980, 1982)
Nicotiana tabacum	36% Cultures underwent differentiation as compared to 53% controls, plant regeneration	Bajaj (1988b)
Oryza sativa × *Pisum sativum*	10.8–13.9% Survival	Bajaj (1983d)
Oryza sativa × *Sorghum bicolor*	10.3 to 13.8% Survival	Bajaj (1983e)
Triticum aestivum	33% Viability, callus	Takeuchi et al. (1982)
Triticum aestivum × *Pisum sativum*	12.2–17.9% Survival	Bajaj (1983d)

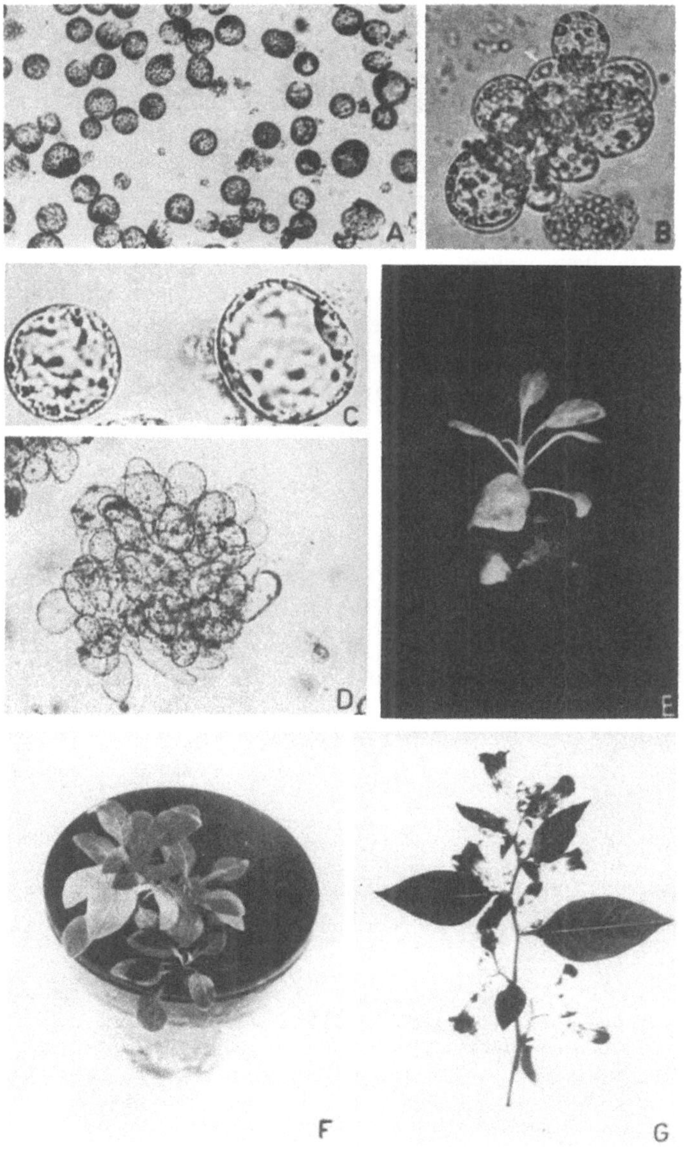

Fig. 7A-G. Regeneration of plants from protoplasts of *Atropa belladonna, Datura innoxia*, and *Nicotiana tabacum* frozen in liquid nitrogen. **A** Mesophyll protoplasts of *Atropa* after freezing and thawing. **B** Same, 4 weeks after culture in a drop of liquid medium. **C** Retrieved callus cells suspension-derived protoplasts of *Datura* 3 weeks after culture. **D** Callus cell colony obtained from frozen mesophyll protoplast of *Nicotiana*. **E** *Nicotiana* plantlet regenerated from frozen protoplast-derived callus. **F,G** Frozen protoplast-derived plant of *Atropa* 4 weeks after transferring to soil (**F**) and showing normal flowering (**G**). (Bajaj 1988b)

4 Conclusions

The maintenance and short-term storage of in vitro cultures do not present many problems, however, methods must still be refined and perfected for long-term conservation of germplasm. The underlying principle is to reduce or completely stop growth and metabolic activity. Various methods, such as minimal media and the use of growth retardants, low atmospheric pressure, mineral oil, desiccation, and low temperature storage, have been investigated. Of these, low temperature storage, both nonfrozen and frozen, have yielded encouraging results. Freeze preservation of cultures in liquid nitrogen has enabled one to retain the biosynthetic and morphogenetic potential of a large number of plant species. Cryopreservation can be of immediate use for (1) the storage of germplasm of vegetatively propagated crops; (2) plants with recalcitrant seeds; and (3) storage of desired and selected somaclonal cell lines.

References

Aitken-Christie J, Singh AP (1987) Cold storage of tissue cultures. In: Bonga JM, Durzan DJ (eds) Cell and tissue culture in forestry, vol 2. Martinus Nijhoff, Dordrecht, pp 285–304

Augereau JM, Courtois D, Petiard V (1986) Long-term storage of callus cultures at low temperatures or under mineral oil layer. Plant Cell Rep 5:372–376

Bajaj YPS (1976a) Regeneration of plants from cell suspensions frozen at –20, –70, and –196 °C. Physiol Plant 37:263–268

Bajaj YPS (1976b) Gene preservation through freeze storage of cell, tissue and organ cultures. Acta Hortic 63:75–84

Bajaj YPS (1977a) Survival of *Atropa* and *Nicotiana* pollen embryos frozen at –196 °C. Curr Sci 46:305–307

Bajaj YPS (1977b) Initiation of shoots and callus from potato tuber-sprouts and axillary buds frozen at –196 °C. Crop Improv 4:48–53

Bajaj YPS (1977c) Clonal propagation and cryopreservation of cassava through tissue culture. Crop Improv 4:198–204

Bajaj YPS (1978a) Tuberization in potato plants regenerated from freeze-preserved meristems. Crop Improve 5:137–141

Bajaj YPS (1978b) Effect of super-low temperature on excised anthers and pollen embryos of *Atropa, Nicotiana* and *Petunia*. Phytomorphology 28:171–176

Bajaj YPS (1979a) Technology and prospects of cryopreservation of germplasm. Euphytica 28:267–285

Bajaj YPS (1979b) Establishment of germplasm banks through freeze storage of plant tissue cultures and their implications in agriculture. In: Sharp WR, Larsen PO, Paddock PO, Raghavan V (eds) Plant cell and tissue culture – principles and applications. Ohio State Univ Press, Columbus, pp 745–774

Bajaj YPS (1979c) Freeze preservation of meristems of *Arachis hypogaea* and *Cicer arietinum*. Indian J Exp Biol 17:1405–1407

Bajaj YPS (1980a) Freeze preservation of plant cells – a novel approach to the conservation of germplasm. In: Gupta AK (ed) Genetics and wheat improvement. Oxford & IBH, New Delhi, pp 141–149

Bajaj YPS (1980b) Induction of androgenesis in rice anthers frozen at –196 °C. Cereal Res Commun 8:365–369

Bajaj YPS (1981a) Growth and morphogenesis in frozen (–196 °C) endosperm and embryos of rice. Curr Sci 50:947–948

Bajaj YPS (1981b) Regeneration of plants from ultra-low frozen anthers of *Primula obconica*. Sci Hortic 14:93–95

Bajaj YPS (1981c) Regeneration of plants from potato meristems freeze – preserved for 24 months. Euphytica 30:141–145

Bajaj YPS (1982a) Cryopreservation of germplasm of potatoes, cassava, peanut and chickpea. In: Fujiwara A (ed) Plant tissue culture 1982. Maruzen, Tokyo pp 799–800
Bajaj YPS (1982b) Induction and cryopreservation of genetic variability in rice. In: Rice tissue culture planning conf. IRRI, Los Baños, pp 99–111
Bajaj YPS (1983a) Cryopreservation and the international exchange of germplasm. In: Sen SK, Giles KL (eds) Plant cell culture in crop improvement. Plenum, New York, pp 19–41
Bajaj YPS (1983b) Production of normal seeds from plants regenerated from meristems of *Arachis hypogaea* and *Cicer arietinum* cryopreserved for 20 months. Euphytica 32:425–430
Bajaj YPS (1983c) Regeneration of plants from pollen embryos of *Arachis*, *Brassica* and *Triticum* spp. cryopreserved for one year. Curr Sci 52:484–485
Bajaj YPS (1983d) Survival of somatic hybrid protoplasts of wheat × pea and rice × pea subjected to –196 °C. Indian J Exp Biol 21:120–122
Bajaj YPS (1983c) Cryopreservation of germplasm of cereals – progress and prospects. In: Sakamoto S (ed) Proc 6th Int Wheat Genet Symp, Kyoto Univ, Kyoto, pp 565–574
Bajaj YPS (1983f) Cassava plants from meristem cultures freeze-preserved for three years. Field Crop Res 7:161–167
Bajaj YPS (1984a) Cryopreservation of germplasm of crops – present status and prospects. Proc Int Symp Plant tissue and cell culture application to crop improvement. Czech Acad Sci, Prague, pp 607–616
Bajaj YPS (1984b) Induction of growth in frozen embryos of coconut and ovules of citrus. Curr Sci 53:1215–1216
Bajaj YPS (1984c) The regeneration of plants from frozen pollen embryos and zygotic embryos of wheat and rice. Theor Appl Genet 67:525–528
Bajaj YPS (1985a) Cryopreservation of embryos. In: Kartha KK (ed) Cryopreservation of plant cells and organs. CRC, Boca Raton, pp 227–242
Bajaj YPS (1985b) Cryopreservation of plant cell cultures and its prospects in agricultural and forest biotechnology. In: Natesh S et al. (ed) Int Worksh Biotechnology in agriculture: evolving a research agenda for the International Centre of Genetic Engineering and Biotechnology. Oxford & IBH, New Delhi, pp 109–131
Bajaj YPS (1985c) Cryopreservation of germplasm of potato and cassava – viability studies on excised meristems cryopreserved up to four years. Indian J Exp Biol 23:285–287
Bajaj YPS (1986a) In vitro preservation of genetic resources – techniques and problems. Int Symp Nuclear techniques and in vitro culture for plant improvement. IAEA/FAO, Vienna 1985, pp 43–57
Bajaj YPS (1986b) Cryopreservation of potato somaclones. In: Semal J (ed) Proc Symp Somaclonal variations and crop improvement. Nijhoff, Dordrecht, pp 244–251
Bajaj YPS (1986c) In vitro regeneration of diverse plants and the cryopreservation of germplasm in wheat (*Triticum aestivum* L.). Cereal Res Commun 14:305–311
Bajaj YPS (1987) Cryopreservation of pollen and pollen embryos, and the establishment of pollen banks. In: Giles KL, Prakash J (eds) Pollen development and cytology. Academic Press, London, pp 397–420
Bajaj YPS (1988a) Cryopreservation and the retention of biosynthetic potential in cell cultures of medicinal and alkaloid-producing plants. In: Bajaj YPS (ed) Biotechnology in agriculture and forestry, vol 4. Medicinal and aromatic plants I. Springer, Berlin Heidelberg New York Tokyo, pp 169–187
Bajaj YPS (1988b) Regeneration of plants from frozen (–196 °C) protoplasts of *Atropa belladonna* L., *Datura innoxia* Mill and *Nicotiana tabacum* L. Indian J Exp Biol 26:289–392
Bajaj YPS (1989a) Induction and cryopreservation of somaclonal variation in wheat and rice. Int Symp Genetic manipulation in crops. CIMMYT, Mexico, pp 195–203
Bajaj YPS (1989b) Cryopreservation of plant protoplasts. In: Bajaj YPS (ed) Biotechnology in agriculture and forestry, vol 8. Plant protoplasts and genetic engineering I. Springer, Berlin Heidelberg New York Tokyo, pp 97–106
Bajaj YPS (1990a) Cryopreservation of germplasm of vegetatively propagated crops. Bull Soc bot France 137:99–114
Bajaj YPS (1990b) Cryopreservation of germplasm of wheat. In: Bajaj YPS (ed) Biotechnology in agriculture and forestry, vol 13. Wheat. Springer, Berlin Heidelberg New York Tokyo, pp 669–681
Bajaj YPS (1990c) Cryopreservation of germplasm of legumes and oilseed crops. In: Bajaj YPS (ed) Biotechnology in agriculture and forestry, vol 10. Legumes and oilseed crops. I. Springer, Berlin Heidelberg New York Tokyo, pp 49–62

Bajaj YPS, Reinert J (1977) Cryobiology of plant cell cultures and establishment of gene banks. In: Reinert J, Bajaj YPS (eds) Applied and fundamental aspects of plant cell, tissue, and organ culture. Springer, Berlin Heidelberg New York, pp 757–777

Bajaj YPS, Sala F (1991) Cryopreservation of germplasm of rice (*Oryza sativa* L.). In: Bajaj YPS (ed) Biotechnology in agriculture and forestry, vol 14. Rice. Springer, Berlin Heidelberg New York Tokyo, pp 553–571

Bajaj YPS, Komeva SB, Gutierrez R, Maribona RH (1987) Freeze preservation of plantlets, excised meristems and cell cultures of sugarcane (*Saccharum officinarum* L.) at –196 °C. In: Bose A, Sengupta P (eds) Proc Int Conf Cryogenics. McGraw-Hill, New Delhi, pp 222–226

Braun A (1988) Cryopreservation of sugarbeet germplasm. Plant Cell Tissue Organ Cult 14:161–168

Bridgen MP, Staby GL (1981) Low pressure and low oxygen storage of plant tissue cultures. Plant Sci Lett 22:177–186

Butenko RG, Popov AS, Volkova LA, Chernyak ND, Nosov AM (1984) Recovery of cell cultures and their biosynthetic-capacity after storage of *Dioscorea deltoides* and *Panax ginseng* cells in liquid nitrogen. Plant Sci Lett 33:285–292

Caplin SM (1959) Mineral-oil overlay for conservation of plant tissue cultures. Am J Bot 46:324–329

Cella R, Colombo R, Galli MG, Nielsen E, Rollo F, Sala F (1982) Freeze-preservation of rice cells: a physiological study of freeze-thawed cells. Physiol Plant 55:279–284

Chen THH, Kartha KK, Leung NL, Kurz WGW, Chatson KB, Constabel F (1984) Cryopreservation of alkaloid-producing cell cultures of periwinkle (*Cathranthus roseus*) Plant Physiol 75:726–731

Chen THH, Kartha KK, Gusta LV (1985) Cryopreservation of wheat suspension culture and regenerable callus. Plant Cell Tissue Organ Cult 4:101–109

Chen WH, Cockburn W, Street HE (1979) Preliminary experiments on freeze preservation of sugarcane cells. Taiwania 24:70–74

Cheyne VA, Dale PJ (1980) Shoot tip culture in forage legumes. Plant Sci Lett 19:303–309

Dale PJ (1980) A method for in vitro storage of *Lolium multiflorum*. Ann Bot 45:497–502

Dereuddre J, Fabre J, Bassaglia C (1988) Resistance to freezing in LN of carnation (*Dianthus caryphyllus* L. var. Eolo) apical and axillary shoot tips. Plant Cell Rep 7:170–173

Diettrich B, Popov AS, Pfeiffer B, Neumann D, Butenko RG, Luckner M (1982) Cryopreservation of *Digitalis lanata* cell cultures. Planta Med 46:82–87

Engelmann F, Duval Y, Dereuddre J (1985) Survie et prolifération d'embryons somatiques de palmier á huile (*Elaeis guineesis* Jacq.) aprés congélation dans l'azote liquide. CR Acad Sci Paris Ser III, 3:111–116

Finkle BJ, Ulrich JM (1979) Effects of cryoprotectants in combination on the survival of frozen sugarcane cells. Plant Physiol 63:598–604

Galzy R (1969) Recherches sur la croissance de *Vitis rupestris* Scheel sans et court noué cultivé in vitro à différentes températures. Ann Phytopathol 1:149–166

Gray D (1987) Quiescence in monocotyledonous and dicotyledonous somatic embryos induced by dehydration. HortSci 22:810–814

Grout BWW, Henshaw GG (1978) Freeze-preservation of potato shoot-tip cultures. Ann Bot 42:1227–1229

Haskins RH, Kartha KK (1980) Freeze-preservation of pea meristems: cell survival. Can J Bot 833–840

Hauptmann RM, Widholm JM (1982) Cryostorage of cloned amino acid analog resistant carrot and tobacco suspension culture. Plant Physiol 70:30–37

Henshaw GG, Stamp JA, Westcott RJ (1978) Tissue culture and germplasm storage. In: Sala F, Parisi B, Cella R, Ciferri O (ed) Plant cell cultures: results and perspectives. Elsevier/North-Holland, Amsterdam, pp 277–282

Henshaw GG, O'Hara JF, Westcott RJ (1980) Tissue culture methods for the storage and utilization of potato germplasm. In: Ingram D, Helgeson J (eds) Tissue culture for plant pathologists. Blackwell, Oxford, pp 71–76

Hiraoka N (1988) Nonfrozen storage of cell cultures and its effect on metabolites. In: Bajaj YPS (ed) Biotechnology in agriculture and forestry, vol 4. Medicinal and aromatic plants I. Springer, Berlin Heidelberg New York, pp 157–168

Hunter CS (1986) In vitro propagation and germplasm storage of *Cinchona*. In: Withers L, Alderson P (eds) Plant tissue culture and its agricultural applications. Butterworth, London, pp 291–301

Jones LH (1974) Long term survival of embryoids of carrot (*Daucus carota* L.). Plant Sci Lett 2:221–224

Kartha KK, Leung NL, Mroginski LA (1982) In vitro growth responses and plant regeneration from cryopreserved meristems of cassava (*Manihot esculenta* Crantz). Z Pflanzenphysiol 107:133–140

Katano M, Ishihara RA, Sakai A (1983) Survival of dormant apple shoot tips after immersion in liquid nitrogen. HortSci 18:707–708
Kobayashi S, Sakai A, Oiyama I (1990) Cryopreservation in liquid nitrogen of cultured navel orange (*Citrus sisnesis* Osb.) nucellar cells and subsequent plant regeneration. Plant Cell Tissue Organ Cult 23:15–20
Lundergen C, Janick J (1979) Low temperature storage of in vitro apple shoots. HortSci 14:514
Maddox AD, Gonsalves F, Shields R (1983) Successful preservation of suspension cultures of three *Nicotiana* species at the temperature of liquid nitrogen. Plant Sci Lett 28:157–162
Marin ML, Duran-Vila N (1988) Survival of somatic embryos and recovery of plants of sweet orange (*Citrus sinensis* L. Osb.) after immersion in liquid nitrogen. Plant Cell Tissue Organ Cult 7:51–57
Mazur RA, Hartmann JX (1979) Freezing of plant protoplasts. In: Sharp WR, Larsen PO, Paddock EF, Raghavan V (eds) Plant cell and tissue culture – principles and applications. Ohio State Univ Press, Columbus, pp 876
Miedema P (1982) A tissue culture technique for vegetative propagation and low temperature preservation of *Beta vulgaris*. Euphytica 31:635–643
Monette PL (1986) Cold storage of kiwifruit shoot tips in vitro. HortSci 21(5):1203–1205
Mullin RH, Schlegel DE (1976) Cold storage maintenance of strawberry meristem plantlets. HortSci 11:100–101
Nag KK, Street HE (1975) Freeze-preservation of cultured plant cells. II. The freezing and thawing phases. Physiol Plant 34:261–265
Nitzsche W (1980) One year storage of dried carrot callus. Z Pflanzenphysiol 100:269–271
Normah MN, Chin HF, Hor YL (1986) Desiccation and cryopreservation of embryonic axes of *Hevea brasiliensis* Muell.-Arg. Pertanika 9(3):299–303
Reed BM (1990) Survival of in vitro-grown apical meristems of *Pyrus* following cryopreservation. HortSci 25(1):111–113
Reed BM, Lagerstedt HB (1987) Freeze preservation of apical meristems of *Rubus* in liquid nitrogen. HortSci 22(2):302–303
Sakai A, Yamakawa M, Sakato D, Harada T, Yakuwa T (1978) Development of a whole plant from an excised strawberry runner apex frozen to –196 °C, Low Temp Sci Ser B, Biol Sci 36:31–38
Sala F, Cella R, Rollo F (1979) Freeze-preservation of rice cells. Physiol Plant 45:170–176
Seibert M (1976) Shoot initiation from carnation shoot apices frozen to –196 °C. Science 191:1178–1179
Seibert M, Wetherbee PJ (1977) Increased survival and differentiation of frozen herbaceous plant organ cultures through cold treatment. Plant Physiol 59:1043–1046
Seitz U, Alfermann W, Reinhard E (1983) Stability of biotransformation capacity in *Digitalis lanata* cell cultures after cryogenic storage. Plant Cell Rep 2:272–276
Smith DR (1985) Radiata pine (*Pinus radiata* D Don). In: Bajaj YPS (ed) Biotechnology in agriculture and forestry. vol 1. Trees 1. Springer, Berlin Heidelberg New York Tokyo, pp 274–291
Takeuchi M, Matasushima H, Sugawara Y (1980) Long-term freeze-preservation of protoplasts of carrot and Marchantia. Crylett 1:519
Takeuchi M, Matasushima H, Sugawara Y (1982) Totipotency and viability of protoplasts after long-term freeze preservation. In: Fujiwara (ed) Plant tissue culture 1982. Maruzen, Tokyo, pp 797–798
Towill LE (1981) *Solanum etuberosum*: a model for studying the cryobiology of shoot-tips in the tuber-bearing *Solanum* species. Plant Sci Lett 20:315–324
Uragami A, Sakai A, Nagai M, Takahashi T (1989) Survival of cultured cells and somatic embryos of *Asparagus officinalis* cryopreserved by vitrification. Plant Cell Rep 8:418–421
Watanabe K, Mitsuda H, Yamada Y (1983) Retention of metabolic and differentiation potential of green *Lavandula vera* callus after freeze-preservation. Plant Cell Physiol 24:119–122
Weber G, Roth EJ, Schweiger HG (1983) Storage of cell suspensions and protoplasts of *Glycine max* (L.) Merr., *Brassica napus* (L.), *Datura innoxia* (Mill.), and *Daucus carota* (L.) by freezing. Z Pflanzenphysiol 109:29–39
Withers LA (1979) Freeze-preservation of somatic embryos and clonal platelets of carrot (*Daucus carota* L.). Plant Physiol 63:460–467
Withers LA (1980) Preservation of germplasm. Int Rev Cytol 11B:101–136
Withers LA (1982) The development of cryopreservation techniques for plant cell, tissue and organ cultures. In: Fujiwara A (ed) Plant tissue culture 1982. Maruzen, Tokyo, pp 793–794
Yakuwa H, Oka S (1988) Plant regeneration through meristem culture from vegetative buds of mulberry (*Morus bombycis* Koidz) stored in liquid nitrogen. Ann Bot 62:79–82
Zheng Guang-zhi, He Jing-bo, Wang Shi-ling (1983) Cryopreservation of calli and their suspension culture cells of *Anisodus acutangulus*. Acta Bot Sin 25:512–517 (in Chinese)

III.2 Quiescence and Dormancy in Somatic Embryos

D.J. Gray and A. Purohit[1]

1 Introduction

Certain organisms avoid stress by entering a state of deep rest or dormancy; an adaptive mechanism that allows colonization of otherwise unfavorable environments. During dormancy, need for nutrients is greatly reduced as is sensitivity to environmental changes, thereby ensuring survival through hostile periods, such as dry or cold seasons. The adaptive range of organisms that possess this ability is greatly increased. For example, in the animal kingdom, a number of different taxa, notably phylum Tardigrada (tardigrades), can become dormant upon dehydration, which allows them to survive dry periods. Others, such as the fairy shrimp (*Eubranchipus*) produce eggs that withstand severe drying (Hickman et al. 1974). They become active and reproduce upon rehydration during moist periods. Such animals can, thus, survive and thrive in areas that are subjected to periodic cycles of drying and wetting. Plants such as *Selaginella lepidophylla* (Hook. and Grev.) Spring behave similarly (Adams et al. 1990); they stop growth and shrivel during unfavorable dry seasons but resume a normal appearance and growth habit upon rehydration. Both of these diverse types of organisms can remain dormant and survive prolonged periods of dehydration.

Perhaps the most significant example of dormancy, when considering its impact on mankind, involves the ascendance of higher plants, most of which produce seed that can be dried and preserved. Seed, which evolutionarily represent simply the survival and dispersal structures of young sporophytes, allowed higher plants to spread into new habitats at an unprecedented rate. Due to a combination of their small size and quiescence, which made storage and handling a simple task, seed have allowed mankind to transport abundant germplasm of valuable crops into new areas. Seed also represents the single most important source of human nutrition worldwide (e.g. Anderson 1976; Roberts 1976) because of their relatively high concentrations of important amino acids, proteins, lipids, and polysaccharides.

In nearly all instances, seed has evolved to favor random dispersal. Man has long sought to change this natural function by developing complex machinery to uniformly plant seed. The development and manipulation of seed has become a major business. However, one practical drawback that has never been overcome is the fact that, with few exceptions, plants that grow from seed represent the meiotic recombinants of two parents and, as such, are not genetically identical. Although this factor has been manageable, since acceptably uniform seed for most crops could be produced by selfing

[1]Central Florida Research and Education Center, Institute of Food and Agricultural Sciences, University of Florida, 5336 University Avenue, Leesburg, FL 34748, USA

Biotechnology in Agriculture and Forestry, Vol. 17
High-Tech and Micropropagation I (ed. by Y.P.S. Bajaj)

inbred parental lines, others, such as fruit and nut crops, have to be propagated vegetatively from cuttings due to self-incompatibility. If genetically identical seeds could be produced, improvement of certain crops, such as forest trees, would be greatly facilitated. The advent of genetically engineered plants, which may not be meiotically stable, adds impetus to produce seeds that germinate into exact duplicates of highly selected, outstanding individuals (Gray 1987a).

Genetically identical seed may be produced via obligate apomixis, where the embryo in seed is produced from cells of maternal origin, or through development of in vitro somatic embryogenesis technology. Although apomictic seed is produced by certain grass species and a few others, this trait has not been transferred for use into any other commercial crops and, therefore, appears to be limited to a small group of species. On the other hand, somatic embryogenesis appears to have more widespread promise since it has been documented for hundreds of species and varieties.

The new field that seeks to utilize somatic embryos for clonal plant production is termed "synthetic (or artificial) seed technology". Broadly defined, synthetic seeds are somatic embryos processed to be of use in commercial plant propagation (Gray 1990). Thus, synthetic seed can take several different forms depending upon application. For example, somatic embryos that produce plants after placement into greenhouse liners would be cost-effective and advantageous in the commercial ornamental plant industry, where many varieties are already produced by in vitro micropropagation. Application of synthetic seed technology in such instances would be preferable to existing micropropagation systems which are typically complex, usually consisting of several shoot proliferation, rooting, and acclimatization stages. Furthermore, micropropagation, while more efficient than traditional vegetative propagation, is relatively time-consuming and labor-intensive such that it cannot be economically utilized for many crops and profit margins are slim for others. Substitution of synthetic seed technology for micropropagation would decrease time and labor required for plant production since the somatic embryo proliferation stage would occur very rapidly in comparison to axillary bud proliferation. A separate rooting stage would not be needed, since somatic embryos germinate bipolarly. Plants produced from somatic embryos germinating in the greenhouse would be acclimated to their environment. Thus, time and labor would be reduced due to the relative rapidity and simplicity of synthetic seed technology when compared to micropropagation.

For crops that are normally field planted, somatic embryos processed to more closely resemble natural seed would be necessary (Gray and Purohit 1991). Two major modifications to existing culture systems are needed in order to produce such somatic embryos: a synthetic seed coating must be provided and a reversible resting phase must be induced. Since somatic embryos typically develop without protective or nutritive tissues that normally enclose zygotic embryos, synthetic seed coats are necessary for protection and ease of handling. An inducible resting phase is needed so that somatic embryos can be conveniently stored and germinated into plants at will.

Progress in synthetic seed coat development has been previously reviewed (Redenbaugh et al. 1988; see also Chap. III.3 this Vol.). A distinct terminology has arisen to categorize and describe various types of resting seeds, based on seed physiology research. This terminology is directly applicable to synthetic seed research. In this presentation, we provide background information from seed physiology and somatic embryogenesis literature as well as discuss research aimed at investigating quiescence and dormancy in somatic embryos.

2 Background and Terminology

A fundamental difference between zygotic and somatic embryos is that somatic embryos tend to mature incompletely while not entering a resting phase (Gray 1987b). Embryos in orthodox seed begin a resting period during late stages of maturation; a major characteristic that allows seed to be conveniently stored (Gray 1990). The antithesis of orthodox, when considering ability to achieve a resting stage, is recalcitrant seededness.

Recalcitrant seed cannot be dehydrated or stored for useful time periods (King and Roberts 1979). This discussion will concentrate on the orthodox type in comparisons of seed with somatic embryos.

Several forms of rest have been distinguished in orthodox seed. "Quiescence" and two forms of "dormancy" are recognized here, based on definitions by Bewley and Black (1985): (1) Quiescence – resting phase is reversed solely by the addition of water. Quiescent seeds typically germinate rapidly after inhibition. (2) Primary dormancy – resting phase is present in the mature seed and is controlled either by internal mechanisms or by tissues external to the embryo such as endosperm, seed coat, pericarp, or other floral tissues. Germination occurs in the presence of water after internal mechanisms are reversed or after removal of inhibitory tissues. (3) Secondary dormancy – dormancy is induced later, after previously favorable germination conditions may have occurred. Secondary dormancy occurs when rest is induced in nondormant seed by environmental factors such as drying, light, dark, and temperature extremes. The occurrence of any resting type is characteristic of a given species or variety.

The resting phase is influenced by many different factors, which can be categorized into three basic groups (Bidwell 1974): (1) Environmental factors which include light, temperature, and water requirements; (2) internal factors such as the presence of seed coats, embryo immaturity, gas (i.e. CO_2, O_2, and ethylene) concentrations, presence of inhibitors (i.e. ABA), absence of promoters (i.e. GA_3); and (3) timing mechanisms, including after-ripening, degradation of inhibitors, synthesis of promoters.

Ability of seed to withstand dehydration is critical to attainment of the resting phase. Maturation of seed is accompanied by loss of water. A concomitant reduction in metabolism during dehydration culminates in the resting phase. After the resting period, an increase in metabolism leads to germination (Kermode et al. 1986). Seed embryos must be at the "desiccation tolerant stage" in order to survive dehydration (Gray 1987b; Koster and Leopold 1988; Neumann et al. 1989; Senaratna et al. 1990). Prior to dehydration, metabolism is oriented toward accumulation of storage compounds, whereas after rehydration, storage reserves are consumed for germination. These observations suggest that dehydration followed by rehydration functions to switch the mode of development from that of storage synthesis to catabolism of storage material (Kermode et al. 1986). Therefore, acquisition of desiccation tolerance appears to be central to normal embryo development. Dehydration appears to cause a basic switch in physiology such that pathways favoring embryo development are replaced by those that lead to plant development. Poor germination of somatic embryos may be due to the absence of these complex developmental mechanisms. However, as described below, their absence may be due more to differences between the in vitro and in vivo environment than to intrinsic problems of somatic embryos. Induction of desiccation tolerance, followed by dehydration and quiescence, may prompt somatic embryos to function in a manner more like seed embryos.

Somatic embryos tend either to grow and germinate without normal maturation, become disorganized into embryogenic tissue, or die. They rarely enter a resting stage (Gray 1990). However, dormancy has been documented in somatic embryos of grape and may occur but be unrecognized in others (see below). Dormancy controlled by the endosperm or seed coat obviously does not function in embryogenic culture systems. Other types of dormancy, such as those regulated by water availability, temperature, light, inhibitors, and promoters (e.g. Tran and Cavanagh 1984; Bewley and Black 1985) may be present at some level in somatic embryos, especially those of species with corresponding zygotic embryo dormancy.

Quiescent or dormant somatic embryos will be needed in order to mimic seed storage and handling characteristics. Although quiescence and dormancy have only recently been documented in somatic embryos, there are many instances where plants cannot be obtained from well-developed somatic embryos. This suggests that unrecognized dormancy may be involved. Traditional methods for inducing and maintaining quiescence in seeds (e.g. Owen 1956; Barton 1961) may be applicable for somatic embryos.

3 Quiescence

Categorization of the type of rest that can be induced in somatic embryos is somewhat different from that of seed. Rest induced in somatic embryos that would have otherwise continued growing could be defined as a type of secondary dormancy (see above). However, we consider drying to be a previously unrecognized step, missing in conventional culture systems, that is needed for proper embryo development. Therefore, if somatic embryos enter a resting phase after drying, "quiescent" would be the best designation. Use of dehydration treatments to induce quiescence in somatic embryos is an obvious choice since dehydration is known to cause arrest of growth in seed embryos.

Rate and degree of dehydration can be finely regulated by use of a controlled relative humidity environment, as illustrated by Fig. 1 which shows the weight loss over time by groups of 100 grape somatic embryos when exposed to 30, 70, or 90% relative humidity air. Embryos in 30% relative humidity dehydrated most rapidly and lost the greatest percentage of weight, whereas the reverse was true for the 90% level. The embryos were all placed in a 60 °C oven after 330 h to remove additional unbound water, resulting in a sharp drop in weight for the 70 and 90% treatments such that all three became equivalent. By comparing fresh weights with those obtained at equilibrium at each relative humidity and with dry weights, we calculated that nondehydrated embryos contained 75% water, whereas those maintained at 30, 70, and 90% relative humidity until weight stabilization occurred contained 22, 30, and 35% water, respectively. This demonstrates that distinct somatic embryo water contents can be produced and maintained with this protocol.

Of several early attempts to obtain plants from dehydrated callus and/or somatic embryos, the first that we are aware of utilized somatic embryos of carrot (Jones 1974). Somatic embryos were stored for 21 days at 80–90% relative humidity (RH), which resulted in a nonquantified amount of drying. Although some of the embryos "survived", plants were not obtained. Another worker dried carrot callus cultures for 7

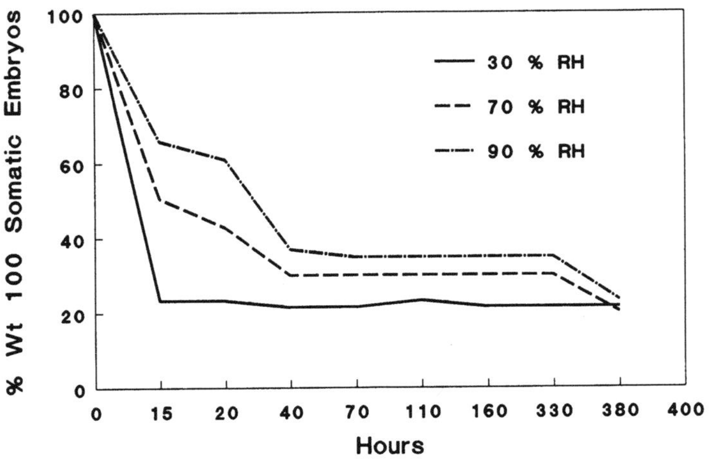

Fig. 1. Weight loss by grape somatic embryos maintained at 25 °C under 30, 70, or 90% relative humidity. Note that weight loss is greater and more rapid at lower relative humidities. Weight loss after 330 h was due to additional drying at 60 °C to remove remaining water. This demonstrates that relative humidity can be used to modulate drying rate and control equilibrated water content in somatic embryos

days at room temperature and unspecified relative humidity (Nitzsche 1978). Growth of callus resumed and plant regeneration occurred after imbibition, but the origin of the plants could not be determined. It was possible that the plants originated from somatic embryos that formed after imbibition. Subsequently, carrot somatic embryos survived 4 days of drying in a water-soluble plastic resin, although recovery of plants was not reported (Kitto and Janick 1985). This report also represented the first demonstration of a dehydrated synthetic seed coat. Pretreatments with ABA and high concentrations of sucrose prior to dehydration were employed in each of these studies. In retrospect, it appears that these pretreatments functioned to induce a state of desiccation tolerance.

The first report of actual plant recovery utilized dehydrated somatic embryos of orchardgrass isolated from callus cultures (Gray and Conger 1985a; Gray et al. 1987). Somatic embryos were exposed to 70% RH air and stored for up to 21 days at 23 °C and the amount of water loss was determined. Use of a regulated relative humidity environment allowed precise control and repeatability of drying rate. When exposed to this relative humidity, embryos dehydrated rapidly, became discolored, decreased in size, and their outer cell walls collapsed (Fig. 2A). Embryo water content dropped from 83 to 13% within 24 h and was maintained over the entire storage period. This water content is similar to that of seeds maintained at 70% RH and is adequate for maintenance of viability during prolonged seed storage. Imbibition was accomplished by placement of embryos on solidified medium. Within 15 min, they swelled, regained their previous appearance and, subsequently, germinated (Fig. 2B). However, only well-developed, white, opaque somatic embryos were responsive. Such somatic embryos were structurally mature and contained starch and lipid storage compounds (Gray and Conger 1985). This demonstrated that only embryos that had achieved a degree of developmental maturity were able to become quiescent and further suggested that such embryos were desiccation tolerant. When these embryos were stored in a dehydrated state for 21 days, 4% germinated and produced plants after imbibition (Table 1).

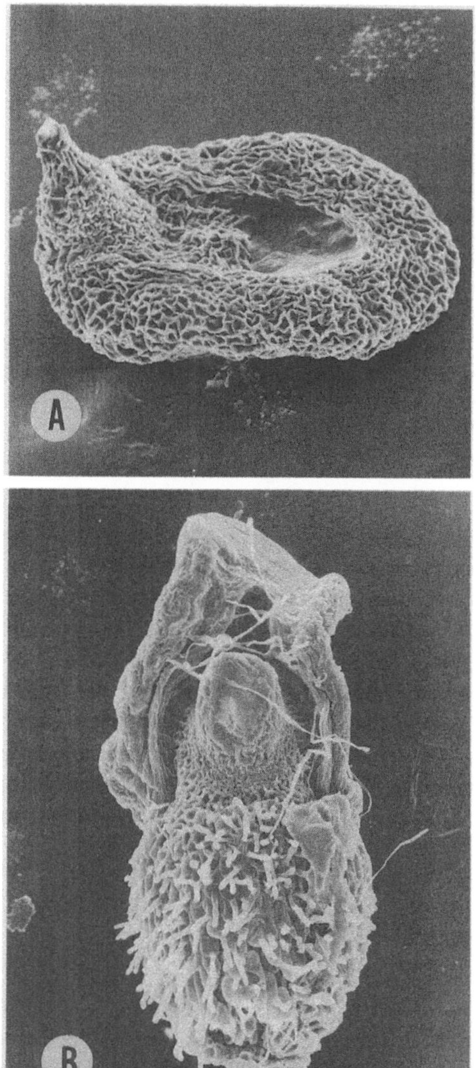

Fig. 2A,B. Scanning electron microscopy of dehydrated and imbibed somatic embryos of orchardgrass. **A** Dehydrated orchardgrass embryo; note collapsed cells. **B** Orchardgrass embryo after imbibition and germination. Development of root hairs is the first sign of germination. (Gray et al. 1987)

That somatic embryos could become reversibly quiescent in the manner of seed embryos was demonstrated.

Plants were also recovered from somatic embryos of grape which were dehydrated for 21 days as will be described in more detail with regard to dormancy in the next section (Gray 1987b, 1987d, 1989). We have also been able to induce quiescence and subsequently obtain plants from somatic embryos of corn (*Zea mays* L.) and soybean (*Glycine max* Merr.) (Gray unpubl.). Most recently, somatic embryos of alfalfa were dehydrated with concomitant plant recovery (McKersie et al. 1989). Approximately 90–100% of dehydrated high quality embryos produced plants when treated with ABA

Table 1. Germination of dehydrated orchardgrass somatic embryos after storage at 23 °C for 0, 7, and 21 days.[a] (Gray et al. 1987)

Response[b]	Days of dehydrated storage		
	0	7	21
No germination	126/28[c]	333/74	396/88
Germination – no further growth	180/40	81/18	36/18
Germination – viable plants	144/32	36/8	18/4
Total	450/100	450/100	450/100

[a]Embryos were imbibed on solidified medium after test storage periods.
[b]Embryos that produced root hairs, roots, coleoptiles, and/or shoots but failed to develop further were scored as germinated – no further growth. Those that produced green leaves and continued to grow were considered to be viable.
[c]Number/percentage of total.

before drying. About 60% of embryos dried for 1 year without humidity control germinated into plants (Senaratna et al. 1990). Pretreatments with ABA were reportedly necessary for alfalfa somatic embryos to survive dehydration. Authors postulated that ABA pretreatment induced a state of desiccation tolerance (Senaratna et al. 1987, 1989). ABA applied at 10 μM to 14-day-old embryos was most effective. Drying rate was also found to be critical such that slow drying was more effective than rapid drying. These treatments approximate certain aspects of the developing seed environment. Our ability to induce desiccation tolerance in somatic embryos clearly demonstrates that they are capable of entering complex developmental pathways normally associated with those of seed embryos when proper environmental conditions are provided.

4 Dormancy

There are few recognized reports of somatic embryo dormancy in the literature. However, it is highly likely that dormancy has been present but overlooked in certain instances. The poor germination rates typical of many embryogenic culture systems have been blamed on inherent problems and abnormalities of somatic embryo development when compared to seed embryos. However, poor germination cannot be attributed to abnormalities in culture systems where well-developed embryos are formed (Gray 1990). Dormancy that is not being alleviated through typical manipulations in vitro may function in these instances. There are two possible causes of dormancy in vitro (Gray and Purohit 1991): (1) dormancy that is typical for zygotic embryos may also be expressed by somatic embryos of the same species in vitro; (2) dormancy may be artificially induced by aspects of the culture environment such as exogenously supplied growth regulators. In either of these instances, dormancy could be broadly categorized as either the primary or secondary type as described above. With well-developed but poorly germinating somatic embryos, dormancy might be detected through application of standard dormancy-breaking treatments such as cold stratification or exposure to dormancy-breaking growth regulators (e.g. cytokinins or gibberellins) (Hepher and Roberts 1985a,b).

The grape embryogenic culture system provides the best documented example of in vitro dormancy. With grape, somatic embryos become very well developed and morphologically faithful to zygotic embryos, but germinate poorly. Grape seed exhibits a type of dormancy that is alleviated by cold stratification. Germination occurs after approximately 4 weeks of treatment at 4 °C (Flemion 1937). Grape somatic embryos mimic seed in that they germinate better following cold treatments, providing evidence to suggest that they are also dormant (Gray and Mortensen 1987). Studies of ABA concentration in grape embryogenic cultures showed a rapid increase during embryo development, which reached a peak at maturation (Rajasekaran et al. 1982). Cold stratification of somatic embryos resulted in a rapid decrease in ABA. Exogenously supplied ABA inhibited somatic embryo germination (Gray 1989). Because ABA is implicated as a controlling factor of dormancy in many types of seeds (e.g. Walton 1980; Bewley and Black 1985; Thevenot et al. 1987), it is plausible to consider that ABA also functions in grape somatic embryos. In contrast to ABA, exogenously supplied gibberellin (GA_3) induced grape somatic embryo germination and the concentration of endogenous GA-like compounds increased during cold stratification (Pearce et al. 1987). Results of these studies suggest a simple endogenous control of embryogenesis and germination, whereby ABA inhibits precocious germination and thus promotes normal development while GA promotes germination. Embryo dormancy appears to be a developmental mechanism that promotes full maturation and allows germination to be timed to the season (spring) or other environmental conditions that are most amenable to successful plant establishment. The close resemblance of grape somatic and zygotic embryo dormancy provides compelling evidence for our theory that dormancy also exists in other embryogenic culture systems with corresponding zygotic embryo dormancy.

In addition to cold stratification, we found that dehydration could also be used to manipulate grape somatic embryo dormancy (Gray 1987b, 1990). During dehydration, grape somatic embryos underwent morphological changes similar to those of orchardgrass (Gray 1987b). Their water content equilibrated to 13% when stored at 70% RH and they resumed a normal appearance after rehydration. Genotypic differences in response were noted; culture lines that produced relatively well-developed somatic embryos were most responsive (Gray 1989). After 21 days of dehydrated storage, 34% of embryos from one grape genotype produced plants following imbibition (Table 2). Dehydrated embryos germinated to produce plants immediately after imbibition,

Table 2. Comparison of dehydration and benzyladenine (BA) for inducing germination in grape somatic embryos.[a] (Gray 1989)

Treatment	Percent germination response[b]			
	Hypocotyl	Root	Cotyledon	Shoot
Dehydration	77	68	65	34
0.5 μM BA	100	92	98	12
Control	76	88	36	0

[a]Well-developed embryos were either dehydrated for 21 days at 70% RH and 27 °C, placed directly on medium with BA, or placed on basal medium (control).

[b]Germination response was based upon either enlargement and greening of hypocotyls and cotyledons or emergence of roots or shoots.

whereas no plants were recovered from nondehydrated controls. Prior to our investigations of dehydration, the best treatment for inducing germination and producing plants from grape somatic embryos was via culture on medium containing BA (Table 2). This study demonstrated that higher germination percentages were possible through dehydration. In considering terminology, an interesting conundrum occurs: Since dehydrated somatic embryos of grape do not require typical dormancy-breaking pretreatments and germinate directly after imbibition, they can be considered to be quiescent, not dormant. Alternatively, dehydration may be viewed as a dormancy-breaking treatment. This latter conclusion is supported by studies of somatic embryogenesis in red oak (*Quercus rubra* L.) (Gingas and Lineberger 1988) and soybean (Parrott et al. 1988). Somatic embryos of red oak germinated and produced shoots only after a period of dehydration or treatment with osmotically active sugars. With soybean, only a few somatic embryos from several test genotypes germinated without dehydration pretreatments but up to 100% germinated after dehydration. These studies suggest that dehydration would increase germination and plant recovery rates from somatic embryos of other species.

One unexpected result of our dehydration experimentation was that the germination pattern of grape somatic embryos could be strikingly manipulated (Gray 1989). Only somatic embryos that had been dehydrated were capable of germinating in a pattern reminiscent of that of a seedling such that shoot and root emergence occurred synchronously (Fig. 3). This was in contrast to germination of nondehydrated control embryos with and without treatment with dormancy-breaking growth regulators. These embryos germinated abnormally such that embryonic organs often became enlarged. Although plants were occasionally obtained, notably with benzyladenine pretreatment, root and shoot emergence was not synchronous.

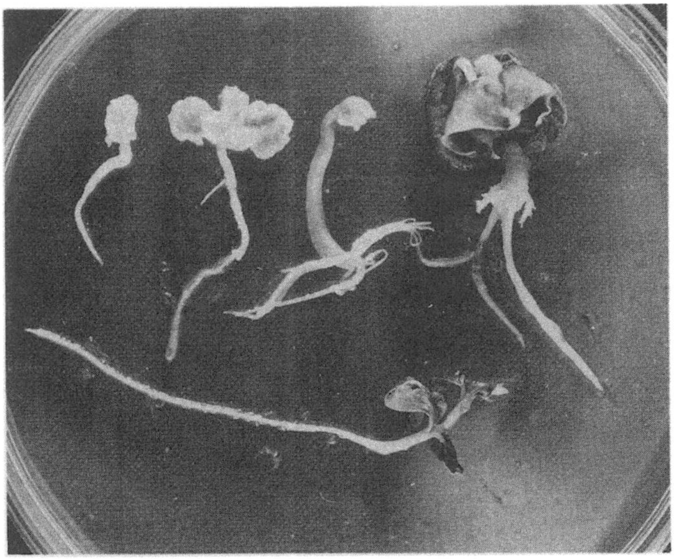

Fig. 3. Comparison of germination treatments for grape somatic embryos. *From top left* control treatment; 1 μM gibberellic acid; 1 μM abscisic acid; 1 μM benzyladenine. *Below* dehydration at 70% RH for 21 days. (Gray 1989)

5 Conclusions

When taken as a whole, these results demonstrate that dehydration exerts a profound effect on somatic embryo maturation (Gray 1990; Gray and Purohit 1991). Since a dehydration step has not been a feature of conventional embryogenic culture systems, its effect has heretofore gone unrecognized. Dehydration of seed to storage water levels is generally considered to be possible only after maturation and the acquisition of desiccation tolerance (Adams et al. 1983). Desiccation tolerance may result from the stabilizing effect of increased sugar concentration on cell membranes (Koster and Leopold 1988). However, one fact often not considered in relation to somatic embryo maturation is that, in seeds, water content drops continuously during most stages of development as relative dry weight increases. Furthermore, developing seed is subjected to diurnal fluxes in relative water content due to daily cycles of temperature and plant metabolic activity. Gradual dehydration and/or fluctuating cycles of dehydration may be key factors involved in regulation of seed (and zygotic embryo) maturation. These types of dehydration are missing or are much less pronounced in somatic embryos due to the hydrated nature of the in vitro environment. Dehydration treatments that mimic the water fluxes in developing seed may lead to better somatic embryo maturation and germination.

Ability of somatic embryos to enter a reversible resting phase has several important ramifications to future research and development. Firstly, embryogenic cultures represent a valuable in vitro system for studying the phenomena of quiescence and dormancy. Asynchronous development, typical of embryogenic systems, allows embryos at various developmental stages to be simultaneously isolated for study (e.g. Gray et al. 1984). This allows the effects of experimental treatments to be evaluated on all stages at one time and differs from equivalent experiments with zygotic embryos (Gray and Conger 1985b,c). With zygotic embryos, treatments must be repeated over time in order to sample embryo stages. However, this lack of synchrony does have deleterious consequences when considering synthetic seed production, where uniform embryo production is required (Gray 1987c). Since somatic embryos develop without seed coat and endosperm, the effect of these tissues on the resting phase can be eliminated and, thus, better understood. Secondly, induction of quiescence in somatic embryos is necessary in order to develop useful synthetic seed for many crops, although nonquiescent embryos may be useful for other crop types (Table 3). Quiescent synthetic seed will be

Table 3. Types of synthetic seed and crop applications. (Gray and Purohit 1991)

Synthetic seed type	Relative development cost	Relative cost per seed	Example of crop application
Naked embryo			
Nonquiescent	Low	High	Ornamentals
Quiescent (dried)	Low	High	Germplasm conservation
Encapsulated embryo			
Nonquiescent	Medium	High	Vegetable transplants
Quiescent (dried)	High	Low	Agronomic and conifers

needed for all crops that must be field planted at high density in order to take advantage of the convenience offered by seed. When available for a broad range of important species, quiescent synthetic seed will revolutionize the existing micropropagation industry by dramatically increasing the number and type of crops that can be economically cloned. Thirdly, use of quiescent somatic embryos for germplasm conservation would eliminate reliance on field gene banks. Field gene banks are expensive to maintain and subject to loss due to environmental disasters. This has prompted exploration of approaches dealing with in vitro culture methodology (e.g. Bajaj 1986; Towill 1988; Withers 1989). The amount of germplasm maintained on several acres in the field could be kept in one or few drawers of a seed repository.

The relevance of somatic embryo quiescence and dormancy to aspects of both fundamental and applied research make further investigation highly compelling.

Acknowledgments. We thank the International Board for Plant Genetic Resources for providing partial support to this project as part of the IBPGR research program. Florida Agricultural Experiment Station Journal Series No. R-01105.

References

Adams CA, Fjerstad MC, Rinne RW (1983) Characteristics of soybean seed maturation: necessity for slow drying. Crop Sci 23:265–267

Adams RP, Kendall E, Kartha KK (1990) Comparison of free sugars in growing and desiccated plants of *Selaginella lepidophylla*. Biochem Syst Ecol 18:107–110

Anderson RG (1976) Prospects for improving production of cereals. In: Scrimshaw NS, Behar M (eds) Nutrition and agricultural development: significance and potential for the tropics. Plenum Press, New York, pp 295–304

Bajaj YPS (1986) In vitro preservation of genetic resources – techniques and problems. Int Symp Nuclear techniques and in-vitro culture for plant improvement. IAEA/FAO, Vienna 1985, pp 43–57

Barton LV (1961) Seed preservation and longevity. Leonard Hill, London, pp 216

Bewley JD, Black M (1985) Seeds: physiology of development and germination. Plenum, New York, 367 pp

Bidwell RGS (1974) Dormancy, senescence and death: plant physiology. Macmillan, New York, pp 466–493

Flemion F (1937) After-ripening at 5 °C favours germination of grape seeds. Contr Boyce Thompson Inst 9:7–15

Gingas VM, Lineberger RD (1988) Plantlet regeneration from asexual embryos of *Quercus rubra* L. HortScience 23:786

Gray DJ (1987a) Introduction to the symposium. In: Proc Symp Synthetic seed technology for the mass cloning of crop plants: problems and perspectives. HortScience 22:796–797

Gray DJ (1987b) Quiescence in monocotyledonous and dicotyledonous somatic embryos induced by dehydration. In: Proc Symp Synthetic seed technology for the mass cloning of crop plants: problems and perspectives. HortScience 22:810–814

Gray DJ (1987c) Concluding remarks. In: Proc Symp Synthetic seed technology for the mass cloning of crop plants: problems and perspectives. HortScience 22:814

Gray DJ (1987d) Effects of dehydration and other environmental factors on dormancy in grape somatic embryos. HortScience 22:1118

Gray DJ (1989) Effects of dehydration and exogenous growth regulators on dormancy, quiescence and germination of grape somatic embryos. In Vitro Cell Dev Biol 25:1173–1178

Gray DJ (1990) Synthetic seed for clonal production of crop plants. In: Taylorson RB (ed) Recent advances in the development and germination of seeds. Plenum Press, New York, pp 29–45

Gray DJ, Conger BV (1985a) Quiescence in somatic embryos of orchardgrass (*Dactylis glomerata*) induced by desiccation. Am J Bot 72:816

Gray DJ, Conger BV (1985b) Somatic embryo ontogeny in tissue cultures of orchardgrass. In: Henke RR, Hughes KW, Constantin MJ, Hollaender A (eds) Tissue culture in forestry and agriculture. Plenum, New York, pp 49–57

Gray DJ, Conger BV (1985c) Time-lapse light photomicrography and scanning electron microscopy of somatic embryo ontogeny from cultured leaves of *Dactylis glomerata* (Gramineae). Trans Am Microsc Soc 104:395–399

Gray J, Mortensen JA (1987) Initiation and maintenance of long term somatic embryogenesis from anthers and ovaries of *Vitis longii* 'Microsperma'. Plant Cell Tissue Organ Cult 9:73–80

Gray DJ, Purohit A (1991) Embryogenesis and development of synthetic seed technology. Crit Rev Plant Sci 10:33–61

Gray DJ, Conger BV, Hanning GE (1984) Somatic embryogenesis in suspension and suspension-derived callus cultures of *Dactylis glomerata*. Protoplasma 122:196–202

Gray DJ, Conger BV, Songstad DD (1987) Desiccated quiescent somatic embryos of orchardgrass for use as synthetic seeds. In Vitro Cell Dev Biol 23:29–33

Hepher A, Roberts JA (1985a) The control of seed germination in *Trollius ledebouri*: the breaking of dormancy. Planta 166:314–320

Hepher A, Roberts JA (1985b) The control of seed germination in *Trollius ledebouri*: a model of seed dormancy. Planta 166:321–328

Hickman CP Sr, Hickman CP Jr, Hickman FM (1974) Integrated principles of zoology, 5th edn. CV Mosby, St Louis, 1025 pp

Jones LH (1974) Long term survival of embryoids of carrot (*Daucus carota* L.). Plant Sci Lett 2:221–224

Kermode AR, Bewley JD, Dasgupta J, Misra S (1986) The transition from seed development to germination: a key role for desiccation? HortScience 21:1113–1118

King MW, Roberts EH (1979) The storage of recalcitrant seeds. Report for the International Board for Plant Genetic Resources Secretariat, Rome, IBPGR, p 96

Kitto SL, Janick J (1985) Hardening treatments increase survival of synthetically-coated asexual embryos of carrot. J Am Soc Hort Sci 110:283–286

Koster KL, Leopold CA (1988) Sugars and desiccation tolerance in seed. Plant Physiol 88:829–832

McKersie BD, Senaratna R, Bowley SR, Brown DCW, Krochko JE, Bewley JD (1989) Application of artificial seed technology in the production of hybrid alfalfa (*Medicago sativa* L.). In Vitro Cell Dev Biol 25:1183–1188

Neumann D, Norer L, Parthier B, Rieger R, Scharf KD, Wollgiehn R, Nieden UZ (1989) Heat shock and other stress response systems of plants. Biol Zentralbl 108:1–146

Nitzsche W (1978) Erhaltung der Lebensfähigkeit in getrocknetem Kallus. Z Pflanzenphysiol 87:469–472

Owen EB (1956) The storage of seeds for maintenance of viability. Commonw Agric Bur Bucks, England, 81 pp

Parrot WA, Dryden G, Vogt S, Hildebrand DF, Collins GB, Williams EG (1988) Optimization of somatic embryogenesis and embryo germination in soybean. In Vitro Cell Dev Biol 24:817–820

Pearce D, Pharis RP, Rajasekaran K, Mullins MG (1987) Effects of chilling and ABA on [^{3}H]gibberellin A_4 metabolism in somatic embryos of grape (*Vitis vinifera* L. × *V. rupestris* Scheele). Plant Physiol 80:381–385

Rajasekaran K, Vine J, Mullins MG (1982) Dormancy in somatic embryos and seeds of *Vitis*: changes in endogenous abscisic acid during embryogeny and germination. Planta 154:139–144

Redenbaugh K, Fujii JA, Slade D (1988) Encapsulated plant embryos. In: Mizrahi A (ed) Biotechnology in agriculture. Liss, New York, pp 225–248

Roberts LM (1976) Improving the production and nutritional quality of food legumes. In: Scrimshaw NS, Behar M (eds) Nutrition and agricultural development: significance for the tropics. Plenum Press, New York, pp 309–317

Senaratna T, McKersie BD, Borochov A (1987) Desiccation and free radical mediated changes in plant membranes. J Exp Bot 38:2005–2014

Senaratna T, McKersie BD, Bowley SR (1989) Desiccation tolerance of alfalfa (*Medicago sativa* L.) somatic embryos – influence of abscisic acid, stress pretreatments and drying rates. Plant Sci 65:253–259

Senaratna T, McKersie BD, Bowley ST (1990) Artificial seeds of alfalfa (*Medicago sativa* L.) induction of desiccation tolerance in somatic embryos. In Vitro Cell Dev Biol 26:85–90

Thevenot C, Ralambosa J, Simond-Cote E (1987) Influence of abscisic acid and oxygen supply on germination of more or less dormant apple embryos. Isr J Bot 36:101–112

Towill LE (1988) Genetic consideration for germplasm preservation of clonal materials. HortScience 23:91–95
Tran VN, Cavanaugh AK (1984) Structural aspects of dormancy. In: Murray DR (ed) Seed physiology, vol 2. Germination and reserve mobilization. Academic Press, Australia, pp 1–44
Walton DC (1980) Biochemistry and physiology of abscisic acid. Annu Rev Plant Physiol 31:453–489
Withers LA (1989) In vitro conservation and germplasm utilisation. In: Brown AHD, Frankel OH, Marshall DR, Williams JT (eds) The use of plant genetic resources. Cambridge Press, New York, pp 309–334

III.3 Artificial Seeds – Encapsulated Somatic Embryos

K. REDENBAUGH[1], J. FUJII[2], D. SLADE[3], P. VISS[4], and M. KOSSLER[1]

1 Introduction

Somatic embryogenesis has been observed for a wide array of species (Ammirato 1983). Despite advances made in the past 30 years with somatic embryogeny, there has been no commercial development. Although oil palm (*Elaeis guineensis*) can be propagated by somatic embryogenesis, the method being commercialized uses in vitro-rooted shoots, a high-cost propagation system (Corley 1982). Unfortunately, for most crops in which the per-unit seed value is low, labor-intensive micropropagation systems are not feasible. Somatic embryogenesis appears to be the only clonal propagation system suitable for crops currently propagated by seeds.

The limitations for a somatic embryo delivery system have been twofold: no suitable coating system was available, and the quality of the somatic embryos for all species was poor. In this chapter our progress over the past 8 years in these two areas is discussed.

2 Artificial Seed Concepts

The concepts for artificial seeds have been developed during the past 10 years. Although somatic embryos were first produced in 1958, it was not until the late 1970's that the idea was presented at a conference (Murashige, *Symposium on Tissue Culture for Horticultural Purposes* in Ghent, Belgium, September 6–9, 1977). Although Murashige discussed the concept of artificial seeds in detail at the meeting, his comments in the proceedings were limited to: "But to be applicable, the cloning method must be extremely rapid, capable of generating several million plants daily, and competitive economically with the seed method" (Murashige 1977). In the following year, the first mention of artificial seeds appeared in the literature (Murashige 1978). Other early efforts for development of artificial seeds were by Keith Walker, at Monsanto Company, working with a *Medicago sativa* L. (alfalfa) line (Regen S) identified by Bingham et al. (1975) and by Robert Lawrence at Union Carbide working with celery and lettuce. Lawrence focused on delivery of somatic embryos using fluid drilling technology (Lawrence 1981) as well as using polyoxyethylene to form seed tapes or

[1]Calgene Inc., 1920 Fifth Street, Davis, CA 95616, USA
[2]The University of Chicago, Chicago, IL
[3]University of Seattle, WA
[4]Plant Research Laboratory, Modesto, CA, USA

Biotechnology in Agriculture and Forestry, Vol. 17
High-Tech and Micropropagation I (ed. by Y.P.S. Bajaj)

sheets (Lawrence pers. commun.). At the symposium workshop, *Advances in Methods of in Vitro Cloning for Large-Scale Propagation of Plants* (September 21–22, 1981 at the W. Alton Jones Cell Science Center in Lake Placid, New York), Walker and Lawrence discussed the concept of a low-cost, high-volume propagation system for vegetable and agronomic crops. The focus was to use somatic embryos delivered by fluid drilling, in a seed tape, or as artificial seeds.

Other early research on developing delivery systems for somatic embryos also focused on these three areas. Drew (1979) suggested delivering carrot somatic embryos using fluid drilling, but did not conduct any experiments, since he was able to produce only three plants from carrot embryos on a carbohydrate-free medium. Baker (1985) was also unsuccessful in using fluid drilling to plant somatic embryos. Kitto and Janick (1982, 1985a,b) coated clumps of carrot embryos, roots, and callus with polyoxyethylene. Some embryos survived the coating process as well as a desiccation step. Styer (1985) suggested using a seed tape, but gave no data. Durzan (1979) mentioned pelleting conifer somatic embryos, but also presented no results.

Recently, we reported the use of hydrogel coatings, such as sodium alginate, for producing single-embryo artificial seeds (Redenbaugh et al. 1984, 1986, 1988; Fujii et al. 1987). To date, we have produced artificial seeds for several crops (Table 1). It is anticipated that the first commercial artificial seeds will be for those crops in which it is technically feasible to produce embryos of sufficient quality and for which there is a high cost per seed. Candidate crops are listed in Table 2. Although there are many crops for which somatic embryos have been produced, only a few have been sufficiently developed with the concept of producing high-quality embryos (Category 1, Table 2). However, commercial application is unlikely. Other crops, because of their high seed costs, would be good candidates for artificial seed technology, but they lack good somatic embryogeny systems (Category 2). Finally, there are a few crops in which both technology and commercial potential are present (Category 3). For commercial development of artificial seeds, the crops in Category 3 will most likely be the first used.

Since results on using hydrogel encapsulation of somatic embryos to form artificial seeds appeared, several laboratories have conducted research with alfalfa, asparagus, carrot, celery, loblolly pine, and Norway spruce (Table 3). However, except for our research and that of Kamada et al. (1988), no consistent embryo-to-plant conversion has occurred from encapsulated embryos. Demarly (pers. commun.) indicated that

Table 1. Crops for which plants have been produced from somatic embryos encapsulated in calcium alginate beads

Species	Common name
Apium graveolens L.	Celery
Brassica sp.	Rapid-cycling *Brassica*
Daucus carota L.	Carrot
Gossypium hirsutum L.	Cotton
Lactuca sativa L.	Lettuce
Medicago sativa L.	Alfalfa
Oryza sativa L.	Rice
Zea mays L.	Corn

Table 2. Potential crops for artificial seeds

Category 1 – Strong technological basis	
Efficient somatic embryogenesis system already exists	
Caraway	*Panicum*
Carrot	*Pennisetum*
Orchardgrass	
Category 2 – Strong commercial basis	
High per-unit value crops	
Asparagus	Impatiens
Begonia	Lettuce
Broccoli	Petunia
Cauliflower	Pineapple
Cucumber	Potato
Cyclamen	Rice
Daylily	Spinach
Garlic	Sugarcane
Geranium	Tobacco
Gerbera daisy	Tomato
Ginsing	Watermelon
Category 3 – Good technological and commercial basis	
Alfalfa	Cotton
Celery	Date palm
Coffee	Grape
Corn	Mango

Table 3. Artificial seeds using hydrogels

Species	Results	Reference
Alfalfa (*Medicago sativa*)	90% in vitro conversion in various hydrogels	Redenbaugh et al. (1984, 1986, 1987a,b)
Asparagus (*Asparagus officinalis*)	Calcium aglinate encapsulation, no results published	Yamakawa (1985) and Nishimura (1986)
Carrot (*Daucus carota*)	Calcium aglinate encapsulation, no results published	Yamakawa (1985) and Nishimura (1986)
Carrot (*Daucus carota*)	Encapsulated somatic embryos, no conversion data or coating composition given	Lutz et al. (1985)
Carrot (*Daucus carota*)	50% conversion using calcium alginate capsules	Kamada et al. (1985, 1988)
Carrot (*Daucus carota*)	Encapsulation using alginate and carrageenan, possible conversion on vermiculite	Hama (1986)
Celery (*Apium graveolens*)	90% in vitro conversion in calcium alginate	Redenbaugh et al. (1984, 1987a)
Loblolly pine	Encapsulation using alginate, no germination or conversion	Gupta and Durzan (1987)
Norway spruce	Encapsulation using alginate, germination, but no conversion	Durzan and Gupta (1988)

his laboratory had encapsulated somatic embryos of alfalfa, carrot, date palm, soybean, and grapevine in calcium alginate, but did not indicate whether any of the artificial seeds germinated. Similarly, Sanada (pers. commun.) reported on celery, lettuce, and rice. Bapat et al. (1987) reported on encapsulation of mulberry shoot buds in calcium alginate beads and in agar blocks, with alginate giving a better response in terms of plant production. Other research on alginate encapsulation was reported by Upadhyaya et al. (1987) and Gautz et al. (1987). Upadhyaya et al. modified and tested machinery for planting hydrated capsules, but presented data only on machinery design. Gautz et al. were able to separate hydrogel capsules using a vibrating box. Capsules that were uncoated did not separate well, whereas wax-coated capsules could be separated using vibration. Although not identified in the article, the wax coating was that described in Redenbaugh et al. (1987a).

Recently, considerable attention has been given to somatic embryo desiccation, beginning with reports by Kitto and Janick (1982, 1985a,b) on carrot. Other researchers have focused on a variety of species (Table 4). Based on most of the results to date, somatic embryos are damaged by the process, despite their ability to survive desiccation. Two possibilities as to the cause of this damage are (1) the dehydration and rehydration protocols damage the embryos, or (2) the somatic embryos have not reached the appropriate maturation stage with the greatest desiccation tolerance. One notable exception in Table 4 is grape somatic embryo dehydration (Gray 1987, 1988), for which an improved conversion frequency over nondehydrated embryos was achieved. However, the frequency declined with storage. Although significant progress has been made with somatic embryo desiccation, efficient coating or encapsulation methods

Table 4. Desiccation of somatic embryos

Species	Results	Reference
Alfalfa (*Medicago sativa*)	33% conversion after desiccation, 21% conversion without desiccation	Fujii and Redenbaugh (unpubl.)
Alfalfa (*Medicago sativa*)	65% conversion after desiccation, 65% conversion without desiccation	McKersie et al. (1988)
Carrot (*Daucus carota*)	3% survival of polyoxyethylene-coated embryos, no survival of uncoated embryos	Kitto and Janick (1982, 1985a,b)
Celery (*Apium graveolens*)	Plants produced after desiccation in polyoxyethylene coating	Janick (1988); Kim and Janick (1987)
Grape (*Vitis longii*)	20% conversion after desiccation and 21 days storage, 34% conversion after desiccation and 7 days storage, 5% conversion after desiccation only	Gray (1987, 1988)
Orchardgrass (*Dactylis glomerata*)	4% conversion after desiccation and 21 days storage, 8% conversion after desiccation and 7 days storage, 32% conversion without desiccation	Gray (1987)
Soybean (*Glycine max*)	30% conversion after desiccation, 60% conversion without desiccation	Obendorf and Slawinska (1986, 1988); Slawinska and Obendorf (1987)
Wheat (*Triticum aestivuum*)	50% conversion after desiccation	Carman et al. (1987, 1988)

have not been developed that allow for high frequency plant recovery. For the remainder of this Chapter, we will focus on hydrated artificial seeds.

3 Artificial Seed Coating

3.1 Requirements for Artificial Seeds

The first requirement for an artificial seed is a somatic embryogeny protocol that results in developmentally normal, phenotypically uniform embryos. The need for high-quality embryos is actually the limiting step for commercialization of artificial seeds and will be discussed in greater detail later. The coating material needs to have several qualities. It must be nondamaging to the embryos. The coating must not only be mild enough to protect the embryo and allow for germination, but it must be sufficiently durable for rough handling during manufacture, storage, transportation, and planting. In addition, the coating must contain nutrients, developmental control agents (such as plant hormones), and other components necessary for germination and plant growth. For many crops, it is necessary that single-embryo artificial seeds be produced. Finally, it is highly desirable that the artificial seeds be plantable using existing farm machinery.

3.2 Encapsulation Methods

Initially, several methods of encapsulation were examined for producing artificial seeds. The methods included gelation, complex coacervation, and interfacial polymerization (Redenbaugh et al. 1987a). Of these, the only useful technique was gelation. The other methods, although common for microencapsulation, did not produce firm enough capsules or else the embryos did not survive. A number of gels were tested for somatic embryo encapsulation (Redenbaugh et al. 1987a) and several were found to produce a sufficiently hard capsule, while maintaining embryo viability (Table 5).

Embryo-to-plant conversion occurred using alginate, alginate with gelatin, and carrageenan with locust bean gum. Because of its gelation properties, ease of use, nontoxicity, and low cost, sodium alginate was chosen for use in further development of artificial seeds.

To form hydrogel capsules, two methods were used, (1) gel complexing via a dropping procedure or (2) molding. Alfalfa somatic embryos were mixed with sodium alginate (2% w/v) and dropped using a plastic pipet into a suitable di- or trivalent metal

Table 5. Useful gels for encapsulation of somatic embryos. (Redenbaugh et al. 1987b)

Gel	Conc. % w/v	Complexing agent	Conc. mM
Sodium alginate	0.5–5.0	Calcium salts	30–100
Sodium alginate with gelatin	2.0 5.0	Calcium salts	30–100
Carrageenan with locust bean gum	0.2–0.8 0.4–1.0	Potassium or ammonium chloride	500
Gelrite	0.25	Temperature lowered	

salt such as calcium nitrate (100 mM). An ion exchange reaction occurred in which calcium replaced the sodium. Surface complexing began immediately and the drops gelled completely in 30 min. Gelation time was determined by incubating the capsules in the calcium salt for various time periods (Table 6). Although 15 min of complexing produced capsules of hardness equal to those produced after 30 min, not all the 15-min capsules were completely gelled. The size of the capsules could be controlled by varying the inside diameter of the pipet tip (Table 7). Based on this information, a pipet tip with an inside diameter of 4.0 mm was selected for making capsules approximately 5 mm in diameter. For producing medium quantities of artificial seeds (up to about 5000), a second, faster method was used. This involved inserting embryos into drops of sodium alginate as the drops fell from a separatory funnel into a solution of calcium nitrate (Fig. 1). Both encapsulation methods were used interchangeably to produce spherical artificial seeds (Fig. 2) without effecting subsequent embryo performance.

Alternatively, the embryos could be mixed in a temperature-dependent gel such as Gelrite or agar, placed in the well of a microtiter plate, and gelled as the temperature was lowered. Similarly, a molding system could be used in which sodium alginate and embryos were placed in wells and calcium nitrate was added to complex the gel.

Table 6. The effects of complexing time on capsule hardness

Minutes in 100 mM calcium chloride	Hardness (kg/capsule)[a]	Standard deviation
5	0.12a	0.10
15	0.75b	0.43
30	0.78b	0.28

[a]Common letter indicates no significant difference (alpha = 0.01) using Duncan's New Multiple range test.

Table 7. Effect of size of pipet tip on capsule diameter

Inside diameter of pipet tip	Mean capsule diameter[a]	Standard deviation
1.5 mm	3.5a	0.26
2.0	4.4b	0.20
2.5	4.7bc	0.34
2.9	4.6cd	0.21
3.6	4.9cde	0.10
4.0	4.9cde	0.35
4.4	5.0def	0.39
5.0	5.2efg	0.33
5.5	5.3fg	0.39
6.5	5.4g	0.37

[a]Common letters indicate no significant difference (alpha = 0.01) using Duncan's New Multiple range test.

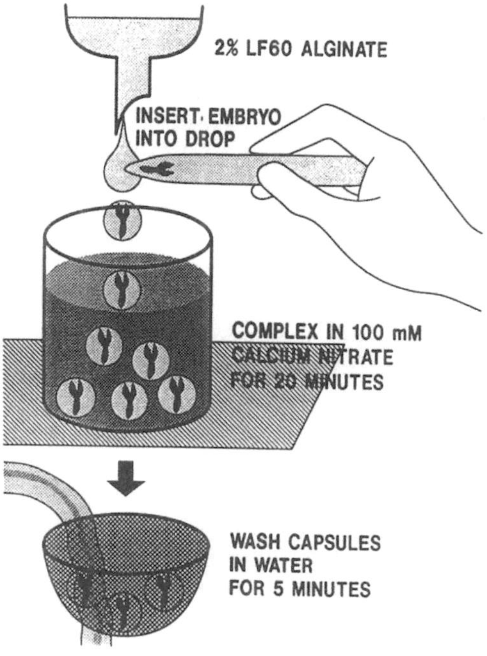

Fig. 1. Production of calcium alginate capsules using a separatory funnel

Fig. 2. Alfalfa somatic embryos encapsulated in calcium alginate beads. Beads are approximately 4 mm in diameter

3.3 Coating Alginate Capsules

The calcium alginate capsules were very wet and consequently tended to stick together slightly. This caused difficulty in handling and determining whether existing machinery could be used to plant the capsules. Furthermore, the capsules would lose water rapidly and dry down to a hard pellet within a few hours when exposed directly to the atmosphere. To overcome these problems, various hydrophobic coatings were tested and an Elvax polymer coating (developed by Dr. Zoila Reyes, SRI International) was found to be suitable for coating alginate capsules. (Redenbaugh et al. 1987a).

The production of a nontacky, water-barrier coating with Elvax first required altering the surface properties of the alginate capsule so that a slightly less hydrophilic interface was formed. This was done by stirring 4 g of calcium alginate capsules in 2 ml of a 50 mM tris buffer solution containing 0.1 g glucose and 0.2 ml glycerol for 1 min.

To apply the coating, the capsules were carefully dipped into the Elvax polymer solution, dried, and then dipped again. Fifty milliliters of a 10% solution of Elvax 4260 (ethylene vinyl acetate acrylic acid terpolymer, Dupont) in cyclohexane was prepared. Five grams stearic acid, 10 g cetyl alcohol, and 25 g spermaceti wax substitute (S. L. Abbot Co., Los Angeles, CA) were mixed together at 40 °C. The resultant solution was combined with Elvax in 295 ml petroleum ether and 155 ml methylene chloride to produce the Elvax polymer solution. Capsules were dipped in the warm solution six times for 10 s per dip. Between dips, the capsules were dried with a warm air stream. The Elvax polymer solution precipitated around the alginate capsules to form a hydrophobic coating. The final step was to rinse the capsules with petroleum ether and then to air-dry the capsules. The coated capsules were dry to the touch.

This coating was found to minimize tackiness such that coated capsules could be planted using a Stanhay seed planter (Redenbaugh et al. 1986). When properly applied, the coating greatly reduced capsule desiccation and the capsules could readily be handled without drying during a 1-day period. Without the coating, the capsules would dry completely during the first day. For most purposes, the 1-day time period will most likely be sufficient, since the coated artificial seeds could be packaged in sealed containers. Finally, encapsulated alfalfa somatic embryos were coated with the Elvax solution without reducing shoot emergence (Fig. 3). The percent shoot emergence from one or two-dip coatings was equal to the control conditions (both replicates for the two-dip experiment had an 80% shoot emergence; hence, there was no error bar).

4 Somatic Embryogeny

4.1 Importance of High-Quality Embryos

Somatic embryogeny consists of all stages of embryo development from the onset of embryogenesis to the production of a fully mature, developmentally arrested embryo. The use of the term somatic embryogeny is important in order to emphasize the necessity of producing high-quality embryos for use as artificial seeds. Most reports in the literature deal primarily with somatic embryogenesis and the production of a population of embryos, some of which will continue development and produce plants, but most of which are developmentally abnormal.

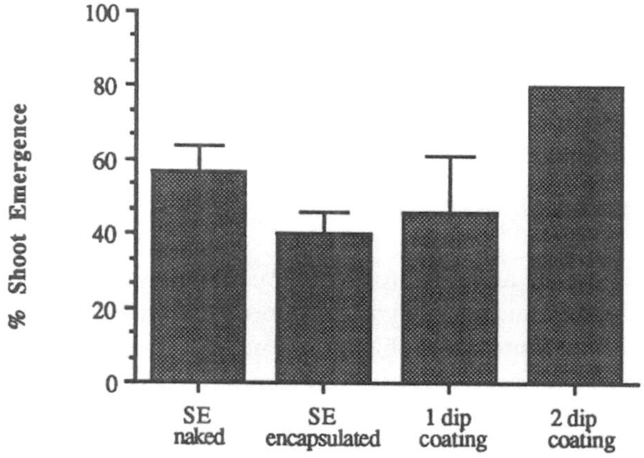

Fig. 3. Shoot emergence from alfalfa alginate capsules coated with Elvax 4260

4.2 Alfalfa Somatic Embryogeny

We focused on alfalfa (*Medicago sativa* L.) for development of high-quality somatic embryos, not only as tools in our breeding program and for development of hybrid alfalfa seed, but also as a model system for research on artificial seeds. Over the past 8 years, we have refined the somatic embryogeny protocol as described by Walker and Sato (1981) into a stepwise procedure in which each step is critical for subsequent performance of the embryo.

Callus was initiated from petioles placed on Schenk and Hildebrandt (SH) medium (1972) containing 25 μM naphthaleneacetic acid, 10 μM kinetin, and 8 g/l agar. The callus was subcultured every 21 days. The callus could be cycled on this maintenance medium for 6 months.

For embryo induction, callus was transferred for 3 days to SH medium containing 50 μM 2,4-dichlorophenoxyacetic acid, 5 μM kinetin, and 8 g/l agar. The callus was removed from this medium and placed on a regeneration medium consisting of SH medium with 10 mM ammonium, 30 mM proline, and 8 g/l agar. After 21 days, 1–5 mm length embryos had formed. For some experiments, a cold maturation treatment was done in which the embryos were stored on the regeneration plates at 4 °C for 7 to 10 days.

For final evaluation, the somatic embryos were placed on a conversion medium consisting of half strength SH medium with 1.5% maltose (instead of sucrose) and 8 g/l agar. The embryos were scored for conversion after 28 days. The term "conversion" was adopted to more adequately describe subsequent development of the embryos. "Germination" of somatic embryo was too nebulous a term, since it could refer to anything from simple radicle elongation (without further development) to complete plant formation. Consequently, in order to better assess the effects of the embryogeny treatments, we rated embryos for conversion frequency. Conversion was defined as including all of the following:

Germination (radicle elongation).
Development of a vigorous root system.
Growth and development of the shoot meristem.
Production of at least two true leaves.

A direct shoot-to-root connection.
Absence of hypertrophy (swelling) of the hypocotyl.
Minimization of callus growth in the hypocotyl.
A green plant with a normal phenotype.

4.3 Improving Conversion Frequencies

Using the above protocol, various components were tested both for increasing production of somatic embryos and improving conversion frequencies. Maltose was discovered to substitute for proline in the regeneration medium (Strickland et al. 1986, 1987). A number of experiments were conducted to examine this carbohydrate.

A grid experiment was conducted to determine the effect of maltose and ammonium on embryo numbers and conversion. Five concentrations of maltose and three of ammonium were used (Figs. 4, 5). Although the standard concentrations of 3% maltose and 15 mM ammonium resulted in the greatest numbers of embryos (Fig. 4), four concentrations of maltose (3 to 9%) and two of ammonium (15 and 30 mM) gave the highest conversion frequencies (Fig. 5). The maltose effect may be useful for regenerating embryos under high osmotic stress to harden the embryos and load them with carbohydrate. In subsequent experiments, maltose was found to produce lower-quality embryos than those produced on a proline regeneration medium. However, it was discovered that when maltose was incorporated into the conversion medium as a substitute for sucrose, the conversion frequency was significantly increased (Redenbaugh et al. 1987b). Other experiments showed that glucose, the monosaccharide component of maltose, gave the same high conversion frequency (Redenbaugh et al. 1987b).

Over the past decade, the in vitro conversion frequency was increased from less than 1% to 70–90% for random-picked embryos (Table 8). Because such high-quality embryos could be produced, other experiments on artificial seeds could be conducted. Of considerable interest was the storage of artificial seeds. Ideally, artificial seeds should have a storage life as long as true seeds. For some crops, particularly where current practice allows for long seed shelf life, a comparable storage period of 6 to 12 months

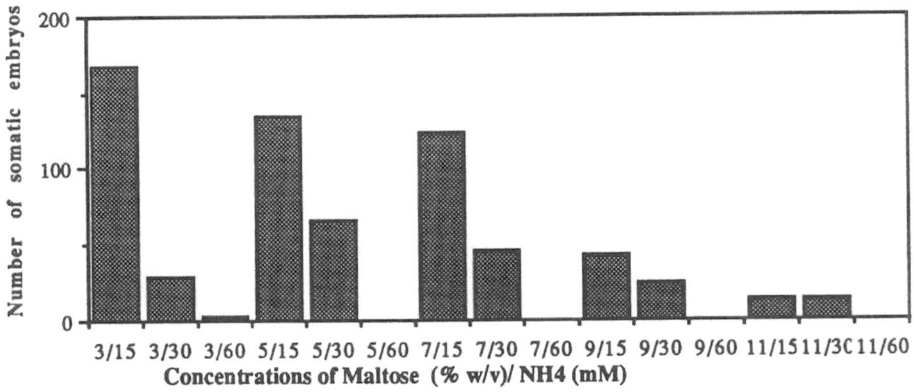

Fig. 4. Effect of maltose and ammonium on somatic embryo numbers

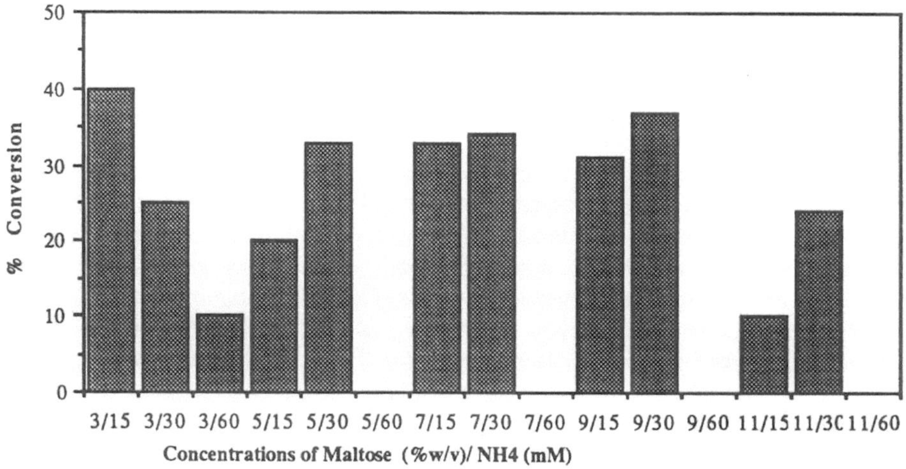

Fig. 5. Effect of maltose and ammonium on somatic embryo conversion

Table 8. Progress in alfalfa somatic embryo conversion

Time	% in vitro conversion of random SE
1982	<0.5%
1983	8.0
1984	30.0
1985	40–50
1986	50–70
1987	60–80
1988	70–90

or longer may be desirable. However, if the value added for a specific crop is sufficiently great, then shorter storage periods can be tolerated, as evidenced by Royal Sluis's Quick Pill, which has a shelf life of 2 weeks (de Vries 1985). In our early experiments, none of the embryos (whether encapsulated or not) survived a cold storage treatment. However, as the embryogeny protocol improved, cool treatment was found to actually mature the embryos. Using the current alginate protocol for alfalfa, the storage life for artificial seeds was found to be quite short. After 7 days there was a significant decrease in subsequent conversion, whereas the conversion frequencies of naked embryos was only slightly diminished over a 49-day period. Preliminary results suggest that embryo respiration may be inhibited inside the alginate capsule.

4.4 Soil Conversion Research

Another critical area for development is conversion of artificial seeds outside an in vitro environment. For these experiments, select embryos were used. The artificial seeds were planted in one of two soilless media (peat or vermiculite in aluminum trays) under mist in the greenhouse, partially covered with Saran, and watered daily. Conversion frequencies ranged between 2 and 30%. As more experiments were conducted, a conversion of 20% was routinely obtained (Fig. 6). This indicated that the quality of embryos was sufficient to achieve greenhouse conversion, and allowed us to redirect our focus from in vitro to soil conversion.

In order to evaluate the nutrients required for conversion of alfalfa somatic embryos, a complex experiment was conducted in which six levels of SH salts were tested against six concentrations of maltose. Although the graph is complicated (the tabular data are presented in Table 9), it is immediately obvious that at least a small level of SH salts was required for conversion to occur (Fig. 7). The minimum level was 1/16 x to achieve a conversion frequency of 40%. In addition, 0.75% w/v was the mininum maltose concentration for optimum conversion. These data were surprising, since it was assumed that the carbohydrate source would be the limiting factor. In fact, 5 to 15% conversion was achieved without any maltose. This experiment was carried a step further to determine which classes of nutrients in the SH salts were critical for conversion (Fig. 8). Without the major salts (potassium nitrate, magnesium sulfate, and calcium chloride) no conversion occurred. Without ammonium phosphate, conversion was reduced to 5 from 30%. Although yet untested, we hypothesize that these critical salts can be applied as fertilizers to alfalfa artificial seeds planted in soil in the greenhouse.

Fig. 6. Alfalfa plants grown from both naked and encapsulated somatic embryos planted directly in the greenhouse

Table 9. Effect of SH salt and maltose concentrations on conversion of alfalfa somatic embryos in vitro

SH level (X)	Maltose level (%)	% Conversion	Standard error (±)
0	0	0.0	0
0	0.09	0.0	0
0	0.19	0.0	0
0	0.375	0.0	0
0	0.75	0.0	0
0	1.5	0.0	0
1/32	0	2.0	2.0
1/32	0.09	14.0	5.1
1/32	0.19	18.0	3.7
1/32	0.375	28.0	5.8
1/32	0.75	19.4	3.2
1/32	1.5	10.0	3.2
1/16	0	6.0	4.0
1/16	0.09	14.0	6.0
1/16	0.19	34.0	6.0
1/16	0.375	29.2	5.7
1/16	0.75	43.2	6.9
1/16	1.5	38.0	10.7
1/8	0	10.0	4.5
1/8	0.09	14.0	4.0
1/8	0.19	23.2	6.8
1/8	0.375	30.0	6.3
1/8	0.75	41.2	8.5
1/8	1.5	45.0	5.5
1/4	0	8.0	5.8
1/4	0.09	24.0	2.4
1/4	0.19	30.0	5.5
1/4	0.375	47.4	7.1
1/4	0.75	39.2	6.1
1/4	1.5	27.6	2.4
1/2	0	16.0	4.0
1/2	0.09	16.0	4.0
1/2	0.19	25.2	4.3
1/2	0.375	35.2	3.9
1/2	0.75	40.0	3.2
1/2	1.5	40.4	5.3

4.5 Maturation Markers

One final area of investigation was to develop biochemical markers that would provide additional assessment for embryo quality and performance. Electrophoretic patterns of isozymes were chosen as potential markers because of extensive use of isozymes to study genetic variation in plant systems and, to some extent, in tissue differentiation and developmental processes (Scandalios 1974). For this research, a rapid cycling clone of *Brassica oleracea* (PHW, Ccc-1) was chosen due to the ease of obtaining both zygotic

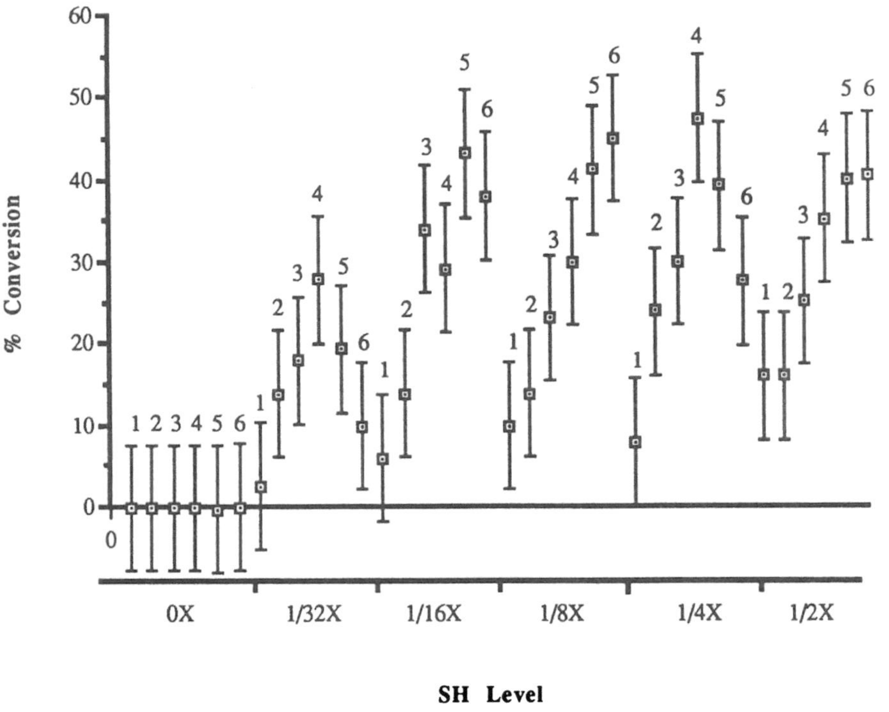

Fig. 7. Effect of SH and maltose on in vitro alfalfa somatic embryo conversion. Maltose Levels: *1* = 0%; *2* = 0.9%; *3* = 0.18%; *4* = 0.375%; *5* = 0.75%; *6* = 1.5%

and somatic embryos. Somatic embryos were produced from immature zygotic embryos using an embryo-off-embryo procedure. The medium was SH salts with 800 mg/l glutamic acid, 100 mg serine, 2% sucrose, 0.5 μM naphthaleneacetic acid, and 0.5 μM 2,4-dichlorophenoxyacetic acid (Loh and Ingram 1982).

Zygotic embryos were isolated and grouped according to time from pollination and according to size for comparison of developmental stages of embryogeny. Larger embryos (3 to 4 mm in length) were extracted individually, while smaller embryos were combined with 1 to 14 additional embryos (depending on size) to ensure sufficient material. All zygotic embryos exhibited some cotyledon expansion. Globular, heart, and early torpedo stage embryos were not used because of the difficulty in obtaining large numbers for analysis. Mature embryos were excised from dry seed.

Individual or grouped embryos were cold extracted in 10 ml of a 0.2 M phosphate buffer containing 2.5 g/l bovine serum albumin, 0.5 g/l dithiothreatol, and 0.5 ml/l 2-mercapthoethanol. Isozymes were separated by starch gel electrophoresis and stained according to procedures adapted from Conkle et al. (1981). Sketches of isozyme patterns were made on all acceptable gels.

Zygotic embryos were initially screened for 25 enzyme systems. Fifteen of these produced acceptable banding patterns (Table 10). No differences were observed among immature zygotic embryos at different ages from pollination. Isozyme differences were observed, however, between immature zygotic and mature embryos (embryos from dry

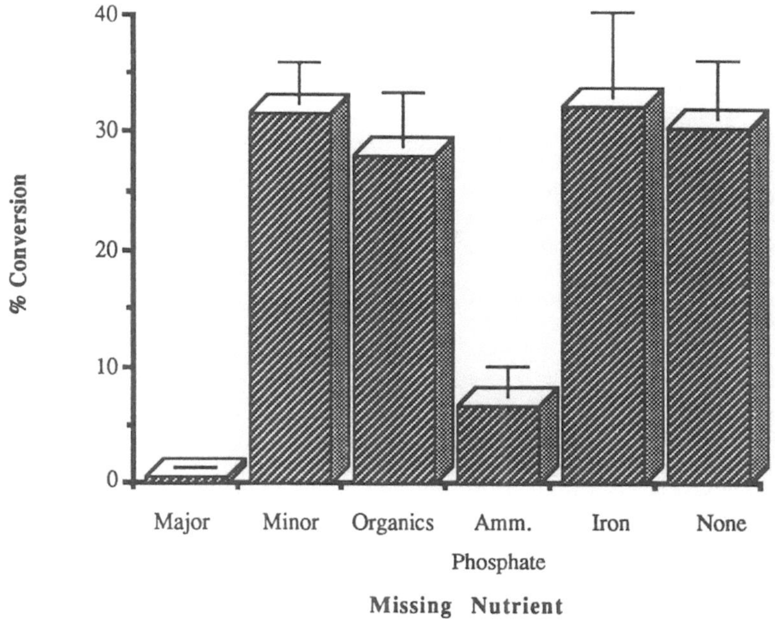

Fig. 8. Effect of deletion of various SH medium components on somatic embryo conversion. *Major* = KNO_3, $MgSO_4$, $CaCl_2$; *Minors* = $MnSO_4$, H_3BO_4, $ZnSO_4$, KI, $CuSO_4$, $NaMoO_4$; *Organics* = Myo-inositol, thiamine HCl, nicotinic acid, pyridoxine HCl; *ammonium phosphate* = $NH_4H_2PO_4$; *Iron* = Na_2EDTA, $FeSO_4$

seeds) for catalase, esterase, and phosphoglucomutase (Table 10 and Figs. 9–11). These three isozymes could potentially be used to assess progression of somatic embryos into a developmentally arrested state.

Brassica somatic embryos of various sizes were compared to zygotic embryos to determine if the somatic embryos followed the same developmental progression in terms of isozyme patterns. Distinct differences between the two embryo types were found for catalase, esterase, acid phosphatase, phosphoglucomutase, and menadione reductase (Table 10 and Fig. 9). We postulate several reasons for the lack of correlation between somatic and zygotic embryos for these isozymes. The somatic embryos could be partially oriented towards germination, although morphologically this was not yet evident. This seems unlikely, since the germinating zygotic embryos (Fig. 9) were dissimilar from the somatic embryos for the five isozymes. A second possibility is that certain isozyme patterns could be induced as a result of the culture environment. This is discounted because zygotic embryos, placed on the same culture medium as the somatic embryos, showed different banding patterns (Fig. 9). A third possibility is that the somatic embryos were poorly developed and resembled somatic tissue more so than zygotic embryos. This does not appear to be the case, since the pattern for pod and leaf tissues were very different from those for somatic embryos. Most likely, the *Brassica* somatic embryos were incomplete in their development and were not of a quality comparable to the zygotic embryos. This assessment is further supported by the low conversion frequencies of the *Brassica* embryos.

Table 10. Enzyme systems in *Brassica oleracea* screened for developmentally related isozyme patterns

Enzyme	Acceptable[a] banding pattern	Different Banding Patterns Between immature ZE[b] and dry seed	Between ZE and SE
Acid phosphatase	+	–	+
Aconitase	+	–	–
Alanine aminopeptidase	+	–	–
Alcohol dehydrogenase	+	–	
Aldolase	–		
Catalase	+	+	+
Diaphorase	–		
Esterase (fluorescent)	+	+	+
Fructose diphosphatase	+		
Fumerase	–		
Glutamate dehydrogenase	+	–	–
Glutamate oxalo acetate transaminase	+		
Glucose-6-phosphate dehydrogenase	–		
Isocitric dehydrogenase	–		
Leucine aminopeptidase	+	–	–
Malic dehydrogenase	+	–	–
Menadione reductase	+	–	+
Peroxidase	–		
Phosphoglucomutase	+	+	+
6-Phosphogluconate dehydrogenase	+	–	–
Phosphoglucose isomerase	+	–	–
Shikimate dehydrogenase	–		

[a]Showed visible band staining and resolution.
[b]Zygotic embryo.

Isozymes were useful for evaluating somatic embryogeny on a comparative basis with zygotic embryos. Although Wetter and Dyck (1983) suggested that isozymes might be useful to study somatic embryogeny, there still is a basic problem in using isozymes as maturation markers. This is that no inherent biochemical information is provided. Because most methods used for assessing somatic embryogeny are time-consuming, other methods that do provide insight into the biochemical nature of the embryos would be more useful. For example, evidence that alfalfa somatic embryos produce seed-specific storage protein (Stuart et al. 1985) is useful because the information supports the argument that somatic embryos are analogous to zygotic embryos.

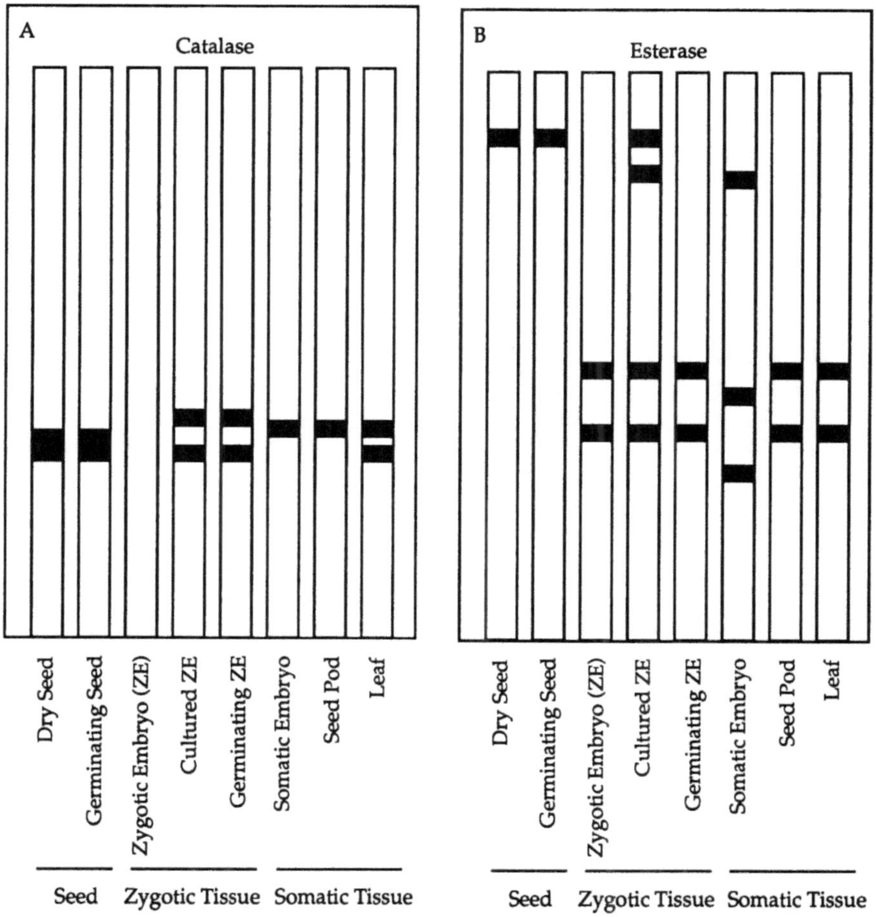

Fig. 9A-E. Isozyme banding patterns for zygotic and somatic embryos and other somatic tissue. ■ Stained enzyme band; □ Destained enzyme band. A blank lane indicates that the specific plant material was not stained for the particular enzyme **C-E** are placed on pp. 412 and 413)

5 Conclusions

The overall value of artificial seeds for research and commercial use has only recently being determined. For research purposes, artificial seeds will be valuable as an analytical tool for studying zygotic embryogeny and for examining the role of the endosperm in embryo development and germination. The primary commercial role will be as a delivery system. Artificial seeds will provide a method to reduce the costs of transplants and provide for a greater stand uniformity. This latter advantage will not be easily attainable because of problems of somaclonal variation, which are still not well understood or controlled. Of much importance will be the ability to provide large-scale delivery of elite genotypes selected from hand-pollinated hybrids or genetically engineered plants. Artificial seeds should provide a method for propagation of sterile and unstable genotypes. Finally, it will be possible to include various adjuvants such as microorganisms, plant growth regulators, and pesticides in the artificial seeds to tailor

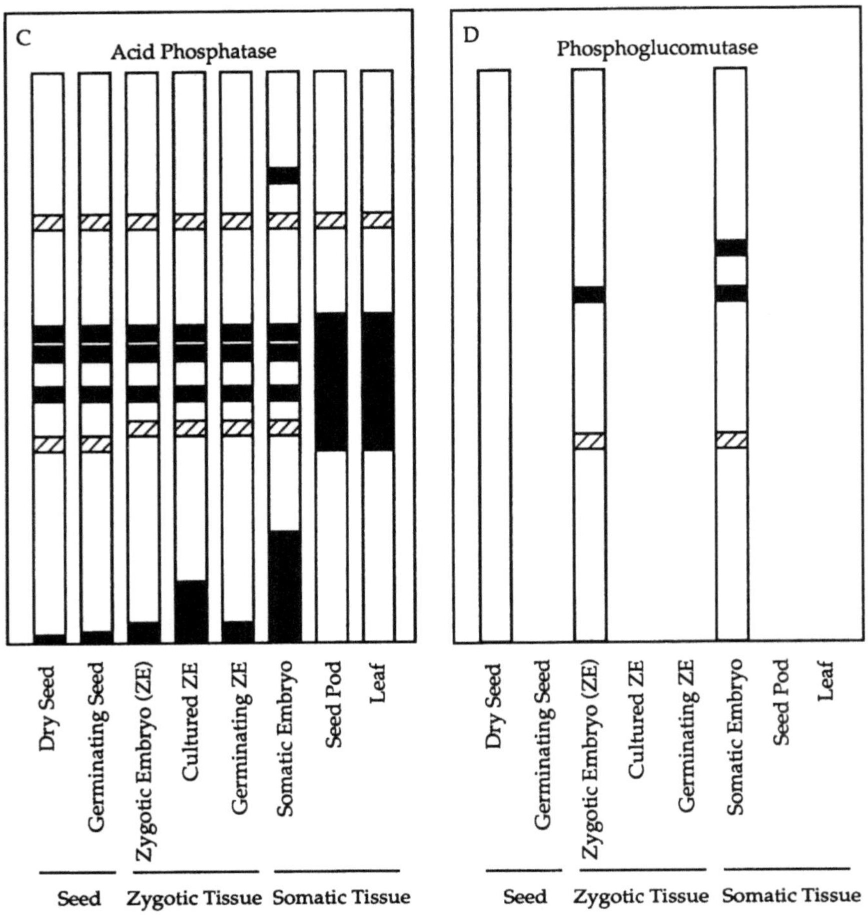

a field-specific, plantable unit for a desired crop. Certainly other uses for artificial seeds will be identified as more progress in made.

Using an optimized alfalfa embryogenesis system and the alginate encapsulation process, we calculated the costs for producing artificial seeds (Table 11). The cost of 0.03¢ per artificial seed compares favorably with seed costs for crops such as hybrid broccoli (0.09¢), cauliflower (0.04¢), hybrid geranium (6.0¢), and petunia (0.03¢). For other crops, such as hybrid celery, seeds are not readily available, and cost comparisons cannot be made.

Currently, the major limitation to commercializing artificial seeds is somatic embryogeny. Although artificial seeds can now be planted directly in the greenhouse, the lack of high-quality embryos continues to be a problem. Street (1979) discussed the reliability in embryogenesis, "morphogenic competence is determined from the time of culture initiation," such that there is "the need to have an initiation medium that will ensure that the competent cells are involved in callus formation." The need for competent cells is absolute. For artificial seeds, competency refers not only to a regeneration ability, but extends beyond to plant recovery, growth habit, and ultimately harvest index. Sunderland (1979) demonstrated the production of hundreds of morphologically

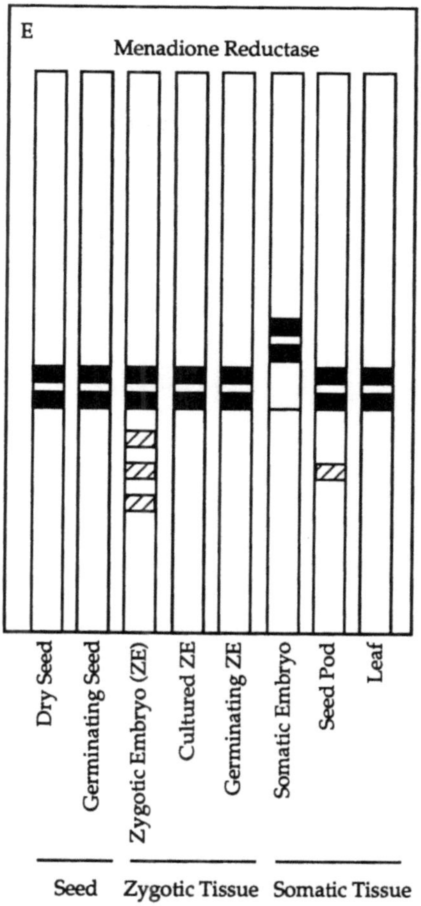

Table 11. Calculated costs for producing one million artificial seeds based on optimized alfalfa model system[a]. (Redenbaugh et al. 1987b)

Production stage	Labor cost (US $)	Materials cost (US $)
Callus growth (21 days)	4.00	1.36
Induction (3 days)	4.00	2.00
Embryo regeneration (28 days)	4.00	2.00
Maturation treatment (21 days)	4.00	2.00
Encapsulation (1 day)	112.00	28.00
Subtotal	128.00	35.36
Combined totals	163.36	
Overhead (60%)	98.00	
Total cost per million artificial seeds	261.36	
Cost per unit	0.03	

[a]Calculations, in part, provided by C. Sluis.

uniform embryos from *Datura* and *Nicotiana* pollen, but little has been done since to assess the quality of somatic embryos. The early viewpoints of Murashige (pers. commun.) and Street (1979) on the difficulty of somatic embryogeny are still valid today. The quality and fidelity of somatic embryos are the limiting factors for development and scale-up of artificial seeds.

Acknowledgments. We wish to acknowledge the laboratory assistance of Kathleen Ave, Jessica Daily, Catharine Dunwoody, Eileen Flynn, Ron Futterman, Polly Goldman, Richard Gumina, Lianna Kirk, Suzanne McCarthy, Steve McEuen, Shelly Mealer, Rebecca Olson, Brian Paasch, Sarah Puerner, Mary Reshwan, Kathy Sullivan, Sarah Viss, Norma Watkins, and Ted Woodruff over the past 10 years.

References

Ammirato P (1983) Embryogenesis. In: Evans DA, Sharp WR, Ammirato PV, Yamada Y (eds) Handbook of Plant Cell Culture, Vol 1. Macmillan, New York, pp 82–123

Baker C (1985) Synchronization and fluid sowing of carrot, *Daucus carota*, somatic embryos. MS Thesis. University Microfilms, Ann Arbor

Bapat V, Mhatre M, Rao PS (1987) Propagation of *Morus indica* L. (mulberry) by encapsulated shoot buds. Plant Cell Rep 6:393–395

Bingham E, Hurley L, Koutz D, Saunders J (1975) Breeding alfalfa which regenerates from callus tissue in culture. Crop Sci 15:719–721

Carman J (1988) Somatic embryogenesis in wheat: factors affecting competence, induction and differentiation. In Vitro Cell Dev Biol 24:71A

Carman J, Jefferson N, Campbell W (1987) ABA, O^2, temperature and relative humidity effects on embryoid maturation in *Triticum aestivum* L. calli. 1987 Annual Agronomy Meeting, Atlanta, GA, Nov 29-Dec 4, p 148

Conkle M, Hodgskiss P, Nunnally L, Hunter S (1981) A manual for starch gel electrophoresis of conifer seeds. USFS, Berkeley, CA

Corley R (1982) Clonal planting material for the oil palm industry. Planter 58:515–528

de Vries J (1985) Industry's view of seed quality and technological developments. Seed Quality Seminar, Royal Sluis, April 11–12, pp 11–15

Drew R (1979) The development of carrot (*Daucus carota* L.) embryoids (derived from cell suspension culture) into plantlets on a sugar-free basal medium. Hortic Res 19:79–84

Durzan D (1979) Progress and promise in forest genetics. Proc 50th Anniversary Conf Paper Science and Technology – the Cutting Edge, The Institute of Paper Chemistry, Appleton, WI, pp 31–60

Durzan D, Gupta P (1988) Somatic embryogenesis and polyembryogenesis in conifers. In: Mizrahi A (ed) Biotechnology in Agriculture. Alan Liss, New York, pp 53–81

Fujii J, Slade D, Redenbaugh K, Walker K (1987) Artificial seeds for plant propagation. Trends Biotech 5:335–339

Gautz L, Upadhyaya S, Garrett R (1987) Vibratory separation of gel encapsulated plant propagules. Trans ASAE 30:652–656

Gray D (1987) Quiescence in monocotyledonous and dicotyledonous somatic embryos induced by dehydration. HortScience 22:810–814

Gray D (1988) Effects of dehydration on quiescence and germination of grape somatic embryos. In Vitro Cell Dev Biol 24:70A

Gupta P, Durzan D (1987) Biotechnology of somatic polyembryogenesis and plantlet regeneration in loblolly pine. Bio/Technology 5:147–151

Hama I (1986) Artificial seeds. Japanese Patent Application No 40708/1986, Feb 27 (in Japanese)

Janick J (1988) Production of synthetic seed via desiccation and encapsulation. In Vitro Cell Dev Biol 24:70A

Kamada H (1985) Artificial seed. In: Tanaka R (ed) Practical Technology on the Mass Production of Clonal Plants. CMC, Tokyo, p 48, 1985 (in Japanese)

Kamada H, Kiyosue T, Harada H (1988) New methods for somatic embryo induction and their use for synthetic seed production. In Vitro Cell Dev Biol 24:71A

Kim Y-H, Janick J (1987) Production of synthetic seeds of celery. HortScience 22:89
Kitto S, Janick J (1982) Polyox as an artificial seed coat for asexual embryos. HortScience 17:448
Kitto S, Janick J (1985a) Production of synthetic seeds by encapsulating asexual embryos of carrot. J Am Soc Hortic Sci 110:277–282
Kitto S, Janick J (1985b) Hardening treatments increase survival of synthetically-coated asexual embryos of carrot. J Am Soc Hortic Sci 110:283–286
Lawrence R Jr (1981) In vitro cloning systems. Environ Exp Bot 21:289–300
Loh C, Ingram D (1982) Production of haploid plants from anther cultures and secondary embryoids of winter oilseed rape, *Brassica napus* ssp. *oleifera*. New Phytol 91:507–516
Lutz J, Wong J, Rowe J, Tricoli D, Lawrence R Jr (1985) Somatic embryogenesis for mass cloning of crop plants. In: Henke R, Hughes K, Constantin M, Hollaender A (eds) Tissue Culture in Forestry and Agriculture. Plenum, New York, pp 105–116
McKersie B, Bowley S, Senaratna T, Brown D, Bewley J (1988) Application of artificial seed technology in the production of hybrid alfalfa (*Medicago sativa* L.). In Vitro Cell Dev Biol 24:71A
Murashige T (1977) Plant cell and organ cultures as horticultural practices. Acta Hortic 78:17–30
Murashige T (1978) The impact of plant tissue culture on agriculture. In: Thorpe T (ed) Frontiers of plant tissue culture 1978. International Association for Plant Tissue Culture, Univ Calgary, Alberta, pp 15–26
Nishimura S (1986) Somatic seeds. In: 4th Tokai Regional Meeting for Biotechnology, Nagoya, Japan, p 12 (in Japanese)
Obendorf R, Slawinska J (1986) Somatic embryogenesis from cotyledonary tissues of soybean and maturation to desiccation tolerant state. In Vitro Cell Dev Biol 22:53
Obendorf R, Slawinska J (1988) Maturation of soybean somatic embryos to a desiccation tolerant state. In Vitro Cell Dev Biol 24:71A
Redenbaugh K, Nichol J, Kossler M, Paasch B (1984) Encapsulation of somatic embryos for artificial seed production. In Vitro 3:256–257
Redenbaugh K, Paasch B, Nichol J, Kossler M, Viss P, Walker K (1986) Somatic seeds: encapsulation of asexual plant embryos. Bio/Technology 4:797–801
Redenbaugh K, Slade D, Viss P, Fujii J (1987a) Encapsulation of somatic embryos in synthetic seed coats. HortScience 22:803–809
Redenbaugh K, Viss P, Slade D, Fujii J (1987b) Scale-up: artificial seeds. In: Green C, Somers D, Hackett W, Biesboer D (eds). Plant Tissue and Cell Culture. Alan Liss, New York, pp 473–493
Redenbaugh K, Fujii J, Slade D (1988) Encapsulated plant embryos. In: Mizrahi A (ed) Biotechnology in Agriculture. Alan Liss, New York, pp 225–248
Scandalios J (1974) Isozymes in development and differentiation. Annu Rev Plant Physiol 25:225–258
Schenk R, Hildebrandt A (1972) Medium and techniques for induction and growth of monocotyledonous and dicotyledonous cell cultures. Can J Bot 50:199–204
Slawinska J, Obendorf R (1987) Soybean somatic embryo composition, respiration and water relations during maturation to a desiccation-tolerant state. American Society of Agronomy, 1987 Annu Agron Meet, Atlanta, GA, Nov 29–Dec 4, p 153
Street H (1979) Embryogenesis and chemically induced organogenesis. In: Sharp WR, Larsen P, Paddock E, Raghavan V (eds) Plant Cell and Tissue Culture – Principles and Applications. Ohio State Univ Press, Columbus, Ohio, pp 123–153
Strickland S, McCall C, Nichol J, Stuart D (1986) Enhanced somatic embryogenesis in *Medicago sativa* by addition of maltose and its higher homologs to the culture medium. In: Somers D, Gengenbach B, Biesboer D, Hackett W, Green L (eds) VI Int Cong Plant Tissue and Cell Culture Abstracts. Univ Minnesota, Minneapolis, MN, p 188
Strickland S, Nichol J, McCall C, Stuart D (1987) Effect of carbohydrate source on alfalfa somatic embryogenesis. Plant Sci 48:113–121
Stuart D, Nelsen J, Strickland S, Nichol J (1985) Factors affecting developmental processes in alfalfa cell cultures. In: Henke R, Hughes K, Constantin M, Hollaender A (eds) Tissue Culture in Forestry and Agriculture. Plenum, New York, pp 59–73
Styer D (1985) Bioreactor technology for plant propagation. In: Henke R, Hughes K, Constantin M, Hollaender A (eds) Tissue culture in forestry and agriculture. Plenum Press, New York, pp 177–130
Sunderland N (1979) Comparative studies of anther and pollen culture. In: Sharp WR, Larsen P, Paddock E, Raghavan V (eds) Plant Cell and Tissue Culture – Principles and Applications. Ohio State Univ Press, Columbus, Ohio, pp 203–219

Upadhyaya S, Gautz L, Garrett R (1987) Retrofitting vegetable planters to seed gel encapsulated propagules. Appl Eng Agric 3:211–214

Walker K, Sato S (1981) Morphogenesis in callus tissue of *Medicago sativa*: the role of ammonium ion in somatic embryogenesis. Plant Cell Tissue Organ Cult 1:109–121

Wetter L, Dyck J (1983) Isoenzyme analysis of cultured cells and somatic hybrids. In: Evans DA, Sharp WR, Ammirato PV, Yamada Y (eds) Handbook of Plant Cell Culture, Vol. 1. Macmillan, New York, pp 607–628

Yamakawa K (1985) Application of artificial seed and its potential. Agric Chem Today 29:68 (in Japanese)

Section IV
Automation, Artificial Intelligence, Robots in Transplanting, Information Management, Bioreactors, Computerized Greenhouses

IV.1 Automated Systems

B. TISSERAT[1]

1 Introduction

1.1 State of Plant Tissue Culture

Murashige (1974) introduced the basic micropropagation plan, which has three distinct stages (Fig. 1). In Stage I, the culture is established; in Stage II, the culture is multiplied; and in Stage III, the culture is prepared for transfer to the free-living environment. Additional stages (e.g., stage 4 or acclimatization of the plantlet in the soil) for the cloning sequence have also been proposed (Anderson 1980). Labor in commercial plant tissue culture is a series of numerous, nontechnical, repetitive tasks linked together in an assembly-line fashion to produce a free-living plant as an end product (Lawrence 1981; Rowe 1986). According to Rowe (1986), the demand for micropropagated plants exceeds production. The number of plants produced by micropropagation is proportional to the labor (Yang 1977; Oglesby 1978; Rowe 1986, Smith 1986). Depending on the species, an average worker can produce about 70 000–200 000 plants a year (Van Uyen and Vander Zaag 1983; Rowe 1986). Labor costs constitute a considerable amount of the expense involved in micropropagation (Yang 1977; Oglesby 1978; Rowe 1986; Smith 1986). Smith (1986) estimates that over 80% of the cost of micropropagating radiata pines in New Zealand is labor. Similarly, about 68% of the cost of producing potato plantlets in Vietnam was attributed to labor (Van Uyen and Vander Zaag 1983).

Current plant tissue culture techniques are manual procedures which have changed little in the last 30–40 years (Murashige and Skoog 1962; Gamborg et al. 1968; Murashige 1974; Rowe 1986). Reculturing is performed as an inherent part of the multi-staged cloning process and may account for 30–40% of the labor involved. Cultures are maintained in individual culture vessels on either a semi-solid gelatinous medium (e.g., agar) or in liquid medium and are transferred every 4 to 8 weeks to fresh medium. Nutrient media compositions have been designed for this method of culture (Murashige and Skoog 1962; Gamborg et al. 1968). Cultures, especially in the multiplication stage, often need several transfers to fresh medium. It remains to be determined exactly why plants require frequent reculturing. Plants may deplete the medium nutrients or release toxins which are detrimental to their survival, or both. Medium

[1]Fruit and Vegetable Chemistry Laboratory, USDA/ARS, 263 South Chester Ave., Pasadena, CA 91106, USA
Mention of a trade name or proprietary product in the paper does not constitute a guarantee or warranty of the product by the US Department of Agriculture and does not imply its approval to the exclusion of other products that may also be suitable.

Biotechnology in Agriculture and Forestry, Vol. 17
High-Tech and Micropropagation I (ed. by Y.P.S. Bajaj)

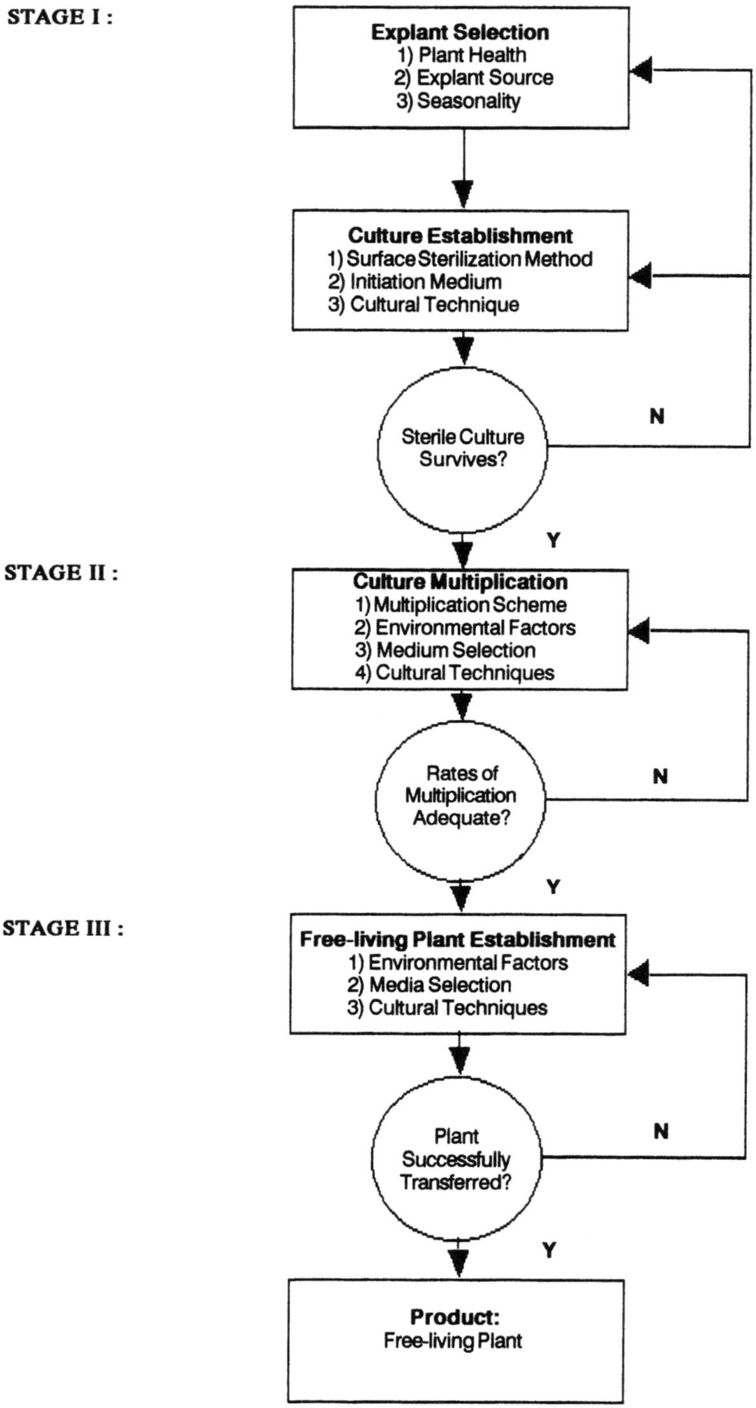

Fig. 1. Flow chart of the traditional plant tissue culture cloning system

replacement also facilitates the advent of particular growth events (e.g., induction of embryoids and plantlets from callus, or rooting of shoots). It should also be noted that sometimes the plant may physically outgrow the culture vessel, thus necessitating re-culture.

Automation of plant tissue culture is sought to decrease production costs and increase micropropagation efficiency (Styer 1985; Rowe 1986). A robotics assembly line system eliminating man-plant interactions has been suggested (Hartney and Kabay 1984; Rowe 1986; De Bry 1986). In such a system, a robot will perform culture transfers and media preparation (Deleplanque et al. 1985; Rowe 1986; De Bry 1986). Automated systems to propagate plants in vitro are in the hypothetical state only, with no marketable prototypes yet available. Automation of commercial plant tissue culture has been slow due to: (1) lack of interest in such activities until recently, (2) over-dependence of commercial industry on university and grant technological developments due to the failure to aggressively institute its own research programs, and (3) lack of expertise of current personnel in the plant tissue culture field to engineer suitable systems. A technology transfer from various disciplines is necessary to effect development of any automated commercial micropropagation system.

1.2 Mechanization Versus Automation

A distinction should be made between mechanization and automation. Mechanization is the use of nonintelligent machinery to replace human activities. Automation is the intelligent control of a system (i.e., a group of procedures) often using machines as tools to increase productivity and reduce costs. Plant tissue culture has been mechanized for several years (e.g., use of automatic pipeters, electronic pH meters, electrical autoclaves). These machines have improved the efficiency of tissue culture by replacing a task formerly necessitating high human involvement with a machine requiring lower human involvement. Fully mechanized tissue culture laboratories, where machines essentially duplicate their human counterparts, have been suggested (Deleplanque et al. 1985; Rowe 1986; De Bry 1986; Simonton and Thai 1988).

In contrast, the institution of an automated tissue culture system first requires a complete appraisal of plant tissue culture activities. Such an appraisal generates new ideas to be employed in propagation procedures. For example, agar can be replaced with liquid medium, propagation steps can be reduced or eliminated, clump dissection minimized, new contaminant detection methods developed, and record-keeping methods to provide a productivity feedback mechanism improved (Hartney and Kabay 1984; Tisserat and Vandercook 1985, 1986; Smith 1986; Aitken-Christie and Jones 1987; Levin et al. 1988; Simonton and Thai 1988). Following such an appraisal, mechanized procedures with suitable controllers are developed to implement these ideas into a workable system.

2 Automated Systems in Plant Tissue Culture

Chemostats or bioreactors are large enclosed chambers containing liquid medium which is usually replenished periodically during culture (Graebe and Novelli 1966;

Wilson 1980; Tanaka et al. 1983; Hagimori et al. 1984; Styer 1985). These systems are usually employed to grow cells or cell clumps. Within the chamber both chemical and physical parameters are often controlled (e.g., agitation, pH and oxygen levels). Often, aeration is provided by bubbling air through the system. Styer (1985) advocated using bioreactors to multiply plants which are amenable to the somatic embryogenesis cloning process. In such a scenario, the explant is induced to produce somatic embryos in one bioreactor which are then transferred to a second, larger bioreactor to produce plants. These plants may then be planted directly in the field or prepared for some type of fluid drilling system (Styer 1985).

Air-lift bioreactors (or air flow systems) have been employed to germinate seeds and proliferate protocorms of orchids (Cheng et al. 1978; Cheng and Chua 1980, 1982). *Digitalis purpurea* shoot production was significantly enhanced in an air-lift bioreactor (i.e., jar fermenter culture) (Hagimori et al. 1984). A nutrient misting system has been found to be satisfactory for the growth of callus, anthers, and shoot meristems of *Brassica, Cinchona, Daucus, Ficus,* and *Lycopersicon.* Also, a method whereby both liquid and agar media are simultaneously employed to culture a wide variety of plants has recently been presented (Molnar 1987). Usually, the use of a submerging medium is adverse to plant foliage (Aitken-Christie and Jones 1987; Tisserat and Vandercook 1985, 1986). In radiata pine, vitrification was common in plantlets grown on or in a liquid medium. However, short immersions in liquid medium, coupled with intermittent dry periods, is an adequate substitute for agar culture for many plants (Figs. 2–6; Table 1) (Tisserat and Vandercook 1985, 1986).

To reduce labor requirements, the number of transfers during a culture's life can be minimized by using an automated plant culture system (APCS) (Tisserat and Vandercook 1985, 1986). The APCS duplicates the features of manual plant tissue culture procedures without the necessity for frequent manual transferring of cultures to fresh medium (Figs. 2, 3). In the APCS, plant cultures are situated and remain in a single vessel during their culture. Plant cultures are thereafter periodically bathed in fresh medium and aerated continuously. Aliquots of the culture medium are periodically discarded, while simultaneously being replenished with fresh medium. In such a system, the necessity for the physical transfer of plant cultures is eliminated. The early APCS was gas sterilized (Fig. 2) (Tisserat and Vandercook 1985, 1986). Subsequent models are completely autoclavable, allowing them to be more easily constructed and employed by tissue culture laboratories (Fig. 3).

3 Automated Plant Culture System Features

3.1 Rates of Growth

Several APCS models have been developed and successfully demonstrated by growing various plant cultures (Figs. 3–6). Results show that plants grew larger and more rapidly in the automated system versus the manual system (Table 1). With carrot particularly, spectacular growth rate increases were obtained by the APCS versus the manual agar procedure (Table 2) (Fig. 4).

Where appropriate (e.g., short-term growth tests) the second pump, second reservoir, valve, and waste reservoir can be eliminated. Use of a single reservoir containing 11 of medium will suffice for short-term growth of most plants (i.e., 30–60 days).

Fig. 2A,B. Components of the automated plant culture system and microcomputer system. **A** Complete automated plant culture and microcomputer system. *1* Sterile culture system composed of culture chamber, pumps, three-way valve, tubing, and reservoirs. *2* interface wiring for sterile culture system to relay board assembly. *3* relay board assembly composed of relay boards, transformers, and input and output strips. *4* black and white monitor to program and monitor operation of the system. *5* Atari 400 (16K RAM) microcomputer. *6* ribbon cables interface relay board assembly to microcomputer. *7* controller jacks used to interface computer to the environment. **B** Side view of relay board assembly. *1* Plug-in relay board which contain switches, timer, and wiring. *2* ribbon cables from four controller jacks. *3* subminiature printed circuit SPDT relay with five VDC coils. *4* forty-four position card-edge socket which holds plug-in relay boards and connects with inlet and outlet wiring. *5* power transformer, capacitor, and bridge rectifier assembly which converts 110 VAC to 5 VDC *6* power transformer and bridge rectifier used to convert 110 VAC to 12 VDC *7* power transformer, capacitor, and bridge rectifier assembly used to convert 110 VAC to 24 VDC *8* input 110 VAC dual barrier strip

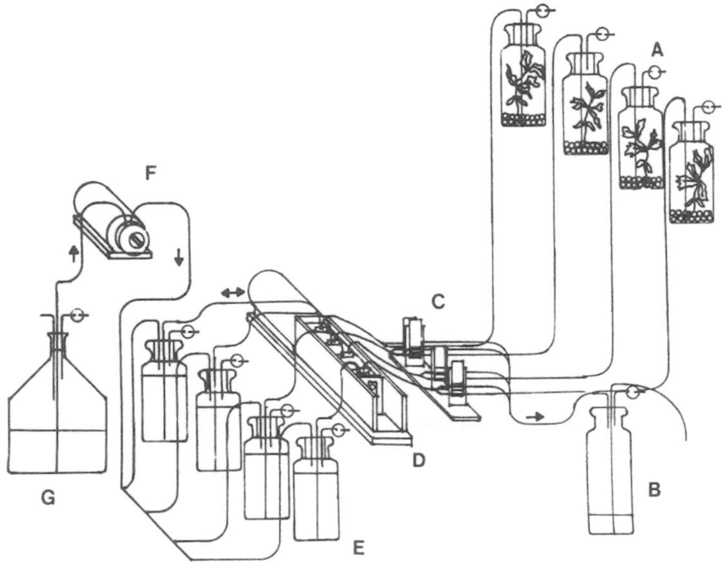

Fig. 3. Diagrammatic representation of the components used in an autoclavable culture system

Fig. 4. Growth response obtained from carrot callus. Callus grown in the APCS (*right*) and callus grown on agar medium (*left*). Note the dramatic increase in size and differentiation using the automated system

Fig. 5. Growth responses obtained from date callus grown in the automated (*right*) and on the traditional agar medium (*left*). Date palm cultures grow substantially larger in the APCS

3.2 Construction Notes

The culture chamber of the APCS is 10 to 100 times larger than most culture vessels commonly employed. Once cultures are freed from the physical constraints of the culture tube wall, culture size and development are enhanced. Also, trays may be introduced into the chamber to hold small cultures in individual compartments. The APCS allows for the co-culture of numerous plantlets in a single chamber, therefore making it useful for a mass plant testing systems. Additional testing of the APCS will further reveal the potential advantages of this system over the manual method.

Numerous types of commercial three way valves and pumps are available which could replace those used in this system. A peristaltic pump costs more than an impeller pump. Medium reservoirs and closure types may be constructed from a variety of sources. For example, an aspirator bottle with a serrated outlet situated at its bottom may be used for the medium exit line; only an air vent and medium inlet line would be required in the cap.

3.3 Computer Control

Several types of microcomputers or programmable controllers can be employed to control the APCS. Atari and Commodore computers are selected to control the APCS because of their low cost, versatility, ease of interfacing, and their ability to accept an

Table 1. Comparison of tissue volume, weight means, and rate of increase for plant cultures grown in the automated and manual culture system. Growth parameters recorded for species may include any of the following measurements: average culture volume (cm^3), culture weight (g), leaf length (cm), or culture length (cm)[a]

Plant tissue growth parameter	Days in culture	Traditional agar culture	Automated plant culture system	Increase (x fold)
Aster shoot tips				
Fresh weight (g):	45	0.9 +/- 0.4	1.8 +/- 0.9	2^{NS}
Carrot callus, clone no. 1				
Fresh weight (g):	45	7.5 +/- 1.2	14.7 +/- 2.0	1.9^{S}
Volume (cm^3):		15.6 +/- 3.7	37.3 +/- 3.7	2.4^{S}
Carrot callus, clone no. 2				
Fresh weight (g):	60	6.7 +/- 1.5	93.9 +/- 32.4	14.1^{S}
Volume (cm^3):		27.6 +/- 1.9	300.0 +/- 23.9	10.8^{S}
Cow tree shoot tips				
Fresh weight (g):	45	0.7 +/- 0.1	1.2 +/- 0.1	1.8^{S}
Volume (cm^3):		30.8 +/- 3.6	90.6 +/- 10.0	1.9^{S}
Date palm callus, clone no. 1				
Fresh weight (g):	45	4.2 +/- 0.5	13.6 +/- 1.4	3.2^{S}
Volume (cm^3):		18.6 +/- 3.2	72.0 +/- 4.5	4^{S}
Lemon shoot tip culture:				
Culture length (cm):	30	3.8 +/- 0.3	7.3 +/- 0.4	1.9^{S}
Leaf length (cm):		21.5 +/- 0.8	96.7 +/- 30.3	4.5^{S}
Orchid shoot tips				
Fresh weight (g):	270	6.9 +/- 0.6	27.8 +/- 1.2	4^{S}
Volume (cm^3):		32.0 +/- 5.8	130.0 +/- 5.0	4.1^{S}

[a] All plants cultured on MS medium except cow tree, which was cultured on B5 medium. Auxin, 0.54 μM NAA was added to all media except for carrot cultures. Treatment means were significantly different at 5% (^{S}) or nonsignificant (^{NS}) using the t-test.

EPROM chip in cartridge form. Once the microcomputer is dedicated with an EPROM cartridge system; rebooting in the event of a power failure is unnecessary. A microcomputer controlled by an EPROM cartridge provides error-free programming and operation. Therefore, again essentially no additional time is wasted in time-consuming programming operations.

It is critical that plants grown in the APCS be given optimum exposure to both nutrient medium and atmosphere. To accomplish this, the operation program can be adjusted to test for the influence of: (1) length of time cultures are immersed in medium, (2) number of cycles per day cultures are exposed to medium, and (3) number of reservoir medium discards and replacements on plant growth.

3.4 Construction Cost Factors

The microcomputer and interface system can be constructed for about US $ 400 ($ 200 for components and $ 200 for labor). Each ethylene oxide sterilized sterile culture system (SCS – culture vessel chamber, media bottles, tubing, and pumps) can be assembled for about $ 100 and requires about 30 min of labor. Ten systems could cost about $ 1100 (i.e., only one microcomputer/interface system is needed). If sterilization with ethylene oxide is not feasible, then chemical sterilization with disinfectants may be performed. The autoclavable system, though requiring the same amount of expertise to construct, is more expensive due to the high cost of the peristalic pump and pinch clamp valves. A five-channel drive peristalic pump with heads costs about $ 770. Pinch clamp valves cost about $ 90 a piece. Therefore to construct ten independent culture systems a cost outlay of $ 4130 is required.

Using one microcomputer, 12–32 SCS's can be independently run; with inclusion of larger amperage transformers and heavy duty relays, hundreds of SCS's can be controlled through a single microcomputer. Each SCS unit occupies 45 cm (h) × 20 cm (w) × 45 cm (d).

4 Remarks on the APCS

Mass propagation on a commercial scale, using the APCS, is accomplished as follows: (1) a shoot cluster or callus piece is introduced into APCS culture chamber to grow and proliferate without any further handling; and (2) after 3–9 months in culture, thousands of suitably sized shoots will be produced which can then be transferred to similar chambers to continue the proliferation processes or be acclimatized for introduction into soil. It is feasible for proliferation and acclimatization of plantlets to occur within the same culture chamber and therefore further reduce labor requirements. A reasonable cost savings program employing the APCS is presented in Table 2.

The APCS is not designed to replace all tissue culture practices but replaces the laborious reculturing. Current methods used in commercial tissue culture propagation are often the result of plant physiology research experiments and may or may not reflect an optimum mass micropropagation method. Manual transferring results in physical injury to cultured plants (e.g., trimming of roots and shoots to accommodate plant introduction and projected growth in vessels). Plants are also inadvertently exposed to physiological stress during culture (nutrient medium is exhausted or altered during culture development). Recultured plants need days and sometimes weeks to recover from such trauma. These traumas are minimized by using the APCS.

The procedure of teasing shoot or callus/plantlet clusters apart is artificial. This practice may not actually increase the plant multiplication rate for many plants versus incubating uninjured plants in the APCS (Tisserat and Vandercook 1985, 1986). Species that produce plantlets from callus or adventitious budding are ideal candidates for the APCS. Perennial plants (such as date palm) that produce shoots through axillary branching but do not exhibit strong apical dominance should also be excellent candidates for the APCS. Plants that require "topping" or removal of the terminal shoot tip to minimize the inhibitory apical dominance factor would probably not benefit as much as those plants not possessing this requirement. However, it remains to be determined if constant "topping" is necessary for mass shoot production if such plants are grown in the APCS.

Table 2. List of activities conducted in a typical plant tissue culture laboratory and how they might be influenced by implementation of APCS. Initial cost to install the APCS would be $15 000

Activity	Traditional tissue culture h	%	Cost of operations ($) Wage earners[a] GS-2	GS-5	Applying APCS h
Hourly rate:			5.03	6.90	
Medium preparation:					
Cleaning glassware					
Medium preparation					
Preparation of stocks					
Distribution of medium					
Preparation of vessels					
Capping vessels					
Autoclaving medium					
Total	8	20	40.24	55.20	6
Culture activities:					
Reculturing activities	6	15	30.18	41.40	2
Establishing cultures	16	40	80.48	110.40	14
Data recording					
Data recording					
Data processing					
Analysis of results					
Total	6	15	30.18	41.40	6
Other					
Culture monitoring					
Miscellaneous activities					
Plant procurement					
Total	4	10	20.12	27.60	4
Weekly total:	40	100	201.20	276.00	32
Yearly total:	2080		10 462.40	13 352.00	
With improved technology					
	Extra Time saved (h):		Extra $ saved:		Percent savings:
Actual $ savings/week:					
+APCS if plants grow 1 x:	8		40.24	55.20	20
+APCS if plants grow 2 x:	16		80.48	110.40	40
+APCS if plants grow 3 x:	24		120.72	165.60	60
+APCS if plants grow 5 x:	40		201.20	276.00	100
Actual $ savings/year:					
+APCS if plants grow 1 x:	416		2092.48	2870.40	20
+APCS if plants grow 2 x:	832		4184.96	5740.80	40
+APCS if plants grow 3 x:	1248		6277.44	8611.20	60
+APCS if plants grow 5 x:	2080		10 462.40	14 352.00	100

[a]Wage earners correspond to USDA Federal GS laboratory technician employees.

5 Future of the APCS

Many novel plant tissue culture systems have been presented in the literature but few have been widely accepted (Graebe and Novelli 1966; Cheng et al. 1978; Wilson 1980; Cheng and Chua 1980, 1981; Tanaka et al. 1983; Deleplanque et al. 1985; Styer 1985; Aitken-Christie and Jones 1987; Molnar 1987; Levin et al. 1988; Weathers and Giles 1988). Acceptance of a new plant culture system is based on construction costs, practicality (i.e., technological expertise required for construction and operation), and versatility (i.e., the ability of the system to be widely used for different plant species and culture types).

One approach to automate the tissue culture process is to use intelligent vision-capable robots which can dexterously remove plants from containers, correctly sever them into smaller cultures to stimulate additional axillary budding and then reintroduce them into the sterile environment (Deleplanque et al. 1985; Rowe 1986; De Bry 1986). However, fulfillment of these tasks may not be necessary for the automated mass propagation of many plants. Also, a robot system may be prone to mechanical breakdowns or faulty decisions due to poor visual recognition, and cause excessive physical damaging of cultured plants. These factors have discouraged the adaptation of robots to the tissue culture field. Lack of an adequate vision system coupled to robots is probably the greatest obstacle to be overcome in order to use this type of automation.

Fig. 6. Growth response of *Amaranthus gangeticus* L. shoot tips cultured in a microecosystem APCS for 2 months. Here, the original nonflowering *Amaranthus* plantlet flowered produced seeds which when shed germinated. These germinated seeds in turn produced flowering plantlets and viable seeds

Microecosystems wherein the plant matures to the flowering stage, undergoes pollination and fertilization, and then proceeds to fruit maturation may occur in the APCS (e.g., *Amaranthus*) (Fig. 6). Modifications of the APCS program, system design, culture medium, and system environment may be necessary to make the system more responsive to the specific needs of a particular plant. The length of time the culture soaks in a medium must be determined empirically to achieve the optimum growth for each plant. Fortunately, the APCS parameters can be easily modified via programming. The essential feature of the APCS is controlled medium removal and replenishment. The APCS may be the basic building block unit for future fully automated sterile plant growth systems. This versatility allows the basic APCS unit to easily interface with remote and direct sensors as well as vision and robotics systems. Integration of such systems with the APCS should enable plant growth to be closely directed to obtaining a desired product and minimizing human intervention to almost nil. Currently available commercial automated plant tissue culture systems include: spin-filter bioreactors (DNA Plant Technology Corp., 2611 Branch Pike, Cinnaminson, NJ 08077), macrocomputerized automation systems (PPS BV, Lage Dijk 26, P.O. Box 3032, 5700 JC Helmond, The Netherlands), and individual automated system units (De Novo, 736 N. Bradish Avenue, San Dimas, CA 91773).

References

Aitken-Christie J, Jones C (1987) Towards automation: radiata pine shoot hedges in vitro. Plant Cell Tissue Organ Cult 8:185–196

Anderson WC (1980) Mass propagation by tissue culture. In: Proc Conf Nursery Production of Fruit Plants through Tissue Culture – Applications and Feasibility. USDA, SEA, Agric Res Results, ARR-NE-11. Beltsville, pp 1–10

Cheng YW, Chua SE (1980) Mass propagation of orchid plants from orchid seeds using the air flow system to accelerate growth. Singapore J Primary Ind 8:57–62

Cheng YW, Chua SE (1982) The use of air flow system in plant tissue and organ culture. In: Rao AN (ed) Tissue culture of economically important plants. Nat Univ Singapore, COSTED, pp 210–212

Cheng YW, Chua SE, Yong-Ho SY (1978) A simple tissue culture method developed to accelerate growth of plant tissues. Singapore J Primary Ind 6:116–123

De Bry L (1986) Robots in plant tissue culture: an insight. IAPTC Newslett 49:2–22

Deleplanque H, Bonnet P, Postaire JG (1985) An intelligent robotic system for in vitro plantlet production. Proc 5th Int Conf Robotic vision and sensory controls, Amsterdam, The Netherlands. Rovisec 5, IFS Publication, pp 305–314

Gamborg OL, Miller RA, Ojima R (1968) Nutrient requirements of suspension cultures of soybean root cells. Exp Cell Res 50:151–158

Graebe JE, Novelli GD (1966) A practical method for large-scale culture. Exp Cell Res 41:509–520

Hagimori MT, Matsumoto T, Mikami Y (1984) Jar fermenter culture of shoot-forming cultures of *Digitalis purpurea* L. using a revised medium. Agric Biol Chem 48:965–970

Hartney VJ, Kabay ED (1984) From tissue culture to forest trees. Comb Proc Int Plant Prop Soc 34:93–99

Lawrence RH (1981) In vitro plant cloning systems. Environ Exp Bot 21:289–300

Levin R, Gaba V, Tal B, Hirsch S, DeNola D, Vasil IK (1988) Automated plant tissue culture for mass propagation. Bio/Tech 6:1035–1040

Molnar G (1987) A new patented method for mass propagation of shoot cultures. Acta Hortic 212:125–130

Murashige T (1974) Plant propagation through tissue cultures. Annu Rev Plant Physiol 25:135–166

Murashige T, Skoog F (1962) A revised medium for rapid growth and bioassays with tobacco tissue culture. Physiol Plant 15:473–497

Oglesby RP (1978) Tissue cultures of ornamentals and flowers: problems and perspectives. In: Hughes KW, Henke R, Constantin M (eds) Propagation of higher plants through tissue culture: a bridge between research and application. Techn Infor Center, USDE, Oak Ridge, pp 59–61
Rowe WJ (1986) New technologies in plant tissue culture. In: Zimmerman RH et al. (eds) Tissue Culture as a plant production system for horticultural crops. Martinus Nijhoff, Dordrecht, pp 35–51
Simonton W, Thai CN (1988) Plant tissue culture production planning using computer simulation/optimization. ASAE Paper No. 86–2351:1616–1622
Smith DR (1986) Radiata Pine (*Pinus radiata* D. Don). In: Bajaj YPS (ed) Biotechnology in agriculture and forestry. Vol I Trees I. Springer, Berlin Heidelberg New York Tokyo, pp 274–291
Styer DJ (1985) Bioreactor technology for plant propagation. Basic Life Sci 32:117–130
Tanaka H, Nishijima F, Suwa M, Iwamota T (1983) Rotating drum fermentor for plant cell suspension cultures. Biotech Bioengin 25:2459–2470
Tisserat B, Vandercook CE (1985) Development of an automated plant culture system. Plant Cell Tissue Organ Cult 5:107–117
Tisserat B, Vandercook CE (1986) Computerized long-term tissue culture for orchids. Am Orchid Soc Bull 55:35–42
Van Uyen N, Vander Zaag P (1983) Vietnamese farmers use tissue culture for commercial potato production. Am Potato J 60:873–879
Weathers PJ, Giles KL (1988) Regeneration of plants using nutrient mist culture. In Vitro Cell Dev Biol 24:727–732
Wilson G (1980) Continuous culture of plant cells using the chemostat principle. In: Fiechter A (ed) Advances in biochemical engineering. Springer, Berlin Heidelberg New York, pp 1–25
Yang HJ (1977) Tissue culture technique developed for asparagus propagation. HortScience 12:140

IV.2 Artificial Intelligence Technology to Control Plant Environments

Y. HARAZONO[1]

1 Introduction

Computers are used to control plant environment because of their flexibility in changing setpoints of controlled variables and their abilities in data acquisition. In the last decade many computer control methods were proposed in order to obtain optimal conditions for plant growth. Most of them use information about the plant/environment relations and/or empirical data base on the greenhouse management (Takakura et al. 1974, 1978; Takakura 1975; Kozai 1975; Takakura and Ohara 1976; Harazono et al. 1984, 1988; Ceulemans 1985; Hashimoto and Fukuyama 1985; Leung et al. 1985; Jones et al. 1988; Kurata 1988a,b).

Plant responses to changes in environmental variables cannot be described uniformly. Usually, the greenhouse climate is regulated according to the set values, although the set values for the control cannot be adjusted flexibly by the computer system even if the plants show different responses to their environment. When the plants show variable responses, the set values are decided according to one feature of the multiform responses of the plant. In this case, the variety of the plant responses to environment could confuse the greenhouse climate, so that the plant growth may not be in favorable condition or the control may become unstable. Therefore, many control methods which decide the set values by their control logic with fixed structure are limited to manipulation of the plant environment.

On the other hand, greenhouse climate controlled by a well-trained grower seldom becomes irregular. The way in which the grower judges set values for plant environment depends on the intuitive feelings acquired from his experience and knowledge. In the decision process on the set values, there is no logical structure but a synthesis of information on the plant and environment. In other words, there is no rigid model or fixed structure in the decision process. This means that the new control method using plant information to stably regulate the plant environment should be able to deal with various responses of plants to the environment and include a concept of synthetic process of information. Therefore, the artificial intelligence built-in control program should be most capable in the decision process of a computer control method.

There are two ways to utilize plant information by artificial intelligence technique in computer-aided control. The first is an expert system which depends on data base on relations between plant response and environment. Characteristics of an expert system are its large knowledge base and its ability for high speed reference. Thus, an expert

[1]National Institute of Agro-Environmental Sciences, Division of Agrometerology, Kannondai 3–1–1, Yatabe-machi Tsukuba-gun, Ibaraki-ken 305, Japan

Biotechnology in Agriculture and Forestry, Vol. 17
High-Tech and Micropropagation I (ed. by Y.P.S. Bajaj)

system can serve as a tool to mimic growers' decision processes for controlling greenhouse environment according to the experiences of the growers, but the final decision of set values for the control remains to be set by the human operator. Recently, there have been several proposals of ideas for controlling greenhouse environment with an expert system. One idea of a greenhouse control system which includes a knowledge base was proposed by Hoshi and Kozai (1984). Jones (1987) presented the system to monitor and diagnose the control system with an expert system operating in real time. Experiments supported by an expert system for controlling plant environment were carried out by Kano et al. (1987) and Kano and Shimaji (1988). A second way to use artificial intelligence is based on the inference engine. The inference engine acts in the same way as the reasoning in human brains and can deal with the synthetic processing of complicated information. Recently, a few commercial software packages of inference engine and hardware systems have appeared on the market and they can be applied to control greenhouse climate and plant environment in the chamber. The inference engine can replace the momentary decision process of a grower with the judgement whereby the computer system mimics the growers' reasoning, referring to the knowledge base of the plant and real-time data measured. However, there are few studies available about real-time control of plant environment by means of the control method with the inference engine.

This chapter introduces a study intended to establish a way of controlling environmental conditions for optimizing crop growth and production by means of artificial intelligence technology, using real-time information on plant activities. The control method proposed here uses a commercial software package as an inference engine to find the reasonable set values of environmental variables instead of the judgement of a grower.

2 The Concept of Control Logic

In the first step, in order to utilize an artificial intelligence technology to control greenhouse environment, we will examine the control habit of well-trained growers. When a grower changes the set values for controlling greenhouse climate, first, he observes outside weather conditions and the plant conditions in the greenhouse, and he recognizes the present situation. Second, he guesses whether the state of the plants will become better or worse if the set values are changed or left as usual. Third, he makes a decision whether the set values should be changed or not. In this judgement, he does not analyze the present situation numerically, but compares the relative situation of the relationship between the plants and their environment. This examining process mainly depends on his knowledge and the instinct trained in long-term experience.

This examination of the judgement process of a grower suggests that the controlling greenhouse environment should be based on the examining process, which is incorporated as the decision process within the inference engine: in other words, the new set values are decided according to the relative comparison of the present situation with the last one, referring to the empirical knowledge and the measured data. If this decision process of a grower is replaced by an artificial intelligence technology, then the control of greenhouse climate and plant environment can be entrusted to the control methods of a computer. Figure 1 shows the schematic diagram of the decision process

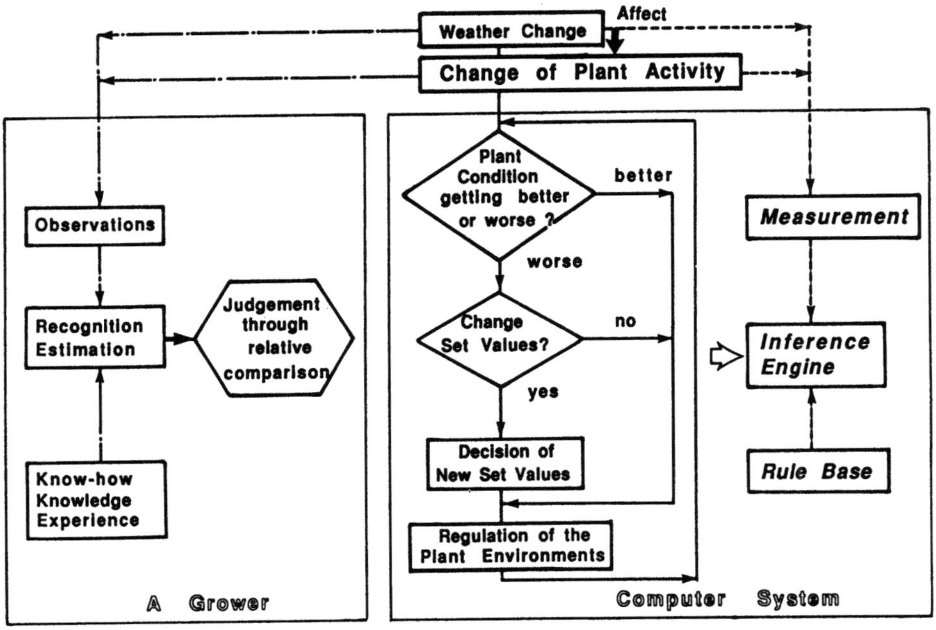

Fig. 1. Schematic diagram of a decision process of new set values by a grower and its analogy for computer simulation (*in italics*). (Harazono et al. 1988)

by a grower and its analogy for computer simulation. Chain lines indicate the flow of information recognized and/or judged by a human operator, and broken lines indicate the flow of data and/or corresponding instructions measured by or resulting from the computer system. The computer control logic proposed here is presented in the right hand side of Fig. 1 as a flow chart, which has some analogy with above concept of the intelligent decision process by a grower.

3 Outline of Control Logic

3.1 Configuration of the Software

In general, an artificial intelligence technique, such as an expert system or an inference engine system, is skillful when dealing with logical statements but unskilled to deal with numerical values. Therefore, the output information from an inference engine is not directly available for the control signal to regulate devices. Moreover, the measured numerical data are not suitably formated for input data of the artificial intelligence system. An interface is thus needed between an artificial intelligence system and the control and measurement system.

The software programs included in the present control method can be grouped into two parts: the inference part, which makes judgements on set values of the plant environment, and the executive part, which controls the plant environment. Figure 2 shows the configuration of the computer software of both parts in the control method. Both

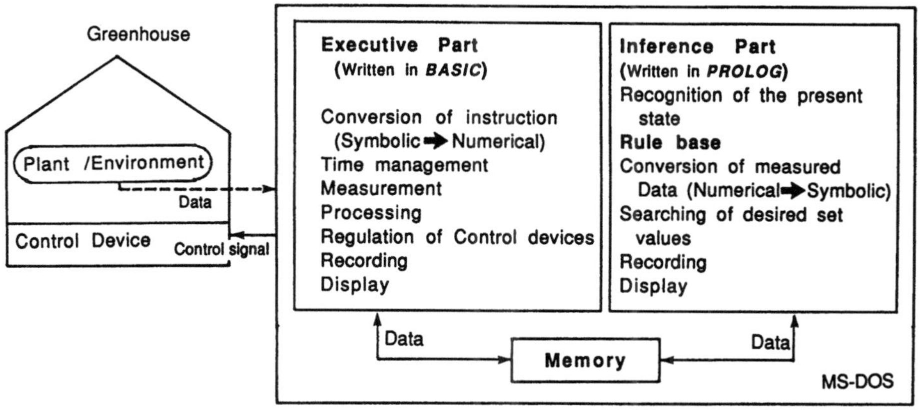

Fig. 2. Schematic diagram of the software configuration of the control method for control of greenhouse climate. (Harazono et al. 1988)

parts include a conversion process program of data flow and several subprograms. The rule base about the plant and its environment is included in the inference part, while the data on plant/environment relations are stored in the executive part.

The inference part, the basic frame within the artificial intelligence technique, decides the new set values of environmental parameters. In this study the commercial software package Prolog KABA (ASTEC, Kyoto Artificial Brain Associations) was used as the inference engine. This Prolog, an artificial intelligence language, runs under the operating system MS-DOS (Version 3.1). Japanese, Chinese, and English alphabets can be used to write programs in this software package. The actual program used in this study was written in Japanese. All subprograms included in this part were written in Prolog. The conversion program is also in this part. The measured numerical data are classified by the conversion program into three groups, namely "too high", "ok", and "too low" for each factor. The rule base, also written in Prolog, includes 19 fundamental plant/environment relations, which will be discussed in later.

The executive part was written in BASIC language which also runs under the operating system MS-DOS (Version 3.1). This part includes programs for time management, measurement, processing, and recording of measured data, device control, and conversion. Measurements of environment and regulation of devices are repeated until the scheduled time or the interruption occurs. At that time, measured data are processed and various information on plant activities evaluated, such as stomatal resistance and photosynthetic rate, then stored to the memory. In this part, numerical data are available and signals for set values sent from the inference part are converted into the numerical one, and are used for the control.

3.2 Execution of Programs and Flow of the Control

At the beginning, the CPU of the computer system runs the executive part. In this part, the present states of plants and environment are measured and several data on plant activities are calculated, then these values are recorded in the memory. The measurements and the regulation of plant environment continue until the scheduled time for examination of the environmental conditions or until the interruption occurs, then the process of the CPU changes from the executive part to the inference part. The interruption occurs when the outside weather condition changes remarkably or an unexpected condition occurs inside the system. The detail of interruption will be discussed later. After the process of the CPU is passed to the inference part, first the previously stored values are moved from the memory to the inference part, and then the measured and calculated data are examined as to whether the plant activities decrease or not and whether the environmental parameters change remarkably or not, compared to the last values. In this process, the numerical data are converted to the symbolic data, and the state of plant environment is examined to know whether plant activities maintain the good conditions or not under the present environment. If it is judged necessary to change the present state, the new set values for environmental variables are searched by the inference engine according to the rule base, and then the corresponding instructions for set values are decided. The corresponding instructions are expressed as symbolic statements and are recorded in the memory, then the process of the CPU moves to that in the executive part. If the present state is judged to be unchanged, the process of the CPU is passed immediately to the executive part again. When the set values are changed, the corresponding instructions are taken from the memory and converted into numerical set values under the CPU process in the executive part, then the environment is regulated according to the new set values.

In this way, the decision process on the new set values is installed in the on-line system for controlling greenhouse environment, using the inference engine with real time information of plant/environment relations.

4 Rule Base and Performance of the Inference Engine

4.1 Relations Among Environmental Variables and Plant Parameters

The rule base is expressed to show the relations among the environmental variables and the plant parameters taken from previous studies. At first we will examine the relation between photosynthetic rate and each environmental variable. It is generally known that the photosynthetic rate increases with light intensity and has a maximum value at an optimum temperature, which depends on the plant species. Moreover, the photosynthetic rate increases with the concentration of carbon dioxide in the atmosphere up to a certain concentration level, and under some conditions, the photosynthetic rate depends on wind speed, humidity, and temperature of the rhizosphere. The optimum conditions for photosynthesis also change with the growth stage of plants. Even if we consider only the photosynthetic rate, we can recognize that there are very complicated relations among the plant parameters and environmental variables.

Some plant parameters affect the photosynthetic rate. These parameters are also affected by environmental variables; for example, the degree of stomatal aperture affects stomatal diffusion resistance directly, then affects the photosynthetic rate strongly, but the degree of the stomatal aperture depends directly on both light intensity and CO_2 concentration and indirectly on humidity, wind speed, and so on. These environmental variables have a complicated effect on the photosynthetic rate, and their

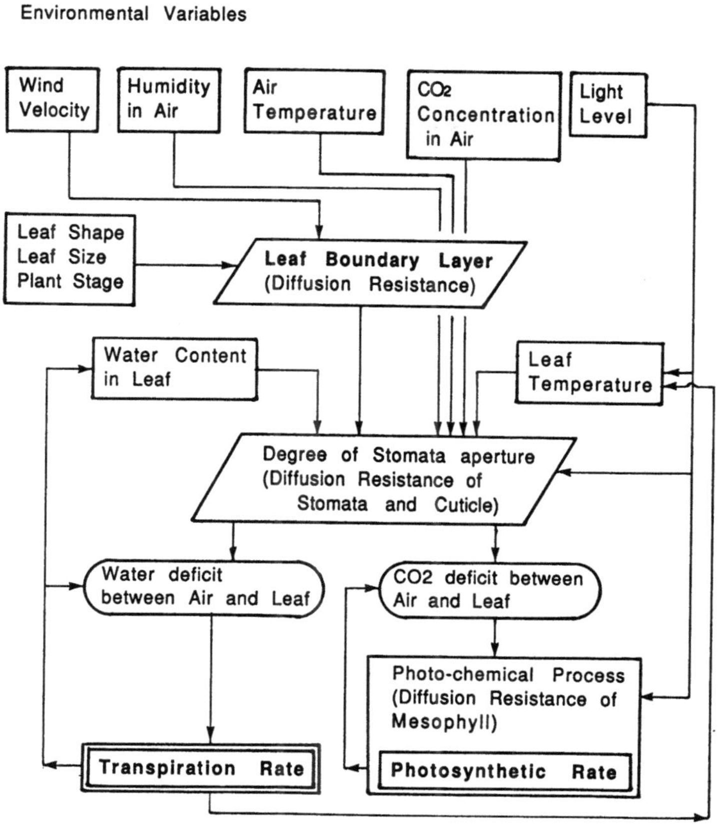

Fig. 3. Block diagram showing the complicated relationships among plant parameters and environmental factors

effects appear compositely and repeatedly. Figure 3 is the illustration of a block diagram showing these complicated relations among the functions and variables, namely, photosynthetic rate, transpiration rate, plant parameters, and environmental variables.

In order to use these experimental results on plant/environment relations as the knowledge in the artificial intelligence system with Prolog language, we must express the above complicated relations as statements written in Prolog, and must construct some consistent rules for the inference engine. Therefore, the complicated relations are divided into individual basic function for each factor or for each plant parameter. When the set values for the newly controlled environment are searched, every function is adapted to the actual states of plant and environment and are examined by the inference engine. In as far as the rule base is strictly expressed correctly to show the plant/environment relations, the decision by the inference engine becomes the right decision referred to the experimental results, and resembles the decision by a well-trained grower.

4.2 Expression of the Rule Base

Now we will try to construct the rule base about the photosynthetic rate, as an example. Some previous studies on the photosynthetic rate and the plant/environment relations are shown in Figs. 4, 5, 6, and 7. The relation shown in Fig. 4 describes the relation between CO_2 transport and diffusion resistances, namely, the model of CO_2 diffusion. Following facts can be derived from the model: in order to increase photosynthetic rate, one should increase the difference in CO_2 concentration between the atmosphere and chloroplast in the leaf or decrease each diffusion resistance. Figure 5 shows that the diffusion resistance of the leaf boundary layer, r_b, decreases with wind speed, U, but stomatal resistance, r_s, increases slightly with U under high wind speed condition (over about 1 m/s). This means that the stomatal aperture decrease when the transpiration rate is too high under low r_b condition. The increase of stomatal aperture means the decrease of diffusion resistance of stomata. Moreover, the stomatal resistance decreases with light level. This means that the high irradiance causes the degree of stomatal aperture to increase. From the above examinations and previous results as shown in Fig. 3, the following facts are derived: in order to decrease stomatal resistance, one should increase light level, and thus increase the stomatal aperture; in order to decrease the boundary layer resistance, one should increase wind speed.

Concerning degree of stomatal aperture, the following can be derived from the previous studies; the decrease of CO_2 concentration and the increase of light level make the stomatal aperture directly increase, and the increase of the humidity in atmosphere, the increase of water content in the leaf, and the decrease of boundary layer resistance make the stomatal aperture indirectly increase, but these secondary effects on stomatal aperture sometimes appear with a certain time lag.

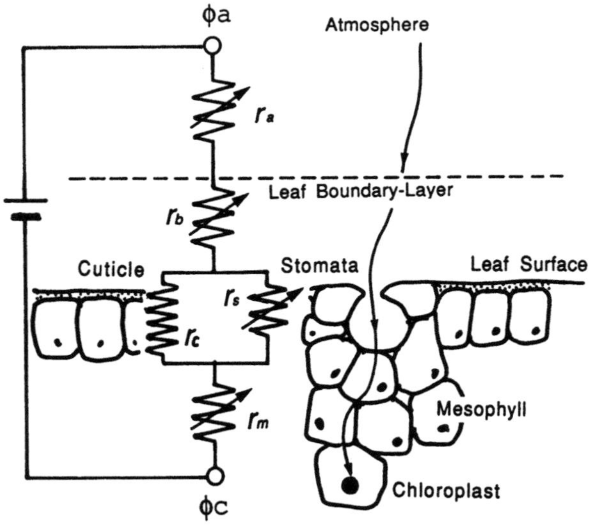

$$Po = \frac{\phi a - \phi c}{\sum r}$$

Fig. 4. Diffusion resistance model for CO_2 transfer between leaf and atmosphere. (After Yabuki 1985). *Po* indicates the photosynthetic rate; ϕ_a and ϕ_c indicate CO_2 concentrations at atmosphere and chloroplast; r_a, r_b, r_c, r_s, r_m indicate the diffusion resistances for atmosphere, leaf boundary layer, cuticle, stomata, and mesophyll

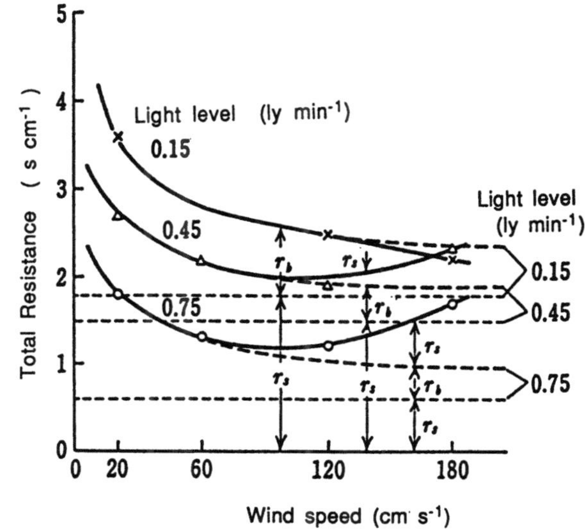

Fig. 5. Relationship among the diffusion resistance for water vapor, light level, and wind speed. Yabuki and Kiyota 1978). r_b and r_s indicate the diffusion resistances for leaf boundary layer and stomata

Figures 6 and 7 show the relation among photosynthetic rate and leaf temperature, CO_2 concentration, and light level. Figure 6 suggests that there are optimum leaf temperatures for photosynthesis, and the optimum leaf temperature slightly increases with the CO_2 concentration. Figure 7 also suggests that the optimum leaf temperature increases with light level, and the saturation level of photosynthetic rate to the light level increases with the CO_2 concentration. These relations from the figures include complicated relations among several factors, so it is difficult to express these whole functions as numerical functions but easy to express them as statements of the rule base.

These relations are rearranged to show that each factor acts as an individual function for controlling the environment. For example, the information about the photosynthetic rate is expressed as the following rule; in order to "increase photosynthetic rate", several instructive functions, including one variable or one parameter, are expressed as the following phrases: "increase light intensity", "increase CO_2 concentration", "increase degree of stomatal aperture", "increase wind speed", "decrease stomatal resistance", "decrease plant respiration", and so on. In this way, every listed item of information about the photosynthetic rate can be translated to the corresponding rules, which constitute the rule base.

Although some relations in the rule base could be expressed as numerical functions, the set values derived from the numerical functions may include inconsistencies, and in such a case, they could not be applied to control plant environment. On the other hand, if there are any inconsistencies within the corresponding instructions obtained from the rule base, they can be logically corrected by the inference engine, and if there is some weighing or ordering among the rules, the correction could be done more reasonably. Therefore, the artificial intelligence technique can be useful in the decision process in the complicated system.

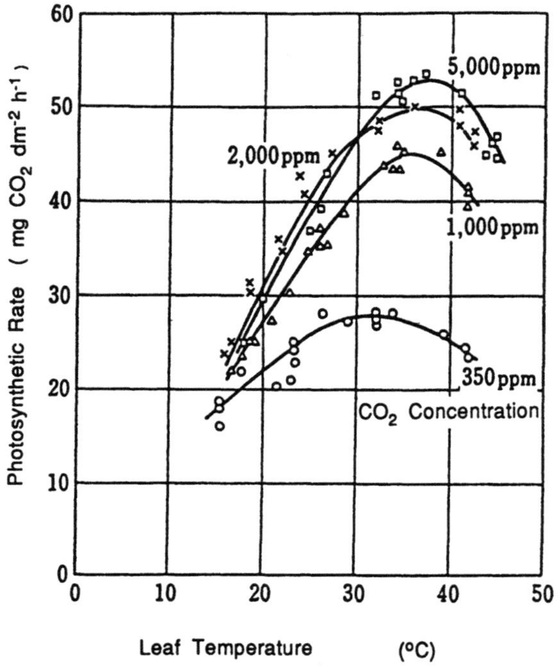

Fig. 6. Relationship among photosynthetic rate, leaf temperature, and CO_2 concentration. (Yabuki 1985)

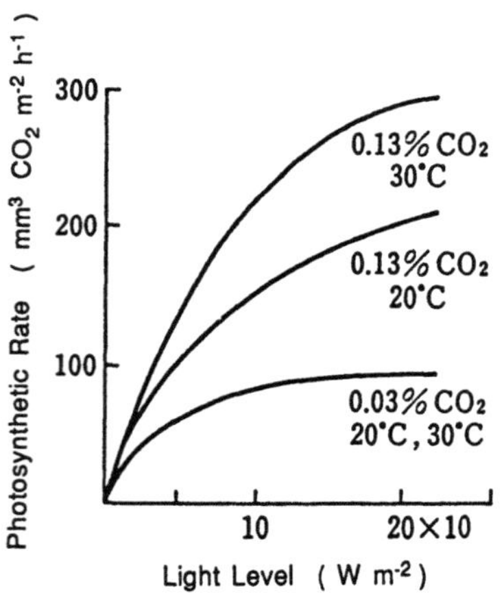

Fig. 7. Relationship among photosynthetic rate, leaf temperature, light level, and CO_2 concentration. (Gaastra 1963)

4.3 Features of the Present Rule Base and Inference Engine

An example of the rule base used in the present experiment is listed in Fig. 8. A large number of rules may reflect the feature of the object more correctly, but it requires a longer time to search for an instruction among many rules. It is better to have a small number of factors and parameters which constitute the rule base for a limited purpose, because the inference engine compares the state of the object with all rules, and picks up all corresponding corrective instructions, so that the presence of unusable parameters or factors could lead to a long interval for searching, and occasionally mislead the corrective instructions which are necessary for actually controlling plant environment, which could confuse the control.

In the present experiment, photosynthetic rate, transpiration rate, leaf temperature, stomatal resistance, and mesophyll resistance are the plant parameters, and air temperature, relative humidity, light level, and CO_2 concentration are the environmental vari-

1. Increase(Photosynthesis, S) :-
   Condition (Increase Photosynthesis, [Increase (CO2 concentration, S), Decrease (Stomatal resistance, S), Decrease (Mesophyllic resistance, S), Fit (Air temperature, S)]).
2. Increase(Transpiration, S) :-
   Condition (Increase Transpiration, [Increase (Leaf temperature, S), Decrease (Relative humidity, S), Open (Stomata, S)]).
3. Decrease(Transpiration, S) :-
   Condition (Decrease Transpiration, [Decrease (Leaf temperature, S), Increase (Relative humidity, S), Close (Stomata,S)]).
4. Decrease(Mesophyll resistance, S) :-
   Condition (Decrease Mesophyll resistance, [Increase (Leaf temperature, S),]).
5. Decrease(Stomatal resistance, S) :-
   Condition (Decrease Stomata resistance, [Decrease (CO2 concentration, S), Increase (Relative humidity, S), Open (Stomata, S)]).
6. Fit (Leaf temperature,S) :-
   Condition (Fit Leaf temperature, [Fit (Air temperature,S)]).
7. Increase(Leaf temperature, S) :-
   Condition (Increase Leaf temperature, [Increase (Air temperature, S)]).
8. Decrease(Leaf temperature, S) :-
   Condition (Decrease Leaf temperature, [Decrease (Air temperature, S)]).
9. Open (Stomata, S) :-
   Condition (Open Stomata, [Decrease (Transpiration, S), Decrease (CO2 concentration, S)]).
10. Close (Stomata, S) :-
    Condition (Close Stomata, [Increase (Transpiration, S), Increase (CO2 concentration, S)]).
11. Increase(Mesophyll resistance, S) :-
    Condition (Increase Mesophyll resistance, [Decrease (Leaf temperature, S),]).
12. Increase(Stomatal resistance, S) :-
    Condition (Increase Stomata resistance, [Increase (CO2 concentration, S), Decrease (Relative humidity, S), Close (Stomata, S)]).
13. Increase(Air temperature, S) :-
    Condition (Increase Air temperature, [Increase (Air temperature, S)]).
14. Decrease(Air temperature, S) :-
    Condition (Decrease Air temperature, [Decrease (Air temperature, S)]).
15. Fit (Air temperature, S) :-
    Condition (Fit Air temperature, [Fit (Air temperature, S),]).
16. Decrease(Relative humidity, S) :-
    Condition (Decrease Relative humidity, [Decrease (Relative humidity, S),]).
17. Increase(Relative humidity, S) :-
    Condition (Increase Relative humidity, [Increase (Relative humidity, S)]).
18. Increase(CO2 concentration, S) :-
    Condition (Increase CO2 concentration, [Increase (CO2 concentration, S)]).
19. Decrease(CO2 concentration, S) :-
    Condition (Decrease CO2 concentration, [Decrease (CO2 concentration, S)]).

Fig. 8. List of the rule base used in the present study

ables. Controlled environmental variables of the chamber are air temperature, CO_2 concentration, and relative humidity. Therefore, there is no use in including instructive functions about the light level and environmental variables outside the greenhouse or chamber. Please note that there are no rules about these in the rule base shown in Fig. 8. When the light level is variably controllable in the system, for example, in the case when supplemental lighting and shading are supplied the instructive functions on light level should be added to the rule base. There are 19 rules in the present system, in which seven rules, from Rule 13 to Rule 19, are expressed as recurrent. The recurrent expressions play very important roles in the inference engine, because the search by the inference engine continues until it finds the recurrent statement.

Rule 1 and several corresponding instructions derived from Rule 1 are shown in Fig. 9, in order to know the role of the rule in the inference engine. Rule 1 shows that there are four ways to increase the photosynthetic rate under the present experimental set up: increasing CO_2 concentration, decreasing stomatal and mesophyll resistances, and controlling the air temperature at the optimal level. Although CO_2 concentration and air temperature are controllable variables, these corresponding instructions can be translated into action in practice, but other instructions on stomatal/mesophyll resistances cannot be translated into action directly. For the factors or parameters which cannot be controlled directly, namely, for the uncontrollable parameters in a narrow sense, other rules on plant parameters, such as Rule 4 and Rule 5 in this case, are adapted at the second searching stage. Rule 5, shown in the bottom half of Fig. 9, is the example of the rule which is adapted at the second search in relation to stomatal resistance. Note that there is an uncontrollable plant parameter on stomata which requires the third search stage. In this way, more search about uncontrollable parameters could be carried out in the inference engine, rules on plant parameters are repeatedly referred to until the instructions on controllable variables are found.

As the optimum air temperature for photosynthesis depends compositely on light level, humidity, and CO_2 concentration, the correcting instruction on air temperature adopted for a case is not uniquely determined, so the corresponding instruction expressed as "fit" is used in the present case. The correcting instruction for set value of air

Rule 1 **Increase(Photosynthesis, S) :-**
Condition (Increase Photosynthesis, [Increase (CO2 concentration, S), decrease(Stomatal resistance, S), decrease (Mesophyll resistance, S), fit (air temperature, S)]).

Increasing	CO2 concentration
fitting	Air temperature suitably
decreasing	***Mesophyll resistance***
	Stomatal resistance

Rule 5 **Decrease(Stomatal resistance, S) :-**
Condition (Decrease Stomatal resistance, [Decrease (CO2 concentration, S), Increase (Relative humidity, S), open (Stomata, S)]).

Increasing	Relative humidity
decreasing	CO2 concentration
making open	***stomata***

Fig. 9. Examples of rules and derived correcting instructions. (Harazono et al. 1988)

temperature is determined more clearly by another rule in the following searching stages. In this rule base we have redundancy for set values by the expression "fitting", which is most advantageous for controlling plant environment by means of the AI technique.

4.4 Decision Process on Set Values

When the new set values are needed for control, the present conditions of the plant and environment are applied to the rule base, then some instructions are derived from the rules for one objective factor. As all conditions of the present state are applied to the rules one by one, some correcting instructions for the environmental variables are found. Sometimes there are inconsistencies among the instructions. In such cases, it is not simple to translate the selected correcting instruction into action. Ordering or weighing of the rules is one possible way to avoid the conflict. In this study, we use the averaging method. First, every correcting instruction has a unit value with a sign which is positive for increasing or negative for decreasing the parameter; in other words, each correcting instruction is the instructive unit value of +1 or –1. Then, correcting instructions found are checked for inconsistencies. For controllable variables, if all the correcting instructions derived from the rule base are not the same, the instructive units are averaged for each controllable variable. For example, if there are two correcting instructions to increase air temperature and one correcting instruction to decrease (note that there is inconsistency) the total number of instructions about air temperature is 3 and sum of the instructive units is +1; therefore, the instructive value is one third. Then the final corrective instruction is decided, that the air temperature should be increased by one third of a unit. The unit means the stepwise value for modification of set points of environmental variables, namely the adjustment value per one unit, whose physical units are °C, %, and ppm for air temperature, relative humidity, and CO_2 concentration, respectively.

In general, both variables to be changed and their adjustment values are required for the control, and both are decided by the rule base in many control methods using artificial intelligence techniques. In the present control method, the variables to be changed are decided readily by the rule base, but the adjustment values to be set are decided as the relative values (instructive values) by the rule base with inference engine, then the actual adjustment values for the control are obtained by multiplying each instructive value by each adjustment value per one unit. New setpoint for a controlled variable is defined as the sum of the last setpoint and the present adjustment value. Therefore, in this control method, the magnitude of the adjustment values per one unit for each variable greatly affects the control manner and the responsibility of the system. Large adjustment values per one unit may lead to the rapid change of the controlled environment, and small adjustment values may effect a steady environment but cannot follow the sudden change of setpoints resulting from change in the outside weather conditions. In other words, in order to modify the responsibility of the system, the adjustment values need only to be changed per one unit. This is the advantage of this control method and may be useful when the responsibility of the complicated system, whose structure is not clear, is to be changed.

5 Measuring Systems for Controlling Plant Environment

In general, the nature of measured variables could influence the basic behavior of the control system; especially the plant parameters to be measured affect the control of greenhouse environment. Therefore, it is important for the control method how the plant parameters are measured.

Parameters indicating plant activities, such as photosynthetic rate, transpiration rate, leaf temperature, and diffusion resistances of stomata and mesophyll, can be determined using an assimilatory box. Figure 10 is a schematic illustration showing the measurement system with a small assimilatory box and the experimental set-up for cucumber plants. Figure 11 is a photograph showing the attachment of the small assimilatory box to a cucumber leaf. An infrared CO_2 gas analyzer (IRGA) of absolute type is used to record CO_2 concentration of atmosphere in the growth chamber, and an IRGA of differential type is used to measure the exchange rate of CO_2 between leaf and atmosphere. According to the formula shown below, photosynthetic rate of the leaf, Po, is calculated from the difference between CO_2 concentrations inside and outside the assimilatory box and the flow rate of the air which passes through the assimilatory box. Data from the differential IRGA are also used to determine total diffusion resistance of CO_2 transport and mesophyll resistance, r_m, for the CO_2 transport. Transpiration rate, Tr, is determined using humidity, AH, and leaf temperature, T_1, measured in the assimilatory box and in the atmosphere. Diffusion resistance of stomata, r_s, is determined from leaf temperature and transpiration rate. Mesophyll resistance, r_m, is determined from total diffusion resistance ($r_s + r_m$) and stomatal resistance. These numerical values about plant activities are calculated by the following formula with measured data,

$$Po = C1 * [CO_2(out) - CO_2(in)] * Q * 273.2 / [A * (273.2 + T_a)]$$

$$Tr = C2 * [AH(in) - AH(out)] * Q / A$$
$$r_s = [SH(T_1) - AH(out)] / Tr$$
$$r_m = C3 * [CO_2(out) - CO_2(in)] / Po - r_s,$$

where $CO_2(out)–CO_2(in)$ and AH(in)–AH(out) are CO_2 concentration difference and absolute humidity difference between inside and outside of the assimilatory box. $SH(T_1)$ is the saturated humidity at leaf temperature. Q, T_a, and A are flow rate of air passing through the assimilatory box, air temperature, and leaf area in the assimilatory box, respectively, C1, C2, and C3 are coefficients.

6 Experiments of Controlling Plant Environment by Means of the A1 Technique

6.1 Hardware Configurations

Experiments were carried out using a growth chamber of the wind channel type which has a measurement space of 1.5 m long, 0.8 m wide, and 0.8 m high, and has artificial light sources. Wind speed near plants in the chamber can be regulated from 0.2 m/s to 2.5 m/s. Light level in the chamber can be regulated by changing the number of metal halide lamps to be turned on. Figure 12 shows the external appearance of the chamber.

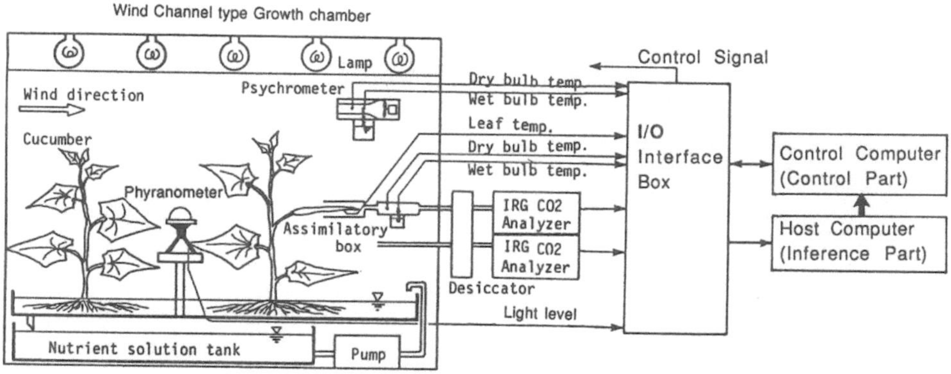

Fig. 10. Schematic illustration of measurement system with small assimilatory box and experimental set-up. (Harazono et al. 1984)

Fig. 11. Photograph showing the attachment of small assimilatory box to a cucumber leaf

The area behind the two small windows shown in the right down of the photo is the measurement section, and other portions constitute the wind channel of vertical rotation. A microcomputer, NEC PC-9801, was used as a host computer of the A1 system and for the measurement of the data on plant activity and the environment. The other microcomputer, Fujitsu FM-7, was used in order to control the growth chamber and measure the atmospheric environment in the chamber. Both computers are widely distributed in Japan. The schematic diagram of the experimental system is shown in Fig. 13. The I/O

Fig. 12. Photograph showing the external appearance of the wind channel type growth chamber which is used in present controlling experiments

interface belonging to the control computer was originally designed for the purpose of controlling the growth chamber, and the interface of the host computer was also originally designed for general measurement purposes.

6.2 Set Conditions of the Experiment

In the present experiment, light level in the chamber was routinely regulated stepwise to stimulate daily variation of outside solar radiation by adjusting the number of lamps to be turned on. Air temperature, relative humidity, and CO_2 concentration in the chamber were controlled according to the set values derived from the logic using the inference engine. In this experiment, the adjustment values per one unit of correcting instructions for control devices were defined as 2 °C for air temperature, 5% for relative humidity, and 20 ppm for CO_2 concentration, respectively. A series of control process in the inference part, such as judgement, searching, and decision to correct instructions, was executed every 10 min. Although whether the set values are actually changed or not is dependent on the state of the plant and environment, regulation and check of the conditions of plant and environment were carried out every 10 s by the control computer under the CPU process of the executive part.

When the present state of the plant and environment was judged as one of the following situations, the new set values were derived by the inference engine through the rule base; (1) the light level changes remarkably, (2) photosynthesis decreases while the light level does not decrease, (3) diffusion resistances increase while the light level

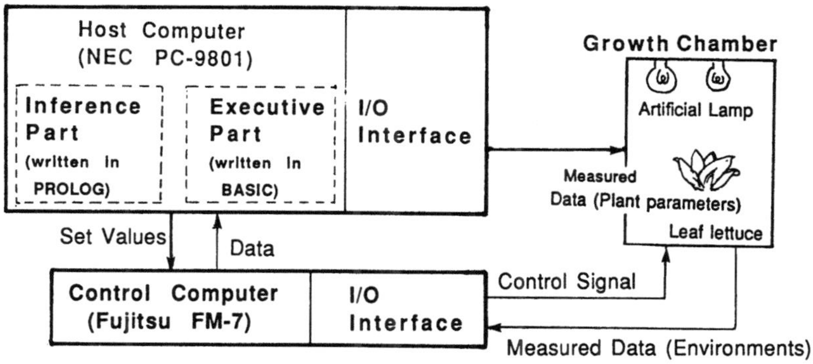

Fig. 13. Schematic diagram of the system tested; control method in plant environments. (Harazono et al. 1988)

does not change. Among the three conditions, only the first condition causes interruption for the movement of the CPU process from the executive part to the inference part, regardless of the scheduled time in the series of the computer control process.

In the experiment, lettuce leaf was used as the test plant. The plant activities were measured using the small assimilatory box (40 mm long, 30 mm wide, 5 mm deep) which covered both sides of a representative leaf of the tested plant. Two infrared CO_2 gas analyzers were used to measure absolute and differential CO_2 concentrations, leaf temperature was measured by a thin thermocouple, light level was measured by a pyranometer at the height of the assimilatory box, and air temperature and humidity were measured by a ventilated psychrometer which measures dry and wet bulb temperatures with thermocouples. Leaf area used in the calculation of plant activities was not directly measured but assumed to be equal to the area of the assimilatory box. Therefore, if the shape of leaf was uneven, as in the case for lettuce leaf the estimated leaf area used in the calculation might be underestimated.

6.3 Experimental Result and Examination of the Control Method

Figure 14 shows an example of the control. Light level at the plant is shown by the broken line, which was controlled independently. The set values for air temperature, relative humidity, and CO_2 concentration are indicated by the thin lines. The thick lines are the measured values of the above.

In the morning, the air temperature and the CO_2 concentration were controlled with a large increase until 12:00 h, except the period between 08:30 and 09:30 h, during which they were decreased. In the afternoon, both the air temperature and the CO_2 concentration were decreased, accompanied by decrease in light level. Set values for the CO_2 concentration derived by the control method were relatively stable, and the air temperature and CO_2 concentration were controlled relatively parallel to the change of light level.

According to the examinations of recorded set values for control devices, it was revealed that the increase of the set value for CO_2 concentration in the morning was mainly attributed to the rules on photosynthesis and transpiration; on the other hand,

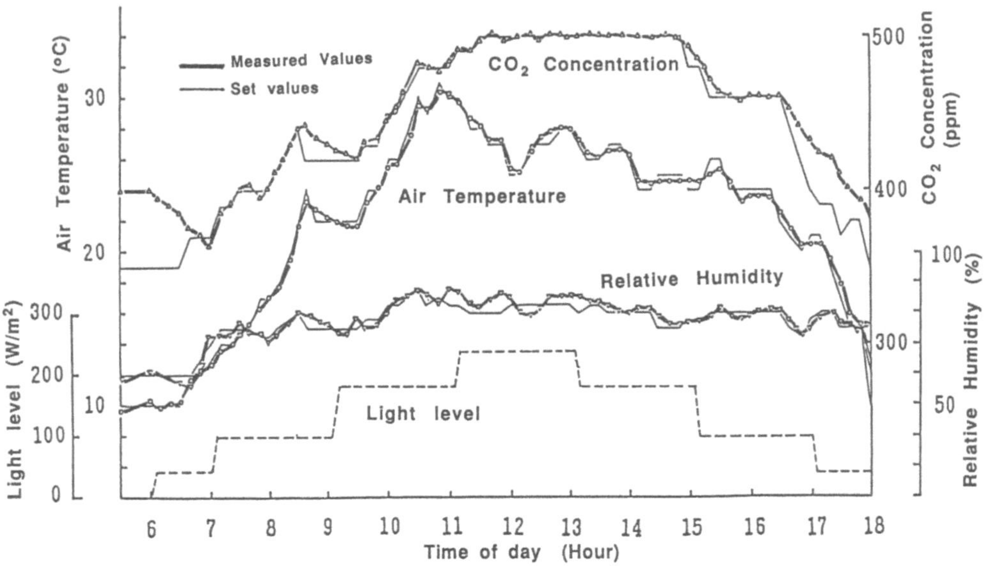

Fig. 14. Simulated daily variation of light level, controlled results of environmental variables, and corresponding set values decided by the inference engine of the A1 control method wind channel type growth chamber. (Harazono et al. 1988)

the decrease of the set values in the afternoon was mainly attributed to the rules on leaf temperature and stomatal and mesophyll resistances. These results show that stomatal resistance becomes smaller with the increase of light level in the morning. The inference engine recognizes this plant response to the environment and decides the above instructive set values. In the afternoon, mesophyll resistance became large because light level decreases; therefore, to decrease mesophyll resistance, the set value for CO_2 concentration was lowered. Note that the set values decided by the method were corrected whenever the activity or response of the plant could not follow the change in the environment. According to the check-through of the recorded correcting instructions, most of the corrections were done logically compared to the conditions, but all controlled environments did not provide good conditions for plant activity, because there was a time lag between the decision on set values and the response. If the adjustment values per one unit for controllable variables are improved, and if the interval time for changing the set values are longer, the controlled environment could maintain better states for plant activities. Although the experiment was only at the first stage of the study, more than half the corrections were successful, and then the manner of decision resembles the decision process by a trained grower, especially for the results in the morning.

Figure 15 shows plant activity under the controlled environment according to the set values shown in Fig. 14. The photosynthetic rate shown in Fig. 15 could be overestimated because of the underestimated leaf area, which could not be measured directly.

Until 08:30 h, the photosynthetic rate increased at a low rate as stomatal resistance decreased. Just after 08:30 h, stomatal resistance increased remarkably, and the photosynthetic rate decreased. These reactions were possibly caused by water stress in the leaf and successive stomata closure, which occurred after a rapid increase in air

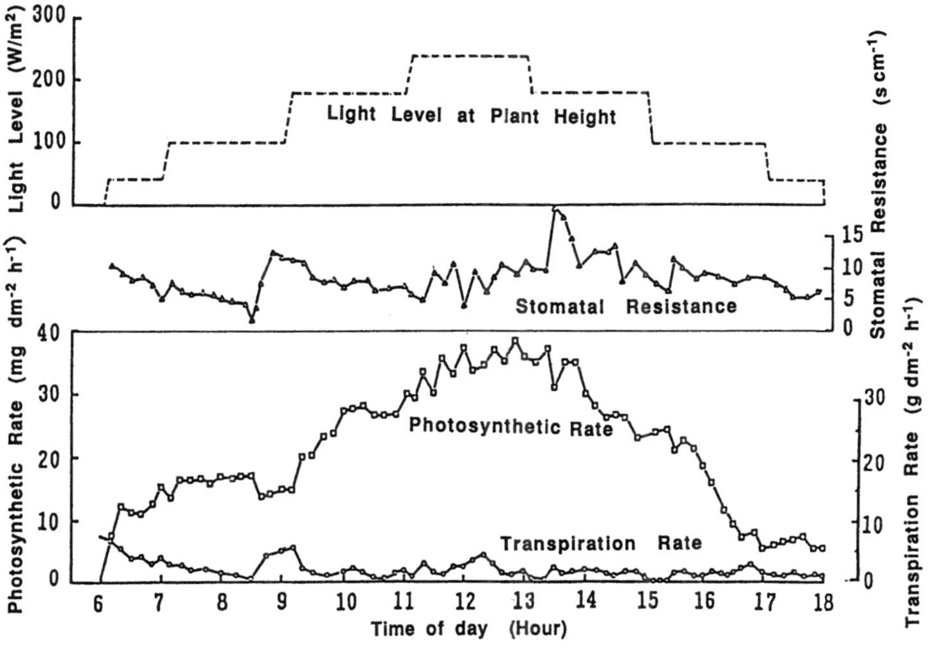

Fig. 15. Simulated daily variation of light level and plant responses to the controlled environments by means of the A1 control method. (Harazono et al. 1988)

temperature. This was recognized by the inference part of the control method; the set values for air temperature and CO_2 concentration were lowered after 08:30 h, then, as a result, the water content in the leaf recovered from the levels of water stress, so the stomatal resistance decreased. Within about 1 h, stomatal resistance and photosynthetic rate recovered to their original levels. At this time the set values of air temperature and CO_2 concentration were increased again to a high level.

As far as we could examine the short-term results of controlled environment and plant responses, the decisions on the set values were reasonable, and the controlled environment relatively stable. The manner of controlling environment was similar to that of a greenhouse grower who regulates the greenhouse environment according to his observations of plant conditions. The measured values of the plant activities included some errors, as stated previously, but the environment was controlled without sudden changes. This is an advantage in the decision process for the actual control of the environment by this control method, because some of the measured values always have some errors and/or range shifts.

7 Conclusions

Environments controlled by the present control method promote plant activities and keep them in good condition, in spite of the small number of rules involved. The manner of control was similar to that of a trained greenhouse grower who watches plant conditions. Therefore, it was confirmed that the control of the plant environment was well done by this method, using an inference engine with a rule base and real-time plant data. Plant environment in the growth chamber was controlled stably, and the change in the environment instructed by this method produced maximal plant activities under the given light level. Although the present method was examined only with the simple system, the validity of the control method using an artificial intelligence technique is confirmed; this method can thus be useful to control actual greenhouse climate and can be adapted to control complicated systems.

Micropropagation is one of the typical production systems in intensive agriculture, and it has led to the rapid production of disease-free plants. Many studies on automated production systems and on the control of their environment in micropropagation have been carried out in the last decade (Harrell and Simonton 1986; Kozai et al. 1986; Aitkin-Christie and Davies 1988; Fujiwara et al. 1988; Hempel and Meszka 1988; Walker et al. 1988). Furthermore, micropropagation is carried out under entirely controlled conditions, but production techniques and their theoretical analysis are not yet completely established, because the micropropagation process includes different production stages in which several environmental and/or plant factors are interrelated. Therefore, the application of the present control method by means of an artificial intelligence technology to the control of environment in micropropagation would seem to be of great benefit. The many data that have been gathered to obtain the descriptions of rules in relationships between the controlled environment and plant responses in vitro can be used to apply the present method to micropropagation.

In this control method, the diffusion resistances of stomata and mesophyll, which were used as the plant parameters in the rule base, acted as if it was a feedback mechanism to increase both CO_2 concentration and air temperature during the reasoning processes in the inference engine. The presence of the compensative relations in the rule base seemed to be one reason that the plant environment were controlled reasonably. Therefore, in order to realize better control or decision for the complicated system, it seems to be necessary to include compensative relationships in the rule base which is incorporated within the artificial intelligence technique.

This study was carried out at the University of Osaka Prefecture, Japan, from 1980 to 1987. At first, the control method was constructed using a pattern-matching technique. The control by the pattern matching sometimes became unstable because the decision process of set values could not respond to the various plant activities. In order to avoid this weakness, the "amendment program", which changes set values empirically according to the relations between limited data base and measured data, was added. Although the programs were written in FORTRAN language, there are limitations in the decision process of set values. The control method presented here was developed from the method described above. The success in controlling environment was obtained by the use of the inference engine included in Prolog language.

In order to apply the present method to the actual control of greenhouse environment, further studies must be undertaken on ways to solve conflicts within the correcting instructions and to construct rules for individual control equipment. The number of inconsistencies among the correcting instructions derived from the inference engine may increase as the number of rules increases, so the reasonable ordering or logical weighing of the rules is required for fast decision on the set values.

Acknowledgments. The author thanks Dr. Kazutoshi Yabuki for his advice, Mr. Toshiyuki Taenaka, Miss Hiromi Kamiya, and Miss Misako Awakura for their assistance, and Dr. Atsushi Kano for his comments and discussions.

References

Aitken-Christie J, Davies HE (1988) Development of a semi-automated micropropagation system. Acta Hortic 230:81–87

Ceulemans RJ (1985) Plant growth optimization by photosynthetic monitoring. Acta Hortic 174:309–312

Fujiwara K, Kozai T, Watanabe I (1988) Development of a photoautotrophic tissue culture system for shoots and/or plantlets at rooting and acclimatization stages. Acta Hortic 230:153–158

Gaastra P (1963) Climatic control of photosynthesis and respiration. In: Evance LT (ed) Environmental control of plant growth. Academic Press, London, pp 113–140

Harazono Y, Taenaka Y, Yabuki K (1984) Optimizing environmental control of greenhouse climate for cucumber growth by means of 'Learning control method'. Acta Hortic 148:259–265

Harazono Y, Kamiya H, Yabuki K (1988) A control method based on an artificial intelligence technique and its application for controlling plant environment. Acta Hortic 230:209–214

Harrell R, Simonton W (1986) Automation opportunities in plant tissue culture operations. Am Soc Agric Eng 86–1596:1–12

Hashimoto Y, Fukuyama T (1985) Some speaking plant approach to the synthesis of control system in the greenhouse. Acta Hortic 174:219–226

Hempel M, Meszka K (1988) Modelling of micropropagation. Acta Hortic 230:137–144

Hoshi T, Kozai T (1984) Knowledge-based and hierarchically distributed on-line control system for greenhouse management. Acta Hortic 148:301–308

Jones P, Jacobson BK, Jones JW (1988) Applying expert systems concepts to real-time greenhouse controls. Acta Hortic 230:201–208

Jones PH (1987) Interfacing an expert diagnostic tool to real-time data. Appl Eng Agric 3(2):233–236

Kano A, Shimaji H (1988) Greenhouse environmental control system with a crop model and an expert system. Acta Hortic 230:229–236

Kano A, Sueyoshi K, Shimaji H (1987) Sky farm, a computerized greenhouse for crop and environmental studies. Yokogawa Tech Rep 31(4):197–200

Kozai T (1976) Optimizing control of plant growth using a digital computer. J Agric Meteorol Jpn 32:41–49

Kozai T, Fujiwara K, Watanabe I (1986) Fundamental studies on environments in plant tissue culture vessels. (2) Effects of stoppers and vessels on gas exchange rates between inside and outside of vessels closed with stoppers. J Agric Meteorol Jpn 42:119–127

Kurata K (1988a) Greenhouse control by machine learning. Acta Hortic 230:195–200

Kurata K (1988b) A fundamental study on automation of greenhouse crop management by machine learning. J Agric Meteorol 44:181–186

Leung CM, Wilson DR, Marquand CJ, Tassou SA (1985) A knowledge-based system for the control of a greenhouse environment. Acta Hortic 174:425–431

Takakura T (1975) Maximizing plant growth using a small computer. Acta Hortic 46:147–156

Takakura T, Ohara G (1976) Direct digital control of plant growth(II). J Agric Meteorol 32:107–115

Takakura T, Kozai T, Tachibana K, Jordan KA (1974) Direct digital control of plant growth(I). Trans ASAE 17:1150–1154

Takakura T, Ohara G, Nakamura Y (1978) Direct digital control of plant growth(III). Acta Hortic 78:257–264

Walker PN, Heuser CW, Heinemann PH (1988) Micropropagation: studies of gaseous environments. Acta Hortic 230:145–152

Yabuki K (1985) Dynamic response of plants to the environment (in Japanese, Japanese title Syokubutu no Douteki Kankyou). Asakura, Tokyo, p 200

Yabuki K, Kiyota M (1978) Studies on the effects of wind speed on photosynthesis (6). J Agric Meteorol 34:59–63

IV.3 Robots in Transplanting

G.E. Miles[1] and L.J. Kutz[2]

1 Introduction

By virtue of their flexibility, which permits them to be used for many different tasks, robots[3] have the potential for lowering production costs. Automated machinery such as combines often go unused for many months each year, and the fixed overhead costs add unnecessary expense. A machine which transplants, irrigates, fertilizes, and harvests many different kinds of crops could be utilized more often during the year, thereby lowering the fixed costs allocated to each crop. Its ability to perform the necessary functions for many different crops would provide additional business options for the farm manager to consider.

A robot has the necessary flexibility, but the variability found in plants and animals requires a modern systems engineering approach to its development. The robot needs sensors to locate objects and measure features. Computer programs must acquire, interpret, and process sensor information, recognize patterns in the data, and reach consistent, intelligent decisions about future movements of the gripper and manipulator. Evaluating improvements in the design of sensors and control algorithms by field testing is virtually impossible because the row to row, field to field, and day to day variations in plants mask the differences. The scatter in the data hides the improvement due to sensor or algorithm design.

A solution to this problem is to follow a systems engineering approach (Kutz and Miles 1986). The first step is to define the gripper and manipulator motions required to perform the desired task. Computer-Aided-Design (CAD) models help organize the working environment (work cell) and plan the motions. Animated simulations with these CAD models illustrate the selected robot's performance for the proposed task. The animation illustrates the dynamic interaction between the gripper and the object being moved and makes it easier to evaluate modifications in the machine or its control algorithms. Laboratory studies are necessary to confirm that the proposed motions satisfactorily perform the intended task. One of the important benefits of the animated CAD simulation is the automatic development of the control program, which can be downloaded to the controller for the laboratory robot. Since this program has been tested during the computer simulations, it should be free of mistakes which could cause

[1]Department of Agricultural Engineering, Purdue University, West Lafayette, IN 47907, USA
[2]Agricultural Engineering Department, Auburn University, Auburn, AL 36849, USA
[3]A robot is a reprogrammable multifunctional manipulator designed to move materials, parts, tools, or specialized devices, through variable programmed motions for the performance of a variety of tasks (Robot Industries Association).

Biotechnology in Agriculture and Forestry, Vol. 17
High-Tech and Micropropagation I (ed. by Y.P.S. Bajaj)

damage in the real-world robot. Once a model of a task has been proven valid by laboratory studies, additional simulations with other, commercial robots will determine the technical feasibility (e.g., can the robot reach all plug and growing flats?) and the production capacity (e.g., plants per hour). These data, along with purchase prices and operational costs, can be used to determine the economic feasibility of using a robot to perform this task. The most important statistic, the per unit cost (i.e., cost per plant), can only be determined through a study of the materials flow with the robot in place as part of the system. Usually there will be multiple solutions: one large robot with a large capacity versus several small robots with lower capacities, and each scenario must be studied to determine the optimum cost/performance ratio. For a given production situation, the daily throughput (i.e., plants per day) determines the necessary capacity of the robot(s), whereas the annual production (i.e., volume of plants produced per year) determines the unit cost (i.e., cost per plant).

This entire process (modeling, simulating, testing, and optimization) is the essence of the systems engineering approach to designing complex machines. Without systems engineering, machine development becomes an endless cycle of design modifications and field testing; with it, machine performance in response to changes in design or crop parameters can be simulated, regardless of the time of the year, crop maturity, or weather conditions. Through simulations, the same row can be harvested again and again, thereby enabling changes in control algorithms or sensors to be evaluated without the interference of plant to plant variation in crop parameters. This process does not eliminate the need to validate machine performance under field conditions, but it does minimize the number of tests because infeasible solutions are identified and discarded before reaching this stage.

As an example of the systems engineering process, consider the study of robotic transplanting of bedding plants by Kutz et al. (1987).

2 Simulation of the Transplanting Task

A geometrical model of the transplanting task was first constructed on a CAD system to determine the feasibility of using Puma 560[4] robot (Unimation, Inc., Danbury, CT) for transplanting, and also to determine where to place the flats in relation to the position of the robot in the work cell (Kutz et al. 1987). The Puma 560 robot is a six degree-of-freedom manipulator (but only five were used for transplanting), and is powered by electric DC servomotor drives. The operating envelope of the robot includes a sphere of radius 0.92 m, and the maximum payload capacity of the end-effector is 58 Newtons. The robot is designed to have a position repeatability of 0.1 mm. This robot was chosen as the basis for these studies because its CAD model already existed and one was available for the laboratory testing with the live seedlings. The objectives of these tests were to determine the motions required to effectively transplant seedlings.

---

[4]Reference to a company or trade name does not imply approval or recommendation of a product by the authors nor their universities, to the exclusion of others.

The CAD facility was a commercially available system marketed by Computervision Co. (Bedford, MA). Three-dimensional wire-frame models of objects were constructed using Computervision's CADDS 4X Revision 3.00-A software. Several components within the robotic transplanting work cell were modeled, including a 392-cell plug flat, a 36-cell growing flat, seedlings and soil plugs, a belt conveyor, table, and a specialized end-effector to grasp the soil plug attached to the seedling. Models of the flats were based on dimensions of the actual plug flat and growing flat used in the validation tests. An example of a complete work cell model is shown in Fig. 1.

Four separate work cell configurations were modeled on the CAD system to evaluate alternative arrangements of the robot and the flats (Fig. 2). The arrangement with flats on opposite sides of the robot was considered because it could be easiest to implement with existing parallel conveyors. The side by side comparison minimizes the distance the manipulator has to travel during a transplant cycle. The arrangements with the sides and ends of the flats tangential to the arc made by the operating circle of the robot were studied to determine how sensitive the cycle times were to which actuator was predominantly used to perform the motions.

Because of limitations which existed in the CAD software when this project began, two separate simulation studies were devised. The first involved simulating the time to individually transplant the first 36 seedlings in the 392-cell plug flat to the 36-cell growing flat. The pattern followed was identical for each of the work cell configurations studied. Seedlings were transplanted row by row from the plug flat to the growing flat, beginning with the row closest to the growing flat. This transplanting order enabled the robot to transplant the 36 seedlings closest to the growing flat, and thus provided an optimistic estimate of the average time to fill one growing flat.

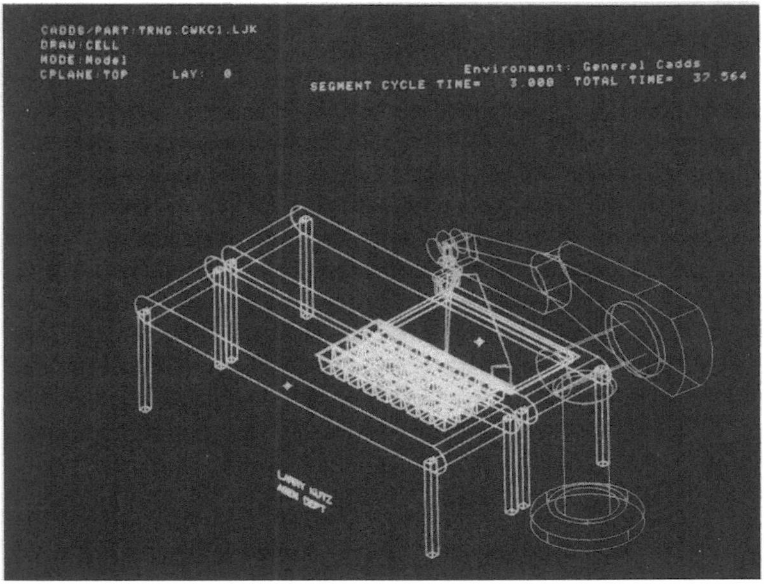

Fig. 1. Typical transplanting work cell model showing Puma 560 robot, plug flat, growing flat, table and belt conveyor. (Kutz et al. 1987)

Therefore, to obtain an estimate of the range of robot cycle times to transplant an individual seedling, a second simulation was devised. In this study, robot cycle times to transplant one seedling were recorded for a maximum and a minimum distance case relative to the locations of the flats in each work cell configuration.

Computervision's robot simulation and off-line programming package called *Robographix* was used to model the Puma 560 and evaluate the four work cell configurations described in Fig. 2. All simulations performed on the CAD system were programmed in the VAL II robot control language using a tool tip speed of 254 mm/s. A simple, straight-down approach path was hypothesized to grasp and release the seedlings.

The VAL II control programs generated by the robot simulation software were downloaded to the actual Puma 560 controller (Fig. 3). The Puma 560 was then used to validate the simulated cycle times and to verify that the robot was capable of reaching each cell in the four work cell arrangements.

3 Simulation Results

The results of the simulation and validation tests are shown in Table 1 and Fig. 4. In general, the validated cycle times agree to within ±10% of the simulated times. The

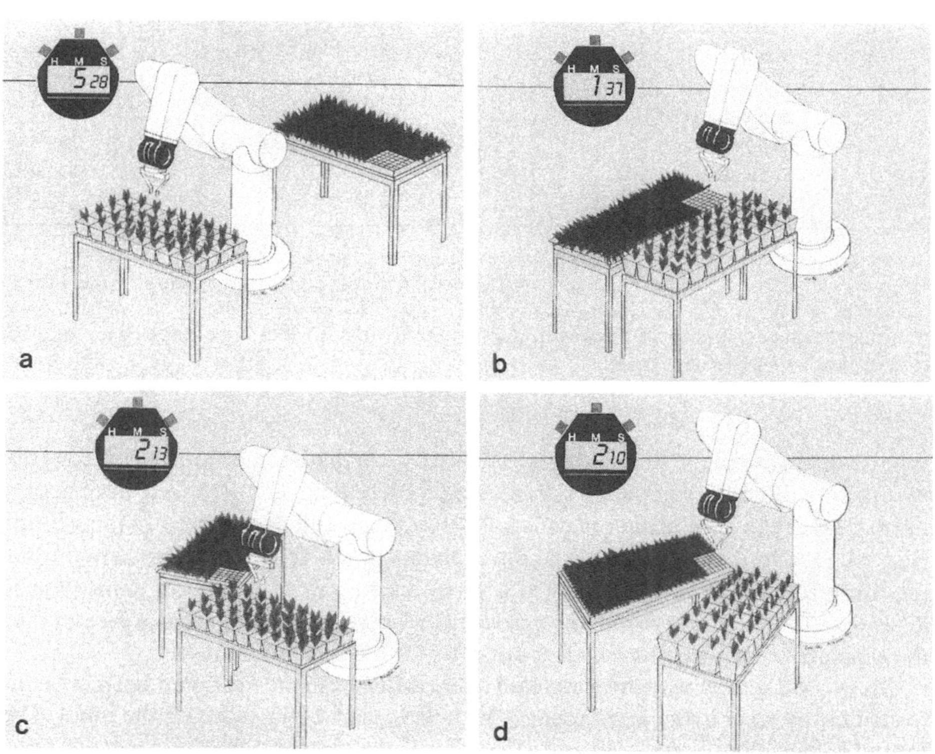

Fig. 2a-d. Artist's illustration of four work cell configurations modeled. Simulated times are shown on the stopwatch, **a** work cell 1, **b** work cell 2, **c** work cell 3, **d** work cell 4

```
SPEED 254.00
RIGHTY
ABOVE
NOFLIP
BASE 0.00, 0.00, 662.00, -0.00
TOOL WKC4000001
FLIP
MOVE WKC4000002
APPRO WKC4000003, 76.20
MOVES WKC4000003
CLOSEI
DEPARTS 76.20
```

Fig. 3. Segment of VAL II control codes for the Puma 560 produced automatically by the off-line programming tool *Robographix*

Table 1. Simulation and validation data summary (Kutz et al. 1987)

Work cell configuration	Procedure[a]	Time to transplant 36 seedlings (s)	Single seedling transplanting study: Minimum distance case: Time (s)	Path length[b] (mm)	Calc. time[c] (s)	Maximum distance case: Time (s)	Path length[b] (mm)	Calc. time[c] (s)
1	Simulation	326.1	7.4	1717	266.4	12.1	2552	435.6
	Validation	303.8	7.1	1621	255.6	11.8	2445	424.8
2	Simulation	96.6	2.8	145	99.4	5.6	1382	201.6
	Validation	98.8	2.4	117	86.4	5.3	1388	190.8
3	Simulation	131.5	2.0	202	72.0	6.7	2257	241.2
	Validation	126.3	2.9	166	104.4	6.9	2207	248.4
4	Simulation	127.2	2.8	201	100.8	4.4	1785	158.4
	Validation	123.1	2.6	175	93.6	4.8	1735	172.8

[a]Simulation data were obtained using the *Robographix* package on the CAD system.
[b]Distances are the computed, point to point distances from the plug cell to the growing cell and back to the next plug cell.
[c]Calculated time to transplant 36 seedlings was computed by multiplying the time observed for the robot to transplant a single seedling by 36.

differences between the simulated and validated cycle times can be attributed to at least two factors. First, in computing the simulated cycle time, the *Robographix* package assumed that the joint actuators on a robot accelerate and decelerate instantaneously (Fig. 5). Due to inertia and friction, the individual joint actuators on an actual robot require a finite time to accelerate to a given speed and a slightly different time to decelerate to a stop. Therefore, the validated cycle time should be slightly greater than the simulated time calculated by *Robographix* for the same robot path.

A second reason why the simulated times differ from the validated times was due to calibration error in the servomotors which drive each of the joints on the robot. This caused errors in positioning the end-effector in three-dimensional (3-D) space. Consequently, actual locations of the flats were not identical to those used in the simulations.

However, the results did show that the side by side arrangement was most desirable in terms of minimizing the time required by the robot to transplant 36 seedlings. The results also determined the magnitude by which flat location affects cycle time.

More importantly, simulation showed where to place the flats so that the robot was capable of reaching every cell in both flats in each work cell configuration. In the side by side arrangement (work cell 2), the robot was required to reach points in the horizontal plane enclosed by a square 500 mm on a side. This plane was positioned approximately 400 mm above the base of the robot for each of the work cell configurations. Vertical motion of the end-effector during the transplanting procedure was confined to a box of height 80 mm above this horizontal plane.

Additional analysis of Table 1 showed that for the single seedling simulation study, work cell 4 had the least variation in the robot cycle times. Although the linear distance between the plug cell and the growing cell for the maximum distance case in the single seedling simulation study was greater for work cell 4 compared to work cell 2, the robot was able to transplant the seedling in less time in work cell 4. Similarly, the cycle time for the minimum distance case for work cell 4 was about equal to the respective time for work cell 2, even though the distance covered in work cell 4 was greater than that of work cell 2.

This was due to the fact that work cells 3 and 4 better utilized the revolute-motion capabilities of the robot. Revolute motion is natural to any robot which utilizes rotary actuators at each of its joints, as does the Puma 560. Linear motion, on the other hand, is natural to robots that utilize linear or slide actuators at the joints, as on gantry-mounted robots (Paul 1981). Since work cells 3 and 4 positioned the flats tangential to the operating circle of the Puma 560, these work cells involved more revolute motions than did work cell 2. Thus, a gantry-mounted robots such as the IBM RS-1 (IBM Corp., Boca Raton, FL) might be better suited to the motions required in work cell 2.

Table 1 also lists the calculated time to transplant 36 seedlings based on the recorded minimum and maximum cycle times for the single seedling study. Note that the cycle times obtained in the first simulation study for work cells 1, 3, and 4 fell between the calculated times based on the minimum and maximum distance cases, and, in fact, lie closer to the calculated time for the minimum distance case. However, the simulated time to transplant 36 seedlings in work cell 2 fell below the calculated cycle time based on the minimum distance case.

Additional analyses showed the simulated time for the robot to move vertically from the approach position of a seedling to the top of the plug cell varied depending on the location of the plug cell. In work cell 2, plug cells closest to the robot required 0.58 s for this motion, while plug cells at the far end of the flat required only 0.30 s. These motions were performed using a robot tool tip speed of 254 mm/s. As a result, the simulated time to transplant the first seedling located at the far end of the plug flat in the first simulation study was 2.08 s, which was less than the time recorded for the minimum distance case in the single seedling simulation study.

This anomaly can be explained as follows. Vertical motion of the end-effector when the robot arm is approaching objects close to the robot base is almost solely due to joint 2 of the Puma robot, or the shoulder joint. For cell locations farther away from the robot base, other joints on the robot are used proportionally more. Since each joint actuator has a different maximum velocity specification, cell locations which require different joint actuators to move the end-effector from the approach position of a seedling to the top of the plug cell will have different times. The maximum velocity for

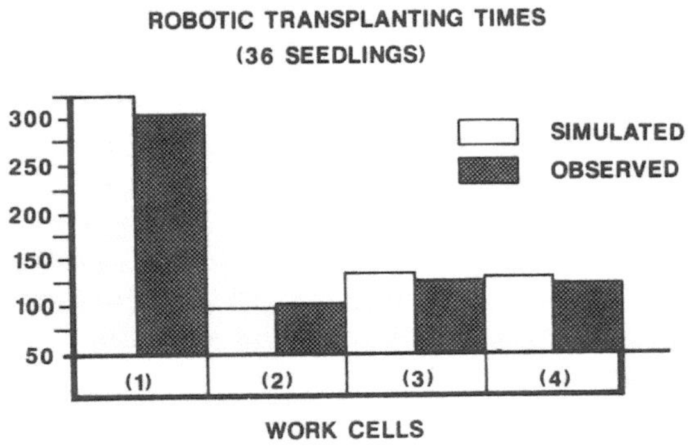

Fig. 4. Bar chart of simulated and observed robot cycle times. (Kutz et al. 1987)

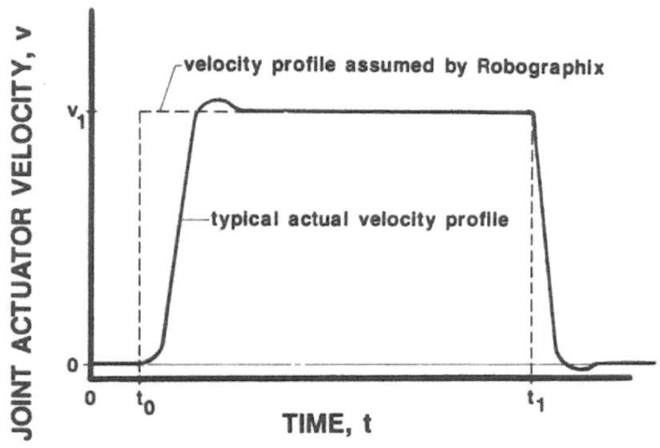

Fig. 5. Actuator speeds differ from the profile assumed by *Robographix*. (Kutz 1985)

joint 2 on the Puma robot is smaller than any of the other joints, and therefore, robot motions which utilize this joint for the majority of the resulting motion of the end-effector will be relatively slower.

The general conclusion which can be drawn from this anomaly is that end-effector or gripper speed is dependent upon where the tip is in the work-envelope. Any model which does not account for this non-linearity is not able to accurately predict the dynamics of motions, or the throughput performance of a robot. Hence animated, CAD simulations are a necessary step toward the successful development of an optimum robotic materials handling system.

4 Transplanting Validation Tests

Validation tests using live seedlings were also conducted using the Puma 560 robot. These tests were performed: (1) to evaluate the performance of the seedling gripper, and (2) to evaluate the ability of the robot to transplant live seedlings using the robot control program generated on the CAD system.

Tomato, *Lycopersicon lycopersicum*, and marigold, *Tagetes erecta*, seedlings 30 to 50 mm tall were used in the transplanting tests (Kutz et al. 1986). The tomato seedlings were flimsy, with a proliferation of leaves at the top of the stem that tended to intertwine with adjacent seedlings. The marigold seedlings, however, had a stiff stem with small, fern-like leaves at the top of the stem that pointed upward. Therefore, the marigold seedlings were better adapted to robotic transplanting.

4.1 Gripper Design

The gripper design utilized major components of a Unimation model 510 pneumatic, parallel-jaw gripper supplied with the robot. Since this gripper was air-actuated, the jaws could only be positioned in the fully open position or the fully closed position. The difference in the distance between the jaws in the open and closed positions was about 20 mm. However, in grasping and releasing a soil plug, the gripper only required about 5 to 10 mm of jaw motion. Therefore, a means of reducing the jaw motion between the open and closed positions was required.

First, the original jaws were cut to a final length of 37 mm, and a jaw adjustment screw was mounted on one of the jaws to provide a means for adjusting the distance between the jaws in the closed position. Secondly, two lightweight fingers constructed from 1.02 mm thick (20 gauge) galvanized sheet metal were mounted just inside each jaw. The beam spring constant of each finger was about 1.3 N/mm. When gripping a soil plug, each finger deflected about 3 mm, resulting in a calculated applied force to the soil plug of about 4 N. This force was found to be adequate to reliably grip the soil plug without damaging the seedling. Figure 6 shows the seedling gripper mounted on the robot.

Fig. 6. Seedling gripper mounted on Puma 560 robot. (Kutz et al. 1987)

4.2 Transplanting Procedure

Flats were located on a table surrounding the robot by entering the 3-D coordinates of the center of each flat, programming the robot to move to the calculated locations of three corners of each flat, and manually moving the flats to the proper locations. Once the flats had been accurately located, the robot was used to dibble the growing flat. A small, wedge-shaped Plexiglas block was grasped by the robot gripper, and the robot was programmed to move each cell in the growing flat. The robot pushed the dibbling block 20 cm into the soil in each cell, thus forming a hole that would later accommodate the transplanted seedling. The robot dibbled a 36-cell flat in 30 s. Dibbling provides a firm foundation for the seedling so that it remains erect after transplanting.

The seedlings were then transplanted by the robot in the same manner as was simulated on the CAD system. Figure 7 shows the Puma 560 with the plug and growing flats before a transplanting test.

4.3 Discussion of Transplanting Tests

Initial transplanting tests with the robot showed that a simple, straight-down approach path to the seedlings in the plug flat was unsatisfactory. As the gripper moved down toward the base of a seedling, the leaves would snag on the tips of the gripper, causing the seedling to bend or buckle. To resolve this problem, the control program was modified so that the robot approached the seedlings using an L-shaped path. The approach point for each seedling was shifted horizontally away from the seedling by one plug cell width, or 18 mm. The robot gripper moved straight down from this approach

Fig. 7. Puma 560 robot and plug and growing flats before a transplanting test. (Kutz et al. 1987)

point until the tips of the fingers of the gripper were at the same height as the top of the plug flat. Then the gripper moved horizontally to the plug cell, and 10 mm down into the cell. The departure path of the gripper from the plug flat was likewise modified to resemble an L-shaped path rather than a simple, straight-line path. In addition, the robot tool tip speed was slowed to 50 mm/s during the approach and departure path segments.

These modifications to the approach and departure path segments greatly increased the success rate of transplanting seedlings. The slower tool tip speed was effective in allowing adjacent seedlings to part as the gripper moved horizontally to a seedling. By moving down into the plug cell, the soil plug attached to the seedling was grasped rather than the stem. As a result of the control program revisions, the cycle time to transplant 36 seedlings in work cell 2 increased to 3.3 min. Using this modified procedure, the average success rate when transplanting tomato and marigold seedlings was 96%.

Initially, a few seedlings were missed by the robot gripper due to low soil levels in some plug cells. In such cases, the gripper sometimes grasped the stem of the seedling and removed the seedling and soil plug successfully, or missed the seedling entirely. Increasing the distance by which the gripper moved down into all plug cells solved this problem but caused other problems, such as compression of the soil plug, which made the release of the soil plug in the growing cell more difficult.

After transplanting, the seedlings were watered and placed in a greenhouse. The water tended to heal the soil around the seedlings and firm the soil naturally. All of the transplanted seedlings continued to grow vigorously in the greenhouse. In all, some 500 seedlings were transplanted by the robot.

5 Machine Vision and Image Processing

Unlike many industrial applications of robots, the variability of agricultural systems requires sensors, means of evaluating and comparing robot performance for different configurations, and means of rapidly reprogramming the machine to perform alternative tasks. Figure 8 illustrates why sensing is necessary for automated transplanting. Practically every plant is a different size and shape. Not all cells in the germination flat have plants in them. The robot must be equipped with sensors to make an intelligent decision about which plants to transplant and which to discard. By measuring features such as leaf area, stem length, and stem diameter which relate to seedling quality, it would be possible to sort the plants into various grades for marketing purposes.

Machine vision and image processing (MVIP) provide the means of sensing many of the desirable features, and the reprogrammability of such systems provides the flexibility to adapt when the robot is used for different tasks. In this text, machine vision refers to the sensor which converts the visual information into an electrical signal (i.e., the camera, lenses, filters, frame grabber, etc.). Image processing refers to the software necessary to convert the data from the camera into useful, features information, such as leaf area, stem diameter, etc. Image processing includes algorithms to enhance the image and emphasize certain features, as well as pattern recognition routines which search the data for combinations which can be grouped into objects such as leaves or stems (Whittaker et al. 1984; Cardenas-Weber et al. 1988). Figure 9 illustrates the relationships between the hardware and software for machine vision and image processing. The following paragraphs describe MVIP sensing of plant features and how these measurements can be used in robotic transplanting.

Sometimes seedlings are grown in pots and the robots must be able to accurately locate the position of each plant, even when the leaves hide portions of the container. In a similar situation, Whittaker et al. (1987), described how machine vision and image processing were used to locate tomatoes when only sections of the fruit were visible in the black and white images (Fig. 10). The centers were computed by applying a circular Hough transform to an image. False centers were caused by round edges of leaves and other objects in the scene having radii of curvature approximately equal to the tomatoes. Whittaker found an inverse relationship between the number of false indications and the number of undetected tomatoes. When approximately 20% of the centers were false (not tomatoes), the number of undetected tomatoes in an image was virtually zero.

Guyer et al. (1986), applied machine vision to plant identification (Fig. 11). In very flexible situations, this ability could be used to recognize that production had shifted to a new species, and that possibly the robot should modify its transplanting task. The routines used to identify plants would also recognize abnormalities such as multiple plants in a cell. Fundamental features such as leaf area which were used to identify species also could be used to grade plants and direct the robot to place them into different containers.

Machine vision and image processing have been used to measure features of bare-root seedlings. Cardenas-Weber et al. (1988), discussed two methods of counting roots on strawberry, *Fragaria xananassa*, plants (Fig. 12). Crown diameter (measured in Fig. 12b)

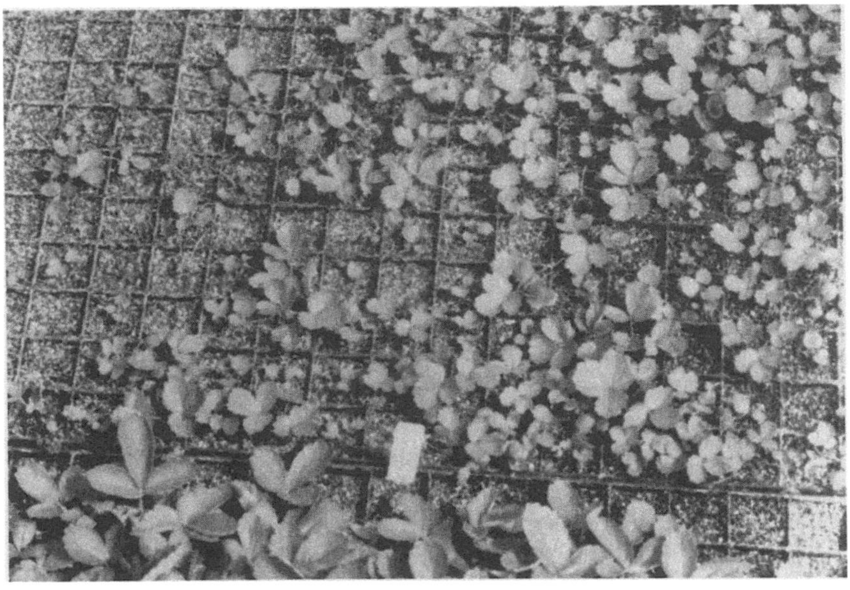

Fig. 8. Machine vision and image processing are needed to locate empty plug cells and to sort plants based on size and conformation

is also a criteria used for grading strawberry plants. Suh and Miles (1988) developed means of measuring stem diameter, shoot height, root volume, and seedling weight for Southern yellow pine, *Pinus palustris*, seedlings (Fig. 13). Figure 14 shows that MVIP measurements of projected area are well correlated to root volume. The accuracy of direct measurements and computed features, such as the ratio of shoot area to root area (sturdiness ratio), indicates that MVIP grading has significant potential for improving the quality of seedlings.

Image processing routines have been written to count evergreen shrubs grown in containers (Fig. 15). Even when foliage extends beyond the container, and the containers are placed "can-tight", the computer vision system is able to correctly count the plants in a scene.

6 Summary and Conclusion

The engineering of intelligent machines (i.e., flexible, reprogrammable machines that sense changes in conditions and respond appropriately) requires a systems engineering approach. In this approach computer-aided engineering is used to model the gripper and work cell, animated simulations are performed to determine the technical feasibility and production capacity of the robotic machine, laboratory studies validate the effectiveness of the proposed gripper motions, and simulations of the materials flow determine an optimum number and size of robot(s) to employ. This approach has been successful in using a robot to transplant bedding plants.

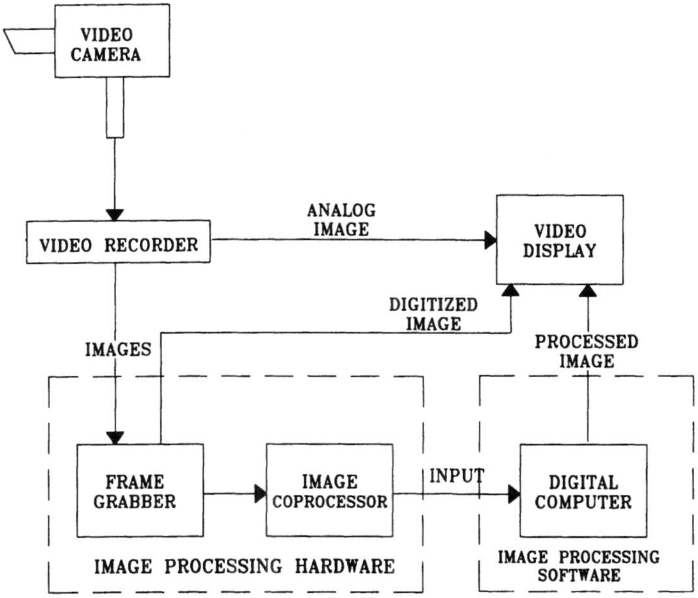

Fig. 9. Block diagram for machine vision hardware and image processing software. (Cardenas-Weber et al. 1988)

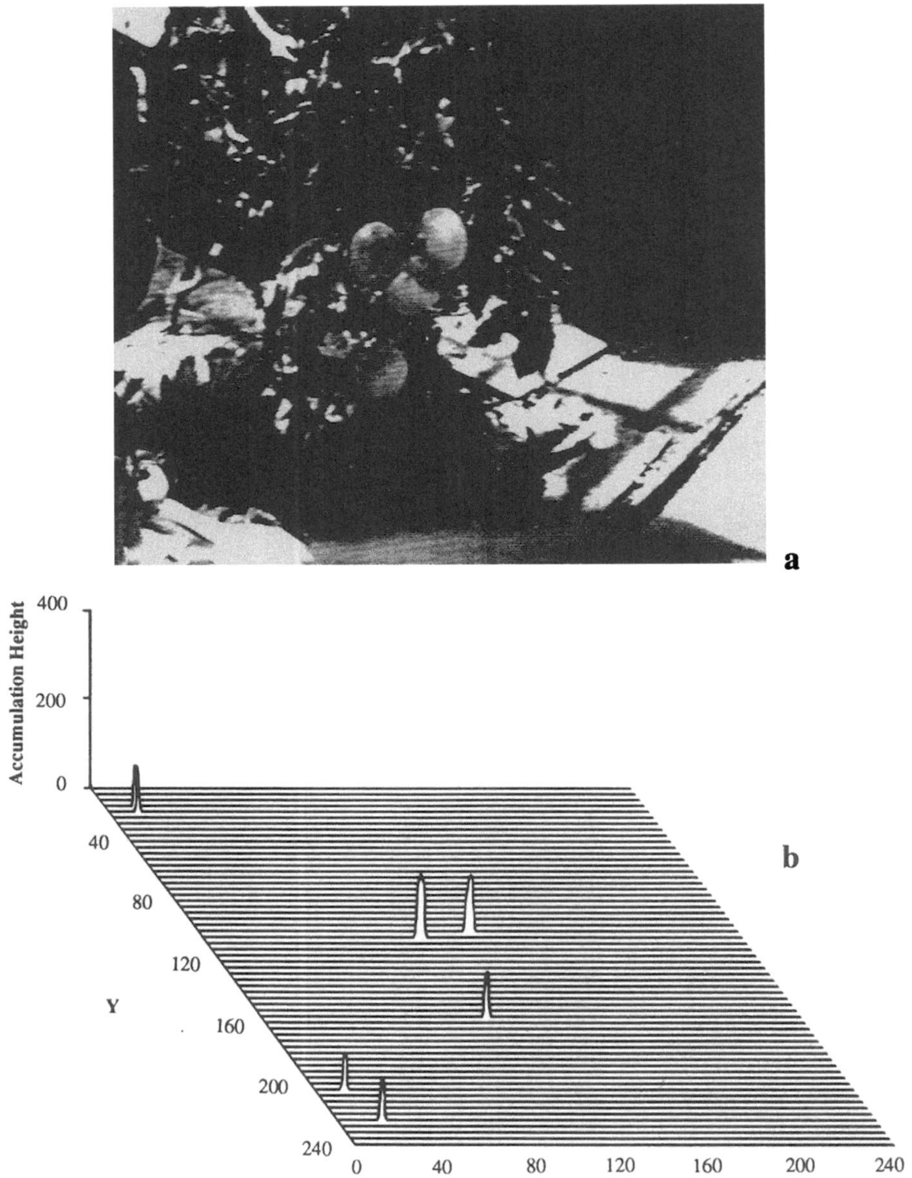

Fig. 10. **a** Grey-level image of tomatoes grown in a greenhouse. **b** Peaks computed by the circular Hough transform are probable locations of tomatoes. (Whittaker et al. 1987)

Machine vision and image processing can sense plant features necessary to automate the transplanting task. The technical feasibility of identifying plant species, locating and counting plants grown in containers, measuring leaf area, stem diameter, shoot length, root volume, and counting the number of roots has been proven.

Commercial implementation of intelligent, automated transplanters based on robotics is the next step.

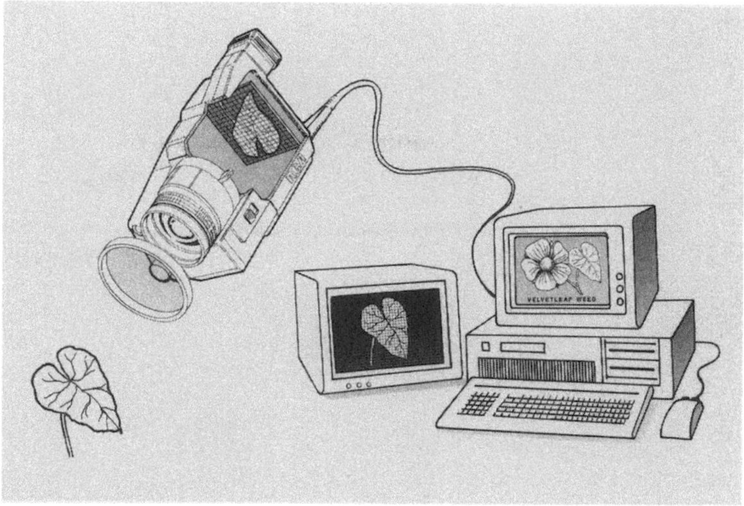

Fig. 11. Leaf area, perimeter, and thickness combine to form features such as complexity and elongation, which are used to identify plant species

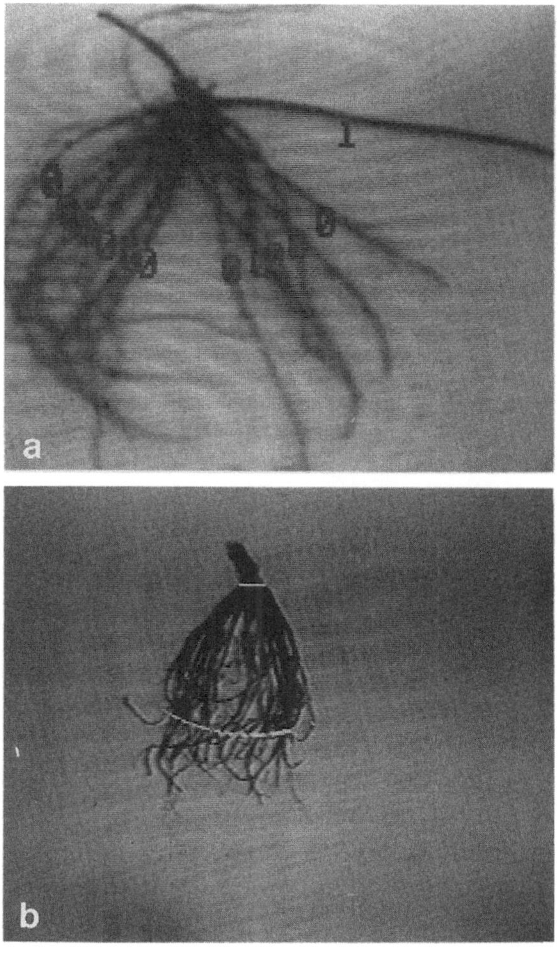

Fig. 12. **a** Primary roots of a strawberry plant are counted by detecting the gray-level changes along an arc about the crown. The software places a "0" on each root and a "1" on the branch to the mother plant. **b** Back lighting was used to capture this image of a bare-root strawberry plant. Image processing located the crown, measured the diameter and counted the roots along the arc shown. (Cardenas-Weber et al. 1988)

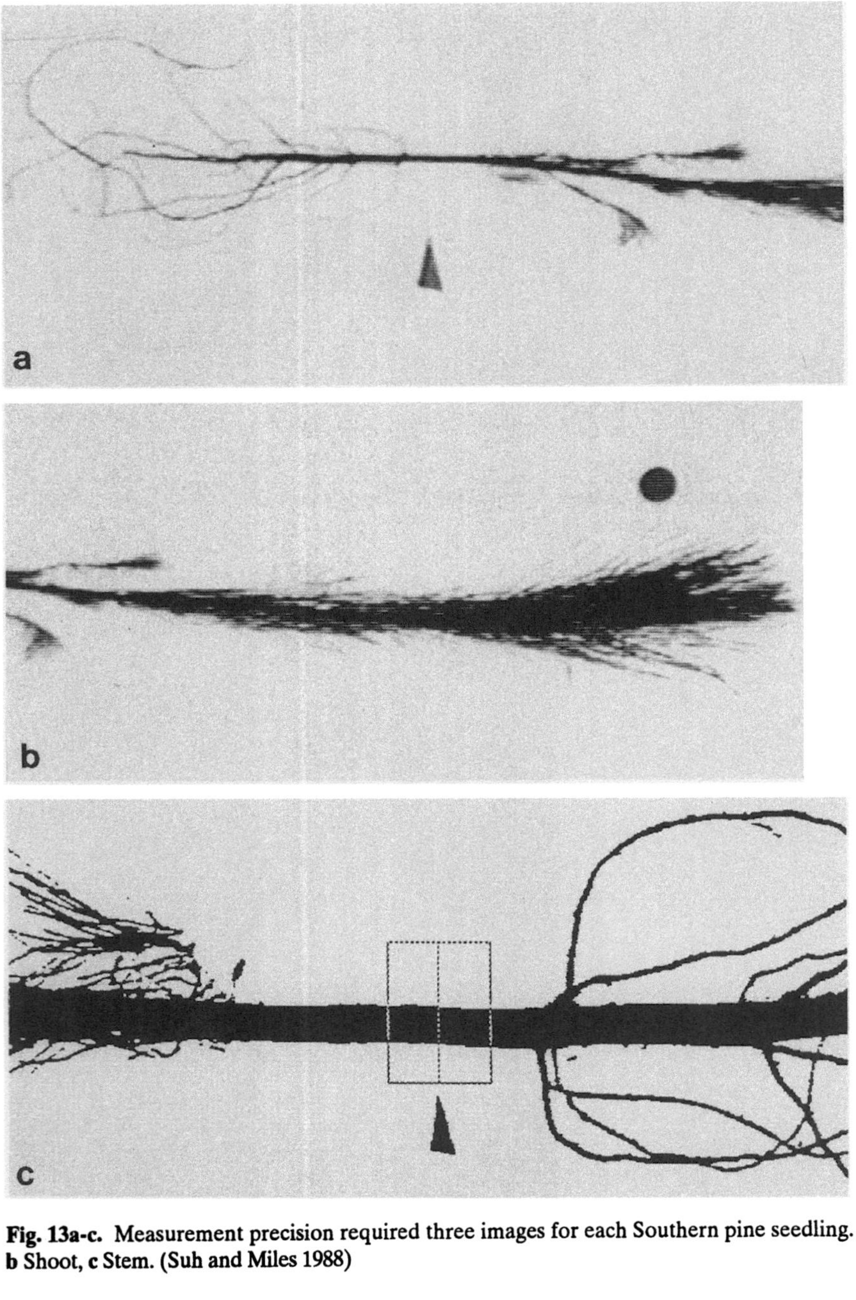

Fig. 13a-c. Measurement precision required three images for each Southern pine seedling. **a** Roots, **b** Shoot, **c** Stem. (Suh and Miles 1988)

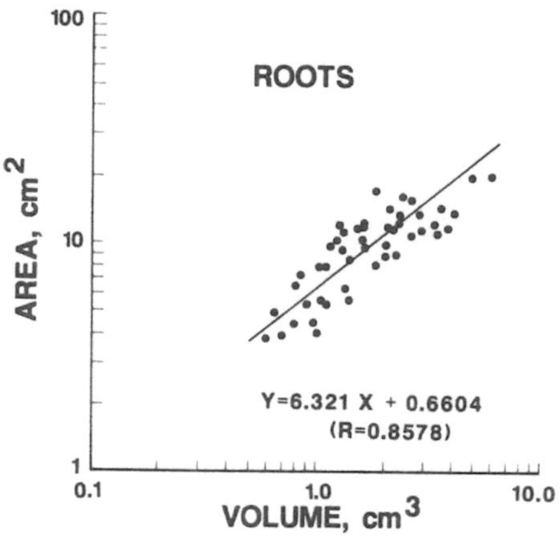

Fig. 14. Machine vision observations of projected root area of Southern pine seedlings are well correlated to water displacement measurements of root volume. (Suh and Miles 1988)

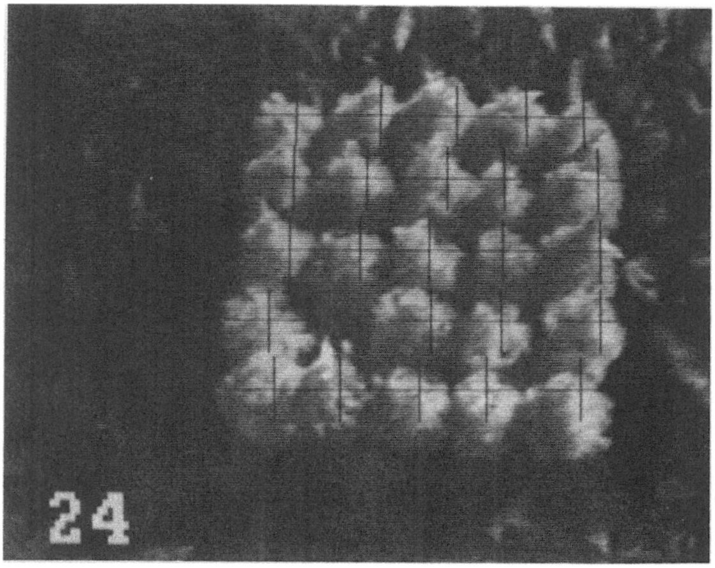

Fig. 15. Evergreen shrubs grown in containers are counted by identifying the peaks in the gray-level histograms for rows and columns. Image processing routines were written by AGEN 690 students: Doug Graham, Seong In Cho, Willie Hart and Hanhua Liang. (Cardenas-Weber et al. 1988)

References

Cardenas-Weber M, Lee FF, Guyer DE, Miles GE (1988) Plant features measurements with machine vision and image processing. Paper 88–1541, ASAE, St Joseph, MI, p 12

Guyer DE, Miles GE, Schreiber MM, Mitchell OR, Vanderbilt VC (1986) Computer vision and image processing for plant identification. Trans ASAE 29(6):1500–1507

Kutz LJ (1985) Robotic transplanting of bedding plants: simulation and validation. PhD Thesis, Purdue University, West Lafayette, IN, 171 pp

Kutz LJ, Miles GE (1986) CAD simulation speeds robotic workcell development. Agric Eng 67(3):20–22

Kutz LJ, Miles GE, Krutz GW, Hammer PA (1986) Robots in the greenhouse. Greenhouse Grower 4(9):14,16,17

Kutz LJ, Miles GE, Hammer PA, Krutz GW (1987) Robotic transplanting of bedding plants. Trans ASAE 30(3):586–590

Paul RP (1981) Robot manipulators. MIT Press. Cambridge, MA, 279 pp

Suh S, Miles GE (1988) Measurement of morphological properties of pine seedlings using machine vision. Paper 88–1542, ASAE, St Joseph, MI, p 18

Whittaker AD, Miles GE, Guyer DE (1984) Introduction of digital image processing. Paper 84–5510, ASAE, St Joseph, MI, p 10

Whittaker AD, Miles OR, Mitchell OR, Gaultney LD (1987) Fruit location in a partially occluded image. Trans ASAE 30(3):591–596

IV.4 Computing and Information Management in Micropropagation

M. HEMPEL[1]

1 Introduction

The word: "computing", which is traditionally understood as calculation, has been broadened to include "... automatic data processing, batch processing, data capture, data processing, data retrieval, electronic data processing, number crunching (colloq.), system analysis, system engineering." (Anonymous 1986). All these activities are connected with the collecting, processing, and storing of information concerning systems.

Analysis, modeling, and the simulation of systems are techniques which have been used in agricultural economics for the last 20–30 years. The aim of these activities is to improve the understanding and optimization of the processes conducted and controlled by human beings and to improve the knowledge of the natural phenomena. The scope of this chapter reaches beyond modeling and simulation. The aim of the author is to discuss the problems of information management in a micropropagation firm. In many places in this chapter, the author expresses his opinion on production matters without citation of the source of this information, which at times could be controversial. The author's knowledge of this subject has been accumulating for 20 years through his involvement in the operation of many micropropagation firms in the positions of research, extension, or as a production worker.

The brief explanation of some terms and graphic marks which may be unfamiliar to tissue culture specialists is necessary before beginning the discussion of micropropagation problems.

2 Basic Concepts

From the point of view of the natural sciences, a system is most often considered as a set of interrelated elements integrated functionally, with the assumption that the function of the whole system cannot be reduced to the functions of its elements (Gutenbaum 1987). The individualization of these elements from within the systems must be done according to the same rules as the individualization of the system from an environment: the functions of the particular elements must be different. A model is a simplified representation of a system (de Wit 1982) where a flow of energy, matter, or information (rate processes) under the influence of driving or auxiliary variables is described (Penning de Vries 1982). Diagrams depicting interrelations between system

[1]Plant World Explorations, Bowral, Australia

Biotechnology in Agriculture and Forestry, Vol. 17
High-Tech and Micropropagation I (ed. by Y.P.S. Bajaj)

elements are the most frequently used forms of model presentation. They facilitate the understanding of a model structure and interrelationships and are usually composed of:

- state variables which are quantities of matter, energy, or information, e.g. plants, shoots, vegetative buds, biomass, production plans, etc.; in biological systems, they change under the influence of auxiliary and/or driving variables;
- driving, auxiliary, and intermediate variables (van Dyne and Abramsky 1975; de Wit 1982) which are the factors influencing state variables; in the case of micropropagation, temperature, light, media and substrate components, management, etc.
- rate processes during which the changes of system state variables occur; they can be described in physiological, physical, biochemical, economical terms as growth, branching, rooting, sales, employment, etc.
- parameters or constants characterizing the intensity of rate processes, e.g., coefficients of multiplication, rate of growth, etc.

The graphic forms listed below will be used to represent model elements (Ferrari 1982):

 — state variables;

 — flow and direction of state variables changes;

 — direction of information flow;

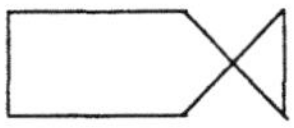 — rate processes (a control of state variables flow); information lines usually connect these symbols with symbols representing state, auxiliary, or driving variables, indicating their influence on a particular process;

 — auxiliary variables which usually mediate flow of information;

 — driving variables (van Dyne and Abramsky 1975);

 — source or sink of matter, energy, or information beyond system boundaries.

To decide which system elements will be treated as state and which as auxiliary or intermediate variables depends only on the objectives and phases of the modeling process. Driving variables are the factors influencing system elements from behind their boundaries. In field production systems, temperature, solar radiation, and precipitation can be treated as driving variables (de Wit 1982). In the case of micropropagation, where most of the physical, chemical, and biological factors are controlled within the system, the trends and situation on a market for materials, equipment, labor, and propagation material can be discussed as the driving variables.

Each system can be discussed on different abstraction levels; therefore, many interdependent models can be built to represent a flow of matter and information within a system (van Dyne and Abramsky 1975). In each production system the only two elements that can be distinguished during modeling on the highest abstraction level are management and direct production environment. Descending in abstraction levels, direct production environment can be broken down into such elements as plant material, labor, equipment, environmental factors, etc.

The abstraction level, i.e., the precision with which a model represents the elements and interdependence within the system depends on the goal of modeling. The model which is aimed to be used as a tool for the management of production processes must be as precise as possible in depicting the relationship between production elements.

To test a system's behavior by analyzing its model operation is called simulation. Simulation enables the correction of assumptions made in the model before their application in practice as management or educational tools.

The system analysis, modeling, and simulation are techniques especially suited for application in intensive, production domains where most of the driving variables are controlled by man. Micropropagation, being the most intensive technology of agriculture, requires very precise advanced management methods based on a sound understanding of the relationship between internal elements of technology as well as the existing technology and the market. Its auxiliary variables, i.e., temperature, light, substrate composition, etc., can be precisely regulated in order to achieve production goals.

In an attempt to bring further discussion closer to tissue culture specialists, such words as system, model, state variables, rate processes, and driving variables will be used interchangeably with their counterparts in the micropropagation expert language such as micropropagation firm, plan, schedule, explants of different kind, growth and organogenetic processes, and factors influencing these processes.

3 The Micropropagation Firm as a System

As mentioned above, the boundaries of the system must be defined before an attempt is made to build its model. Discussion presented in this chapter will be limited to the characteristics of the micropropagation firm operation with such elements as laboratory and ex vitro production of young plants, with their sales, and materials, equipment, and labor supply included (Fig. 1). Modeling activities will be focused on describing organogenetic and growth processes of plant material. Sales and supply of materials, equipment, and labor will be treated with less attention. Market, with customers and suppliers as its elements, will be treated as the environment of the system; but its situation and trends will be considered as the system's driving variables.

The model presented in Fig. 1 depicts the flow of matter and energy (solid lines) with the source in the market for materials, equipment, and labor and the sink in the market for micropropagation products. On this general abstraction level the three state variables are: plant stock, materials, equipment and labor, products and management. The flow is forced by such rate variables (processes) as supply, propagation, and sales.

The orders are distinguished as the auxiliary variables because they mediate the influence of market demand on the rate processes of the system. The role of the management is to gather all this information flowing from the system elements and after its processing to control the processes being the parts of the system (Fig. 1). In practice, the role of management is much broader and includes formation of the general policy

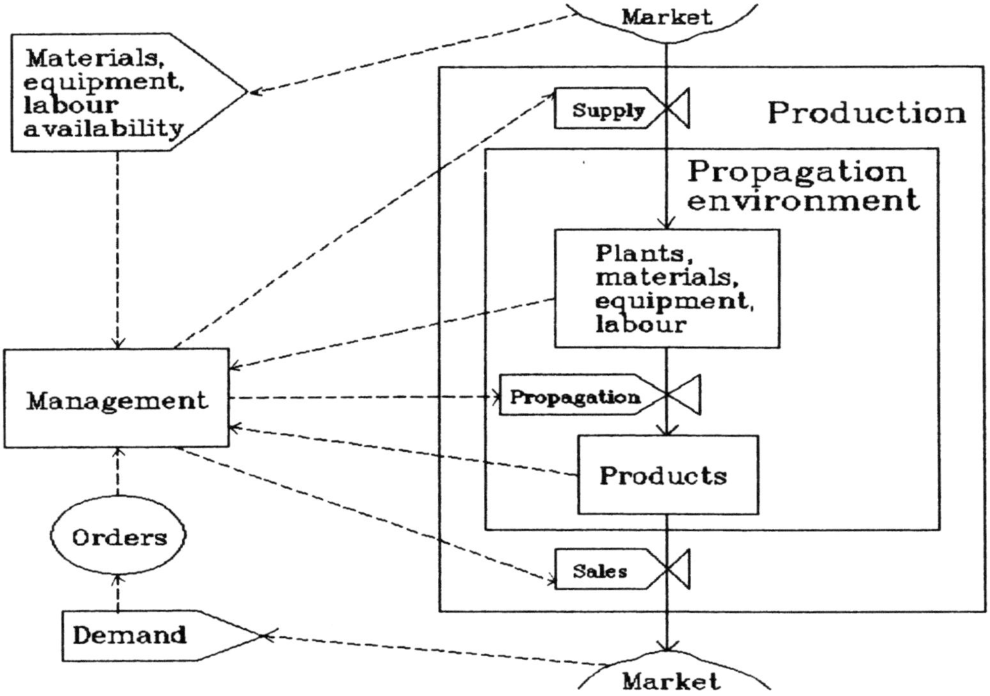

Fig. 1. Graphic representation of a micropropagation firm as a system

of a firm, employee training, market influencing by advertising, etc., but these functions will not be discussed in this text.

Information flowing into the management serves as the foundation for the analysis of the firm (system) momentary situation. This analysis is usually followed by construction of a production schedule, verification, and sometimes optimization of this schedule and its application to control the rate processes. Through the feedback mechanisms, information about the results of the model application flows back to management. If the results vary from those predicted, the plans are reconstructed. Such constant approximations of constants (coefficients) or functions of rate processes lead to the perfectioning of management skills. The shorter the time intervals in which the feedback mechanism occurs, the faster the functions or coefficients will be refined. The character of micropropagation makes this technology especially suited for the application of modern management techniques because of its repetitive multiplication passages and very short (few weeks) other technology stages.

The occurrence of many short passages has its source in the character of propagation techniques. Since plant tissues are grown in closed, small containers on a fixed amount of medium, they use their limited resources very quickly and must be transferred each 3–12 weeks into new conditions.

The time during which tissues stay in the same conditions is so short that it does not allow for very complex organogenetic or physiological changes. Therefore, during modeling and simulation each of the passages can be treated as a "black box". It facilitates the analysis of the whole system. This situation may change in the near future because new technologies have been proposed in the last few years, that are focused on the more fluent environmental manipulations by using sterile (Tisserat and Vandercook 1985; Farrell 1987) or unsterile (Kozai 1988a) liquid medium systems, and "mistifiers" (Weathers and Giles 1987). This technology will bring micropropagation closer to greenhouse production methods but it has not yet been widely accepted in practice.

Two phases can be distinguished in management operation:

— preparation of plans,
— application of plans in production.

In Fig. 2 the model of information management during planning is presented. This model can be treated as a submodel of the model shown in Fig. 1, which depicts infor-

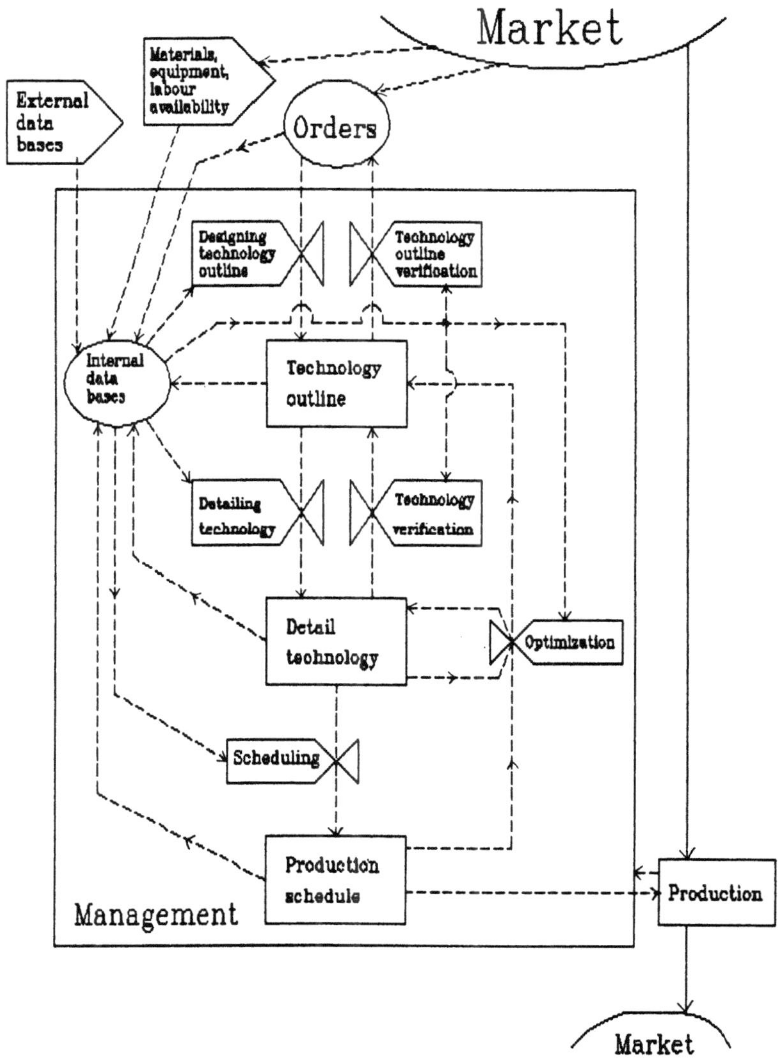

Fig. 2. Graphic representation of managerial activities during the planning phase

mation flow. The elements dispensable at this stage of discussion were simplified or omitted. The plans prepared on different abstraction levels and production are treated here as the state variables, i.e., the subjects of changes under the influence of the internal knowledge base content (manager's memory, traditional and computer files, etc.). In the case of this model, external knowledge bases are distinguished as the driving variable and the market as the environment.

4 The Planning Phase

The order placed by a customer or a need of the management to increase plant stock are stimuli to start the construction of the production plan (model). The specification of an order of a customer or management consists usually of: the type of product, its amount and quality, and the date by which it has to be delivered or produced. Within one genotype several products can be ordered, e.g., young plants grown in multipots, plants rooted in vitro, microcuttings, etc. The technology chosen must allow for quantitative and qualitative changes of the plant stock available into the product desired in the time fixed by the order and delivery dates.

If the ordered product has not been produced in that particular laboratory, the first step undertaken by the management is to search for the technology which will enable obtaining the product from the available stock through a chain of qualitative transformations. In other words, the basic model of technology must be defined.

4.1 Designing Technology Outline

To begin the technology outline, one must first investigate the contents of the internal and/or external data bases in order to determine the technology which will provide the desired organogenetic transformations. In the case of many species which have been propagated in vitro for many years, the management has a choice of at least a few alternative pathways (Fig. 3).

The diagram presented in Fig. 3 contains many variants of a process which in Fig. 1 was described as propagation. Plant tissues of different levels of organization are treated here as the state variables. The auxiliary variables are divided into two sets according to the processes they influence. The first of these sets contains such variables as personnel skills, tools, equipment, etc. These variables affect the rate processes such as division of branched shoots or callus, isolation of initial explants, planting, etc. by direct mechanical intervention. The variables acting upon the rates of organogenesis and growth, i.e., temperature, light, media and substrate components, etc. were numbered in the second set. The influences of explant (plant) quality on their transformations and some less important state variables were not presented, in order to keep the diagram clear.

As shown in Fig. 3, the same product, e.g., young plants grown in a greenhouse, can be obtained by at least three pathways. It does not exhaust all possibilities because the presented diagram is a very general one and does not include the potential of specific genotypes. By the process of initial explant isolation, fragments of leaves can be obtained and then, in order to obtain shoots, not only the growth process must be stimu-

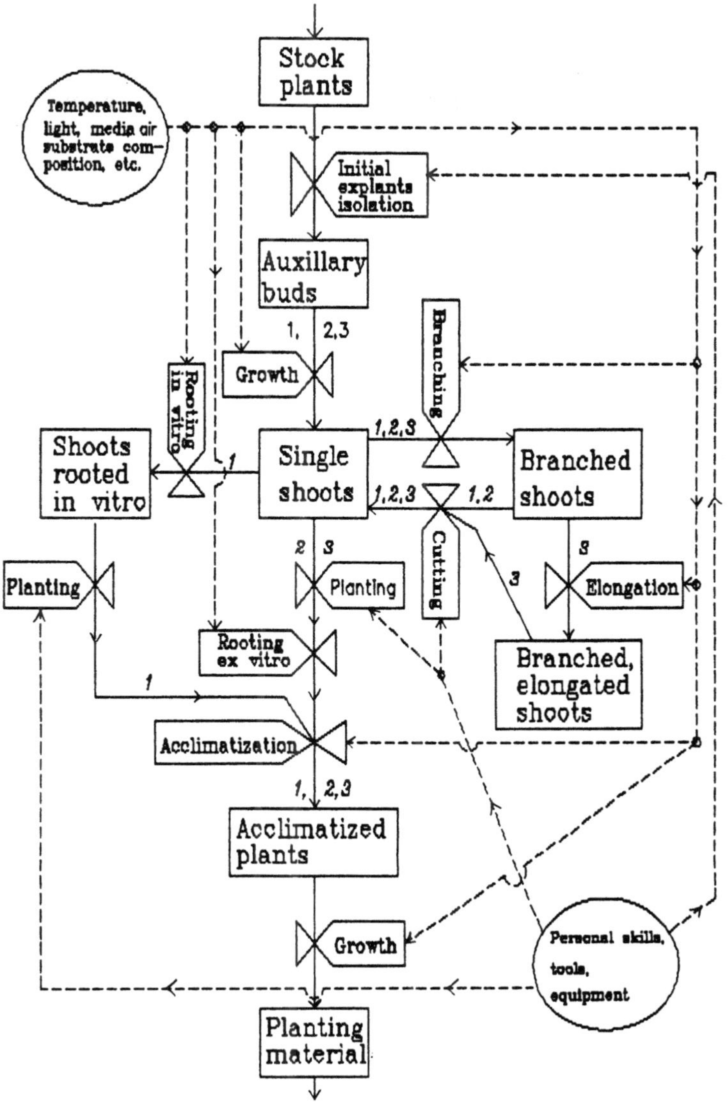

Fig. 3. Schematic representation of three micropropagation variants

lated but also, firstly, shoot primordia formation (Raman 1977; Henny 1978; Peck and Cumming 1984). In the case of some plants, the loop leading to the increase in the amount of plant tissue in a propagation system (single shoots-branched shoots) can be replaced by the loops composed of protocorms-protocorm aggregates (*Cymbidium* – Fonnesbech 1972; Kusumoto and Furukawa 1977) or callus tissue (*Anthurium* – Pierik 1975; Pierik and Steegmans 1976). The transformations shown in Fig. 3 are commonly accepted in production.

Since the time when Murashige (1974) structured a micropropagation as a process composed of three stages, namely initial, multiplication, and pretransplant, many additions to this traditional scheme have been proposed. The rooting of microcuttings directly in a horticultural substrate instead of in vitro was suggested by several authors (Wolfe et al. 1983; Hildebrandt and Harney 1983; Econoumu and Read 1984; Norton and Norton 1988). Maene and Debergh (1985) developed the system in which the shoot elongation process is included with the aim of preparing microcuttings for ex vitro rooting. Special treatments of stock plants before the initiation of cultures seem to be beneficial in the case of some species (Simmonds and Nelson 1988). High temperature treatments are necessary for the appropriate development of some plants (*Lilium* – Stimart et al. 1982).

At this stage of construction of the plan, the basic criterion chosen is the high probability that the desired qualitative changes will occur. The risk of introducing phenotype changes into the plant material during the induction of certain morphogenetic processes (callus formation, shoot regeneration from callus) is especially high in the case of chimera propagation (Hempel 1985).

The need for the construction of a new technology outline can also take place when the results of the technology used are not satisfactory in qualitative (e.g., phenotype changes in a progeny) or quantitative (e.g., to slow rates of organogenetic processes) terms (Fig. 2).

After the general technology has been chosen, the level of variables affecting earlier chosen processes should be defined.

4.2 Detailing a Technology

The goal of this step of production planning is to adjust the auxiliary variables, i.e., temperature, media and substrate composition, and labor skills, to the levels at which the intensity of the processes will enable changes in the stock material into the ordered products. These transformations must be completed within the time which is left until the delivery dates. Besides qualitative transformations, quantitative changes must be taken into account. Therefore, calculations must be made to check if there is enough stock to produce the ordered number of products.

To meet the quantitative demand for products, the intensity of rate processes such as shoot branching or cutting efficiency can be increased by adjustments of levels of auxiliary variables or a multiplication loop can be repeated for a desired number of times.

The most common production practice is to maintain a certain amount of stock by the use of a multiplication loop and to draw from this stock the samples of plant material which are necessary to fulfill the particular orders. The management of this process is easy in the case of plants where all of the products obtained by cutting of explants on smaller pieces are functionally the same, i.e., they can be used both in a multiplication loop or for other processes, e.g., rooting. Examples are *Gerbera* (Murashige et al. 1974) *Rosa* (Hasegawa 1979), and *Spatiphyllum* (Fonnesbech and Fonnesbech 1979), where single shoots obtained after the division of branched shoots can be used to invoke both branching and rooting processes.

However, in the case of such genera as *Cordyline* (Maene and Debergh 1985), *Dieffenbachia* (Taylor and Knauss 1978), *Dracaena* (Chua et al. 1981), and many others, the products of branched shoot cutting can be shoot tips, stem sections containing single or a few nodes, and enlarged, united shoot bases. Development of the proper algorithm describing efficient flow of plant material is much more difficult for such genotypes because the explants mentioned above cannot be used equally for different purposes, e.g., multiplication and rooting.

Although this text is written for tissue culture specialists and the over abundance of mathematical formula can diminish its readability, the presentation of some fundamental model elements in a more formal way is necessary. This will facilitate a more detailed study of the simulation and optimization problems for the reader.

From a mathematical point of view, transformations of state variables during rate processes under the influence of auxiliary variables can be presented as the family of functions (f_i):

where $O < i < k, \quad i \in N$,

f_i: (explants, expl. no., auxiliary var. $_f$) → (products, prod. no.),

explants, products ε STVAR,
expl. no., pro. no. ε N,
auxiliary var. $_f$ ε AUVAR.

STVAR represents the set of all possible state variables which can occur during the micropropagation process. This can be defined both in qualitative terms as buds, shoot tips, leaf fragments, callus tissue, single shoots, branched shoots, rooted shoots, small plants growing in a horticultural substrate, etc., and quantitatively by their height, weight, width, etc.

AUVAR represents the set of auxiliary variables, i.e., light, temperature, media and substrate composition, mode of plant material cutting, persons explanting tissues, tools used during explant manipulation, etc., which can be further defined by their intensity, level, concentration, skills, etc.

When deciding to apply some mathematical techniques at this stage of planning, a management could meet problems. This is because the functions presented above are hardly known for any particular genotype, its explants, growth or organogenetic processes, and factors influencing these processes. The knowledge of functional dependencies between these quantities is in an early stage of development. Many papers have been published showing qualitative interdependence between organogenesis and growth of specific explants and factors such as temperature, light, media and substrate components, etc. There is a lack of publications containing discussion of functional, quantitative, and qualitative interdependence between these variables and rates of state variable changes. This fact visualizes the importance of modeling in finding the failing points of human knowledge systems.

Because micropropagation is composed of passages which are usually treated as "black boxes", the multiple regression analysis is especially suited to describe its behaviour. However, if this technology is to progress in the direction proposed by Weathers and Giles (1987); Fujiwara et al. (1988) and Kozai (1988b), as discussed earlier in this chapter, the functions similar to those developed by Liebig and Lederle (1985), Lederle (1985), and Krug (1985), for radish grown in a greenhouse, could be more suitable.

At present, micropropagation managers have two different approaches to calculate a material flow within a micropropagation model. They are similar to those proposed by van Dyne and Abramsky (1975) for general agriculture.

The first, the constant coefficient approach, is very often used by beginners. Using such a method of approximation of process intensity, one assumes that it is constant for all possible sets of auxiliary and driving variables influencing the explant development. For example, rates of branching or rooting are assumed to be the same in all possible combinations of factors influencing explants (temperature, light, media and substrate composition, etc.). It is obvious that by using such a method of flow calculation, one cannot expect an accurate prediction of the systems behavior.

The second method, which is most often used by micropropagation management, relies on the construction of the matrices of coefficients characterizing the intensity of changes of the particular genotype and explants under the influence of specific sets of auxiliary variables (particular levels of media components, temperature, etc.). These coefficient matrices usually result from laboratory management experience and they change constantly through an adaptation mechanism during the development of laboratory knowledge. In both of the above-mentioned situations, models will have deterministic characters because they will not reflect the stochastic character of living systems.

Besides the lack of knowledge of functional relationships within micropropagation elements, another common weakness of production planning is to neglect the quality of the products of the particular processes. These products are usually the explants used to induce subsequent processes, and their quality can have profound influence on their further growth and development. One must remember that very often the high multiplication rate leads to the lowering of the quality of propagated tissues (vitrification, excessive branching, etc.) (Hempel 1985). Therefore, the proper balance in priorities for quantity and quality of products must be established.

These priorities depend on the customers' demand, which is the result of market trends. Horticulturists will accept a product of inferior quality when it is scarce on the market. However, overproduction of a particular product imposes the necessity for its high quality if it intends to remain competitive. The long-term policy is also important when decisions concerning the quantity-quality balance are made.

Going back to mathematics, the micropropagation technology can be described as function superposition:

$$p_1 * p_2 * p_3 * \ldots\ldots p_{k-1},$$

where p_i represents the functions of state variable transformations, $p_i = f_i$, and for the remaining transformations (not occurring in the particular technology) $p_i = \mathbf{id}$ at assumption that $i \in N$, $O < i < k$.

Since these functions are not known at present, they can be replaced by coefficients developed for the sets of particular state, auxiliary, and driving variables.

The algorithm presented below is an example of simplified quantification of micropropagation and enables the calculation of a number of product items which can be produced in a fixed time.

$$\text{fin. no.} = \text{st. no.} * \text{in. co.} * \text{el. co.} * \text{inr. co.} * \text{exr. co.} * \text{acl. co.} * \text{mul. co.}^{[(\text{w. no.} - \text{in. t.} - \text{el. t.} - \text{inr. t.} - \text{exr. t.} - \text{acl. t.})/\text{mul. t}]},$$

where:

fin. no. is the final number of product items;
st. no. initial amount of stock items;

in. co., el. co., inr. co., exr. co., acl. co., mul. co. are coefficients of initiation, elongation, in vitro rooting, ex vitro rooting, acclimatization and multiplication processes, respectively;
w. no., is the period of time (in weeks) from the beginning of production to its end;
in. t., el. t., inr. t., exr. t., acl. t., mul. t. are the time in weeks of initiation, elongation, in vitro rooting, ex vitro rooting, acclimatization and multiplication processes, respectively.

The above formula contains elements of the most popular technological processes (Fig. 3). To simplify the representation of plant material flow, mechanically induced changes were coupled with organogenetic ones. In production, such couples are named initial, multiplication, rooting, etc. stages. In the formula the intensity of joined processes is represented by one coefficient, e.g., coefficient of multiplication, rooting, etc.

In the case where some of the processes are not planned for the particular technology, their coefficients must be replaced by 1 and time by 0. The duration of the technological processes can be expressed in weeks, which is habitual production practice or, if necessary, in days. The results obtained by calculations differ from real ones in situations when the coefficient of multiplication is expressed by a real number (the number of explants obtained after cutting has always integer value). This difference increases as the number of multiplication passages increases. The algorithm can be used only for the propagation procedures where all products of transformations are functionally the same, so they can be used equally as state variables of subsequent transformations (*Gerbera, Rosa, Nephrolepis*, etc.).

4.3 Technology Validation

Before the detailed technology plan (model) can be used in a further planning process, its validity must be confirmed (Dent and Blackie 1979) (Fig. 2).

In the situation when production modeling with stochastic elements included is conducted by economists or system analysts, the technology model validation becomes a laborious task including discussions of its assumptions with experts of the particular production profile. However, in former discussion production modeling was assumed to be conducted by the micropropagation firm management and to be equal to production planning. Therefore, technology correctness is taken into account during all these processes. The determinism of the technology model and the fact that it is usually based on the results (coefficients) obtained previously in the particular laboratory diminish the probability that it will not be valid. However, after planning is completed, the management usually runs additional tests (simulate the system behavior) to assure itself that previous calculations were correct.

During simulation, the levels of the state and auxiliary variables can be arbitrarily changed and the behavior of the system can be imitated. The analysis of the consequences of introduced changes enables better understanding of the system and necessary amendments in plan assumptions.

When technological plans have been prepared by a management member who is not directly involved in the production process, their correctness must be verified by specialists engaged directly in propagation.

One technology model seldom fits within the constraints of a particular order. It is common that the management must choose between several options using common sense (heuristic) problem-solving rules or one of the optimization techniques. However, the best time for optimization is after the production schedule has been constructed, when such variables as materials, equipment, and labor availability (Fig. 1) can be taken into account.

4.4 Production Scheduling

The production schedule is the set of tasks which must be performed in order to achieve planned changes. These tasks must be described in detail from an organizational point of view and placed within a particular period of time. The description must contain specification of:

— the way in which the operations have to be conducted,
— material, equipment, and personnel necessary to carry them out.

All processes occurring during micropropagation can be divided into two groups with respect to their rapidity. The first group contains mechanical manipulations, e.g., division of branched shoots on single shoots, planting of rooted shoots in a horticultural substrate, plant transplantation, etc. These operations are of short duration and are planned for a particular day.

All processes which are the result of environmental manipulations, i.e., simulation or restriction of branching, rooting, growth, etc. can be placed into the second group. They last a certain time, usually not shorter than 3, and not longer than 12 weeks. The processes belonging to the first group are influenced by labor skills, equipment, tools, etc.; those of the second group by temperature, light, media, substrate and air components, etc. (Fig. 3).

The production schedule must also contain the specifications of the tasks which do not directly affect plant material changes but auxiliary variables of these processes, e.g., media and substrate preparation, supply, etc. (Fig. 4). These second- and further-order tasks must precede in time the first-order manipulations. The higher the task order, the less important is its time dependence on the first order to which it is related. For example, the supply of chemicals and substrate components need not be as strictly co-ordinated in time with cutting or planting as media and substrate preparation. Therefore, the tasks of further orders can be grouped with the aim of bettering exploit equipment and labor potential. During these clustering operations, optimization techniques can also be applied.

However, the stability of some auxiliary variables must be taken into account when their action is transferred in time. As an example, changes occurring during storage of sterile vessels filled with media can be given. The longer they are stored, the higher the probability that they will be contaminated or the chemical and physical properties of media will change. The problem can also occur when certain organic substances are stored beyond their storage time limit or when a soil mix which was pasteurized is used at an appropriate time after this process had been completed.

In practice, production schedules of different time precision are usually constructed. The most common procedure is to build up daily and weekly plans. The purpose is to enable the fast reconstruction of plans within a week when the specific day

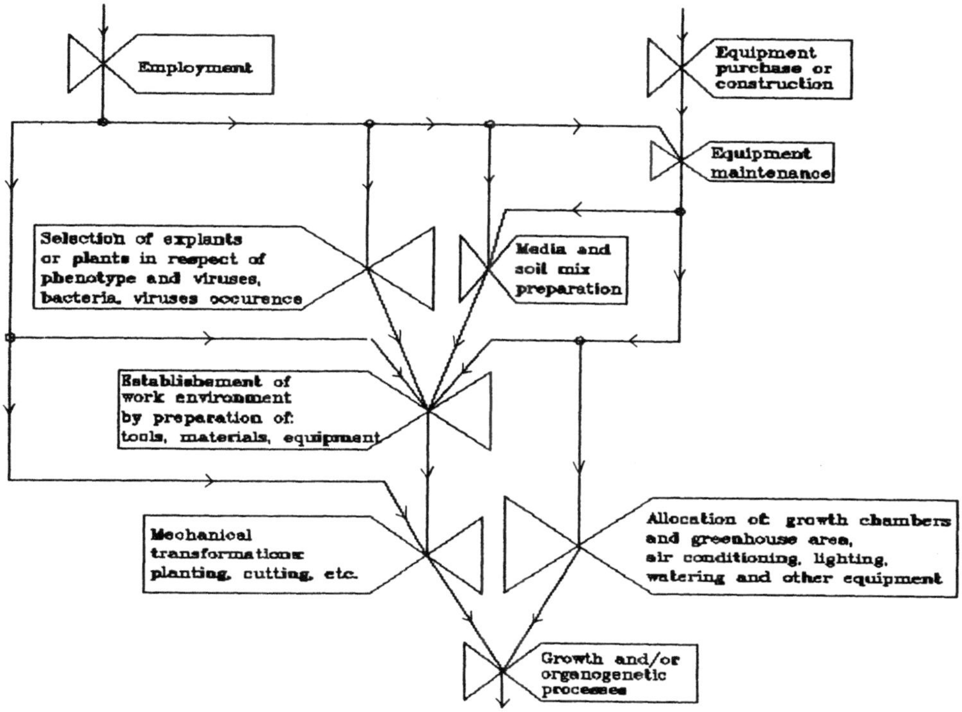

Fig. 4. Schematic representation of processes antecedent to mechanical or environmental manipulations

plan was not fulfilled. In most laboratories, propagation processes are discrete in time within a 1-week accuracy, so the weekly operational plan to be used is developed first and later the daily plan is elaborated. In certain situations, the fortnightly and monthly plans can be elaborated to enable some planning buffering over longer periods of time.

4.5 Evaluation of Production Constraints

Previous discussion was focused on modeling and scheduling of a production technology of an arbitrary genotype in respect of one order. In practice, the situation is far more complicated. An average in vitro laboratory produces tens of genotypes to meet thousands of orders, therefore hundreds, even thousands, of technology plans must be developed simultaneously. In such circumstances, the probability that actions planned in respect to plant tissue reactions and total production time will coincide is very high. Therefore, the management must check in advance the reality of the plans in respect to such variables as materials, equipment, and labor. If necessary, additional equipment and materials must be purchased or workers employed. When one of these constraints is fixed (usually equipment – flow cabinets, growth chambers, etc.) the technology plan must be changed to fit into these restrictions. These restrictions must also be defined before the application of optimization techniques.

4.6 Optimization Techniques

The demand for perfection can have esthetic or economic sources. When trying to improve the economy of one's activities, one looks for optimal or near-optimal solutions to the problem. Optimal means giving the best results in respect to the goals posed by a person seeking improvement. These goals are usually defined in terms of the profitability of the production process, but can also consider the product quality or any other restrictions. The goals of optimization often change in the course of a micropropagation firm operation and depend on the present policy of the firm.

In the past, many optimization techniques have been elaborated for different agricultural branches of production. Their usefulness in a particular situation depends on the structure of the system which has to be optimized and on the knowledge of the interdependence between its variables. Discussion of optimization techniques available can be a topic for a separate book. Here, the future prospects of only the most popular techniques are summarized.

Simulation, critical path analysis, linear programming, and dynamic programming are the optimization techniques most often used in agriculture. Simulation, besides mimicking a systems behavior, can also be applied for seeking the optimal set of state and auxiliary variables. It can be used for the optimization of any system and its effectiveness depends only on the correctness of the algorithm(-s) describing the system. Simulation is the least efficient method of finding the optimal solution (van Elderen 1987).

Optimization by systems simulation differs, depending on the optimum-seeking design applied. Dent and Blackie (1979) proposed such methods as: steepest ascent, Simplex, conjugate direction, random search, and random search with learning. The other optimization procedure, the Monte Carlo method (Carlsson et al. 1969; Hales 1969) is the simulation procedure which is based on random choice of the input variable levels and selection of the best output results. It is equivalent to the random search method proposed by Dent and Blackie (1979).

Critical path analysis is suited for situations where the proper scheduling of tasks in time is the object of optimization (Cooke-Yarborough 1964; Maddison 1969). Duration of tasks may be defined and arranged in the network in which they logically follow each other. The goal of procedure is to find an optimal path through the network in respect to time. The cost of each task can be equated with time; therefore, the solution can also be the most cost-effective one.

Linear programming has standard formulation of its technique (linear equations) and standard optimum seeking design (Simplex algorithm). It gives one solution which is optimal for the set of given variables, assumed to be stable in time (Hales 1969; Meijaard 1969; Schumacher and Weston 1983). Management seldom looks only for the optimal solution but for the set of near-optimal solutions from which the most suitable can be chosen. Another drawback of this method is built-in linearity of interdependence, which rarely occurs in practice. To overcome these inconveniences, the versions of these techniques have been proposed as integer linear programming (Butterworth 1972) and dynamic linear programming (Davis 1979).

Dynamic programming is the technique especially suited for the systems composed of many subsequent stages (Kennedy 1986). Any arbitrary algorithm (including linear programming and critical path algorithms) can be included to seek the optimum with this technique. The solution is unique and derived in many backward recursive steps (stages of a model).

Theoretically, all of these methods can be used for the optimization of micropropagation plans. However, the dynamic programming technique seems to be the best for this purpose because it was elaborated for systems composed of many subsequent stages and enables the incorporation of other optimization procedures (Kennedy 1986).

5 Production, Monitoring and Control

All production processes can be divided into two groups: those which are the subject of mechanical manipulations and those which are under the influence of environmental variables. Production monitoring and control intends to maintain all these driving variables (worker skills, equipment efficiency, temperature, light, etc.) on the planned levels. The way it is conducted depends on the membership of the variables in one of the above groups.

Brief deviations in the control of mechanical manipulations i.e., cutting of plant material, have much more profound consequences than in environmental ones. The whole process can then be completed without the possibility of its compensation by proper adjustment of auxiliary variables levels.

Short-term fluctuations of environmental variables, i.e., temperature and light conditions, have a much weaker effect on plant tissue growth and development because of buffering and compensation mechanisms built in living tissues.

Unfortunately, the variables involved in the control of mechanical manipulations (workers' skills) are much more difficult to control and monitor than, any other. Proper training of staff, increasing employees' job motivation, and appropriate methods of description of tasks can be counted among the procedures the management can use to keep this variable on the proper level. Many of these problems can be eliminated by mechanization and robotization.

6 Planning-Production Feedback Mechanism

Two different reasons for the failure of plans can be distinguished:

— improper monitoring and control procedures;
— inappropriate estimation of process coefficients or functions during plan construction.

In both situations, the plan must be reconstructed in order to enable the production of desired products. In the first case, only the current or subsequent week plan must be changed. In the second case, the new coefficients or functions must be calculated based on the results of particular manipulations and information stored in the data base (Fig. 5). Subsequently, all plan elements in which these quantities are included have to be reconstructed. Through this feedback mechanism the estimation of rate process intensity can be perfected day by day during laboratory operation. In production, the permanent loop, including planning and propagation procedures, should exist permanently in order to improve a laboratory's operation.

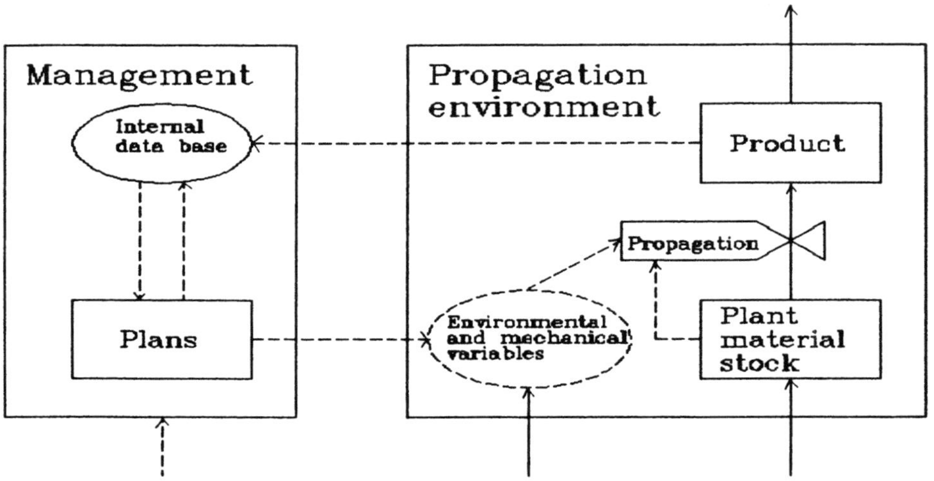

Fig. 5. The feedback loop between planning and propagation effects

7 Technical Aspects of Computing and Information Management

Information management is composed of information collection, processing, output, and storage. Since computing in greenhouse production has been examined in many publications, the discussion of computing details in production monitoring and control will be limited here to the in vitro phase only. The technical aspects of these processes during the planning phase differ from those used during production monitoring and control.

7.1 Production Monitoring and Control

During the production phase, the role of the management is to fulfill plans by keeping all controllable variables on the planned levels. These variables have to be monitored (information collection) and if found undesirable (information processing), corrected (information output). The way this is conducted depends on the type of the variable; environmental variables are handled differently from the variables influencing mechanical transformations.

7.1.1 Environmental Manipulations

7.1.1.1 Temperature and Air Composition

The laboratory levels of some environmental variables (temperature, air composition) have a tendency to fluctuate as the outside levels of these variables change. They have to be considered as continuous variables, constantly monitored and, if necessary, adjusted to the desired levels.

This can be done manually by using monitoring devices such as thermometers, hydrometers, etc., and, when necessary, adjustments can be made through heating, cooling, and humidifying equipment. Human management, however, can easily be replaced by mechanical and electronic equipment, which is far more precise in operation.

Although these devices are accurate in keeping variables at the stable levels, they are not flexible in operation. Often variable levels will fluctuate in different places of a growth chamber or they must be changed periodically. Increased flexibility of variable control can be achieved by the application of computers. To enable this, the analog signals from the sensors must be first digitized and, after processing and transformation back to analog signals, sent to the controlling equipment. This method of control enables the application of very complicated software which can be easily accessible and arbitrarily manipulated during the production planning phase. Contary to greenhouse production, such control techniques have rarely been applied to commercial in vitro laboratories. Such situations exist because the greenhouse production systems are much less insulated from environmental influences and are, therefore, much more difficult to control than in vitro systems.

7.1.1.2 Light

Light intensity, spectrum, and duration are discrete, totally controllable variables. Their levels are usually adjusted to the desired values only once – at the beginning of environmental manipulation (passage). The light intensity and spectrum change continuously during lighting equipment operation, but these changes are so slow and minor that they are usually neglected during micropropagation. The mechanical light clocks ordinarily used in micropropagation laboratories can be replaced by the internal computer clock, but then digital signals must be converted to analog ones.

7.1.1.3 Media Components

With standard micropropagation technology and closed vessel environment, the concentration of media components must be adjusted to the correct levels during media preparation before the beginning of environmental manipulation. These levels, once set, cannot be changed later during the planned manipulations (they change interacting with plant tissues). To control their values during media preparation, equipment such as balances, pH meters, and autoclaves are used. During the operation of most of this equipment, mechanical, electrical, and/or electronic signals between the sensors and controlling devices are monitored by workers who decide on the actions which must be undertaken. These actions are usually manual manipulations, i.e., adding substances during weighing or a base/acid during pH adjustment.

Human involvement in information transformation during these operations increases the risk of errors. Mechanization, automation and/or use of robots diminish this risk as well as increasing work efficiency. The line for automatic media dispensing and manipulation with growth containers proposed by Fari (1987) and the robot for media preparation in Petri dishes and vials reported by Vial (1987) are the first reported steps toward automation and computer control of media preparation.

7.1.2 Mechanical Manipulations

Mechanical manipulations of plant tissues have the effects on their structure or location. They are usually performed with the use of tools which can range from very simple (scalpels, forceps) to complicated (robots).

In the standard micropropagation laboratory, workers cutting plant material evaluate its quality by sight. They then process this information concerning the desired product, i.e., single shoots of defined quality, and decide how the specific plant tissue must be cut to achieve the desired effect. Since each piece of plant tissue to be cut is different, the automation and computerization of information collection and processing are far more difficult than in the case of environmental variables control.

A challenging problem is to substitute human sight by computerized image analysis. The structure of living, dynamically changing three-dimensional tissues is much more complicated to analyze than shapes of standardized parts moving on assembly lines in factories. However, the time is nearing when the latest developments in this field will replace humans by robots in manipulations with plant material cultured in vitro (Smith and Spomer 1987). Grand d'Esnon et al. (1987) and Chee and Cantliffe (1988) reported application of image analysis in propagule sorting during artificial seed production. Zryd (1987) used two-dimensional image analysis for selection of plant cells and protoplasts in transformation experiments. During the last meeting of the American Society for Horticultural Sciences, Meyer and co-workers (1988) presented the results of the experiments in which they compared the effectiveness of the image analysis and visual screening methods for quality evaluation of micropropagated plants. They found high conformity of shoot and root length measurements by the application of both methods. However, precision of internode length and leaf area evaluation by image analysis was higher when tissues were grown in tubes or flat vessels.

Research into applications of computerized image analysis in micropropagation has shown the necessity for adapting propagation methods to the requirements of this equipment. Less dense cultures with elongated shoots, small leaf blades, and visible internodes are much more traceable by image analysis than dense plant tissues with many short shoots. Therefore, until the image analysis techniques are improved sufficiently to manage all plant tissue culture objects, the attention of micropropagation developers will be focused on such methods as the hedging system proposed by Aitken-Christie and Davies (1987) for *Pinus radiata* and the stem elongation method elaborated by Maene and Debergh (1985).

Another fascinating problem in introducing the robotization of mechanical manipulations is the automatic processing of information gathered by video cameras. Stochastisity of living systems and learning capabilities must be included in the software designed for this purpose. The importance of machine learning grows parallel with the complexity of systems which are intended to be automatized. On the lowest abstraction level, i.e., sorting damaged or live protoplasts, robots must be taught to distinguish only two kinds of objects. When the managed system grows in complexity, more objects and relations between them (concepts) are added. The basic concepts through the generalization process lead to the origin of new concepts on the higher abstraction level. Automation of gathering of information, learning, and inferring are the objects of interest of artificial intelligence studies. However, before progressing here toward the significance of these studies for micropropagation, the achievements in mechanization and automation of mechanical manipulations will be discussed.

The studies on robot application in commercial tissue culture began a few years ago and are still in the early stages of development. During the conference organized by Moet-Hennesy in Monaco, Miwa (1987) presented the robot adapted to transplantation of in vitro-grown plants and seedlings. The robot was able to transplant (without cutting) an in vitro-grown plantlet in 60 s, using specially devised shape memory alloy actuators. During the same meeting, Rowe and co-workers (1987) presented a paper on the robot work station which was designed based on industrial manipulators and a vision system interfaced by specialized software. Manipulators, which were invented to enable cutting of delicate, in vitro-grown tissues, could allocate, cut, and plant a potato microcutting in less than 3 s.

It is evident that such advanced and expensive equipment can be applied only in large-scale production. Sluis and Kahn (1987) defined the lowest number of units at which application of robot lines could be economic as 500 million per year.

Before automated production lines are introduced in micropropagation, many individual and simpler operations can be mechanized and/or automatized. As an example, the automated plant surface sterilization system can be given (Maurice et al. 1985). This enables the replacement of humans in the disinfection of plant material before initiation of cultures with no significant differences from manual operation.

7.2 Production Planning

7.2.1 Information Collection

At present, in the majority of micropropagation laboratories, information about production results is gathered on the forms designed by the particular laboratory for the particular process(-es). These forms are manually filled with data by production supervisors or workers directly involved in production processes. The detail of information collected and the periods of time when the forms are sent to management for processing depends on the managerial rules adopted by a laboratory. As mentioned before, the most common micropropagation practice is to work on a weekly basis.

The forms can be easily replaced by computer spreadsheets which may have the same format as the forms, but enable the automatic gathering of information by the proper software interface with production equipment. A number of universal spreadsheets and integrated packages such as SuperCalc, visiCal, Lotus 1–2–3, Symphony, etc. can be used for information collection with their manual input. If information has to be gathered automatically, interfacing of electronic spreadsheets with production devices must be developed for that particular set of equipment.

The application of computers in information collection enables automatic transfer through an internal network to an electronic data base, where it can be stored until processing. Although the electronic information gathering-storage process is optimal from a managerial point of view, at present many micropropagation firms still use the traditional paper form-file cabinet style of information management.

7.2.2 Information Processing

Information processing methods depend on the complexity and the abstraction level of the planning goal. Operations used at the end of the scheduling stage, i.e., calculations of the number of explants to be cut by particular workers or the amount of a substance to be weighed in order to achieve the concentration needed can easily be done with a pencil or an electronic calculator. Similarly, the calculation of the average efficiencies of production processes does not require any other tools. However, as algorithms grow in complexity, the probability of an error increases. The replacement of humans by computers in all these uphill processes not only eliminates an error risk but also dramatically increases their speed. Electronic data manipulation enables an automatic search for the information needed and its incorporation into the calculations. Necessary data can be retrieved not only from an internal data base but also from distant data bases from other continents.

Information processing is impossible without an earlier definition of the processing procedure – an algorithm. Algorithms can contain a range of arithmetic and logical operations translated by a programmer in one of the computer languages during software development. In writing computer software, a programmer tries to mimic the human use of facts and rules in problems solving. In most of fourth and earlier generation computer languages (Basic, C. Forth, Modula, Pascal), facts are represented by assignments of a numerical value to variables and rules serve to manipulate these numerical variables by arithmetic and/or logical operations. Such an approach is sufficient to resolve numerical problems of defined facts and rules on the lower abstraction levels.

However, if human reasoning (complex logical operations) and learning (building up new rules based on old and new facts) are to be simulated, the fifth generation languages, LISP, Prolog, SmallTalk, etc., have to be used for software development. These languages are object-orientated and incorporate tools which enable the imitation of human intelligent behavior. Besides the above-mentioned reasoning and learning, the speech and image recognition processes which interface humans with their environment can be imitated.

In last few years, Prolog and its versions (Turbo Prolog, Arity Prolog, IF/ Prolog, ALS Prolog, etc.) have attracted software developers because of their universality and low cost. These languages enable the simulation of deduction and induction processes, natural language analysis (Walker et al. 1987; Sterling and Shapiro 1987), and the incorporation of programs written in other languages which are better suited for complex arithmetic calculations. At present, the fifth generation languages hold the most hopeful prospects for the future in the automation of planning on the highest abstraction levels, i.e., designing technology outlines and details and the integration of planning with production by robotization of cutting, media preparation, etc.

However, such computerized integration of micropropagation can only be achieved in progressive steps by starting from the automation of simple information manipulations. Although many papers have been published on this subject in relation to other domains of horticulture (Beck et al. 1983a,b; Stanley et al. 1983; Ingram and Yeager 1984; Yeager and Ingram 1984, 1987; Gladwin 1985; Halsey 1985; Hammer et al. 1985; McPherson et al. 1985), the only ones available in relation to micropropagation are those of Hempel (1988) and Hempel and Meszka (1988). In the first paper, the author discusses the foundations of the program which is aimed at facilitating many micropropagation planning procedures. The second is devoted to the formal description of micropropagation as a system. Some ideas from the second article were presented in Section 4.2 of this chapter.

7.2.3 Information Ouput

The products of information processing are reports, plans, and orders. They are usually prepared on the different forms and serve as templates for production or management manipulations. In automatized and robotized micropropagation systems of the future, the planning output will be sent directly to the equipment performing production manipulations.

7.3 Computer Hardware

The basic characteristics of all computer hardware are:

- amount of internal randomly accessible memory (RAM);
- amount of external storage memory;
- speed of computation;
- possibility of communication with external devices (additional information storage devices, printers, another computer micropropagation equipment, etc.).

RAM memory is used by programs and temporary data storage during the program's operation. Most of the standard programs written for microcomputers as spreadsheets, data bases, etc. do not require more than 640 kilobytes (kb) of RAM memory. If intending to implement more sophisticated programs written in the fifth generation languages (Prolog, LISP), one must consider buying the equipment with 1 or more megabytes (Mb) RAM available.

External information carriers (floppy, hard, and optical discs and magnetic tapes) can be used for long-term storage of data gathered from production or obtained after processing and needed for later use. The cheapest of them, floppy discs, can hold from 180 kb to 1.4 Mb; the most expensive, optical discs and magnetic tapes, many gigabytes (Gb) of information (an average page of text consumes about 3 kb of memory). Besides storage capacity, the average access time to recorded information is important in the real time applications, where computing time (if necessary) must follow the events occurring in automatized and/or robotized production or when a customer question concerning the availability of a product must be answered quickly.

The high speed of computation must be taken into consideration in applications where complicated mathematical and logical calculations are necessary. In micropropagation management it can take place when optimization methods are to be applied during production planning. Fast microprocessors are also important in real time systems when complex decisions must be made during equipment operation.

When planning integration of many computers and production devices in a network, one must look for hardware which will support its operation. A few parallel and serial ports and the possibility of the installation of digital-analog card is then important. Another solution is to purchase specialized hardware and software which enables data acquisition, and output converting them from analog to digital format and vice versa (Anonymous 1988).

When choosing appropriate computer hardware for a business, one must consider also its compatibility with the standard hardware and software of the world. Standardization facilitates the gradual development of very advanced computer systems. At present, IBM microcomputers and DOS or UNIX operation systems are such standards, to which many of hardware and software producers adjust their products.

The price of computer equipment is just as important during selection. For economic reasons, most micropropagation enterprises will restrict their choice to microcomputers which, presently, cost between ca. 1000 and 10 000 American dollars.

8 The Micropropagation Firm as a Whole – Prospects of Information Management

In the previous discussion, the planning and production phases were treated and detailed separately. Such an approach was necessary to facilitate the understanding of micropropagation as a system and information flow within it. However, in reality decision making and execution of inferred actions are much more closely interwoven by the feedback mechanism. When the decisions concern simple physical entities, e.g., temperature, the decision-making process (higher, lower, or leave at the same level) is simple, fast, and does not need to be spatially and temporally separated from the object handled. In this case, the decision is the effect of simple arithmetic comparisons such as equal to (=), higher than (>) or lower than (<). The more complicated the decisions are, the more spatially and temporally they are separated from the object of concern. With the growing complication of the planning process, the need for separate information-processing units – individual persons (managers) and/or group of people (management) – increases.

The decision-making tools used on the higher abstraction levels are far more complicated and are based not only on arithmetic operations and comparisons, but also on heuristic rules with learning process included. The selection of a proper technology is much more prone to errors and more time-consuming than decisions on lower abstraction level, i.e., who is to cut particular explants. Complication of management methods leads to an increase in bureaucracy and, subsequently, in production costs.

Mechanization, automation, robotization, and computerization are used to lower the probability of errors, relieve humans of boring and burdensome activities, and reduce production costs. The introduction of these means in managerial activities was not possible until recent developments in artificial intelligence and computer technology. Although the full imitation of human intelligence is still not possible, the expert system techniques which have emerged from artificial intelligence studies (Forsyth 1986; Townsend 1987; Cross 1988) enable programming and the decision-making process as the sequence of logical decisions based on earlier defined rules. The programs can be easily customized to the particular decision-making process by a programmer or even a common user with the help of expert system shells.

However useful in management activities, expert systems support only a few aspects of human intelligence. The main obstacle in computer simulation of processes such as learning with new concept formation intuition (heuristics), natural language analysis, etc., is their lack of full understanding and low speed of computation. Microprocessors of traditional computers perform operations serially one after another. Before each operation, data must be imported and, after it is completed sent back to a memory. Such an approach to computing is a bottleneck in human behavior simulation, where many processes are performed in parallel, i.e., image and speech analysis and reasoning. Computers based on microprocessors operating in parallel are the recent solution to low computing speed (Fox and Messina 1988).

It is difficult to predict when such machines will be applied to micropropagation, because at the present the adequate programming techniques have not been developed and this equipment is very expensive. However, taking into account the speed of computer industry development, it will take place in the near future. It is worth remembering that traditional microcomputers were introduced only in the early 1980's and are now present in almost every office and home.

9 Summary

In this chapter, the author has applied systems analysis to discuss information flow within a micropropagation firm. The model of a firm was created to facilitate the analysis of multidirectional relationships occurring during plant propagation in vitro. Two basic components of the general model, the management and direct propagation environment, were distinguished, and several more detailed elements were described. The information flow between micropropagation elements and micropropagation elements and external variables was also included in the discussion. Besides the theoretical background of information management, its technical aspects, including computer application and the prospects for the future, were outlined. It is obvious that the fast progress in computer technology which has been taking place in recent years will soon enable the automatization and robotization of some micropropagation processes. Automatic management of information is critical in such advanced production systems.

However, before the trials of micropropagation integration in a fully automatized "intelligent" system can be undertaken, the detailed, standardized description of its elements, processes, and interdependencies is necessary. Standardization enables the modular development of software and its compatibility not only within one firm but also within this branch of production and external data bases fed with information by research institutions. The discussion presented in this chapter is the first attempt to outline the foundations of such standardization and is aimed to provoke further discussion on this subject.

Acknowledgments. The editorial help of Ms. Tanya Greenwell during the preparation of the manuscript is gratefully acknowledged.

References

Aitken-Christie J, Davies H (1987) Automation of a new hedging system for micropropagation. Materials of Symposium: Innovations and Plant Production, Monaco, September 1987, p 12
Anonymous (1986) The Penguin Macquarie Thesaurus. Penguin Books Ringwood, Australia
Anonymous (1988) Data acquisition and control. MetraByte Corporation, Tauton, USA
Beck HW, Fasulo TR, Knapp JL, Tucker DPH (1983a) Florida agricultural information retrieval system: a computerized crop management program with application to citrus. Proc Fla State Hortic Soc 96:3–5
Beck HW, Marlowe GA, Pohronezny KL, Johnson FA (1983b) Florida agricultural information retrieval system: a computerized crop management program with application to tomatoes. Proc Fla State Hortic Soc 96:128–130
Butterworth K (1972) Practical application of integer programming to farm planning. Farm Management 2:151–157
Carlsson M, Hovmark B, Lindgren I (1969) A Monte Carlo method for the study of farm planning problems. Rev Market Agric Econ 37:80–103
Chee RP, Cantliffe DJ (1988) Propagule sorting by computerized vision. HortScience 23(3):173
Chua BU, Kunisaki JT, Sagawa Y (1981) In vitro propagation of *Dracaena marginata* Tricolor. HortScience 16(4):494
Cooke-Yarborough RE (1964) Critical path planning and scheduling. Rev Market Agric Econ 32:36–48
Cross TB (1988) Knowledge engineering. Brady Books, New York
Davis J (1979) Planning techniques and models for horticultural business management. Acta Hortic 97:7–23
Dent JB, Blackie MJ (1979) System simulation in agriculture. Applied Science Publishers, Barking, Essex, England
de Wit CT (1982) Simulation of living systems. In: Penning de Vries FWT, van Laar HH (eds) Simulation of plant growth and development. Centre for Agricultural Publishing and Documentation, Wageningen
Economou AS, Read PE (1984) In vitro shoot proliferation of Minnesota deciduous azaleas. HortScience 19(1):60–61
Farell MA (1987) Liquid medium system for plant micropropagation. Materials of Symposium: Innovations and Plant Production, Monaco, September 1987, p 32
Fari M (1987) Development of technical and biological basis of micropropagation toward automation and industrialisation. Materials of Symposium: Innovations and Plant Production, Monaco, September 1987, pp 55–56
Ferrari Th J (1982) Introduction to dynamic simulation. In: Penning de Vries FWT, van Laar HH (eds) Simulation of plant growth and development. Centre for Agricultural Publishing and Documentation, Wageningen
Fonnesbech M (1972) Organic nutrients in the media for propagation of *Cymbidium* in vitro. Physiol Plant 27:360–364
Fonnesbech A, Fonnesbech M (1979) In vitro propagation of *Spathiphyllum*. Sci Hortic 10(1):21–25
Forsyth R (1986) The anatomy of expert systems. In: Yazdani M (ed) Artificial intelligence – principles and applications. Chapman and Hall, London, pp 186–199
Fox GC, Messina PC (1988) Advanced computer architecture. Trends Comput 1 (special issue): 16–23
Fujiwara K, Kozai T, Watanabe I (1988) Development of a photoautotrophic tissue culture system for shoots and/or plantlets at rooting and acclimatization stages. Acta Hortic 230:153–158
Gladwin CH (1985) Budgeting garden costs for the individual garden using GARDBUD. Proc Fla State Hortic Soc 98:331–334
Grand d'Esnon, Harrell R, Chee R (1987) Propagule sorting by computerized vision during artificial seed production. Materials of Symposium: Innovations and Plant Production, Monaco, September 1987, p 22
Gutenbaum J (1987) Modelowanie matematyczne systemow. PWN, Harszawa, Poland
Hales AW (1969) Linear programming in horticulture. Sci Hortic 21:55–62
Halsey LA (1985) Electronic spreadsheet for woody ornamental nursery budgeting. Proc Fla State Hortic Soc 98:134–136
Hammer PA, Sullivan G, Wolford JA (1985) FLORALMART: Computer-assisted floricultural marketing. HortScience 20(20):179–182
Hasegawa PM (1979) In vitro propagation of rose. HortScience 14(5):610–612

Hempel M (1985) The influence of micropropagation on progeny plants. Acta Hortic 167:263–272
Hempel M (1988) Microcomputers in micropropagation. Acta Hortic 226:699–702
Hempel M, Meszka K (1988) Modelling of micropropagation. Acta Hortic 230:137–143
Henny RJ (1978) In vitro propagation of Peperomia Red Ripple from leaf discs. HortScience 13(2):150–151
Hildebrandt V, Harney PM (1983) In vitro propagation of *Syringa vulgaris* Vesper. HortScience 18(4):432–434
Ingram DL, Yeager TH (1984) Computer applications in landscape plant propagation. Proc Fla State Hortic Soc 97:282–284
Kennedy JOS (1986) Dynamic programming. Elsevier Applied Science, Barking, Essex
Kozai T (1988a) High technology in protected cultivation – from environment control engineering point of view. Special lectures: Horticulture in High Technology Era, May 10–11, 1988, Tokyo, Japan, pp 1–43
Kozai T (1988b) Autotrophic (sugar-free) tissue culture for promoting the growth of plantlets in vitro and for reducing biological contamination. Materials of International Symposium on Application of Biotechnology for Small Industries Development in Developing Countries, Bangkok, Thailand, 21–24 September 1988, pp 1–21
Krug H (1985) Growth processes and crop growth modelling. Acta Hortic 174:193–204
Kusumoto M, Furukawa J (1977) Effect of organic matter on the growth of *Cymbidium* protocorms cultured in vitro. J Jpn Soc Hortic Sci 45(4):421–426
Lederle E (1985) Simulation of plant growth based on methods of plant growth analysis exemplary for radish. Acta Hortic 174:247–250
Liebig HP, Lederle E (1985) Strategy for modelling plant growth. Acta Hortic 174:168–190
Maddison AE (1969) Critical path analysis. Sci Hortic 21:43–54
Maene L, Debergh P (1985) Liquid medium additions to established cultures to improve elongation and rooting in vivo. Plant Cell Tissue Organ Cult 5(1):23–33
Maurice V, Vandercock CE (1985) Automated plant surface sterilization system. Physiol Veg 23(1): 127–133
McPherson EG, McCarter J, Baker F (1985) A microcomputer-based park tree inventory system. J Arboric 11(6):177–181
Meijaard D (1969) The importance of farm comparison and linear programming in farm management research of glasshouse crops. Acta Hortic 13:104–112
Meyer MJ, McClelland MT, Smith MAL, Spomer LA (1988) Image analysis of growth and development parameters in microcultured whole plants. HortScience 23(3):173
Miwa Y (1987) Plant tissue culture robot operated by a shape memory alloy actuator and a new type of sensing. Materials of Symposium: Innovations and Plant Production, Monaco, September 1987, p 52
Murashige T (1974) Propagation through tissue cultures. HortScience 9(3):170
Murashige T, Serpa M, Jones JB (1974) Clonal multiplication of gerbera through tissue culture. HortScience 9(3):175–180
Norton ME, Norton CR (1988) A comparative study of in and ex vitro rhizogenesis on micropropagated shoots in the plant families Rosacae and Ericacae. HortScience 23(3):739 (Abstr)
Peck DE, Cumming BG (1984) In vitro propagation of *Begonia* × *tuberhybrida* from leaf sections. HortScience 19(3):395–397
Penning de Vries FWT (1982) System analysis and models of crop growth. In: Penning de Vries FWT, van Laar HH (eds) Simulation of plant growth and development. Centre for Agricultural Publishing and Documentation, Wageningen
Pierik RLM (1975) Callus multiplication of *Anthurium andreanum* Lind. in liquid media. Neth J Agric Sci 23:290–302
Pierik RLM, Steegmans HHM (1976) Vegetative propagation of *Anthurium scherzerianum* shoots through callus cultures. Sci Hortic 4:291–292
Raman K (1977) Rapid multiplication of *Streptocarpus* and *Gloxinia* from in vitro cultured pedicel segments. Z Pflanzenphysiol 83(5):411–418
Rowe WJ, Rowe NM, Roepe HP (1987) The application of robotics for an automated tissue transfer system. Materials of Symposium: Innovations and Plant Production, Monaco, September 1987, p 41
Schumacher S, Weston FC (1983) Linear programming: a tool for production planning in glasshouse floriculture. Acta Hortic 147:83–88

Simmonds J, Nelson SD (1988) Adventitious bud production on explants of *Begonia × hiemalis* depends on the development state of the donor plant. Physiol Plant 73(3):360–367

Sluis C, Kahn R (1987) Economic parameters governing application of robotics to micropropagation. Materials of Symposium: Innovations and Plant Production, Monaco, September 1987, p 6

Smith MAL, Spomer LA (1987) Electronic imaging for objective analyses of plant experiments in vitro and in vivo. Materials of Symposium: Innovations and Plant Production, Monaco, September 1987, p 13

Stanley CD, Price JF, Short DE (1983) Discussion of a microcomputer program for performing pesticidal spray calculations in horticultural crops. Proc Fla State Hortic Soc 96:292–294

Sterling L, Shapiro E (1986) The art of Prolog. MIT Press, Cambridge, MA

Stimart DP, Ascher PD, Wilkins HF (1982) Overcoming dormancy in *Lilium longiflorum* bulblets produced in tissue culture. J Am Soc Hortic Sci 107:1004–1007

Taylor ME, Knauss JF (1978) Tissue culture multiplication and subsequent handling of known pathogen-free *Dieffenbachia maculata* cv. Perfection. Proc Fla State Hortic Soc 89:233–235

Tisserat B, Vandercook CE (1985) Development of an automated plant culture system. Plant Cell Tissue Organ Cult 5(2):107–117

Townsend C (1987) Mastering expert systems with Turbo Prolog. Howard W. Sams, Indianapolis, USA

van Dyne GM, Abramsky Z (1975) Agricultural systems models and modelling: an overview. In: Dalton GE (ed) Study of agricultural systems. Applied Science Publishers, Barking, England

van Elderen E (1987) Scheduling farm operations: a simulation model. Pudoc, Wageningen

Vial J (1987) Robotization of an “in vitro screening”. Materials of Symposium: Innovations and Plant Production, Monaco, September 1987, p 44

Walker A, McCord M, Sowa JF, Wilson WG (1987) Knowledge systems and Prolog. Addison-Wesley, Reading, MA

Weathers P, Giles K (1987) A novel method of in vitro culture adaptable to many varieties of plants. Materials of Symposium: Innovations and Plant Production, Monaco, September 1987, p 45

Wolfe DE, Eck P, Chin CK (1983) Evaluation of seven media for micropropagation of highbush blueberry. HortScience 18(5):703–705

Yeager TH, Ingram DL (1984) Computer-assisted fertigation dilution calculations. Proc Fla State Hortic Soc 97:294–297

Yeager TH, Ingram DL (1987) Use of a microcomputer to compare container fertilization costs. HortScience 22(1):158

Zyrd J-P (1987) Image analysis: its uses in the analysis of plant cells and protoplasts. Materials of Symposium: Innovations and Plant Production, Monaco, September 1987, p 21

IV.5 Mass Propagation of Plants Through Shake- and Bioreactor-Culture Techniques

S. TAKAYAMA[1]

1 Introduction

From the point of view of the industrial application of plant tissue culture, clonal propagation and secondary metabolite production are the most important fields of research (Damiano 1980, Fujita 1987, Sagawa 1989). Today, there are many commercial tissue culture nurseries throughout the world. Most of these nurseries use flasks or bottles containing agar medium for commercial propagation, but efficiency is not high because of labor and time costs. In order to improve efficiency, the use of shake culture techniques for the mass propagation of plants was previously reported (Takayama and Misawa 1982e). Recently, the author has tried to scale up the techniques using bioreactors (Takayama 1984, 1986). Using a small bioreactor (4 to 20 l), over 4000 to 20 000 plantlets were produced within 1 to 2 months. The results are described here.

2 Advantages and Disadvantages of the Present in Vitro Propagation Techniques

The in vitro propagation technique has become routine procedure in many plant species including flowers, ornamentals, vegetables, trees, etc., and has been applied to commercial propagation. At present, in vitro processes are performed mostly using agar media (Table 1). These methods have their advantages compared to conventional propagation methods.

1. Plants propagated in vitro are genetically homogenous.
2. Plants are sterilized and cultured aseptically in vitro; they are healthy. Moreover, plants can be freed from viruses by thermotherapy or by shoot tip culture.
3. Plants can be produced independently of season or environmental conditions, which facilitates production and shipping.
4. Because the propagation efficiency is extraordinarily high, new cultivars or strains can be easily established and distributed to commercial growers.
5. Preservation of a large number of stock plants in vitro is possible in a small area, such as a culture room at 15 to 30 °C or a cold room at 0 to 5 °C. Serial subculture is required before senescence.

[1]Department of Biological Science and Technology, Tokai University, 317 Nishino, Numazu, Shizuoka 410–03, Japan

Biotechnology in Agriculture and Forestry, Vol. 17
High-Tech and Micropropagation I (ed. by Y.P.S. Bajaj)

Table 1. Plants propagated by tissue culture

Plant name	Propagation method[a]				Plant name	Propagation method[a]			
	A	B	C	D		A	B	C	D
Fruit trees					*Phalaenopsis*	o	o	o	
Malus pumila	o	o	o		*Miltonia*	o	o		
Pyrus communis	o	o			*Dendrobium*	o	o		
Prunus avium	o	o							
Prunus persica	o	o			Trees				
Vitis vinifera	o	o			*Eucalyptus*	o	o		
Citrus sp.	o	o			*Populus*	o	o		
Vegetables/Fruits					Ornamentals				
Solanum tuberosum	o	o	o	o	*Ficus benjamina*	o	o		
Asparagus officinalis	o	o	o	o	*Poinsettia*	o	o		
Fragaria ananassa	o	o	o	o	*Anthurium*	o	o	o	o
Zingiber officinale	o	o	o		*Dieffenbachia*	o	o	o	o
Amorphophalus	o	o	o	o	*Monstera*	o	o		
Dioscorea	o	o	o	o	*Philodendron*	o	o		
Colocasia	o	o	o	o	*Syndapsis*	o	o		
Cucumis	o	o			*Caladium*	o	o	o	o
Lycopersicum esculentum	o	o			Alocasia	o	o		
					Ananas	o	o	o	
Flowers					*Neoregelia*	o	o		
Lilium	o	o	o	o	*Doracaena*	o	o		
Begonia	o	o	o	o	*Nandina*	o	o		
Chrysanthemum	o	o	o		*Adiantum*	o	o	o	
Saintpaulia	o	o	o		*Nephrolepis*	o	o		
Gerbera	o	o			*Asplenium*	o	o		
Dianthus	o	o	o		*Cyathea*	o	o		
Geranium	o	o			*Nepenthes*	o	o		
Rosa	o	o			*Spathiphyllum*	o	o	o	o
Rhododendron	o	o			*Hosta*	o	o		
Gloxinia	o	o	o		*Ophiopogon*	o	o		
Streptocarpus	o	o	o						
Paeonia	o	o			Plantation crops				
Limonium	o	o			*Coffea*	o	o	o	
Hyacinthus	o	o	o		*Camellia sinensis*	o	o		
Allium	o	o			*Musa* sp.	o	o		
Colchicum	o	o			*Ananas comosus*	o	o	o	
Hyppeastrum	o	o	o	o	*Eraeis guineensis*	o	o	o	
Nerine	o	o	o		*Manihot esculenta*	o	o		
Lycoris	o	o							
Narcissus	o	o			Medicinal plants				
Crocus	o	o			*Scopolia japonica*	o	o	o	o
Freesia	o	o			*Atropa belladonna*	o	o	o	o
Gladiolus	o	o	o	o	*Pinellia ternata*	o	o	o	o
Cattleya	o	o							
Cymbidium	o	o	o						

[a]Propagation methods are; A = virus-free plants produced by apical meristem culture, B = agar culture, C = shake culture, D = bioreactor culture.

In spite of these merits, the technique has the following disadvantages.

1. Micropropagation on agar medium is labor- and time-consuming.
2. Little or no merits of large-scale production, and difficult to scale-up.
3. Production of a large number of plants from a small number of stock plants is risky because of somaclonal variation, especially when multiplied through adventitious bud formation from calli.

In order to overcome these problems and to establish large-scale mass propagation, the author adopted the shake- and bioreactor-culture techniques. The merits of these techniques are:

1. Time- and labor-saving in transfer of the cultures.
2. Easy scale-up to large-scale mass propagation.
3. Growth and multiplication rates are enhanced by forced aeration.
4. Nutrient uptake is enhanced and results in the stimulation of growth and multiplication rate.
5. Apical dominance of the culture disappears by continuous moving of the cultures in the medium, and induces the favorable proliferation of numerous multiple buds.

Until now, many plant species have been subjected to shake- and bioreactor cultures, as shown in Table 1. Most of these results reveal the benefit of the techniques in the scale-up of cultures.

3 Present State of Development of Mass Propagation Techniques

At present, the following techniques have been developed as new strategies to scale up of mass propagation of plants.

3.1 Bioreactor Culture Techniques (Fig. 1A)

This technique was developed by the author, and the details are described here. The main trait is the use of shaker and bioreactor, which are usually applied to the large-scale cultivation of microbial, animal, or plant cells.

3.2 Simple Scale-Up from Agar Culture Techniques

Nagaoka (1989) developed a "Bionursery System" which aims to produce the plants directly transplantable to soil by effective hardening of the plants in vitro. The process is fundamentally to simply scale up the conventional tissue culture processes, which comprise three main steps; (1) multiplication of propagules under controlled culture environment and by automatized sequences, (2) rapid growth and acclimatization until the plants become resistant to outside conditions, (3) additional benefit of injection of VA mycorrhizas or attenuated viruses. The equipment (Fig. 1B) was designed to control both the oxygen and carbon dioxide concentrations and also the supply of the culture medium. The use of rockwool or ceramic fiber as the adsorbent of the medium results

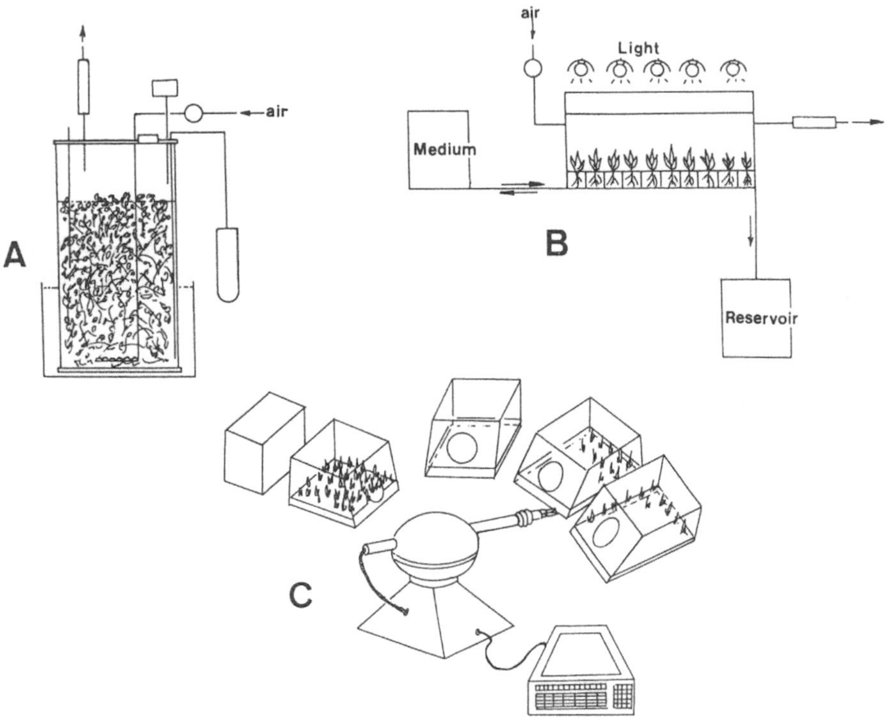

Fig. 1A-C. Various scale-up methods for mass propagation of plants. **A** Bioreactor. **B** Bio-nursery system (Nagaoka 1989). **C** Mericlone robot (Miwa 1989)

in increased transplantation efficiency. The merits are; (2) the labor and time for washing out the agar medium are saved, (2) roots are not damaged and so growth after transplant into soil is stimulated.

3.3 Automatization of Tissue Culture Process Using a "Mericlone Robot" (Fig. 1C)

Miwa (1989) developed a "Mericlone Robot" which aseptically transfers the plant tissues automatically. The robot consists of an actuator and a control system. A Ti-Ni shape memory alloy (SMA) is used as an actuator (fingers and a wrist).

The tissue transfer speed for one explant by man is 30 s. Although the transfer speed with the Mericlone Robot needs 1 min or more, the continuous running of the robot saves labor. Although the price of the robot is very high, this type of robot should be valuable for commercial propagation in the near future.

4 Schematic Outline of Mass Propagation of Plants Using Shake- and Bioreactor-Culture Techniques

It is the schematic outline of the mass propagation of plants using shake- and bioreactor-culture techniques is as follows (Fig. 2).

Stage I. Establishment of aseptic culture.
Stage IIa. Induction of multiple buds on the surface of a tissue segment.
Stage IIb. Rapid growth of buds by shake culture.
Stabe IIc. Rapid growth of buds by bioreactor techniques.
Stage III. Root formation and hardening before transplanting into soil.

Stage I is the process for establishment of aseptic cultures. In Stage IIa (induction of multiple adventitious or lateral buds on the tissue segment by adding cytokinins into the medium), the sizes of the buds multiplied were too small to divide and transfer to Stage III or transplant directly to soil, so tissue segments with multiple buds should be transferred to Stage IIb (shake culture) or to Stage IIc (bioreactor culture). In Stage IIb or in Stage IIc, buds grew rapidly to plantlets. Stage III (root formation and hardening before transplant into soil) is usually necessary, but not in all plant species.

In this scheme, an ideal process for mass propagation starts from Stage I, then multiplication of buds in Stage IIa, then rapid growth in Stage IIc, and finally propagated plantlets are transplanted directly to soil.

5 Characteristics of Cultures in the Bioreactor

In the author's experience, typical cultures are classified into three types according to their culture characteristics in the bioreactor (Fig. 3).

1. Floating just beneath the medium surface (strawberry plantlets).
2. Moving in the medium (begonia, gloxinia).
3. Sinking to the bottom of the bioreactor (lilies, gladiolus, potatoes).

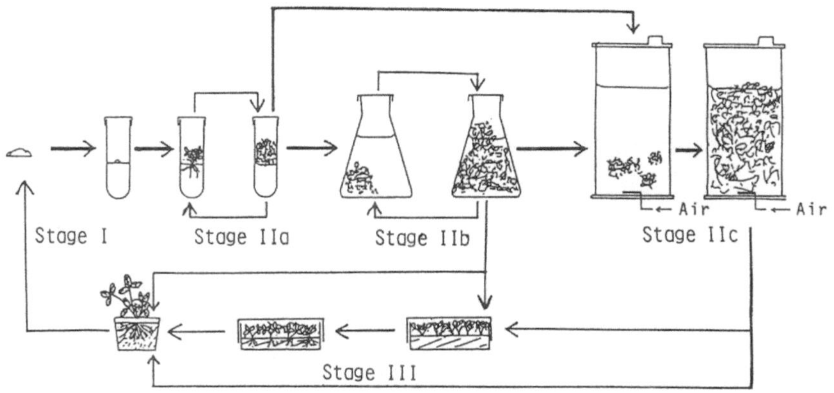

Fig. 2. Schematic outline of mass propagation of plants by shake- and bioreactor-culture techniques

Very good growth was observed in the case of floating and sinking cultures (Fig. 3), but when the cultures are moving in the medium, they suffer severe damage by shear stresses because they crush each other, and consequently growth is severely repressed.

5.1 Types of Bioreactors

Various types of bioreactors have been used in plant tissue cultures (Fig. 4). The hardware configurations of the bioreactors for plant tissue cultures are fundamentally the same as for microbial and animal cell cultures. Followings are the typical examples of the bioreactors.

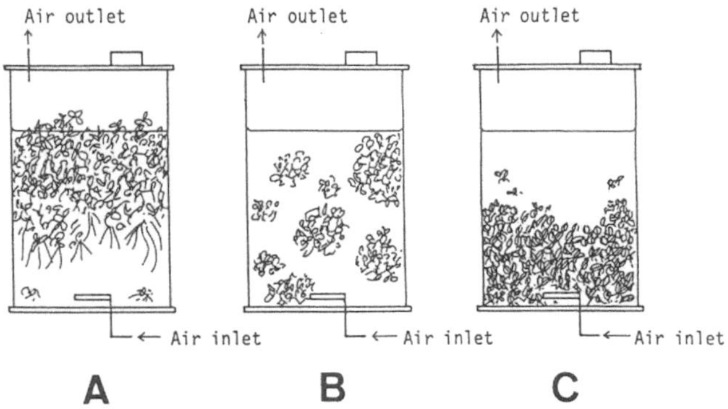

Fig. 3A-C. Characteristics of plant organ cultures in a bioreactor. **A** Floating just beneath the medium surface. **B** Moving in the medium. **C** Sinking to the bottom of the bioreactor

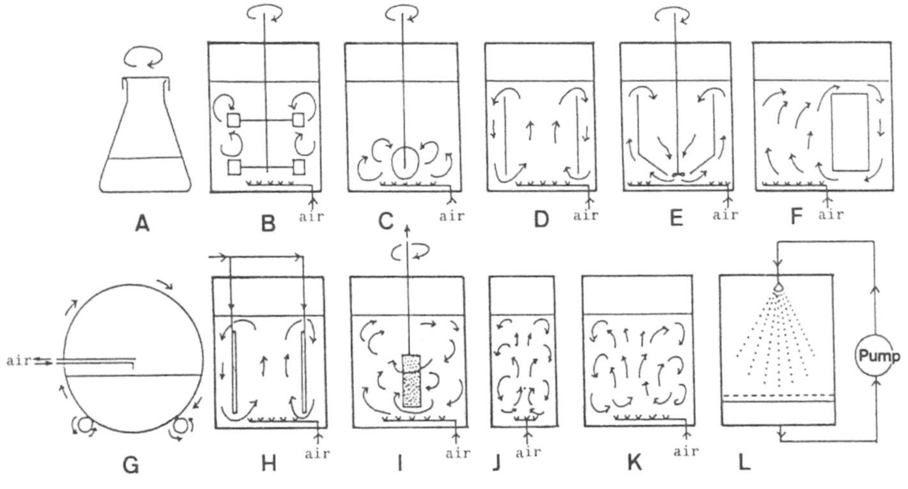

Fig. 4A-L. Different types of bioreactors for plant cells, tissues and organs. **A** Shake Flask. **B** Aeration-agitation. **C** Percolated impeller. **D** Draught tube airlift. **E** Draft tube with Kaplan turbin. **F** Airlift loop. **G** Rotating drum. **H** Light-emitting draught tube. **I** Spin filger. **J** Bubble column. **K** Aeration. **L** Gaseous phase

5.1.1 Aeration-Agitation Bioreactor

The most popular and fundamental bioreactor having an agitator shaft connected to a motor (Fig. 5). Various types of impellers such as paddle impellers are attached to a shaft, and most frequently used for plant cell cultures as well as microbial and animal cell cultures. However, when the bioreactor is applied to differentiated plant organ cultures, many problems arise, especially severe repression of shoot growth and multiplication. So, in the case of plant propagation, this type of bioreactor should be avoided.

5.1.2 Air-Driven Bioreactors

Air-driven bioreactors are the simplest design which is equipped with only a sparger at the bottom of the vessel (Fig. 6). This type of bioreactor has proved excellent for plant cell cultures producing secondary metabolites (Wagner and Vogelmann 1977) and also for plant organ cultures especially for propagation of plants (Takayama 1986). Most of the propagation experiments reported hitherto were performed using this type of bioreactors.

5.1.3 Rotating Drum Bioreactor (Fig. 7)

The reactor vessel rolls on a roller and oxygen is transferred from the membranous medium spread on the inner surface of the reactor vessel (Tanaka et al. 1983). The oxygen supply mechanism is different from either the aeration-agitation or the airlift

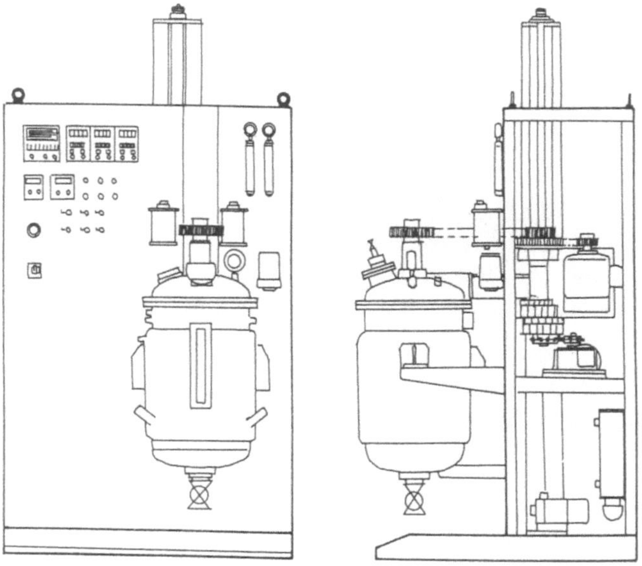

Fig. 5. Aeration-agitation bioreactor

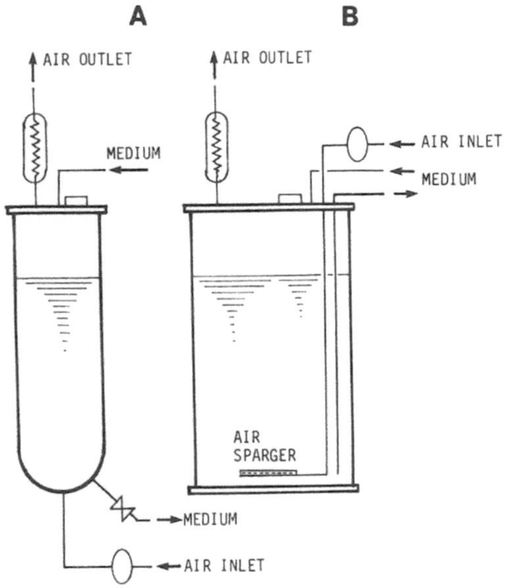

Fig. 6A,B. Two types of aeration bioreactor. **A** Bubble column. **B** Aeration

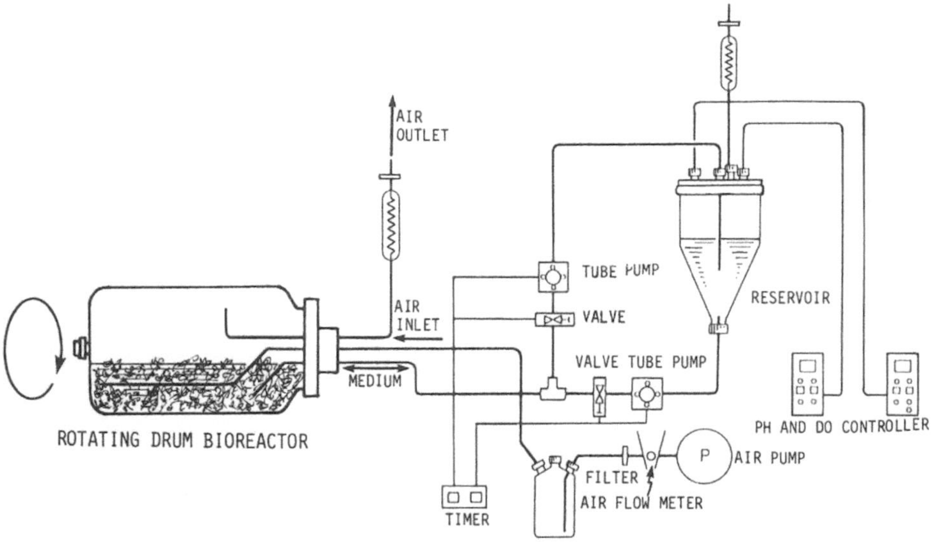

Fig. 7. Rotating drum bioreactor. (Courtesy of Shibata Hario Glass Industries Co. Ltd. Tokyo)

bioreactor. Tanaka et al. (1983) reported that the oxygen transfer coefficient is affected by a change of airflow rate under all rotational speeds. A reactor vessel which has no baffle plate can be used for mass propagation of plants.

5.1.4 **Spin Filter Bioreactor** (Fig. 8)

The spin filter bioreactor (Styer 1985) contains a filter in the bioreactor that spins via magnetic coupling to a stir plate. The spinning of the filter causes agitation of the medium without generating the shear observed with blade stirrers. The spinning also prevents plugging of the filter. The filter agitates the medium and allows withdrawal of spent medium without cell loss. The spin filter bioreactor is suitable not only for continuous culture of plant cells but also for enhancing the growth rate and propagation efficiency of the plants. The production of plants via somatic embryogenesis has been recognized as the most suitable for mass propagation using a spin filter bioreactor.

5.1.5 **Gaseous Phase Bioreactor** (Fig. 9)

In this type of bioreactor, cultures are inoculated onto filters which support plant tissues, and the culture medium is sprayed intermittently from a shower nozzle (Ushiyama et al. 1984a,b; Ushiyama 1988). This type of bioreactor supplies gaseous oxygen and has no agitation mechanism, so its merit is the supply of ample oxygen without generating shear. The growth of *Panax ginseng* or *Wasabia japonica* calli was stimulated remarkably (more than 150% comparing to the cultures in an aeration-agitation bioreactor). The scale-up of the bioreactor was designed using a belt conveyor installed within the culture vessel. The practical use of this type of bioreactor in plant propagation is expected.

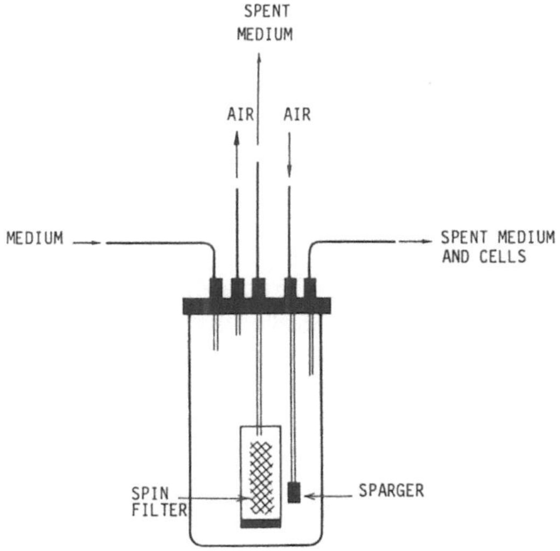

Fig. 8. Spin filter bioreactor. (Styer 1985)

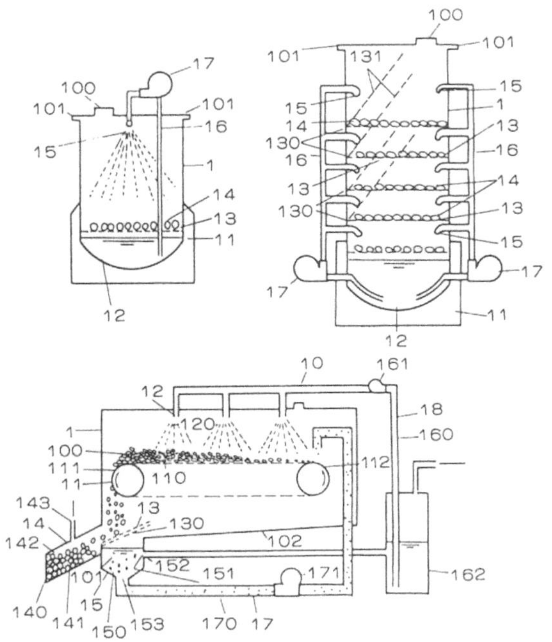

Fig. 9. Gas phase bioreactor. (Ushiyama et al. 1984a,b)

5.1.6 Light-Introducing Bioreactor (Fig. 10)

Introduction of light into a bioreactor can be used to induce important physiological processes such as photosynthesis, photomorphogenesis, or activation of some metabolic processes. Various types of light-introducing bioreactors have been reported. Inoue (1984), and Takada and Oota (1988) reported bioreactors equipped with transparent pipes from which light was emitted from the surface of the pipe into the bioreactor. Ikeda (1985) reported an airlift bioreactor equipped with a light-introducing draft tube (Fig. 10). The draft tube was constructed as an air-tight tube which consisted of a transparent inner and outer tube. Within the center of the draft tube was a light-introducing optical fiber.

The introduction of light into the bioreactor will become an important technique for the mass propagation of plants using a bioreactor, because the growth and hardening of the plants is stimulated by irradiation and results in increased propagation efficiency.

In plant tissues, the physiological and morphological characteristics are quite different from those of microbial or animal cells, and especially highly susceptible to shear stress. Shoots are severely damaged by shear stress, so amongst the various bioreactors mentioned above, the aeration-agitation type bioreactor should be avoided. In order to succeed in large-scale mass propagation, bioreactors should have the following characteristics.

1. Supply of ample oxygen and nutrients (the oxygen transfer coefficient 1/h is 10–50)
2. Low shear stress.

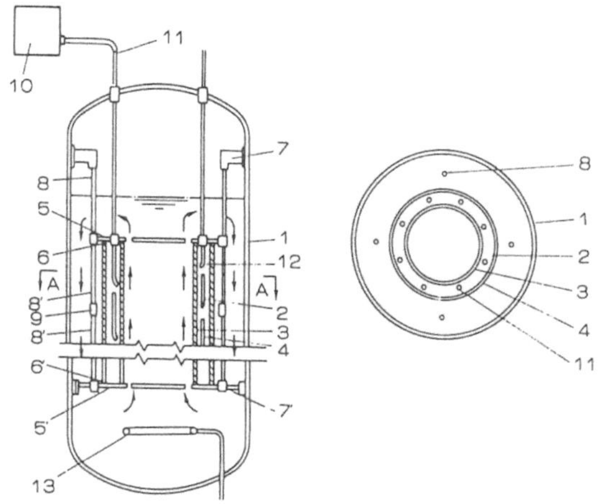

Fig. 10. Light-introducing bioreactor. (Ikeda 1985)

Of the various bioreactors described above, the author used two simple air- driven bioreactors for clonal propagation, i.e., a bubble column bioreactor and a bubbling bioreactor (Takayama and Misawa 1981). Other bioreactors valuable in plant organ cultures are the rotating drum bioreactor (Tanaka et al. 1983), spin filter bioreactor (Styer 1985), and gaseous phase bioreactor (Ushiyama et al. 1984a).

6 Scale-Up Technique

The culture conditions are precisely examined by shake culture then scale-up to bioreactor culture. The general protocol for scale-up is as follows.

6.1 Establishment of Aseptic Culture

At first, shoot cultures are established mostly from shoot tips. This process is performed routinely in order to free them from microbes or viruses. Established shoot cultures are serially subcultured on agar medium or in the liquid medium.

6.2 Multiplication of Shoot Buds (Fig. 11)

Generally, shoot buds are multiplied effectively by; (1) Stimulation of axillary bud formation by cytokinins, (2) in vitro cutting, (3) adventitious bud formation on callus or on vegetative tissue segment, (4) asexual embryogenesis. Among these, to attain high propagation rate and to avoid somaclonal variation, (a) axillary bud formation and (b) in vitro cutting are adopted in our methods. The multiplied shoot buds thus obtained are usually too small, and their morphological features are the aggregates of numerous buds. These buds were grown rapidly after transfer to liquid medium and shake- or bioreactor culturing.

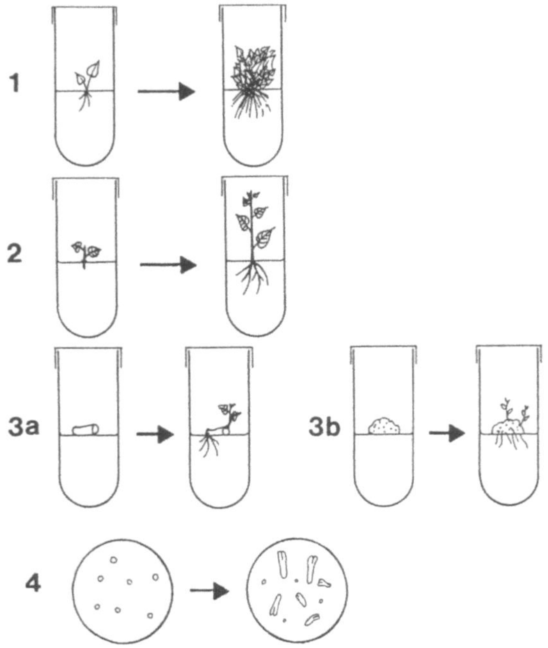

Fig. 11 (1–4). Various types of shoot multiplication methods. **1** Axillary bud formation. **2** In vitro cutting. **3** Adventitious bud formation from **3a** tissue segment or from **3b** callus. **4** Asexual embryogenesis

6.3 Optimization of Culture Conditions in a Liquid Medium by Shake Culture

When shoot buds were transferred and shake cultured in the liquid medium, growth responses were quite different between species or genera. Some plant species grow fairly well in shake culture, but others are inhibited. The growth characteristics depend on the inherited nature of the genera, species, or cultivars. For example, *Begonia × hiemalis* is well adapted to liquid medium, but saintpaulia and strawberry grow fairly well on agar medium and growth was repressed in the liquid medium. These characteristics should be considered for the success of liquid cultures. The culture medium composition is also important for the success of plant propagation in the liquid medium. The optimum composition of the liquid medium is sometimes different from that of the agar medium. Hormonal or mineral composition affect the growth of tissues more severely in the liquid medium than on agar medium. This is because the tissues are completely submerged in the liquid medium, so the contact area of cultures to the medium is wider in the liquid than on the agar medium. In order to select the optimum medium composition, the effects of phytohormone combinations, strength of the MS medium (especially ammonium ions), and the sucrose levels on growth should be examined.

6.4 Scale-Up to Bioreactor

After the optimum medium composition is selected by shake culture, then scale-up to bioreactor culture. Although the medium composition and culture conditions are different between species or genera, in the optimum condition, growth occurs fairly well in a bioreactor in the best cases. Examples are described in the next section.

7 Shake- and Bioreactor-Culture Techniques Applied to Various Plants

7.1 Begonias, Saintpaulias, Gloxinias

In the case of the propagation of begonias, saintpaulias, and gloxinias in a bioreactor, cultures move in the medium according to medium movement by aeration. In this situation, cultures can crush each other in the medium and become severely damaged by shear stress, as a result of which growth is severely repressed. However, when the movement of culture is inhibited, begonias easily grow into plantlets (Takayama and Misawa 1981, 1982a, 1983).

An established mass propagation scheme for *Begonia* × *hiemalis* is presented in Fig. 12. The explants from young leaves (A) were isolated in 7 × 7 mm segments, and cultured on a bud-inducing medium (B). As soon as numerous buds were formed (C), they were transferred into a fresh liquid medium and cultured on a rotary shaker operated at 180 rpm (D) or in a bioreactor (F). After 20 days, the buds grew into plantlets (G), which were divided and transferred to an agar medium (H) for root formation (I). Plantlets thus obtained were transferred into soil (J) and grown to flowering (K). In this scheme, most parts of the processes are mass production of plantlets without roots (B, C, D, E, F, G). These processes are divided into two steps, i.e., differentiation on agar medium (B and C) and growth in a liquid culture (D, E, F, and G), which could be recycled easily at 50-day intervals. A further ca. 15 days are required for root formation on agar medium. Saintpaulia and gloxinia were also propagated easily using this scheme.

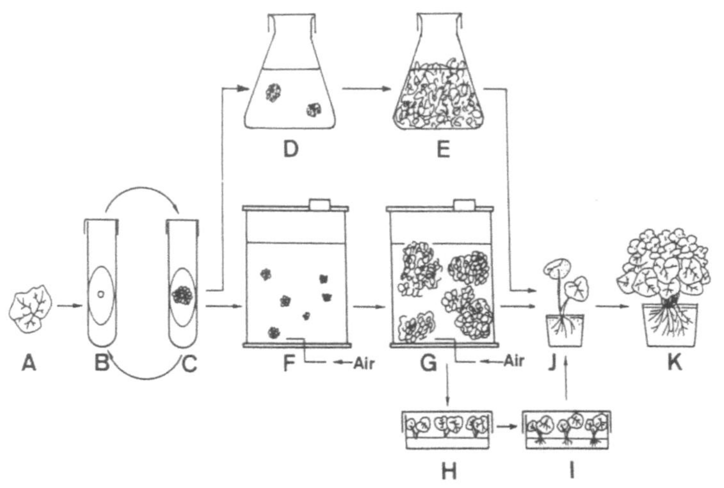

Fig. 12A-I. Schematic outline of mass propagation of *Begonia* × *hiemalis* by shake- and bioreactor-culture methods. For explanation see text

7.2 Lilies

The schematic outline of the propagation of lilies using a bioreactor is shown in Fig. 13. In this scheme, a bulblet established in vitro (A) was serially subcultured and multiplied in (B) and (C), then bulbscales were dissected and further transferred to the medium containing 3 to 10 mg/l cytokinins (D) and regenerated numerous primordia-like multiple adventitious bulbscales (E). Then bulbscales were rapidly grown by shake culture (F, G), and yielded numerous secondary bulbscales (H), which were used as the material for bulblet formation in the bioreactor (I, J). If the bulblets propagated in the bioreactor were removed aseptically from the bioreactor and the bulbscales dissected and transferred back to the bioreactor, the bulblets can be propagated only by the bioreactor methods. In the propagated bulblets (K) dormancy was broken by low temperature treatment at 5 °C, then they were transplanted into soil (L).

In the case of lilies, although the cultures (bulbscales) sink to the bottom of the bioreactor, their growth was reasonably good. In practice, in order to propagate lily bulblets successfully in the bioreactor, selection of the medium should be a main prerequisite; for example, MS medium strength affects bulblet formation of *Lilium longiflorum* (Easter lily). A half (× 1/2) strength MS medium stimulated bulblet formation, but normal (× 1) and second (× 2) strength MS medium repressed bulblet formation, and stimulated callus formation. Accordingly, we selected half (× 1/2) strength MS medium for *Lilium longiflorum* propagation in the bioreactor.

At half (× 1/2) strength MS medium, bulblet formation was stimulated at sucrose concentrations from 0 to 9%. Under these conditions, no callus formation was observed except at 9% sucrose. The number of bulblets produced on one bulbscale was stimulated at 1 to 3% sucrose, but the growth of bulblets was stimulated most effectively at 6 to 9% sucrose.

The size distribution of the bulbs propagated in a 2-l bioreactor was investigated, and a total of 634 bulblets were produced from 250 bulbscales within 2 months. About

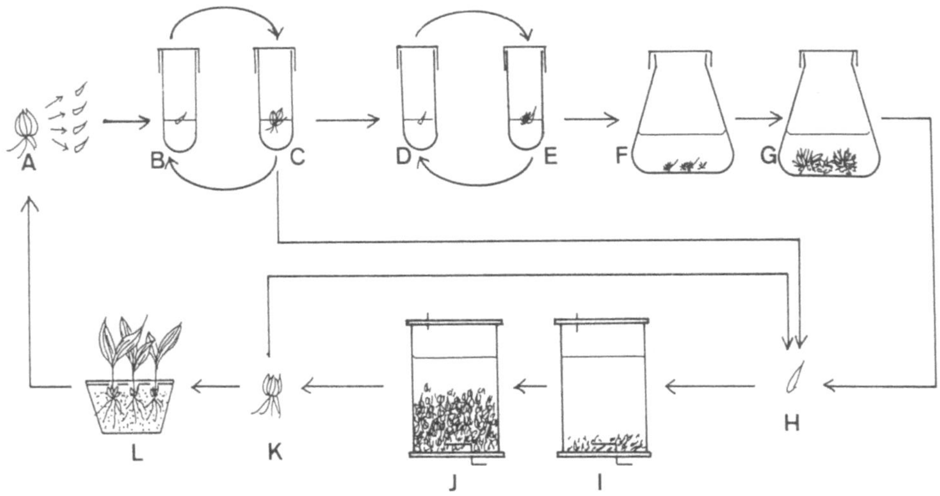

Fig. 13A-H. Schematic outline of mass propagation of *Lilium* by bioreactor-culture methods. For explanation see text

70% of these exceeded 0.3 g, and these flowered during the first year of cultivation in soil (Takayama and Misawa 1979, 1980, 1982b,c,d, 1983).

7.3 *Hyacinthus*

Hyacinthus is frequently used as a garden or pot flower or as cut flowers. The multiplication rate of the bulb is very slow by conventional methods such as scooping or scoring (Hartmann and Kester 1975). To overcome these problems, tissue culture technique has been applied to *Hyacinthus* propagation on agar medium (Hussey 1975, 1977; Tamura 1978; Paek and Thorpe 1989), but propagation efficiency is not high. The main obstacle is the extraordinarily slow growth rate and the difficulty of corm formation. The author has applied shake culture techniques and established a process producing microcorms as shown in Fig. 14 (Takayama et al. 1991). A bulb (A) was sliced aseptically into bulbscale segments with basal part (disc) (B), or apical meristem (C) was dissected aseptically from a bulb. The Bulbscale segment or apical meristem was transferred onto MS agar medium containing combinations of 0.1 mg/l NAA and 1 to 10 mg/l BA (D). Bulbs were regenerated on the surface of regenerative callus (E), and after dividing the regenerative callus (F), transferred to the medium with the same composition and subcultured serially (recycling F and G) and multiplied the regenerative callus on which the small multiple buds were regenerated (G). The cultures were then transferred into liquid medium and shake-cultured (H). The bulbs grew rapidly from buds in shake culture (H, I). The bulbs thus obtained were transplanted into soil (J) and grown to flowering.

7.4 Gladiolus

Most of the important gladiolus cultivars are infected with viruses that severely damage the quality and productivity of the flowers. Virus-free stocks were produced by shoot tip culture, and propagation on agar medium was established (Takatsu 1978; Bajaj et al. 1982). The authors established a jar fermentor culture technique for mass production of gladiolus microcorms, as shown in Fig. 15 (Takayama et al. 1987). In vitro plantlets were established by meristem culture, then transferred to MS liquid medium supplemented with 10 g/l sucrose to induce multiple shoots. The cultures were then transferred

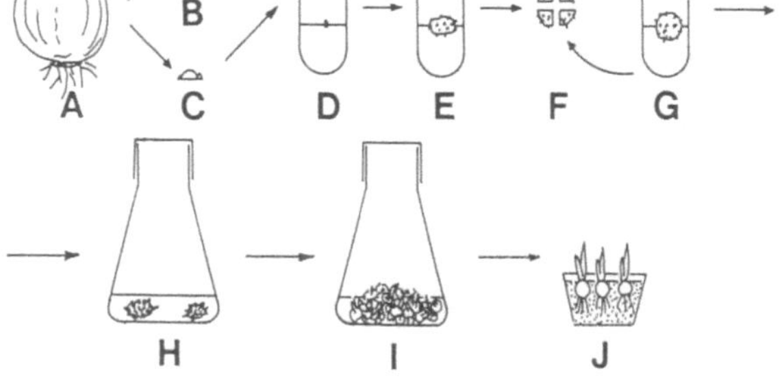

Fig. 14A-J. Schematic outline of the propagation of *Hyacinthus* propagation using shake culture. For explanation see text

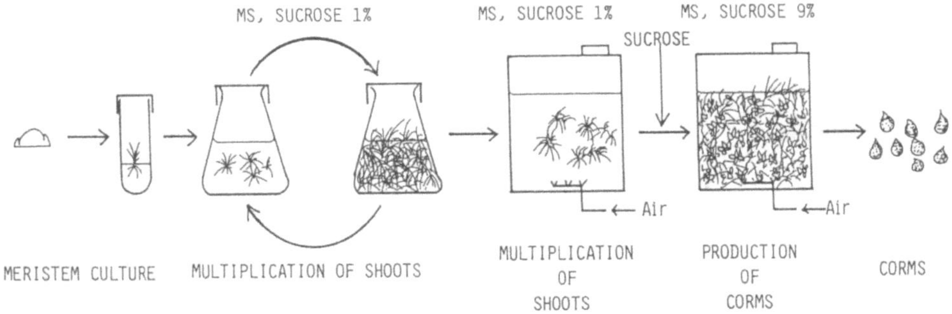

Fig. 15. Schematic outline of the mass propagation of gladiolus by shake- and bioreactor-culture techniques

Table 2. Effect of sucrose and NAA concentrations on corm formation of gladiolus in shake culture

Suc. conc. (mg/l)	NAA conc. (mg/l) 0	1	3	10
0	–	–	–	–
10	–	–	–	–
30	+	+	–	–
60	+	–	–	–
90	+++	–	–	–

to a bioreactor filled with MS liquid medium supplemented with 10 g/l sucrose. After shoots were multiplied, in order to induce corm formation, sucrose was aseptically added to the bioreactor at a final concentration of 90 g/l (Table 2). Further cultivation resulted in the formation of corms at the basal part of the shoots. About 300 corms were harvested from a 2-l bioreactor.

7.5 Amaryllis

Amaryllis (*Hyppeastrum hybridum*) is an important garden or cut flower. Its propagation by conventional methods such as twin scaling is problematic because of low multiplication rate and virus transmission. Tissue culture methods on agar medium have been developed (Mii et al. 1974; Yanagawa and Sakanishi 1977), but bulb formation is limited. The use of shake culture to increase bulb growth, and the established propagation method is shown in Fig. 16.

A bulb (A) was sliced aseptically into twin scale segments with basal part (disc) (B) and transferred onto MS agar medium (C), and bulbs were regenerated (D). Shoot tip culture (D, E and F) was initiated from the in vitro-produced bulblet (D), and again a bulblet was produced (F). The proximal end of the bulb was then sectioned (G) and transferred to liquid medium and shake cultured at 100 to 180 rpm (H). After 1 to 3 months, bulblets were regenerated (I) and removed from the culture vessel and transplanted into soil and cultivated to flowering (J).

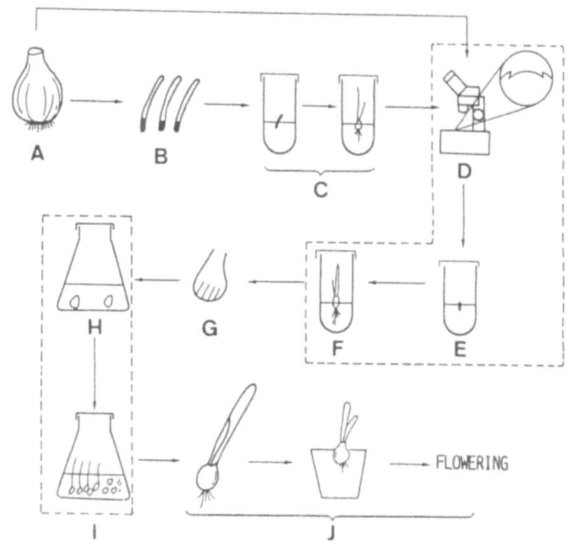

Fig. 16. Schematic outline of the mass propagation of amaryllis by shake culture. For explanation see text

7.6 Strawberries (Fig. 17)

An aseptic culture was established from meristems, then transferred to the medium containing cytokinins to induce multiple axillary buds. This process should be recycled every 30 days to increase the number of axillary buds [the cultures with multiple axillary buds were transferred every 30 days (Takayama et al. 1985a)]. The cultures with multiple axillary buds were transferred to the bioreactor and cultured for 1 month. During this period, buds grew rapidly to plantlets and after being taken out of the bioreactor, these plantlets were divided and transplanted directly into soil.

The growth of plantlets in a bioreactor (16.9 ± 4.8 g/explant) was almost the same as that of shake culture (12.9 ± 2.8 g/explant), and is superior to agar cultures (3.6 ± 0.8 g/explant).

Although the growth of *Begonia* plantlets is repressed severely by medium movement because of aeration, the growth of strawberries was not inhibited in bioreactor cultures (Takayama et al. 1985b).

7.7 Potato Microtubers

Potato microtubers can be propagated effectively by shake- and bioreactor-culture techniques (Akita and Takayama 1988). The process for microtuber production is shown in Fig. 18. Potato plantlets were maintained in vitro successfully by transferring the nodal segments onto MS agar medium. Using these plantlets, the process of microtuber production consists of two phases, Phase 1: shoot multiplication, and Phase 2: tuber formation. In Phase 1, shoots were multiplied in MS liquid medium containing 30 g/l sucrose under continuous irradiation. In Phase 2, microtubers were induced in MS liquid medium containing 90 g/l sucrose under complete darkness.

The culture conditions in Phase 1 severely affected microtuber induction. When shoots were grown in the aerial part of a bioreactor in Phase 1, microtubers were

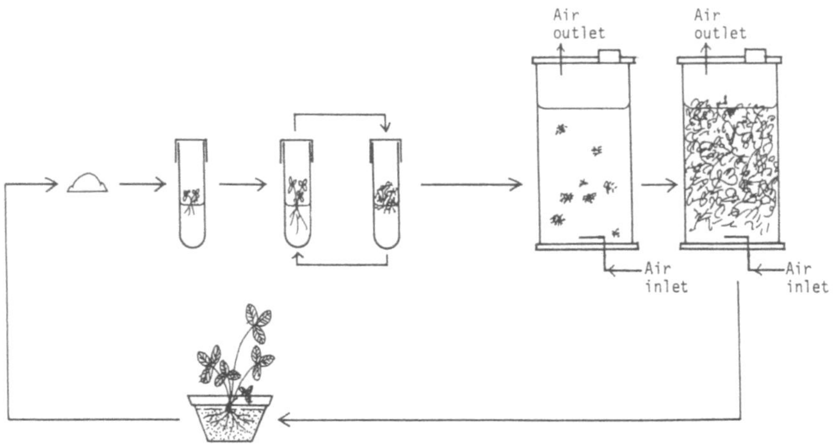

Fig. 17. Schematic outline of mass propagation of strawberry plants through bioreactor culture techniques

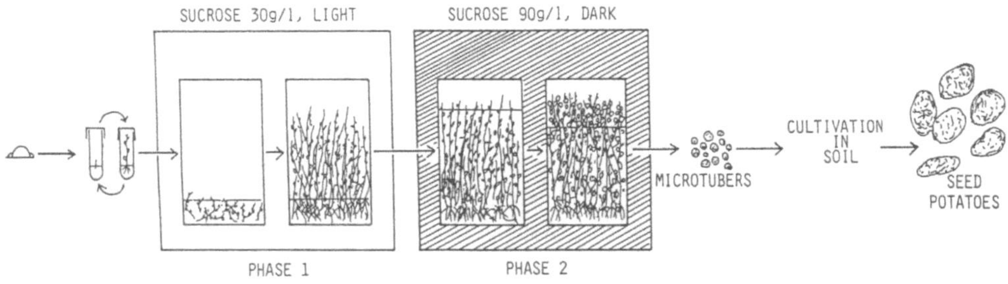

Fig. 18. Schematic outline of the propagation of potato microtubers in a bioreactor

induced efficiently in Phase 2. In contrast, when shoots were grown in the completely submerged conditions in the medium, microtuber induction was repressed in Phase 2.

After about 2 months of cultivation in a bioreactor, about 50% of the total fresh weight of the cultures was harvested as microtubers. About 400 microtubers were harvested using a bioreactor containing a 4-l liquid medium (Akita and Takayama 1988).

7.8 *Colocasia esculenta, Pinellia ternata*

Some araceous plants such as *Colocasia esculenta* and *Pinellia ternata* produce corms on the basal part of shoots. Although corm formation by tissue culture is usually quite difficult, the author has succeeded in producing microcorms of these plant in a bioreactor. The merits of microcorms are; (1) harvesting directly from bioreactor and handling of microcorms are quite easy, (2) saving on labor and cost on transplanting into soil, (3) microcorms are resistant to desiccation, so the hardening process can be abbreviated, (4) preservation of the microcorms is easy. In order to induce microcorm formation in vitro, two factors are most important: (1) increase of the sucrose concentration in the medium up to 90 g/l, (2) stimulation of evaporation of the medium by strong forced aeration to dry up the medium in the bioreactor. The schematic outline of the process is as follows. (1) Shoot cultures established from shoot tips were transferred to half strength MS liquid medium containing 30 g/l sucrose and were shake-cultured. In this process, by repeated subculture, shoots were multiplied quite effectively. (2) Shoots were further multiplied in a bioreactor with the same composition medium. (3) When the shoots were fully multiplied and grown in a bioreactor (Fig. 19), concentrated sucrose solution was added until the final concentration becomes 90 g/l, and at the same time the aeration rate is increased. Through these processes, a large number of microcorms were produced.

8 Concluding Remarks

In order to develop bioreactor techniques for plant propagation, the following problems must be overcome.

1. Application of bioreactor techniques to other plant species.
2. Selection of the optimum medium for the culture of plant organs in a bioreactor.
3. Introduction of light into the bioreactor.
4. Development of low cost bioreactor specialized for plant propagation.
5. Scale-up of bioreactor techniques.

Fig. 19. Shoot cultures of *Colocasia esculenta* in a 20-l bioreactor

The author is currently engaged in trials to apply bioreactor techniques to propagation for other species, with good results, and believes that plant propagation through bioreactor techniques will become an important industrial process in the near future.

References

Akita M, Takayama S (1988) Mass propagation of potato tubers using jar fermentor techniques. Acta Hortic 230:55–61

Bajaj YPS, Sidhu MMS, Gill APS (1982) Some factors affecting the in vitro propagation of gladiolus. Sci Hortic 18:269–275

Damiano C (1980) Planning and building a tissue culture laboratory. In: Proc Conference on Nursery Production of Fruit Plants Through Tissue Culture-Applications and Feasibility. USDA, Beltville, pp 93–101

Fujita Y, Tabata M (1987) Secondary metabolites from plant cells. Pharmaceutical applications and progress in commercial production. In: Green C et al. (ed) Plant Tissue and Cell Culture. Alan R Liss, New York, pp 169– 185

Hartmann HT, Kester DE (1975) Plant propagation: principles and practices, 3rd edn. Prentice-Hall, Englewood Cliffs, NJ

Hussey G (1975) Propagation of *Hyacinthus* by tissue culture. Sci Hortic 3:21–28

Hussey G (1977) In vitro propagation of some members of the Liliaceae, Iridaceae and Amaryllidaceae. Acta Hortic 78:303–309

Ikeda H (1985) Culture vessel for photoautotrophic culture. Japanese Patent, Kokai Tokyo Koho 60–237984 (in Japanese)

Inoue M (1984) Culture instrument for photoautotrophic organisms. Japanese Patent, Kokai 59–21682

Mii M, Mori T, Iwase N (1974) Organ formation from excised bulb scales of *Hyppeastrum hybridum* in vitro. J Hortic Sci 49:241–244

Miwa Y (1989) Robot in tissue culture nursery (in Japanese). In: ICMSA Committee (eds) Measurement, control and information in production of plants. Keisoku jidou Seigyo Gakkai, Tokyo, pp 202–210

Nagaoka M (1989) Bio-Nursery System. In: Takatsuji M (ed) Syokubutu Seisan Sisutemu Jituyou Jiten. Fuji Technosystem, Tokyo, pp 1271–1272

Paek KY, Thorpe TA (1989) Hyacinth. In: Ammirato PV, Evans DA, Sharp RW, Bajaj YPS (eds) Handbook of plant cell culture Vol 5. McGraw-Hill, New York, pp 479–508

Sagawa Y, Kunisaki JT (1989) Micropropagation of floricultural crops. In: Ammirato PV, Evans DA, Sharp WR, Bajaj YPS (eds) Handbook of plant cell culture, Vol 5. McGraw-Hill, New York, pp 25–56

Styer DJ (1985) Bioreactor technology for plant propagation. In: Henke RR, Hughes KW (eds) Tissue culture in forestry and agriculture. Plenum, New York, pp 117–130

Takada T, Oota Y (1988) Culture of plant cells under illumination. In: Plant tissue culture for bioengineering. The Japanese Association for Plant Tissue and Cell Culture, Tokyo, pp 94–95

Takatsu I (1978) Practical methods for meristem culture of gladiolus and the production of virus-free plants (in Japanese). Agric Hortic 53:797–800

Takayama S (1984) Mass propagation of plants through shake culture techniques. Plant Tissue Cult Lett 1:8–13

Takayama S (1986) Mass propagation of plants by fermentor culture techniques, Abst VI Intern Congr Plant Tissue and Cell Culture, Minnesota University, pp 449

Takayama S, Misawa M (1979) Differentiation in *Lilium* bulbscales grown in vitro. Effect of various cultural conditions. Physiol Plant 46:184–190

Takayama S, Misawa M (1980) Differentiation in *Lilium* bulbscales grown in vitro. Effects of activated charcoal, physiological age of bulbs and sucrose concentration on differentiation and scale leaf formation in vitro. Physiol Plant 48:121–125

Takayama S, Misawa M (1981) Mass propagation of *Begonia* × *hiemalis* plantlets by shake culture. Plant Cell Physiol 22:461–467

Takayama S, Misawa M (1982a) Factors affecting differentiation and growth in vitro, and a mass propagation scheme for *Begonia* × *hiemalis*. Sci Hortic 16:65–75

Takayama S, Misawa M (1982b) Regulation of organ formation by cytokinin and auxin in *Lilium* bulbscales grown in vitro. Plant Cell Physiol 23:67–74
Takayama S, Misawa M (1982c) A scheme for mass propagation of *Lilium* in vitro. Sci Hortic 18:353–362
Takayama S, Misawa M (1982d) Cultivation of in vitro propagated *Lilium* bulbs in soil. J Am Soc Hortic Sci 107:830–834
Takayama S, Misawa M (1982e) Mass propagation of ornamental plants through tissue culture. In: Fujiwara A (ed) Plant Tissue Culture 1982, Maruzen, Tokyo, pp 681–682
Takayama S, Misawa M (1983) The mass propagation of *Lilium* in vitro through stimulation of multiple adventitious bulbscale formation and through shake culture. Can J Bot 61:224–228
Takayama S, Amo T, Fukano M, Oosawa K (1985a) Mass propagation of strawberries by jar fermentor culture. (1) Establishment of aseptic culture and the long-term storage conditions in vitro under low temperature (in Japanese). Abst 1985 Spring Meeting of Japanese Society for Horticultural Science, Tokyo, April 3, 1985, Japanese Society for Horticultural Science, pp 218–219
Takayama S, Amo T, Fukano M, Oosawa K (1985b) Mass propagation of strawberries by jar fermentor culture. (2) Studies on the optimum conditions in a liquid medium and the establishment of mass propagation scheme using a jar fermentor (in Japanese). Abst 1985 Spring Meeting of Japanese Society for Horticultural Science, Tokyo, April 3, 1985, Japanese Society for Horticultural Science, pp 220–221
Takayama S, Amo T, Fukano M, Akita M (1987) Mass propagation of gladiolus by liquid culture methods (in Japanese). Abst 10th Meeting Japanese Assoc Plant Tissue Culture, Sendai, July 14, 1987, Japanese Association for Plant Tissue Culture, 88 pp
Takayama S, Amo T, Fukano M (1991) Rapid clonal propagation of *Hyacinthus orientalis* bulbs by shake culture. Scientia Hortic 45:315–321
Tamura S (1978) Adventitious bud formation from excised hyacinth bulbscale in vitro. J Jpn Soc Hortic Sci 46:501–508
Tanaka H, Nishijima F, Suwa M, Iwamoto T (1983) Rotating drum fermentor for plant cell suspension cultures. Biotechnol Bioeng 25:2359–2370
Ushiyama K (1988) Mass culture technology for plant cells and the production of secondary metabolites (In Japanese). BioScience and Industry 46:7–11
Ushiyama K, Miyamoto Y, Oda H, Ishida Y (1984a) Gaseous phase culture equipment for continuous culture of plant tissues (in Japanese). Japanese Patent, Kokai Tokyo Koho, 59–45873
Ushiyama K, Oda H, Miyamoto Y, Ishida Y (1984b) Gaseous phase culture method and the equipment for plant tissues. Japanese Patent, Kokai Tokyo Koho, 59–45879, (in Japanese)
Ushiyama K, Oda H, Miyamoto Y (1986) Large-scale tissue culture of *Panax ginseng* root. Abstr VI Intern Congr Plant Tissue and Cell Culture, Minnesota University, pp 252
Ushiyama K (1988) Large scale culture techniques of plant cell and the production of secondary metabolites. BioSci Industry 46:7–11
Wagner F, Vogelmann H (1977) Cultivation of plant tissue in bioreactors and formation of secondary metabolites. In: Barz W, Reinhard E, Zenk MH (eds) Plant formation of secondary metabolites. Plant tissue culture and its bio-technological application. Springer, Berlin Heidelberg New York, pp 245–252
Yanagawa T, Sakanishi H (1977) Regeneration of bulblets on *Hippeastrum hybridum* bulb segments excised from various parts of a parent bulb. J Jpn Soc Hort Sci 46:250–260

IV.6 Climate Control of Greenhouses and Crop Growth

G.P.A. Bot[1] and H. Challa[2]

1 Introduction

A greenhouse is a light-transparent structure, constructed to modify the environmental conditions for crop production. This structure forms a physical barrier between the interior and the ambient environment and is therefore responsible for the modification of some of the major growth factors: compared to the outside there is less light, less air movement, and a lower CO_2 concentration during the light period; on the other hand, water vapor pressure and air temperature are higher than outside. Whereas in the past temperature increase was the first objective of the use of greenhouses and related structures, it became gradually obvious that control of temperature, although an important aspect, should be considered as part of the control of the complex of growth factors.

Greenhouse climate control is a complex subject for a number of reasons: the physics of the greenhouse climate is rather complex, one has to deal with a control system in which no simple relation exists between the control actions and the processes to be controlled, and last but not the least the reactions of crops to the environment are extremely complex and still insufficiently understood.

Greenhouse cultivation, although expensive, has the great advantage of allowing out-of-season production, cultivation of crops not adapted to the local climatic conditions, and furthermore the improved control may give rise to products of improved quality. This is being practiced in the Netherlands in large areas (Table 1). Moreover, the enveloping of air is favorable for disease control and diminishes the stress of pathogens and pests from the outside environment. For the propagation of many crops, a protected environment is required in order to obtain good results. Besides these advantages, disadvantages may occur. Particularly light conditions, both in day light integral and day length, may differ greatly from the natural conditions for the crop involved. The principle of improving growth conditions of crops by manipulation of the microclimate was already known in ancient times. Sheltering is an example of this general principle; already in the ancient Egyptian and Roman civilizations crops were protected against a hostile environment for out-of-season production. In Europe the evolution was from orangeries and winter gardens to the commercial greenhouses in use nowadays (Van de Muijzenberg 1980).

Parallel to the evolution of the greenhouse structure, the equipment developed. At first a ventilation system had to prevent excess temperatures at high irradiation. With unheated, cold greenhouses the potentials for production are limited. Therefore heating

[1,2]Institute of Agricultural Engineering (JMAG), Dept. of Horticulture, Wageningen Agricultural University, 6700 AA Wageningen, The Netherlands

Biotechnology in Agriculture and Forestry, Vol. 17
High-Tech and Micropropagation I (ed. by Y.P.S. Bajaj)

Table 1. Some statistics of glasshouse culture in The Netherlands (situation of 1987). 1 ha (hectare) = 10 000 m^2; 1 Dfl (Dutch guilder) ≈ US$ 0.45 (1989)

Crops	Area (ha)	Estimated annual average yield (Dfl/m^2)	Estimated annual production value ($Dfl \times 10^9$)
Vegetables	4452	58	2.3
Cut flowers	3569	70	2.6
Pot plants	806	114	1.4
Bedding plants	218	–	–
Total	9045	68	6.3

systems were introduced to improve environmental control and to extend the growing season to year-round production. Again, Van de Muijzenberg (1980) describes this evolution in detail. Originally, ventilation and heating were controlled by hand. Especially in regions with strongly fluctuating outside weather conditions, as in northwestern Europe, this was a cumbersome operation requiring much labor and attention. Therefore automatic climate control was introduced, at first by simple on-off thermostats, which were subsequently replaced by electronic controllers. The control algorithms were based on the practical procedures developed in manual control (Strijbosch and Van de Vooren 1975). The greater potentials of electronic controllers and improved knowledge of the physiology of greenhouse crops (Gaastra 1959) stimulated the development of more scientifically based adaptations of control algorithms (Bowman and Weaving 1970; Bokhorst et al. 1972; Heijna 1975) and of methods to control crop growth and production directly (Takakura et al. 1974). The practical significance of these developments, however, was limited due to the limitations of the hardware and the complexity of the problems that had to be solved.

The introduction of computers for greenhouse climate control was a challenge to exploit its potentials by implementation of horticultural and plant physiological knowledge (Challa 1976). For this purpose, however, information from the various disciplines had to be available in a coherent and structured form and a systems approach had to make clear what kind of information was needed (Udink ten Cate et al. 1978). Especially information on short- and long-term crop growth and development was missing. Present research in greenhouse climate control is aimed at optimization, following the systems approach (Challa et al. 1988), applying operation research (Bailey and Seginer 1989), or following the "speaking plant" approach (Hashimoto et al. 1989). Besides the objective of improving production, there was an increased interest in energy saving after the sharp rise of energy costs in the 1970's and early 1980's (Germing 1978). This was achieved by improvement of the greenhouse equipment, e.g., thermal screens and heating systems, improved greenhouse constructions and greenhouse materials, adapted greenhouse crops, better climate control strategies, and improved horticultural operation. Although the price of fuel decreased in the late 1980's, this development has triggered an approach which aims at production at a high level of quantity and quality against minimal costs.

This short and global survey of the history of greenhouse operation shows that a development has taken place where the emphasis has changed from protection from adverse conditions to highly sophisticated production. Optimization of the production process became the first objective, accompanied by an increasing demand for scientific knowledge.

2 A Greenhouse as a Production Facility

Growing a crop in a greenhouse allows manipulation of the growing conditions and thereby management of the production processes. In fact, a greenhouse can be considered as a biotechnical reactor in which a crop converts carbon dioxide, water, and fertilizer to a marketable product with the aid of radiative energy. Environmental conditions in this reactor have to be controlled in such a way that the conversion, i.e., the exploitation of available light, is optimal. However, this conversion process is very complex, with often highly crop-specific reaction patterns and with subprocesses with quite different response times ranging from seconds to weeks or even to more than a year.

A discussion on the greenhouse as a production facility has to include two aspects.

1. How is the plant environment modified by the greenhouse and how can the environmental factors be manipulated? (Sect. 2.1).
2. What are the mechanisms of crop production and how is the production process affected by the environmental factors? (Sect. 2.2).

It should, however, be noted that although it is useful to analyze the individual growth factors and the crop reactions to these factors, it is practically impossible to control these factors independently in normal greenhouses (Challa 1990) and as a consequence one has always to deal with combined effects of growth factors.

2.1 Greenhouse Climate

Greenhouses are developed in various shapes and sizes and with a great variety of equipment for different conditions (Hanan et al. 1978; Von Zabeltitz 1978). In this short survey only some basic characteristics are given, not specific for a particular greenhouse type.

The set of momentary environmental factors in the greenhouse is referred to as the greenhouse climate. The modification of the outside weather to the greenhouse climate by the greenhouse is mainly achieved by two mechanisms:

— The cover is (partly) transparent for incoming shortwave radiation and absorbs (completely or partly) the backward thermal radiation (energy trap).
— The air is enclosed in the envelope and therefore stagnant. Thus both the turbulent mixing with outside air and the local air velocities are strongly decreased.

The last effect is by far the most important; however, the first effect has to be taken into account for a proper understanding of greenhouse climate.

Although the basic mechanisms are simple, the various physical processes have to be studied both separately and in their mutual interaction to understand their impact on the greenhouse climate and therefore on the growing conditions for the crop. The

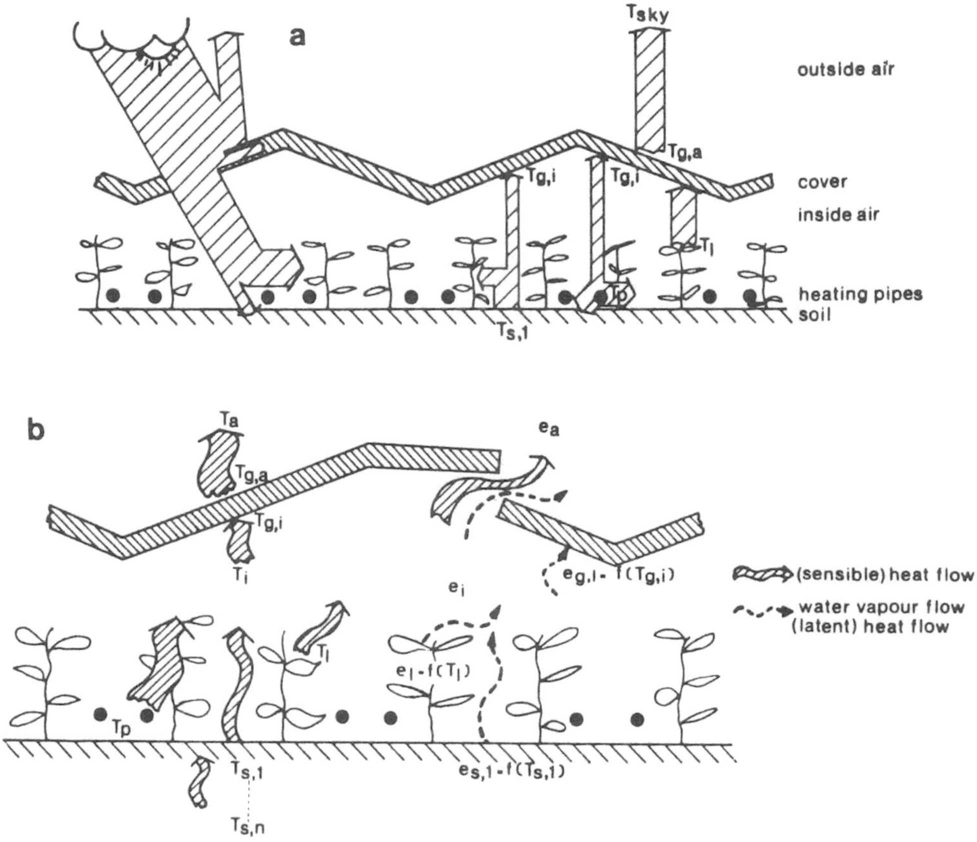

Fig. 1a,b. Scheme of the major physical processes in greenhouse climate. **a** Radiative short- and longwave energy transfer. **b** Convective and conductive energy and mass transfer

main physical processes can be considered schematically according to Fig. 1a,b, following the above-mentioned mechanisms.

2.1.1 Shortwave Radiation

Incoming shortwave global radiation (direct solar plus diffuse hemispherical) is partly reflected, absorbed, and transmitted by the transparent and opaque components of the cover (Fig. 1a). The transmission can be calculated, knowing the optical properties of the materials, the greenhouse construction, geometry, and orientation, and the directional distribution of the radiation (Stoffers 1971; Kozai et al. 1978; Bot 1983; Critten 1989). The shortwave radiation regime at crop level differs from that outside as a result of the light interception of the cover: the shortwave radiation is decreased and its composition changed due to shadowing of the opaque parts and the diffusing effect of the transparent materials. Not only photosynthesis of the crop is affected by the altered shortwave radiation level and composition (Challa and Schapendonk 1984), but also the energy budget of the greenhouse (Bot 1983).

2.1.2 Longwave Radiation

The longwave radiation between the various parts of the greenhouse and the outside environment is intercepted by the cover (Fig. 1a). Glass is almost completely opaque for radiation in the thermal wavelength region of about 5–50 μm. Most plastic covering materials are partly transparent to longwave radiation but due to film condensation at the inner surface, also these covers act in almost all cases as ideal absorbers. The apparent temperature at which the complete atmosphere emits and absorbs thermal radiation, the so-called sky temperature, is of crucial importance for the thermal radiation exchange between the greenhouse and the environment. This sky temperature can be calculated from meteorological data as an average for long periods only with restricted accuracy (Wartena et al. 1973); momentary values can only be determined experimentally, e.g., applying a pyrgeometer. This is not included in standard meteorological observations.

Especially the cover temperature is strongly affected by the thermal radiation exchange, and this in turn has its impact on the greenhouse climate.

2.1.3 Ventilation

The enveloping of air strongly affects the greenhouse climate through the decreased air exchange with the environment. This means that allowing more or less air exchange by ventilation is a powerful tool in the control of environmental conditions. The rate of air exchange directly affects temperature and composition (concentration of water vapor, carbon dioxide etc.) of greenhouse air.

Almost all greenhouses are naturally ventilated through one-side-mounted windows in the cover and the side walls. Only few greenhouses, operating under extreme conditions, are provided with forced ventilation systems or even cooled by pad and fan transpiration. Forced ventilation gives rise to a substantial increase of operational costs, but an advantage is that ventilation can be well controlled.

Natural ventilation is driven by pressure differences over the ventilation openings, which are due to the wind field around the greenhouse and to the temperature difference between inside and outside. The rate of ventilation depends on the flow characteristics of the openings. A quantitative model of ventilation in small greenhouses, validated in a wind tunnel, was presented by Kozai et al. (1980). For large Venlo-type greenhouse blocks, Bot (1983) presented a model with validation in situ. Recently, De Jong (1990) generalized this model.

2.1.4 Crop Transpiration

From a physical point of view, the crop is an energy converter, converting net radiative energy influx to a flux of latent (transpiration) and sensible heat to the air. Only a negligible part of the absorbed shortwave radiation is exploited for photosynthesis. The net radiative energy input (Fig. 1a) results in a crop temperature at which the input is balanced by the energy losses due to transpiration and sensible heat (Fig. 1b). Crop transpiration is the major vapor flow to the greenhouse air, which gives rise to an increased humidity. For the greenhouse climate, crop transpiration is crucial; without a crop the climate is hot and dry, with a crop it is cool and humid (Businger 1963). Transpiration is not only determined by the vapor pressure difference between leaf tissue and air as driving force, but also by the resistance to vapor transport between leaf tissue and air. This resistance cannot be altered directly. The total resistance is primarily determined by the stomatal resistance, and to a lesser extent by the boundary layer resistance. Boundary layer resistance is a function of local air velocity, which can be marginally controlled by the use of fans, and is affected by the temperature of the heating pipes. Stomatal resistance cannot be controlled directly, but it is affected by irradiation, CO_2 concentration, air humidity, and crop water status (Zeiger 1983). For outdoor crops transpiration is investigated extensively as a function of relevant parameters and reviewed in handbooks (e.g., Monteith 1975, 1976); for greenhouse crops few fundamental publications are available. Stanghellini (1987) reviewed literature on this item and presented a validated model.

2.1.5 Energy Transfer

The major convective energy flows between the greenhouse air and its environment are that from or to the heating pipes, the soil surface, and the cover (Fig. 1b). They directly affect greenhouse air temperature. Due to the low local air velocities in the greenhouse (0.05–0.2 m/s), the exchange mechanism is mainly natural convection. However, the exchange differs from literature data on pure natural convection in an undisturbed laboratory environment and has to be determined experimentally in situ (Bot 1983). The exchange of sensible heat to the cover is directly connected to the exchange of latent heat due to condensation of water vapor to the cover. The condensation flow directly affects the humidity in the greenhouse, especially when the ventilation rate is low. The condensation is driven by the vapor pressure difference between greenhouse air and the cover. This difference increases strongly with decreasing cover temperature through the strong temperature dependency of the saturated vapor pressure on the cover. So the cover temperature not only has a direct effect on greenhouse air temperature, but also has a strong effect on the greenhouse air humidity. Therefore the cover temperature plays a key role in greenhouse climate. The thermal properties of the cover and the exchange to the outside environment directly affect the cover temperature. The thermal properties of the cover can be deduced from the composition of the cover. The exchange to outside is due to forced convection at normal wind speeds and, of course, dependent on the outside wind speed. Also in this case, general literature data on forced convection about flat plates are not applicable; the exchange has to be determined about the saw-tooth surface (Bot 1983).

2.2 Crop Reactions and Climate Control

2.2.1 Radiation

Shortwave radiation provides the energy required for photosynthesis, the process that generates the primary building blocks and the biochemical energy equivalents for crop growth (Penning de Vries 1974; Challa 1976). As such, photosynthesis is to be considered as one of the key processes in crop production. Especially under winter conditions at the higher latitudes of northwestern Europe, the available amount of radiation poses a severe limit on crop growth and the potentials for adequate production in a qualitative and quantitative sense. Many crops under these extreme light conditions are not able to produce at all, or give rise to products of poor quality (Cockshull 1988). However, during the last few decades, plant breeding has provided varieties that are particularly adapted to these specific conditions (Smeets 1984). For other crops, additional lighting is required for cultivation during the winter season, but the expenses are high (Sect. 2.2.1.3).

A special category of crops is formed by the group of mainly heterotrophically growing crops, that use stored assimilates for growth. For these crops, light has mainly a control function and energy supply is of little significance. Examples are the forcing of tulips, narcissus and hyacinth (bulb crops), lilac, *Prunus*, and *Forsythia* (woody ornamentals).

At lower latitudes and during the summer season, high irradiation, often in combination with high outside temperature, forms the major problem for the growth of many crops. Diminished yield and damage to the crop may result from the direct effect of radiation on sensitive leaves, but more often negative effects are also caused by overheating and water stress. Reduction of light is obtained by shading of the greenhouse cover or application of shade screens.

Daylength is an important factor for the development of many crops. Crops cultivated in a greenhouse are often exposed to a daylength that may differ from their natural daylength. Therefore, in some cases it is necessary to control daylength for good crop performance. In addition, daylength control can be used to plan the production cycle, which is important in relation to the market and labor requirements. In order to shorten the natural daylength, darkening screens are used, whereas extension of the natural day is obtained by artificial lighting (Sect. 2.2.1.3).

2.2.1.1 Photosynthesis and Crop Growth

The relation between radiation and photosynthesis of individual leaves is well documented, but often it is not fully realized that the response of crop photosynthesis may deviate considerably from that of individual leaves (Fig. 2). In general, a crop has a better ability to utilize high radiation levels, because the light is distributed over leaves with different orientations and at different positions within the canopy, and consequently only a small fraction of the leaves will receive full irradiance (Goudriaan 1977). An irradiance of 300 W/m^2 (photosynthetically active radiation, PAR) is saturating at a CO_2 pressure of 34 Pa (~340 ppm) and close to saturation at 100 Pa (~1000 ppm) for a single leaf, whereas no saturation is observed with a closed canopy at an irradiance of 500 W/m^2.

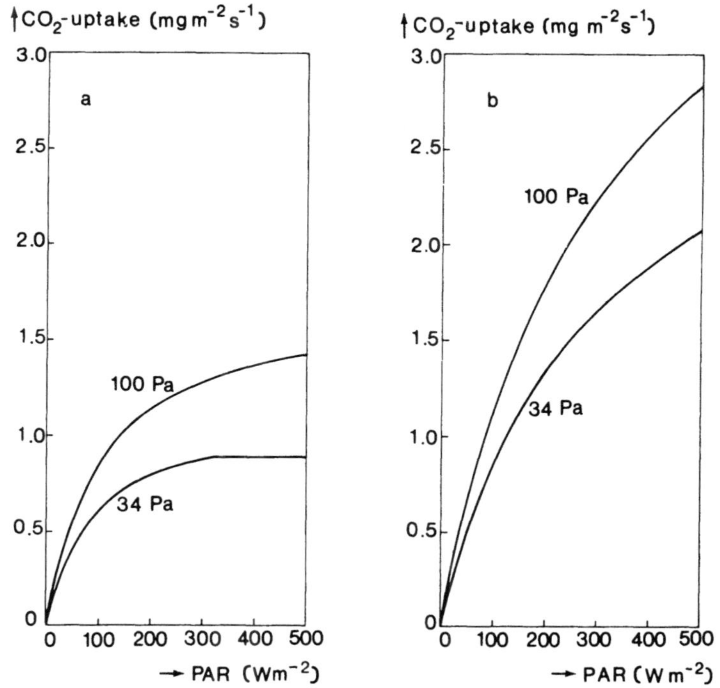

Fig. 2a,b. Light response curves of the rate of gross CO_2 uptake of a single leaf (**a**) and of a crop (**b**) with leaves with photosynthetic properties as in (**a**). Radiation is photosynthetic active radiation (*PAR*) at the top of the leaf or the canopy. Response curves are calculated according to Challa (1990), using the leaf photosynthesis model of Farquhar and Von Caemmerer (1982). Conditions for **b**: diffuse radiation, leaf area index (LAI) = 3, otherwise see Challa (1990)

Dry matter production of crops is linearly related to crop photosynthesis according to Penning de Vries and Van Laar (1982):

$$Y = \alpha (P - R_m), \tag{1}$$

where Y = dry matter production (g/m^2/day), α = conversion efficiency (g dry weight/g CH_2O), P = gross photosynthesis (g CH_2O/m^2/day), R_m = maintenance respiration (g CH_2O/m^2/day).

Equation (1) makes clear that under low light conditions increase of Y due to increase of P will be more than proportional, especially at high R_m and low P (Challa and Schapendonk 1984). High values of R_m may be expected at high temperature and in particular when the biomass per unit area is high (Penning de Vries 1975).

With young isolated growing plants, a situation characteristic for the initial phase of the production cycle of many crops, the rate of photosynthesis depends strongly on the interception of light. This interception is dependent on the crop characteristics according to Goudriaan (1982):

$$I_a = I_0(1 - 4)(1 - e^{-k\,LAI}), \tag{2}$$

where I_a = total amount of radiation absorbed by the crop, I_O = radiation level at top of the canopy, r = reflection coefficient of the crop, k = extinction coefficient of the canopy, LAI = leaf area index, the leaf area per unit ground area.

At low LAI, I_a, and consequently P at not too high radiation, and Y [Eq. (1)] are almost linearly related to LAI, and as a result plants grow approximately exponentially. In this situation besides P also the reaction of the leaf area ratio (LAR, the ratio of leaf area over total plant dry weight) to radiation is important. LAR increases with decreasing radiation, compensating partly for the loss of photosynthesis per unit leaf area by increasing that area. As a result, a less than proportional reaction of relative growth rate of young plants to variations in radiation is usually observed (Challa and Schapendonk 1984).

2.2.1.2 Morphological Reactions

Radiation has a pronounced influence on the morphology of crops. Low radiation in many cases results in an increase of the length of internodes and a decrease of the size of leaves, flowers, leaf thickness, stem diameter, etc. (Björkman 1981; Pieters 1986). It is difficult to generalize the consequences for different crops, because each crop has its own requirements and problems. For young plants, the decrease of leaf thickness is the reason for an increase of the specific leaf area (leaf area/leaf weight), which is the most important determinant of LAR, which in turn is, as has been discussed previously, an important growth factor with young plants of most species.

With vegetable crops, morphological reactions of the crop to light are generally less important than with ornamentals, but a well-known example is the shape of leaves with lettuce, where low radiation causes the formation of narrow and long leaves, giving rise to a low-quality grade of lettuce (Bensink 1971). With ornamentals the morphology is, of course, of crucial significance, because the ornamental value is partly determined by the shape of the plant, such as plant height, plant width, leaf size, internodal length, etc.

2.2.1.3 Artificial Light

In horticultural practice artificial light is either used to increase crop photosynthesis, when natural light is a limiting factor for production or product quality, or for growth control. The first type of lighting may be referred to as assimilation light (AL), in the second case light is a qualitative factor (e.g., daylength control), and in that case we may use the term control light (CL).

AL is quite expensive, both with respect to the capital investment and the running cost, and an efficient use of AL therefore is essential. From the relation between radiation and crop photosynthesis it follows that best efficiency of AL is obtained when natural light is absent or insignificant (Fig. 3). Usually, however, a given minimum daily light integral is desired when using AL, which can be obtained by a short period of high radiation or a long period at low radiation. In the first case, the capital investment in a heavier installation may outweigh the benefit of maximum efficiency obtained with a shorter lighting period. In practice, a compromise is obtained by using AL also during a part of the day, when natural light is below a certain threshold level. It should be mentioned here that, although the purpose of AL is different from that of CL, the effects

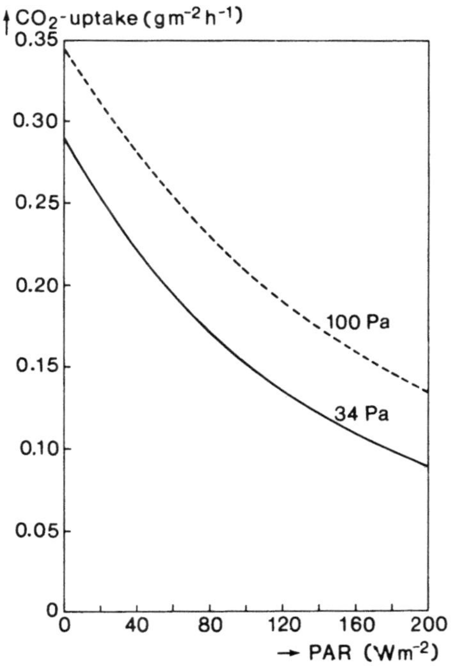

Fig. 3. Response of gross CO_2 uptake of a crop to a radiation increase of 7.36 W/m^2 (PAR) at a CO_2 pressure of 34 (~340 ppm) and 100 Pa (~1000 ppm), as a function of natural radiation (*PAR*) available. The data are obtained from Fig. 2b

cannot always be separated, and it may well be that in some cases where AL is used, its role as CL is also important (e.g., Craker et al. 1983).

At present, the main use of CL is for daylength control. Because the response of crops to daylength is a low light reaction, only low light levels are required to obtain the desired effect. Often it is even not necessary to extend the day with continuous light: a night-break may suffice for this purpose. This has resulted in the system of cyclic lighting, where the glasshouse area is divided into sections and each section is irradiated in turn. The quality of light, however, does play a role and therefore the light source deserves attention. In practice, incandescent light is used in general, because the installation is cheap and the energy requirements are small, but other light sources also play a role, such as fluorescent lamps (mainly with pot plants) and compact gas-discharge lamps. In cases where increase of photosynthesis is required besides daylength control, assimilation light is also used for daylength control.

2.2.1.4 High Radiation

High radiation levels cannot be used efficiently by all crops. A great number of pot plants, but also certain cut flowers have to be protected from too high irradiation, because the crop may be damaged or quality loss may occur. In order to reduce the radiation level inside the greenhouse, the greenhouse cover in those cases is shaded, or a shading screen is applied. The advantage of the latter solution is that the screen can be opened when the radiation level is low, and that less labor is required to apply and to remove the shading material. However, the capital investment in screens may be substantial.

2.2.2 CO_2 Concentration

The greenhouse cover is a barrier between the interior and the outside environment. With respect to CO_2 concentration inside the greenhouse this has two consequences: when the air exchange rate is small during the light period, the CO_2 concentration may decrease to values close to the compensation point (Schapendonk and Gaastra 1984), due to the photosynthetic activity of the crop. On the other hand, the CO_2 concentration may be increased to values far above the natural concentration by CO_2 enrichment, and this gives rise to substantial increase of the efficiency of light utilization for photosynthesis (Bravdo 1986) and consequently yield (Kimball 1986).

The economics of CO_2 enrichment depend on the conditions (loss of CO_2), the price of CO_2, which in turn is related to the source of CO_2, and the resulting yield increase (in economic terms) (Challa and Schapendonk 1986; Seginer et al. 1986). CO_2 enrichment is profitable in many cases because the value of the yield increase is usually high compared to the cost of CO_2. Therefore CO_2 enrichment is particularly important at high irradiation, but then the practical problem arises that there is a surplus of heat in the greenhouse, requiring ventilation, and thus extra CO_2 supply (Bellamy and Kimball 1986). An extra practical problem is that when CO_2 is obtained from the combustion of gas or oil, as is the most common procedure, also heat is generated during the periods without energy demand. In the case of CO_2 from flue gases of the central boiler, a separation of heat use and CO_2 generation in time may be achieved by storing the resulting hot water in storage tanks until required for heating. This solution is common practice in The Netherlands.

When gas or oil are combusted, besides CO_2 also other products are formed, such as water vapor, NO_x, and other contaminants (Hand 1986; Van Berkel 1986). The increase of water vapor pressure, especially at low ventilation rates, is often undesirable, but the presence of air contaminants may result in a substantial decrease of the positive effect of CO_2 enrichment. This negative effect may be further enhanced by air pollution due to the industry. It is important to adjust the CO_2 burners properly to reduce the formation of air contaminants to a minimum, but total elimination is impossible. Particularly when sensitive crops are grown in areas where air is polluted, CO_2 enrichment has to be limited to values lower than would be profitable when considering only photosynthesis (Wolting et al. 1985).

2.2.2.1 Photosynthesis

The primary effect of CO_2 concentration of the air inside the greenhouse is on photosynthesis. In textbooks on crop growth it is often stated that CO_2 concentration affects only the maximum rate of leaf photosynthesis. However, due to competition of CO_2 and O_2 for binding sites of ribulosebiphosphate-carboxylase-oxygenase, there is also a marked effect of CO_2 concentration on the rate of leaf photosynthesis at low irradiance (Schapendonk and Gaastra 1984). CO_2 enrichment is therefore not only important at high radiation, but may contribute substantially to enhance crop photosynthesis at intermediate and low irradiance (Fig. 4). In view of present knowledge on the reaction mechanisms involved, the Blackman theory, where only one factor can be limiting at any moment, does not provide a proper description of the real situation.

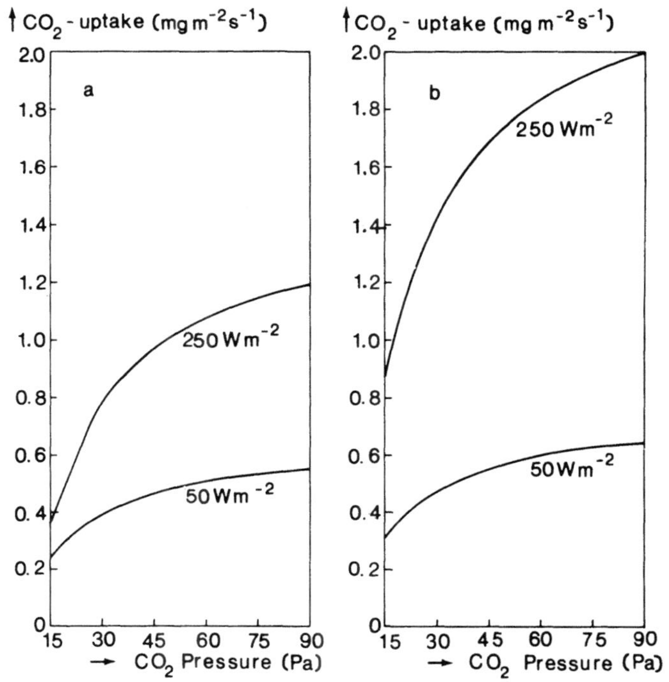

Fig. 4a,b. Response of the rate of gross CO_2 uptake of a leaf (**a**) and a crop (**b**) to CO_2 (partial pressure in the greenhouse air) at a PAR of 50 or 250 W/m^2 at the top of the leaf or the canopy. 1 Pa CO_2 pressure is approximately equivalent to a relative CO_2 concentration of 10 ppm under normal conditions. Other details according to Fig. 2

2.2.2.2 Stomatal Control

It is well known from plant physiological literature that CO_2 concentration in the air affects the stomatal opening of leaves. Raschke (1986) stated that in a number of plants stomatal control is such that the ratio between internal and external CO_2 concentration of the leaf is maintained constant, but in many plants, especially when they have not been exposed to stress, there is no sensitivity of stomata to CO_2 concentration at all. The result of a decreased stomatal conductivity due to high CO_2 concentration is that transpiration will be diminished and that CO_2 enrichment will be less effective than may be anticipated if there were no effect on stomatal resistance. The mechanism of stomatal control, however, is so complicated that it is not possible to generalize stomatal behavior. The decrease of transpiration due to CO_2 enrichment is a problem with sensitive crops and in that case growers are very careful to limit the CO_2 concentration to safe values.

2.2.3 Air Temperature

As has been discussed before, temperature inside a greenhouse can be controlled to a certain extent, primarily by heating and ventilation. The reactions of crops to temperature are quite complex, because temperature interferes in a great number of growth processes. A first distinction can be made in stress- and normal reactions. Stress conditions occur when temperature reaches low or high extremes where normal crop growth is disturbed and the crop may be damaged (Steponkus 1981). At low temperature, membranes may be malfunctioning and phenomena known as chilling and photobleaching may occur (Larcher and Bauer 1981). At high temperature, vital enzymes may be damaged and critical processes may be hampered (Kappen 1981). In addition, high temperatures usually are accompanied by high radiation, resulting in high transpiration rate and water stress, which in turn lead to stomatal closure and a further rise of temperature of the crop. Although it is clear that extreme temperatures should be prevented, this is not always possible. In the case of too low temperatures this can be due to a lack of heating capacity. In the case of too high temperatures this can be due to a combination of high irradiation and high outside temperature. Within the "normal" temperature range, however, limits for temperature have to be respected also in order to keep the crop in a good condition and to obtain a good product quality. These limits are known from horticultural practice, but are sometimes poorly understood. Especially with ornamentals, a fine tuning of temperature is often required in order to obtain the desired crop.

2.2.3.1 Dry Matter Production

The general relation between gross photosynthesis and dry matter production was given by Eq. (1). In this equation photosynthesis (P) (Berry and Björkman 1980) and maintenance respiration (R_m) (Penning de Vries 1975) are sensitive to temperature. In contrast, the conversion efficiency of carbohydrates to structural dry weight is generally supposed to be independent of temperature (De Wit et al. 1978). The relation between gross photosynthesis and temperature is affected by radiation and CO_2 concentration (Fig. 5). The reaction of a crop resembles that of an individual leaf, though it is slightly less pronounced due to the averaging of radiation over the leaf area of the crop. The reaction of R_m to temperature is less well documented due to the poor understanding of the nature of this process and its control. In general, its behavior is described by means of a Q_{10} with a value of approximately 2.0 (Penning de Vries and Van Laar 1982). Especially under conditions of low radiation and a high biomass per unit ground area, the effect of temperature on R_m will dominate the response of dry matter production.

2.2.3.2 Development

In general, crop development is stimulated by high temperature. Leaf plastochron (the time between initiation of successive leaves) is decreased (Milthorpe 1959) and the period from initiation until maturity of organs is generally decreased. A general description of this phenomenon is the well-known concept of heatsums (Hesketh et al. 1980; Vos et al. 1982), where the developmental stage of crops or organs D_t is quantified by referring to a total number of degree-days required to obtain the desired stage:

$$D_t = \Sigma\{MAX[0, (T - T_{min})]/HSUM\}, \quad (3)$$

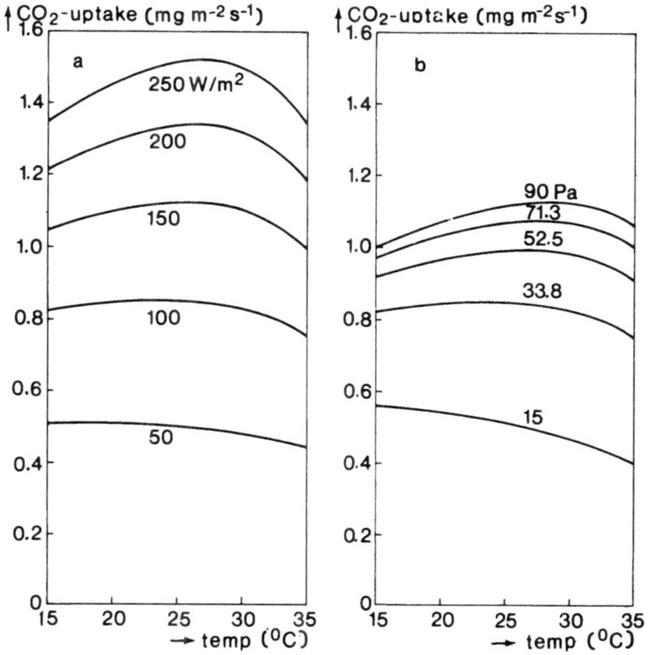

Fig. 5a,b. Effect of temperature on the rate of gross CO_2 uptake of a crop at 34 Pa CO_2 pressure and different PAR at top of the canopy (**a**) and at a PAR of 100 W/m^2 and different CO_2 pressure (**b**). 1 Pa CO_2 pressure is approximately equivalent to a concentration of 10 ppm under normal conditions. Other details according to Fig. 2

where D_t = developmental stage at day t, T = average diurnal temperature, T_{min} = minimum temperature for development, HSUM = the number of degree-days required to reach the stage desired.

The term MAX[0, $(T - T_{min})$] is used to indicate that at $T < T_{min}$ the rate of development is 0.

The heatsum approach has its benefits to describe development within a given stage. It should be stressed that transitions of crops from one stage to another, such as the transition from vegetative to generative growth, or break of dormancy, are in general under the control of far more complex mechanisms. Temperature may be one of the factors involved and in general the heatsum concept alone does not provide an adequate description. In addition, it should be realized that sensitivity to temperature may change in the course of a 24-h period (Cockshull 1988). In the context of this chapter it is not possible to treat these phenomena in detail, because they are mostly quite specific and can be understood to a certain extent by referring to the ecology of the site from where the species originated.

2.2.3.3 Low Temperature

Low temperatures normally should not occur in greenhouses, where the heating system should have sufficient capacity to cover the energy demand under almost all situations.

Low temperatures, however, do occur in low-cost greenhouses that are commonly used in mild climates (Bot 1987; Pardossi et al. 1988). Chilling injury, photo-bleaching, and malformation of products may be the result. In modern, intensively managed greenhouses, too low temperatures may occur locally, due to a not uniform distribution of temperature inside the greenhouse, a problem that is hard to prevent under all circumstances with the present greenhouse technology.

Some thermophyllic crops, such as, e.g., cucumber, are particular sensitive to low root temperature (Karlsen 1980), and for those crops control of root temperature is essential, a condition that is particularly easy to accomplish in soilless cultures. Other crops, such as freesia, require a low soil temperature for the transition from the vegetative to the generative growth and in that case soil cooling is required to obtain a good crop (Dijkhuizen and Van Holsteijn 1975).

2.2.3.4 High Temperature

High temperatures of the greenhouse air under conditions of high radiation and high outside temperature have to be prevented for the cultivation of many crops by some way of cooling (Quan and Takakura 1988). Sprinkling of water on the roof gives some relief. Mechanical cooling of greenhouses is only feasible in research facilities. In dry climates, evaporative cooling may contribute considerably to reduce temperatures and simultaneously crop transpiration. The other possibility to decrease temperature under these extreme conditions is to use screens or to shade the roof in order to diminish the radiative energy entering the greenhouse (Sect. 2.2.1.4).

High temperatures give rise to many unwanted effects. With respect to dry matter production of crops, it is clear that negative effects on photosynthesis do occur. These effects are partially direct, partially indirect. A high temperature results in a decrease of the rate of potential photosynthesis, an increase of photorespiration and an increase of the compensation point for CO_2, and each of these will contribute to a decrease of the rate of photosynthesis (Berry and Björkman 1980). In addition, stomatal resistance will tend to increase, due to direct effects and to water stress.

Besides the effects of high temperature on dry matter production, many, more crop-specific, negative effects are known, such as damage to the crop, quality loss of the product, or even total loss of production.

2.2.3 Air Humidity, Transpiration, and Crop Water Status

The significance of water for crop growth is manifold:

- water is the medium required for the living structures and for many biochemical reactions,
- an adequate water supply is required to maintain turgor pressure, which in turn is responsible for providing firmness to the plant and which is the motor for extension growth of cells,
- circulation of water through the plant enables transport of nutrients, assimilates, and hormones from root to shoot and vice versa,
- transpiration of leaves contributes to the reduction of leaf temperature under conditions of high radiation and high air temperature.

The great significance of water to plant life has resulted in extensive literature on this topic, particularly with respect to the problem of drought. Yet it is curious to notice that in the case of greenhouse culture the theoretical background for control of crop water status and transpiration is still very weak. Conditions of high humidity and low transpiration cause a good deal of worry to most growers. The plant physiological understanding of the effect of these conditions, however, is still very limited (Van de Sanden 1985).

A direct control of crop water status does not yet exist in practice for some simple reasons. The desired levels are unknown, reliable sensors to measure water status in practice are not available, and, as has been indicated previously, air humidity is not controlled directly: in practice its control is always connected to the control of other factors like temperature or radiation.

2.2.3.1 Transpiration

Inside the greenhouse, energy input from solar radiation is lower and water vapor pressure is higher than outside. As a result, transpiration of greenhouse crops usually will be smaller than that of crops growing in the open. This can be an advantage in areas where water is expensive, but more often low transpiration is considered as the cause of many physiological disorders, particularly in situations of low radiation and relatively high outside temperature and therefore high inside relative humidity (Hand 1988). Crops growing under these conditions often are weak plants that are sensitive to pests and diseases and to sudden changes in climatic conditions. Under conditions of high radiation and particularly in the early stages of crop growth, when the crop has little influence on greenhouse climate, the temperature of the crop will rise to high values and consequently the vapor pressure difference between leaves and air will also become large. In that case the rate of transpiration on a leaf area basis may attain high values, water stress may occur, and crops may be damaged.

The indirect effect of controlled air temperature and humidity on the transpiration process makes it hard to control its rate directly. It may be enabled by the implementation of transpiration models (Stanghellini and Van Meurs 1989). A more passive way of controlling transpiration at high radiation is realized by shading of the greenhouse roof, and use of screens or evaporative cooling. At low radiation, transpiration is stimulated by extra ventilation and heating.

2.2.3.2 Water Status and Crop Growth

The control of the water status of greenhouse crops is an important problem: it affects crop growth quantitatively as well as qualitatively. Crop growth and production are affected indirectly because of the relation between stomatal opening and photosynthesis. In addition, there is a direct effect via the turgor pressure which provides the driving force for cell extension growth (Cosgrove 1986). Under unfavorable conditions, cell extension may be limited by the availability of water and, as a result, growth may diminish. The background of many qualitative effects of an unfavorable water status on crop quality in many cases is poorly understood. A rather general principle that growers try to follow is the rule that sudden great changes in water status have to be prevented. There are many examples in horticultural practice where such changes give rise to physiological disorders in the crop and the product.

3 Greenhouse Climate Control

3.1 The Control Problem

From the overview of relevant physical factors and their effect on the crop (Sect. 2), it can be pointed out how the greenhouse climate and thereby crop growth and production can be manipulated. This can be depicted schematically according to Fig. 6. The main actuators are the heating system and the ventilation system, both affecting directly and indirectly greenhouse air temperature and humidity, and thereby crop performance. Besides these major actuators, some others are installed in modern greenhouses, such as:

— Carbon dioxide is supplied as pure gas from storage tanks or it is obtained from the exhaust gases of the central heating system or of special burners.
— Thermal screens to reduce heat losses during night time, shading screens to reduce too strong irradiation in the day time, and darkening screens for daylength control.
— Artificial lighting systems to increase crop photosynthesis or for daylength control
— A great variation in crop watering systems adapted to specific situations.

Heating and ventilation are controlled continuously in feedback or feedforward control loops. In a feedback loop this means that greenhouse air temperature and humidity are measured and interpreted by a controller to regulate the actuator. The dynamics of the greenhouse climate and the feedback loop determine the control scheme that is employed, where accuracy and stability are important criteria. When the dynamics are known beforehand, feedforward control can be applied to compensate for fast disturbances. Udink ten Cate (1983) reviewed literature and introduced adaptive greenhouse climate control. He stressed the complexity of practical control systems as a mixture of control algorithms and horticultural knowledge which can be disentangled by a systems approach.

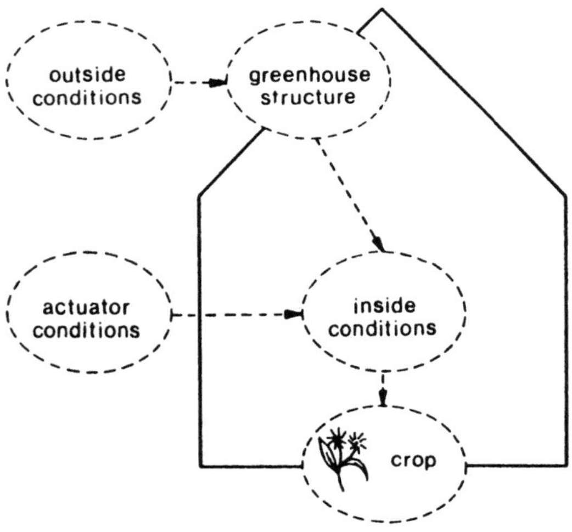

Fig. 6. Scheme of greenhouse control. (After Udink ten Cate 1983)

Table 2. Summary of factors controlled in greenhouses

Factor	Setpoint based on:
Air temperature	Day/night; season; irradiation; crop stage
Air humidity	No extremes/condensation/rapid changes
CO_2 Concentration	Day/night; CO_2 source; ventilation rate
Screens	Day/night; irradiation; energy demand
Artificial lighting	Period; irradiation; daily light integral
Water supply	Crop transpiration (measurements/model)
EC[a] of nutr. solution	Season; weather conditions; crop stage

[a]EC = Electrical Conductivity.

The other above-mentioned actuators are not controlled in feedback or feedforward control loops, but just on a time base or evaluating some systems information, e.g., measured irradiation in the watering system. The operation of these actuators is separated from the control of heating and ventilation. In the present systems, for each actuator there is a separate setpoint. The basic regimes used are summarized in guidelines which are specified for different phases and seasons in the so-called blueprint. In practice, of course, the grower does not follow these blue-prints blindly, but adjusts the setpoints according to his own experience and to his observations on the appearance and the performance of the crop. In Table 2 an overview of the various climate factors is given together with the factors governing the setpoints.

In digital controllers in use nowadays a great number of parameters have to be set. In most commercial greenhouse climate control systems some hundreds of settings have to be installed. The tuning of the settings is the responsibility of the grower (Anonymous 1987), so if the control system fails, the grower is blamed for a wrong setting. Because systems are developed by private companies, no literature is available on the structure and algorithms. Through the settings, growers try to optimize crop production both on a short and a long term, based on experience and intuition. The basic problem with present control systems is that there are strong interactions between the various control actions, that the number of parameter settings is large, that the significance of most settings is based only on practical experience, and that precise information about both the short- and the long-term reactions of the crop is lacking.

3.2 Optimal Control

Modern climate control systems should exploit in a better way the potentials of advanced computer technology and the available knowledge in the field of crop science. Recent investigations demonstrated that for the main environmental factors a rigid control in accordance to the blue-prints is by no means required (Hurd and Graves 1984; Bakker and Van Uffelen 1988; De Koning 1988a,b). Moreover, there is clear evidence that optimal setpoints for crop performance may vary depending on the outside weather conditions and the crop status over short periods of time (Challa 1976, 1978; Challa and Brouwer 1985; Challa and Schapendonk 1986; Challa 1990). Furthermore, it should be realised that the cost of maintaining setpoints for temperature, air humidity, and CO_2 concentration may vary considerably from situation to situation, depending on the weather conditions, ventilation rate, use of thermal screens, etc.

The previous discussion illustrates that present climate control could be improved fundamentally:

- setpoints for various climatic factors should be based on an instantaneous evaluation of the balance between costs and benefits,
- the control system should be able to realize setpoints for climatic factors rather than for the individual actuators, while taking into account the multiple effects they exert on greenhouse climate,
- the control system should perform adequately in a dynamic sense.

Research based on these concepts has been reported by various authors. Seginer and Albright (1980) formulated optimization criteria with respect to the operation of thermal screens. Hurd and Graves (1984) showed that substantial savings could be obtained when temperature control was optimized by taking into account the variations in energy loss from greenhouses in relation to wind speed. With respect to CO_2 control, it has been pointed out theoretically (Challa and Schapendonk 1986) and demonstrated experimentally (Nederhoff 1988) that optimization may give rise to substantial increase in economic returns. Challa et al. (1988) presented an interdisciplinary research program, aimed at application of dynamic optimization in practical greenhouse climate control. Bailey and Seginer (1989) focused on optimum control of greenhouse temperature, and Seginer (1989) tried to determine best management policy.

To evaluate costs and benefits in a dynamic way, quantitative knowledge on these items should be available. The momentary production costs depend on the desired greenhouse climate, the physical greenhouse properties, and the outside weather conditions. This dependency can be described by a dynamic greenhouse model (Bot 1989), so the momentary production costs can be calculated. The dynamic evaluation of the benefits is of a more complex nature. Momentary crop growth rate can be derived from the simulated rate of crop photosynthesis with the environmental factors as boundary conditions (Gourdriaan 1982; Spitters et al. 1988), but the resulting production of harvestable crop is obtained only after some time, at harvesting time. So both a transformation of crop growth to marketable product and an expected economic yield have to be considered in the instantaneous evaluation of the benefits. Moreover, the present crop growth models do not yet predict all relevant aspects of crop behavior adequately (e.g., crop development) and besides, in the control, aspects other than instantaneous crop performance only have to be taken into account (e.g., crop adaptation, crop morphology, product quality, disease and pest control, labor environment, interference with crop management). This additional information is available as practical experience and has to be implemented in optimization schemes. It is to be expected that this information can be implemented in an expert system or a knowledge system.

In this way the structure of an optimizing control system can be composed from subsystems, as depicted in Fig. 7. According to this scheme, the setpoints for the climate factors are set dynamically from the dynamic cost-benefit analysis, implementing dynamic systems information and practical experience. The realization of the setpoints by the controllers can also be optimized by means of "modern control theory" (Van Henten 1989).

In the complete control system, the realization of the setpoints, the levels of the setpoints, and the long-term adjustments can be distinguished according to Fig. 8. The

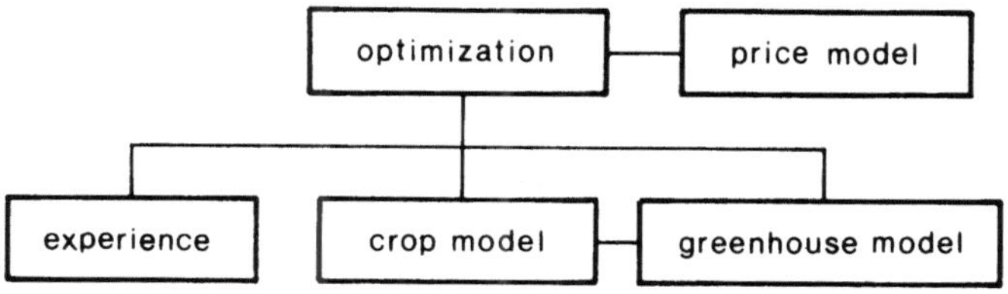

Fig. 7. Structure of an optimizing climate control system

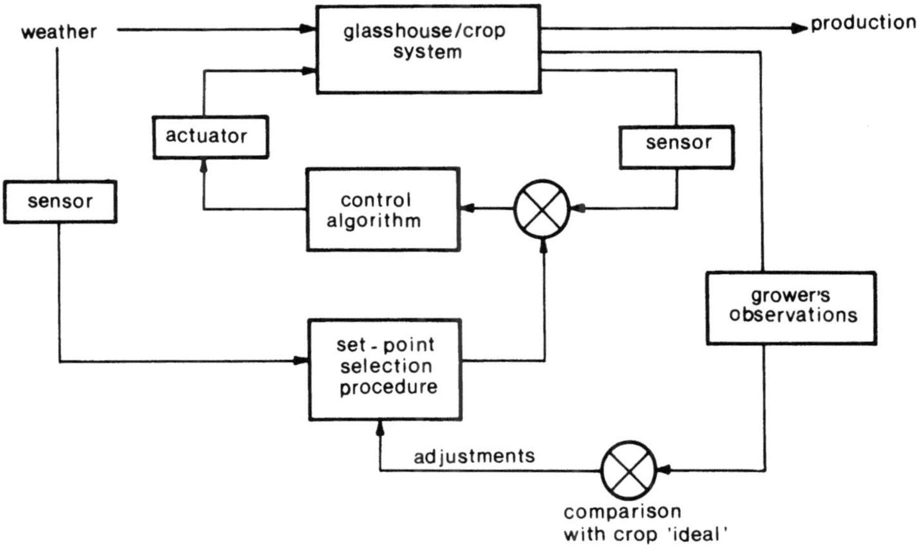

Fig. 8. Lay-out of a climate control system with various levels of control

grower has to deal with long-term objectives only, short-term control is completely performed by the control system. In this way, the grower can manage his holding to realize his long-term objectives, supported by a control system that provides him with information on the systems behavior (Challa et al. 1988; Challa 1990).

4 Summary and Conclusions

Greenhouse climate is the result of the action of the outside weather conditions on the interacting greenhouse- and crop systems. The greenhouse conditions will be modified as a result of changes in the outside weather conditions and of manipulations of the greenhouse actuators: ventilator opening, heating pipe temperature, CO_2 supply, etc.

For a proper understanding of greenhouse climate control in relation to crop performance, it is essential to take into account all the relevant physical and physiological processes that play a role in the establishment of the climate inside the greenhouse, as well as the response of the crop to its environment. These basic processes are described in this chapter. Essential processes for the greenhouse environment are those that contribute to the balances of energy, humidity, and CO_2. Generic crop processes are photosynthesis, respiration, and transpiration. In addition, a number of more crop-specific reactions are mentioned that have to be taken into account when considering the response of crops to the greenhouse environment.

The present survey makes clear that the problem of climate control in greenhouses is not only an engineering problem, but certainly also to a large extent a problem of evaluation of the effects on crop performance. The engineering problem is that the control system is principally inadequate to deal with disturbances caused by the outside weather conditions and that, in addition, the controls of temperature, humidity, and CO_2 are largely interacting. Furthermore, one has to deal with a system where multiple goals with conflicting requirements have to be satisfied. The evaluation of crop performance is partly hindered by a lack of basic plant physiological knowledge and further complicated by the largely divergent response times of often strongly interacting crop reactions.

A fundamental improvement of present climate control systems requires a systems approach where all relevant processes are described in models. Instead of setpoints for the environmental conditions, evaluation of the relevant production processes and the required inputs should form the basis for the control of the actuators. The fast development of hardware and of models of crop growth and greenhouse climate makes such developments feasible within the next 10 years.

References

Anonymous (1987) Terminologie geautomatiseerde kasklimaatregeling. [in Dutch] Stichting Mechanisatie Centrum, Wageningen, The Netherlands

Anonymous (1988a) Kwantitatieve informatie voor de glastuinbouw 1988–1989, PTG and CADG, Naaldwijk; PBN and CADB, Aalsmeer, The Netherlands

Anonymous (1988b) Landbouweconomisch bericht 1988, LEI, The Hague, The Netherlands

Bailey BJ, Seginer I (1989) Optimum control of greenhouse heating. Acta Hortic 245:512–518

Bakker JC, van Uffelen JAM (1988) The effects of diurnal temperature regimes on growth and yield of glasshouse sweet pepper. Neth J Agric Sci 36(19):36–88

Bellamy LA, Kimball BA (1986) CO_2 enrichment duration and heating credit as determined by climate. In: Enoch HZ, Kimball BA (eds) Carbon dioxide enrichment of greenhouse crops, Vol II. CRC Boca Raton, Fla, pp 167–197

Bensink J (1971) On morphogenesis of lettuce leaves in relation to light and temperature. Meded Landbouwhogesch 71(15):93

Berry JA, Björkman O (1980) Photosynthetic response and adaptation to temperature in higher plants. Annu Rev Plant Physiol 31:491–543

Björkman O (1981) Responses to different quantum flux densities. In: Pirson A, Zimmermann MH (eds) Physiological plant ecology I, Encyclopedia of plant physiology new series vol. 12A. Springer, Berlin Heidelberg New York, pp 57–107

Boers A, van der Velden NJA (1987) Raming van de bedrijfsuitkomsten in 1987 van de glastuinbouwen champigonbedrijven. PR No 24-87, LEI, The Hague, The Netherlands

Bokhorst D, Drenth A, Holsteijn GPA (1972) Lichtafhankelijke klimaatregeling voor kassen. [in Dutch] ITT publ 74, IMAG Wageningen

Bot GPA (1983) Greenhouse climate: from physical processes to a dynamic model. PhD Thesis, Wageningen Agric Univ, 240 pp

Bot GPA (1987) Analysis of the micro climate in greenhouses for flower production in Costa Rica. WMO, Geneva

Bot GPA (1989) A validated physical model of greenhouse climate. Acta Hortic 245:389–396

Bowman GE, Weaving GS (1970) A light-modulated greenhouse control system. J Agric Eng Res 15(2):100–105
Bravdo B (1986) Effect of CO_2 enrichment on photosynthesis of C_3 plants. In: Enoch HZ, Kimball BA (eds) Carbon dioxide enrichment of greenhouse crops, Vol II. CRC, Boca Raton, Fla, pp 14–27
Businger JA (1963) The glasshouse (greenhouse) climate. In: van Wijk WR (ed) Physics of plant environment. North Holland, Amsterdam
Challa H (1976) An analysis of the diurnal course of growth, carbon dioxide exchange and carbohydrate reserve content of a cucumber. Agric Res Rep 861. Pudoc, Wageningen, 88 pp
Challa H (1978) Programming of night temperature in relation to the diurnal pattern of the physiological status of the plant. Acta Hortic 76:147–150
Challa H (1990) Crop growth models for greenhouse climate control. In: Rabbinge R, Goudriaan J, van Keulen H, Penning de Vries FWT, van Laar HH (eds) Theoretical production ecology: reflections and prospects. Simulation monographs 34, Pudoc, Wageningen, pp 125–145
Challa H, Brouwer P (1985) Growth of young cucumber plants under different diurnal temperature patterns. Acta Hortic 174:211–217
Challa H, Schapendonk AHCM (1984) Quantification of effects of light reduction in greenhouses on yield. Acta Hortic 148:501–510
Challa H, Schapendonk AHCM (1986) Dynamic optimalization of CO_2 concentration in relation to climate control in greenhouses. In: Enoch EC, Kimball BA (eds) Carbon dioxide enrichment of greenhouse crops. CRC, Boca Raton, Fla
Challa H, Bot GPA, Nederhoff EM, Van de Braak NJ (1988) Greenhouse climate control in the nineties. Acta Hortic 230:459–470
Cockshull KE (1988) The integration of plant physiology with physical changes in the greenhouse climate. Acta Hortic 229:113–123
Cosgrove DJ (1986) Biophysical control of plant cell growth. Annu Rev Plant Physiol 37:377–405
Craker LE, Seibert M, Clifford JT (1983) Growth and development of radish (*Raphanus sativus* L.) under selected light environments. Ann Bot 51:59–64
Critten DL (1989) Light transmission into greenhouses. Acta Hortic 245:377–383
de Jong T (1990) Greenhouse ventilation. PhD Thesis, Wageningen Agric University, 116 pp
de Koning ANM (1988a) The effect of different day/night temperature regimes on growth, development and yield of glasshouse tomatoes. J Hortic Sci 63(3):97–103
de Koning ANM (1988b) An algorithm for controlling the average 24-hour temperature in glasshouses. J Hortic Sci 63(3):105–109
de Wit CT et al. (1978) Simulation of assimilation, respiration and transpiration of crops. Simulation Monographs. Pudoc, Wageningen, The Netherlands, pp 141
Dijkhuizen T, van Holsteijn GPA (1975) Soil cooling for freesia. Acta Hortic 51:122–130
Farquhar GD, von Caemmerer S (1982) Modelling of photosynthetic response to environmental conditions. In: Lange OL, Nobel PS, Osmond CB, Ziegler H (eds) Physiological Plant Ecology II, Encyclopedia of plant physiology new series vol 12B. Springer, Berlin Heidelberg New York, pp 549–582
Gaastra P (1959) Photosynthesis of crop plants are influenced by light, carbon dioxide, temperature and stomatal diffusion resistance. Meded Landbouwhogesch, Wageningen 59(13):1–68
Germing GH (1978) Energy and protected cultivation in The Netherlands. Acta Hortic 87:363–369
Goudriaan J (1977) Crop micrometeorology: a simulation study. Simulation monographs. Pudoc, Wageningen, pp 249
Goudriaan J (1982) Potential production processes. In: Penning de Vries FWT, Van Laar HH (eds) Simulation of plant growth and crop production. Simulation Monographs. Pudoc, Wageningen, pp 98–113
Hand DW (1986) CO_2 sources and problems in burning hydrocarbon fuels for CO_2 enrichment. In: Enoch HZ, Kimball BA (eds) Carbon dioxide enrichment of greenhouse crops, Vol I. CRC, Boca Raton, Fla, pp 99–121
Hand DW (1988) Effects of atmospheric humidity on greenhouse crops. Acta Hortic 299:143–158
Hanan JJ, Holley WD, Goldsberry KL (1978) Greenhouse management. Springer, Berlin Heidelberg New York
Hashimoto T, Nishina H, Nonami H, Youlin Yi (1989) Dynamical approach to physiological plant ecology – vegetable cultivation under controlled environment. LAEE Ehime University, Matsuyama, Japan
Heijna BJ (1975) Delta X – a new control system. Acta Hortic 46:13–19

Hesketh JD, Elmore CD, Jones JW (1980) Predicting flowering and subsequent leaf expansion. In: Heketh JD, Jones JW (eds) Predicting photosynthesis for ecosystem models. CRC, Boca Raton, Fla, pp 123–131
Hurd RG, Graves CJ (1984) The influence of different temperature patterns having the same integral on the earliness and yield of tomatoes. Acta Hortic 148:547–554
Kappen L (1981) Ecological significance of resistance to high temperature. In: Pirson A, Zimmermann MH (eds) Physiological plant ecology I, Encyclopedia of plant physiology new series vol 12A. Springer, Berlin Heidelberg New York, pp 439–474
Karlsen P (1980) The influence of root and air temperature on young cucumber plants. Acta Hortic 118:95–103
Kimball BA (1986) Influence of elevated CO_2 on crop yield. In: Enoch HZ, Kimball BA (eds) Carbon dioxide enrichment of greenhouse crops, Vol II. CRC, Boca Raton, Fla, pp 105–115
Kozai T, Goudriaan J, Kimura M (1978) Light transmission and photosynthesis in greenhouses. Simulation Monographs. Pudoc, Wageningen
Kozai T, Sase S, Nara M (1980) A modelling approach to greenhouse ventilation control. Acta Hortic 106:125–136
Larcher W, Bauer H (1981) Ecological significance of resistance to low temperature. In: Pirson A, Zimmermann MH (eds) Physiological plant ecology I, Encyclopedia of plant physiology new series vol 12A. Springer, Berlin Heidelberg New York, pp 403–437
Milthorpe FL (1959) Studies on the expansion of the leaf surface. I. The influence of temperature. J Exp Bot 10:233–249
Monteith JL (1975) Principles of environmental physics. Edward Arnold, London
Monteith JL (ed) (1976) Vegetation and the atmosphere I, II. Academic Press, London, UK
Nederhoff EM (1988) Dynamic optimization of the CO_2 concentration in greenhouses: an experiment with cucumber (*Cucumus sativus* L.). Acta Hortic 229:341–348
Pardossi A, Tognoni F, Lovemore SS (1988) The effect of different hardening treatments on tomato seedling growth, chilling resistance and crop production in cold greenhouse. Acta Hortic 229:371–378
Penning de Vries FWT (1974) Substrate utilization and respiration in relation to growth and maintenance in higher plants. Neth J Agric Sci 22:40–44
Penning de Vries FWT (1975) The cost of maintenance processes in plant cells. Ann Bot 39:77–92
Penning de Vries FWT, van Laar HH (1982) Simulation of growth processes and the model BACROS. In: Penning de Vries FWT, van Laar HH (eds) Simulation of plant growth and crop production, Simulation Monographs. Pudoc, Wageningen, pp 114–135
Pieters GA (1986) Relation between the diameter of the growing shoot and the mature size of leaves and flower head of sunflower. Physiol Plant 68:113–118
Quan Z, Takakura T (1988) A strategy for calculating the maximum cooling load of a greenhouse. Acta Hortic 230:519–526
Raschke K (1986) The influence of the CO_2 content of the ambient air of stomatal conductance and the CO_2 concentration in leaves. In: Enoch HZ, Kimball BA (eds) Carbon dioxide enrichment of greenhouse crops, Vol II. CRC, Boca Raton, Fla, pp 87–102
Schapendonk AHCM, Gaastra P (1984) A simulation study on CO_2 concentration in protected cultivation. Sci Hortic 23:217–229
Seginer I (1989) Optimal greenhouse production under economic constraints. Agric Syst 29:67–80
Seginer I, Albright L (1980) Rational operation of greenhouse thermal-curtains. Trans ASAE 23(5):1240–1245
Seginer I, Angel A, Gal S, Kantz D (1986) Optimal CO_2 enrichment strategy for greenhouses: a simulation study. J Agric Eng Res 34:285–304
Smeets L (1984) Breeding for low energy requirements in glasshouse crops. Acta Hortic 148:37–42
Spitters CJT, van Keulen H, van Kraalingen DWG (1988) A simple and universal crop growth simulator: SUCROS87. In: Rabbinge R, Ward SA, van Laar HH (eds) Simulation and systems management in crop protection. Simulation Monographs. Pudoc, Wageningen, The Netherlands
Stanghellini C (1987) Transpiration of greenhouse crops. An aid to climate management. PhD Thesis, Wageningen Agric University
Stanghellini C, van Meurs W (1989) Crop transpiration: a greenhouse climate control parameter. Acta Hortic 245:384–388
Steponkus PL (1981) Response to extreme temperatures. Cellular and sub-cellular basis. In: Pirson A, Zimmermann MH (eds) Physiological plant ecology I, Encyclopedia of plant physiology new series vol 12A. Springer, Berlin Heidelberg New York, pp 372–402

Stoffers JA (1971) Lichtdurchlässigkeit von Gewächshäusern in Blockbauweise. [in German] ITT publ 39. IMAG, Wageningen, The Netherlands

Strijbosch Th, van de Vooren J (1975) Developments in climate control. Acta Hortic 46:21–22

Takakura T, Kozai T, Tachibana K, Jordan KA (1974) Direct digital control of plant growth. I. Design and operation of the system. Trans ASAE 17(6):1150–1154

Udink ten Cate AJ (1983) Modeling and (adaptive) control of greenhouse climates. PhD Thesis, Wageningen Agric University

Udink ten Cate AJ, Bot GPA, van Dixhoorn J (1978) Computer control of greenhouse climates. Acta Hortic 87:265–272

van Berkel N (1986) CO_2 enrichments in The Netherlands. In: Enoch HZ, Kimball BA (eds) Carbon dioxide enrichment of greenhouse crops, Vol I. CRC, Boca Raton, Fla, pp 17–33

van de Muijzenberg EWB (1980) A history of greenhouses. IMAG, Wageningen

van de Sanden PACM (1985) Effect of air humidity on growth and water exchange of cucumber seedlings. Acta Hortic 174:259–267

van Henten EJ (1989) Model based design of optimal multivariable climate control systems. Acta Hortic 248:301–306

von Zabeltitz Chr (1978) Gewächschäuser, Planung und Bau. [in German] Eugen Ulmer, Stuttgart

Vos J, Drees EM, Penning de Vries FWT (1982) Grain formation and assimilate partitioning in wheat. Part II. Modelling of post floral growth of wheat. In: Penning de Vries FWT, van Laar H (eds) Simulation of plant growth and crop production. Simulation Monographs, Pudoc, Wageningen, pp 144–151

Wartena L, Palland CL, Vossen GHL (1973) Checking of some formulae for the calculation of longwave radiation from clear skies. Arch Meteorol Geophys Bioklimatol B 21:335–348

Wolting HG, van Remortel EAM, van Berkel N (1985) Air quality in greenhouses with and without CO_2 enrichment. Acta Hortic 174:351–357

Zeiger E (1983) The biology of stomatal guard cells. Annu Rev Plant Physiol 34:441–475

IV.7 Computer-Controlled Greenhouses in Japan

K. DATE[1]

1 Introduction

Application of microcomputers to environmental control for commercial greenhouses in Japan is said to date back to 1979. The Japan Greenhouse Horticulture Association's statistic data for 1990 (investigated during the period of July 1988 to June 1989) gives a total facility area of 2074 ha for glasshouses and 46737 ha for plastic houses, with a combined total area amounting to 48811 ha (covering greenhouses providing enough height for persons working in them to stand upright). Among these greenhouses, 74 ha of well-equipped glasshouses (totaling 571 units) and 114 ha of plastic houses (totaling 470 units) employ microcomputers for environmental control. About ten manufacturers produce microcomputers used for these purposes. The present statistics does not include equipment used in soilless culture.

Microcomputers used for greenhouse environmental control are roughly classified into two types. One of them is the type used for greenhouse control normally having control programs written in the ROM's. Microcomputers of this type are relatively inexpensive. Many of them, however, pose difficulties in modifying control logics or take a long time for the modifications to be carried out. Another type comes from general-purpose microcomputers that have control programs RAM-written, allowing control logics to be changed.

Our company is a greenhouse manufacturer, designing and erecting greenhouses to specifically meet grower's requirements. Today, more and more growers want to use microcomputers to automatize their greenhouses. A few examples of the experiences we have had are provided below.

2 Examples of Using a Microcomputer for Greenhouse Control

2.1 Description of Facility

At the time Nakayama started growing orchids (*Cattleya* and *Oncidium*) in 1985, he introduced the use of microcomputers to achieve labor saving in environmental control. His growing facility consists of glasshouses amounting to 1020 m^2. Among them, a double-span, even-span type glasshouse, width 8 m and length 35.75 m, is controlled

[1]Yamamoto Agricultural Corporation, 532 Yatoutajima Toyoda-Chou, Iwata-Gun Shizuoka-Prefecture, Japan

Biotechnology in Agriculture and Forestry, Vol. 17
High-Tech and Micropropagation I (ed. by Y.P.S. Bajaj)

by a microcomputer. The two sections of the glasshouse are controlled by settings common to both (Fig. 1a,b). The glasshouse is equipped with four motors for opening and closing the roof ventilator (operated in pairs), two motors for opening and closing the side ventilator, and two movable internal curtain units. To provide heating in the glasshouse, hot water produced by a boiler is fed to heating pipes in the glasshouse by line pumps.

2.2 Input into and Output from the Control System

This glasshouse is controlled by a microcomputer for greenhouse control (made by ESD, GREEN MICOM-8). Unlike a personal computer, this microcomputer is not provided with a CRT and a keyboard. Instead of these the microcomputer has the input typing device and numeric display. Pressing these keys allows setting points, such

Fig. 1a,b. Orchid culture. **a** View of the entire facilities. **b** Inside view of a glasshouse

as setpoint of room temperature for ventilation, to be input. The numeric display gives the values of setting points and current environmental conditions including the amount of solar radiation outdoors and temperature inside the glasshouse. On the left side of the microcomputer it is displaying the operating condition of microcomputer-controlled equipment by flashing lamps.

Analog signals from a solar radiation sensor, a temperature sensor, and an anemometer, all installed outdoors, and a temperature sensor installed in the glasshouse, and digital signals from a rain sensor are input into the microcomputer. Provided for control outputs are two pairs of contacts (two contacts each for opening and closing) for roof ventilator control, two pairs for side ventilator control, one for upper curtain control, one for lower curtain control, one for heater ON/OFF control, and a few for the alarm circuit.

2.3 Description of the Control

The operation of the principal devices installed in the glasshouse is as follows:

When the air temperature in the glasshouse exceeds the setpoint of room temperature for ventilation, the roof ventilators open first. If the air temperature is still higher than the setpoint of room temperature for ventilation with the roof ventilators fully opened (opened to the full opening limit), then the side ventilators are opened. When the air temperature in the glasshouse drops below the setpoint of room temperature for ventilation, the ventilators are closed in reverse order. The displacement of each ventilator is proportional to the difference between the setpoint of room temperature for ventilation and the air temperature in the glasshouse and is corrected for the difference between the outdoor air temperature and air temperature in the glasshouse. If a wind velocity exceeding the wind velocity setting point is detected, all the ventilators are closed to protect the glasshouse. If rainfall is sensed, the roof ventilators are totally closed. However, the roof ventilators can be kept slightly opened under a rainfall condition by using an aperture setting for rainfall.

When the air temperature in the glasshouse drops below the setpoint of room temperature for heating that has been set beforehand, the line pumps are operated for a set time duration. The line pumps repeat starting and stopping until the air temperature in the glasshouse reaches the setpoint of room temperature for heating. However, as the air temperature in the glasshouse comes closer to the setpoint of room temperature for heating, the operating time of the line pumps is shortened to avoid overshooting.

The movable internal curtains are operated as follows: the upper curtains, used for shading of sunlight, are opened and closed according to the amount of solar radiation outdoors. The curtains are opened when the amount of solar radiation outdoors exceeds the setting point. Since the amount of solar radiation changes from moment to moment, controlling the curtain system by instantaneous values of the solar radiation could lead to too frequent opening and closing of the curtains. To avoid this problem, the curtains are so controlled as to open or close depending on the total integrated radiation measured over a definite period of time after the preceding opening or closing operation. The transparent lower curtains are opened or closed depending on the air temperature in the glasshouse and the hour of the day. Opening the curtain requires that the setting conditions for both the air temperature in the glasshouse and the hour be met, while closing the curtains requires that either the setting condition for the air temperature in the glasshouse or that for the hour be met.

The glasshouse is managed in the following manner.

In order to manage the air temperature in the glasshouse, a day is divided into six time periods. The user sets the starting hour of each time period, the set-point of room temperature for ventilation (temperatures when undesirable to exceed from the standpoint of plant growth), and the setpoint of room temperature for heating (lowest temperatures desired from the standpoint of plant growth). When the air temperature in the glasshouse exceeds the setpoint of room temperature for ventilation, the ventilators are actuated; when the air temperature in the glasshouse drops below the setpoint of room temperature for heating, the line pumps are started. On a day with fine weather, therefore, the air temperature in the glasshouse is kept approximately at the setpoint of room temperature for ventilation during the daytime and at the setpoint of room temperature for heating during the nighttime.

The first through third time periods cover the daytime and the other three the nighttime. The setting points are so determined that photosynthesis is intensified during the former time periods and the photosynthate produced during the daytime is quickly translocated and the consumption of the photosynthate caused by excessive respiration after translocation is suppressed during the nighttime. For the first through third time periods, the user usually sets the setpoint of room temperature for ventilation for each time period at the temperature suitable to the photosynthesis process for the particular crop being grown, and the setpoint of room temperature for heating at the temperature desired to be kept on a cloudy day for growing plants. There are, however, cases in which solar radiation in the early morning does not cause the air temperature in the glasshouse to rise to the temperature suitable for photosynthesis. To cope with this problem, it is a frequent practice to set the setpoint of room temperature for heating for the first time period at the temperature suitable for photosynthesis and operate the line pumps. In winter, when the air temperature in the glasshouse becomes low, heating the glasshouse during the fourth through sixth time periods is a common practice. In order to accelerate translocation of photosynthate and suppress excessive respiration, the setpoint of room temperature for heating for each of the fourth through sixth time periods is set at a lower value than for the preceding time period. In the translocation accelerating time periods (the fourth and fifth time periods), a quantity proportional to the day's total integrated solar radiation is automatically added to the setpoint of room temperature for heating, causing the air temperature in the glasshouse to be kept at a temperature a few degrees higher than the setpoint of room temperature for heating setting. Naturally, the addition is not done on cloudy days when the total integrated solar radiation is small.

Nakayama has introduced the above-mentioned microcomputer and is satisfied with the microcomputer-controlled temperature management. When introducing microcomputers of this type, growers should thoroughly examine the details of control before choosing a manufacture, because to change control logic is difficult once it has been determined.

3 Examples of Using a General-Purpose Microcomputer

3.1 Growing Musk Melons

3.1.1 Description of the Facility

Kobayashi has been growing musk melons for about 15 years, operating seven glasshouses with his wife (Fig. 2a). Musk melons are difficult to grow in Japan, which makes them a rather expensive fruit. Four of the Kobayashis' seven glasshouses have a span of 7 m and a length of 22 m, and the remaining three a span of 5.8 m and length of 22 m. All of the glasshouses are east-west-oriented three-quarter type glasshouses (Fig. 2,b). These glasshouses are individually controlled by a microcomputer. In order to reduce the temperature difference in the east-west direction in each glasshouse, the ventilators are arranged in the following manner. The roof ventilators and side ventilators facing the south are divided into two halves on the east and west sides. Each side is provided with a motor for opening and closing (a thermometer is installed on each side). In contrast, the roof ventilators facing the north are opened or closed by a single motor. In addition to these motors, two motors for opening and closing the movable internal curtains are provided in each glasshouse. In order to heat each glasshouse, hot water produced by the boiler is fed into the heating pipes of each glasshouse by seven line pumps.

The following situation led Kobayashi to introducing a microcomputer for his melon-growing business. Melons in different glasshouses should differ in their stage of growth, and melons are harvested every 10 days. The individual glasshouses therefore are kept at different temperatures. In addition, planned shipment of melons requires that the temperature distribution in a glasshouse be made as uniform as possible for planned harvesting. These considerations make management of glasshouses very painstaking work, to say nothing of taking care of the ventilators during the daytime; management of up to five glasshouses has been the limit in the case of manually controlled greenhouses.

3.1.2 Input into and Output from Control System

On the Kobayashis' farm, seven glasshouses are individually controlled by an industrial general-purpose microcomputer (made by Yokogawa Electric Corporation, Model YEWMAC300). Like common personal computers, this computer is provided with a keyboard and CRT. The programming language used is BASIC. The user follows the instructions on each setting display appearing on the CRT and interlocutorily enters setting values in the microcomputer. Each setting display is divided into a portion for setting parameters of an individual glasshouse and another for setting parameters common to the entire system. Environmental conditions inside and outside the glasshouses and the operating conditions of all the devices are constantly displayed on the CRT. Mere operation of the function keys enables the user to see graphically represented information of short-term (2 days) changes in environmental conditions inside and outside the glasshouses and operating conditions of devices, and of the items in the grower's diary. In the hierarchy of the present system, a microprocessor-based computer commanding input, output, and control, named line controller, works under

Fig. 2a,b. Musk melon culture. **a** View of the entire facilities. **b** Three-quarter glasshouse

the YEWMAC300. This allows programs to be modified without disrupting the control of the glasshouses. Further, simply adding a line controller allows the control system to be expanded without changing the basic software. The use of an analog recorder as an A/D converter allows data coming from the input sensor to be stored for a long time.

Inputs to this control system as outdoor environment information are analog signals from the solar radiation sensor, the ventilated psychrometer and the wind vane and anemometer, and digital signals from the rain sensor. Inputs as environmental information of the glasshouses are analog signals from the ventilated thermometers (installed on the west side of a glasshouse) and the ventilated psychrometers (installed on the east side of a glasshouse), installed in each glasshouse, and digital signals from the limit switches informing the control system of whether the roof ventilators, side ventilators, upper curtains, and lower curtains are fully opened or closed. Additionally, signals giving the boiler water temperature and alarm signals for failure of boiler ignition are input. Provided as control outputs are five pairs of contacts for opening and closing the ventilators in each glasshouse, two pairs of contacts for opening and closing the upper curtains and the lower curtains each. In addition, a contact for each of the kerosene-fired CO_2 generators, the dehumidifiers, the air conditioners, and the boiler ignition systems are provided, plus seven contacts for the line pumps supplying hot water to each glasshouse.

3.1.3 Description of the Control

The devices installed in the glasshouses operate in the following manner.

When the air temperature in the glasshouse exceeds the setpoint of room temperature for ventilation, the roof ventilators facing the south are opened. If the total of each aperture traveled reaches the maximum aperture setting, then the roof ventilators on the north side are opened to the maximum aperture. If the air temperature of the glasshouse continues to rise, then the side ventilators on the south side are opened. When the air temperature in the glasshouse drops below the setpoint of room temperature for ventilation, the ventilators are closed in reverse order to the opening procedure. The roof ventilators and side ventilators facing the south are divided into the east and west halves, and are actuated upon indication of the thermometers installed on the east and west sides of a glasshouse. The ventilators facing the north are actuated as a result of the mean value of the indications by the thermometers installed on the east and west sides of a glasshouse. Each ventilator is controlled by PID. When a wind velocity greater than the setting point is detected, all the ventilators are totally closed. Then, if the air temperature in the glasshouse exceeds the setpoint of room temperature for ventilation, it is possible to detect the predominant wind direction and control the air temperature in the glasshouse by opening or closing the ventilators that happen to be leeward. When rainfall is detected, the roof ventilators are closed for a time; depending on the amount of solar radiation afterwards, they can be opened to the maximum aperture set for rainfall.

The line pumps for heating start and stop within the preset operational limits when the air temperature in the glasshouse drops below the setpoint of room temperature for heating.

The boiler is so operated as to produce hot water having the temperature calculated by an equation taking the difference between the outdoor air temperature and the setpoint of room temperature for heating into account.

The movable internal curtains are drawn open when the amount of solar radiation becomes greater than the setting point, the time period set comes in and the average air temperature in the glasshouse becomes higher than the setpoint of room temperature for ventilation. The movable internal curtains are closed, depending on the value calculated by means of an expression taking the outdoor air temperature, the setpoint of room temperature for heating, the daily amount of solar radiation, and the difference between the anticipated outdoor air temperature and air temperature in the glasshouse.

Actually, the glasshouses are managed as follows. As in the preceding example, a day is divided into six time periods. The user sets the setpoint of room temperature for ventilation and setpoint of room temperature for heating for each time period, drawing a curve for the day's temperature management. The setpoint of room temperature for both ventilation and heating, of course, are so set that photosynthesis is intensified during the daytime and, during the nighttime, translocation of photosynthate is accelerated and excessive respiration after translocation is suppressed. The air temperature in the glasshouse during the translocation accelerating time is automatically set at a higher value than the setpoint of room temperature for heating, depending on the day's total integrated solar radiation. If the relative humidity in a glasshouse exceeds the relative humidity set for the time period, the line pumps are actuated to maintain the setpoint of room temperature for heating, with the ventilators being kept open, to remove any moisture unnecessary for plant growth from the glasshouse.

In addition to the functions described above, this control program contains others specially designed for growing musk melons. As explained above, a microcomputer of this type, though expensive, has the advantage of being capable of taking the user's experience into the functional design. In addition, a microcomputer of this type is advantageous in controlling several glasshouses separately.

Since Kobayashi's present hardware allows up to 15 glasshouses to be separately controlled, he is planning to add more glasshouses to expand his scale of business. He holds regular meetings with musk melon growers who have introduced microcomputers of this type and the manufacturer, modifying his control programs.

3.2 Example of Growing *Phalaenopsis*

3.2.1 Description of the Facility

Yamagoshi, after having grown musk melons for about 20 years, started growing *Phalaenopsis* in November 1987. He owns an even-span, double-span glasshouse having a width of 10 m and length of 43.05 m (each glasshouse section is separately controlled) and three-quarter type glasshouse having a span of 7 m and length of 22 m (used for growing seedlings) (Fig. 3a). Both are microcomputer-controlled. Each glasshouse is equipped with two motors for opening and closing the roof ventilators, two motors for opening and closing the side ventilators, a motor for opening and closing the external shading curtains, a motor for opening and closing the internal shading curtains, two motors for opening and closing the movable internal curtains, and two air conditioners (the seedling glasshouse equipped with an air conditioner). To heat the glasshouses, hot water produced by the boiler is supplied to heating pipes in the glasshouses by three line pumps.

Yamagoshi introduced a microcomputer to his farm business for the following reasons: to automatize and save on labor for such simple jobs as ventilator management, to improve the quality of his products by taking care of them meticulously, and to save on energy. In addition, because he had never grown *Phalaenopsis* before, he wanted to quickly obtain a cultivation management method by collecting and analyzing data related to the cultivation environment to put environment control into practice.

3.2.2 Input into and Output from the Control System

This is another example of an industrial general-purpose microcomputer (made by Yokogawa Electric Works, Corporation, Model YEWMAC300) being used to control glasshouses, in this case, three glasshouses. The hardware is the same as in the example of growing musk melons. Various setting values are interlocutorily put into the microcomputer in the same way as described in the preceding example.

Fig. 3a-d. *Phalaenopsis* culture. **a** View of the entire facilities. **b** Input devices for external environmental factors. **c** Ventilated psychrometer and CO_2 sensor. **d** Air conditioners

The inputs into the microcomputer are the same as in the case of the musk melon culture, with the addition of signals from the CO_2 gas sensors and solar radiation sensors installed in each glasshouse, and the pH meter and EC meter installed in the watering tank (Fig. 3b,c).

Provided as control outputs for each glasshouse are two pairs of contacts for opening and closing the roof ventilators, two pairs of contacts for opening and closing the side ventilators (the even-span glasshouse is provided with upper side ventilators and lower side ventilators, while the seedling glasshouse is provided with single-frame side ventilators on both sides of the house), a pair of contacts for opening and closing the external shading curtains, a pair of contacts for opening and closing the internal shading curtains, and a pair of contacts for opening and closing the upper curtains and the lower curtains, respectively. Provided in addition to these contacts are three contacts for switching the electromagnetic valve controlling the supply of CO_2 gas (liquefied carbon dioxide gas being used), a contact for boiler starting, a contact for boiler stopping, and three contacts for the line pumps supplying hot water to the glasshouses. For the air conditioners (Fig. 3d), contacts for cooling operation and contacts for operating only fans in the air conditioners are provided.

3.2.3 Description of the Control

The ventilators, internal curtain, boiler, and line pumps operate in the same way as in the case of growing musk melons. Other devices operate in the following manner.

The external shading curtains and internal shading curtains are opened and closed independently of each other. When the amount of solar radiation in the glasshouse exceeds the setting point for solar radiation for each curtain, they are closed. The setting point for the amount of solar radiation is so set that the closing of the external curtains is followed by that of the internal curtains.

When the CO_2 concentration in the glasshouse drops below the CO_2 concentration setting point for that particular time period, the electromagnetic valve is opened; when the CO_2 concentration in the glasshouse exceeds the setting point, the valve is closed. When the valve is opened, the fans in the air conditioners are driven to diffuse the supplied CO_2 gas quickly and uniformly.

When the air temperature in the glasshouse exceeds the temperature setting point for that particular time period, the air conditioners are actuated. However, the output signals are given after the control system has determined whether one air conditioner or two be operated, depending on the cooling load.

Since these devices are interlinked with each other, the control system is so programmed that they operate efficiently. For example, the control system is programmed not to operate the line pumps and air conditioners and supply CO_2 gas with the ventilators opened.

In order to leave the environmental control of the glasshouses to a microcomputer, setting points suitable for growing *Phalaenopsis* must be input. How Yamagoshi came to obtain these setting points is described below.

Phalaenopsis was very difficult to propagate and very expensive. Generally it has been grown under weak light (to avoid leaf scorching) and with a small amount of applied nutrients and water. Even today, when the established technology of nonsymbiotic germination permits mass propagation of the seedlings, *Phalaenopsis* is often produced by the method previously described.

At first, Yamagoshi grew *Phalaenopsis* under weak light. Analysis of the data on the glasshouse environment in which he made the first culture revealed the following facts: the gradient at which the CO_2 concentration in the glasshouses decreases during the nighttime is steeper on a day having a greater total integrated solar radiation than on a day having a smaller total integrated solar radiation and the CO_2 concentration is lower on a day of intense solar radiation (the ventilators are usually totally closed during the nighttime). He then grew *Phalaenopsis*, gradually increasing total integrated solar radiation in the glasshouse, and noted that the tendency became more remarkable. Based on this observation, he decided on a basic line of increasing the amount of solar radiation in the glasshouse to such an extent that leaf scorching does not occur and maintaining a higher CO_2 concentration during the nighttime. The *Phalaenopsis* he grew under these conditions was favorably evaluated by other growers, who commented that Yamagoshi's *Phalaenopsis* grew fast and the leaves were stronger and more vigorous. It is because he introduced a microcomputer-based control system with an instrumentation system, he says, that he was able to put the new culture management method into practice with confidence.

Whether the effect of CO_2 enrichment or a greater amount of solar radiation compared with the conventional culture method has brought the good results described above cannot be determined with certainty, because he experimented without a control. However, the introduction of a computer worked favorably for his business. He is planning to double his farm area this year.

4 Conclusion

We have presented two types of microcomputers for greenhouse environment control in Japan. Our company established the following guideline to be referred to when introducing microcomputers to growers for controlling their greenhouse environment. For a crop for which culture management has been well studied, an inexpensive microcomputer for glasshouse control is used when only one greenhouse is to be controlled. However, a general-purpose microcomputer, which though expensive, allows easy modification of control programs, is used even for such a crop when: a number of greenhouses are to be separately controlled; the crop to be grown is expensive and the grower's experience is to be taken into programming the control; or, the crop to be grown is expensive and culture management methods are not established or culture methods involved are open to question. In the event that more information is desired about the development of microcomputer-based control systems and control logic, refer to the papers by Takakura et al. (1979) and Kozai et al. (1982) for further details.

The environmental factors controlled by these microcomputers are mainly such aboveground environmental elements as temperature in the greenhouse, CO_2 gas concentration, and humidity. Special devices are used in the management of the temperature in a greenhouse. In a greenhouse that is not computer-controlled, ventilator-controlling devices, heating-controlling devices, internal curtain-controlling devices and the like are separately installed. This makes it difficult laborwise for a grower to make settings manually and create an environment suitable for growing a crop, with a potential loss of energy. The introduction of microcomputers has made it a reality to meticulously control each device, decrease man hours for environment management, and attain the goal of saving energy. The effect of saving labor by introducing microcomputers is difficult to express as reduction in the operating expenses, because of the small scale of this type of business that often does not need labor outside a family group. As a future task, development of favorable software including not only the environmental control of aboveground parts but also that of belowground parts is desired.

In addition, more and more interest has been shown for soilless culture such as rockwool (Fig. 4) and NFT culture. Manufacturers of soilless culture systems have developed microcomputers for soilless culture for automatize the management of pH and EC of nutrients. Particularly in rockwool culture, in which not only pH and EC of nutrients but also the amount of drip-watering is controlled depending on the amount of solar radiation, there are many cases of introducing microcomputers. In this way,

Fig. 4a,b. Rockwool culture. **a** View of the entire facilities. **b** Rockwool bed

microcomputers are finding applications not only in the environmental control of greenhouses but also in the field of soilless culture. Soilless culture allows easier control of underground parts than soil culture does. Soilless culture offers the possibility of increased harvest based on the progress of future cultivation studies.

In the field of protected cultivation in Japan, studies have recently been underway on support systems for culture environment control and culture management involving crops, with the help of engineering, artificial intelligence, and expert systems (Harazono et al. 1988; Hirafuji et al. 1988; Kano and Shimaji 1988; Kurata 1988). Technological innovation in computer hardware being remarkable, the author believes that control systems embodying the new ideas described above will be supplied to growers at reasonable prices in the near future.

References

Harazono Y, Kamiya H, Yabiki K (1988) A control method based on an artificial intelligence technique and its application for controlling plant environment. Acta Hortic 230:209–214

Hirafuji M, Ono Y, Kobayashi K (1988) A knowledge-based system and a neurocomputer for agricultural information processing. Acta Hortic 230:253–259

Japan Greenhouse Horticulture Association (1990) Area of glasshouses, plastic houses, etc. for horticulture. Vegetable Production and Marketing Division, Food and Marketing Bureau, Ministry of Agriculture, Forestry and Fisheries (ed) Jpn Greenhouse Hortic Assoc, Tokyo

Kano A, Shimaji H (1988) Greenhouse environmental control system with a crop model and expert system. Acta Hortic 230:229–236

Kozai T, Itou T, Inayama M, Kobayashi H (1982) Development of a microcomputer-based environment control and data-logging system for greenhouses and its field test. J Agric Meteorol 38:45–50

Kurata K (1988) Greenhouse control by machine learning. Acta Hortic 230:195–200

Takakura T, Okada M, Shimaji H, Nara M (1979) Development of a microprocessor-based multi-variable control system for greenhouses. J Agric Meteorol 35:97–102

Subject Index

MIX
Papier aus verantwortungsvollen Quellen
Paper from responsible sources
FSC® C105338

If you have any concerns about our products,
you can contact us on
ProductSafety@springernature.com

In case Publisher is established outside the EU,
the EU authorized representative is:
Springer Nature Customer Service Center GmbH
Europaplatz 3, 69115 Heidelberg, Germany

Printed by Libri Plureos GmbH
in Hamburg, Germany